D1700168

The chemistry of
Organomagnesium Compounds

Patai Series: The Chemistry of Functional Groups

A series of advanced treatises founded by Professor Saul Patai and under the general editorship of Professor Zvi Rappoport

The **Patai Series** publishes comprehensive reviews on all aspects of specific functional groups. Each volume contains outstanding surveys on theoretical and computational aspects, NMR, MS, other spectroscopic methods and analytical chemistry, structural aspects, thermochemistry, photochemistry, synthetic approaches and strategies, synthetic uses and applications in chemical and pharmaceutical industries, biological, biochemical and environmental aspects.
To date, 118 volumes have been published in the series.

Recently Published Titles

The chemistry of the Cyclopropyl Group (Volume 2)
The chemistry of the Hydrazo, Azo and Azoxy Groups (Volume 2, 2 parts)
The chemistry of Double-Bonded Functional Groups (Volume 3, 2 parts)
The chemistry of Organophosphorus Compounds (Volume 4)
The chemistry of Halides, Pseudo-Halides and Azides (Volume 2, 2 parts)
The chemistry of the Amino, Nitro and Nitroso Groups (2 volumes, 2 parts)
The chemistry of Dienes and Polyenes (2 volumes)
The chemistry of Organic Derivatives of Gold and Silver
The chemistry of Organic Silicon Compounds (2 volumes, 4 parts)
The chemistry of Organic Germanium, Tin and Lead Compounds (Volume 2, 2 parts)
The chemistry of Phenols (2 parts)
The chemistry of Organolithium Compounds (2 volumes, 3 parts)
The chemistry of Cyclobutanes (2 parts)
The chemistry of Peroxides (Volume 2, 2 parts)
The chemistry of Organozinc Compounds (2 parts)
The chemistry of Anilines (2 parts)

Forthcoming Titles

The Chemistry of Hydroxylamines, Oximes and Hydroxamic Acids
The Chemistry of Metal Enolates

The Patai Series Online

Starting in 2003 the **Patai Series** is available in electronic format on Wiley InterScience. All new titles will be published as online books and a growing list of older titles will be added every year. It is the ultimate goal that all titles published in the **Patai Series** will be available in electronic format.
For more information see under **Online Books** on:

www.interscience.wiley.com

R−Mg

The chemistry of
Organomagnesium Compounds

Part 2

Edited by

ZVI RAPPOPORT

The Hebrew University, Jerusalem

and

ILAN MAREK

Technion-Israel Institute of Technology, Haifa

2008

John Wiley & Sons, Ltd

An Interscience® Publication

Copyright © 2008   John Wiley & Sons Ltd, The Atrium, Southern Gate, Chichester,
West Sussex PO19 8SQ, England

Telephone (+44) 1243 779777

Email (for orders and customer service enquiries): cs-books@wiley.co.uk
Visit our Home Page on www.wileyeurope.com or www.wiley.com

All Rights Reserved. No part of this publication may be reproduced, stored in a retrieval system or transmitted in any form or by any means, electronic, mechanical, photocopying, recording, scanning or otherwise, except under the terms of the Copyright, Designs and Patents Act 1988 or under the terms of a licence issued by the Copyright Licensing Agency Ltd, 90 Tottenham Court Road, London W1T 4LP, UK, without the permission in writing of the Publisher. Requests to the Publisher should be addressed to the Permissions Department, John Wiley & Sons Ltd, The Atrium, Southern Gate, Chichester, West Sussex PO19 8SQ, England, or emailed to permreq@wiley.co.uk, or faxed to (+44) 1243 770620.

Designations used by companies to distinguish their products are often claimed as trademarks. All brand names and product names used in this book are trade names, service marks, trademarks or registered trademarks of their respective owners. The Publisher is not associated with any product or vendor mentioned in this book. This publication is designed to provide accurate and authoritative information in regard to the subject matter covered. It is sold on the understanding that the Publisher is not engaged in rendering professional services. If professional advice or other expert assistance is required, the services of a competent professional should be sought.

Other Wiley Editorial Offices

John Wiley & Sons Inc., 111 River Street, Hoboken, NJ 07030, USA

Jossey-Bass, 989 Market Street, San Francisco, CA 94103-1741, USA

Wiley-VCH Verlag GmbH, Boschstr. 12, D-69469 Weinheim, Germany

John Wiley & Sons Australia Ltd, 42 McDougall Street, Milton, Queensland 4064, Australia

John Wiley & Sons (Asia) Pte Ltd, 2 Clementi Loop #02-01, Jin Xing Distripark, Singapore 129809

John Wiley & Sons Canada Ltd, 6045 Freemont Blvd, Mississauga, Ontario, L5R 4J3, Canada

Wiley also publishes its books in a variety of electronic formats. Some content that appears in print may not be available in electronic books.

British Library Cataloguing in Publication Data

A catalogue record for this book is available from the British Library

ISBN 978-0-470-05719-3

Typeset in 9/10pt Times by Laserwords Private Limited, Chennai, India
Printed and bound in Great Britain by Biddles Ltd, King's Lynn
This book is printed on acid-free paper responsibly manufactured from sustainable forestry in which at least two trees are planted for each one used for paper production.

Dedicated to the memory of

Yair Avni

Contributing authors

Jaap Boersma	Chemical Biology & Organic Chemistry, Faculty of Science, Utrecht University, Padualaan 8, 3584 CH Utrecht, The Netherlands
Katja Brade	Department Chemie und Biochemie, Ludwig-Maximilians-Universität München, Butenandtstr., 5-13, D-81377 München, Germany. Fax: +49-89-2180-77680
Gérard Cahiez	Laboratoire de Synthèse Organique Sélective et de Chimie Organométallique (SOSCO), UMR 8123 CNRS-ESCOM-UCP, 5 Mail Gay Lussac, Neuville ˢ/Oise, F-95092 Cergy-Pontoise, France. Fax: +3-313-425-7383; e-mail: g.cahiez@escom.fr
Christophe Duplais	Laboratoire de Synthèse Organique Sélective et de Chimie Organométallique (SOSCO), UMR 8123 CNRS-ESCOM-UCP, 5 Mail Gay Lussac, Neuville ˢ/Oise, F-95092 Cergy-Pontoise, France. Fax: +3-313-425-7383
Ben L. Feringa	Stratingh Institute for Chemistry, University of Groningen, Nijenborgh 4, 9747 AG, Groningen, The Netherlands. Fax: ++3-150-363-4296; e-mail: B.L.Feringa@rug.nl
Andrey Gavryushin	Department Chemie und Biochemie, Ludwig-Maximilians-Universität München, Butenandtstr., 5-13, D-81377 München, Germany. Fax: +49-89-2180-77680
Claude Grison	UMR CNRS-Université de Montpellier 2 5032, ENSCM, 8 rue de l'Ecole Normale, F-34296 Montpellier, France. Fax: +3-346-714-4342; e-mail: cgrison@univ-montp2.fr
Peter J. Heard	School of Science and Technology, North East Wales Institute, Mold Road, Wrexham LL112AW, UK; e-mail: p.heard@newi.ac.uk
Kenneth W. Henderson	Department of Chemistry and Biochemistry, 251 Nieuwland Science Hall, University of Notre Dame, Notre Dame, IN 46556, USA. Fax: +1-574-631-6652; e-mail: khenders@nd.edu
Katherine L. Hull	Department of Chemistry and Biochemistry, 251 Nieuwland Science Hall, University of Notre Dame, Notre Dame, IN 46556, USA. Fax: +1-574-631-6652

Torkil Holm	Department of Chemistry, Technical University of Denmark, Building 201, DK-2800, Lyngby, Denmark. Fax: +45-4-593-3968; e-mail: th@kemi.dtu.dk
Kenichiro Itami	Department of Chemistry and Research Center for Materials Science, Nagoya University, Chikusa-ku, Nagoya 464-8602, Japan. Fax: +8-152-788-6098; e-mail: itami@chem.nagoya-u.ac.jp
Johann T. B. H. Jastrzebski	Chemical Biology & Organic Chemistry, Faculty of Science, Utrecht University, Padualaan 8, 3584 CH Utrecht, The Netherlands. Fax: +3-130-252-3615; e-mail: j.t.b.h.jastrzebski@uu.nl
Jan S. Jaworski	Faculty of Chemistry, Warsaw University, 02-093 Warszawa, Poland. Fax: +4-822-822-5996; e-mail: jaworski@chem.uw.edu.pl
Gerard van Koten	Chemical Biology & Organic Chemistry, Faculty of Science, Utrecht University, Padualaan 8, 3584 CH Utrecht, The Netherlands. Fax: +3-130-252-3615; email: g.vankoten@uu.nl
Paul Knochel	Department Chemie und Biochemie, Ludwig-Maximilians-Universität München, Butenandtstr., 5-13, D-81377 München, Germany. Fax: +49-89-2180-77680; e-mail: paul.knochel@cup.uni-muenchen.de
Joel F. Liebman	Department of Chemistry and Biochemistry, University of Maryland, Baltimore County, 1000 Hilltop Circle, Baltimore, Maryland 21250, USA. Fax: +1-410-455-2608; e-mail: jliebman@umbc.edu
Fernando López	Departamento de Química Orgánica, Facultad de Química, Universidad de Santiago de Compostela, Avda. de las ciencias, s/n, 15782, Santiago de Compostela, Spain; e-mail: qofer@usc.es
Adriaan J. Minnaard	Stratingh Institute for Chemistry, University of Groningen, Nijenborgh 4, 9747 AG, Groningen, The Netherlands. Fax: ++3-150-363-4296; e-mail: A.J.Minnaard@rug.nl
Richard A. J. O'Hair	School of Chemistry, The University of Melbourne, Victoria 3010, Australia; Bio21, Molecular Science and Biotechnology Institute, The University of Melbourne, Victoria, 3010, Australia; ARC Centre of Excellence for Free Radical Chemistry and Biotechnology, Australia. Fax: +6-139-347-5180; e-mail: rohair@unimelb.edu.au
Koichiro Oshima	Department of Material Chemistry, Graduate School of Engineering, Kyoto University, Kyoto-daigaku Katsura, Nishikyo, Kyoto 615-8510, Japan. Fax: +8-175-383-2438; e-mail: oshima@orgrxn.mbox.media.kyoto-u.ac.jp
Mathias O. Senge	School of Chemistry, SFI Tetrapyrrole Laboratory, Trinity College Dublin, Dublin 2, Ireland. Fax: +3-531-896-8536; e-mail: sengem@tcd.ie.

Natalia N. Sergeeva	School of Chemistry, SFI Tetrapyrrole Laboratory, Trinity College Dublin, Dublin 2, Ireland. Fax: +3-531-896-8536
Tsuyoshi Satoh	Department of Chemistry, Faculty of Science, Tokyo University of Science; Ichigayafunagawara-machi 12, Shinjuku-ku, Tokyo 162-0826, Japan. Fax: 8-135-261-4631; e-mail: tsatoh@rs.kagu.tus.ac.jp
Suzanne W. Slayden	Department of Chemistry, George Mason University, 4400 University Drive, Fairfax, Virginia 22030, USA. Fax: +1-703-993-1055; e-mail: sslayden@gmu.edu
James Weston	Institut für Organische Chemie und Makromolekulare Chemie, Friedrich-Schiller-Universität, Humboldtstraße 10, D-07743 Jena, Germany. Fax: +49(0)-36-419-48212: e-mail: c9weje@uni-jena.de
Shinichi Yamabe	Department of Chemistry, Nara University of Education, Takabatake-cho, Nara, 630-8528, Japan. Fax: +81-742-27-9208; e-mail: yamabes@nara-edu.ac.jp
Shoko Yamazaki	Department of Chemistry, Nara University of Education, Takabatake-cho, Nara, 630–8528, Japan. Fax: +81-742-27-9289; e-mail: yamazaks@nara-edu.ac.jp
Hideki Yorimitsu	Department of Material Chemistry, Graduate School of Engineering, Kyoto University, Kyoto-daigaku Katsura, Nishikyo, Kyoto 615-8510, Japan. Fax: +81-75-383-2438; e-mail: yori@orgrxn.mbox.media.kyoto-u.ac.jp
Jun-Ichi Yoshida	Department of Synthetic Chemistry and Biological Chemistry, Graduate School of Engineering, Kyoto University, Nishikyo-ku, Kyoto 615-8510, Japan. Fax: +81-75-383-2727; e-mail: yoshida@sbchem.kyoto-u.ac.jp
Jacob Zabicky	Department of Chemical Engineering, Ben-Gurion University of the Negev, P. O. Box 653, Beer-Sheva 84105, Israel. Fax: +9-72-8647-2969; e-mail: zabicky@bgu.ac.il

Foreword

The present book, *The Chemistry of Organomagnesium Compounds*, is a continuation of the sub-group of volumes in 'The Chemistry of Functional Groups' series that deals with organometallic derivatives. Closely related to it are the two volumes, *The Chemistry of Organolithium Compounds* (Zvi Rappoport and Ilan Marek, Eds., 2003 and 2005) in three parts and the two parts of *The Chemistry of Organozinc Compounds* (Zvi Rappoport and Ilan Marek, Eds., 2006). Organomagnesium (or Grignard) reagents play a key role in organic chemistry. Although considered as one of the oldest organometallic reagents in synthesis, there have been a complete renaissance of the field in the last decade.

The two parts of the present volume contain 17 chapters written by experts from 11 countries. They include chapters dealing with structural chemistry, thermochemistry and NMR of organomagnesium compounds, formation of organomagnesium compounds in solvent-free environment, photochemistry of magnesium derivatives of porphyrins and phthalocyanines, and electrochemistry, analysis and biochemistry of organomagnesium derivatives. Special chapters are devoted to special families of compounds, such as magnesium enolates, ate-complexes, carbenoids and bonded-complexes with groups 15 and 16 compounds. Processes such as enantioselective copper-catalyzed 1,4-addition of organomagnesium halides, the iron-catalyzed reactions of Grignard reagents, and theoretical aspects of their addition to carbonyl compounds as well as carbomagnesiation reactions are covered in separate chapters. Both synthesis and reactivities of organomagnesium compounds are extensively discussed.

Unfortunately, the planned chapter on 'Theoretical Aspects of Organomagnesium Compounds' was not delivered. However, some theoretical aspects are covered in other chapters, especially Chapter 9. Another chapter on 'Mechanisms of Reactions of Organomagnesium Compounds' was not included after it was found that recent material on the topic was meager as compared with the coverage of the topic in Richey's book *Grignard Reagents, New Developments*, published in 2000. We gratefully acknowledge the contributions of all the authors of these chapters.

The literature coverage is mostly up to and sometimes including 2007.

We will be grateful to readers who draw our attention to any mistakes in the present volume or to omissions, and to new topics which deserve to be included in a future volume on organomagnesium compounds.

Jerusalem and Haifa
November 2007

Zvi Rappoport
Ilan Marek

The Chemistry of Functional Groups
Preface to the series

The series 'The Chemistry of Functional Groups' was originally planned to cover in each volume all aspects of the chemistry of one of the important functional groups in organic chemistry. The emphasis is laid on the preparation, properties and reactions of the functional group treated and on the effects which it exerts both in the immediate vicinity of the group in question and in the whole molecule.

A voluntary restriction on the treatment of the various functional groups in these volumes is that material included in easily and generally available secondary or tertiary sources, such as Chemical Reviews, Quarterly Reviews, Organic Reactions, various 'Advances' and 'Progress' series and in textbooks (i.e. in books which are usually found in the chemical libraries of most universities and research institutes), should not, as a rule, be repeated in detail, unless it is necessary for the balanced treatment of the topic. Therefore each of the authors is asked not to give an encyclopaedic coverage of his subject, but to concentrate on the most important recent developments and mainly on material that has not been adequately covered by reviews or other secondary sources by the time of writing of the chapter, and to address himself to a reader who is assumed to be at a fairly advanced postgraduate level.

It is realized that no plan can be devised for a volume that would give a complete coverage of the field with no overlap between chapters, while at the same time preserving the readability of the text. The Editors set themselves the goal of attaining reasonable coverage with moderate overlap, with a minimum of cross-references between the chapters. In this manner, sufficient freedom is given to the authors to produce readable quasi-monographic chapters.

The general plan of each volume includes the following main sections:

(a) An introductory chapter deals with the general and theoretical aspects of the group.

(b) Chapters discuss the characterization and characteristics of the functional groups, i.e. qualitative and quantitative methods of determination including chemical and physical methods, MS, UV, IR, NMR, ESR and PES—as well as activating and directive effects exerted by the group, and its basicity, acidity and complex-forming ability.

(c) One or more chapters deal with the formation of the functional group in question, either from other groups already present in the molecule or by introducing the new group directly or indirectly. This is usually followed by a description of the synthetic uses of the group, including its reactions, transformations and rearrangements.

(d) Additional chapters deal with special topics such as electrochemistry, photochemistry, radiation chemistry, thermochemistry, syntheses and uses of isotopically labeled compounds, as well as with biochemistry, pharmacology and toxicology. Whenever applicable, unique chapters relevant only to single functional groups are also included (e.g. 'Polyethers', 'Tetraaminoethylenes' or 'Siloxanes').

This plan entails that the breadth, depth and thought-provoking nature of each chapter will differ with the views and inclinations of the authors and the presentation will necessarily be somewhat uneven. Moreover, a serious problem is caused by authors who deliver their manuscript late or not at all. In order to overcome this problem at least to some extent, some volumes may be published without giving consideration to the originally planned logical order of the chapters.

Since the beginning of the Series in 1964, two main developments have occurred. The first of these is the publication of supplementary volumes which contain material relating to several kindred functional groups (Supplements A, B, C, D, E, F and S). The second ramification is the publication of a series of 'Updates', which contain in each volume selected and related chapters, reprinted in the original form in which they were published, together with an extensive updating of the subjects, if possible, by the authors of the original chapters. Unfortunately, the publication of the 'Updates' has been discontinued for economic reasons.

Advice or criticism regarding the plan and execution of this series will be welcomed by the Editors.

The publication of this series would never have been started, let alone continued, without the support of many persons in Israel and overseas, including colleagues, friends and family. The efficient and patient co-operation of staff-members of the Publisher also rendered us invaluable aid. Our sincere thanks are due to all of them.

The Hebrew University SAUL PATAI
Jerusalem, Israel ZVI RAPPOPORT

Sadly, Saul Patai who founded 'The Chemistry of Functional Groups' series died in 1998, just after we started to work on the 100th volume of the series. As a long-term collaborator and co-editor of many volumes of the series, I undertook the editorship and I plan to continue editing the series along the same lines that served for the preceding volumes. I hope that the continuing series will be a living memorial to its founder.

The Hebrew University ZVI RAPPOPORT
Jerusalem, Israel
May 2000

Contents

1. Structural organomagnesium chemistry 1
   Johann T. B. H. Jastrzebski, Jaap Boersma and Gerard van Koten

2. The thermochemistry of organomagnesium compounds 101
   Joel F. Liebman, Torkil Holm and Suzanne W. Slayden

3. NMR of organomagnesium compounds 131
   Peter J. Heard

4. Formation, chemistry and structure of organomagnesium species in solvent-free environments 155
   Richard A. J. O'Hair

5. Photochemical transformations involving magnesium porphyrins and phthalocyanines 189
   Natalia N. Sergeeva and Mathias O. Senge

6. Electrochemistry of organomagnesium compounds 219
   Jan S. Jaworski

7. Analytical aspects of organomagnesium compounds 265
   Jacob Zabicky

8. Biochemistry of magnesium 315
   James Weston

9. Theoretical studies of the addition of RMgX to carbonyl compounds 369
   Shinichi Yamabe and Shoko Yamazaki

10. Organomagnesium-group 15- and Organomagnesium-group 16-bonded complexes 403
    Katherine L. Hull and Kenneth W. Henderson

11. Preparation and reactivity of magnesium enolates 437
    Claude Grison

12. Functionalized organomagnesium compounds: Synthesis and reactivity 511
    Paul Knochel, Andrey Gavryushin and Katja Brade

13.	Iron-Catalyzed Reactions of Grignard Reagents **Gérard Cahiez and Christophe Duplais**	595
14.	Carbomagnesiation reactions **Kenichiro Itami and Jun-ichi Yoshida**	631
15.	The chemistry of organomagnesium ate complexes **Hideki Yorimitsu and Koichiro Oshima**	681
16.	The chemistry of magnesium carbenoids **Tsuyoshi Satoh**	717
17.	Catalytic enantioselective conjugate addition and allylic alkylation reactions using Grignard reagents **Fernando López, Adriaan J. Minnaard and Ben L. Feringa**	771
Author index		803
Subject index		855

List of abbreviations used

Ac	acetyl (MeCO)
acac	acetylacetone
Ad	adamantyl
AIBN	azoisobutyronitrile
Alk	alkyl
All	allyl
An	anisyl
Ar	aryl
Bn	benzyl (PhCH$_2$)
Bu	butyl (C$_4$H$_9$)
Bz	benzoyl (C$_6$H$_5$CO)
c-	cyclo
CD	circular dichroism
CI	chemical ionization
CIDNP	chemically induced dynamic nuclear polarization
CNDO	complete neglect of differential overlap
Cp	η^5-cyclopentadienyl (C$_5$H$_5$)
Cp*	η^5-pentamethylcyclopentadienyl (C$_5$Me$_5$)
DABCO	1,4-diazabicyclo[2.2.2]octane
DBN	1,5-diazabicyclo[4.3.0]non-5-ene
DBU	1,8-diazabicyclo[5.4.0]undec-7-ene
DIBAH	diisobutylaluminium hydride
DME	1,2-dimethoxyethane
DMF	N,N-dimethylformamide
DMSO	dimethyl sulfoxide
E-	entgegen
ee	enantiomeric excess
EI	electron impact
ESCA	electron spectroscopy for chemical analysis
ESR	electron spin resonance
Et	ethyl (C$_2$H$_5$)
eV	electron volt

List of abbreviations used

Fc	ferrocenyl
FD	field desorption
FI	field ionization
FT	Fourier transform
Fu	furyl (OC_4H_3)
GLC	gas liquid chromatography
Hex	hexyl (C_6H_{13})
c-Hex	cyclohexyl (c-C_6H_{11})
HMPA	hexamethylphosphortriamide
HOMO	highest occupied molecular orbital
HPLC	high performance liquid chromatography
i-	iso
ICR	ion cyclotron resonance
Ip	ionization potential
IR	infrared
LAH	lithium aluminium hydride
LCAO	linear combination of atomic orbitals
LDA	lithium diisopropylamide
LUMO	lowest unoccupied molecular orbital
M	metal
M	parent molecule
MCPBA	m-chloroperbenzoic acid
Me	methyl (CH_3)
Mes	mesityl (2,4,6-$Me_3C_6H_2$)
MNDO	modified neglect of diatomic overlap
MS	mass spectrum
n-	normal
Naph	naphthyl
NBS	N-bromosuccinimide
NCS	N-chlorosuccinimide
NMR	nuclear magnetic resonance
Pen	pentyl (C_5H_{11})
Ph	phenyl
Pip	piperidyl ($C_5H_{10}N$)
ppm	parts per million
Pr	propyl (C_3H_7)
PTC	phase transfer catalysis or phase transfer conditions
Py	pyridine (C_5H_5N)
Pyr	pyridyl (C_5H_4N)

R	any radical
RT	room temperature
s-	secondary
SET	single electron transfer
SOMO	singly occupied molecular orbital
t-	tertiary
TCNE	tetracyanoethylene
TFA	trifluoroacetic acid
TFE	2,2,2-trifluoroethanol
THF	tetrahydrofuran
Thi	thienyl (SC_4H_3)
TLC	thin layer chromatography
TMEDA	tetramethylethylene diamine
TMS	trimethylsilyl or tetramethylsilane
Tol	tolyl (MeC_6H_4)
Tos or Ts	tosyl (*p*-toluenesulphonyl)
Trityl	triphenylmethyl(Ph_3C)
Vi	vinyl
XRD	X-ray diffraction
Xyl	xylyl ($Me_2C_6H_3$)
Z-	zusammen

In addition, entries in the 'List of Radical Names' in *IUPAC Nomenclature of Organic Chemistry*, 1979 Edition, Pergamon Press, Oxford, 1979, p. 305–322, will also be used in their unabbreviated forms, both in the text and in formulae instead of explicitly drawn structures.

CHAPTER 11

Preparation and reactivity of magnesium enolates

CLAUDE GRISON

UMR CNRS-Université de Montpellier 2 5032, ENSCM, 8 rue de l'Ecole Normale, F-34296 Montpellier, France
Fax: +33-4-67-14-43-42; e-mail: cgrison@univ-montp2.fr

I. INTRODUCTION	438
II. PREPARATION OF MAGNESIUM ENOLATES	438
A. Reductive Metal Insertion into Carbon–Halogen Bonds	438
B. Permutation Heteroatom/Metal	441
C. Permutational Metal/Metal Salts Interconversions (Transmetallations)	445
D. Conjugate Addition	450
E. Permutational Hydrogen/Metal Interconversions (Metallations)	457
F. Miscellaneous Methods	471
III. REACTIVITY OF ENOLATES	472
A. Introduction	472
B. Reactions of Magnesium Ketone Enolates with Electrophiles	472
C. Reactions of Magnesium Ester Enolates and Magnesium Lactone Enolates with Electrophiles	484
D. Reactions of Magnesium Dicarbonyl Enolates with Electrophiles	489
1. Reactions of magnesium α-ketoester enolates	489
2. Reactions of magnesium chelates of β-ketoesters or β-diketones	493
3. Reactions of magnesium dialkyl malonate or magnesium hydrogen alkyl malonate	494
E. Reactions of Magnesium Amide and Lactam Enolates with Electrophiles	499
F. Reactions of Magnesium Thioesters and Thioamide Enolates with Electrophiles	500
G. Reactions of Carboxylic Acid Dianions with Electrophiles	503
H. Reactions with Anions of Chiral Oxazolidinones and Derivatives with Electrophiles	503
I. Reactions of Miscellaneous Magnesium Chiral Enolates	505
IV. REFERENCES	506

The chemistry of organomagnesium compounds
Edited by Z. Rappoport and I. Marek © 2008 John Wiley & Sons, Ltd

I. INTRODUCTION

Enolate anions are among the most important synthetic intermediates, largely because of their great utility to form carbon–carbon bonds. In this field, lithium enolates have been used extensively in modern synthetic organic chemistry and asymmetric synthesis. Recently, increasing interest has been directed toward new developments of magnesium enolates that may be advantageous in chemio-, regio- and stereoselective transformations. Magnesium enolates are highly suitable metal synthons. Using a divalent metal reagent allows one to formally bond two chiral ligands to the Mg center. Moreover, this higher degree of covalency leads to simpler and more stable systems than lithium analogues. Because they are good candidates in stereoselective carbon–carbon bond formation, there is a need for a comprehensive survey covering new developments of magnesium enolates.

The chapter is organized under the headings *Preparation* and *Reactivity* of magnesium enolates.

II. PREPARATION OF MAGNESIUM ENOLATES
A. Reductive Metal Insertion into Carbon–Halogen Bonds

Grignard reagents are usually prepared from the corresponding halides by reaction with metallic magnesium. This method has been used to prepare magnesium enolates of ketones. The problem with the reductive metal insertion in an α-halo carbonyl compound is the presence of the electrophilic carbonyl function. Preparation of magnesium enolate by this route can lead to the formation of an intractable mixture of products, including addition, reduction or coupling of the substrate. Thus, the preparation of magnesium enolates with elemental magnesium is often described with highly sterically hindered bromo ketones, therefore minimizing the possibility of side reactions with the carbonyl group.

The action of magnesium on α-halo ketone has been used by Malmgren, for the synthesis of bromomagnesium enolate **2** derived from 3-bromocamphor **1** (equation 1)[1].

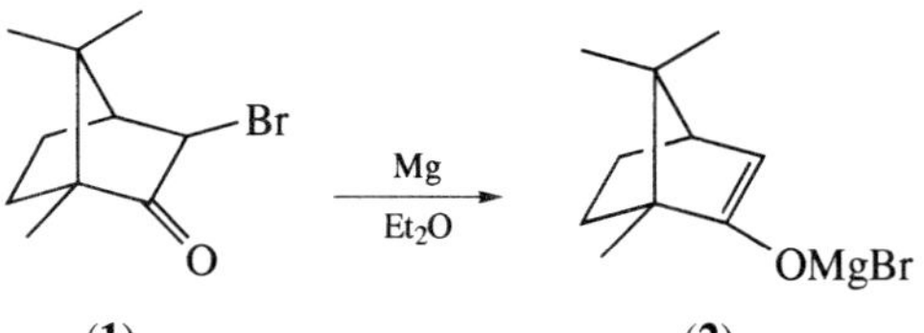

(1)

Similarly, the bromo magnesium enolate **4** of 2,2-diphenylcyclohexanone has been prepared by the action of magnesium on 6-bromo-2,2-diphenylcyclohexanone **3** (equation 2). A small amount of iodine is added to initiate the reaction. The enolate **4** is obtained in 79% yield after 15 min at reflux[2].

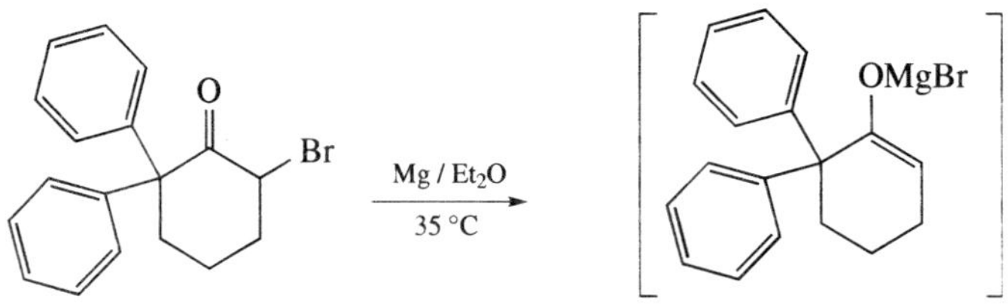

(2)

Colonge and Grenet have reported that this type of reaction may be utilized in the preparation of magnesium enolates from simple aliphatic α-bromo ketones[3].

The treatment of α-bromo *t*-butyl alkyl ketone **5** with magnesium gives almost exclusively the (*Z*)-enolate **6**. The process can be extended to cyclic systems, such as 2-bromo-2,5-trimethyl cyclopentanone **7** that leads to the enolate **8** (equation 3)[4].

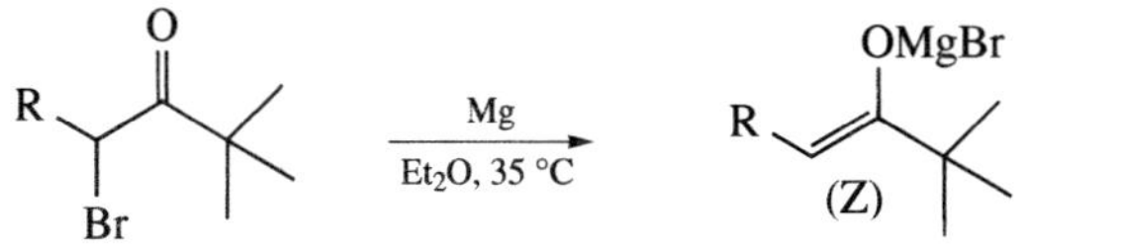

(5) → (6) R = Me, Et, *i*-Pr, *i*-Bu

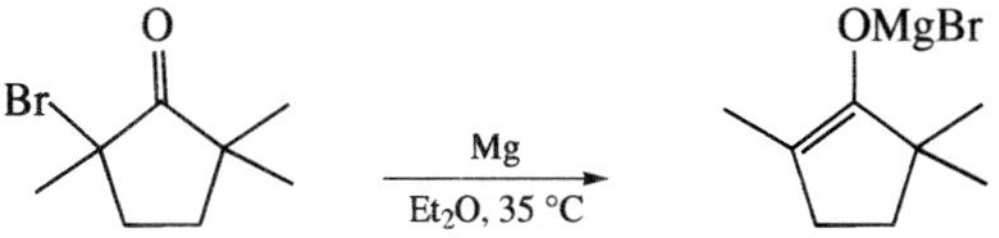

(7) → (8)

(3)

In diethyl ether, the bromo magnesium enolate derived from *t*-butyl ethyl ketone has been characterized as the dimer **9** with bridging enolate residues[5].

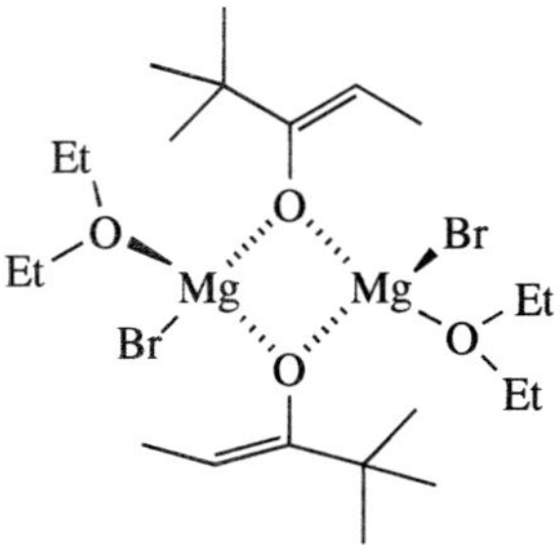

(9)

The procedure for the preparation of magnesium enolates from α-bromo ketones does not need a particular purity or nature of the metal. Magnesium turnings are often convenient to use and sufficiently reactive in many cases. Thus, magnesium enolates **11** of 2,6-dimethyl- and 2,2,6-trimethylcyclohexanones are obtained by treatment of the corresponding 2-bromo ketones **10** with magnesium turnings in a benzene/ether mixture (equation 4). These magnesium reagents are prepared and used at 0 °C, whereas the lithium and titanium analogues are obtained at −78 °C. These experimental conditions illustrate the higher thermal stability of the magnesium species[6].

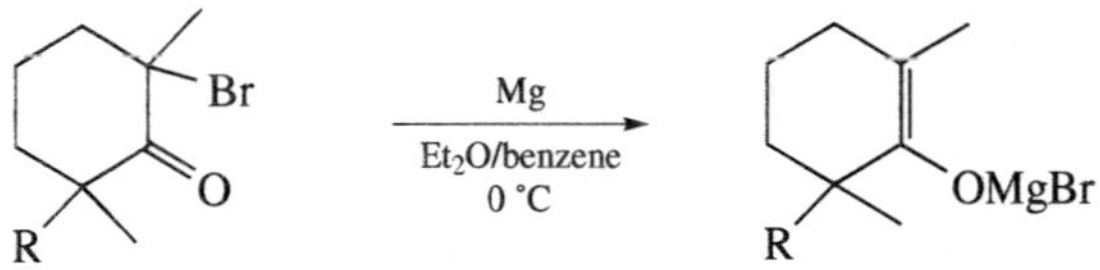

(10) R = H, Me → (11)

(4)

Side-reactions can sometimes be avoided by preparing the magnesium enolate in the presence of the electrophile.

Recently, Altarejos and coworkers have described the direct coupling of α-campholenic aldehyde **12** with α-bromoketone by a magnesium-mediated aldol-type reaction[7]. Thus, under conditions similar to a classical Grignard reaction, the reaction of 3-bromo-3-methyl-2-butanone with magnesium in refluxing diethyl ether generates the corresponding bromomagnesium enolate and subsequent coupling with α-campholenic aldehyde **12** gives the β-hydroxyketone **13** (equation 5). In contrast to the direct aldolisation reaction between the α-campholenic aldehyde and aliphatic ketones, the process allows the regioselective preparation of disubstituted keto-enolate and the aldol condensation through the sole carbon C-3. It limits the other aldol condensation product through the terminal carbon C-1, and opens an interesting route to the synthesis of the sandalwood-type odorant Polysantol **14**.

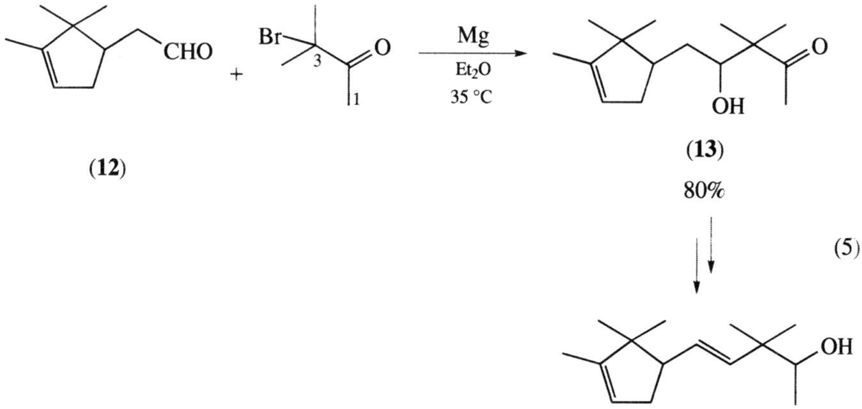

(5)

(**12**)   (**13**) 80%

(**14**) Polysantol

The authors have extended this methodology to other α-bromoketones in order to determine the scope of the reaction and prepared several Polysantol structurally related compounds. Noteworthy is the excellent chemioselectivity without any side reactions on the aldehyde moiety.

Recently, a reductive magnesium insertion into a carbon–iodine bond of a β-iodo-α-ketoester has been described[8]. The preparation of the iodomagnesium enolate **17** derived from an α-ketoester is the first preparation of such metallic species in this series. It was obtained from the reaction between the β-iodo-α-ketoester precursor **16** and magnesium. In this case, the form of the metal is critical and magnesium powder with a large surface area is necessary (equation 6).

The β-iodo-α-ketoester precursor **16** is previously obtained from the reaction between the α-chloroglycidic ester **15** and MgI_2 in ether. The β-iodo-α-ketoester **16** is not isolated and is *in situ* transformed into the iodomagnesium enolate **17** by the presence of the active magnesium produced during the preparation of MgI_2 (the 2/1 magnesium–iodine ratio was used intentionally for the preparation of MgI_2)[9].

The enolate structure of **17** is deduced from the IR data of the reaction medium as a result of the presence of absorption bands at 1490 cm^{-1} for the C=C bond and 1665 cm^{-1} for the C=O bond of the ester group, characteristic for an internal coordination of the enolate magnesium atom with the ester C=O[9].

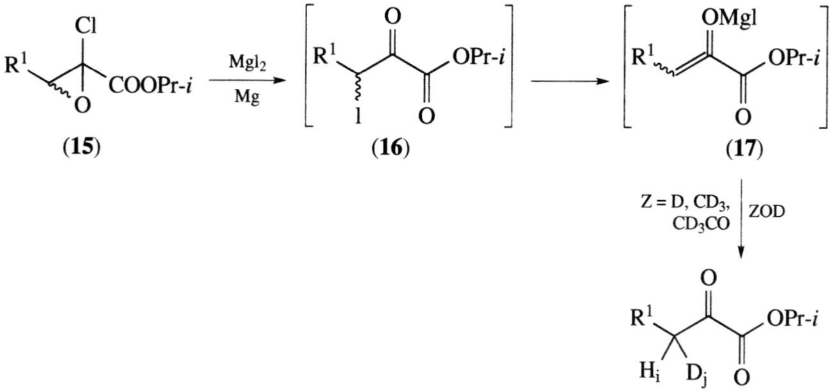

(a) R¹ = i-Pr
(b) R¹ = Et
(c) R¹ = Ph(CH)Me

(18) : i = 2, j = 0
(18)-d_1 : i = j = 1
(18)-d_2 : i = 0, j = 2

(6)

Z = D, CD₃, CD₃CO | ZOD

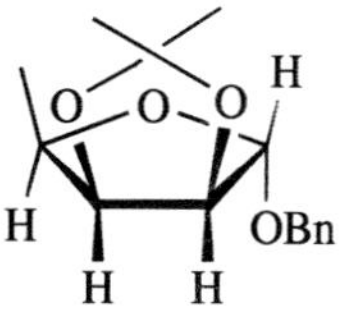

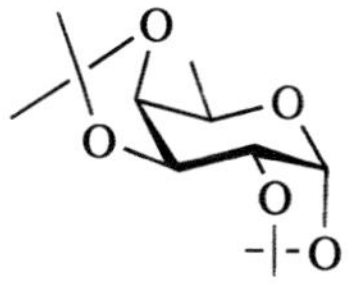

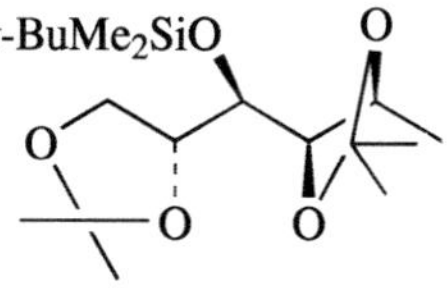

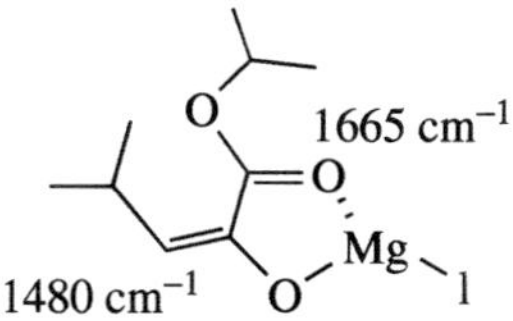

(17a)

All prepared magnesium enolates **17** are stable in refluxing diethyl ether. Deuteriation, and reactions with various electrophiles confirm their structure (see section III). It is noteworthy that the lithiated carbanion-enolate analogue, directly obtained by deprotonation of an α-ketoester **18** with lithiated bases (LDA, for example), is not stable and immediately degrades in the medium, whatever the temperature. Comparatively, the magnesium chelate **17** shows a higher stability, which allows its preparation and synthetic applications.

B. Permutation Heteroatom/Metal

Metal–halogen interconversions are primarily used to prepare organolithium compounds from alkyl and aryl halides. Particular Grignard reagents can be prepared from this methodology. They often give better yields than reactions between halides and magnesium metal due to less side reactions. This method has found applications in the preparation of magnesium enolates, even in the presence of functional groups.

Several years ago, Castro, Villieras and coworkers described the preparation of the magnesium enolate derived from an alkyl α,α-dihaloacetate by halogen–metal exchange between isopropylmagnesium chloride and alkyl trihaloacetate. THF is required as solvent (equation 7)[10].

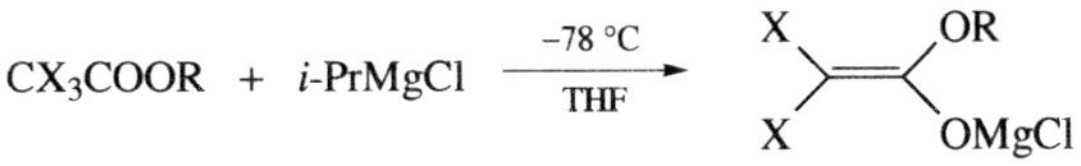

$$CX_3COOR + i\text{-PrMgCl} \xrightarrow[\text{THF}]{-78\,°C} \begin{array}{c} X \\ \diagup \\ X \end{array} = \begin{array}{c} OR \\ \diagdown \\ OMgCl \end{array}$$

(7)

R = Me, i-Pr

X = Cl, Br

Different protocols have been tested to prepare enolates from β-aryl-α-iodoketones. Reactions using Et$_3$B/Ph$_3$SnH in benzene, Et$_3$B in benzene or ether and n-BuLi in ether failed to provide the corresponding enolates. Alternatively, the use of EtMgBr succeeds in generating reactive magnesium enolates from α-iodo ketones. In most cases, the formation of the enolate in THF is cleaner than that in Et$_2$O[11, 12] (equation 8).

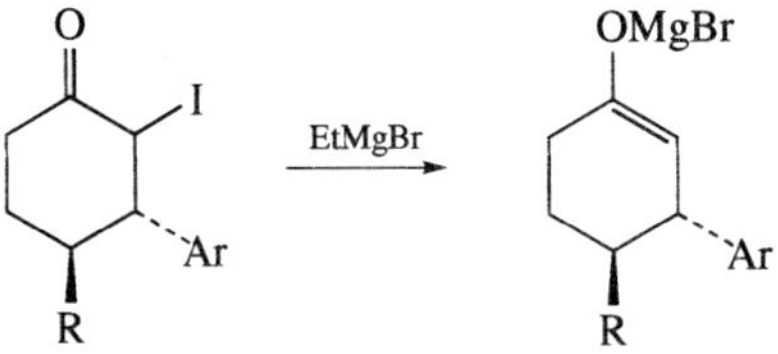

(8)

Ar = bulky aryl groups

R = H, t-Bu, CMe$_2$(OTES)

This method has been used to prepare the antibiotic thienamycin. The magnesium enolate of the β-lactam was prepared from the 6-iodo derivative in THF (equation 9)[4b].

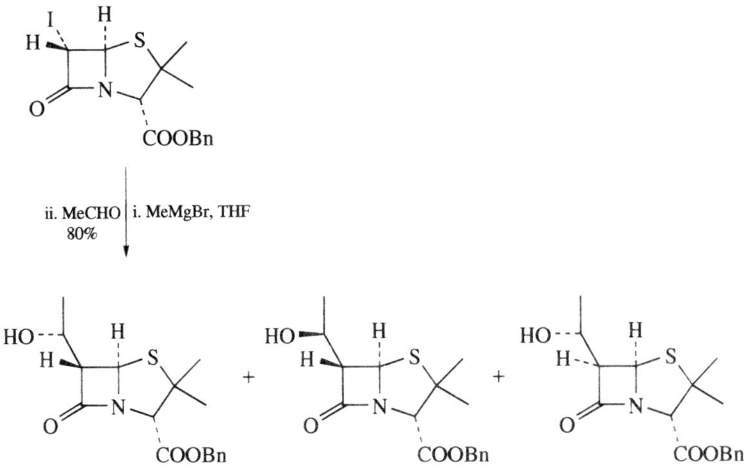

(9)

ii. MeCHO | i. MeMgBr, THF
80%

24 / 49 / 27

11. Preparation and reactivity of magnesium enolates

To increase the stereoselectivity of the aldol reaction, successful reaction of dibromo-β-lactams with methylmagnesium bromide followed by addition of acetaldehyde has been studied (equation 10)[4b].

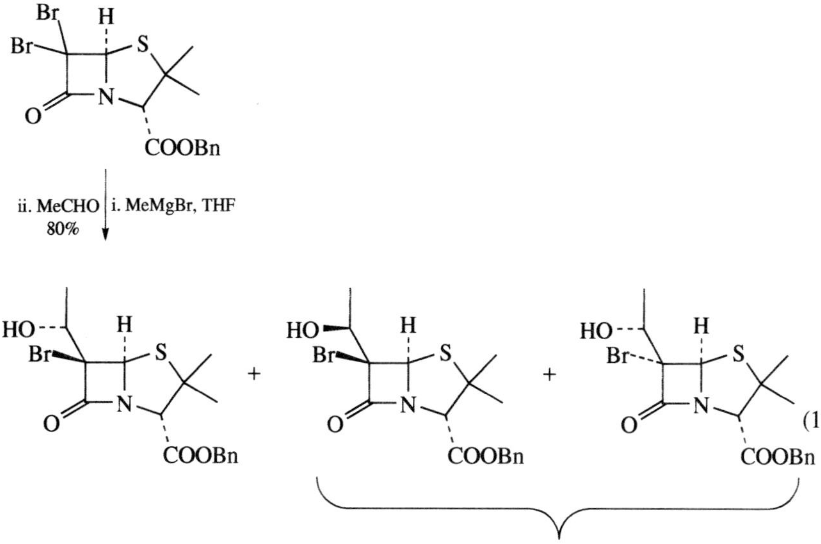

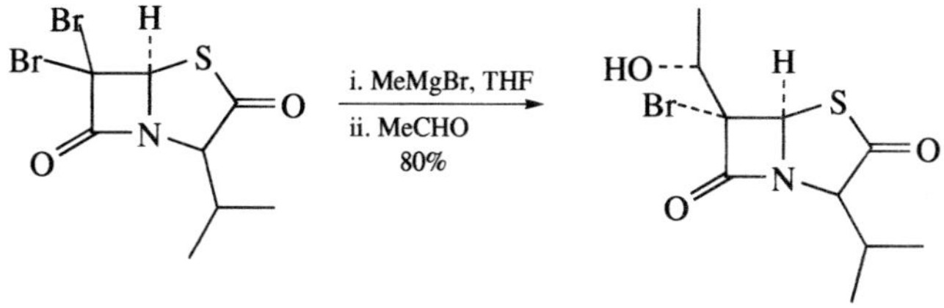

If the metal–halogen interconversions were originally used to prepare magnesium enolates, other heteroatoms than halogen can undergo the exchange. Sulfur atom is often employed.

For instance, Trost, Mao and coworkers use α,α-disulfenylated lactones as enolate precursors[13]. Reaction of ethylmagnesium bromide and α,α-di-(phenylthio)-γ-butyrolactone provides such an enolate that is quenched by ethanal (equation 11).

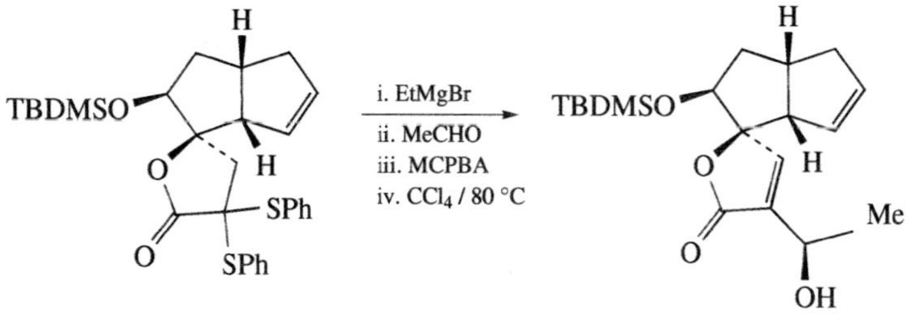

The process has been employed with α,α-disulfenylated ketones, but it needs a catalytic amount of copper(I) bromide in the reaction mixture.

Ligand exchange reaction of sulfoxides has been reported as a novel method for the generation of magnesium enolates. Reaction of a sulfinyl group of β-keto-sulfoxydes, sulfinylamides, sulfinylesters and sulfinylcarboxylic acids with ethylmagnesium bromide gives the corresponding magnesium enolates at low temperature[14–17] (equation 12). The results for EtMgBr-promoted desulfinylation of α-halo-α-sulfinyl derivatives are summarized in Table 1 after quenching the magnesium enolates with aqueous NH_4Cl.

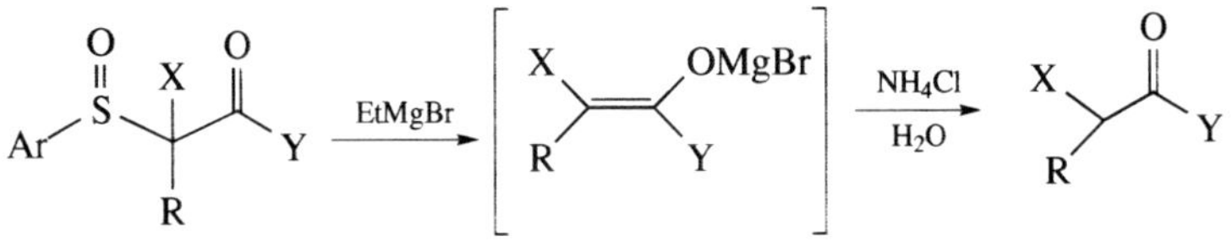

$$\underset{Ar}{\overset{O}{\underset{\|}{S}}}\underset{R}{\overset{X}{\underset{|}{C}}}\underset{}{\overset{O}{\underset{\|}{C}}}Y \xrightarrow{EtMgBr} \left[\underset{R}{\overset{X}{C}}=\underset{Y}{\overset{OMgBr}{C}}\right] \xrightarrow[H_2O]{NH_4Cl} \underset{R}{\overset{X}{\underset{|}{C}}}\underset{}{\overset{O}{\underset{\|}{C}}}Y \quad (12)$$

It should be noted that this desulfinylation is totally regioselective and gives the trisubstituted enolate. In the cases of α-halo α-sulfinyl ketones, reactions are carried out at $-78\,°C$ in Et_2O with 1.1 equivalent of ethylmagnesium bromide. Interesting results have been obtained with α-fluoro-, α-bromo- and α-chloro α-sulfinyl ketones. Because of the

TABLE 1. Synthesis of α-haloketones from α-halosulfoxides

X	Ar	R	Y	Yield (%)
F	Ph	$(CH_2)_3CH_3$	$(CH_2)_2Ph$	79
F	Ph	$(CH_2)_3CH_3$	$(CH_2)_8CH_3$	90
F	Ph	$(CH_2)_3CH_3$	Ph	82
F	Ph	$(CH_2)_3CH_3$	c-Hex	85
Cl	p-Tol	$(CH_2)_3CH_3$	$(CH_2)_2Ph$	91
Cl	p-Tol	$(CH_2)_3CH_3$	$(CH_2)_8CH_3$	93
Cl	p-Tol	$(CH_2)_3CH_3$	Ph	78
Cl	Ph	$(CH_2)_3CH_3$	c-Hex	95
Br	p-Tol	$(CH_2)_3CH_3$	$(CH_2)_2Ph$	68
Br	Ph	$(CH_2)_3CH_3$	$(CH_2)_8CH_3$	70
Br	Ph	$(CH_2)_3CH_3$	Ph	56
Br	Ph	$(CH_2)_3CH_3$	c-Hex	59
F	p-Tol	$(CH_2)_2Ph$	H	82
F	p-Tol	$(CH_2)_9CH_3$	H	95
Cl	Ph	$(CH_2)_2Ph$	H	73
Cl	Ph	$(CH_2)_9CH_3$	H	91
Cl	p-Tol	$(CH_2)_2CH_3$	CON⟨⟩ 	78
Cl	p-Tol	$(CH_2)_2CH_3$	$CONH\,(CH_2)_5CH_3$	73
Cl	p-Tol	$(CH_2)_2CH_3$	$CONHCH_2Ph$	80
Cl	p-Tol	$(CH_2)_2CH_3$	$CONH_2$	78
F	p-Tol	CH_2-Ph	COOEt	60
F	p-Tol	$(CH_2)_9CH_3$	COOEt	58
Cl	p-Tol	CH_2-Ph	COOEt	72
Cl	p-Tol	$(CH_2)_9CH_3$	COOEt	91
Cl	p-Tol	CH_2-Ph	COOBu-i	93
Cl	p-Tol	$(CH_2)_9CH_3$	COOBu-i	67
Cl	p-Tol	CH_3	COOH	67

unstable nature of the products containing bromine, the yields for the synthesis of α-bromoketones are lower than for other α-haloketones.

With α-chloro α-sulfinylcarboxylic acids and their derivatives, the optimal conditions are different. Treatment of primary or secondary α-chloro α-sulfinylamides with 3 equivalents of EtMgBr in THF at −78 °C for 10 min gives cleanly the desulfinylated α-chloroamides in good yields. An excess of Grignard reagent is necessary for completion of the reaction. Similar conditions (EtMgBr, 2 equiv/−78 °C/THF) for the α-chloro α-sulfinylesters give α-chloroesters with better yields. It is worth noting that even α-chloro α-sulfinylcarboxylic acid gives α-chloropropionic acid.

The reductive removal of a sulfinyl group to prepare magnesium enolates is also mentioned from α-chloro-β-hydroxy sulfoxides **20**. These latter are obtained by reaction between lithium carbanions derived from 1-chloroalkyl p-tolyl sulfoxides **19** and carbonyl compounds. These substrates are treated with t-BuMgCl to give magnesium alkoxides, which reacted with i-PrMgCl to afford the magnesium enolates via the rearrangement of β-oxido carbenoids[18] (equation 13).

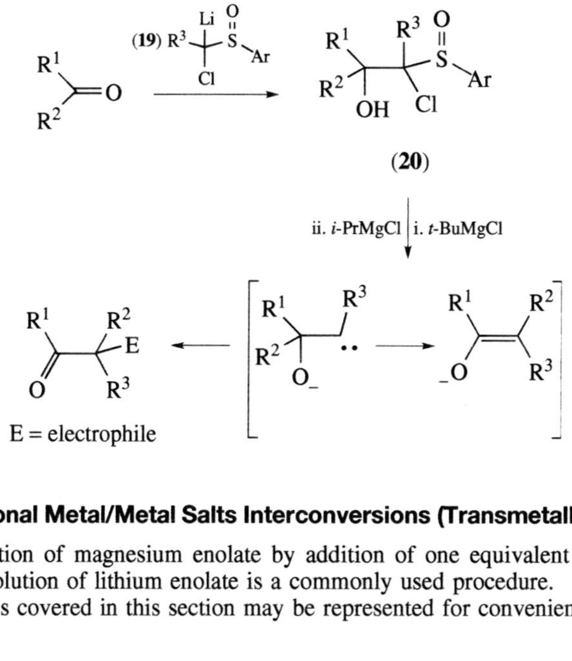

C. Permutational Metal/Metal Salts Interconversions (Transmetallations)

The preparation of magnesium enolate by addition of one equivalent of magnesium dihalide to a solution of lithium enolate is a commonly used procedure.

The reactions covered in this section may be represented for convenience by equation 14.

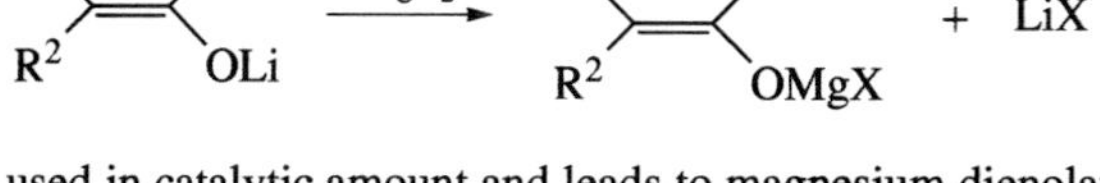

MgX_2 can be used in catalytic amount and leads to magnesium dienolate (equation 15).

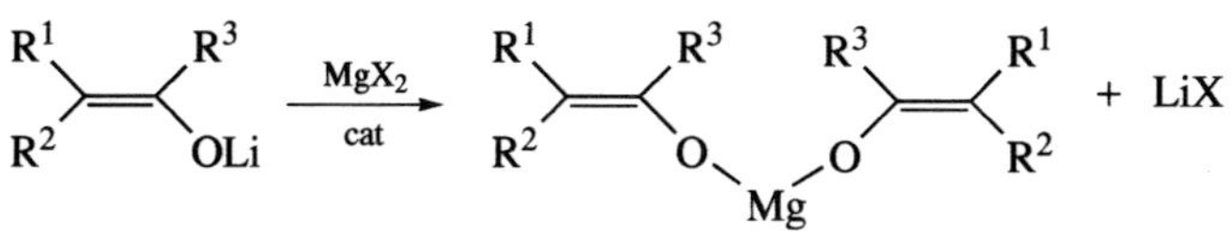

It should be emphasized that, in many cases, these equations are oversimplified. In many of the reported examples, the produced enolate is used *in situ*, and its real structure and nature of the counterion (Mg or Li) are obscured.

However, the formation of a conformationaly rigid chelate with a fixed geometry is generally admitted in the transmetallation of organolithiums with magnesium halides. That explains the stability of the enolate and the stereoselectivity of its reactions.

Some synthetic applications of this methodology are the synthesis of sterically high demanding α-alkylated unsaturated amino acids **21**. Deprotonation of allylic *N*-protected aminoesters with LDA at $-78\,^\circ$C and subsequent addition of $MgCl_2$ result in the formation of a chelated magnesium enolate (equation 16). In contrast to the corresponding lithium enolates, which decompose by warming, the chelate magnesium enolates are more stable and can be used for stereoselective ester enolate Claisen rearrangements[19].

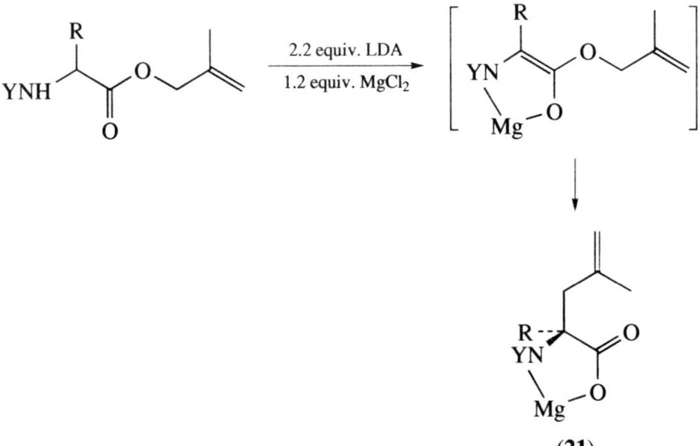

(16)

The method has been extended to other polyfunctionnal systems, such as *O*-ethyl *S*-(tetrahydro-2-oxo-3-furanyl)dithiocarbonate. Treatment of γ-butyrolactone with bis[methoxy(thiocarbonyl)]disulfide in the presence of 2.2 equivalents of lithium diisopropylamide at $-78\,^\circ$C in THF provides the lithium enolate which reacts with $MgCl_2$ to furnish the magnesium enolate[20] (equation 17).

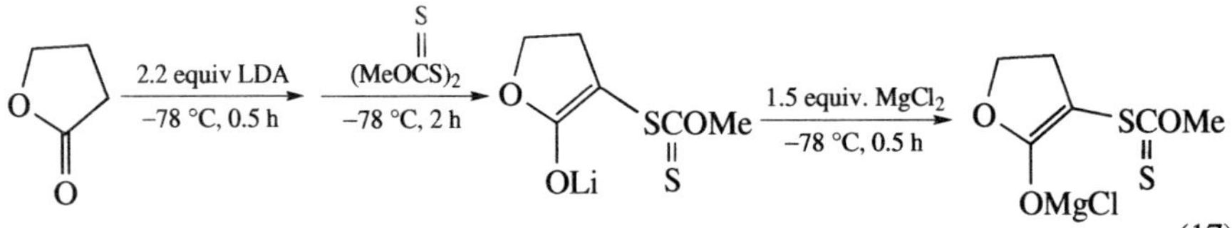

(17)

Reactions presented below show that this method is used frequently to prepare magnesium enolates derived from functionalized carboxylic esters or lactones[21–23,4] (equations 18–24, Tables 2 and 3).

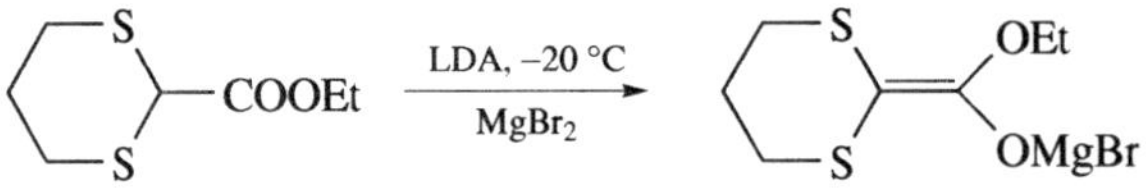

(18)

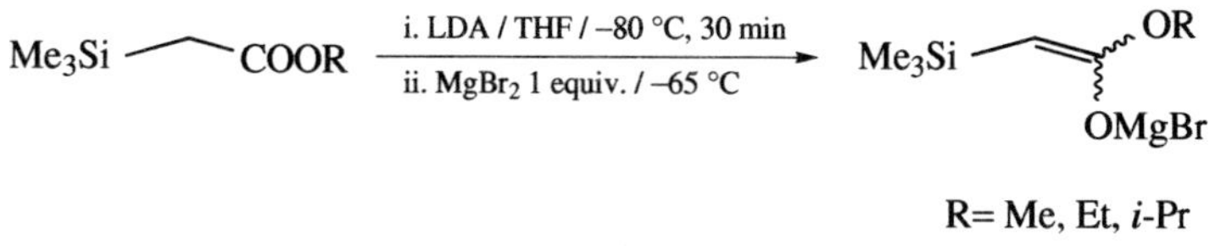

The chelation between a Boc group and Mg(II) is often used to control the stereochemistry in aldol reactions. For instance, Donohoe and House have reported the diastereoselective reductive aldol reactions of Boc-protected electron-deficient pyrroles. The key step of the synthesis is the preparation of an exocyclic magnesium enolate of Boc-protected 2-substituted pyrroles[24].

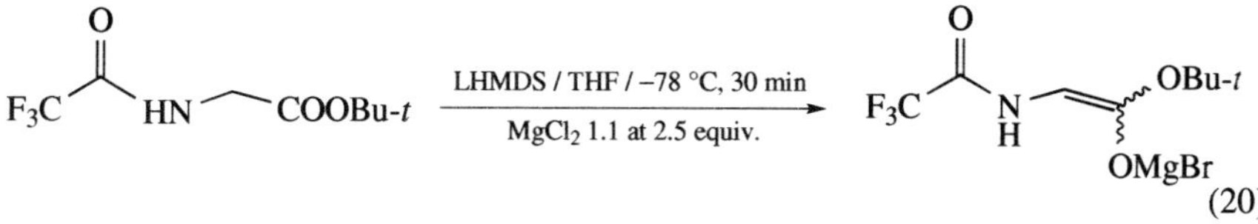

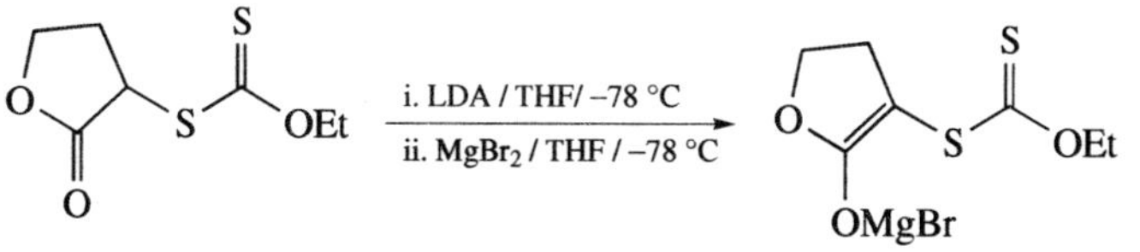

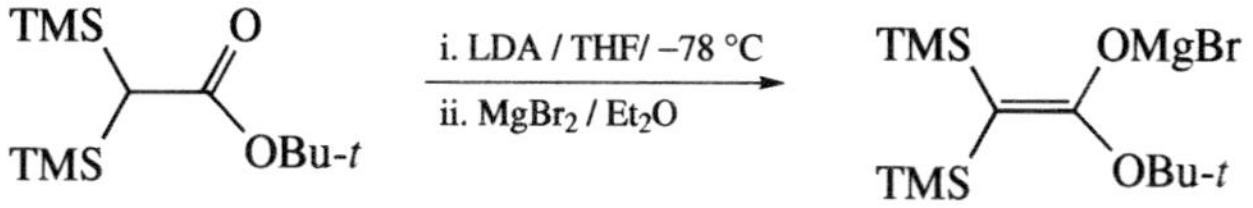

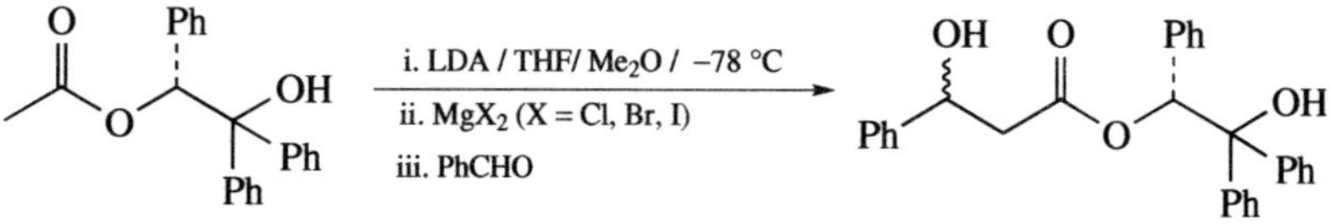

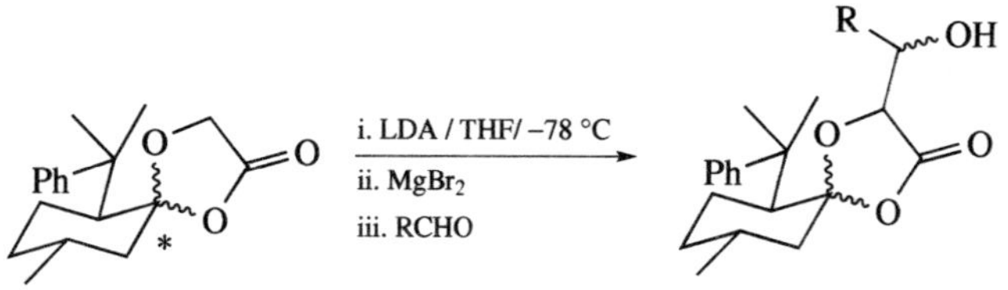

TABLE 2. Aldol stereochemistry of chiral acetates

X	dr
Cl	96/4
Br	97/3
I	98/2

TABLE 3. Yields and aldol stereochemistry of chiral dioxolones

Dioxolone	R	Yield (%)	dr
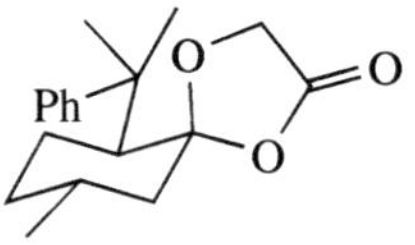 (Ph, cyclohexyl dioxolone, ester)	Ph	92	83/17
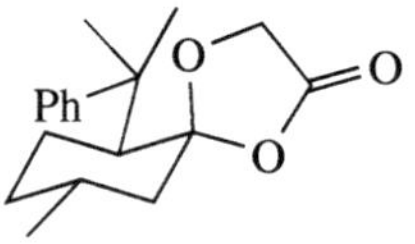 (Ph, cyclohexyl dioxolone, ester)	n-Pr	86	90/10
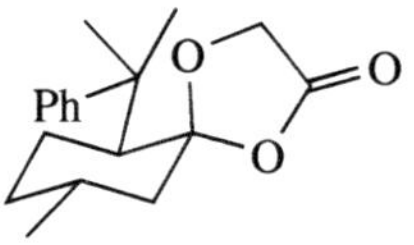 (Ph, cyclohexyl dioxolone, ester)	i-Pr	86	73/27
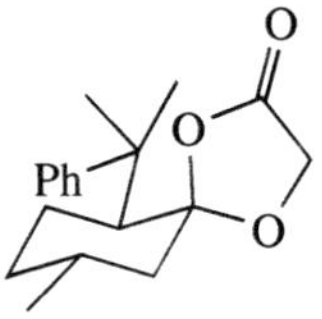 (Ph, cyclohexyl dioxolone, ketone)	Ph	89	75/25
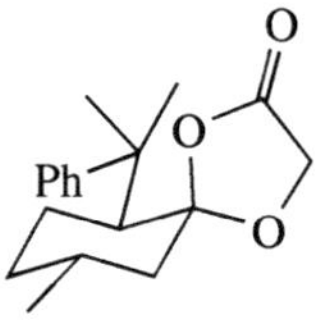 (Ph, cyclohexyl dioxolone, ketone)	n-Pr	92	74/26
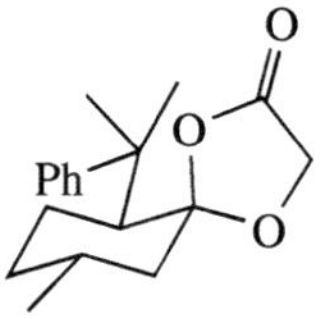 (Ph, cyclohexyl dioxolone, ketone)	i-Pr	89	89/11

Once the substrate is reduced by a LiDBB (4,4'-di-*tert*-butyl-biphenyllithium) solution, the transmetallation of the produced lithium enolate is performed with 1 equiv. of the complex magnesium bromide-diethyl ether (MgBr$_2$, Et$_2$O) leading to the chelated (Z)-magnesium enolate (equation 25).

Enantiomerically pure sulfoxides have been investigated as precursors of α-sulfinyl carbanions for asymmetric synthesis. The results are rationalized in terms of chelated intermediates. In the case of the addition of β-sulfinyl ester enolates on benzaldimines, interesting and surprising results were obtained with magnesium enolates[25]. The preparation of these magnesium enolates is based on the transmetallation of the corresponding lithium species (equation 26). The reaction is carried out by a successive deprotonation of the β-sulfinyl ester with LDA, and a subsequent addition of MgX$_2$ (MgBr$_2$·OEt$_2$ or MgI$_2$) to the lithium enolate. The imine is added to the resulting magnesium species at −78 °C.

Bromo- and iodomagnesium enolates lead to different diastereoselections (Table 4). The discrepancy results from the difference of the Lewis acidity between magnesium bromide and magnesium iodide, and the bulkiness of the halogen atoms. The possible transition states of this asymmetric addition are presented in Section III. Most probably the reaction proceeds through a chelated model with bromomagnesium enolates, and through a nonchelated model with two equivalents of magnesiumiodide[25].

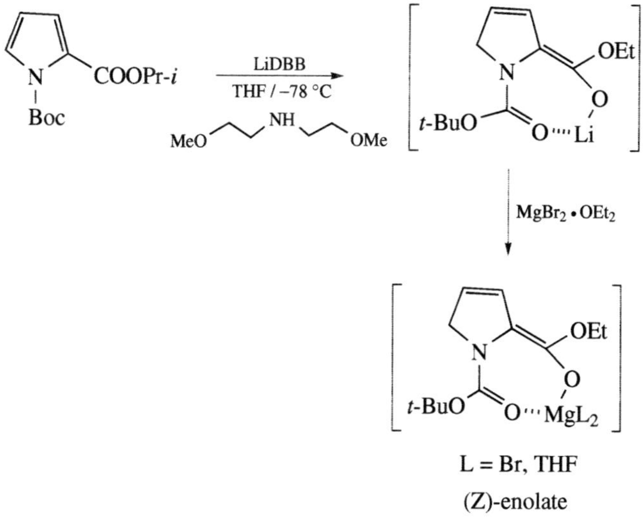

(25)

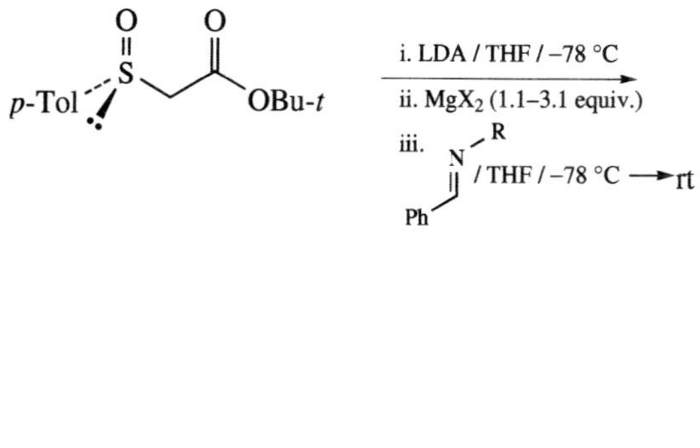

(26)

Transfer of chirality in aldol reactions has been attempted using β-allenyl ester enolates. These ambident nucleophiles have an axis of chirality, and such compounds have been less utilized in stereoselective reactions. They are prepared by transmetallation of the

TABLE 4. Yields and selectivities of asymmetric addition of α-sulfinyl ester enolate with benzaldimines

X	R	Yield (%)	dr
Br	Ts, COOMe	99	91/9
I	Ts, COOMe	81	30/70

lithium enolate with magnesium bromide etherate and are used in stereoselective aldol reaction (equation 27)[26].

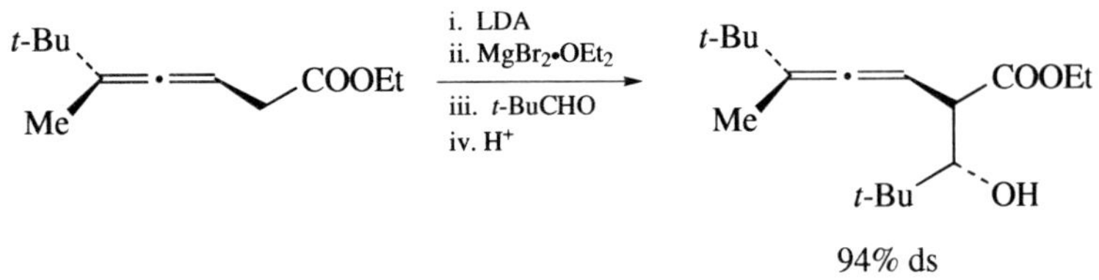

$$\text{t-Bu}\diagdown\diagup\text{COOEt} \quad \xrightarrow[\text{iii. t-BuCHO}]{\substack{\text{i. LDA}\\\text{ii. MgBr}_2\cdot\text{OEt}_2}} \quad \text{t-Bu}\diagdown\diagup\text{COOEt} \atop \text{t-Bu}\text{OH}$$

(27)

94% ds

Other magnesium allenyl enolates, such as **22**, obtained by transmetallation of the lithium species have been used successfully in the preparation of α,β-unsaturated acyl silanes (equation 28)[27].

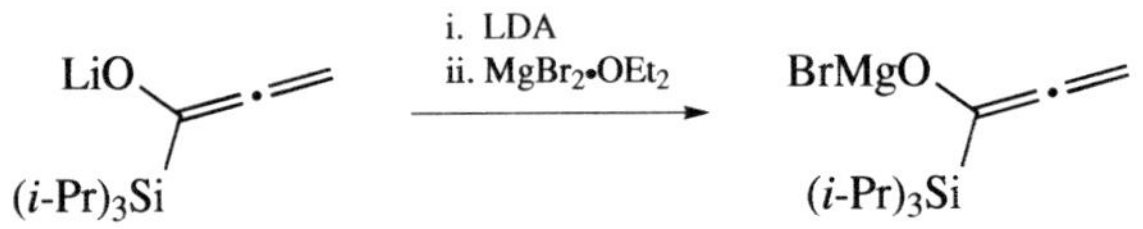

$$\text{LiO}\diagdown\diagup \atop (i\text{-Pr})_3\text{Si} \quad \xrightarrow[\text{ii. MgBr}_2\cdot\text{OEt}_2]{\text{i. LDA}} \quad \text{BrMgO}\diagdown\diagup \atop (i\text{-Pr})_3\text{Si}$$

(28)

(**22**)

It should be noted that the sequence deprotonation/reverse Brook rearrangement between the triisopropylsilyloxy allene and isopropylmagnesium chloride in THF does not provide the magnesium enolate.

Permutational silicium/magnesium salts interconversion can occur with silylenol ethers. Indeed, several metal salts such as $TiCl_4$ and Bu_2BOTf are known to promote the corresponding trialkylsilyl/metal salt exchange with silylenol ether, and consequently this reaction is called a transmetallation reaction. This type of reaction with magnesium salts has been reported recently by Mukaiyama (equation 29)[28]. The reaction is carried out with $MgBr_2\cdot OEt_2$ in toluene at $-19\,°C$. It has to be noted that the magnesium enolate is not formed if the silylenol ether is treated with MgI_2, $MgCl_2$, $Mg(OTf)_2$ or $Mg(ClO_4)_2$.

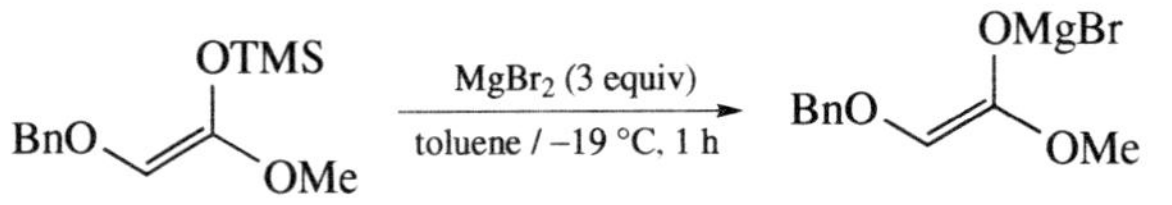

$$\text{BnO}\diagup\!\!\!\diagdown_{\text{OMe}}^{\text{OTMS}} \quad \xrightarrow[\text{toluene}/-19\,°\text{C, 1 h}]{\text{MgBr}_2\,(3\text{ equiv})} \quad \text{BnO}\diagup\!\!\!\diagdown_{\text{OMe}}^{\text{OMgBr}}$$

(29)

An additional silicium/magnesium exchange reaction of silylenol ether with alkyl Grignard reagents was reported. Such reaction is generally employed for the regiospecific generation of magnesium enolates from the corresponding silyl enolate (equation 30)[4a].

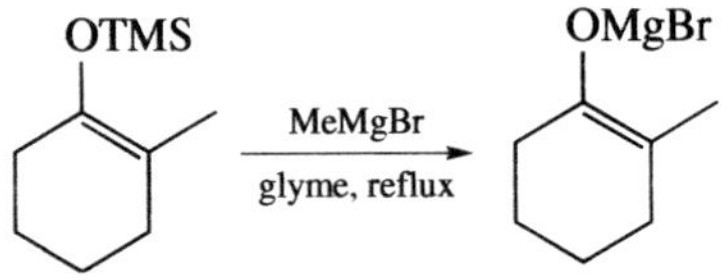

$$\text{OTMS} \quad \xrightarrow[\text{glyme, reflux}]{\text{MeMgBr}} \quad \text{OMgBr}$$

(30)

D. Conjugate Addition

The conjugate addition of organometallic reagents to an electron-deficient carbon–carbon double bond is one of the most widely used synthetic methods to generate enolate. It is well known that Grignard reagents usually give a mixture of 1,2- and 1,4-addition products in which the proportions vary depending upon the steric constraints imposed and the reaction conditions. Thus, by careful choice of the reagent (tertiary organomagnesium reagent or magnesium-ate complexes) and the Michael acceptor (bulky substituents on the carbonyl group, electron-withdrawing group), it is possible to achieve either 1,2- or 1,4-addition[29]. As a result, the Michael addition of magnesium anions has been developed as a new source of magnesium enolates, and several different types are now available[30, 31].

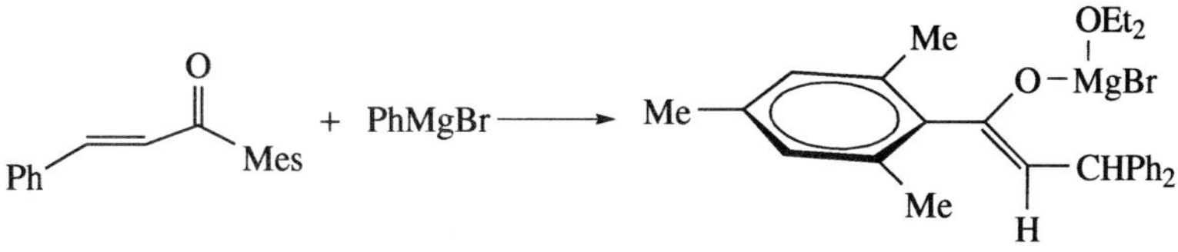

$$\text{Ph}\diagdown\!\!\!\diagup\!\!\!\diagdown\!\!\!\text{C(O)Mes} + \text{PhMgBr} \longrightarrow \text{(Z)-magnesium enolate} \quad (31)$$

One of the simplest methods is the addition of Grignard reagent onto a sterically hindered enone[32]. For example, the (Z)-magnesium enolate is formed from the reaction of benzalacetomesitylene and phenylmagnesium bromide (equation 31).

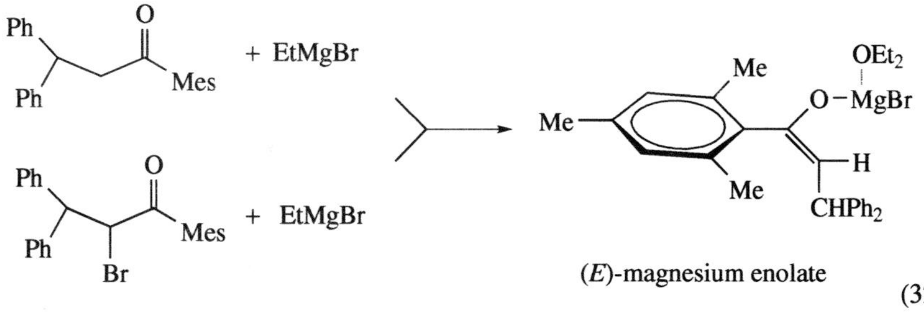

$$\text{Ph}_2\text{CH-C(O)Mes} + \text{EtMgBr}$$
$$\text{Ph}_2\text{CBr-C(O)Mes} + \text{EtMgBr} \longrightarrow (E)\text{-magnesium enolate} \quad (32)$$

It should be noted that the reaction of EtMgBr either on the 2,2-diphenyl 2,4,6-trimethylphenyl ketone or on 2,2-diphenyl-1-bromoethyl 2,4,6-trimethylphenylketone gave the (E)-magnesium enolate (equation 32). Cryoscopic studies in naphthalene show that (Z)- and (E)-magnesium enolates are both monomeric; they contain tricoordinate magnesium analogous to the Grignard compound derived from isopropyl mesityl ketone. IR spectra confirm the enolic nature of the magnesium adducts[32].

The mechanism of these reactions has been extensively studied. A recent book summarizes this literature[30]. A review refers to many other examples involving conjugate addition of organomagnesium reagents to α,β-unsaturated ketones. It includes the stereochemistry of the reaction and copper-induced conjugate additions[33].

The Michael acceptor can be an α,β-unsaturated carboxylic ester. This possibility is illustrated with the conjugate addition of phenylmagnesium bromide to ethyl arecaidinate, which leads to 4-phenylnipecotic acid[34] (equation 33).

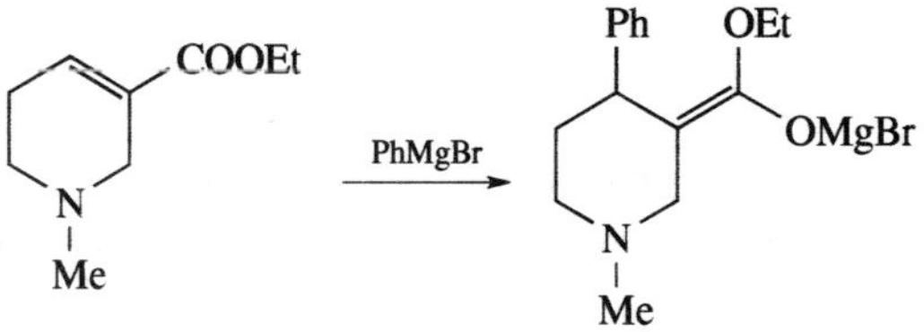

$$\text{(N-Me tetrahydropyridine-COOEt)} \xrightarrow{\text{PhMgBr}} \text{(N-Me piperidine with Ph and C(OEt)=OMgBr)} \quad (33)$$

Methyl 2-(trimethylsilyl)propenoate serves as an excellent Michael acceptor in the reactions with Grignard reagents (equation 34). The resulting adduct anions can be applied to the subsequent condensation with a variety of carbonyl compounds. Such Michael addition is sensitive to the reaction conditions and to the nature of the organomagnesium species. With a stoichiometric amount of PhMgBr, the Michael adduct is the major product, whereas the copper-catalyzed addition of MeMgI and CH_2=CHMgBr leads to a double reaction resulting from the reaction of magnesium enolate with the Michael acceptor[35].

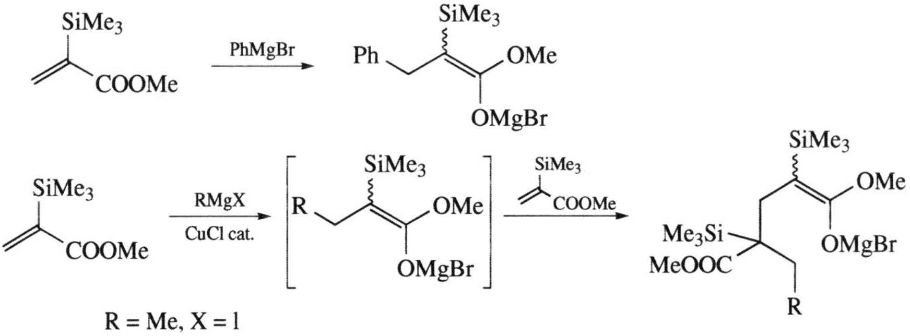

$$R = Me, X = I$$
$$R = CH_2=CH, X = Br$$

(34)

Interestingly, even methyl 2-bromo- or 2-chloro-2-cyclopropylideneacetate reacts cleanly with various Grignard reagents in a 1,4-addition without any chlorine–metal exchange side reactions (equation 35). These *in situ* generated magnesium enolates are particularly reactive with aromatic aldehydes[36].

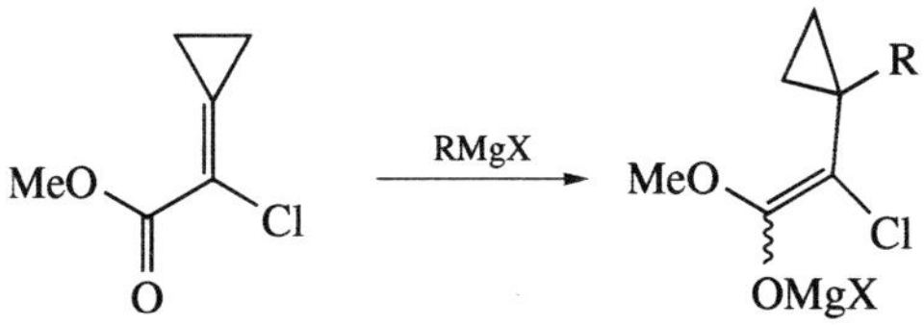

(35)

$$R = i\text{-Pr}, c\text{-Hex}, Ar, CH_2=CH, PhCH_2, -C\equiv C$$

N,N-Dialkyl-α-methacrylthioamides are other examples of Michael acceptors that generate magnesium enolates and as such they undergo conjugate addition reactions with Grignard reagents[37]. It is believed that the conjugate addition provides (Z)-enolates, via a cyclic, six-centered transition state (equation 36)[4b].

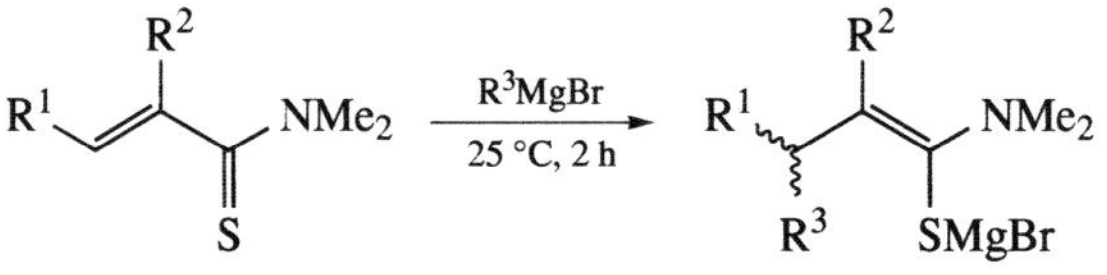

$$R^1 = H, Me, MeCH=CH, Ph$$
$$R^2 = H, Me$$
$$R^3 = Me, Et, i\text{-Pr}, n\text{-Bu}, Ph$$

(36)

Surprisingly, hard magnesium nucleophiles can serve as Michael donors. For example, magnesium amide, derived from deprotonation of **23** with magnesium diisopropylamide, adds to an α,β-unsaturated ester to give the corresponding magnesium enolate **24** via a conjugate addition (equation 37)[38]. Subsequent internal cyclic addition of the magnesium enolate onto the next nitrile yields 4-amino-1,2-dihydro-3-quinolinecarboxylic acid derivatives **25**.

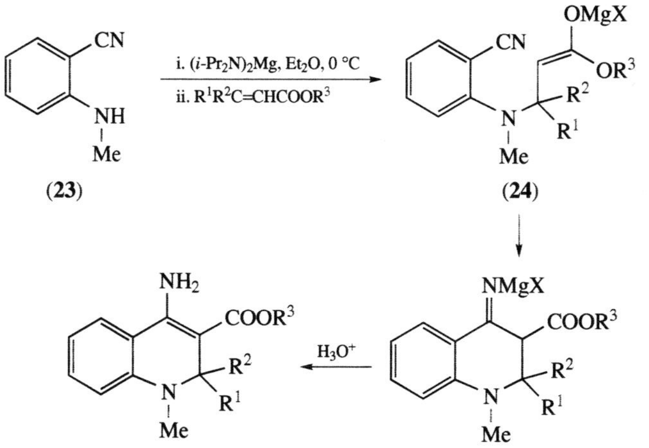

It can be assumed that the bivalent magnesium ion, which probably stabilizes the cyclic anionic intermediate, promotes further coupling reactions.

The same type of reaction has been applied to the preparation of 1,2-dihydroquinoline-3- and 2H-1-benzothiopyran-3-carboxylic acid derivatives (equation 38) via a magnesium amide-induced sequential conjugate addition–aldol condensation reaction between 2-(alkylamino)phenylketones or 2-mercaptobenzophenones[39].

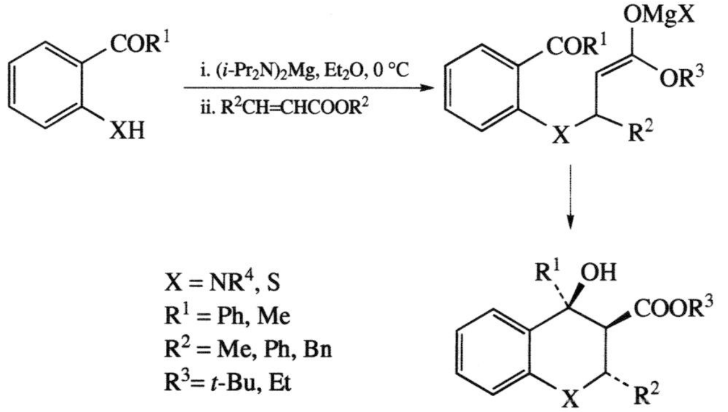

2,3-Dihydroquinoline-3-carbonitriles can also be prepared by the tandem reaction between 2-(methylamino)-benzophenone and α,β-unsaturated carbonitriles in the presence of magnesium amide under the same conditions as for the reaction with α,β-unsaturated carboxylates (equation 39).

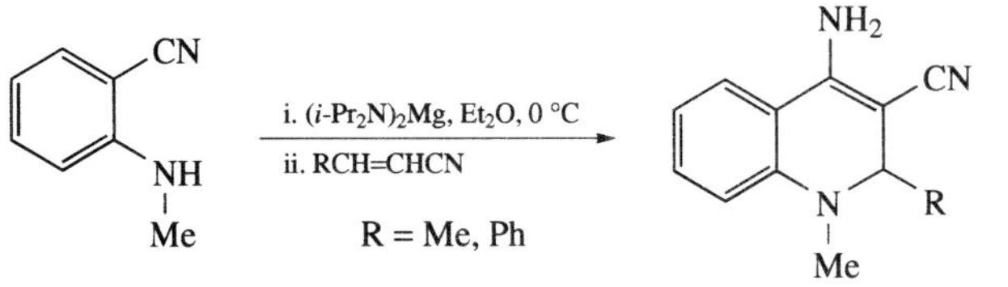

[Equation 39 scheme]

(39)

The propensity of magnesium amide to undergo Michael additions with α,β-unsaturated esters has been developed into a general protocol using homochiral amide to give magnesium β-aminoenolate intermediates[40] (equation 40).

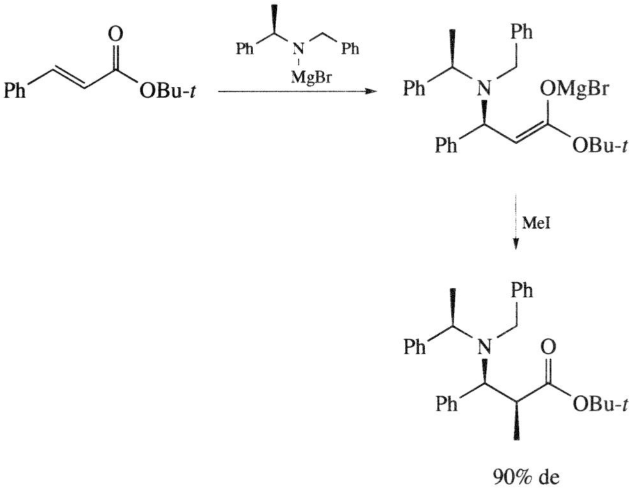

[Equation 40 scheme]

90% de

(40)

Chiral organomagnesium amides form an efficient method to realize enantioselective conjugate addition. Sibi and Asano have reported the Michael addition of σ-bound magnesium reagents derived from bisoxazolines to enamidomalonates (equation 41). The enantioselectivity of the addition is discussed in Section III. This method allows the preparation of chiral β-amino acid derivatives[41].

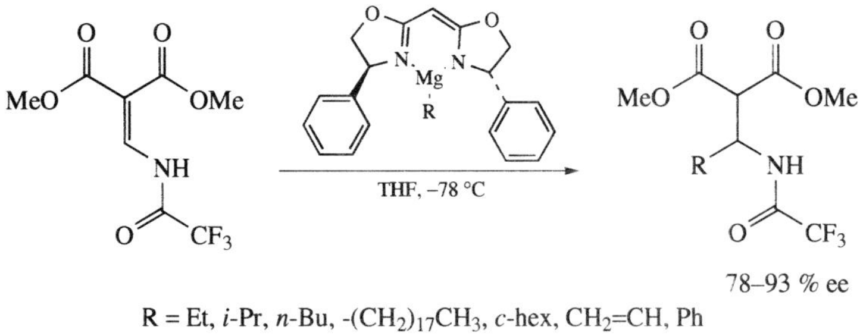

[Equation 41 scheme]

THF, −78 °C

78–93 % ee

(41)

R = Et, i-Pr, n-Bu, -(CH$_2$)$_{17}$CH$_3$, c-hex, CH$_2$=CH, Ph

The introduction of asymmetry in conjugate additions can be promoted by chiral acceptors. The preparation of regio-defined magnesium enolates from copper(I)-mediated conjugate addition of Grignard reagents has been extensively used in many important

11. Preparation and reactivity of magnesium enolates

syntheses. The understanding of these reactions is important to improve the reaction conditions[42].

A convenient and interesting way to control the absolute stereochemistry is to use the Evans-type 4-phenyloxazolidinone auxiliary. Studies of the mechanism for such conjugate addition have been made by Hruby and coworkers (equation 42)[43].

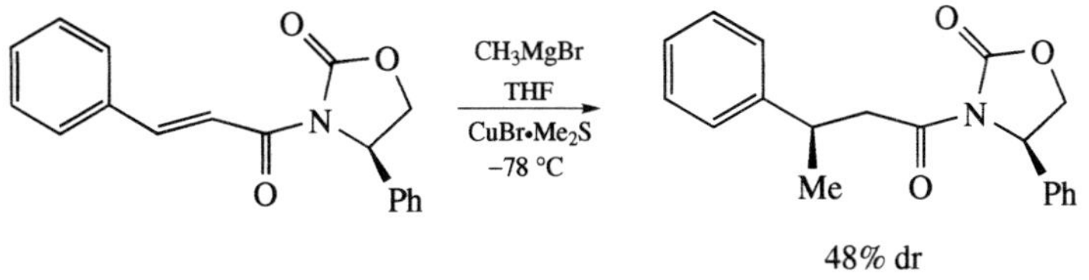

(42)

48% dr

Three chiral intermediates are observed directly by ^{1}H and ^{13}C NMR spectroscopy: one olefin–copper(I) complex **26** and two magnesium enolates **27** and **28** (equation 43).

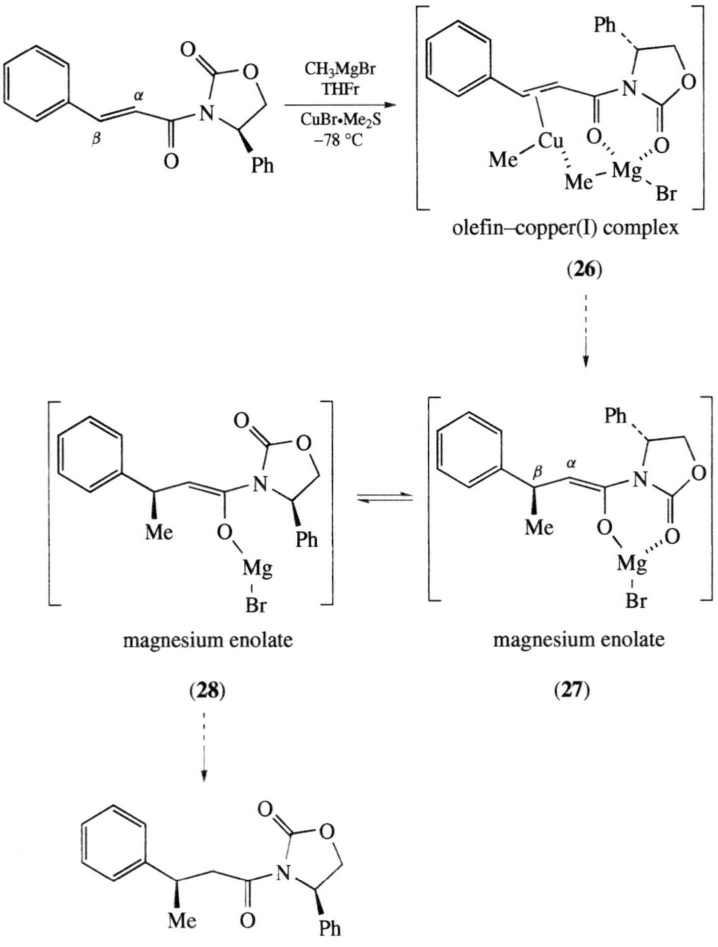

olefin–copper(I) complex

(**26**)

(43)

magnesium enolate

(**28**)

magnesium enolate

(**27**)

456 Claude Grison

The two methyl groups in the olefin–copper(I) complex **26** are crucial for asymmetric induction. The 150° dihedral angle between the α- and β-protons of the magnesium enolate **27** provides valuable information to determine the stereochemical effects on the α center. The two magnesium enolates **27** and **28** are reversibly temperature-dependent. Enolate **27** is the major component at 253 K, while enolate **28** becomes the major component at 293 K. Therefore, temperature lower than *ca* 256 K is required to obtain high stereoselectivity.

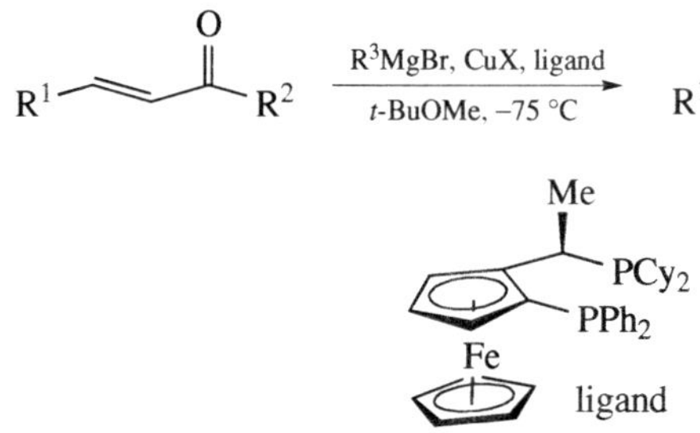

(44)

R^1 = Me, Et, *n*-Pr, *n*-Bu, *n*-Pent, BnO(CH$_2$)$_3$, Ph

R^2 = SMe, SEt

R^3 = Me, Et, *n*-Pr, *i*-Pr, *n*-Bu, *i*-Bu

X = Br·Me$_2$S

R^1 = Me, Et, *n*-Pr, *n*-Bu, *n*-Pent

R^2 = Me, *n*-Bu, *t*-Bu

R^3 = Me, Et, *n*-Pr, *i*-Pr, *n*-Bu, *i*-Bu, Cl-Bu, *i*-Pr(CH$_2$)$_2$, Ph

X = Cl, Br·Me$_2$S, I

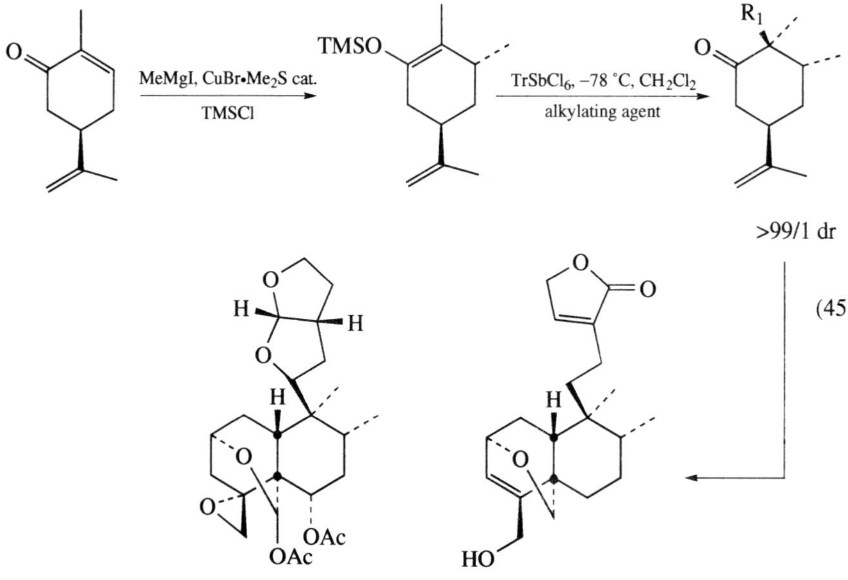

(45)

cledoranes

11. Preparation and reactivity of magnesium enolates

Numerous other protocols have been developed to prepare magnesium enolates by asymmetric 1,4-addition of Grignard reagents to electron-deficient alkenes. Recently, an enantioselective metal-catalyzed version of this key reaction has been studied with enones and α,β-unsaturated thioesters[44–46]. Using chiral ferrocenyl-based diphosphines leads to interesting results (equation 44).

The copper-catalyzed conjugate addition of methylmagnesium iodide to cyclohexenone and trapping of the resulting enolate as its trimethylsilyl enolate, followed by $TrSbCl_6$-catalyzed Mukaiyama reaction, are the first steps of an elegant synthesis of enantiomerically pure clerodanes[47] (equation 45).

Tandem reactions attract significant research interest as they can lead to new methodologies to generate metal enolates. For example, the reaction of ethyl 2-methyl-2-(diphenylmethylsilyl)propionate with vinylmagnesium bromide (or 2-methyl-1-propenylmagnesium bromide) results in the addition of two equivalents of the Grignard reagent. The first addition gives the enone intermediate while the second undergoes a Michael reaction. This sequence allows the synthesis of β-ketosilanes and tetrasubstituted ketones[48] (equation 46).

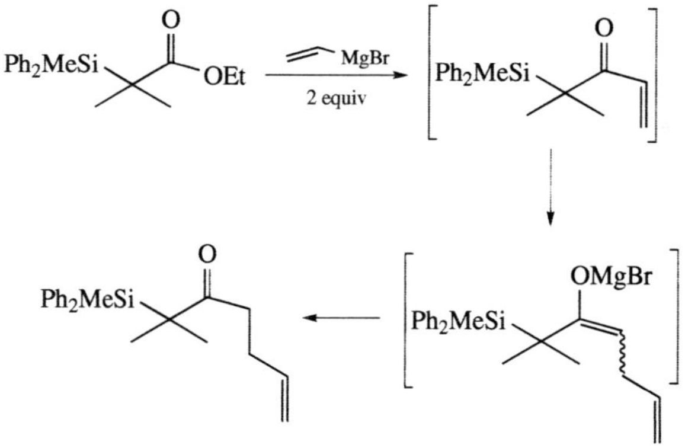

(46)

E. Permutational Hydrogen/Metal Interconversions (Metallations)

As in the preceding section, it should be emphasized that the exact constitution of magnesium enolates is still undetermined, since they are generally produced and used *in situ*. Consequently, the general equations may be oversimplified, since they do not take into account solvatation, association or interactions between different species.

The preparation of magnesium enolates by metallation competes with nucleophilic addition. Thus, until recently, this strategy was only valuable for sterically hindered or relatively acidic substrates, which were metallated by Grignard reagents or magnesium dialkoxydes.

However, it has now been reported that sterically hindered magnesium amides analogous to LDA or LTMP are effective and selective metallating reagents. By comparison with the lithium reagents, they have distinctive stability (even in boiling THF), reactivity (they are compatible with a number of functional groups) and selectivity (they are useful in regio- and stereoselective formation of magnesium enolates) properties. Finally, they are also good candidates for enantioselective deprotonation.

In view of the importance currently attached to the usefulness and synthesis of magnesium enolates, these different aspects are discussed in this section.

Alkylmagnesium halides and dialkylmagnesium compounds are efficient metallating agents toward alkyl mesityl ketones such as Kohler's ketone (2,2-diphenylethyl 2,4,6-trimethylphenyl ketone) and hindered carboxylic esters such as *t*-butyl acetate.

A study of the enolisation mechanism with Kohler's ketone has shown the formation of an (E)-magnesium enolate[49] (equation 47).

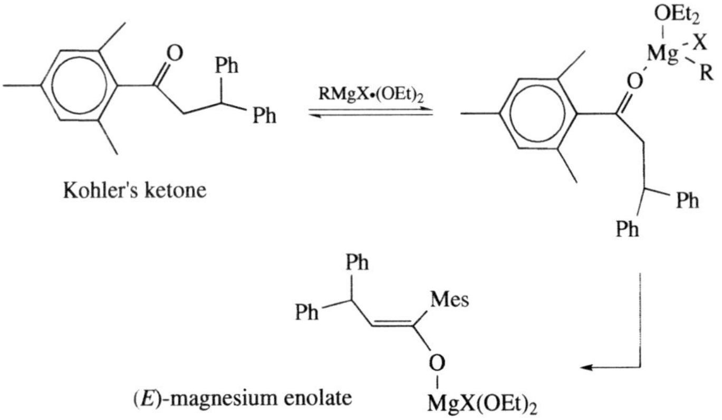

$$\text{Kohler's ketone} \xrightarrow{\text{RMgX·(OEt)}_2} \text{[intermediate]} \longrightarrow (E)\text{-magnesium enolate} + \text{MgX(OEt)}_2 \quad (47)$$

Bertrand and coworkers have studied the complexity of magnesium enolates in solution. The NMR and IR of the products generated from alkyl- and dialkylmagnesium reagents with enolizable ketones indicate that enolates as well as solvent-separated carbonyl conjugated carbanion ion pairs are formed. The analysis of the reaction mixture is complicated by the occurrence of ketone reduction[50].

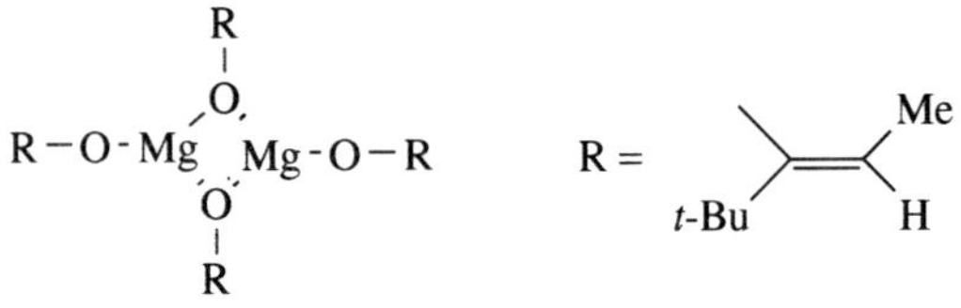

(29)

Fellmann and Dubois[51] have described the structure of the enolate **29** derived from the reaction of *t*-butyl acetate with $(MeO)_2Mg$. The ^{13}C NMR spectrum reveals two *O*-metallated species, which should be symmetric enolates as proposed by Pinkus and Wu for the bromomagnesium enolate of methylmesityl ketone (metal is tricoordinated)[32].

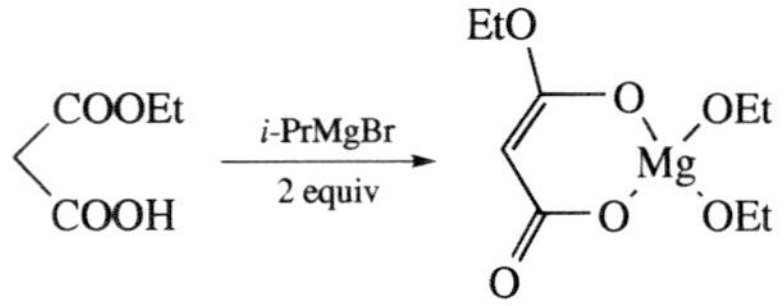

$$\begin{array}{c}\text{COOEt}\\\text{COOH}\end{array} \xrightarrow[\text{2 equiv}]{i\text{-PrMgBr}} \text{[cyclic Mg chelate]} \quad (48)$$

Magnesium enolates derived from β-dicarbonyl compounds can be easily obtained by metallation with *i*-PrMgBr. A stable cyclic chelate is obtained[52]. As example, the magnesium enolate of mixed malonate is shown in equation 48.

$$\text{ArCH}_2\text{COOH} + 2\text{RMgX} \longrightarrow \text{ArCH=C(OMgX)}_2 \quad (49)$$

The bis-deprotonation of arylacetic acids by Grignard reagents is known[53] and the resulting bis(bromomagnesium) salts (equation 49) have been used for preparing β-hydroxy acids (Ivanov reaction).

Extensive studies have been performed on the metallation of thioamides, sulfones and sulfoxides.

Thioamides of secondary amines are deprotonated with isopropylmagnesium to give (Z)-enolates. Thioamides of primary amines react with two equivalents of i-PrMgBr to afford dianions that have been shown to have the (Z)-configuration. These magnesium species are versatile intermediates in stereoselective aldol reaction (equation 50, Table 5; see Section III).

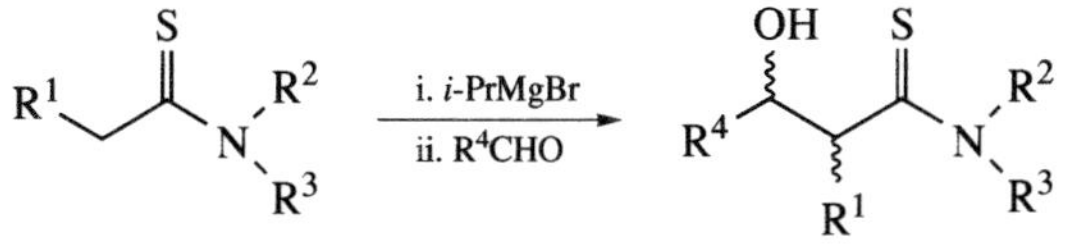

$$R^1 = Me, Ph, SPh, i\text{-}Pr$$
$$R^2 = H, Me$$
$$R^3 = Me, Ph, CH_2CH_2OMe$$
$$R^4 = Me, i\text{-}Pr, Ph, Et, CH_2=CMe$$

(50)

α-Sulfinyl magnesium carbanion enolates have been also investigated as reagents for asymmetric aldol reactions. The first example was introduced by Solladié[54]. The magnesium enolate is prepared by reaction of the sulfinyl ester with t-butylmagnesium bromide. It reacts with aldehydes and ketones with high levels of asymmetric induction. The results are rationalized in terms of rigid chelated intermediates (see Section III). This method has been extended to acetamides. It is illustrated by the preparation of the magnesium enolate derived from N,N-diethyl 2-acyl-1,3-dithiolane-S-oxide (equation 51). This metal anion is an interesting chiral synthon, which has found applications in asymmetric aldol-type addition[55].

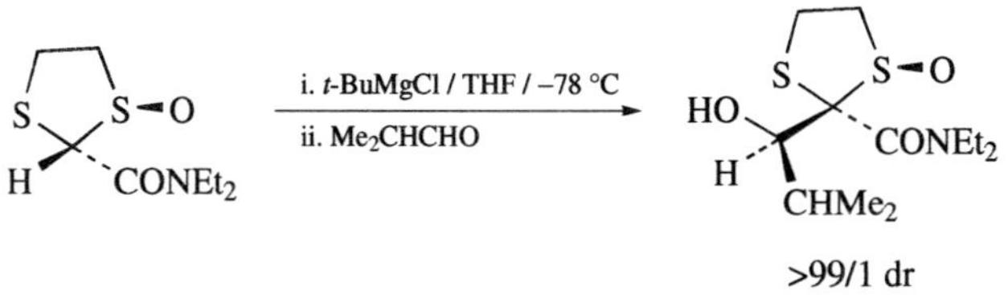

(51)

>99/1 dr

TABLE 5. Stereochemistry of aldol reaction of magnesium enolates derived from thioamides with aldehydes

R^1	R^2	R^3	R^4	dr
Me	Me	Me	i-Pr	95/5
Me	Me	Me	Ph	93/7
Me	Me	Me	$CH_2=CH$	89/11
Ph	Me	Me	i-Pr	72/28
Ph	Me	Me	$CH_2=CH$	73/27
PhS	Me	Me	i-Pr	66/34
i-Pr	Me	Me	Ph	34/66
Ph	Me	H	i-Pr	98/2
Ph	Me	H	Ph	94/6
Ph	CH_2CH_2OMe	H	Ph	97/3

To find a synergic mixed-metal reagent, Hevia and coworkers have investigated the reactivity of variant monosodium–monomagnesium trialkyl NaMgBu$_3$ and disodium–monomagnesium Na$_2$MgBu$_4$ complexes toward the sterically demanding 2,4,6-trimethylacetophenone[56].

Reactions occur smoothly at room temperature, affording the enolate products **30–32** as isolated crystalline solids (in 35–78% yields). In these reactions, 2,4,6-trimethylacetophenone is selectively deprotonated at the methyl position and no nucleophilic side-reactions take place.

The solid-state structures of the homoanionic magnesium enolates **30** and **31** and heteroanionic enolate **33** have been determined by X-ray crystallography, establishing as a common motif a polymetallic chain of four members for **30** and **32** or three members for **31**, with the anionic ligands in each case bridging the different metals together (equations 52–55).

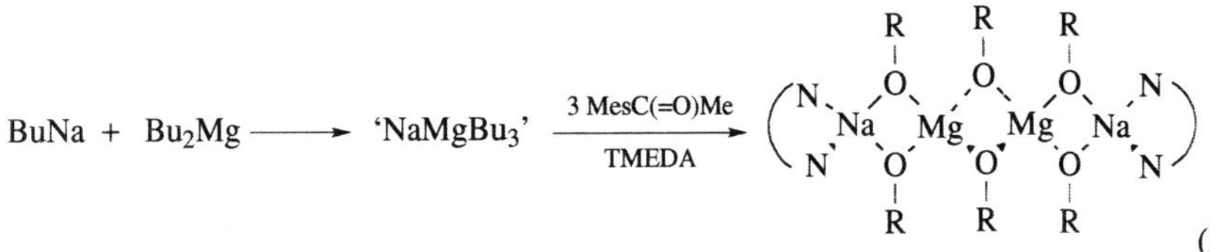

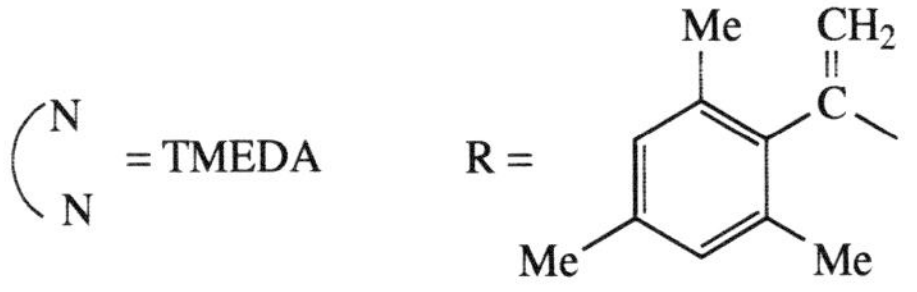

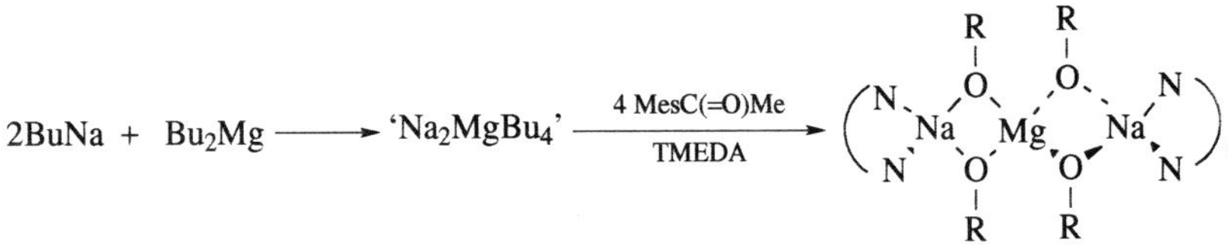

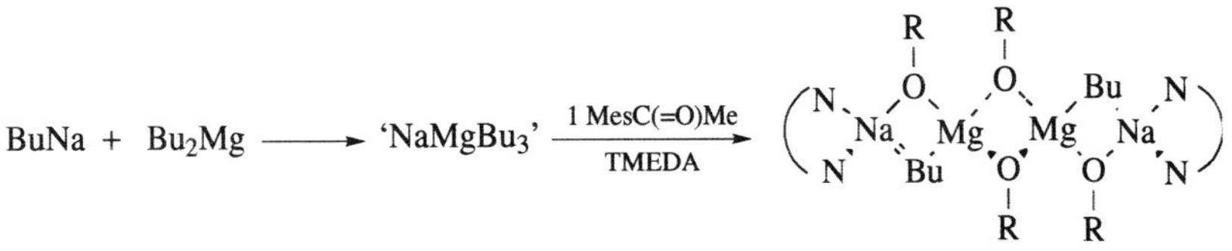

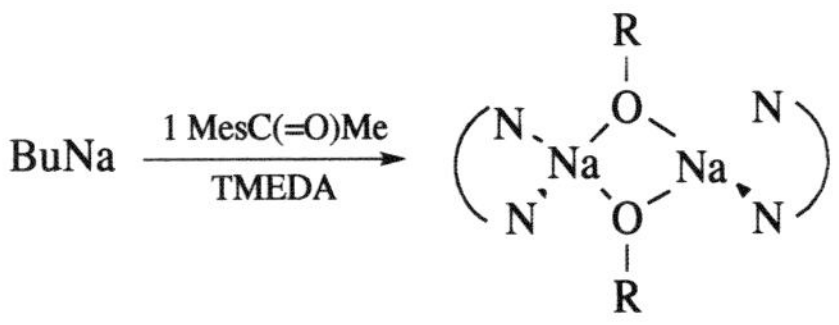

A problem inherent in metallation reactions with Grignard reagents is the poor chemoselectivity of the reactions. The most common side-reactions are the competing nucleophile addition and the reduction of the carbonyl compounds. An interesting alternative would be to use the high electrophilicity of the Mg^{2+} cation and its tendency to form a multicoordinate complex. The preformation of a Mg(II) complex with a carbonyl compound or a carboxylic acid derivative enhances the acidity of the substrate to the point where a relatively mild base can be used.

Two simple basic systems are described: $Mg(OR)_2$ and MgX_2/Et_3N.

Magnesium alkoxide, simply prepared by dissolving magnesium in alcohol in the presence of a crystal of iodine, can be used to prepare magnesium enolates of bifunctional compounds.

For example, the chelated magnesium enolate of a β-diketone or a β-ketoester **34** (formed via **33**) can be easily prepared using $(MeO)_2Mg/MeOH$ as a base. It is stable in a refluxing solution of the reagent, in contrast with the sodium enolate analogue, which is unstable in these conditions.

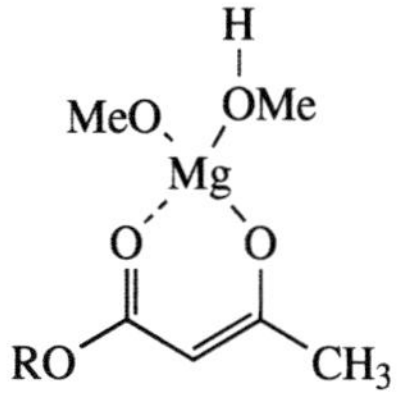

(34)

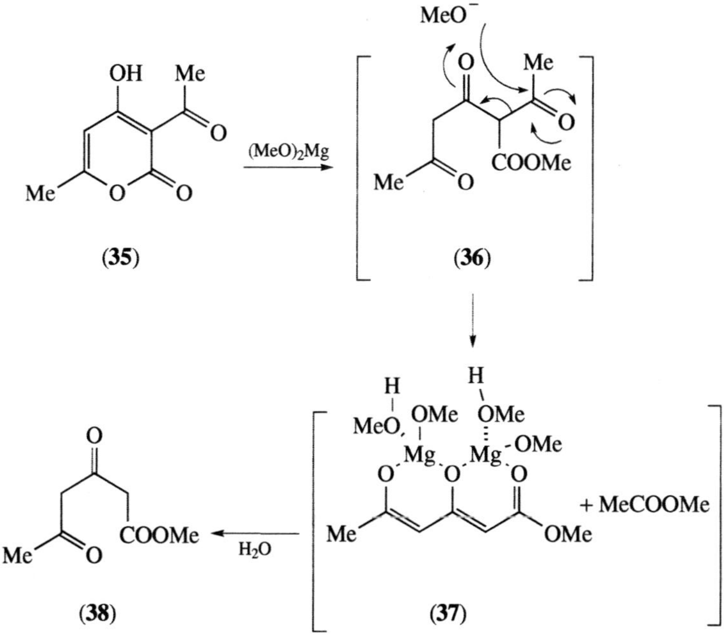

(35) → (MeO)₂Mg → (36) → (37) → (38) + MeCOOMe

(56)

Thus, the methanolysis of the lactone **35** by MeONa or (MeO)$_2$Mg produces initially the triketoester **36**, which is subsequently deacylated *in situ* to give after hydrolysis the diketone ester **38**. The yield of **38** is considerably better with (MeO)$_2$Mg, as the result of the formation in the reaction medium of a chelated bis-enolate **37** which protects **38** from degradation. Extensive degradation is observed with the sodium enolate analogue of **37** which is unstable in these conditions (equation 56)[57].

Rathke and Cowan have used the combination MgCl$_2$/R$_3$N for the metallation of β-dicarbonyl compounds[58]. In the presence of magnesium chloride and 2 equivalents of Et$_3$N, diethyl malonate is easily metallated. The C-acylation of the resulting anion with acid chloride gives excellent yields (equation 57). Other metal chlorides (ZnCl$_2$, CuCl$_2$, FeCl$_3$, LiCl, TiCl$_4$, AlCl$_3$) are ineffective. Similarly, ethyl acetoacetate is C-acylated by acid chlorides in the presence of magnesium chloride and 2 equivalents of pyridine. The reaction of diethyl malonate or ethyl acetoacetate with tertiary amine and MgCl$_2$ provides a remarkably simple entry into the enolate chemistry derived from dialkyl malonate, which will be described in Section III.

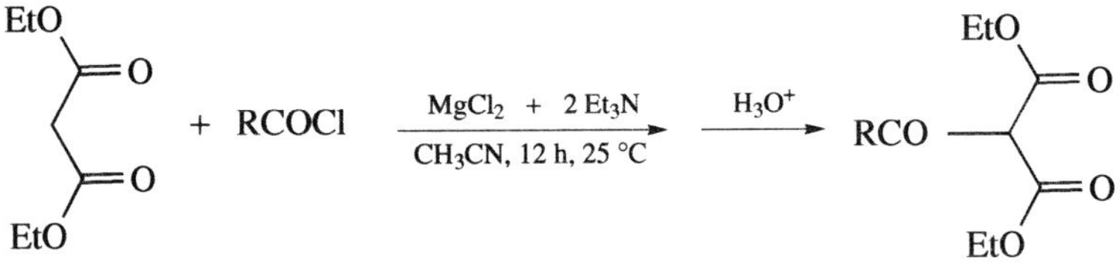

$$\text{EtO-CO-CH}_2\text{-CO-OEt} + \text{RCOCl} \xrightarrow[\text{CH}_3\text{CN, 12 h, 25 °C}]{\text{MgCl}_2 + 2\text{Et}_3\text{N}} \xrightarrow{\text{H}_3\text{O}^+} \text{EtO-CO-CH(COR)-CO-OEt} \quad (57)$$

R = Me, *n*-Pr, *i*-Pr, *t*-Bu, Ph
85–92% isolated yields

The same process has been extended to trialkylphosphonoacetates. Acylation with acid chlorides of the magnesium enolates derived from trimethyl and triethyl phosphonoacetates using a MgCl$_2$/Et$_3$N system provides 2-acyl dialkylphosphonoacetates. Further decarboxylation of these latter compounds affords β-ketophosphonates[59].

A similar procedure for the synthesis of α-acyl aminoesters has been proposed using a MgCl$_2$/R$_3$N base system to generate the magnesium enolates of a series of α-carboxy aminoesters. These reagents react smoothly at 0 °C with a variety of acid chlorides to give α-acyl aminoesters in good to excellent yields[60] (equation 58).

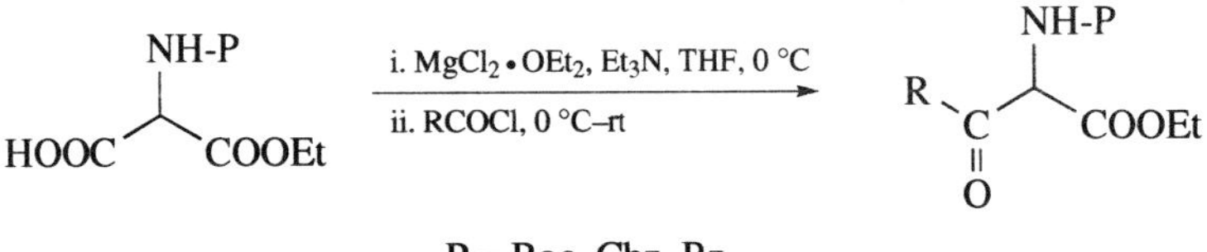

$$\text{HOOC-CH(NH-P)-COOEt} \xrightarrow[\text{ii. RCOCl, 0 °C–rt}]{\text{i. MgCl}_2 \cdot \text{OEt}_2, \text{Et}_3\text{N, THF, 0 °C}} \text{R-C(=O)-CH(NH-P)-COOEt} \quad (58)$$

P = Boc, Cbz, Bz
R = Ar, Me, *n*-Pr, *i*-Pr
77–92%

The MgX$_2$/R$_3$N systems offer another useful synthetic interest. Recently, Evans and coworkers have demonstrated that substoichiometric amounts of magnesium halides in the presence of an amine and chlorotrimethylsilane catalyze the direct aldol reaction of *N*-acyloxazolidinones and *N*-acylthiazolidininethiones with high diastereoselectivity[61,62] (equation 59).

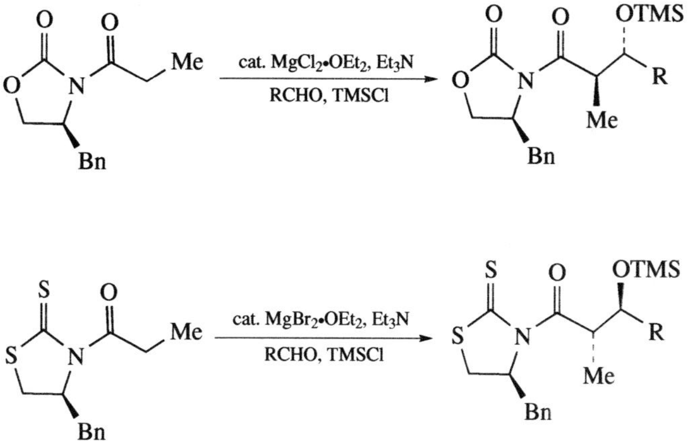

11. Preparation and reactivity of magnesium enolates

(59)

Interestingly, another Mg(II)-mediated aldol-type reaction has been investigated by using $MgBr_2-Et_3N$ with bislactim ethers and aliphatic aldehydes. The aldol products are converted in α-substituted serines[63] (equation 60, Table 6).

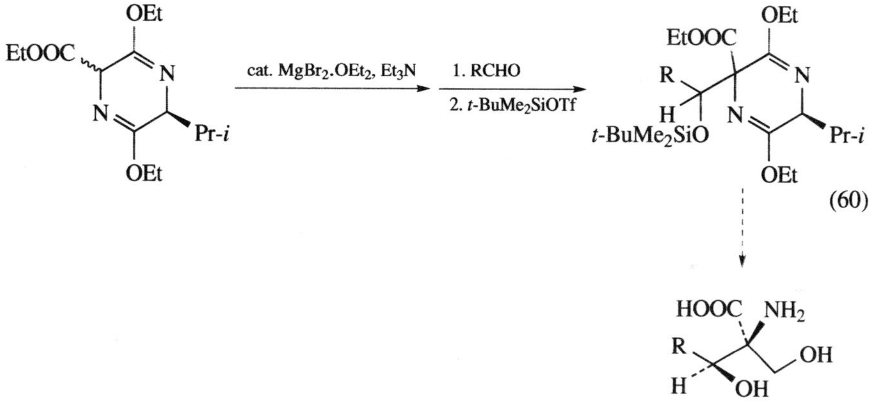

(60)

Stile has developed an alternative way to generate magnesium enolate under mild basic conditions[64]. Ketones react with magnesium methyl carbonate (MMC or Stile's reagent)

TABLE 6. Diastereoselective aldol-type reaction of bislactim ether with aldehydes

R	Yield (%)	dr
i-Pr	86	7/83/1/9
n-Bu	75	7/75/2/16
Ph	70	18/72/0/10
$Me_2C=CH$	74	16/64/3/17

to give stable chelated adducts, which are converted to β-keto carboxylic acids or α-substituted ketones after alkylation and subsequent decarboxylation (equation 61). The process is compatible with carboxylic acid derivatives (nitrile, ester and amide)[65].

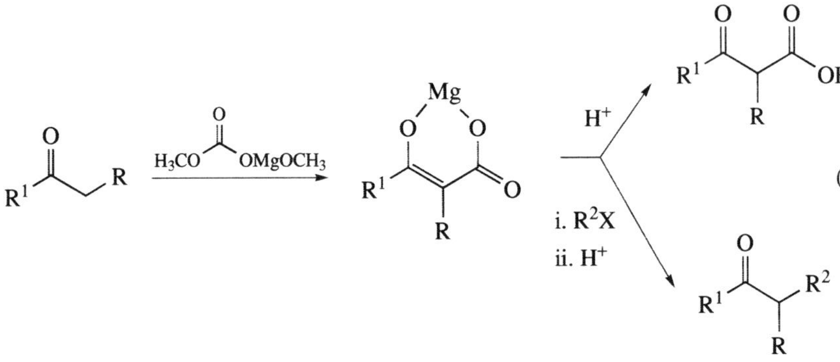

(61)

In recent years, a variety of hindered magnesium amides have been used to produce magnesium enolates. The versatility of these bases is now well recognized. Some typical examples are presented below.

The magnesium amides may be prepared either by reaction of lithium amide and magnesium bromide, by reaction of DIBAL-H with R_2Mg, or by reaction of the corresponding amine with a Grignard reagent[66-68].

For instance, BMDA **39**, the LDA-analogue, is generated by treatment of diisopropylamine with ethylmagnesium bromide at 0 °C in diethyl ether and the mixture is stirred for 1 h at the same temperature.

The magnesium amides and diamides are more thermally stable and less reactive than their lithium analogues, leading to different selectivities.

Examples for the preparation of magnesium enolates (or further subsequent reactions of the enolates) by magnesium amides are listed in Table 7.

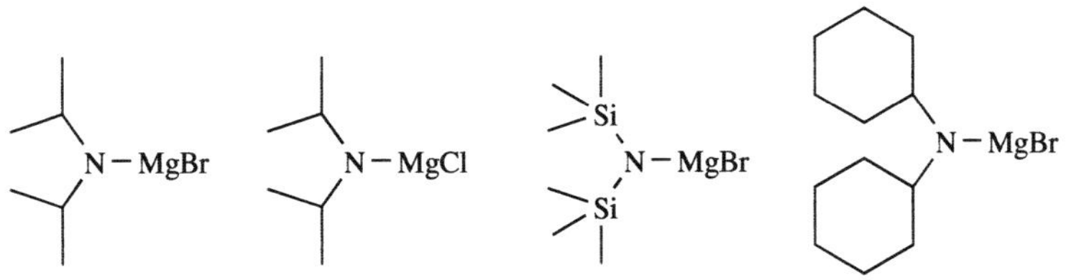

BMDA or DAMgBr (bromomagnesium diisopropylamide)	(chloromagnesium diisopropylamide)	BMHMDS (bromomagnesium hexamethyldisilyamide)	BMDCA (bromomagnesium dicyclohexylmide)
(39)	**(40)**	**(41)**	**(42)**

The magnesium amides of choice for the preparation of magnesium enolates via metallation are the Hauser bases, such as **39** and **40**, or (bis)amidomagnesium reagents, such as **46** and **47**. The reaction has been successfully applied to the preparation of enolates derived from cyclic, acyclic and α-siloxyketones, benzylic ketones, aldehydes, carboxylic esters and amides, even with the less hindered Hauser bases.

However, these reagents show significant differences in their reactivity and in their selectivity as compared to their lithium analogues. It is possible to control the regioselective formation of enolate. By a careful choice of base, (bis)amidomagnesium reagent such

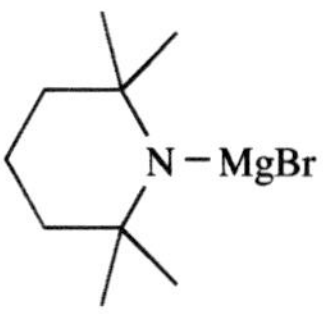

11. Preparation and reactivity of magnesium enolates

MTMP
(bromomagnesium
2,2,6,6-tetramethylpiperidide)
(43)

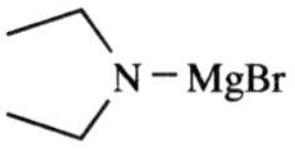

(bromomagnesium
diethylamide)
(44)

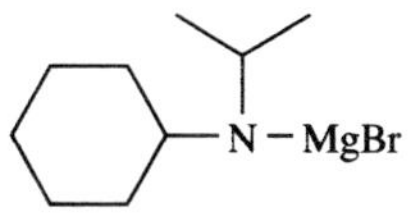

(bromomagnesium
isopropylcyclohexylamide)
(45)

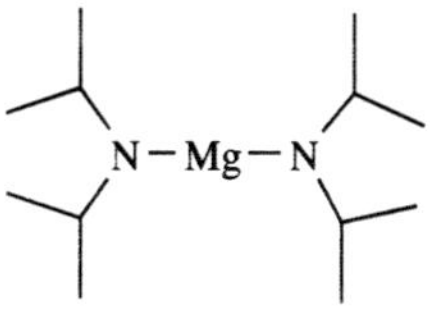

MBDA or (DA)$_2$Mg
[magnesium
bis(diisopropylamide)]
(46)

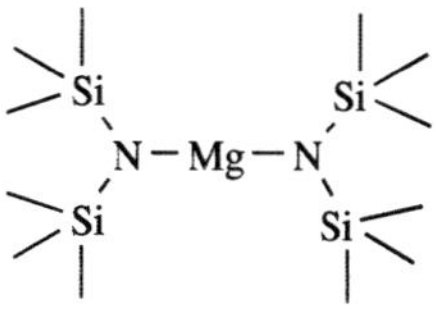

(magnesium salt of
hexamethyldisilylazane)
(47)

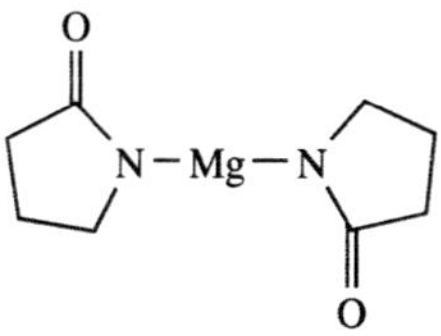

(2-pyrr)$_2$Mg
(2-pyrrolidinone
magnesium salt)
(48)

as **46** [(DA)$_2$Mg] in THF/heptane or the electrogenerated base **47** in DME/HMPA leads to kinetic enolates of cyclic and acyclic ketones via an *in situ* reaction with TMSCl. Yields are high and the regioselectivity is similar to the one obtained with LDA/DME at $-78\,°C$, although the reaction with **46** is performed at room temperature. It should be noted that the high *E*-enol stereoselectivity for benzylic ketones is opposite to the one obtained with LDA. This result can be rationalized considering steric interactions in Ireland's transition state model[74] (equation 62).

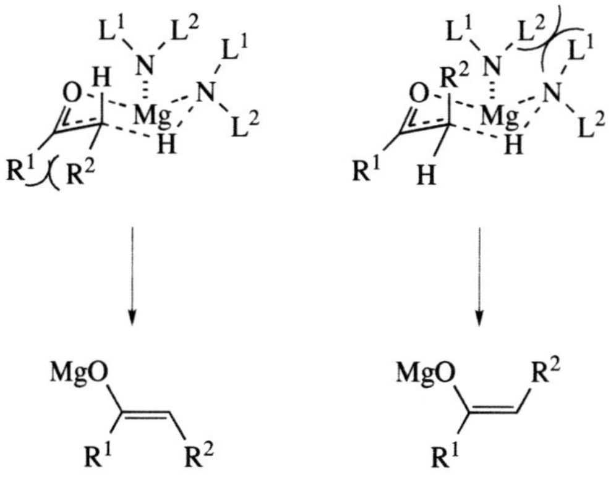

(62)

(*E*)-magnesium enolate   (*Z*)-magnesium enolate

The use of DAMgBr **39** in Et$_2$O/HMPA/TMSCl/Et$_3$N leads to the thermodynamic enolates or silyl enol ethers. This methodology is one of the best direct regiospecific preparations of thermodynamic silylenol ethers from unsymmetric cyclic ketones.

TABLE 7. Examples for the preparation of magnesium enolates (or reaction products after trapping) by magnesium amides

Substrate	Magnesium amide	Magnesium enolate or reaction product after trapping procedure of enolates	Reference
CH_3COOBu-t	46	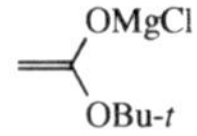 $CH_2=C(OMgCl)(OBu$-$t)$	66, 68
CH_3COOR* ($R* = (-)$-menthyl, $(+)$-bornyl)	44	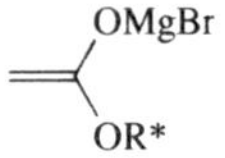 $CH_2=C(OMgBr)(OR*)$	69
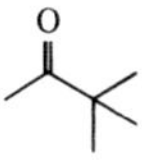 t-Bu−C(=O)−CH$_3$	39	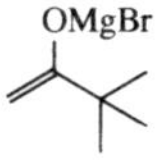 $CH_2=C(OMgBr)(t$-Bu$)$	70
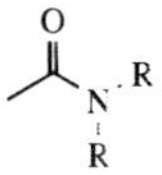 $CH_3-C(=O)-NR_2$ (R = Me, piperidin-1-yl, morpholin-4-yl, 4-methylpiperazin-1-yl)	46	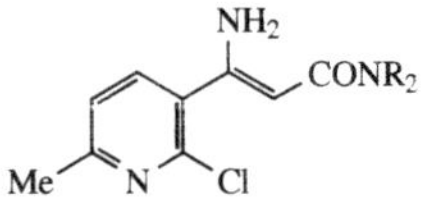 3-(1-amino-2-CONR$_2$-vinyl)-2-chloro-6-methylpyridine	71
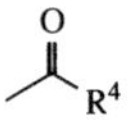 $CH_3-C(=O)-R^4$ (R^4 = OMe, OEt, OPr-n, OBu-n, OBu-t, NMe$_2$)	46	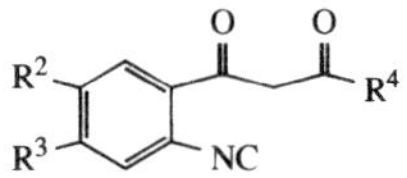 2-NC-4-R^3-5-R^2-C$_6$H$_2$−C(=O)−CH$_2$−C(=O)−R^4	68, 72
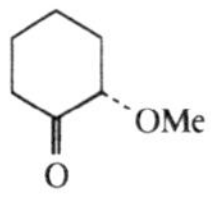 2-methoxycyclohexanone	40	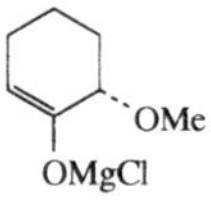 2-methoxy-1-(OMgCl)-cyclohexene	73
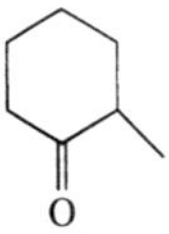 2-methylcyclohexanone	39/ TMSCl/Et$_3$N/Et$_2$O/HMPA	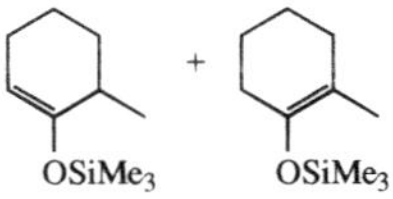 2-methyl-1-OSiMe$_3$-cyclohexene + 6-methyl-1-OSiMe$_3$-cyclohexene (3/97)	74
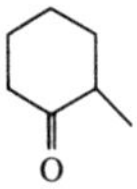 2-methylcyclohexanone	46/ TMSCl/THF/ heptane	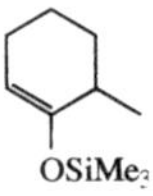 2-methyl-1-OSiMe$_3$-cyclohexene	74
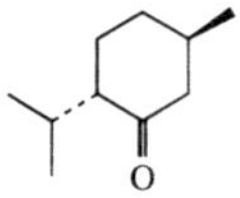 menthone	33/ TMSCl/Et$_3$N/Et$_2$O/ HMPA	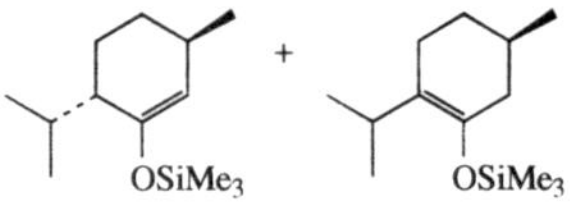 OSiMe$_3$ menthene isomers (3/97)	75

TABLE 7. (continued)

Substrate	Magnesium amide	Magnesium enolate or reaction product after trapping procedure of enolates	Reference
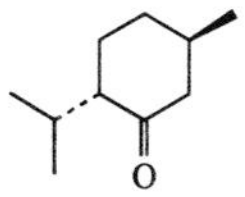 (2-isopropyl-5-methylcyclohexanone)	46/ TMSCl/THF/ heptane	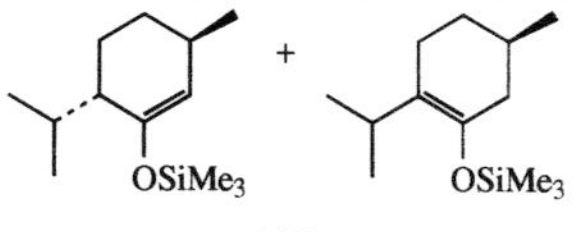 (two trimethylsilyl enol ethers) 98/2	74
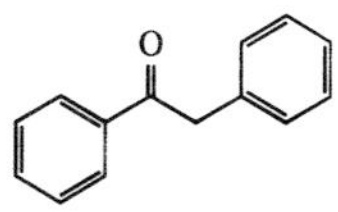 (deoxybenzoin, PhCOCH₂Ph)	46/ TMSCl/THF/ heptane	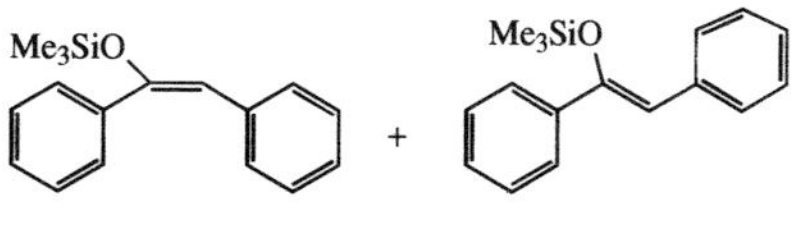 (Z/E silyl enol ethers of stilbene type) 80/20	74
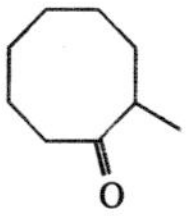 (2-methylcyclooctanone)	39/ TMSCl/Et₃N/Et₂O/ HMPA	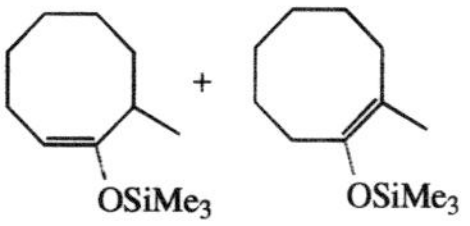 (two silyl enol ethers) 5/95	75
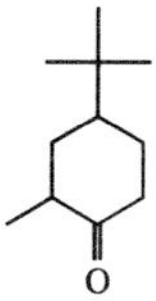 (4-tert-butyl-2-methylcyclohexanone)	39/ TMSCl/Et₃N/Et₂O/ HMPA	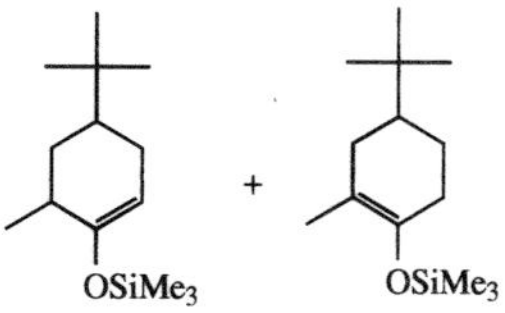 (two silyl enol ethers) 3/97	75
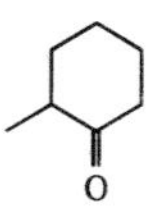 (2-methylcyclohexanone)	47 electrogenerated base DME/ HMPA	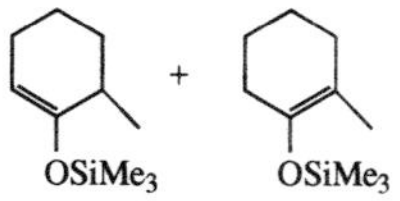 (two silyl enol ethers) 95/5	76
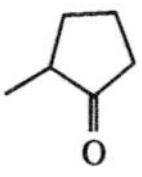 (2-methylcyclopentanone)	47 electrogenerated base DME/ HMPA	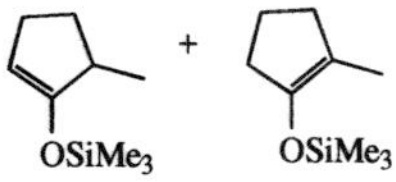 (two silyl enol ethers) 91/9	76
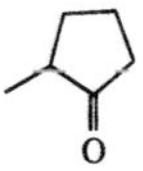 (2-methylcyclopentanone)	48 electrogenerated base DME/ HMPA	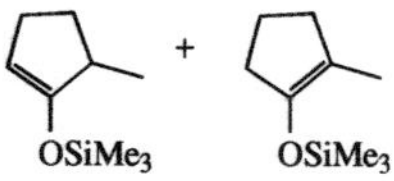 (two silyl enol ethers) 2/98	77

(*continued overleaf*)

TABLE 7. (continued)

Substrate	Magnesium amide	Magnesium enolate or reaction product after trapping procedure of enolates	Reference
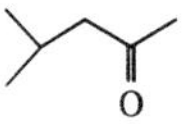 (CH3)2CHCH2C(O)CH3	47 electrogenerated base DME/HMPA	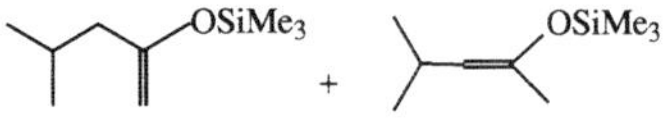 iPr-CH2-C(=CH2)-OSiMe3 + (CH3)2C=C(CH3)-OSiMe3 97/3	76
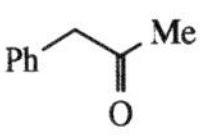 PhCH2C(O)Me	47 electrogenerated base DME/HMPA	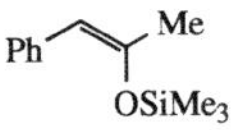 Ph-CH=C(Me)-OSiMe3 (Z)/(E) = 96/4	76
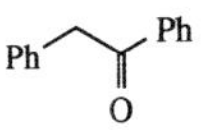 PhCH2C(O)Ph	47 electrogenerated base DME/HMPA	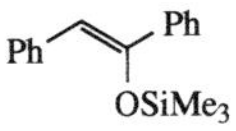 Ph-CH=C(Ph)-OSiMe3 100 (Z)	76
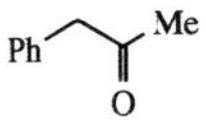 PhCH2C(O)Me	47 (2 equiv.) toluene	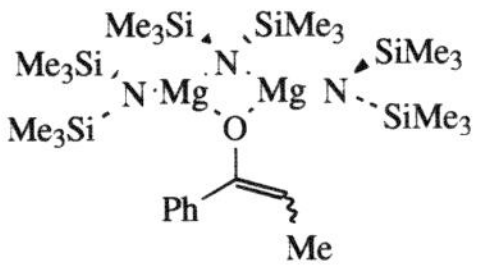 Me3Si\N/Mg-O-Mg\N/SiMe3 cluster with PhC(=CHMe)O (E)/(Z) = 74/26	78, 79
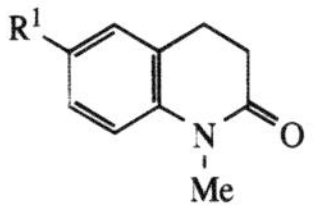 6-R¹-1-Me-3,4-dihydroquinolin-2(1H)-one	39	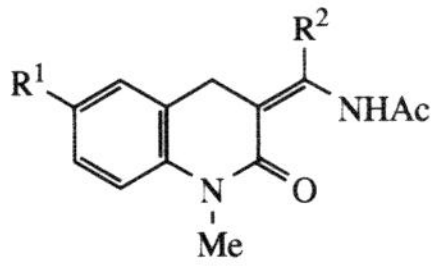 3-(R²C=NHAc)-6-R¹-1-Me-3,4-dihydroquinolin-2(1H)-one	80
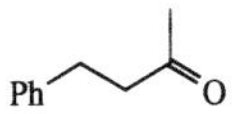 PhCH2CH2C(O)H (wrong – PhCH2CH2C(O)CH3)	46	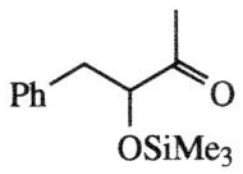 Ph-CH2-C(OSiMe3)=... (α-silyloxy ketone)	81
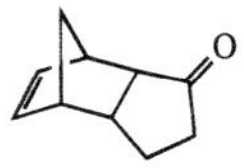 bicyclic enone (norbornene-fused cyclopentanone)	39	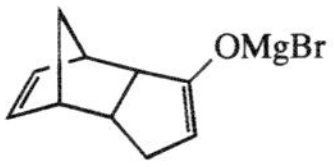 corresponding enol OMgBr	82
 n-hexyl-CHO	46	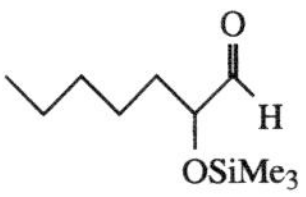 n-pentyl-CH(OSiMe3)-CHO	81
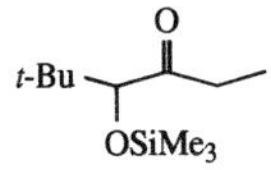 t-Bu-C(O)-CH(OSiMe3)-CH3	43	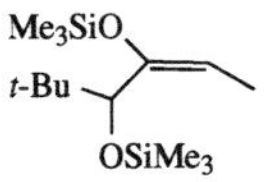 t-Bu-C(OSiMe3)=CH-CH3 with Me3SiO	83

TABLE 7. (continued)

Substrate	Magnesium amide	Magnesium enolate or reaction product after trapping procedure of enolates	Ref.
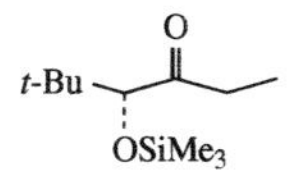 t-Bu—C(=O)—CH(OSiMe₃)—CH₃	45	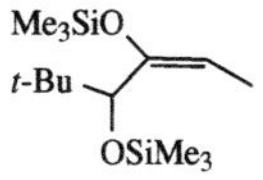 Me₃SiO—C(t-Bu)=CH—OSiMe₃ (E)/(Z) = 91/9	84

The increasing interest in enolization reactions mediated by magnesium amides led to new investigations for structural features of these reagents[78,79].

Magnesium amides have also found good utility in enantioselective deprotonation processes. A range of chiral amines has been prepared by Henderson and coworkers and it was found after conversion to their Mg-bisamide derivatives that it react with 4- and 2,6-substituted cyclohexanones with good to excellent selectivities[85–89] (see Section III). Structures of some chiral magnesium amides are given in Chart 1.

The concept of chiral magnesium amides for the preparation of magnesium enolates has been extended to chiral magnesium bis(sulfonamide) complexes as catalysts for the enolization of N-acyloxazolidines[90] (equation 63).

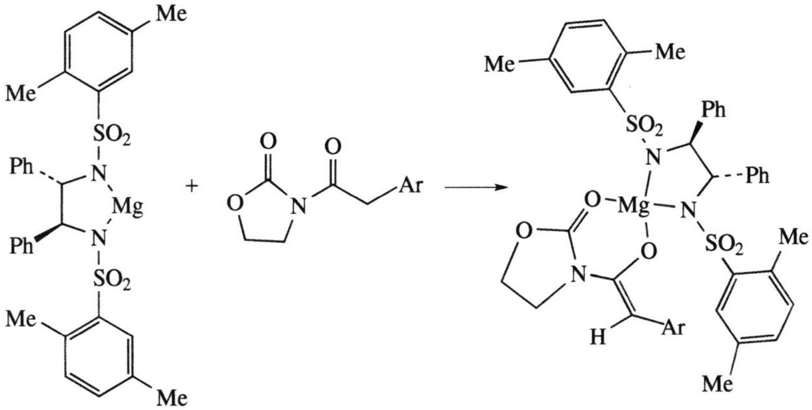

10% (63)

The metallation should proceed via the formation of a chelated tetrahedral magnesium enolate complex, with a (Z)-geometry. The conformational rigidity would be enforced by chelation of both the imide enolate and bis(sulfonamide) ligand to the tetrahedral magnesium ion.

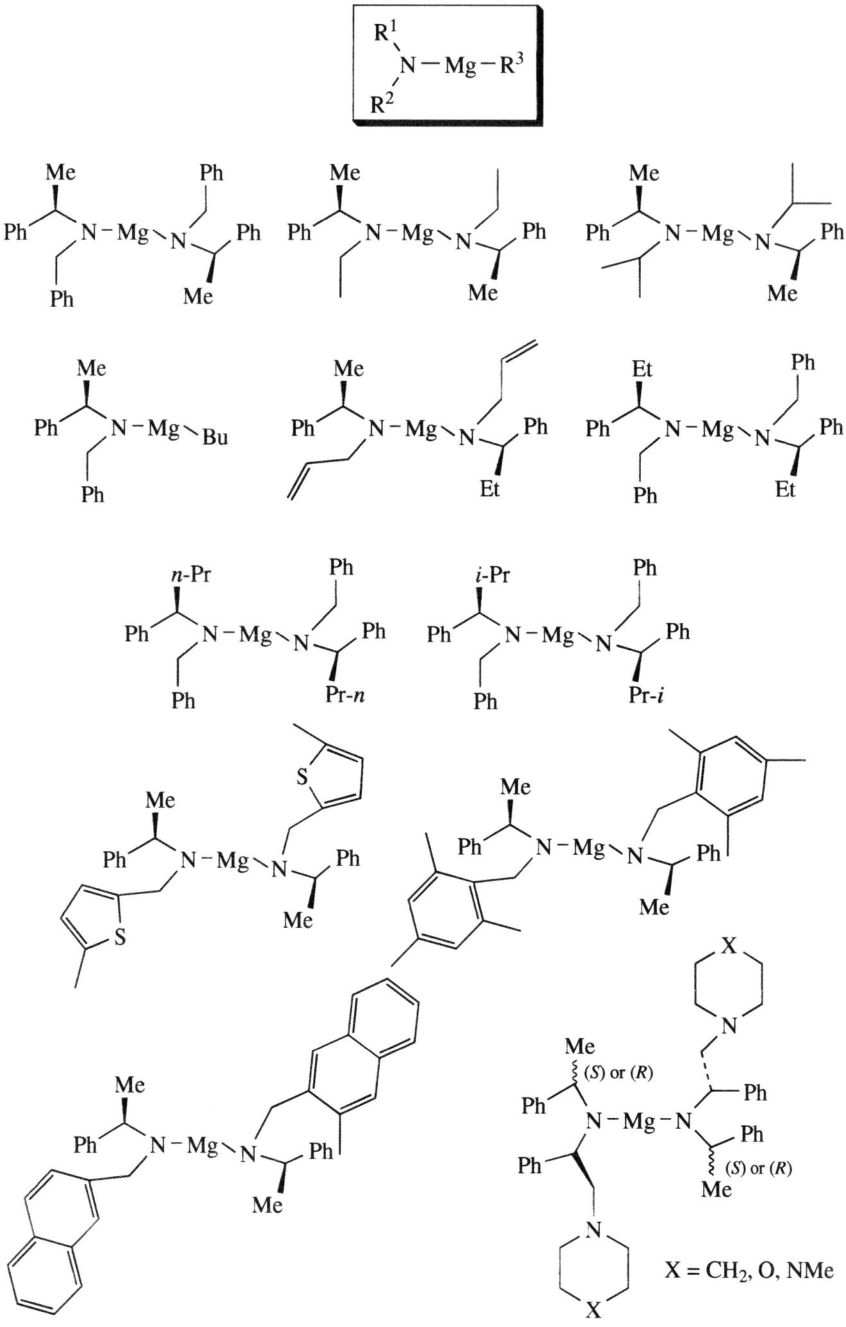

CHART 1

F. Miscellaneous Methods

Other methods to prepare magnesium enolates were also reported. They involve the addition to carbon–carbon multiple bonds. Two different mechanisms are possible: (a) the addition–elimination sequence to a carbon–carbon double bond, (b) the addition to a carbon–carbon of a ketene.

Uncatalyzed addition reactions of Grignard reagents with nonconjugated alkenes and alkynes are of limited use in synthesis. However, carbon–carbon double bonds substituted by a leaving group, such as an acetate, are susceptible to be displaced by organomagnesium compounds presumably by an addition–elimination pathway. A few examples have been reported[91,92] (equation 64).

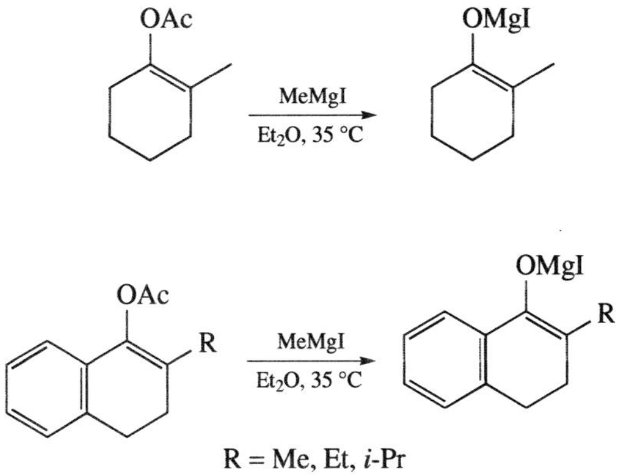

$$(64)$$

Early investigations of reactions of organomagnesium compounds with ketenes are described, as illustrated by the example of mesitylketene described by Rappoport and coworkers[93] (equation 65).

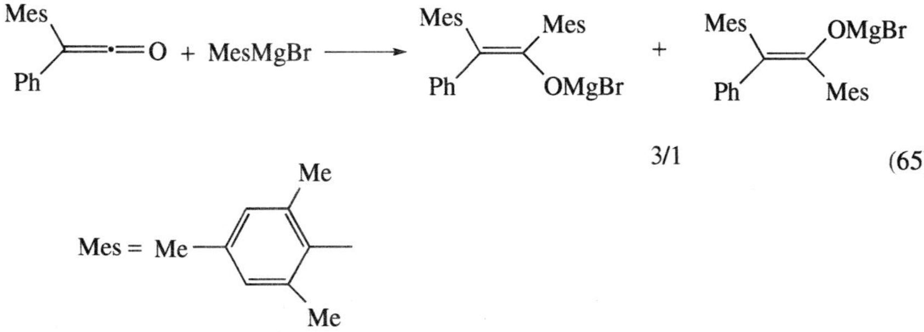

$$(65)$$

The preference for the formation of the major (E)-enolate indicates that the attack occurs preferentially from the side of the formally bulkier mesityl ring. *Ab initio* calculations allow one to rationalize this result. The ketene adopts a conformation with a planar Ph–C=C moiety while the mesityl is nearly perpendicular to this plane. Since the attack onto the C=O group occurs in the plane of the C=C double bond, the coplanar Ph is effectively bulkier and the preferred attack is from the mesityl side.

Recently, Verkade and coworkers have reported the successful synthesis of β-hydroxynitriles from carbonyl compounds in a reaction promoted by strong nonionic bases, such as proazaphosphatrane types. The reaction occurs in the presence of magnesium sulfate, which activates the carbonyl group and stabilizes the enolate thus produced. However, the nature of the counterion of enolate is questionable[94] (equation 66).

$$CH_3CN \ + \ \underset{\substack{(10\%) \\ R = Me, \ i\text{-}Pr}}{\text{[bicyclic guanidine base]}} \ + \ \underset{(2 \text{ equiv.})}{MgSO_4} \ \xrightarrow{R^1 \diagup R = O} \ \underset{R^1}{\overset{R}{\diagup}}\underset{CN}{\overset{OH}{\diagdown}} \quad (66)$$

III. REACTIVITY OF ENOLATES

A. Introduction

This chapter focuses on the reactivity of the magnesium enolates and the potential that they offer in the development of synthetic strategies. These reactive intermediates are useful tools and were found to be trapped with a range of electrophiles, allowing the introduction of numerous functionalities. The purpose of this section is to survey the most important applications of magnesium enolates derived from different functional groups: ketone enolates, ester and lactone enolates, dicarbonyl enolates, amide and lactam enolates, thioester, thioamide enolates, carboxylic acid dianions, chiral oxazilidinones and derivatives and miscelleneous chiral magnesium enolates.

B. Reactions of Magnesium Ketone Enolates with Electrophiles

These reactions are divided into two sections. In the former, representative examples of organic electrophiles, which can be used in reactions with magnesium ketone enolates, are summarized. The second section shows that magnesium ketone enolates can be employed as interesting alternatives to their more known lithium counterparts in aldol addition reactions. This part is discussed in terms of regio- and stereoselectivity.

Magnesium enolates are usually produced as an intermediate in the transformation of ketones into the corresponding silylenol ethers. These reagents are more stable; they may be isolated, purified and characterized using standard organic procedures. The silylation step allows an internal quench of the formed enolates and an analysis of the distribution of magnesium enolates present prior to silylation. Although magnesium enolates are often less reactive than their lithio-analogues, Mekelburger has reported that Li and Mg enolates have a similar reactivity toward Me_3SiCl[4a]. However, there is a controversy on the nonequilibrating silylation conditions of magnesium enolates. The reaction conditions should depend on the method of formation of the enolates (for example, choice and conditions of preparation of the base), the solubility of the magnesium enolates, the nature of the solvent and the composition of the solvent mixture[76]. The range of reagents and substrates is too large to allow a procedure to be given for each reaction. Instead, some representative examples for the most important applications are given.

Krafft and Holton[75] have found that bromomagnesium diisopropylamide (BMDA) in an ethereal solution may be used in conjunction with the system $TMSCl/Et_3N/HMPA$ to prepare thermodynamic silylenol ethers. Reaction times of 8–12 h at 25 °C are required for the complete conversion to trimethylsilyl enolates (equation 67).

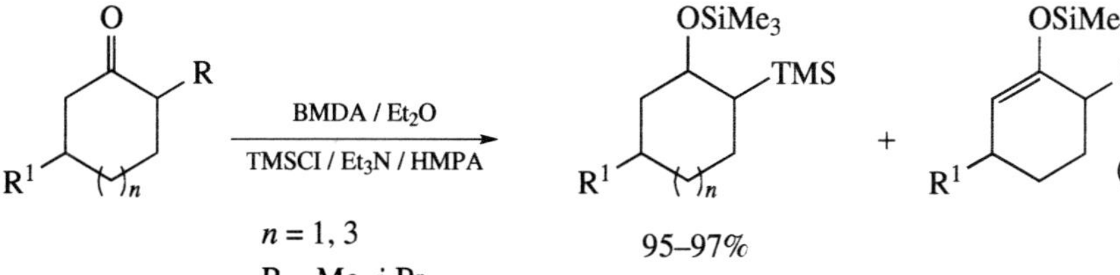

11. Preparation and reactivity of magnesium enolates

$$\text{ketone} \xrightarrow[\text{TMSCl / Et}_3\text{N / HMPA}]{\text{BMDA / Et}_2\text{O}} \text{silyl enol ether (TMS) + silyl enol ether} \quad (67)$$

$n = 1, 3$
$R = \text{Me}, i\text{-Pr}$
$R^1 = \text{H, Me}, t\text{-Bu}$

95–97%

The effectiveness of magnesium enolates as nucleophilic agents limits the interest of the reaction. With less substituted substrates (R = H), the aldol reaction is faster than the silylation. Moreover, due to solubility limitations, the authors are unable to determine whether the high thermodynamic:kinetic ratio of silylenol ethers obtained accurately represents the magnesium enolate composition. Nonetheless, this method is an excellent procedure to selectively prepare the thermodynamic silylenol ether from an unsymmetrical ketone[75].

Bordeau and coworkers have described an efficient and stereoselective synthesis of kinetic silylenol ethers[74]. Less highly substituted silylenolates are regiospecifically prepared in high yield, around room temperature under kinetic conditions, from unsymmetric cyclic ketones and [(DA)$_2$Mg] in THF/heptane (equation 68).

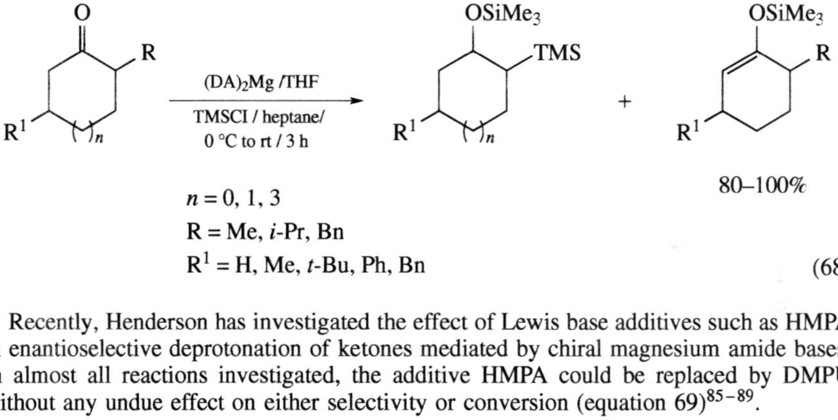

$$\text{ketone} \xrightarrow[\substack{\text{TMSCl / heptane/} \\ 0\,°\text{C to rt / 3 h}}]{(\text{DA})_2\text{Mg /THF}} \text{silyl enol ether (TMS) + silyl enol ether}$$

$n = 0, 1, 3$
$R = \text{Me}, i\text{-Pr}, \text{Bn}$
$R^1 = \text{H, Me}, t\text{-Bu}, \text{Ph, Bn}$

80–100%

(68)

Recently, Henderson has investigated the effect of Lewis base additives such as HMPA in enantioselective deprotonation of ketones mediated by chiral magnesium amide bases. In almost all reactions investigated, the additive HMPA could be replaced by DMPU without any undue effect on either selectivity or conversion (equation 69)[85–89].

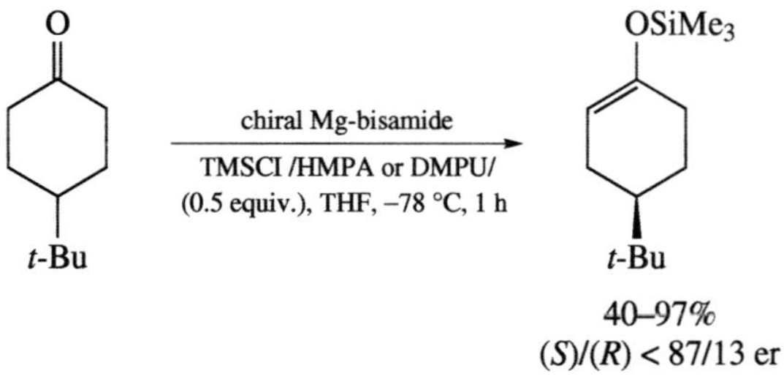

$$\text{4-}t\text{-Bu-cyclohexanone} \xrightarrow[\substack{\text{TMSCl /HMPA or DMPU/} \\ (0.5 \text{ equiv.}), \text{THF}, -78\,°\text{C, 1 h}}]{\text{chiral Mg-bisamide}} \text{silyl enol ether} \quad (69)$$

40–97%
$(S)/(R) < 87/13$ er

An important reaction of silylenol ethers is their use as enolate equivalent in Mukaiyama aldol additions. An example of the synthetic utility of this reaction with a magnesium enolate as starting reagent is shown below.

The copper-catalyzed conjugate addition of methyl magnesium iodide to cyclohexenone and trapping the enolate as its trimethylsilyl enol ether, followed by a trityl hexachloroantinomate-catalyzed Mukaiyama reaction, is applied to R-$(-)$carvone. C-2, C-3 functionalized chiral cyclohexanones are converted into their α-cyano ketones, which are submitted to Robinson annulation with methyl vinyl ketone. Highly functionalized chiral decalones are obtained that can be used as starting compounds in the total synthesis of enantiomerically pure clerodanes[47] (equation 70).

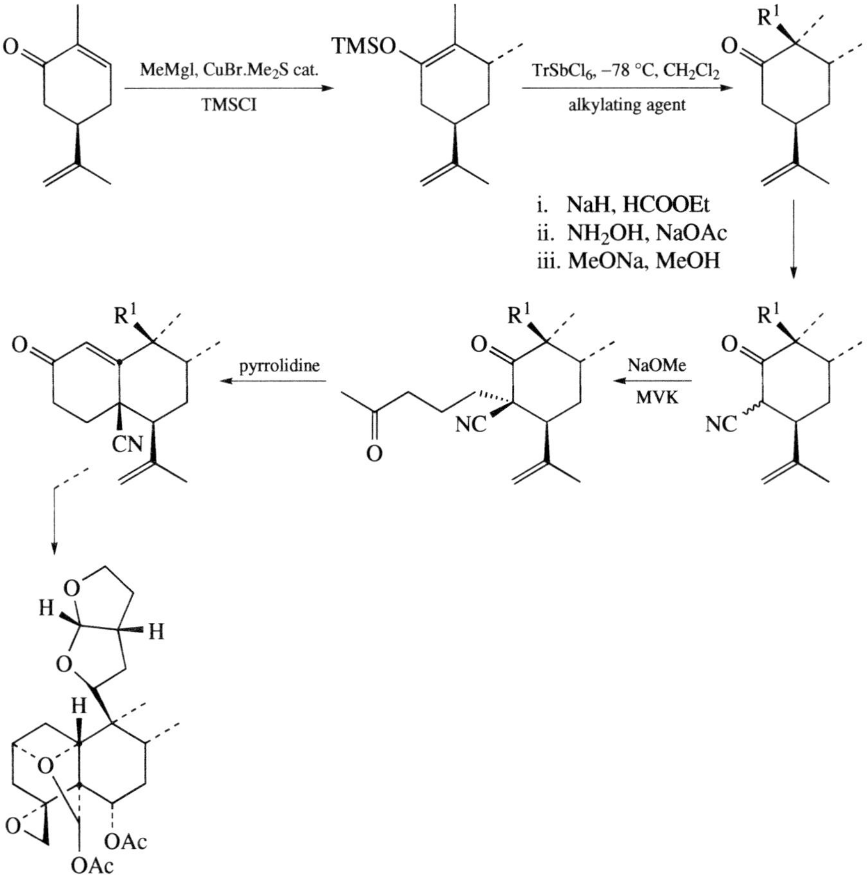

$$\text{cledorane} \tag{70}$$

The conjugate addition of a Grignard reagent (often copper-induced) to an enone, followed by reaction of the resulting enolate with an electrophile, provides numerous examples of tandem vicinal functionalizations[95]. For example, equation 71 depicts the generation of an α-magnesium enolate by the addition of methylmagnesium iodide to 3-trimethylsilylbut-3-en-2-one; the subsequent addition of benzaldehyde generates an alkene via a Peterson olefination[96].

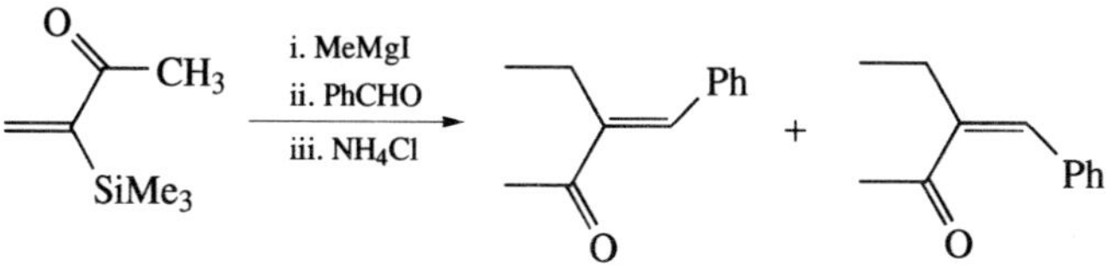

11. Preparation and reactivity of magnesium enolates

$$\text{(eq. 71)}$$

87 / 23

To develop new electrophilic reagents, Ricci and coworkers have described the synthesis of trimethylsilyloxy and hydroxy compounds from magnesium enolates and bis(trimethylsilyl)peroxide. Magnesium enolates, generated using magnesium diisopropylamide, $(DA)_2Mg$, give the hydroxycarbonyl compounds in excellent yields[81] (equation 72, Table 8).

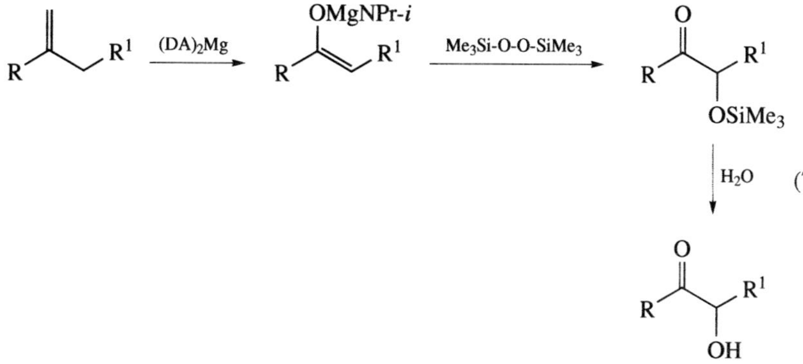

$$\text{(eq. 72)}$$

Magnesium ketone enolates are capable of C-alkylation. In general O-alkylated compounds are not observed. Matsumoto and coworkers have reported a diastereoselective synthesis for the preparation of tricyclic β-lactam antibiotics[73]. The key step is the reaction between magnesium enolate of (2S)-2-methoxycyclohexanone and 4-acetoxyazetidinone (equation 73). The direct coupling reaction between the magnesium enolate and the acetoxyazetidinone proceeds with high yield, regio- and diastereoselectively. Several similar methods are reported with tin and lithium enolates but, among the various enolates screened, the magnesium enolate is found to be the most simple and efficient.

TABLE 8. Synthesis of trimethylsiloxy and hydroxy compounds from magnesium enolates and bis(trimethylsilyl)peroxide

R	R^1	Yield of trimethylsilyloxy derivatives (%)	Yield of hydroxyl derivatives (%)
n-Hex	H	100	46
$PhCH_2CH_2$	H	100	42
$(CH_2)_6$		100	61
$(CH-CH_3)(CH_2)_6$		100	40

Surprisingly, the magnesium enolate of 2-methylcyclohexen-1-one reacts with chloroacetone to give an unexpected product via a cyclohexane ring contraction[91] (equation 74).

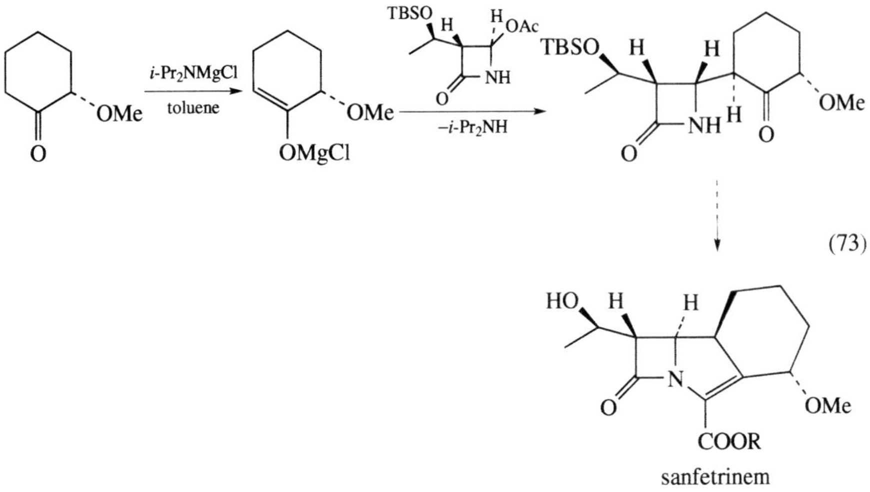

(73)

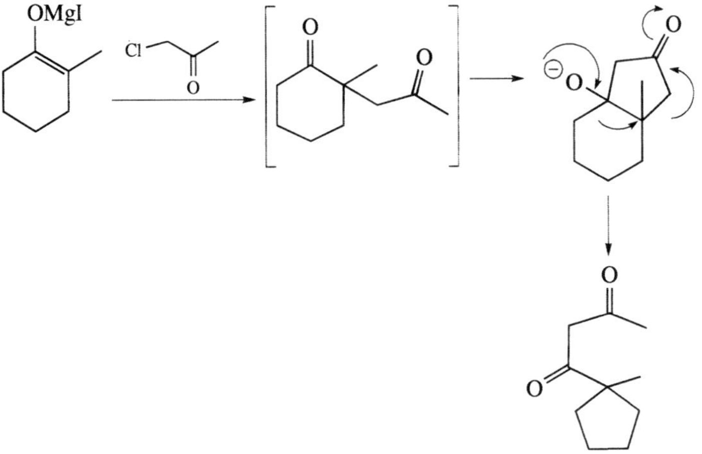

sanfetrinem

(74)

Stile, then Baker and coworkers[64, 65a], have shown that certain magnesium ketone enolates react with magnesium methyl carbonate (MMC: Stiles reagent) to give stable chelated adducts, which are either converted to β-keto carboxylic acids by treatment with aqueous HCl, or to the methyl esters by reaction with methanolic HCl. MMC adducts can be alkylated *in situ* with various alkyl halides. For example, the MMC adduct of 8-[1-(*t*-butyldimethylsiloxy)2-phenylethyl]-2-(1-oxoethyl)dibenzofuran reacts with ω-halo compounds bearing nitrile, ester or amide groups (equation 75, Table 9). The obtained dibenzofuranic derivatives are important intermediates in the synthesis of a series of leukotriene B_4 antagonists. Good to moderate yields (40–86%) of monoalkylated products are formed. In contrast to the unsubstituted β-ketoacids, all α-alkyl β-ketoacids intermediates decarboxylate during the reaction or workup.

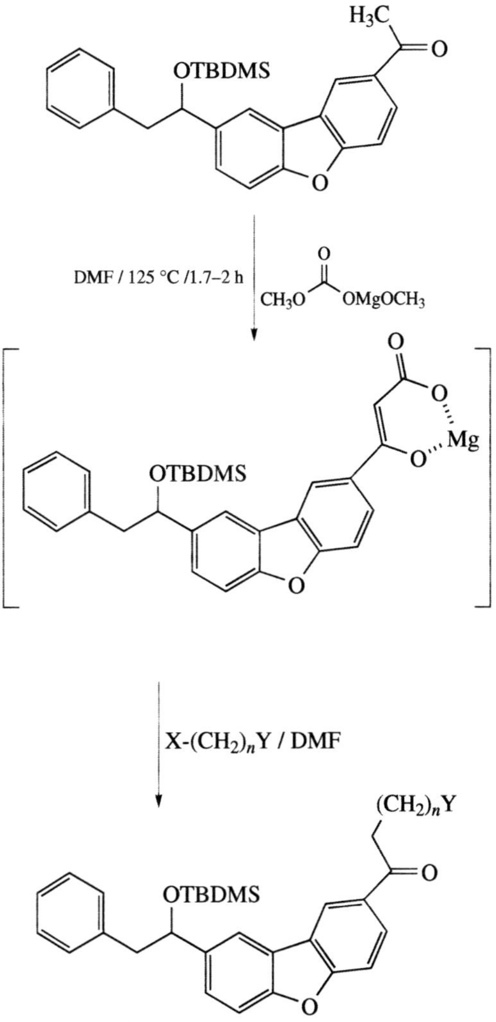

11. Preparation and reactivity of magnesium enolates

(75)

TABLE 9. Reactants and products of MMC-activated substitution reactions

X	Y	n
Br	COOMe	1
Br	CN	1
Br	CONMe$_2$	1
Br	CON(Pr-i)$_2$	1
Br	CON(CH$_2$)$_4$	1
Br	CON(CH$_2$)$_5$	2
I	COOMe	2
Br	COOMe	3

A similar procedure has been applied to the preparation of tetracyclic intermediates having the Bruceantin tetrahydrofuran ring[65b]. The enolic β-ketoesters are isolated with excellent yields (85–95%) (equation 76).

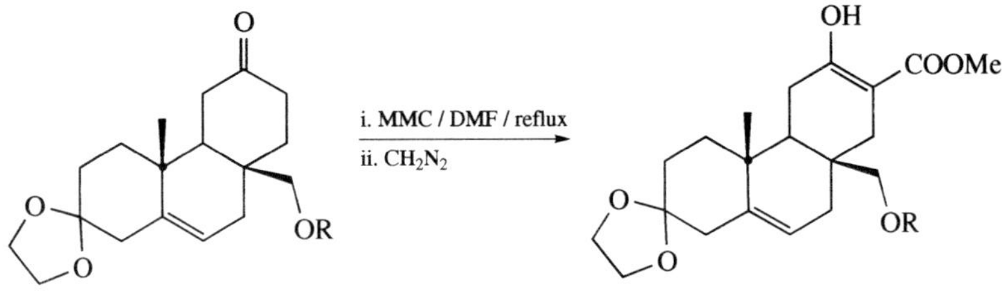

$$R = t\text{-BuMe}_2\text{Si, THP} \tag{76}$$

Magnesium enolates derived from hindered ketones are also possible Michael donors. For example, enolization of t-butyl alkylketones with $(i\text{-Pr})_2$Mg allows the 1,4-addition on the chalcone. A long reaction time (>3 h) limits the competing 1,2-addition and increases the proportion of the threo isomer[97] (equation 77).

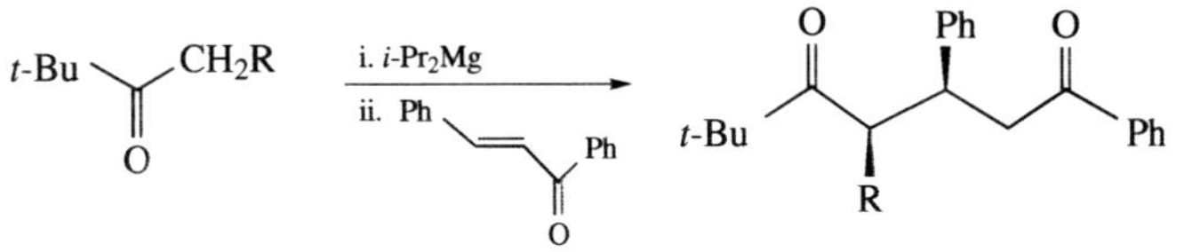

$$R = H, Me \tag{77}$$

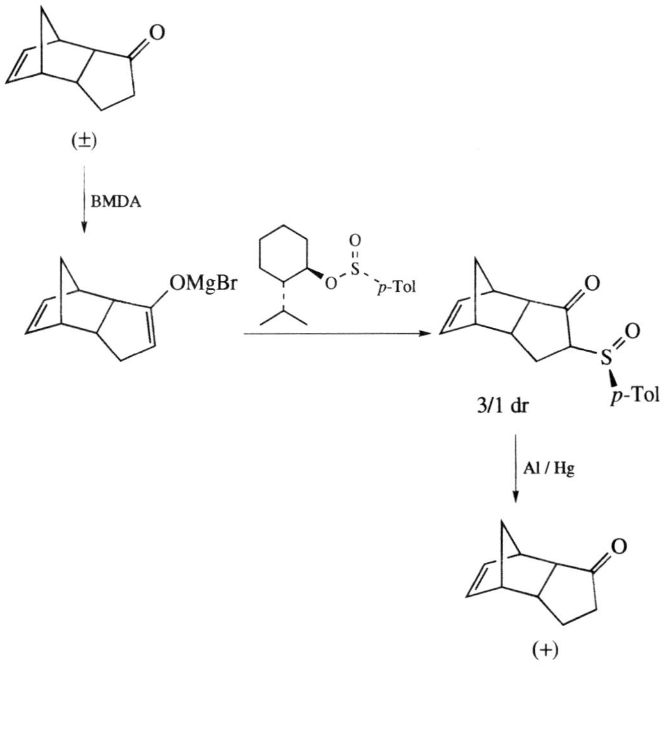

$$\tag{78}$$

The main access to pure enantiomer of chiral sulfoxides is the reaction of Grignard reagents with sulfinate esters. The reaction has been extended to magnesium enolates. It proceeds with clean inversion of configuration at the sulfur atom to yield the β-ketosulfoxide as a single diastereomer. An interesting example of kinetic resolution has been observed by Childs and Edwards[82] for the preparation of β-ketosulfoxide from a racemic ketone and the S_S-menthyl-p-toluensulfinate (equation 78). The rate of reaction of the (+)- and (−)-magnesium enolates with chiral sulfinate differs markedly and leads to the formation of two diastereomers in a 3/1 ratio. The two diastereomers cannot be separated by column chromatography, but the parent ketone is regenerated in optically active form by reductive desulfination.

Due to their inherent polarizability, α-halo-β-ketosulfoxides may be used as electrophilic partners in desulfination reaction to generate metal enolate. Therefore, treatment of α-halo-β-ketosulfoxides with EtMgBr gives magnesium enolates. Trapping these reagents with various electrophiles allows the preparation of α-haloketones[14, 16] (equation 79, Table 10).

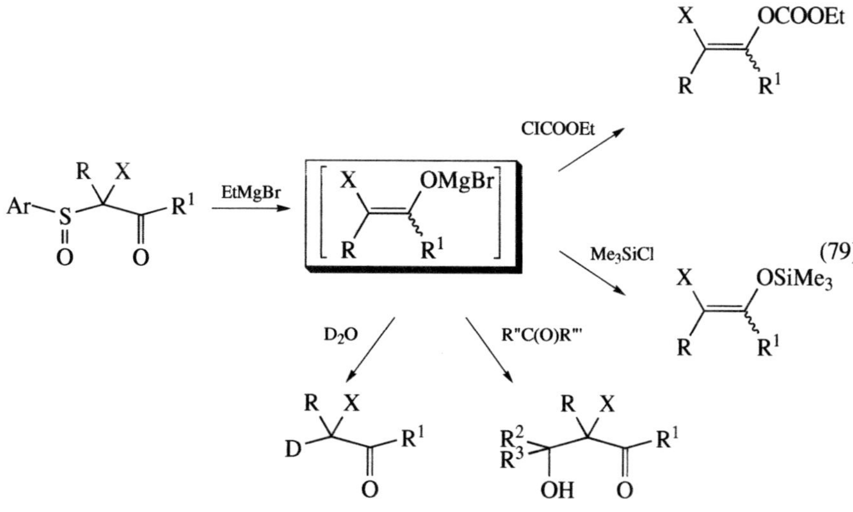

(79)

In constrat to α-halo-β-ketosulfoxides, 1-chloroalkyl aryl sulfoxides react with Grignard reagents to give β-oxido carbenoids. The rearrangement of these intermediates leads

TABLE 10. Substrates and reagents for synthesis of α-halocarbonyl compounds

Ar	X	R	R^1	R^2	R^3
Ph	F	CH_3	Ph	CH_3	CH_3
Tol	Cl	$CH_3(CH_2)_2$	$PhCH_2CH_2$	Ph	H
	Br	$CH_3(CH_2)_7$	$CH_3(CH_2)_8$	$PhCH_2CH_2$	
			c-Hex	CH_3	
				$CH_3(CH_2)_8$	
				c-Hex	
				$(CH_2)_5$	

to magnesium enolates, which can be trapped with various electrophiles to give α,α-disubstituted carbonyl compounds in moderate to good yields[18] (equation 80).

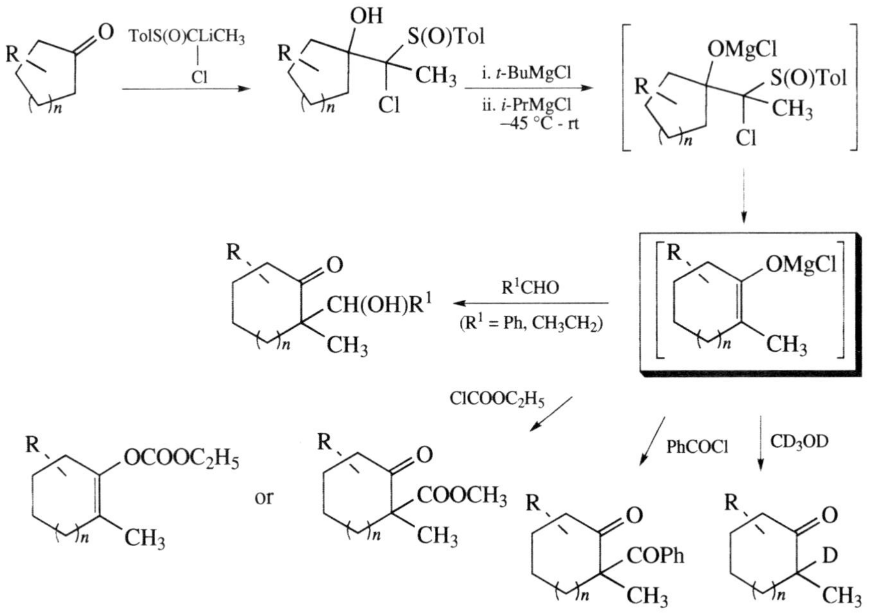

(80)

Magnesium enolates generated by interconversion metal/halogen show a high reactivity toward diethylphosphorochloridate to furnish enol phosphates. This reaction has been used in the synthesis of tetrahydrocannabinols[11,12] as illustrated in equation 81. Reaction of the α-iodo-β-aryl-cyclohexanone **49** with EtMgBr gives the magnesium enolate **50** by metal–halogen exchange, which upon reaction with ClP(O)(OEt)$_2$ provides the enol diethyl phosphate **51**. The enol phosphate thus formed is converted in silyl derivative **52** by treatment with ClMgCH$_2$SiMe$_2$(OPr-i) in the presence of Ni(acac)$_2$ as a catalyst in THF. The product **52** is the direct precursor of Δ_9-THC metabolites.

It should be noted that the reaction of boron and lithium enolates analogues of **50** with ClP(O)(OEt)$_2$ are unsuccessful.

Enantioselective protonation of ketone metal enolates constitutes an important method for the preparation of optically active ketones. Fuji and coworkers[92] have shown interest in the magnesium countercation in the enantioselective protonation of such enolates. Pertinent results are obtained with protonation of Mg(II) enolates of 2-alkyltetralones and carbamates derived from 1,1′-binaphtalene-2,2′-diol as chiral proton sources, as indicated in equation 82 and Table 11.

Magnesium enolates react with aldehydes and ketones to give aldol products after hydrolysis. The reaction proceeds both regio- and stereoselectively and has found many applications in the synthesis of natural products.

By a careful choice of the base and of the experimental conditions, either the kinetic or the thermodynamic magnesium enolate could be prepared (see Section II).

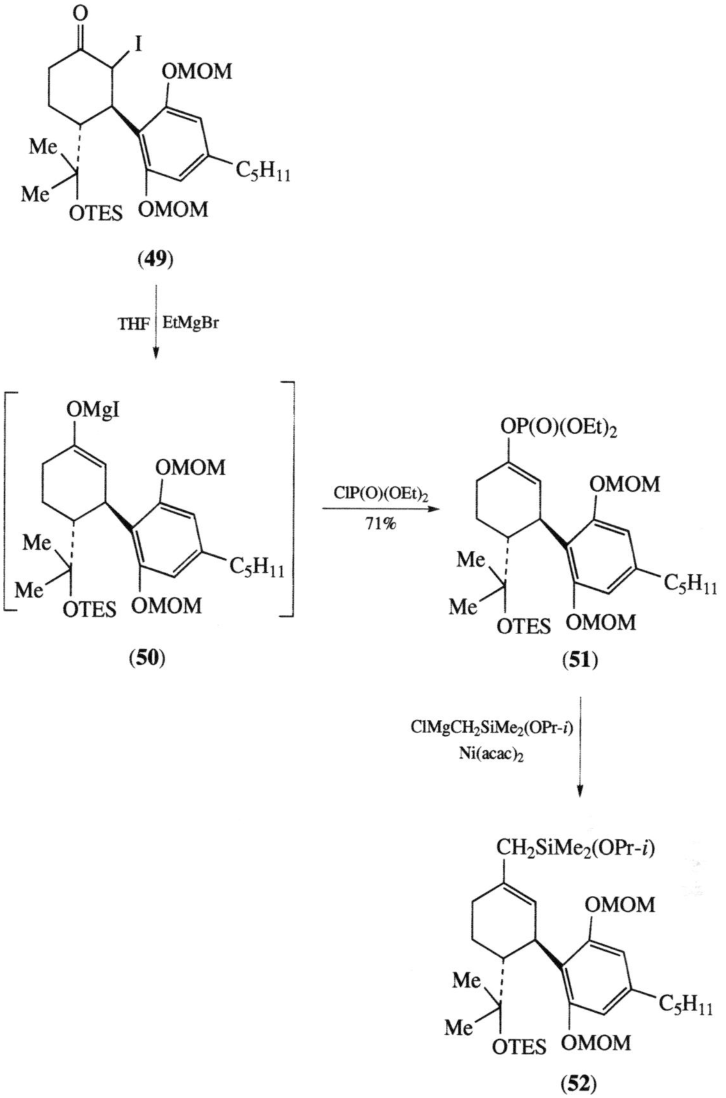

11. Preparation and reactivity of magnesium enolates

(49) → [THF, EtMgBr] → (50) → [ClP(O)(OEt)₂, 71%] → (51) → [ClMgCH₂SiMe₂(OPr-*i*), Ni(acac)₂] → (52)   (81)

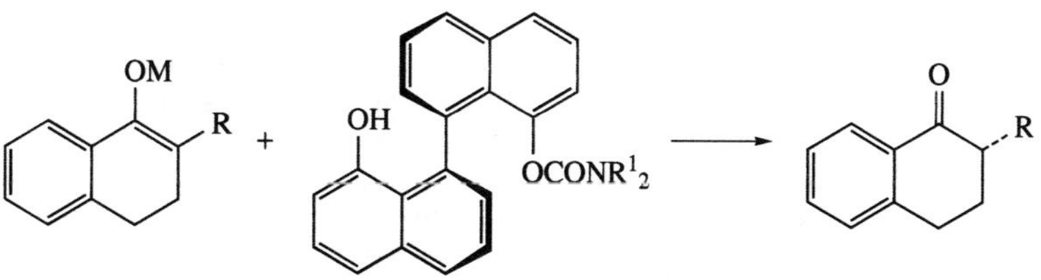

R = Et, *i*-Pr   (82)

482   Claude Grison

TABLE 11. Enantioselective protonation
of magnesiumenolates

M	R	R^1	ee (%)
MgI	i-Pr	i-Pr	9
Li	i-Pr	i-Pr	92
MgI	i-Pr	Et	15
Li	i-Pr	Et	93

Aldol reactions of magnesium enolates are frequently more diastereoselective than the corresponding reactions of lithium enolates. The aldol condensation proceeds via a cyclic transition state in agreement with the Zimmerman–Traxler chelated model[53b].

A few years ago, Fellmann and Dubois[98] studied the aldol reaction of magnesium enolates of trialkyl-substituted α-bromocyclopentanones with different aldehydes, as exemplified in equation 83 and Table 12. Upon addition of aldehydes, 2-unsubstituted cyclopentanone magnesium enolates (obtained by magnesium insertion in the C—Br bond) are converted to the *threo* aldol products via a *lk* approach. A chair-like transition state, in which the R^2 substituent of the aldehyde is placed in an equatorial position to prevent unfavorable 1,3-diaxial interactions with the cyclopentane ring, explains the stereochemical result. The *threo* isomer is also observed with a small alkyl group R^1 for the same reasons. When R^1 is larger, the *gauche* R^1/R^2 interactions become important and disfavor **53** favoring **54**. This decreases the energy difference between both transition states and smaller selectivity is therefore observed.

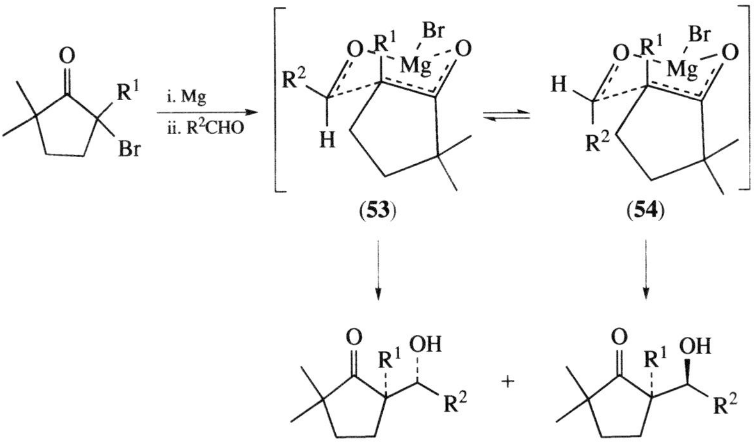

(83)

The study of Mateos and Fuente Blanco[6] on the aldol condensation between magnesium enolate of 2,2,6-trimethylcyclohexanone and 3-furaldehyde is in accord with the preceding stereochemical results. Application to the preparation of model compounds of limonoid, such as pyroangelensolide, is described (equation 84).

In addition to the structural effects due to the geometry of a substituted magnesium enolate, the stereochemistry of the reaction with a chiral aldehyde can be controlled, as described in equation 85. The aldol reaction based on the addition of magnesium enolate **56** to aldehyde **55** has been applied to the synthesis of monensin. The chiral center in the aldehyde induces the preferential approach of one diastereotopic face of the aldehyde by

11. Preparation and reactivity of magnesium enolates

TABLE 12. Stereochemistry of cyclopentanones

R^1	R^2	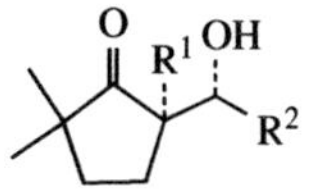 (2R*,1'S*)-isomer	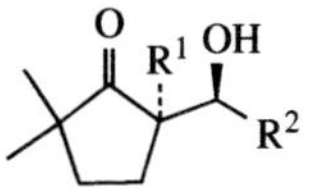 (2R*,1'R*)-isomer
H	Me	93.5	6.5
H	Et	94	6
H	i-Bu	93.5	6.5
H	neo-Pe	94	6
H	i-Pr	97	3
H	t-Bu	>99	<1
Me	Me	93.5	6.5
Et	Me	87.5	12.5
i-Bu	Me	80	20
i-Pr	Me	46	54
i-Bu	Me	29	71

the magnesium enolate; the aldol product formed is converted into a carboxylic acid (cf. **57**) by H_5IO_6 with a facial preference of 5/1.

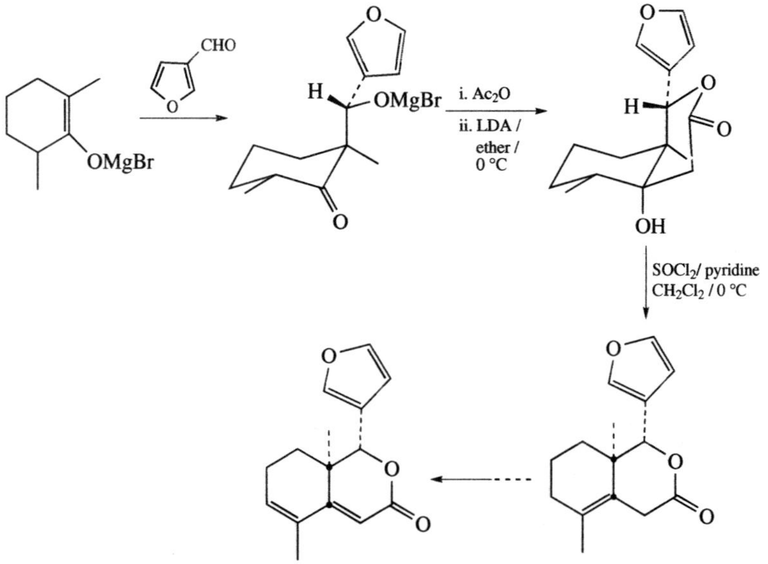

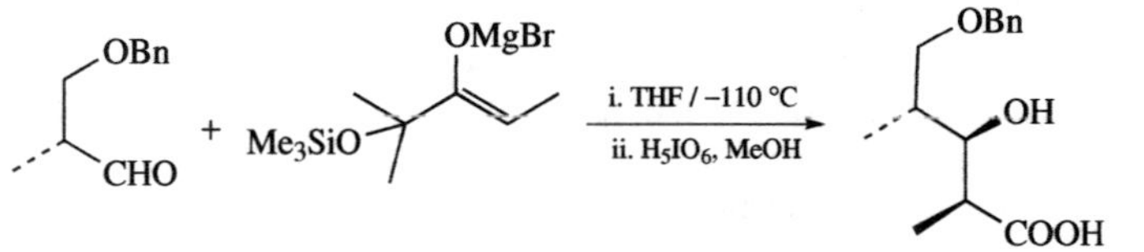

Magnesium enolates derived from hindered ketones are able to initiate polymerization. For example, addition of 2′,4′,6′4-trimethylacetophenone in toluene to a suspension of $(DA)_2Mg$ results in the isolation of $(DA)Mg(OC(=CH_2)-2,4,6-Me_3C_6H_2)$, which is found to be an excellent initiator for the living syndioselective ($\sigma_r > 0.95$) polymerization of methyl methacrylate[99] (equation 86).

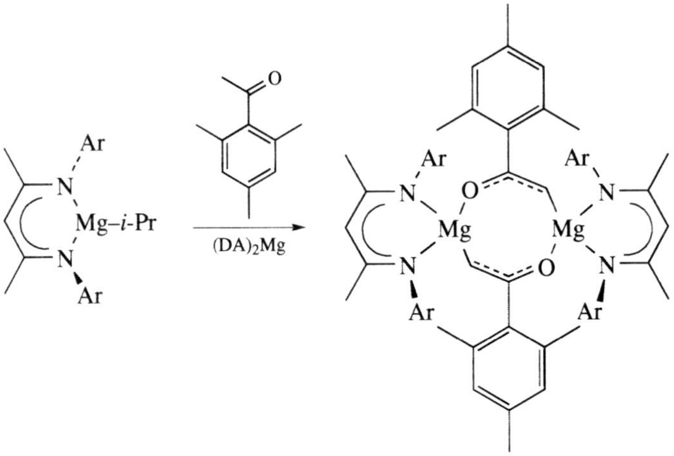

C. Reactions of Magnesium Ester Enolates and Magnesium Lactone Enolates with Electrophiles

Different approaches for the stereoselective transformations and applications of magnesium ester and lactone enolates to organic synthesis have been reported. In most cases, a second functional group is present in the β position. This can be explained by an easy preparation and by a greater stability of the enolate. They are useful reagents in aldol reactions and are attractive for the construction of biologically active products. Equation 87 describes a few types of bifunctional compounds involving magnesium ester and lactone enolates and typical examples of synthetic utility.

Asymmetric synthesis in aldol-type reaction involving magnesium ester or lactone enolates has also been reported. Enolate of (−)-menthyl or (+)-bornyl acetate reacts with substituted benzophenones or α-naphtophenones to yield, upon hydrolysis of the resulting esters, optically active β-hydroxyacids. Although these results are interpreted in terms of a steric factor, Prelog's rules are not applicable to these reactions[69] (equation 88).

The reaction of alkyl dihalogenoacetate magnesium enolates with 2,3-isopropylidene-D-glyceraldehyde affords the expected β-hydroxy-α-dihalogenoesters[100]. The *erythro* isomer is obtained with isopropyl dichloroacetate magnesium enolate. This result is in agreement with theoretical models. 2-Deoxy-pentono-1,4-lactones are obtained after removal of the halogen atom by either Raney nickel or tributyltin hydride reduction (equation 89).

Magnesium-halide-mediated aldol reactions are often reported using magnesium salts simply as Lewis acids[101–103]. More recently, Mukaiyama and coworkers[28] have described a highly diastereoselective aldol reaction between chiral alkoxy aldehydes and magnesium enolate, formed by transmetallation from a silylenol ether and $MgBr_2 \cdot OEt_2$ via a six-membered chelated cyclic transition state. High yield and excellent diastereoselectivity are observed (equation 90).

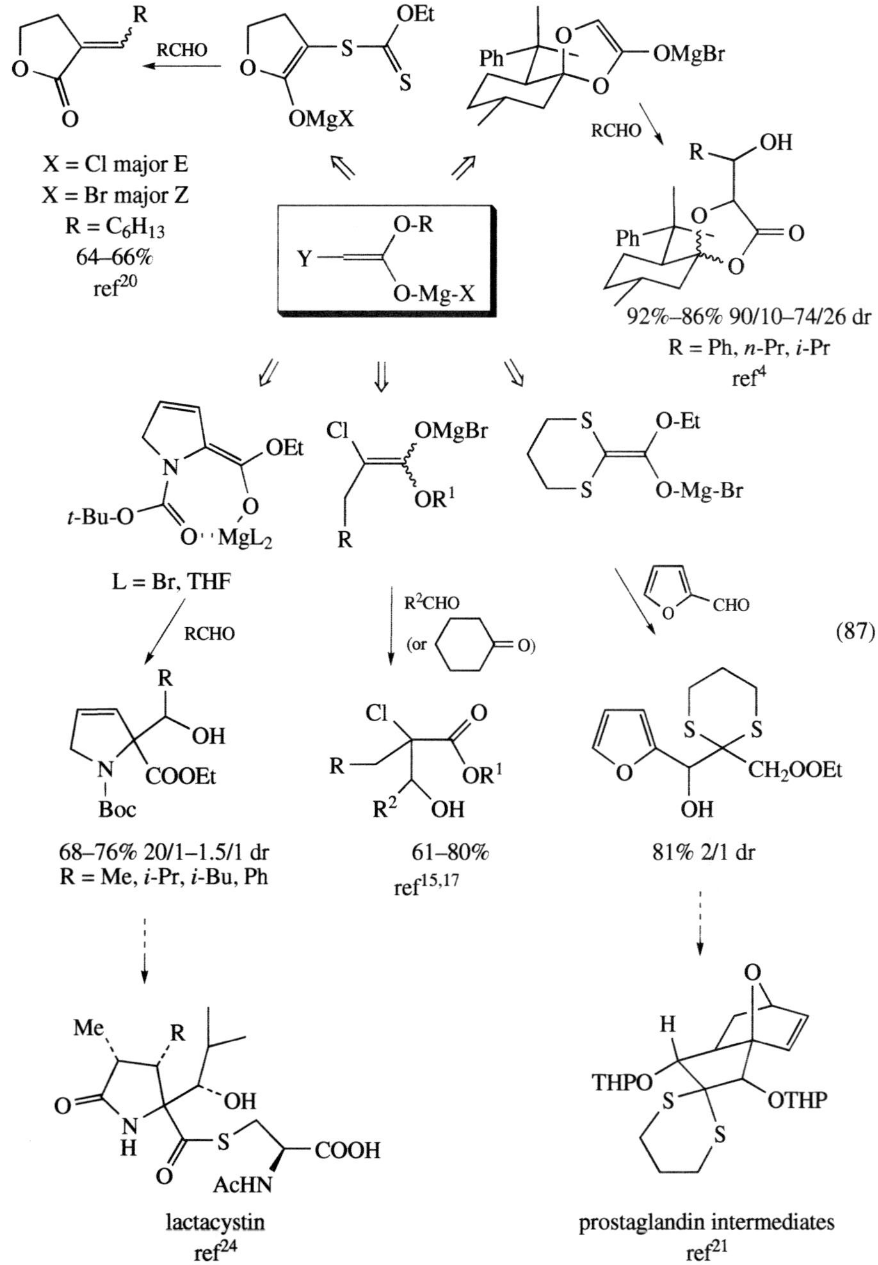

11. Preparation and reactivity of magnesium enolates

X = Cl major E
X = Br major Z
R = C_6H_{13}
64–66%
ref[20]

92%–86% 90/10–74/26 dr
R = Ph, n-Pr, i-Pr
ref[4]

L = Br, THF

68–76% 20/1–1.5/1 dr
R = Me, i-Pr, i-Bu, Ph

61–80%
ref[15,17]

81% 2/1 dr

(87)

lactacystin
ref[24]

prostaglandin intermediates
ref[21]

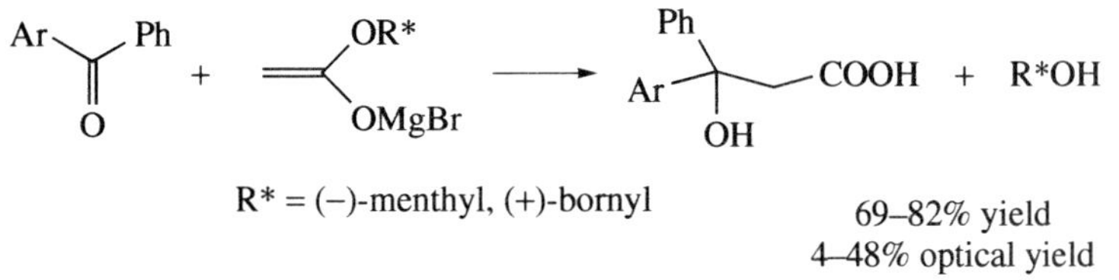

$$\text{Ar}\underset{O}{\overset{\|}{-}}\text{Ph} + \underset{\text{OMgBr}}{\overset{\text{OR*}}{=}} \longrightarrow \text{Ar}\underset{\text{OH}}{\overset{\text{Ph}}{-}}\text{COOH} + \text{R*OH} \qquad (88)$$

R* = (−)-menthyl, (+)-bornyl

69–82% yield
4–48% optical yield

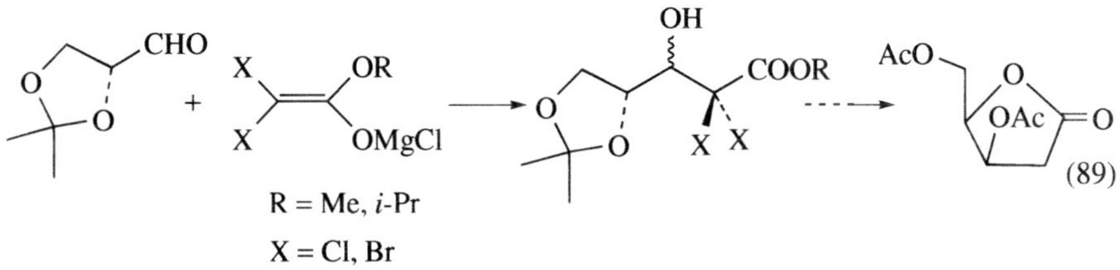

(89) R = Me, i-Pr; X = Cl, Br

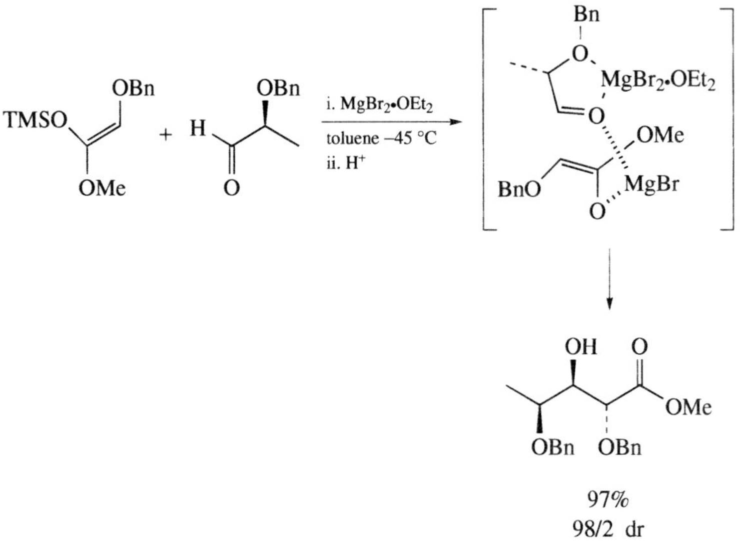

(90)

97%
98/2 dr

Carbanions of enantiomerically pure sulfoxides have been investigated as precursors for asymmetric synthesis. However, they react with carbonyl compounds with modest selectivity. This selectivity is increased by the incorporation of an ester group adjacent to the carbanionic center. The observed stereochemistry in this reaction is consistent with chelated intermediates, where magnesium chelate is particularly efficient, as illustrated in equation 91[54,55].

Fujisawa and coworkers[25] have studied the reaction of pure α-sulfinyl ester enolate **58** with benzaldimines possessing an electron-withdrawing group at the nitrogen atom. The reaction gives β-aminoester in both enantiomeric forms in satisfactory yields (67–99%), in which the changeover of the diastereofacial selectivity was induced by the choice of the protecting group at the nitrogen and the use of additives. Two transition states are proposed. With the magnesium bromide, the reaction probably proceeds through a

11. Preparation and reactivity of magnesium enolates 487

chelated model, whereas by using the less reactive magnesium iodide, it proceeds through a nonchelated model (equation 92).

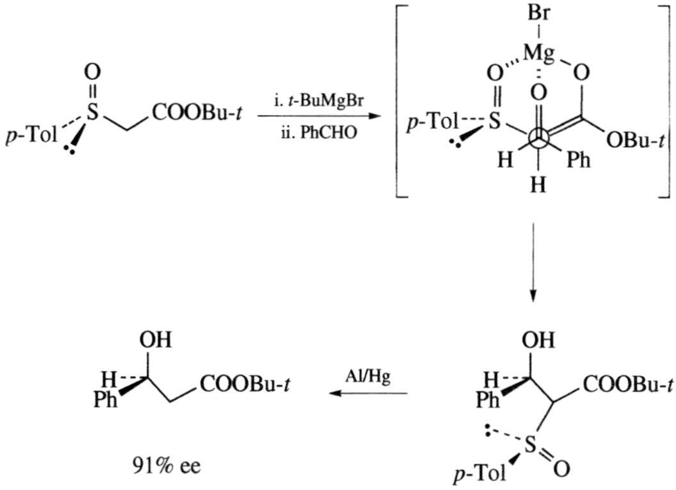

(91)

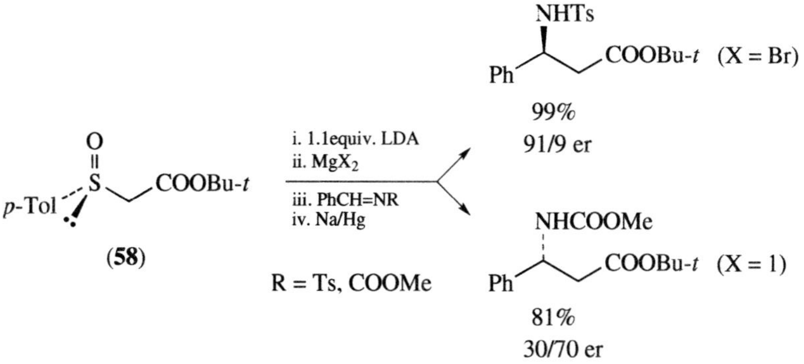

(92)

Hiyama and Kobayashi have studied the reaction between magnesium enolates of *t*-butyl (or ethyl) acetate or *t*-butyl propionate and nitriles. The reaction furnishes 3-amino-2-alkenoates **59** having Z configuration[68]. It has been successfully extended to isonitriles. The addition of magnesium enolate of alkyl acetate to isonitriles **60** affords 4-hydroxy-3-quinolinecarboxylic esters or amides **61** by a tandem Claisen-type condensation/cyclization sequence[72]. This approach has also been used for the synthesis of heterocycles[67] and aminosugar such as L-acosamine **62**[104] (equation 93).

The transmetallation of α-silylated ester lithium enolates by magnesium bromide and subsequent Peterson olefination provides a stereoselective synthesis of α,β-unsaturated esters[22] (equation 94).

Magnesium aminoester enolates are of much interest for the synthesis of complex aminoacids and peptides. These chelate enolates have been used as nucleophiles for a wide range of stereoselective transformations, as diastereoselective Michael additions[23] and Claisen rearrangement[19]. Two typical examples are presented in equation 95.

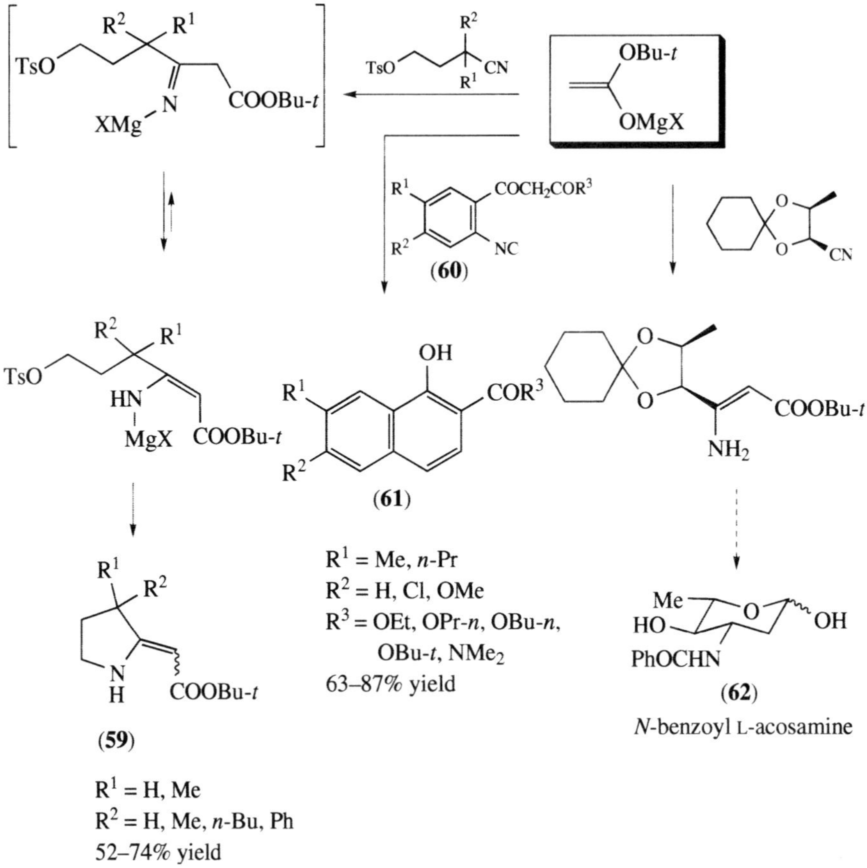

(93)

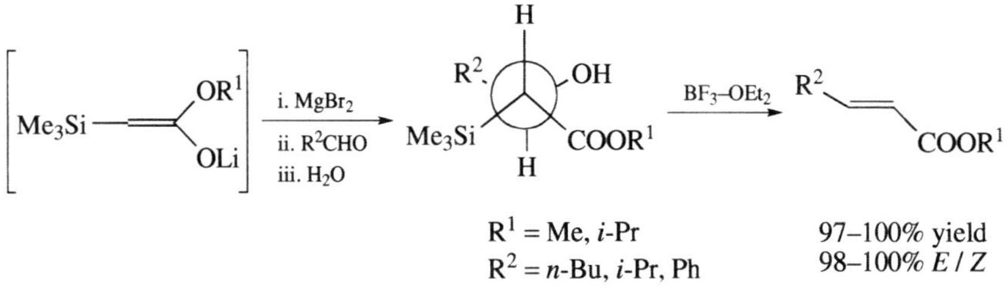

(94)

The formation of chelated magnesium enolate presents three advantages: the enolate is stable and doesn't decompose during the rearrangement; the chelation accelerates the Claisen rearrangement; the fixed enolate geometry that results from the chelation leads to interesting degrees of diastereoselectivity.

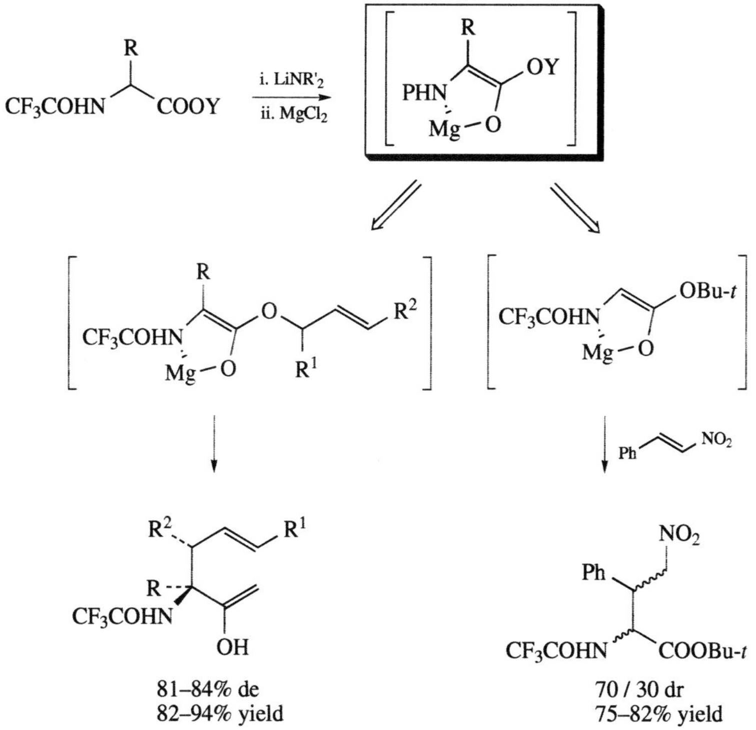

11. Preparation and reactivity of magnesium enolates 489

$$\text{CF}_3\text{COHN}\overset{R}{-}\text{COOY} \xrightarrow[\text{ii. MgCl}_2]{\text{i. LiNR}'_2} \left[\begin{array}{c} \underset{\text{PHN}}{\overset{R}{\diagdown}}\diagup^{\text{OY}} \\ \text{Mg}-\text{O} \end{array}\right]$$

$$\left[\begin{array}{c} \text{CF}_3\text{COHN}\underset{\text{Mg}-\text{O}}{\overset{R}{\diagdown}}\diagup^{\text{O}}\diagdown_{R^1}\diagup^{R^2} \end{array}\right] \quad \left[\begin{array}{c} \text{CF}_3\text{COHN}\underset{\text{Mg}-\text{O}}{\diagdown}\diagup^{\text{OBu-}t} \end{array}\right] \quad (95)$$

$$\downarrow \qquad\qquad\qquad\qquad \downarrow \text{Ph}\diagup\diagdown\text{NO}_2$$

R²⟋⟍R¹
R--
CF₃COHN⟋⟍OH

81–84% de
82–94% yield

Ph⟋⟍NO₂
CF₃COHN⟋⟍COOBu-t

70 / 30 dr
75–82% yield

D. Reactions of Magnesium Dicarbonyl Enolates with Electrophiles

1. Reactions of magnesium α-ketoester enolates

Iodomagnesium enolates **17** derived from α-ketoesters are obtained as indicated in Section II from α-chloro glycidic esters **15**. They can react with different electrophiles. Hydrolysis of the magnesium enolates **17** yielded the α-ketoesters **18**. A detailed study on deuteriolysis reports the regioselective C-mono **18-d_1** and C-dideuteriation **18-d_2** products using [D₄] acetic acid as deuterium donor (94–99% yield) (see equation 6, Section II). This constitutes an efficient method to introduce the pyruvic moiety, deuteriated or not deuteriated, into aliphatic or glucidic substrates[8]. Knowing the biological interest of α-ketoacid group, this first example of deuterium incorporation in such moiety appears especially important as it opens the route to radioactive tracers in this series.

However, due to the high stability of these magnesium enolates, it is necessary to add an excess of HMPA to increase their nucleophilicity toward other electrophiles. Under these conditions, the aliphatic and glucidic magnesium enolates **17** react with hard alkylating reagents such as chloromethylmethyl ether, dimethyl sulfate, diethyl phosphorochloridate and chlorotrimethylsilane to provide O-alkylation products **63–66** in fair to good yields[9] (equation 96, Table 13). Enol ether moieties of different resistance to acid-catalyzed hydrolysis are therefore obtained. It is noteworthy that the steric hindrance of a glucidic residue is not the limiting factor and the phosphorylation of enolates is even more efficient with derivatives of D-galactose and D-lyxose. This procedure constitutes an interesting

TABLE 13. Reactions of magnesium enolates 17

R²	X		Yield (%)	E/Z		Yield (%)	E/Z		Yield (%)	E/Z		Yield (%)	E/Z		Yield (%)	E/Z
H[a]	OH[a]	18a	98	—	18b	95	—	18c	100	—	18d	99	—	18e	98	—
MeOCH₂	Cl	63a	75	13/87	63b	58	10/90	63c	53	19/81	63d	47	0/100	63e	62	0/100
Me	SO₄Me	64a	76	0/100	64b	34	0/100	64c	54	0/100	64d	20	7/93	64e	44	0/100
(EtO)₂P(O)	Cl	65a	28	93/7	65b	34	95/5	65c	28	92/8	65d	45	100/0	65e	43	100/0
Me₃Si	Cl	66a	78	15/85	66b	86	7/93	66c	62	19/81	66d	90	10/90	66e	50	0/100

[a] See equation 6, Section II.

alternative to the Perkow reaction for the preparation of glucidic phosphoenolpyruvic acid derivatives **65d–e**.

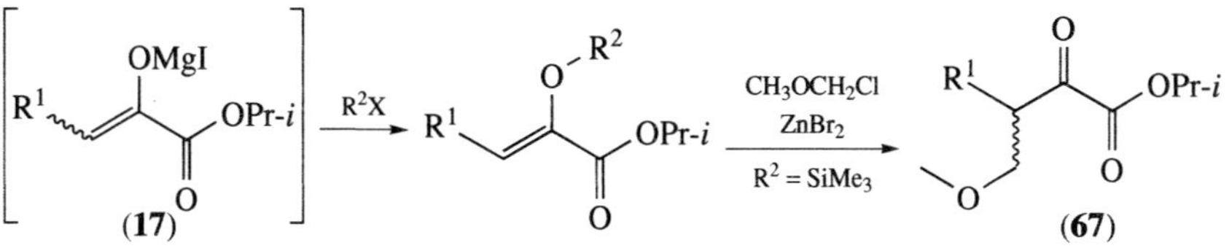

(96)

R²:
(**63**) MeOCH₂
(**64**) Me
(**65**) (EtO)₂P(O)
(**66**) Me₃Si

R¹: (**a**) *i*-Pr, (**b**) Et, (**c**) Ph(CH)Me, (**d**) ... , (**e**) ...

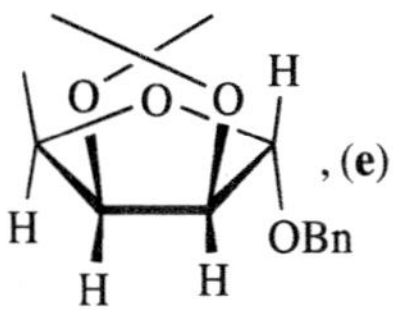

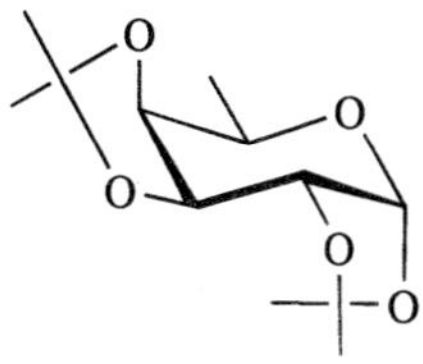

The reaction is stereoselective giving the *Z*-isomer as the major product with alkyl halides and silyl halides. Interestingly, *E* selectivity is observed with enolphosphates.

In the aliphatic series, the C-alkylation of enolates **17** is achieved through their O-silylated derivatives **66**. In the presence of a catalytic amount of ZnBr₂, the silyl enol ether **66** reacted with ClCH₂OCH₃ to give only the C-alkylated α-ketoester **67**. The alkylation is regiospecific but the ketoesters (**67a, c**) are obtained with modest yields (about 50% yield).

An interesting application in the glucidic series concerns the one-pot magnesium dihalide-catalyzed Claisen rearrangement of 2-alkoxycarbonyl allyl vinyl ethers obtained from the iodomagnesium enolate **17e**. This represents the first synthetic application for the stereoselective construction of a disaccharide analogue **70** including a galactosyl and an ulosonic isopropyl ester moiety[105]. Thus, the reaction of the magnesium enolate **17e** with allyl bromide led to the α-ketoester enol precursor **68** (equation 97). The reaction is rather slow, due to the poor reactivity of enolate **17e**, so that, as the slow O-allylation proceeds, a mild Claisen rearrangement occurs, catalyzed by the magnesium salts present in the reaction medium. Finally, after 48 h at 33 °C, the C-allylation product **69** is obtained in 50% yield accompanied by α-ketoester **18e**. The latter compound **18e** results from the hydrolysis of the unreacted starting magnesium enolate **17e** and is easily separated from **69** (37% yield after a chromatographic purification).

The reaction is 100% stereoselective and affords **69** as a unique stereomer with the (*R*)-configuration at C-6, assigned by X-ray analysis. Consequently, the configuration at C-6 results from the well-known chair transition state model for the (*Z*)-O-allyl enol **68** in which the allyl unit is on the opposite side to the isopropylidene ketal at C3–C4 and reacts on the *Si* face of the trisubstituted carbon–carbon double bond. Such a transition state leads to the sole (*R*)-configuration in **69**.

The asymmetric inducting effect of the chiral pyranic ring placed outside of the six centers of the chair transition state has to be noted. In the present case, magnesium dihalide seems to play a crucial role in terms of stereocontrol as an efficient tool to fix a defined and single transition state conformation and to maximize asymmetric induction. It allows one also to carry out the rearrangement under mild reaction conditions at low temperature.

The C-allyl ketoester **69** represents a key precursor for the stereoselective synthesis of the disaccharide analogue **70** in which an ulosonic residue could be installed via the dihydroxylation of the double bond of **69** (equation 98). These glucidic α-ketoacids are involved in biosynthetic pathways of bacteria and constitute important targets for the design of new antibacterial agents.

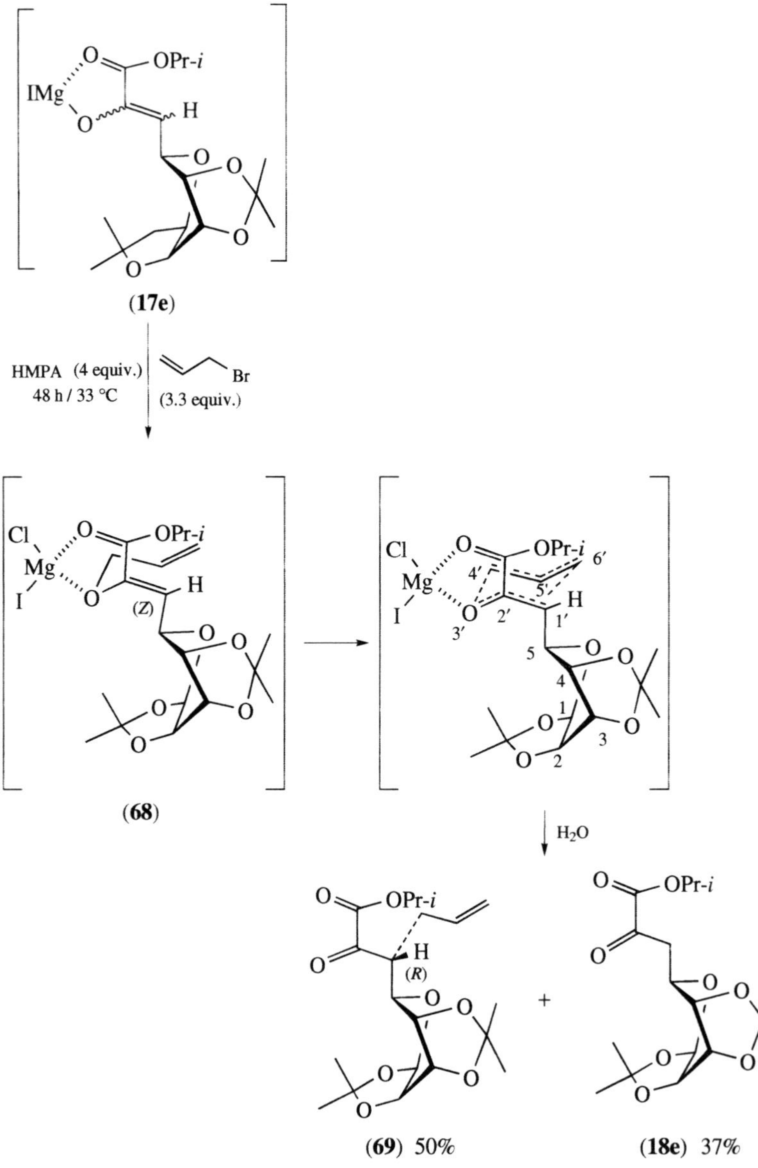

(97)

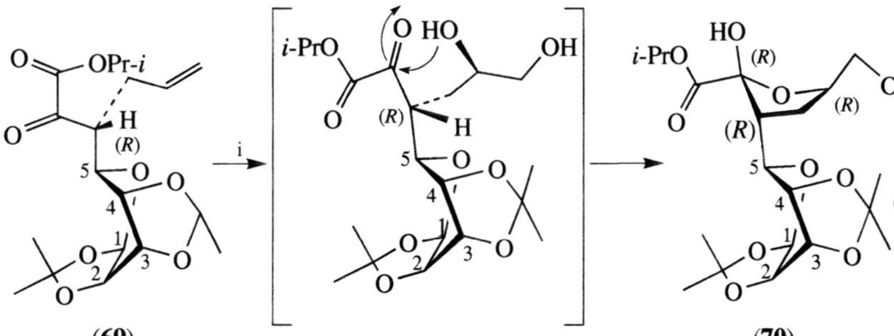

i = OsO$_4$ / NMO, 24 h (95% yield, de: 3/2) or AD-mix-β, 48 h (90% yield, 9/1 dr)

2. Reactions of magnesium chelates of β-ketoesters or β-diketones

Prepared from (MeO)$_2$Mg/MeOH, the magnesium chelates of β-ketoesters or β-diketones show high stability and sometimes a particular reactivity[106,107] (see Section II).

Base-catalyzed transformations can be carried out elsewhere on a complex molecule in the presence of such protected β-dicarbonyl magnesium chelate. For example, the chelated magnesium enolate of a β-ketoester such as **71** prevents the carbonyl keto group becoming an acceptor in aldol condensations. However, in the presence of excess of magnesium methanolate, exchange of the acetyl methyl protons can occur via a carbanion **72** stabilized by delocalization into the adjacent chelate system (equation 99).

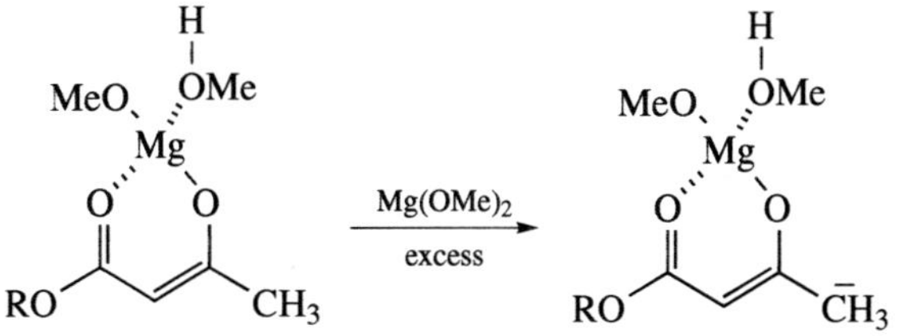

Consequently, the magnesium chelate **71** can also react as a nucleophilic donor in aldol reactions. In the chemistry involving magnesium chelates, these two aspects model their mode of action as nucleophilic partners in aldol condensations. This is exemplified in aldol condensations of γ-diketones[106]. Thus, sodium hydroxyde catalyzed cyclization of diketone **73** to give a mixture of 3,5,5-trimethyl-cyclopent-2-enone **74** and 3,4,4-trimethyl-cyclopent-2-enone **75** in a 2.2/1 isomeric ratio (equation 100). When treated with magnesium methanolate, the insertion of a α-methoxy carbonyl group as control element, as in **76**, allows the formation of a chelated magnesium enolate **77**, and the major product is now mainly the aldol **78**. This latter treated with aqueous NaOH provides the trimethylcyclopent-2-enones **74** and **75** in a 1/49 ratio.

Alternatively, the displacement of alkoxide by attack of a nucleophile at the carbonyl unit of complexed magnesium β-ketoester can proceed smoothly according to a Claisen-type reaction, as described in the transformation of xantophanic enol **79** into resorcinol **80** (equation 101)[107].

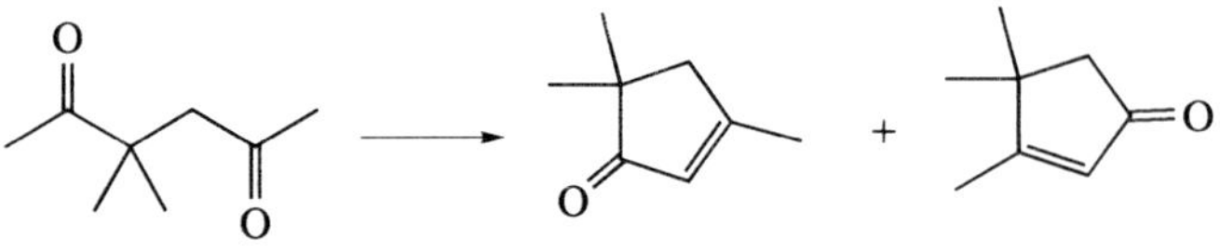

(73) → (74) + (75)

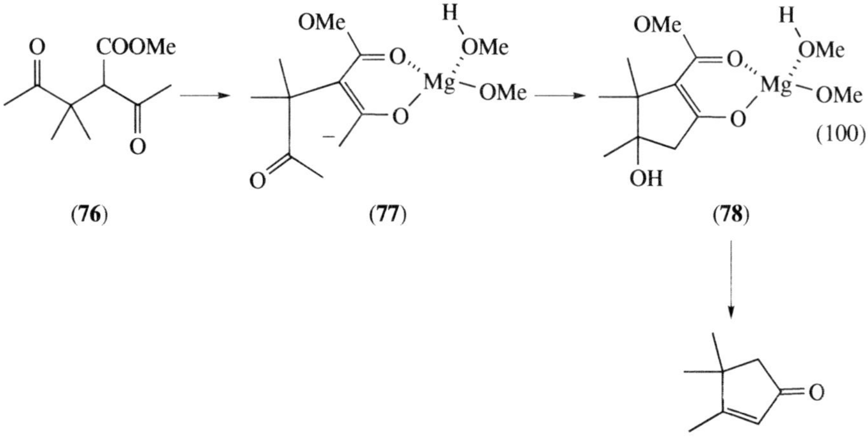

(76) → (77) → (78) (100)

↓

(75)

Other examples including the compared reactivity of sodium and magnesium enolates of β-ketoester, especially in xanthyrone and glaucyrone chemistry, are detailed in the review of L. Crombie[108].

3. Reactions of magnesium dialkyl malonate or magnesium hydrogen alkyl malonate

Anions derived from malonates are ambident nucleophiles, which can react at the carbon or oxygen atom. Therefore, carbon–carbon bond-forming reactions by alkylation or acylation of enolates have been encountered with difficulties. Side reactions which may cause problems are the above-mentioned competing O-reaction and dialkylation[109].

Ireland and Marshall[110] have shown that the use of magnesium malonate enolates offered several advantages: (a) magnesium stabilizes the malonate anion and allows some selective C-alkylation and acylation; (b) magnesium malonate is more soluble and more stable than the corresponding lithium malonate.

Thus, magnesium malonate enolates are useful tools for C-acylation reactions in the synthesis of β-ketoesters. Two reagents could be used for such purposes: magnesium malonate derived from hydrogen methylmalonate, and magnesium malonate derived from dialkylmalonate.

They are easily prepared from Grignard reagents, magnesium ethoxide or with the complex magnesium chloride-trialkylamine (see Section II where the generation of these reagents is discussed). The acylation of anions derived from malonates can be achieved with acyl chlorides, acyl imidazoles, alkoxycarbonylimidazoles or mixed anhydrides.

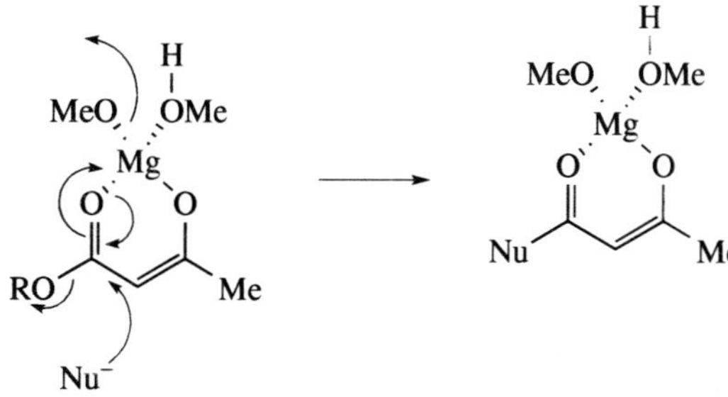

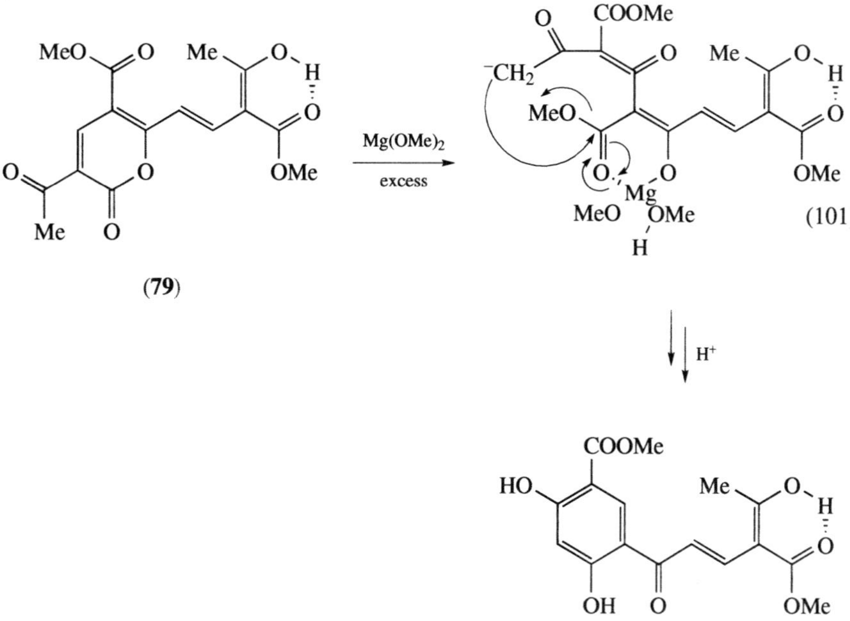

This acylation procedure is very useful in the synthesis of biologically active products, such as steroids and aminoacids.

Recently, Francis and coworkers[111] have reported a scalable stereoselective synthesis of Scymnol. Scymnol is a derivative of cholic acid, which finds applications in the treatment of various skin problems and liver dysfunction. Triformyloxycholic acid chloride **81** is treated with the magnesium enolate of diethylmalonate to afford the β-ketodiester diethyl 3α,7α,12α-triformyloxy-24-oxo−5β-cholestane-26,27-dioate (**82**). The stereoselective hydrogenation of **82** gives the corresponding β-hydroxy diester **83** using a BINAP ruthenium(II) catalyst. Subsequent reduction of the diester moiety and deprotection of the hydroxyl groups afforded Scymnol, **84** (equation 102).

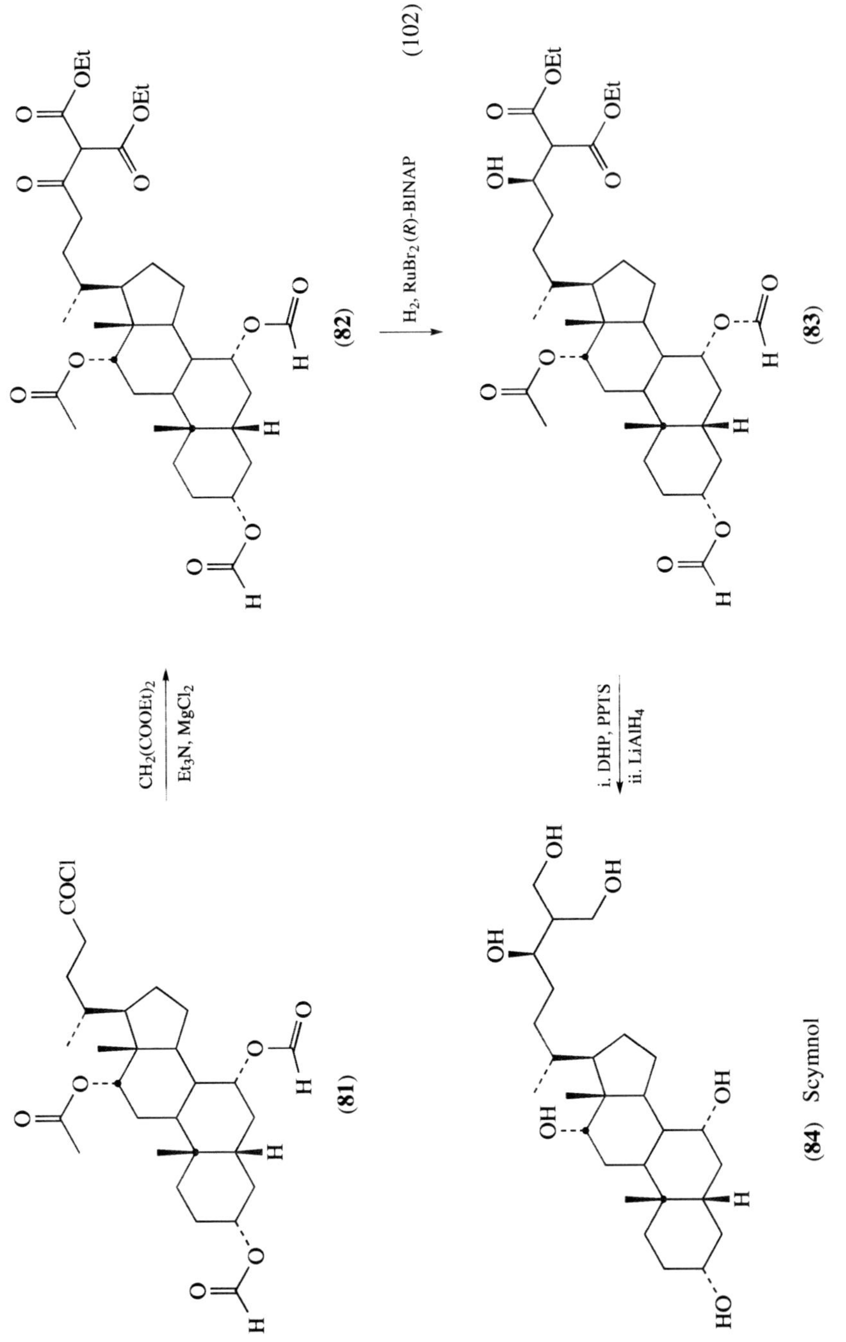

(81) →[CH₂(COOEt)₂ / Et₃N, MgCl₂]→ (82) →[H₂, RuBr₂(R)-BINAP]→ (83)   (102)

(83) →[i. DHP, PPTS; ii. LiAlH₄]→ (84) Scymnol

11. Preparation and reactivity of magnesium enolates

Different syntheses of γ-amino-β-ketoeter derivatives from N-protected L-aminoacids by N,N'-carbonyldiimidazole activation and treatment with the magnesium enolate of hydrogen ethyl malonate are described. These compounds are useful intermediates in the preparation of active amino acid analogues, as illustrated and summarized in equation 103.

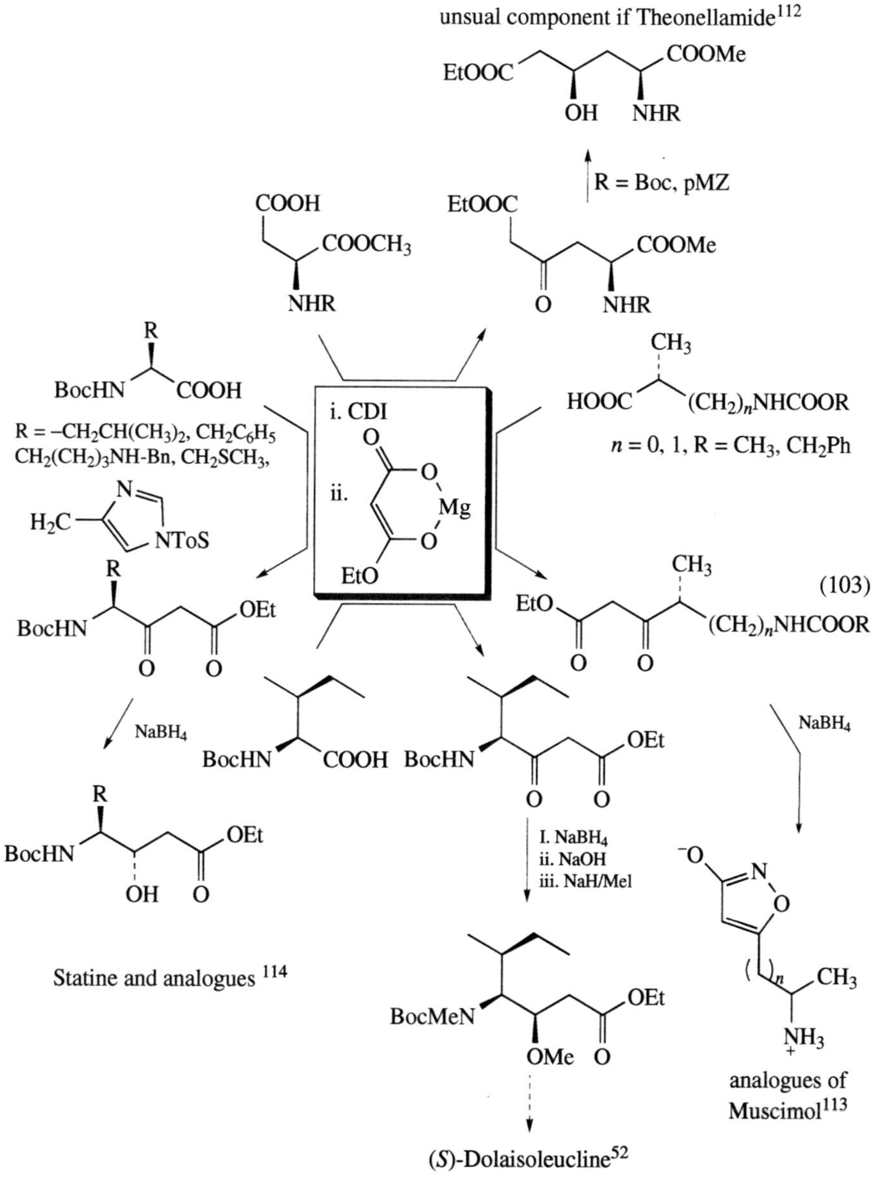

(103)

With a similar procedure, Krysan[60] has described a practical synthesis of α-acylamino-β-ketoesters. The combination of $MgCl_2$ and R_3N is used to metallate a series of alkyl hydrogen (acylamino)malonates at 0 °C. The resulting magnesium enolates react with a wide range of acid chlorides to give the corresponding α-acylamino-β-keto esters in good to excellent yields (equation 104).

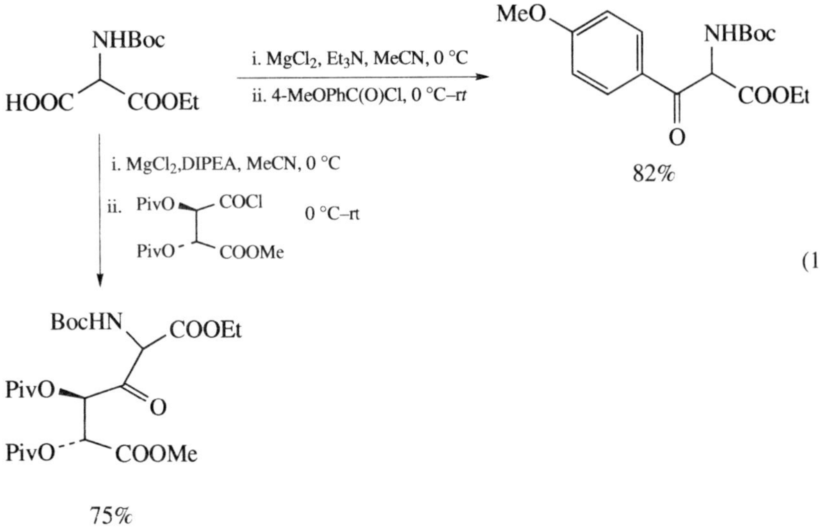

(104)

75%

A general methodology for the preparation of β-amino-α,β-diesters has been developed by Sibi and Asano[41]. It is based on an enantioselective conjugate addition of chiral organomagnesium amides to enamidomalonates. Addition of a large variety of nucleophiles gives adducts in good yields and enantioselectivities. Alkyl, vinyl and aryl nucleophiles are compatible in this method. It should be noted that a simple change of the chiral ligand allows one to prepare products with an opposite configuration (equations 105 and 106, Table 14).

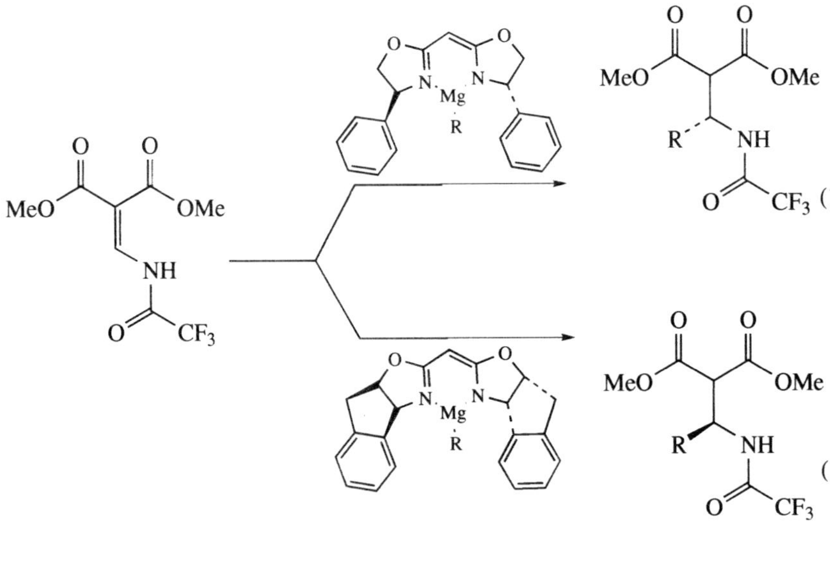

11. Preparation and reactivity of magnesium enolates 499

The simplicity and convenience of this chemistry led to an industrial application on a large scale, as illustrated by the synthesis of Danofloxacin, an animal health quinolone antibiotic[115] (equation 107).

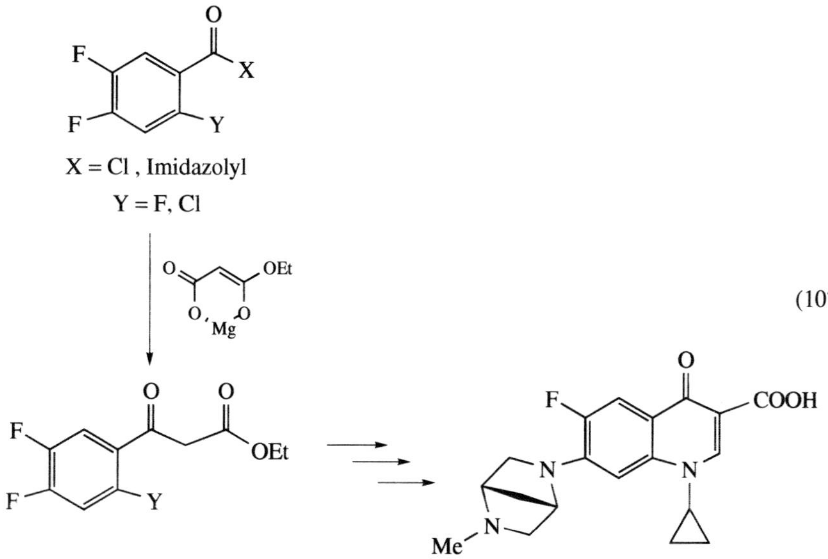

X = Cl, Imidazolyl
Y = F, Cl

(107)

Danofloxacin

E. Reactions of Magnesium Amide and Lactam Enolates with Electrophiles

Aldol reactions of simple amide enolates give poor stereoselection. Stimulated by the interest in β-lactams, the stereochemistry of aldol reactions of chiral magnesium enolates of β-lactams has been studied[116, 117]. The best results have been obtained with 6,6-dibromopenams **85** (equation 108). After bromine–magnesium exchange with MeMgBr,

TABLE 14. Conjugate addition of different nucleophiles to enamidomalonates

R	Yield % (ee%) 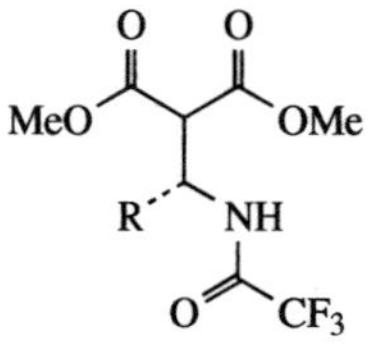	Yield % (ee%) 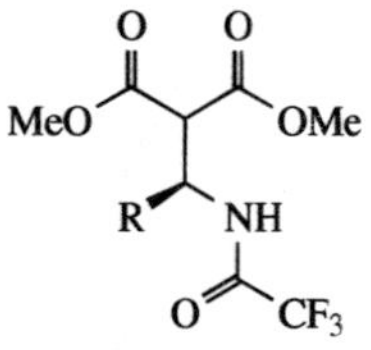
Et	79 (78)	65 (−78)
i-Pr	72 (83)	62 (−80)
n-Bu	72 (81)	65 (−70)
(CH$_2$)$_{17}$CH$_3$	71 (83)	68 (−70)
Cyclohexyl	69 (93)	64 (−74)
CH$_2$=CH	65 (86)	58 (−59)
phenyl	70 (81)	65 (−56)

then addition of ethanal, a single isomer **86** is obtained in excellent yield. It results from the attack of the concave face of the bicyclic system[118].

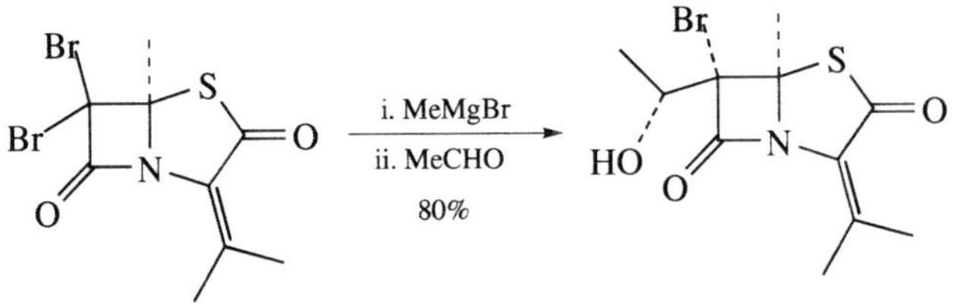

(85) → i. MeMgBr; ii. MeCHO, 80% → (86)   (108)

Other chiral magnesium enolates derived from amides are known to react with aldehydes. For example, the aldol-type reaction of magnesium enolate of (−)-*trans*-2-*N*,*N*-diethylacetamide-1,3-dithiolanes-*S*-oxide with isobutyraldehyde affords a single diastereomer in 82%. The relative stereochemistry of the adduct originates from a rigid transition state **87** where the oxygen atoms of the enolate and the aldehyde are coordinated to the magnesium atom[55].

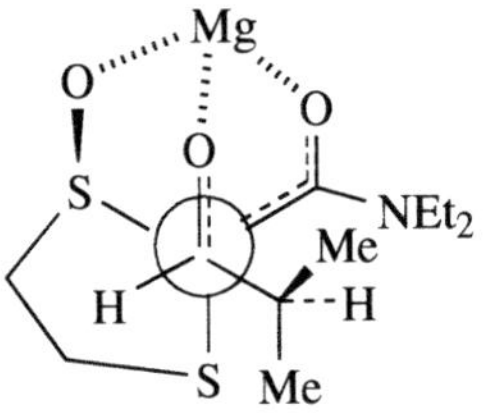

(87)

F. Reactions of Magnesium Thioesters and Thioamide Enolates with Electrophiles

Recently, Feringa and coworkers[46] have described a very interesting protocol for the catalytic asymmetric 1,4-addition of Grignard nucleophiles to α,β-unsaturated thioesters. These conjugate thioesters may be also used to perform some tandem 1,4-addition-aldol reactions via the copper-catalyzed 1,4-addition of organomagnesium derivatives, in the presence of a catalytic amount of JOSIPHOS, in high yields and level of stereocontrol. It was found that the magnesium enolate intermediate readily undergoes aldol reaction with benzaldehyde after 1 min at −75 °C (equation 109).

The utility and efficiency of this methodology is demonstrated by the first catalytic asymmetric synthesis of (−)-phaseolinic acid[46], a natural product displaying useful antifungal, antitumor and antibacterial properties, as illustrated in equation 110.

This elegant tandem protocol allows the synthesis of this natural product in a good 54% overall yield.

Thioamide enolates are also interesting substrates for the stereoselective aldol-type reactions. The aldol stereochemistry is very sensitive to the conditions of preparation of magnesium thioamide enolates and it generally gives different results depending on the procedure used. Illustrations of some aspects of the reactivity are provided in the examples presented below.

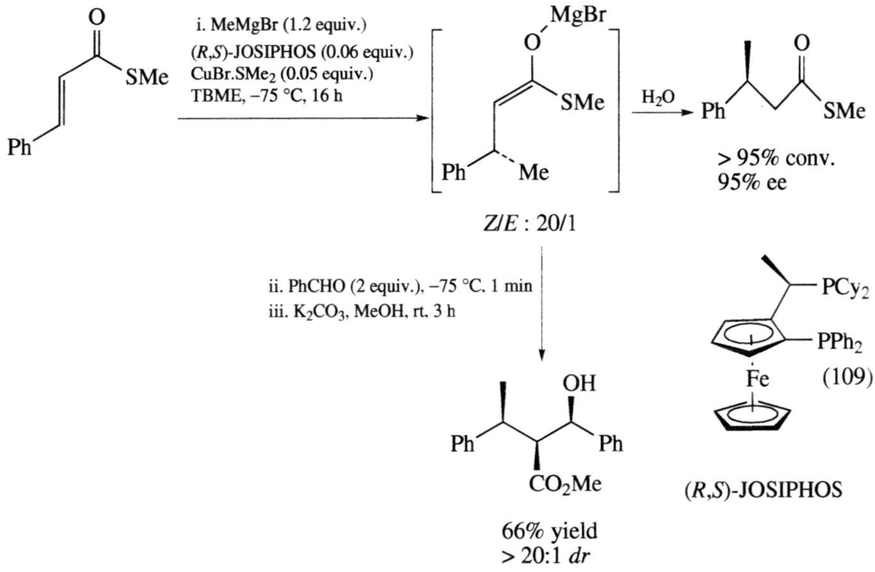

11. Preparation and reactivity of magnesium enolates

(109)

(R,S)-JOSIPHOS

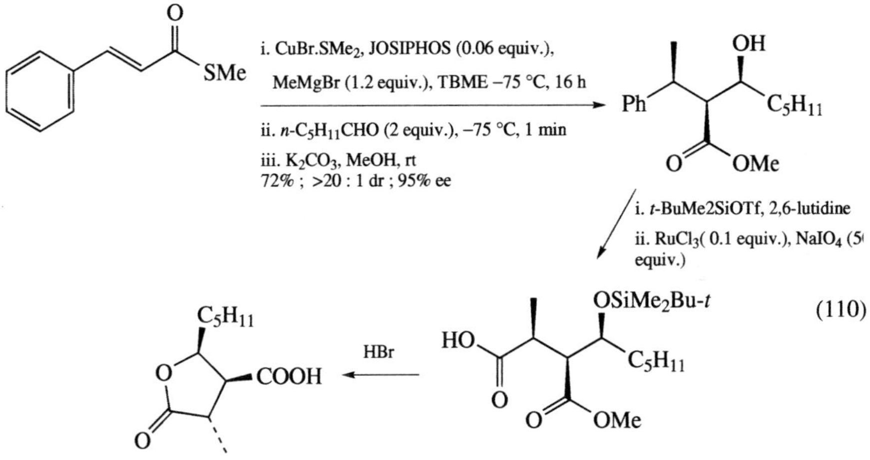

(110)

Phaseolinic acid

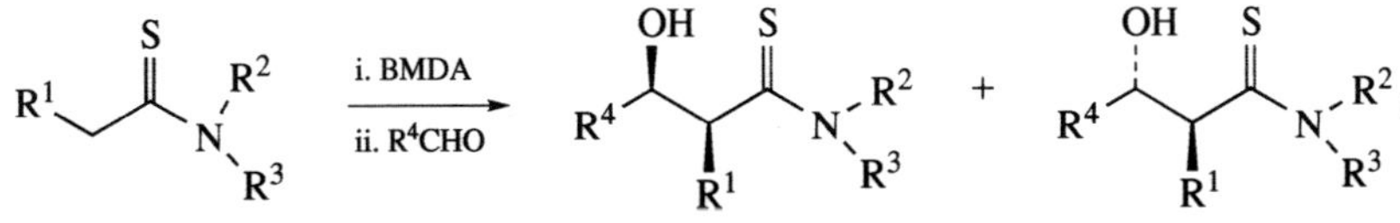

(111)

The products obtained after treatment of thioamides with magnesium bromide diisopropylamide (BMDA) and subsequent aldol condensation are presented in equation 111. The difference in the stereochemistry between thioamides of secondary amines and primary amines should be noted (Table 15). Although, in each case, the metallation leads to

TABLE 15. Aldol stereochemistry of thioamides

R^1	R^2	R^3	R^4	syn/anti
Me	Me	Me	i-Pr	95/5
Me	Me	Me	Ph	93/7
Me	Me	Me	$CH_2=CH$	89/11
Ph	Me	Me	i-Pr	72/28
Ph	Me	H	i-Pr	2/98
Ph	Me	H	Ph	6/94

TABLE 16. Aldol stereochemistry of α,β-unsaturated thioamides

R^1	R^2	R^3	X	syn/anti
Me	Me	Me	I	4/96
Me	Me	i-Pr	I	<1/99
Me	Et	Ph	Br	15/85
Ph	i-Pr	Me	Br	<1/99
Ph	i-Pr	i-Pr	Br	<1/99
Ph	i-Pr	H	Br	18/82

the (Z)-enolates, dianions of thioamides derived from primary amines react with aldehydes to afford predominantly the *anti* aldols whereas enolates of thioamides derived from secondary amines give the *syn* aldols.

If thioamide enolates are prepared by conjugate addition of Grignard reagents to α,β-unsaturated thioamides of secondary amines, the reaction of these enolates with aldehydes affords *anti* aldols. These results are rationalized by the formation of a boat-like, chelate transition state[119,120]. Representative examples are provided in equation 112 and Table 16.

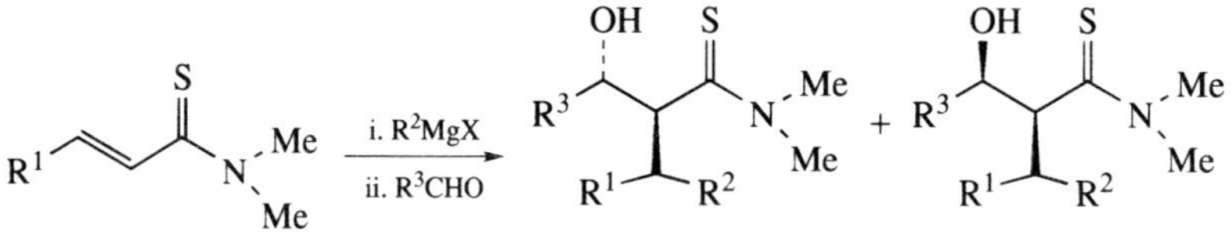

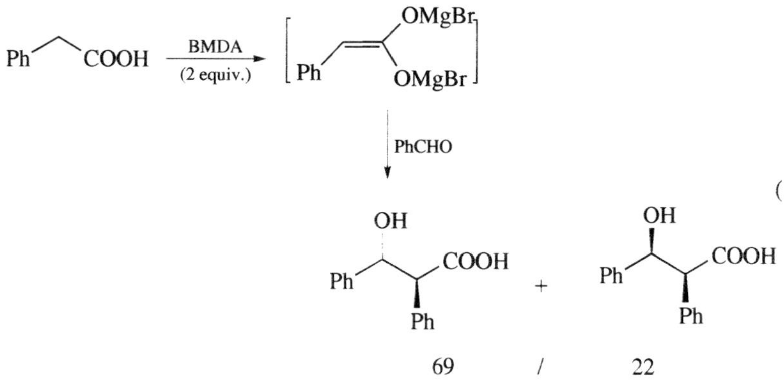

G. Reactions of Carboxylic Acid Dianions with Electrophiles

A few years ago, Blagouev and Ivanov described the bis-deprotonation of aryl acetic acids by Grignard reagents[121]. These magnesium dianions, known as Ivanov reagents, react with aldehydes and ketones. Reaction between dianions of phenylacetic acid and benzaldehyde yields the *anti* β-hydroxy acid as the major diastereomer (*anti/syn* 69/22) (equation 113). This result is in agreement with the formation of a cyclic chair-like transition state according to the model of Zimmerman–Traxler[53b].

H. Reactions with Anions of Chiral Oxazolidinones and Derivatives with Electrophiles

The oxazolidinones have been used as chiral auxiliaries for enolate alkylation and aldol reactions in enantioselective and total syntheses[122–124]. The interest in these substrates is largely known for *syn*-diastereoselective aldol reactions with chlorotitanium or dialkylboron oxazilidinone enolates[123,124] (equation 114).

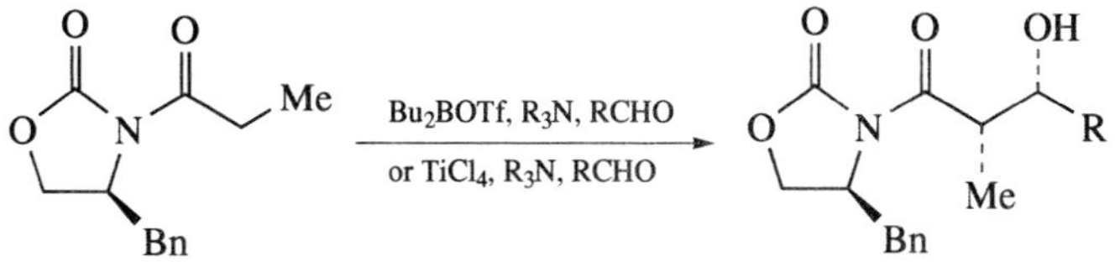

Recently, Evans has reported diastereoselective magnesium halide-catalyzed *anti*-aldol reactions of chiral *N*-acyloxazolidinones **88** (equation 115)[61] and *N*-acylthiazolidine-thiones **90** (equation 116)[62].

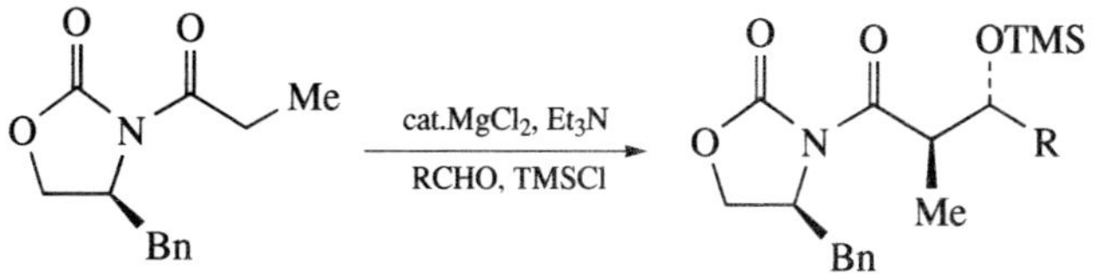

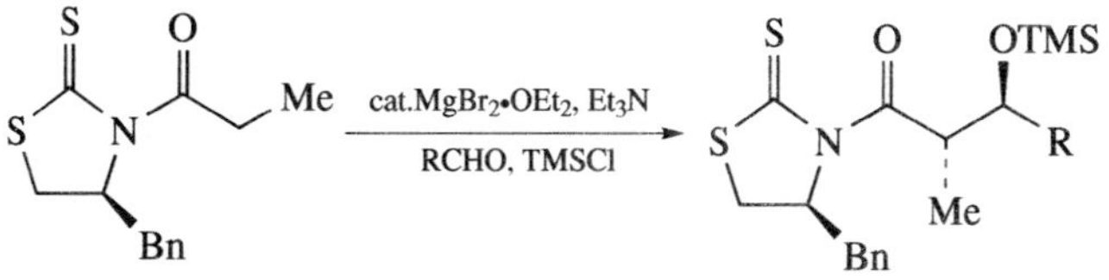

Although *N*-acyloxazolidinones **88** and *N*-acylthiazolidinethiones **90** lead to an *anti* aldol, the respective products **89** and **91** present a different *anti* configuration. Consequently, the corresponding derived magnesium enolates exhibit the opposite face selection in these reactions. On the basis of previous results involving enolates of various metal complexes such as boron, titanium, lithium or sodium enolates, the (Z)-metal enolate is always formed. Evans suggests that these reactions proceed via either a five- or six-coordinate magnesium chair-like transition state **92** for *N*-acyloxazolidinones **88** and a boat-like transition state **93** for *N*-acylthiazolidinethiones **90**. The observed stereochemical results negate the possibility that the thione C=S moiety is coordinated to the magnesium center in the aldol transition state **93**, whereas a chair Zimmerman–Traxler transition state is assumed in the case of magnesium enolate **92**. Semiempirical calculations support such hypotheses.

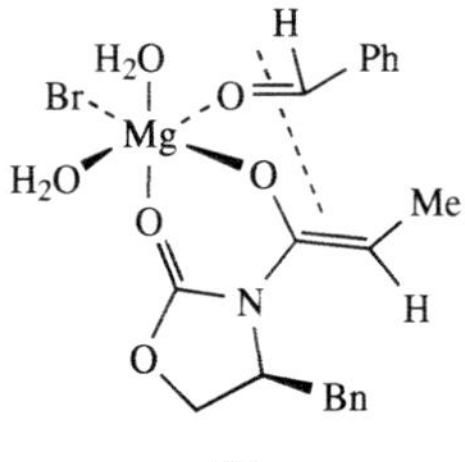

(92)

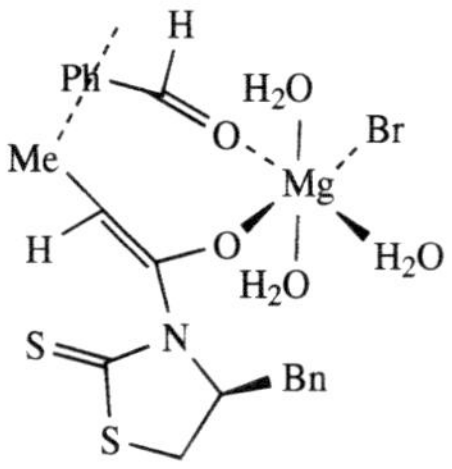

(93)

The enantioselective amination of *N*-acyl oxazolidinones has been studied as part of a general approach to the synthesis of arylglycines. In this case, the enolization is initiated by a chiral magnesium bis(sulfonamide) complex. The oxazolidinone imide enolates are generated using catalytic conditions (10 mol% of magnesium complex) and treated *in situ* with BocN=NBoc to provide the corresponding hydrazide. 20 mol% of *N*-methyl-*p*-toluensulfonamide are added to accelerate the reaction[90] (equation 117).

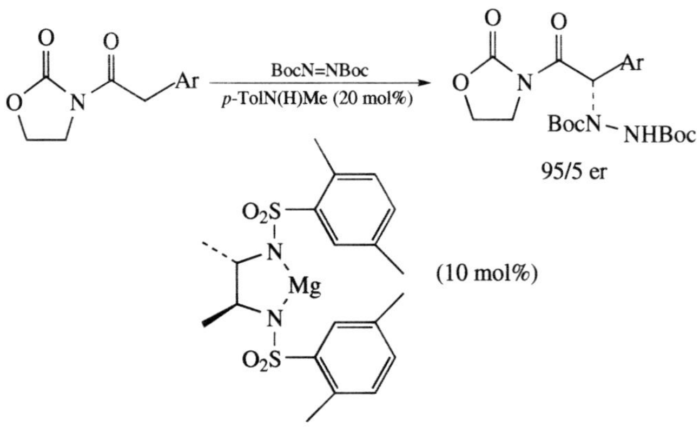

(117)

The procedure is applicable to a variety of aryl-substituted imides incorporating either electron-withdrawing or electron-donating substituents. The reaction leads to high enantiomeric ratio (95/5 er) and the products are easily purified by recrystallization (84–97% yields).

I. Reactions of Miscellaneous Magnesium Chiral Enolates

A straightforward example of asymmetric synthesis is the enantioselective synthesis of α-amino acid which uses (S)-valine ethyl ester **95** as a chiral auxiliary[61] (equation 118). In the process, the chiral auxiliary **95** is condensed with σ-asymmetric α-amino diethylmalonate **94**. Treatment of the product with trimethyloxonium tetrafluroroborate provides the bis-imino ether **96**, which is enolized with MgBr$_2$/triethylamine. The geometry of this enolate is fixed by a six-membered ring, and the isopropyl group of the valine residue imposes a strong facial bias by hindering the 'lower' face of the enolate. The aldehyde is directed to the 'upper' face. Silylation, reduction and hydrolysis provide the acids **97** or esters **98** of alkylated serine derivatives. Four diastereomers are obtained and separated by chromatography (70–86% yields).

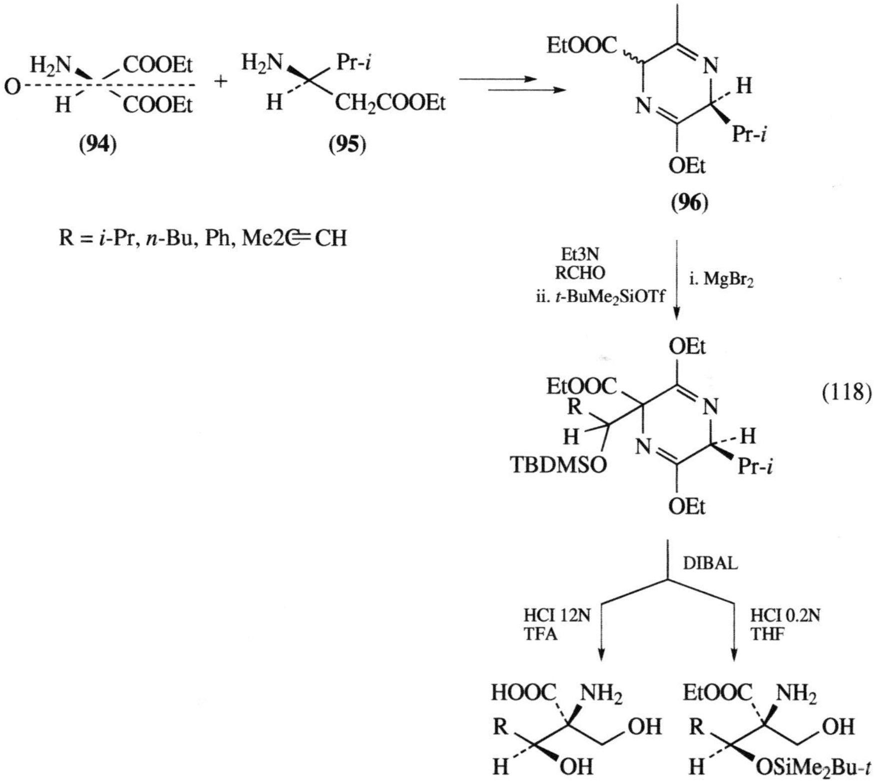

An interesting class of chiral enolates are allenyl enolates. These ambident nucleophiles bear an axis of chirality. Krause and coworkers have found that an axis to center chirality transfer takes place in the aldol reaction of chiral magnesium allenyl enolate with pivalic aldehyde[26]. The aldol reaction proceeds with good diastereofacial selectivity if the lithium enolate is transmetallated with magnesium bromide etherate. The traditional Zimmerman–Traxler chair-like transition state doesn't explain the u-configuration of the aldol adduct. However, the stereochemical course of this axis to center chirality transfer can be rationalized by assuming a boat transition state, as illustrated in equation 119.

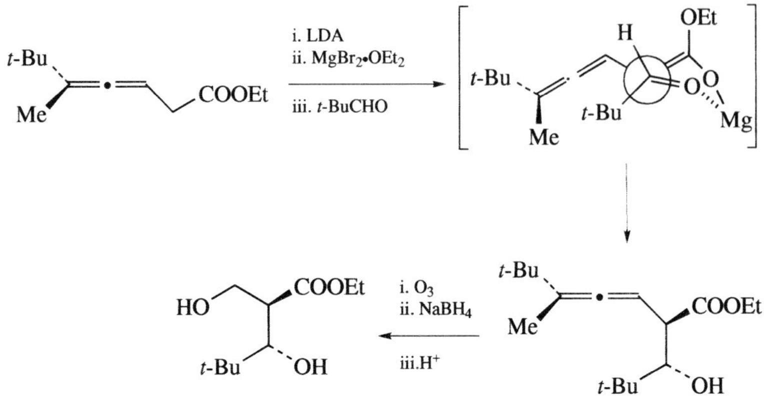

(119)

This reaction was used to prepare enantiomerically pure ethyl (2S,3S)-4,4-dimethyl-3-hydroxy-2-hydroxymethylpentanoate.

IV. REFERENCES

1. S. M. Malmgren, *Chem. Ber.*, **35**, 3910 (1902).
2. J. P. Collman, *J. Org. Chem.*, **26**, 3162 (1961).
3. J. Colonge and S. Grenet, *Bull. Soc. Chim. Fr.*, **10**, C41 (1953).
4. (a) H. B. Mekelburger and C. S. Wilcox, in *Comprehensive Organic Synthesis* (Eds. B. M. Trost and I. Fleming), Vol. 2, Pergamon Press, Oxford, 1991, pp. 99–128.
   (b) C. H. Heathcock, in *Comprehensive Organic Synthesis* (Eds. B. M. Trost and I. Fleming), Vol. 2, Pergamon Press, Oxford, 1991, pp. 181–235.
5. P. G. Williard, *Comprehensive Organic Synthesis* (Eds. B. M. Trost and I. Fleming), Vol. 1, Pergamon Press, Oxford, 1991, pp. 1–42.
6. A. F. Mateos and J. A. de la Fuente Blanco, *J. Org. Chem.*, **56**, 7084 (1991).
7. J. M. Castro, P. J. Linares-Palomino, S. Salido, J. Altarejos, M. Nogueras and A. Sànchez, *Tetrahedron Lett.*, **45**, 2619 (2004).
8. C. Grison, S. Petek and P. Coutrot, *Tetrahedron*, **61**, 7193 (2005).
9. C. Grison, S. Petek and P. Coutrot, *Synlett*, 331 (2005).
10. J. Villieras, B. Castro and N. Ferracutti, *C. R. Acad. Sci., Ser. C*, **267**, 915 (1968).
11. A. D. William and Y. Kobayashi, *Org. Lett.*, **3**, 2017 (2001).
12. A. D. William and Y. Kobayashi, *J. Org. Chem.*, **67**, 8771 (2002).
13. B. M. Trost, M. K. T. Mao, J. M. Balkovec and P. Buhlmayer, *J. Am. Chem. Soc.*, **108**, 4965 (1986).
14. T. Satoh, K-I. Onda, N. Itoh and K. Yamakawa, *Tetrahedron Lett.*, **32**, 5599 (1991).
15. T. Satoh, Y. Kitoh, K-I. Onda and K. Yamakawa, *Tetrahedron Lett.*, **34**, 2331 (1993).
16. T. Satoh, N. Itoh, K-I. Onda, Y. Kitoh and K. Yamakawa, *Bull. Chem. Soc. Jpn.*, **65**, 2800 (1992).
17. T. Satoh, Y. Kitoh, K-I. Onda, K. Takano and K. Yamakawa, *Tetrahedron*, **50**, 4957 (1994).
18. T. Satoh and K. Miyashita, *Tetrahedron Lett.*, **45**, 4859 (2004).
19. U. Kazmaier and S. Maier, *Chem. Commun.*, 1991 (1995).
20. S. Matsui, *Bull. Chem. Soc. Jpn.*, **60**, 1853 (1987).
21. D. D. Sternbach, D. M. Rossana and K. D. Onan, *J. Org. Chem.*, **49**, 3427 (1984).
22. M. Larchevêque and A. Debal, *J. Chem. Soc., Chem. Commun.*, 877, (1981).

23. B. Mendler and U. Kazmaier, *Synthesis*, 2239 (2005).
24. T. J. Donohoe and D. House, *Tetrahedron Lett.*, **44**, 1095 (2003).
25. M. Shimizu, Y. Kooriyama and T. Fujisawa, *Chem. Lett.*, 2419 (1994).
26. M. Laux, N. Krause and U. Koop, *Synlett*, 87 (1996).
27. I. A. Stergiades and M. A. Tius, *J. Org. Chem.*, **64**, 7547 (1999).
28. H. Fujisawa, Y. Sasaki and T. Mukaiyama, *Chem. Lett.*, 190 (2001).
29. T. Eicher, in *The Chemistry of the Carbonyl Group* (Ed. S. Patai), Interscience, New York, 1966, p. 638.
30. T. Holm and L. Crossland, *Grignard Reagents New Developments* (Ed. H. G. Richey, Jr.), Wiley, Chichester, 2000, p. 19.
31. B. J. Wakefield, *Organomagnesium Methods in Organic Synthesis*, Academic Press, London, 1995, p. 124.
32. A. G. Pinkus and A-B. Wu, *J. Org. Chem.*, **40**, 2816 (1975).
33. M. J. Chapdelaine and M. Hulce, *Org. React.*, **38**, 225 (1990).
34. D. D. Weller, R. D. Gless and H. Rapoport, *J. Org. Chem.*, **42**, 1485 (1977).
35. J. Tanaka, S. Kanemasa, Y. Ninomiya and O. Tsuge, *Bull. Chem. Soc. Jpn.*, **63**, 466 (1990).
36. S. Dalai, M. Limbach, L. Zhao, M. Tamm, M. Sevvana, V. V. Sokolov and A. de Meijere, *Synthesis*, 471 (2006).
37. Y. Tamaru, T. Hioki, S.-I. Kawamura, H. Satomi and Z.-I. Yoshida, *J. Am. Chem. Soc.*, **106**, 3876 (1984).
38. K. Kobayashi, H. Takabatake, T. Kitamura, O. Morikawa and H. Konishi, *Bull. Chem. Soc. Jpn.*, **70**, 1697 (1997).
39. K. Kobayashi, R. Nakahashi, A. Shimizu, T. Kitamura, O. Morikawa and H. Konishi, *J. Chem. Soc., Perkin Trans. 1*, 1547 (1999).
40. M. E. Bunnage, S. G. Davies, C. J. Goodwin and I. A. S. Walters, *Tetrahedron: Asymmetry*, **5**, 35 (1994).
41. M. P. Sibi and Y. Asano, *J. Am. Chem. Soc.*, **123**, 9708 (2001).
42. B. Breit and P. Demel, *Modern Organic Chemistry* (Ed. N. Krause), Wiley-VCH, Weinheim, 2002, p. 188.
43. B-S. Lou, G. Li, F-D. Lung and V. J. Hruby, *J. Org. Chem.*, **60**, 5509 (1995).
44. F. Lopez, S. R. Harutyunyan, A. J. Minnaard and B. L. Feringa, *J. Am. Chem. Soc.*, **126**, 12784 (2004).
45. R. Des Mazery, M. Pullez, F. Lopez, S. R. Harutyunyan, A. J. Minnaard and B. L. Feringa, *J. Am. Chem. Soc.*, **127**, 9966 (2005).
46. G. P. Howell, S. P. Fletcher, K. Geurts, B. ter Horst and B. L. Feringa, *J. Am. Chem. Soc.*, **128**, 14977 (2006).
47. B. J. M. Jansen, C. J. Hendrikx, N. Masalov, G. A. Stork, T. M. Meulemans, F. Z. Macaev and A. de Groot, *Tetrahedron*, **56**, 2075 (2000).
48. G. L. Larson, D. Hernandez, I. Montes de Lopez-Cepero and L. E. Torres, *J. Org. Chem.*, **50**, 5260 (1985).
49. A. G. Pinkus, D. F. Mullica, W. O. Milligan, D. A. Grossie and P. W. Hurd, *Tetrahedron*, **40**, 4829 (1984).
50. Y. Maroni-Barnaud, J. Bertrand, F. Ghozland, L. Gorrichon-Guigon, Y. Koudsi, P. Maroni and R. Meyer, *C. R. Acad. Sci. Paris*, **280**, C-221 (1975).
51. P. Fellmann and J-E. Dubois, *Tetrahedron Lett.*, **3**, 247 (1977).
52. (a) T. Shioiri, K. Hayashi and Y. Hamada, *Tetrahedron*, **49**, 1913 (1993).
    (b) Y. Hamada, K. Hayashi and T. Shioiri, *Tetrahedron Lett.*, **32**, 931 (1991).
53. (a) D. Ivanov and N. Nicoloff, *Bull. Soc. Chim. Fr.*, **51**, 1325 (1932).
    (b) H. E. Zimmerman and M. D. Traxler, *J. Am. Chem. Soc.*, **79**, 1920 (1957).
54. G. Solladié, in *Asymmetric Synthesis* (Ed. J. D. Morrison), Vol. 2, Academic Press, New York, 1983, p. 157.
55. M. Corich, F. Di Furia, G. Licini and G. Modena, *Tetrahedron Lett.*, **33**, 3043 (1992).
56. E. Hevia, K. W. Henderson, A. R. Kennedy and R. E. Mulvey, *Organometallics*, **25**, 1778 (2006).
57. L. Crombie, D. F. Games and A. W. G. James, *J. Chem. Soc., Perkin Trans. 1*, 2715 (1996).
58. M. W. Rathke and P. J. Cowan, *J. Org. Chem.*, **50**, 2622 (1985).
59. B. Corbel, I. L'Hostis-Kervella and J. P. Haelters, *Synth. Commun.*, **26**, 2561 (1996).
60. D. J. Krysan, *Tetrahedron Lett.*, **37**, 3303 (1996).

61. D. A. Evans, J. S. Tedrow, J. T. Shaw and C. Wade Downey, *J. Am. Chem. Soc.*, **124**, 392 (2002).
62. D. A. Evans, C. Wade Downey, J. T. Shaw and J. S. Tedrow, *Org. Lett.*, **4**, 1127 (2002).
63. S. Sano, T. Miwa, X-K. Liu, T. Ishii, T. Takehisa, M. Shiro and Y. Nagao, *Tetrahedron: Asymmetry*, **9**, 3615 (1998).
64. M. Stile, *J. Am. Chem. Soc.*, **81**, 2598 (1959).
65. (a) E. S. Hand, S. C. Johnson and D. C. Baker, *J. Org. Chem.*, **62**, 1348 (1997).
    (b) S. M. Kerwin, A. G. Paul and C. H. Heathcock, *J. Org. Chem.*, **52**, 1686 (1987).
66. P. E. Eaton, C. H. Lee and Y. Xiong, *J. Am. Chem. Soc.*, **111**, 6016 (1989).
67. K. Kobayashi and H. Suginome, *Bull. Chem. Soc. Jpn.*, **59**, 2635 (1986).
68. T. Hiyama and K. Kobayashi, *Tetrahedron Lett.*, **23**, 1597 (1982).
69. K. Sisido, K. Kumazawa and H. Nozaki, *J. Am. Chem. Soc.*, **82**, 125 (1960).
70. H. Yamataka, M. Shimizu and M. Mishima, *Bull. Chem. Soc. Jpn.*, **75**, 127 (2002).
71. K. Kobayashi, D. Iitsuka, O. Morikawa and H. Konishi, *Synthesis*, 51 (2007).
72. K. Kobayashi, T. Nakashima, M. Mano, O. Morikawa and H. Konishi, *Chem. Lett.*, 602 (2001).
73. T. Matsumoto, T. Murayama, S. Mitsuhashi and T. Miura, *Tetrahedron Lett.*, **40**, 5043 (1999).
74. G. Lessène, R. Tripoli, P. Cazeau, C. Biran and M. Bordeau, *Tetrahedron Lett.*, **40**, 4037 (1999).
75. M. E. Krafft and R. A. Holton, *Tetrahedron Lett.*, **24**, 1345 (1983).
76. D. Bonafoux, M. Bordeau, C. Biran, P. Cazeau and J. Dunogues, *J. Org. Chem.*, **61**, 5532 (1996).
77. D. Bonafoux, M. Bordeau, C. Biran and J. Dunogues, *Synth. Commun.*, **28**, 93 (1998).
78. X. He, J. Jacob Morris, B. C. Noll, S. N. Brown and K. W. Henderson, *J. Am. Chem. Soc.*, **128**, 13599 (2006).
79. X. He, J. F. Allan, B. C. Noll, A. R. Kennedy and K. W. Henderson, *J. Am. Chem. Soc.*, **127**, 6920 (2005).
80. K. Kobayashi, Y. Fuchimoto, K. Hayashi, M. Mano, M. Tanmatsu, O. Morikawa and H. Konishi, *Synthesis*, 2673 (2005).
81. L. Camici, P. Dembech, A. Ricci, G. Seconi and M. Taddei, *Tetrahedron*, **44**, 4197 (1988).
82. B. J. Childs and G. L. Edwards, *Tetrahedron Lett.*, **34**, 5341 (1993).
83. N. A. Van Draanen, S. Arseniyadis, M. T. Crimmins and C. H. Heathcock, *J. Org. Chem.*, **56**, 2499 (1991).
84. C. H. Heathcock and S. Arseniyadis, *Tetrahedron Lett.*, **49**, 6009 (1985).
85. J. D. Anderson, P. Garcia Garcia, D. Hayes, K. W. Henderson, W. J. Kerr, J. H. Moir and K. P. Fondekar, *Tetrahedron Lett.*, **42**, 7111 (2001).
86. M. J. Bassindale, J. J. Crawford, K. W. Henderson and W. J. Kerr, *Tetrahedron Lett.*, **45**, 4175 (2004).
87. K. W. Henderson, W. J. Kerr and J. H. Moir, *Chem. Commun.*, 479 (2000).
88. K. W. Henderson, W. J. Kerr and J. H. Moir, *Synlett*, 1253 (2001).
89. E. L. Carswell, D. Hayes, K. W. Henderson, W. J. Kerr and C. J. Russell, *Synlett*, 1017 (2003).
90. D. A. Evans and S. G. Nelson, *J. Am. Chem. Soc.*, **119**, 6452 (1997).
91. C. J. R. Adderley, G. V. Baddeley and F. R. Hewgill, *Tetrahedron*, **23**, 4143 (1967).
92. K. Fuji, T. Kawabata and A. Kuroda, *J. Org. Chem.*, **60**, 1914 (1995).
93. H. Yamataka, O. Aleksiuk, S. E. Biali and Z. Rappoport, *J. Am. Chem. Soc.*, **118**, 12580 (1996).
94. P. Kisanga, D. McLeod, B. D' Sa and J. Verkade, *J. Org. Chem.*, **64**, 3090 (1999).
95. R. K. Dieter, in *Modern Organic Chemistry* (Ed. N. Krause), Wiley-VCH, Weinheim, 2002, p. 79.
96. J. Tanaka, H. Kobayashi, S. Kanemasa and O. Tsuge, *Bull. Chem. Soc. Jpn.*, **62**, 1193 (1989).
97. J. Bertrand, N. Cabrol, L. Gorrichon-Guigon and Y. Maroni-Barnaud, *Tetrahedron Lett.*, **47**, 4683 (1973).
98. P. Fellmann and J. E. Dubois, *Tetrahedron*, **34**, 1349 (1978).
99. A. P. Dove, V. C. Gibson, E. L. Marshall, A. J. P. White and D. J. Williams, *J. Chem. Soc., Chem. Commun.*, 1208 (2002).
100. B. Rague, Y. Chapleur and B. Castro, *J. Chem. Soc., Perkin Trans. 1*, 2063 (1982).

101. A. Bernardi, S. Cardani, L. Colombo, G. Poli, G. Schimperna and C. Scolastico, *J. Org. Chem.*, **52**, 888 (1987).
102. E. J. Corey, W. Li and G. A. Reichard, *J. Am. Chem. Soc.*, **120**, 2330 (1998).
103. K. Takai and C. H. Heathcock, *J. Org. Chem.*, **50**, 3247 (1985).
104. T. Hiyama, K. Nishide and K. Kobayashi, *Tetrahedron Lett.*, **25**, 569 (1984).
105. C. Grison, T. K. Olszewski, C. Crauste, A. Fruchier, C. Didierjean and P. Coutrot, *Tetrahedron Lett.*, **47**, 6583 (2006).
106. M. L. F. Cadman, L. Crombie, S. Freeman and J. Mistry, *J. Chem. Soc., Perkin Trans. 1*, 1397 (1995).
107. L. Crombie, M. Eskins, D. E. Games and C. Loader, *J. Chem. Soc., Perkin Trans. 1*, 472 (1979).
108. L. Crombie, *Tetrahedron*, **54**, 8243 (1998).
109. M. W. Rathke and P. J. Cowan, *J. Org. Chem.*, **50**, 2622 (1985).
110. R. E. Ireland and J. A. Marshall, *J. Am. Chem. Soc.*, **81**, 2907 (1959).
111. R. Adhikari, D. J. Cundy, C. L. Francis, M. Gebara-Coghlan, B. Krywult, C. Lubin, G. W. Simpson and Qi Yang, *Aust. J. Chem.*, **58**, 34 (2005).
112. K. Tohdo, Y. Hamada and T. Shioiri, *Synlett*, 105 (1994).
113. P. Krogsgaard-Larsen, A. L. Nordahl Larsen and K. Thyssen, *Acta Chem. Scand., Ser. B*, **32**, 469 (1978).
114. J. Maibaum and D. H. Rich, *J. Org. Chem.*, **53**, 869 (1988).
115. F. R. Busch, R. S. Lehner and B. T. O'Neil, *U.S. Patent* **5**, 380,860 (1995).
116. F. Di Ninno, T. Beattie and B. G. Christensen, *J. Org. Chem.*, **42**, 2960 (1977).
117. J. A. Aimetti and M. S. Kellog, *Tetrahedron Lett.*, 3805 (1979).
118. A. Martel, J. P. Daris, C. Bachand and M. Menard, *Can. J. Chem.*, **65**, 2179 (1987).
119. Y. Tamura, T. Hioki, S. Nishi and Z. I. Yoshida, *Tetrahedron Lett.*, **23**, 2383 (1982).
120. Y. Tamura, T. Hioki and Z. I. Yoshida, *Tetrahedron Lett.*, **25**, 5793 (1984).
121. B. Blagouev and D. Ivanov, *Synthesis*, 615 (1970).
122. G. Procter, *Asymmetric Synthesis*, Chap 5, Oxford University Press, New York, 1999.
123. D. A. Evans, D. L. Rieger, M. T. Bilodeau and F. J. Urpi, *J. Am. Chem. Soc.*, **113**, 111 (1991).
124. J. R. Gaje and D. A. Evans, *Org. Synth.*, **68**, 83 (1990).

CHAPTER 12

Functionalized organomagnesium compounds: Synthesis and reactivity

PAUL KNOCHEL, ANDREY GAVRYUSHIN and KATJA BRADE

Department Chemie und Biochemie, Ludwig-Maximilians-Universität München, Butenandtstr. 5-13, D-81377 München, Germany
Fax: +49-89-2180-77680; e-mail: paul.knochel@cup.uni-muenchen.de

I. INTRODUCTION	512
II. PREPARATION OF FUNCTIONALIZED GRIGNARD REAGENTS	512
A. Direct Oxidative Addition of Magnesium to Organic Halides	512
B. The Halogen–Magnesium Exchange Reaction	515
1. Early studies	515
2. Application of the halogen–magnesium exchange reaction for the synthesis of functionalized Grignard reagents	517
a. Scope and limitations	517
b. Functional group tolerance	530
C. Metalation Reactions with Magnesium Amide Bases	537
D. Miscellaneous Methods	540
III. REACTIVITY OF GRIGNARD REAGENTS	543
A. Substitutions at an sp^3-center	543
1. Transition-metal-catalyzed cross-coupling reactions	543
2. Ring-opening of small cycles	547
3. Diverse reactions	547
B. Substitutions at an sp^2-center	550
1. Transition-metal-catalyzed cross-coupling reactions	550
2. Transition-metal-free cross-coupling reactions	556
3. Allylic substitution reactions	557
4. Synthesis of carbonyl compounds	559
C. Substitutions at an sp-center	559
D. Addition of Organomagnesium Reagents to Multiple Bonds	559
1. Addition to carbon–carbon bonds	559

The chemistry of organomagnesium compounds
Edited by Z. Rappoport and I. Marek   © 2008 John Wiley & Sons, Ltd

a. Catalyzed addition to non-activated C=C bonds 559
b. 1,4-Addition to Michael acceptors 563
c. Addition to other activated alkenes 565
d. Addition to carbon–carbon triple bonds 567
2. Addition to carbon–oxygen bonds 569
3. Addition to carbon–nitrogen bonds 571
E. Diverse Reactions 575
1. Amination reactions 575
a. Electrophilic amination reactions 575
b. Oxidative coupling of polyfunctional aryl and heteroaryl
amidocuprates 580
2. Synthesis of cyclopropanes 582
3. Synthesis of chiral sulfoxides 582
IV. REFERENCES 583

I. INTRODUCTION

Organomagnesium reagents now play one of the key roles in organic synthesis. Since their discovery at the beginning of the last century by Victor Grignard[1], the development of new methods for their preparation opened a way toward highly functionalized organomagnesium reagents. Their practical synthesis, good stability and excellent reactivity make organomagnesium compounds indispensable organometallic intermediates for industry and academic laboratory. Transmetalation to less reactive but more chemoselective organometallic species (zinc, copper[2], titanium etc.) allows additional fine-tuning of their reactivity pattern.

Several comprehensive reviews and books have been published, encompassing the preparation and use of Grignard reagents[3], their chemical and physical properties[4], mechanistic investigations of their formation[3b,5] and studies of their structures in solution and in solid state[5d,6]. In the present chapter, emphasis will be placed on synthetic methods for the preparation of *functionalized* organomagnesium compounds as well as their applications in organic synthesis.

II. PREPARATION OF FUNCTIONALIZED GRIGNARD REAGENTS

A. Direct Oxidative Addition of Magnesium to Organic Halides

A widely used route for the synthesis of Grignard reagents is the oxidative addition of magnesium metal to organic halides in a polar, aprotic solvent like THF or diethyl ether (equation 1).

$$RX \xrightarrow{\text{Mg}}_{\text{THF or Et}_2\text{O}} RMgX \quad (1)$$

Nevertheless, these solvents represent a safety hazard for large-scale industrial processes[7]. They can be substituted by less flammable high boiling glycol ethers like 'butyl diglyme' ($C_4H_9O(CH_2)_2OC_4H_9$). Additional possibilities might arise from the use of non-ethereal solvents like toluene[8], though the presence of one or two equivalents of diethyl ether or THF was found to be beneficial[9]. Controlling the reactant feed rate by on-line concentration monitoring using near-IR spectroscopy can help to further improve the safety of especially large scale processes[10]. Using this method, excessive reactant accumulations leading to unsafe situations like overpressure in the reactor system can be avoided.

12. Functionalized organomagnesium compounds: Synthesis and reactivity 513

Usually, magnesium metal is covered with an 'oxide layer' which mainly consists of $Mg(OH)_2$[11]. The nature of this metal surface plays a pivotal role in the oxidative addition reaction[12]. Thus, to shorten the induction period and obtain a better reproducibility of the reaction time, activation of the magnesium surface with agents as 1,2-dibromoethane prior to reaction is normally desired[13]. A radical mechanism for this reaction is widely accepted[14], though details are still being discussed[15]. Another method of activation of Mg metal using diisobutylaluminium hydride (DIBAH) allows lowering of the reaction temperature (0–10 °C), which can be crucial for the preparation of less stable compounds like **1–3** (equation 2)[16].

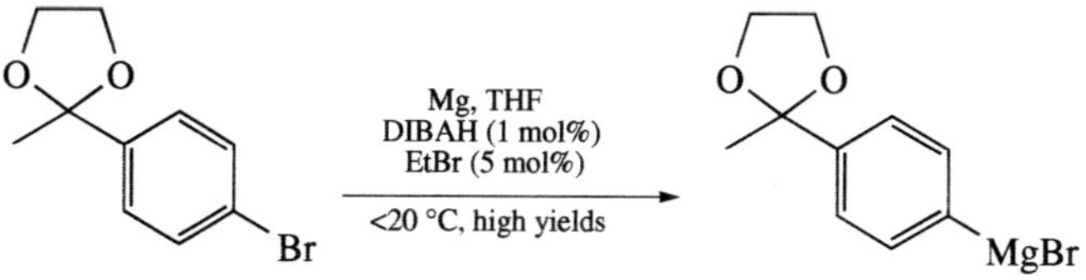

$$\text{ArBr} \xrightarrow[\text{<20 °C, high yields}]{\substack{\text{Mg, THF}\\\text{DIBAH (1 mol%)}\\\text{EtBr (5 mol%)}}} \text{(1)} \qquad (2)$$

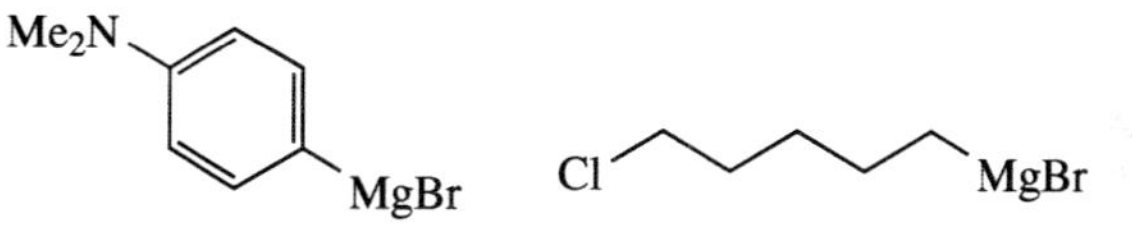

(**2**) (**3**)

Low reactive aryl chlorides are converted to the respective organomagnesium species in excellent yields through transition metal catalysis using 2 mol% $FeCl_2$ (**4–6**, equation 3)[17]. Alternatively, a safe and reproducible method for activation of aryl chlorides or bromides **7** uses microwave irradiation (equation 4). In a synthesis of a novel HIV-1 protease inhibitor, microwave irradiation was essential to generate the starting arylmagnesium halide as well as to promote the subsequent Kumada coupling reaction[18].

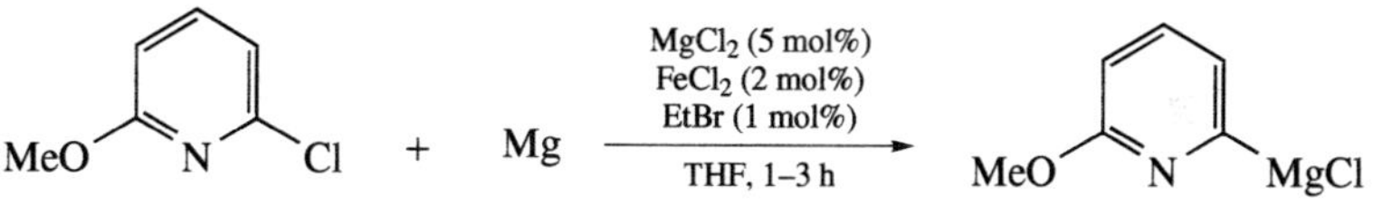

$$\text{MeO-Py-Cl} + \text{Mg} \xrightarrow[\text{THF, 1-3 h}]{\substack{MgCl_2\ (5\ \text{mol\%})\\FeCl_2\ (2\ \text{mol\%})\\\text{EtBr (1 mol\%)}}} \text{MeO-Py-MgCl}$$

(**4**) 96% (3)

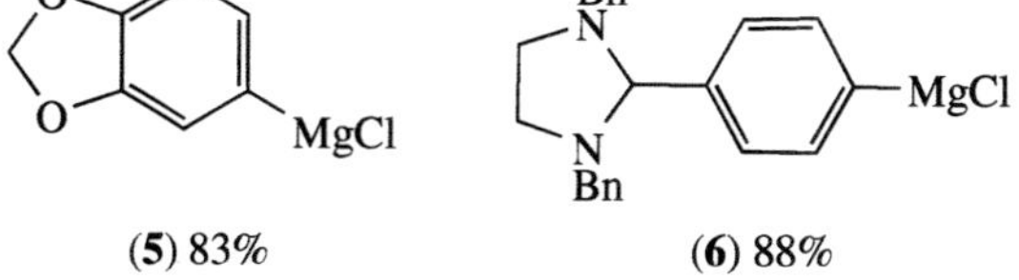

(**5**) 83%   (**6**) 88%

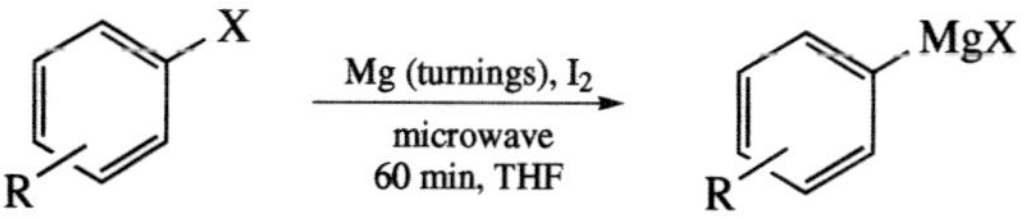

$$\text{(7) X = Cl, Br} \xrightarrow[\substack{\text{microwave}\\\text{60 min, THF}}]{\text{Mg (turnings), } I_2} \quad \text{ArMgX} \qquad (4)$$

For the synthesis of functionalized Grignard reagents bearing sensitive functional groups, highly reactive Rieke magnesium (Mg*) can be used[19]. The low reaction temperatures ($-78\,°C$) allow the presence of a number of functional groups, which are not compatible with the usual methodology (equation 5)[20]. Direct trapping of the resulting Grignard reagents with an electrophile (Barbier conditions), e.g. in the formation of **8**, sometimes improves these results even further (equation 6)[21].

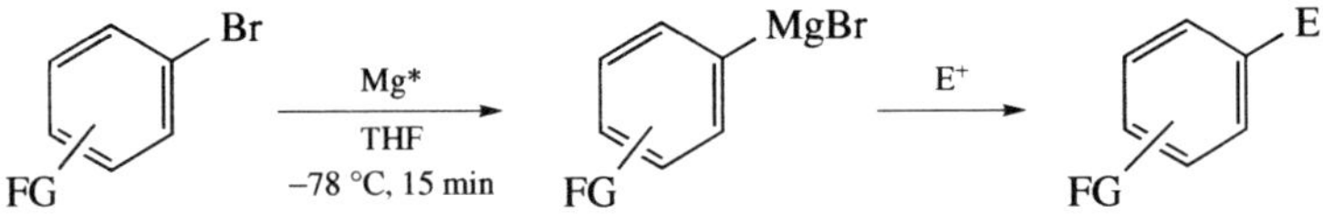

FG = CO_2Bu-t, $OCOBu$-t, CN etc.
E$^+$ = PhCHO, PhCOCl, allyl iodide
E = PhCH(OH), PhCO, $CH_2=CHCH_2$

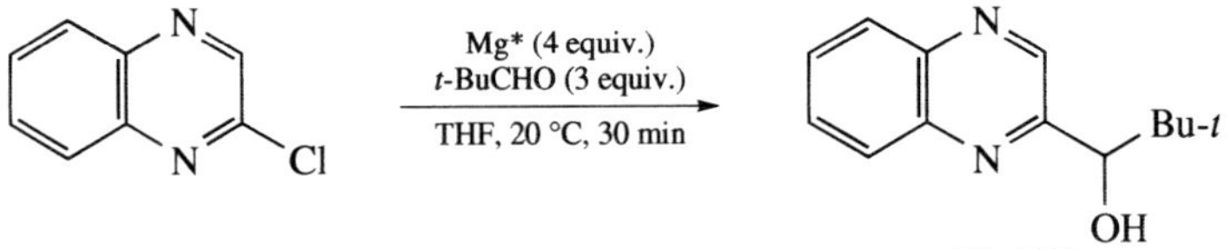

Another source for highly reactive magnesium is soluble Mg-anthracene **9**[22]. It can be used for the reductive metalation of allylic phenyl sulfides **10** (equation 7)[23].

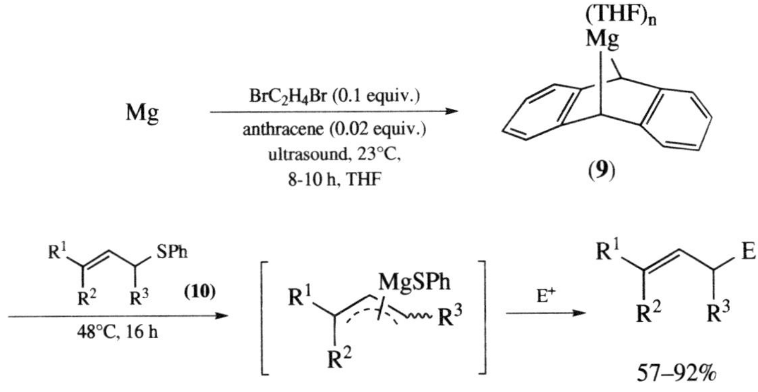

Interesting information on their configurational stability was obtained in the direct synthesis of Grignard reagents from enantiomerically pure alkyl halides like (−)-menthyl chloride (MenCl, **10**)[24]. The corresponding Grignard reagents MenMgCl (**11**) and NeomMgCl (**12**) are formed in a 1:1 ratio, with no equilibrium between them. If the reactivity of the two epimers differs significantly, as in the present case, it is possible to selectively use each component in reactions with electrophiles by using kinetic or thermodynamic reaction control. Thus, reaction with diphenylphosphine chloride leads to full consumption of the more reactive epimer MenMgCl (**11**), while the pure chiral Grignard reagent

12. Functionalized organomagnesium compounds: Synthesis and reactivity   515

12 (NeomMgCl) remains in the reaction mixture (equation 8).

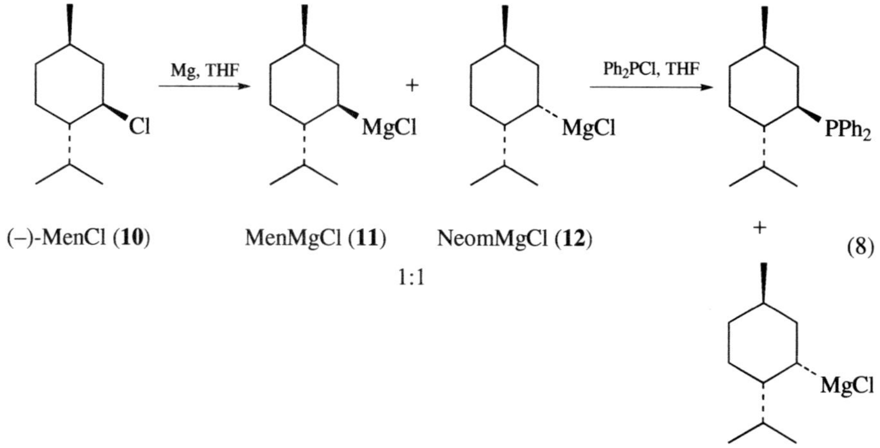

(−)-MenCl (**10**)   MenMgCl (**11**)   NeomMgCl (**12**)   (8)
                              1:1

(**12**)

Grignard reagents can be generated from organic iodides and magnesium in ionic liquids[25] like n-butylpyridinium tetrafluoroborate ([bpy][BF$_4$])[26]. The resulting Grignard reagents show a different reactivity in this solvent (equation 9).

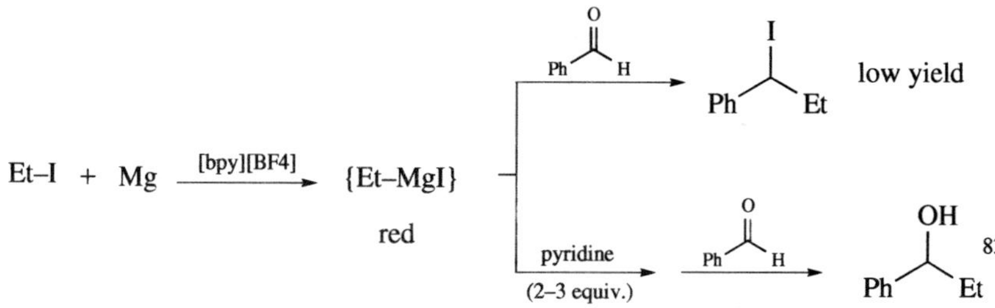

(9)

Unfortunately, the preparation of functionalized Grignard reagents via direct oxidative addition of magnesium metal to organic halides still suffers from severe limitations. This is mainly due to the intrinsic high reducing potential of magnesium metal.

B. The Halogen–Magnesium Exchange Reaction

1. Early studies

In 1931 Prévost reported the reaction of cinnamyl bromide (**13**) to cinnamylmagnesium bromide (**14**), which was the first example of a bromine–magnesium exchange reaction (equation 10)[27].

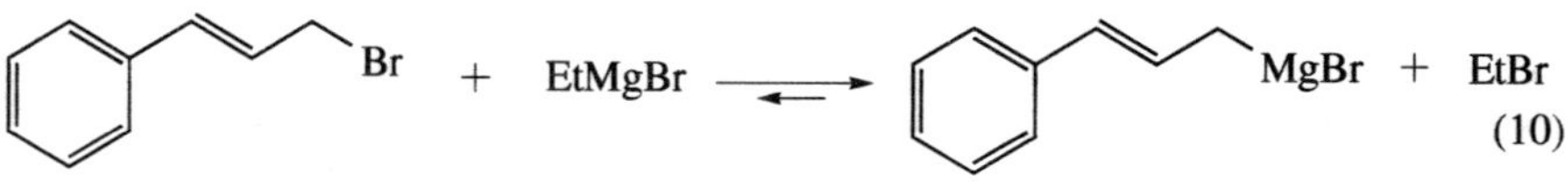

(**13**)   (**14**) 14%   (10)

The halogen–magnesium exchange is an equilibrium process, where the formation of the most stable organomagnesium compound is favored (sp>sp^2(vinyl)>sp^2(aryl)>sp^3 (prim.)>sp^3(sec.)). The mechanism of the exchange reaction is not yet fully clarified, but calculations show that it proceeds via a concerted 4-centered mechanism, in contrast to the halogen–lithium exchange that goes via the formation of a halogenated complex[28].

One of the first synthetically useful procedures, employing a halogen–magnesium exchange reaction, is the synthesis of perfluoroalkylmagnesium halides of type **15** starting from the perfluorinated iodide **16** (equation 11)[29]. This procedure showed significant advantages compared to the oxidative addition reaction, such as higher yields and less side reactions. It is one of the best methods for the synthesis of perfluorinated Grignard reagents[30].

$$C_3F_7-I \quad \xrightarrow[\text{Et}_2\text{O, 15 min} \\ -40 \text{ to } -50\,°C]{\text{PhMgBr}} \quad C_3F_7-MgBr + Ph-I \quad \xrightarrow{\text{cyclohexene oxide}} \quad C_3F_7-\text{cyclohexanol} \quad 90\% \quad (11)$$

(**16**) (**15**)

The halogen–magnesium exchange reaction was later used as a general approach to magnesium carbenoids[31]. Reaction of i-PrMgCl with CHBr$_3$ at $-78\,°C$ furnishes the corresponding magnesium carbenoid **17** which is trapped with chlorotrimethylsilane, leading to (dibromomethyl)trimethylsilane **18** in 90% yield (equation 12).

$$CHBr_3 \quad \xrightarrow[-78\,°C]{i\text{-PrMgCl}} \quad Br_2CHMgCl + i\text{-PrBr} \quad \xrightarrow{Me_3SiCl} \quad Br_2CHSiMe_3 \quad (12)$$

(**17**) (**18**) 90%

This pioneering work paved the way to the systematic study of magnesium carbenoids[32]. Furthermore, it demonstrated that the halogen–magnesium exchange rate is enhanced by the presence of electronegative substituents. A few years later, it could be shown that the formation rate of the new Grignard reagent does not only depend on the electronic properties of the organic molecule, but on the halogen atom as well[33]. The reactivity order (I>Br>Cl≫F) is influenced by the bond strength of the carbon–halogen bond, the halide electronegativity and polarizability. Only for very electron-poor systems, such as the tetra- or pentafluorobenzenes, is the exchange of a chlorine possible, requiring elevated temperatures and longer reaction times. For instance, the exchange reaction of 1-chloro-2,3,4,5,6-pentafluorobenzene (**19a**) with EtMgBr requires 1 h at room temperature to reach complete conversion to the Grignard reagent. The corresponding bromo- and iodoperfluorobenzenes **19b** and **19c** react already at $0\,°C$, leading to perfluorophenylmagnesium bromide **20** within 1 min (equation 13).

$$C_6F_5-X \quad \xrightarrow[\substack{X=Cl;\ 25\,°C,\ 1\,h \\ X=Br;\ 0\,°C,\ 1\,min \\ X=I;\ 0\,°C,\ 1\,min}]{\text{EtMgBr}} \quad C_6F_5-MgBr \quad (13)$$

(**19a**) X = Cl (**20**)
(**19b**) X = Br
(**19c**) X = I

The strong dependency of the reactivity of carbon–magnesium bonds on the reaction temperature, as well as the fact that only reactive electrophiles like aldehydes and most ketones react rapidly at temperatures below 0 °C, turned the halogen–magnesium exchange reaction into a first choice method for the preparation of magnesium organometallics bearing reactive functional groups[34].

2. Application of the halogen–magnesium exchange reaction for the synthesis of functionalized Grignard reagents

a. Scope and limitations. Functionalized iodoarenes react readily with i-PrMgCl or i-PrMgBr in THF at temperatures below 0 °C, sometimes even at −78 °C, affording a range of functionalized arylmagnesium compounds. Sensitive carbonyl group derivatives like nitriles, esters or amides are well tolerated under such conditions[35].

Thus, treatment of methyl 4-iodobenzoate (**21**) with i-PrMgBr in THF at −20 °C provides the corresponding Grignard reagent **22** after 30 min. The magnesium reagent **22** reacts smoothly with aldehydes at this temperature (equation 14)[36].

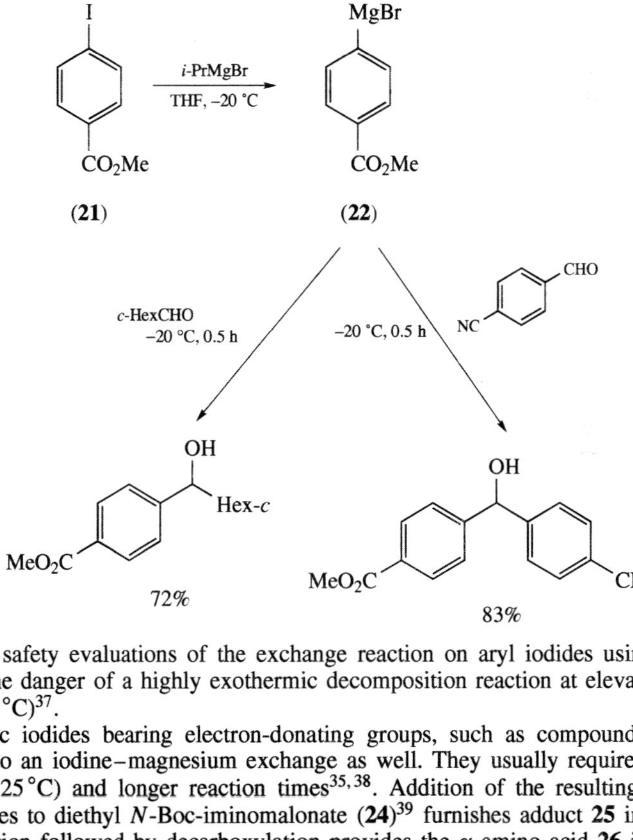

Process safety evaluations of the exchange reaction on aryl iodides using i-PrMgCl revealed the danger of a highly exothermic decomposition reaction at elevated temperatures (>80 °C)[37].

Aromatic iodides bearing electron-donating groups, such as compound **23**, can be subjected to an iodine–magnesium exchange as well. They usually require higher temperatures (25 °C) and longer reaction times[35, 38]. Addition of the resulting arylmagnesium species to diethyl N-Boc-iminomalonate (**24**)[39] furnishes adduct **25** in 79% yield. Saponification followed by decarboxylation provides the α-amino acid **26** in 81% yield (equation 15)[38]. Even a highly electron-rich system like **27** possessing three methoxy groups undergoes the iodine–magnesium reaction as shown in a total synthesis of colchicines (equation 16)[40].

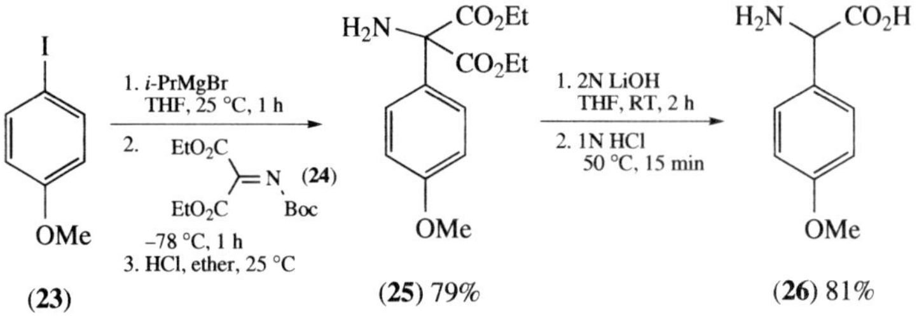

Likewise, heteroaryl iodides react with *i*-PrMgCl in THF giving the corresponding magnesium compounds in excellent yields (Scheme 1)[36, 41].

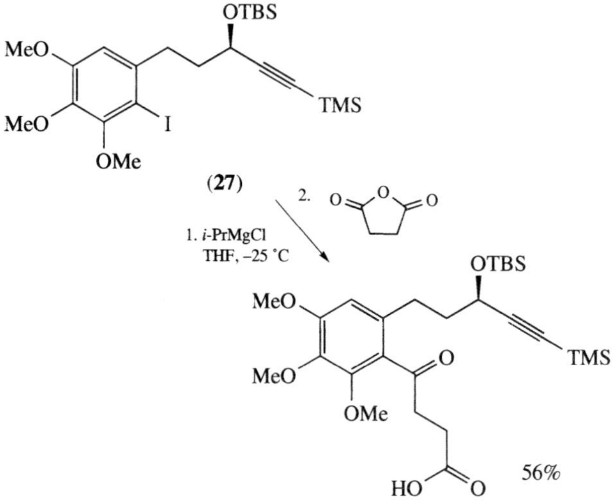

The I/Mg-exchange reaction can be extended to the use of iodo-substituted pyridines[42, 43], uracils[42, 44], purines[45], imidazoles[42, 46], quinolines[47], imidazo[1,2-a]-pyridines[48], pyrroles[49]

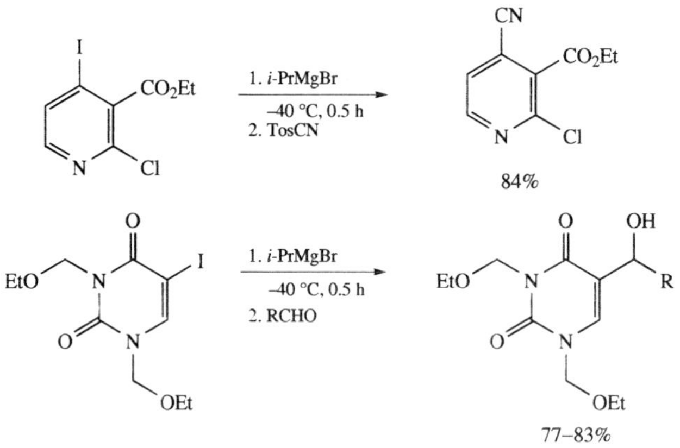

SCHEME 1

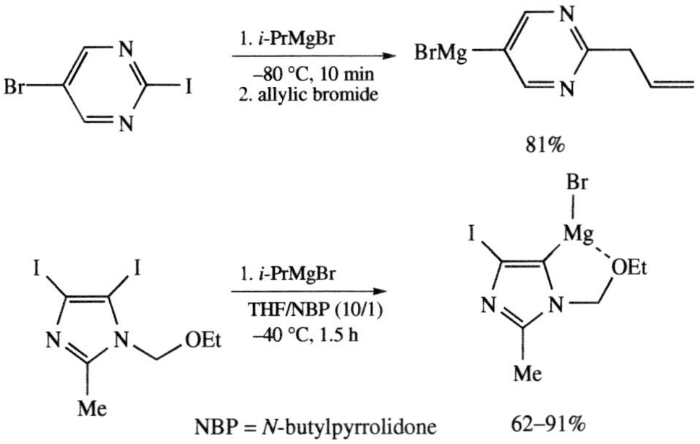

SCHEME 1. (continued)

and isoxazoles[50]. Functional groups like an ester or a nitrile are well tolerated at the temperatures which are required for the exchange.

The Br/Mg-exchange reaction is significantly slower than the I/Mg exchange. Using *i*-PrMgCl or *i*-PrMgBr, a fast exchange at temperatures below 0 °C can only be achieved for systems bearing strong electron-withdrawing groups[41a, 51]. For example, Grignard reagent **28**, a valuable building block for the synthesis of a neurokin 1 receptor agonist, is obtained from the readily available aryl bromide **29** using *i*-PrMgBr (equation 17)[52].

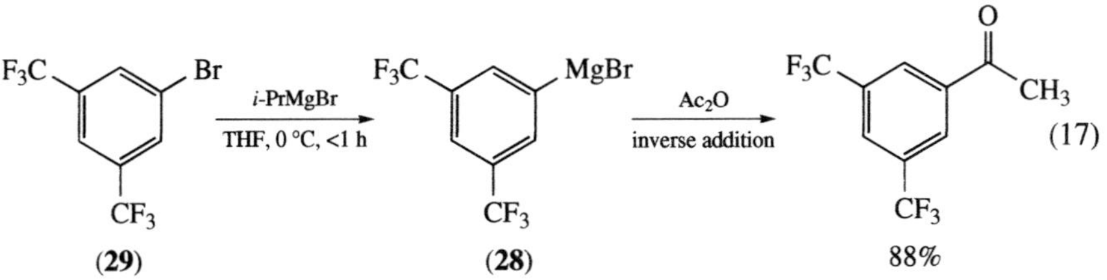

A low reaction temperature (0 °C) allows a safe synthesis of the highly useful, but explosive class of trifluoromethylphenyl Grignard reagents[53]. Polyfunctional aryl bromides **30** and **31**, bearing a chelating function at the *ortho*-position to the bromine, undergo a bromine–magnesium exchange much easier (equations 18 and 19)[41a, 51]. Even the less effective methoxy group directs and facilitates the exchange as in the case of 2,4-dibromoanisole (**32**) (equation 20)[54], although its electron-releasing nature requires a higher reaction temperature.

Inactivated aryl bromides do not react with *i*-PrMgCl in a sufficient rate even at temperatures as high as room temperature. However, the presence of 1 equivalent of LiCl in the reaction mixture enhances the rate of the exchange reaction tremendously, thus even allowing the use of electron-rich aryl bromides (equation 21)[55].

The addition of a stoichiometric amount of LiCl breaks the aggregates of the otherwise dimeric *i*-PrMgCl producing a highly reactive Grignard reagent *i*-PrMgCl•LiCl (equation 22). Commercially available *i*-PrMgCl•LiCl can be used for the preparation of a variety of substrates bearing functional groups. Thus, 3-bromobenzonitrile (**33**) undergoes a fast Br/Mg exchange at −10 °C, leading to the 3-magnesiated species **34** which reacts upon transmetalation with CuCN•2LiCl using benzoyl chloride furnishing ketone

35 (equation 23). The exchange in the presence of LiCl can further be successfully used for the functionalization of *ortho*-dibromo- and tribromobenzenes like **36**, since low reaction temperatures permit the generation of unstable *o*-bromoarylmagnesium species **37** (equation 24)[55]. The addition of LiCl also proved to be beneficial for industrial-scale Grignard reactions, since it avoids the formation of gaseous side products[56].

(**30**) 68% (18)

(**31**) 80% (19)

(**32**) 90% (20)

(21)

FG = F, Cl, Br, CN, CO_2R, OMe

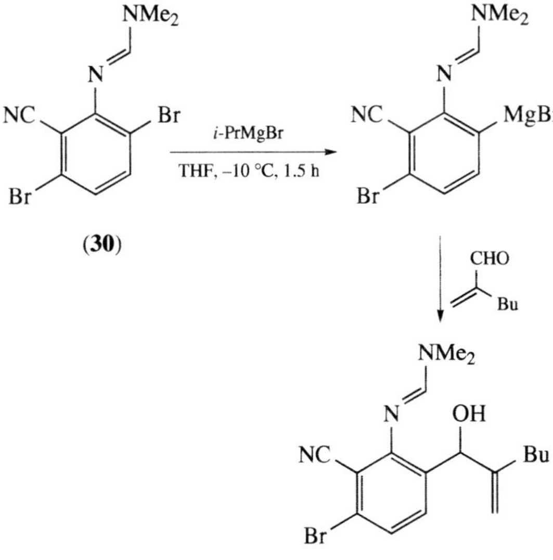

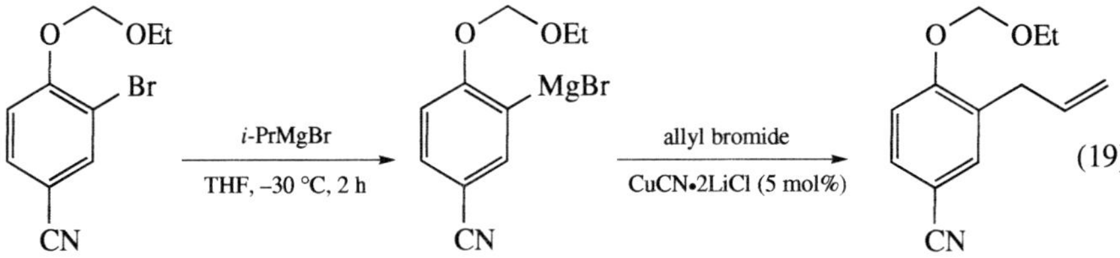

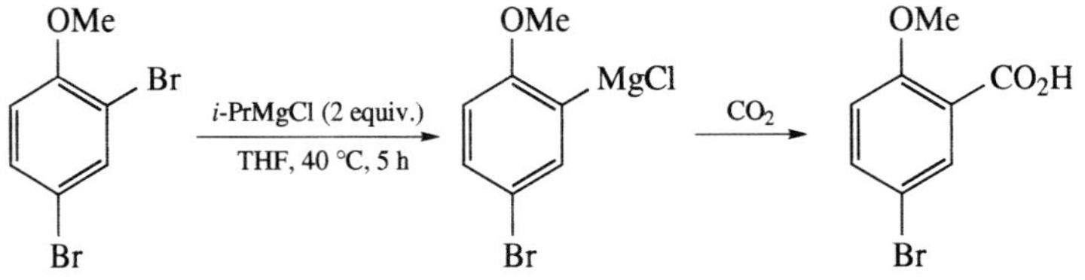

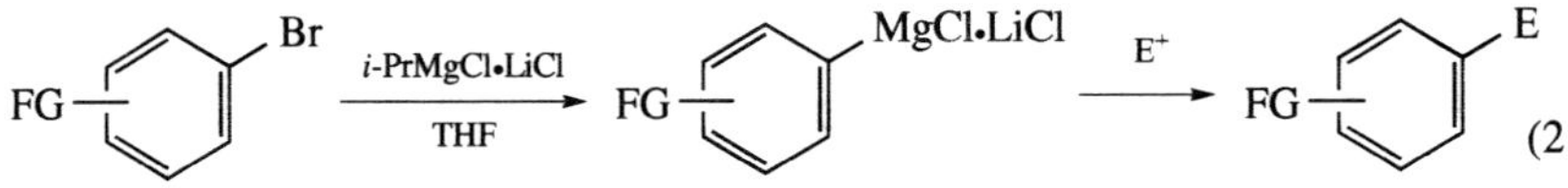

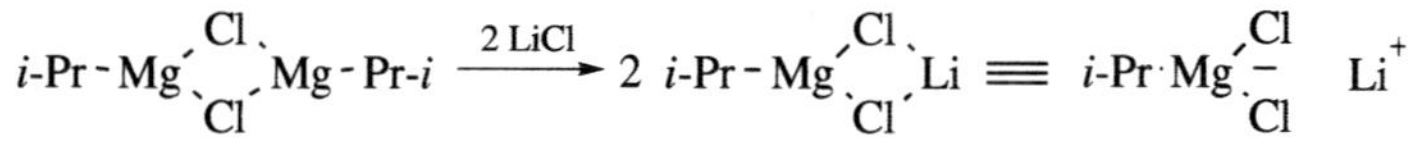

Additionally, the yields of the reactions of arylmagnesium reagents with electrophiles are higher in the presence of LiCl. The ionic salt LiCl ensures a good solubility of the reaction products, such as magnesium alcoholates, in the reaction mixture, making the whole process much easier to handle. For example, the addition of MeMgCl in the presence of LiCl followed by the addition of *i*-PrMgCl to aromatic and heteroaromatic substrates **38** bearing a hydroxy function leads to the corresponding THF-soluble dimagnesiated species, which can be reacted with electrophiles to give *ortho*-functionalized phenols **39a–c** (equation 25)[57].

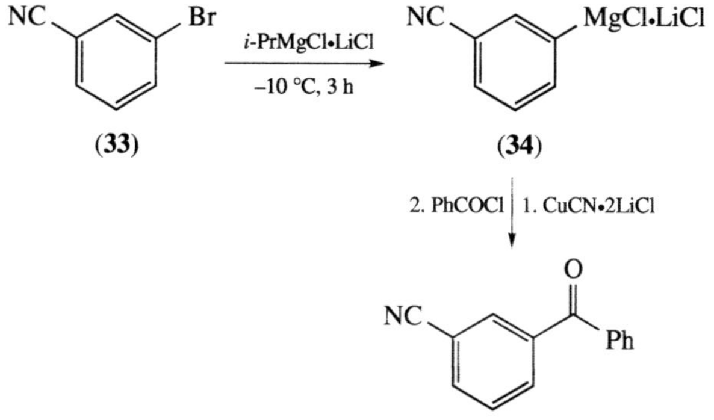

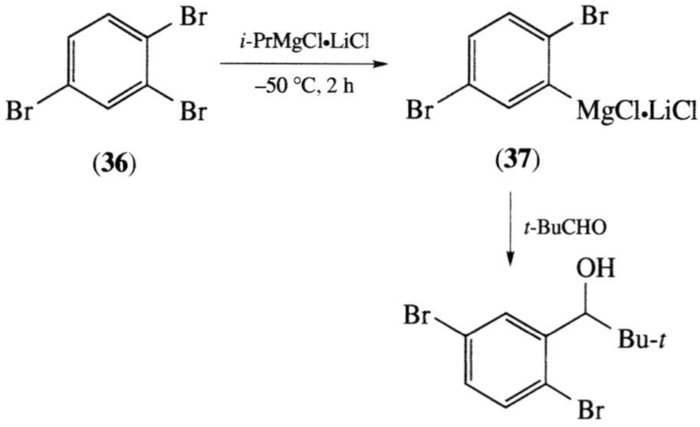

In an extension to this method, lithium magnesiate **40** was prepared (equation 26). This new class of highly reactive exchange reagents allows fast conversion of the electron rich aryl bromide **41** to the corresponding diarylmagnesium reagent **42**, which is trapped with benzaldehyde to give alcohol **43** in 40% yield (equation 27)[28d]. The exchange reagent **40** further allows a facile preparation of the polymeric aryl-bis-magnesium reagent **44** (*n* not determined) starting from diiodobenzene (**45**) (equation 28).

The presence of bis[2-(*N*,*N*-dimethylamino)ethyl]ether allows a selective halogen–magnesium exchange of iodo- and bromoaromatics at ambient temperature using isopropylmagnesium chloride. Sensitive carboxylic ester and cyano groups are well tolerated (equations 29 and 30)[58].

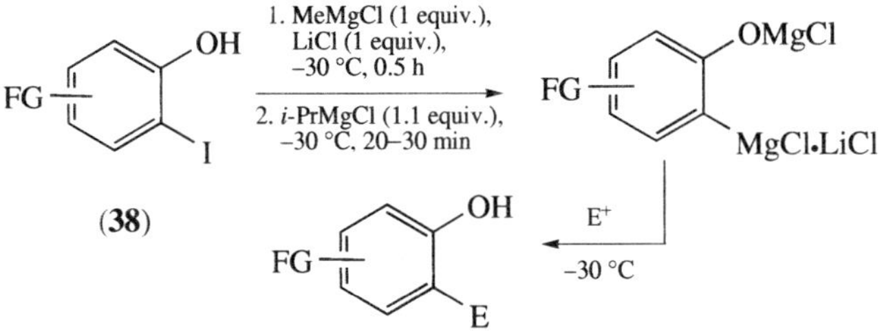

522 Paul Knochel, Andrey Gavryushin and Katja Brade

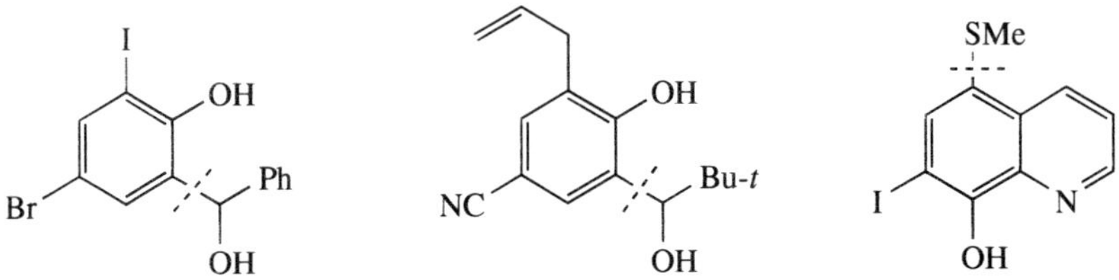

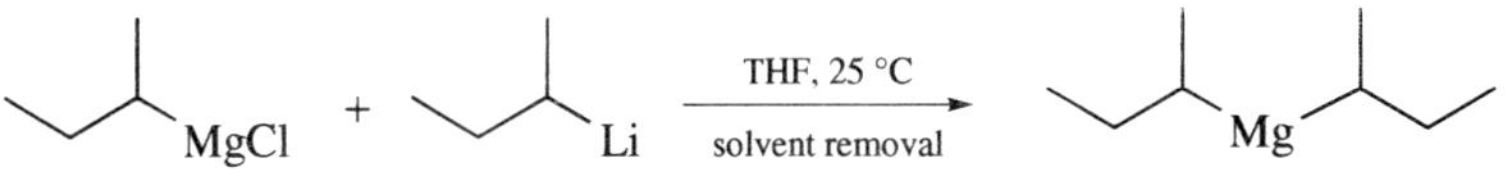

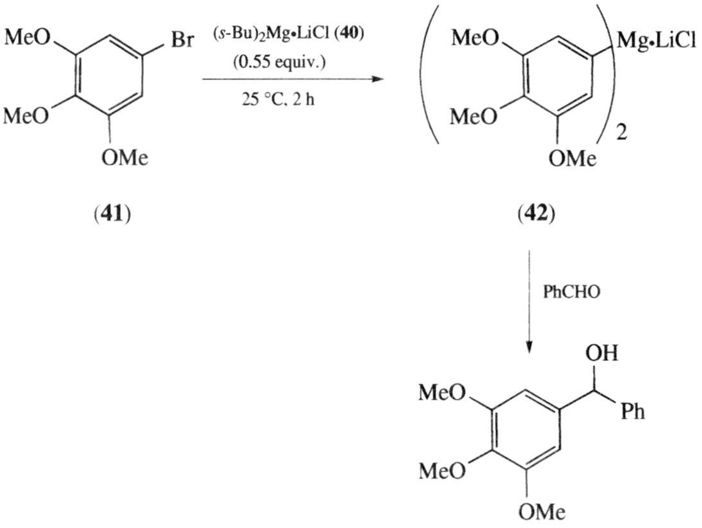

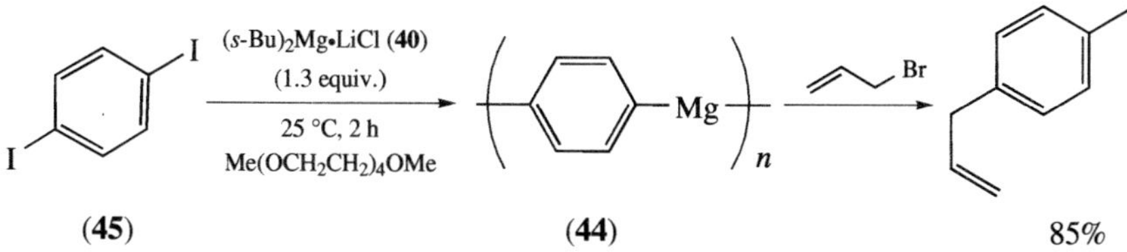

Lithium magnesiates are another class of highly reactive exchange reagents. They are prepared by reacting an organolithium reagent (2 equiv.) with an alkylmagnesium halide (1 equiv.). These lithium magnesiates are substantially more reactive than usual Grignard reagents and undergo a Br/Mg exchange on various aryl bromides[59]. Even 0.5 equiv. of the lithium dibutylmagnesiate relative to the aromatic halide can be sufficient to achieve complete conversion (Scheme 2).

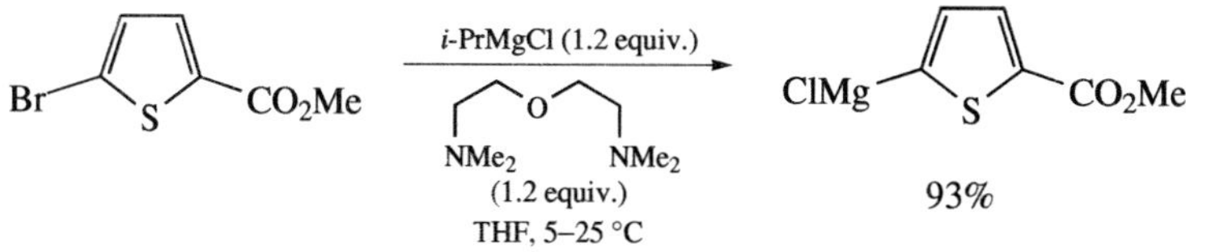

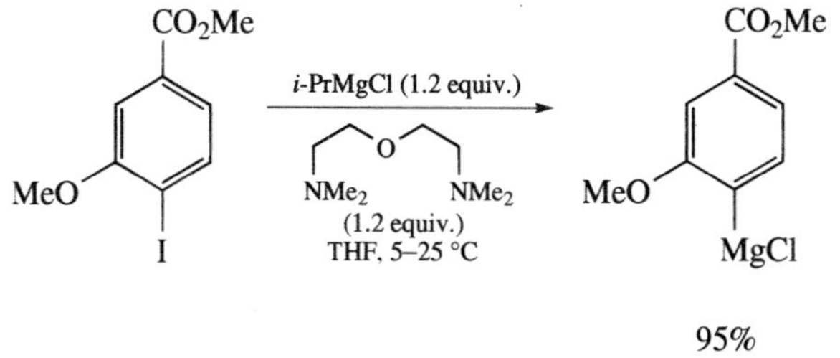

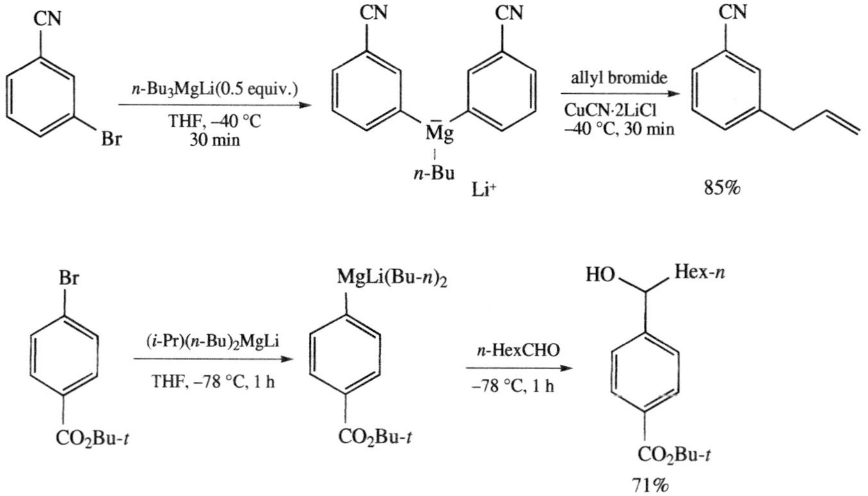

SCHEME 2

Although bromopyridines like **46** are reactive enough to undergo the exchange with *i*-PrMgCl at room temperature, higher yields are obtained using *i*-PrMgCl•LiCl (equation 31)[60]. A tosyloxy substituent in position 2 allows regioselective Br/Mg exchange on position 3 of 3,5-dibromopyridine derivatives **47** (equation 32). The resulting pyridylmagnesium species **48** reacts readily with various electrophiles[61]. Even functionalized uracil derivatives **49a–c** can be obtained via Br/Mg exchange using *i*-PrMgCl•LiCl allowing an efficient synthesis of the HIV replication inhibitor Emivirine (MKC-442, equation 33)[62]. Also, lithium magnesiates can be successfully applied for the bromine–magnesium exchange of heteroaryl bromides like **50** (equation 34)[63].

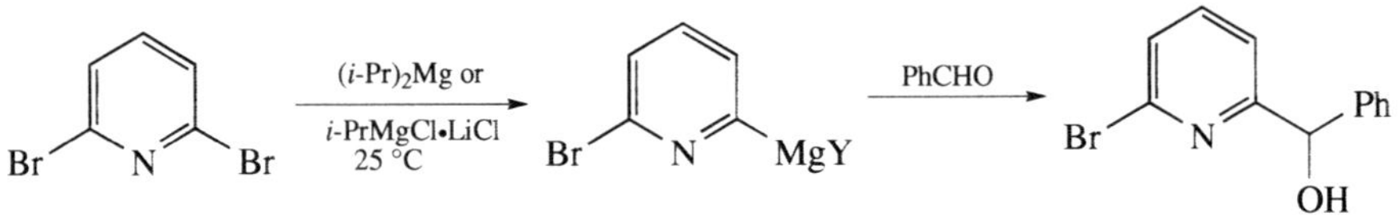

(**46**) Y = *i*-Pr, Cl•LiCl

with $(i\text{-Pr})_2$Mg: 50%
with *i*-PrMgCl•LiCl:85%

(31)

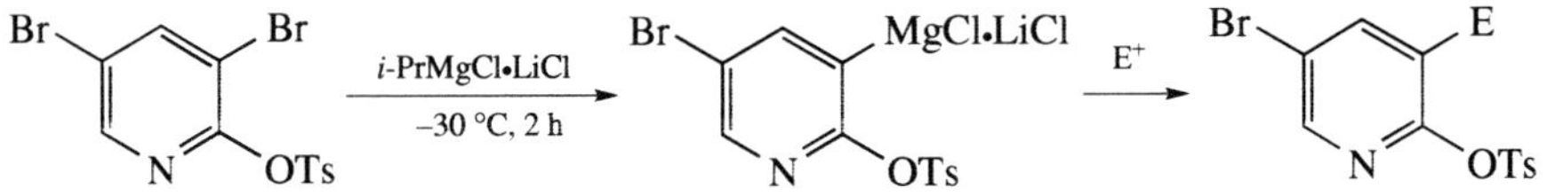

(**47**) (**48**)

E = CHO: 88%
E = COPh: 89%
E = CN: 71%

(32)

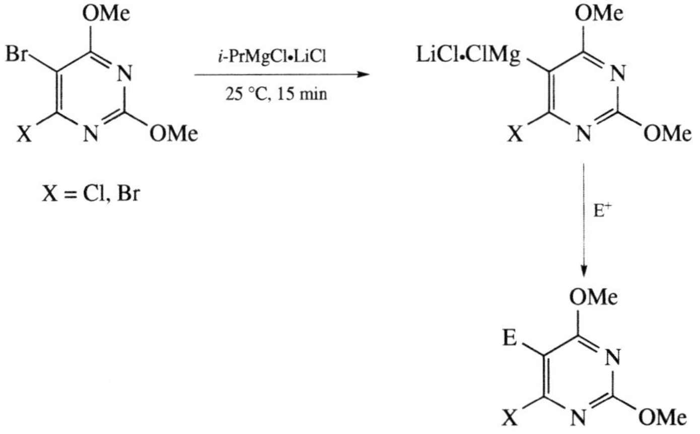

X = Cl, Br

(33)

(**49a**) X = Cl, E = CH(OH)Ph: **91%**
(**49b**) X = Br, E = CH(OH)Ph: **95%**
(**49c**) X = Cl, E = CO_2Et: **91%**
(**49d**) X = Br, E = CO_2Et: **81%**

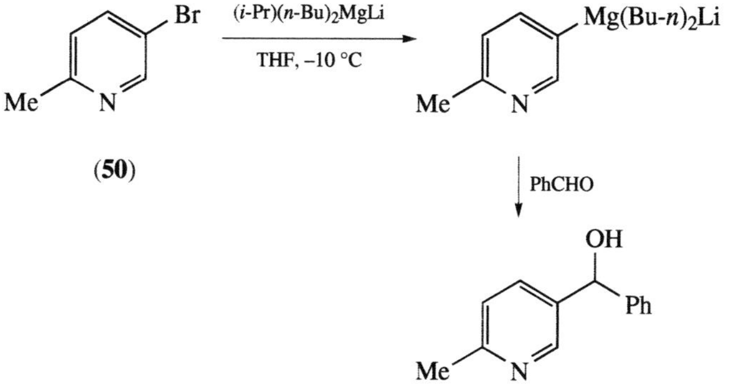

12. Functionalized organomagnesium compounds: Synthesis and reactivity

(34)

Selective formation of 2- or 3-substituted bromothiophenes can be achieved via halogen–magnesium exchange using EtMgBr. Thus, treatment of 2,3-dibromothiophene (**51**) gives solely the 2-magnesiated product **52** (equation 35), whereas 2-bromo-3-iodothiophene (**53**) selectively exchanges the iodine atom (equation 36)[41a, 64].

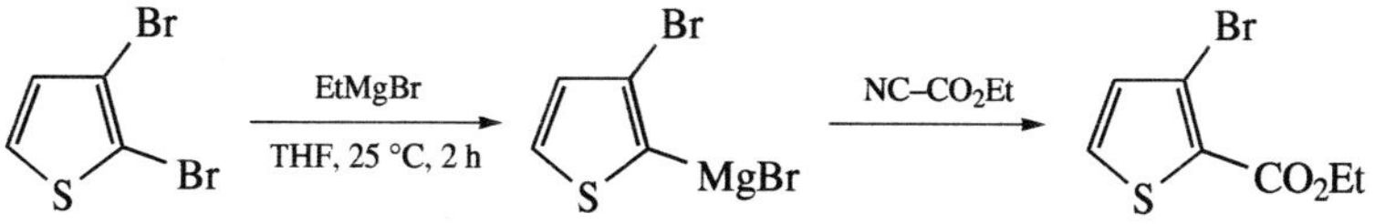

(35)

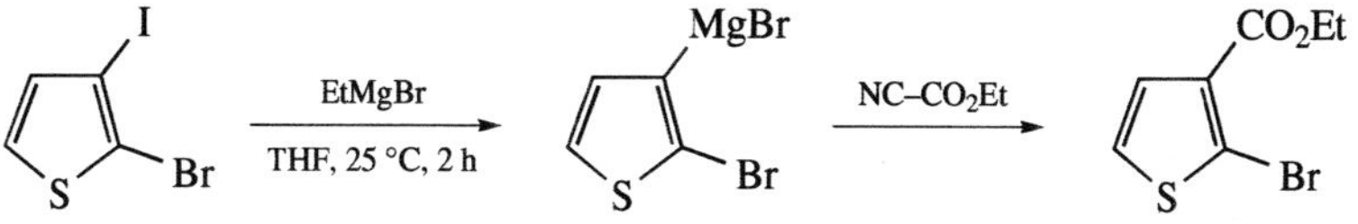

(36)

A selective exchange reaction was observed on 2,4-dibromothiazoles allowing the synthesis of substituted 4-bromothiazoles[65]. The use of functionalized organomagnesium compounds as intermediates for the synthesis of polyfunctionalized heterocycles has been reviewed recently[66].

For a long time, the exchange reaction on functionalized alkenyl halides was limited to reactions on systems either bearing an electron-withdrawing group in α-position[67] or a coordinating substituent in β-position (equation 37)[68]. In the case of β-dibromoacrylic esters like **54**, only the halogen placed *cis-* to the ester function is exchanged due to the strong intramolecular coordination (equation 38)[69].

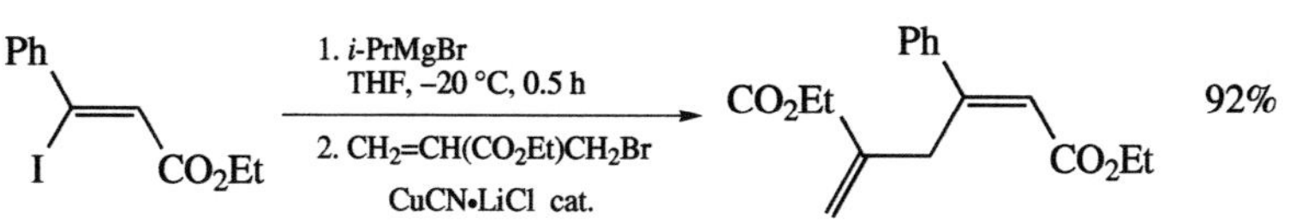

(37)

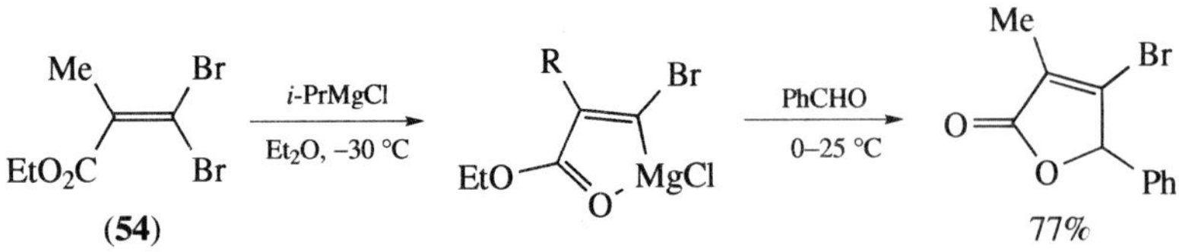

The aryl sulfoxide moiety may serve as a good leaving group in the exchange reaction. Thus, 1-haloalkenyl sulfoxide **55** undergo the exchange at $-78\,°C$ to give carbenoid compounds **56** which can be trapped by electrophiles or converted to acetylenes **57** (equation 39)[70]. Reaction of carbenoid **58** with lithium acetylides leads to the formation of enynes **59** (equation 40)[71].

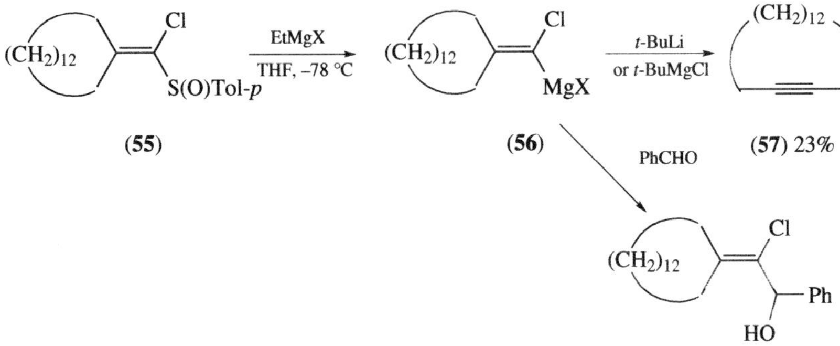

Unfunctionalized alkenyl iodides like **60** react with i-PrMgCl or i-Pr$_2$Mg (Scheme 3). Unfortunately, the exchange reaction requires high temperatures like room temperature or higher to take place, thus precluding the presence of sensitive groups. By using i-PrMgCl·LiCl as exchange reagent, the reaction of unactivated, but functionalized alkenyl iodides **61–63** and iododienes **64** and **65** as substrates can be realized. Complete retention of the double bond configuration is observed (Scheme 3)[72].

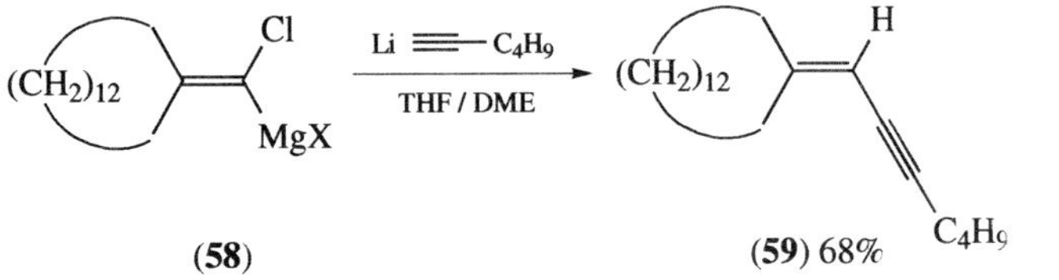

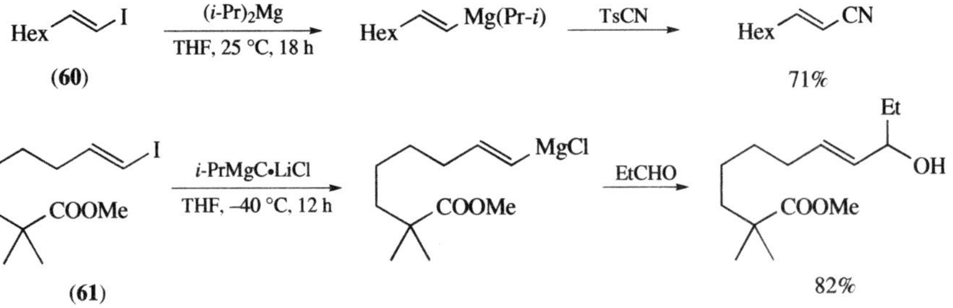

SCHEME 3

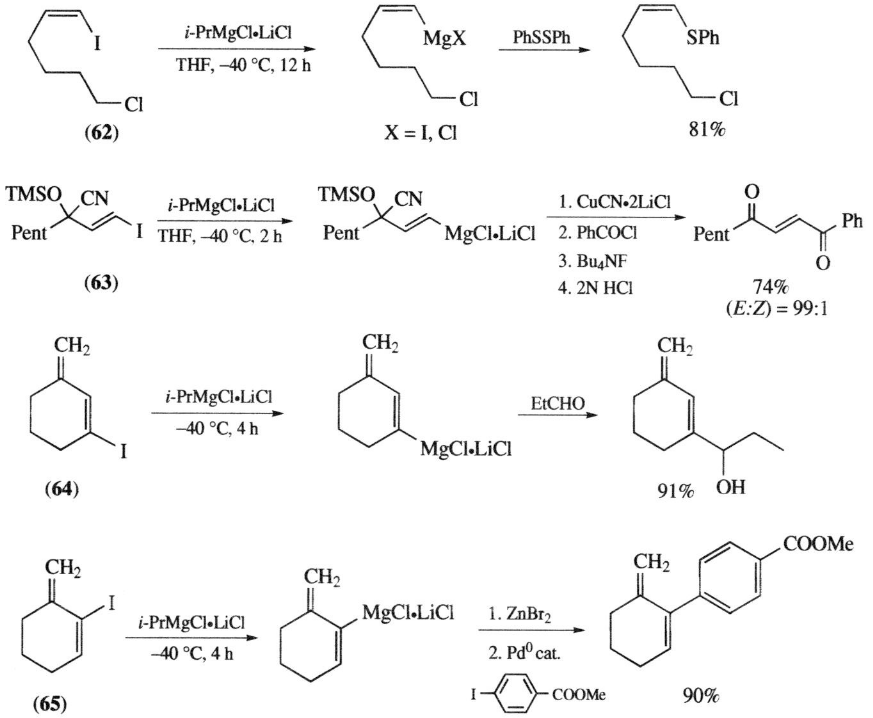

SCHEME 3. (*continued*)

The Br/Mg-exchange reactions on alkenyl bromides are very sluggish. This problem was overcome by the use of s-Bu$_2$Mg•LiCl[28d]. For example, Grignard reagent **66** is obtained by reaction of α-bromostyrene (**67**) with the complex **40** for 1 h at 25 °C. Quenching with benzaldehyde gives the allylic alcohol **68** in 93% yield (equation 41).

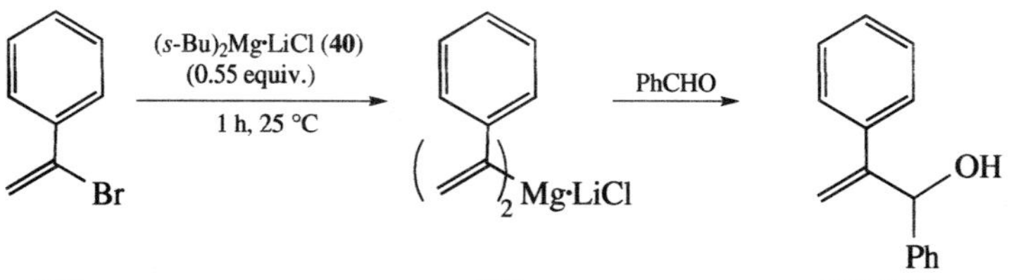

Iodine–magnesium exchange of allenyl iodides takes place on reaction with *i*-PrMgBr in ether. Subsequent reaction with aldehydes or ketones provides homopropargylic alcohols with high regioselectivity[73]. Exchange on chloroalkyl phenyl sulfoxides **69** can also be performed successfully[74–76]. It can be applied for the synthesis of olefins[77] as well as for the preparation of chiral Grignard reagents starting from a chiral sulfoxide as described in equation 42[78,79].

The aspects of the preparation and reactions of chiral Grignard reagents were reviewed[80]. Exchange of alkyl halides is synthetically useful mostly if α-halogen or α-acyloxy substituents are present. The resulting Grignard reagents react smoothly with various electrophiles (equation 43)[77, 81].

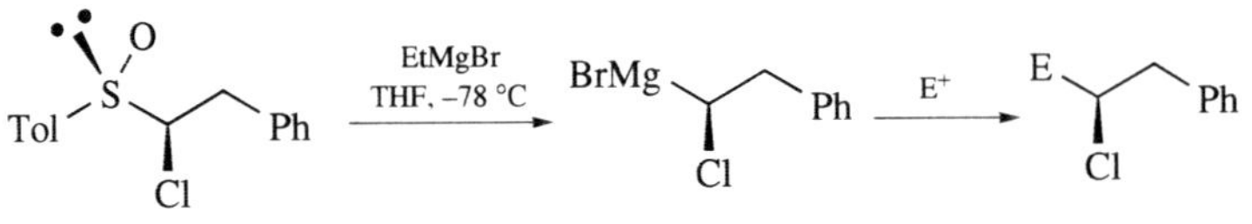

$$\text{Tol-S(O)-CH(Cl)-CH}_2\text{-Ph} \xrightarrow{\text{EtMgBr, THF, -78 °C}} \text{BrMg-CH(Cl)-CH}_2\text{-Ph} \xrightarrow{E^+} \text{E-CH(Cl)-CH}_2\text{-Ph} \quad (42)$$

(**69**) 97% *ee*, dr = 6.4:1 ; 93% *ee*

$$\text{R-C(O)-O-CH}_2\text{-I} \xrightarrow[\text{2. } E^+]{\begin{array}{c}\text{1. }i\text{-PrMgCl (1.1 equiv.)}\\\text{THF/NBP (5/1)}\\-78\,°\text{C, 15 min}\end{array}} \text{R-C(O)-O-CH}_2\text{-E} \quad (43)$$

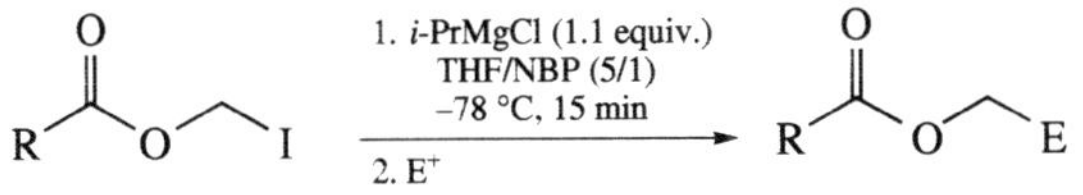

74–82%

R = *c*-Hex, *t*-Bu

E^+ = ArCHO, (PhS)$_2$, Ph$_2$PCl, CH$_2$=C(CO$_2$Et)CH$_2$Br, etc.

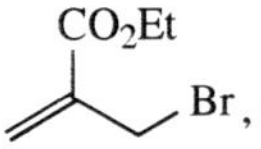

NBP: *N*-butyl-2-pyrrolidinone

Cyclopropyl iodides like **70** and bromides are good substrates for the exchange reaction (equation 44)[77, 82, 83]. The reaction is stereoselective and sufficiently fast at low temperatures, thus allowing the preparation of functionalized compounds. If a coordinating group like an ester is present in a *gem*-dihalocyclopropane like **71**, the *cis*-halogen substituent is exchanged selectively in ether (equation 45)[84].

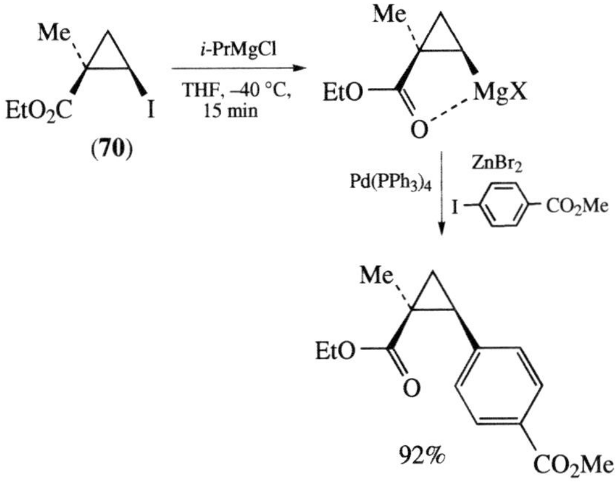

$$\text{(70): Me, EtO}_2\text{C, I on cyclopropane} \xrightarrow{\begin{array}{c}i\text{-PrMgCl}\\\text{THF, -40 °C,}\\15\text{ min}\end{array}} \text{Me, EtO-C(=O)···MgX on cyclopropane}$$

$$\downarrow \text{Pd(PPh}_3)_4, \text{ZnBr}_2, \text{I-C}_6\text{H}_4\text{-CO}_2\text{Me}$$

$$\text{Me, EtO-C(=O), (4-CO}_2\text{Me-C}_6\text{H}_4) \text{ on cyclopropane} \quad (44)$$

92%

α-Metalated nitriles are versatile nucleophiles[85]. They combine a high nucleophilicity with a small steric hindrance of the CN unit, thus allowing sterically demanding alkylation reactions. Afterwards, the nitriles are converted into a large variety of other functional groups[86]. α-Magnesiated nitriles can be obtained via Br/Mg exchange starting from α-bromo nitriles like **72**[87]. The rapid exchange at low temperature (−78 °C) allows an *in situ* reaction protocol, where the exchange reaction selectively takes place in the presence of reactive electrophiles including aldehydes (equation 46).

(71)

61%

(45)

(72)

65% 73% 70%

(46)

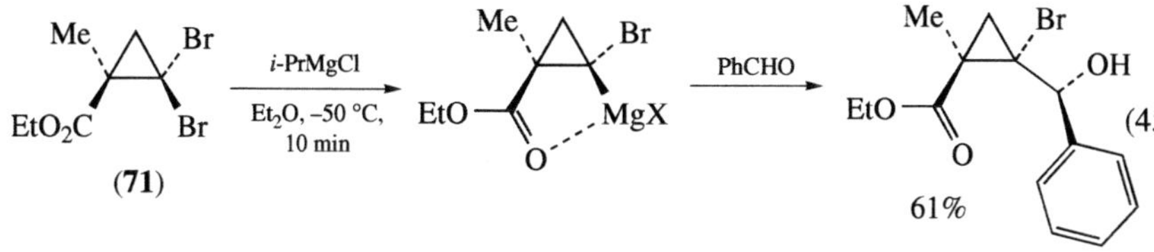

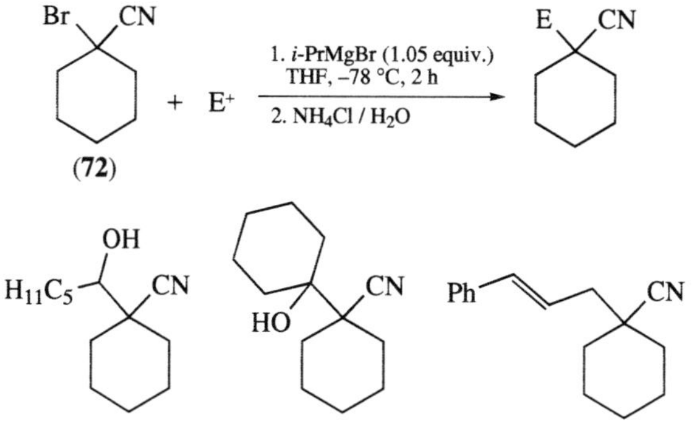

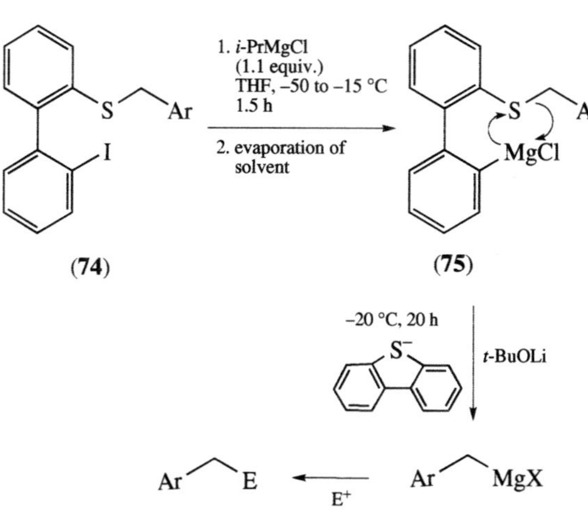

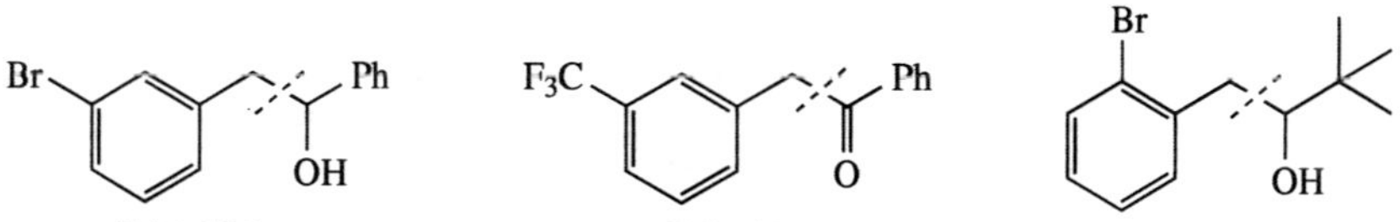

Recently, the preparation of functionalized benzylic magnesium reagents **73** could be realized by using a new sulfur–magnesium exchange reaction (equation 47)[88]. I/Mg

(74) (75)

(76) (73) X = Cl or OBu-*t*

(47)

(**76a**) 89% (**76b**) 89% (**76c**) 81%

exchange on *o*-(*o*-iodophenyl)phenylthio derivative **74** affords the expected exchange product **75** which undergoes an intramolecular sulfur–magnesium exchange reaction after treatment with *t*-BuOLi leading to the desired Grignard reagent **73**. It can be trapped with electrophiles leading to functionalized benzylic compounds **76a–c**.

b. Functional group tolerance. The magnesium exchange reaction tolerates an impressive number of functional groups. For example, the amino function is tolerated after protection as amidine **77** or as diallyl derivative **78** (Scheme 4). Imines like **79** and **80** are suitable protecting groups of anilines and aromatic aldehydes during exchange reactions (Scheme 5)[89].

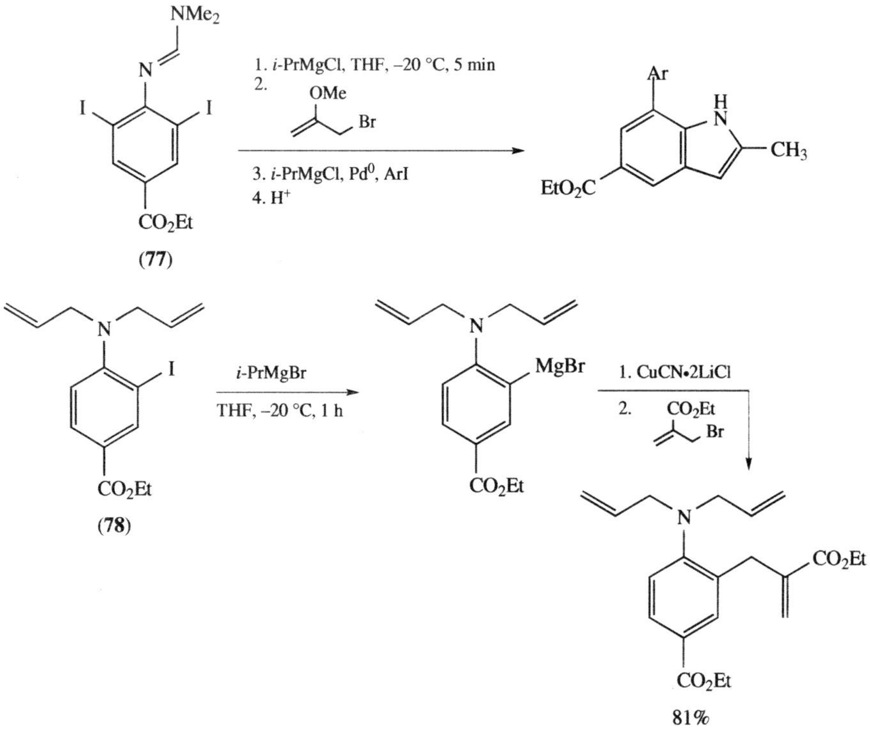

SCHEME 4

While aryl iodides bearing an aldehyde group preferentially react with the aldehyde function during attempted iodine–magnesium exchange, the corresponding imine **80** undergoes a smooth exchange reaction leading to the Grignard reagent **81** (Scheme 5).

The tedious protection and deprotection sequence of anilines can be avoided through the formation of magnesium amides. Halogen–lithium exchange reactions on aryl halides bearing acidic protons have been successfully conducted with alkyllithium reagents. But the necessity of low reaction temperatures ($-78\,°C$) and the considerable amounts of side products make this methodology less attractive, while the formation of unprotected functionalized Grignard reagents can be easily accomplished (Scheme 6). The acidic amine proton of the functionalized aniline **82** is first abstracted with methyl- or phenylmagnesium chloride. These two Grignard reagents only reluctantly undergo exchange reactions

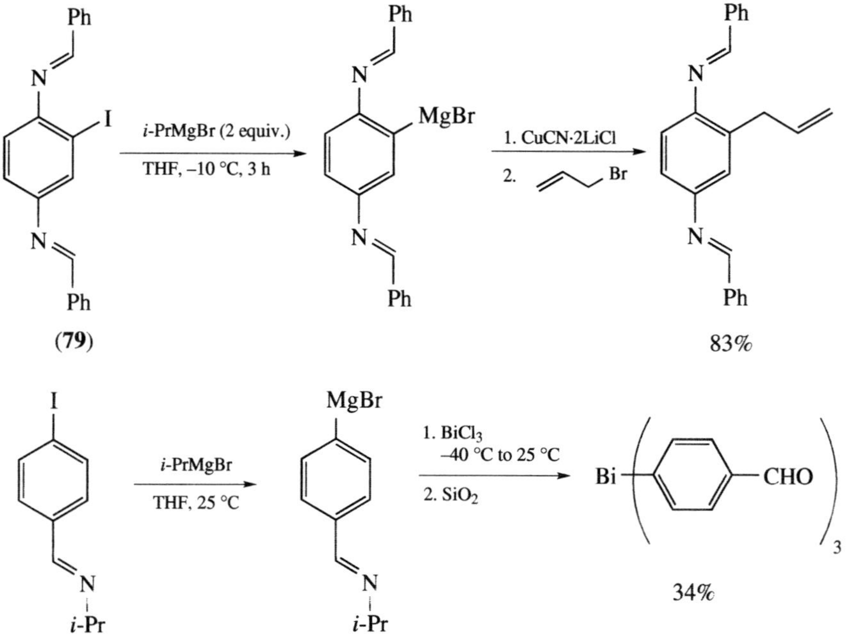

SCHEME 5

and lead to an intermediate of type **83**. In a second step, the actual I/Mg-exchange reaction is carried out with *i*-PrMgCl, leading to the desired Grignard reagent **84** (Scheme 6).[89]

Thus, the functionalized anilines **85** and **86** are obtained in 71–89% yield starting from diiodoanilines **87** and **88** (Scheme 6). Other proton-donating groups like hydroxy groups (equation 25)[57], acids, amides or benzylic alcohols[90] are also compatible with this approach.

Nitro compounds are key intermediates in organic synthesis[91] and are rather reactive toward organomagnesium reagents. They are in general believed not to be compatible with organometallic functionalities. The I/Mg-exchange reaction proceeds readily in the case of *ortho*-nitroaryl iodides such as **89** by using the less reactive PhMgCl instead of *i*-PrMgCl[92]. For nitro-containing substrates, bearing an additional coordinating group such as **90** and **91**, or *ortho*-disubstituted substrates like **92**, the exchange on *meta*- and *para*-aryl iodides is also possible (Scheme 7).

The triazene group is a convenient synthetic equivalent of a diazonium salt and is readily converted to an iodine functionality[93]. It reacts with *i*-PrMgCl when an I/Mg exchange with an iodoarene bearing a triazene functionality is attempted. By using the more reactive *i*-PrMgCl•LiCl, the exchange reaction can be realized at lower temperatures (−40 °C) allowing an excellent compatibility with a triazene group. Thus, the reaction of the iodotriazene **93** with *i*-PrMgCl•LiCl provides the desired Grignard reagent which undergoes a smooth addition–elimination with 3-iodo-2-cyclohexenone in the presence of CuCN•2LiCl. The resulting enone **94** is readily converted to aryl iodide **95** by treatment with CH₃I (equation 48)[94]. This iodine can be subjected to further functionalizations[95].

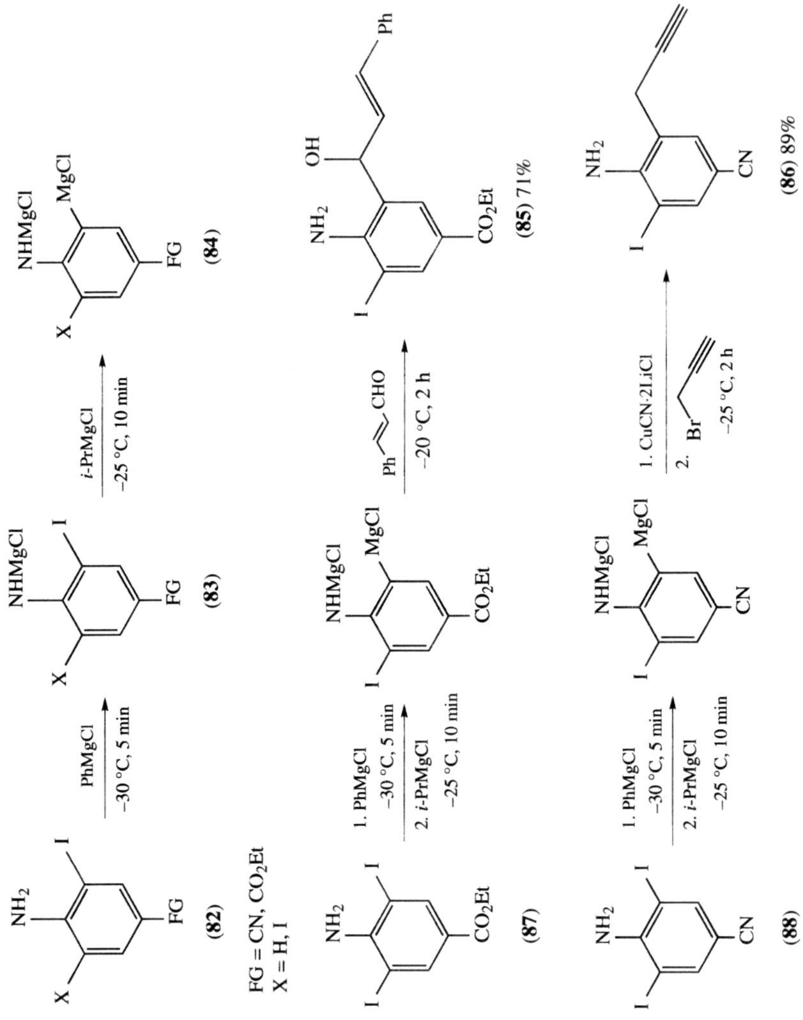
NH2
FG
(82)
FG = CN, CO2Et
X = H, I
PhMgCl
−30 °C, 5 min
NHMgCl
(83)
i-PrMgCl
−25 °C, 10 min
MgCl
(84)
NH2
CO2Et
(87)
1. PhMgCl
−30 °C, 5 min
2. i-PrMgCl
−25 °C, 10 min
Ph
CHO
−20 °C, 2 h
OH
(85) 71%
CN
(88)
1. CuCN·2LiCl
2.
Br
−25 °C, 2 h
(86) 89%

SCHEME 6

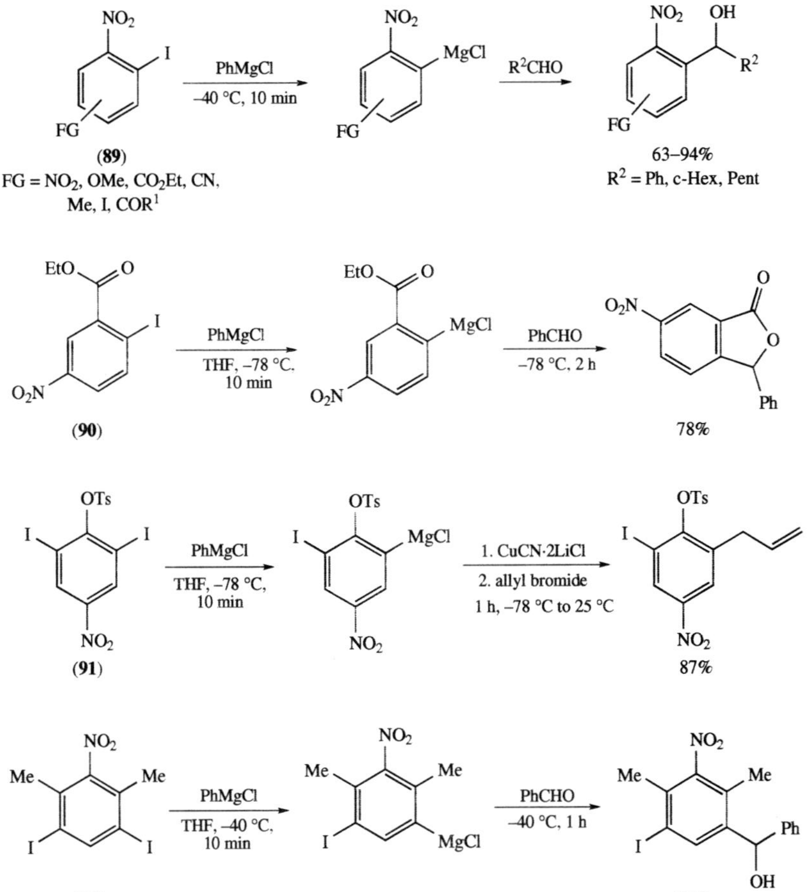

12. Functionalized organomagnesium compounds: Synthesis and reactivity 533

SCHEME 7

Magnesiated triazene derivatives like **96** can further be used for the preparation of functionalized carbazoles **97** (equation 49)[95].

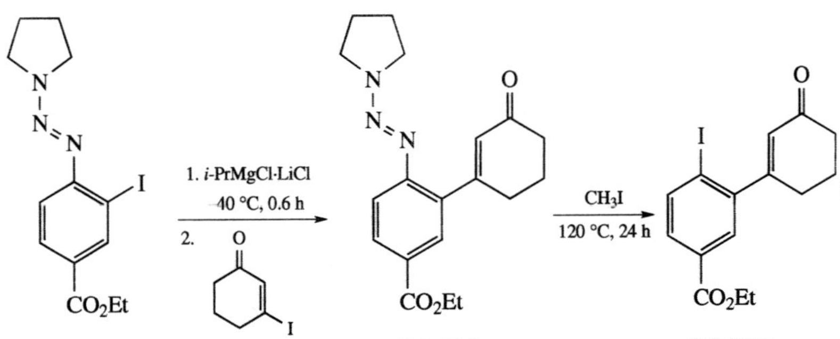

(48)

By carefully tuning the reaction conditions, the preparation of ketone group-containing arylmagnesium species can be achieved. To avoid side reactions, neopentylmagnesium bromide (NpMgBr) **98**, a sterically hindered but reactive Grignard reagent, is used. In conjunction with N-methylpyrrolidinone (NMP) as a polar cosolvent, complete conversion to the organomagnesium species **99** is observed at $-30\,^\circ$C within 1 h (equation 50)[96]. The *ortho*-keto function actually facilitates the formation of the Grignard reagent by pre-coordination of NpMgBr **98** and stabilizes the resulting arylmagnesium species **99** by chelation. Iodo-substituted aryl or heteroaryl ketones or cycloalkenyl ketones can alternatively be protected as cyanohydrins **100** and **101** allowing I/Mg-exchange reactions with i-PrMgCl·LiCl[97]. After deprotection the functionalized products **102** and **103** are obtained in 76–87% yield (equations 51 and 52). This protocol can also be applied to aromatic iodoaldehydes[97].

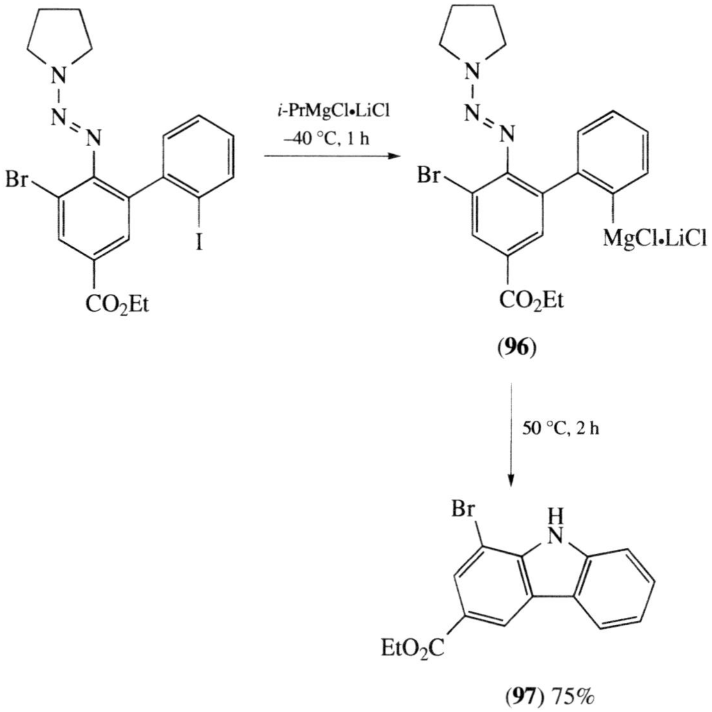

(49)

Electrophilic functional groups in *ortho*-position to the carbon–magnesium bond allow two sequential alkylations. Starting from *ortho*-iodobenzyl chloride **104**, the benzannulated heterocycles **105** and **106** are obtained after the reaction with appropriate electrophiles (Scheme 8)[98].

The high activity of i-PrMgCl·LiCl in exchange reactions allows the preparation of highly functionalized aryl and hetaryl pinacolborates **107**, e.g. **107a, b** (bimetallic reagents, equation 53)[99]. The halogen–magnesium exchange reaction can be easily applied to solid-phase synthesis[81b, 100, 101], affording polymer-bound Grignard reagent **108** (equation 54).

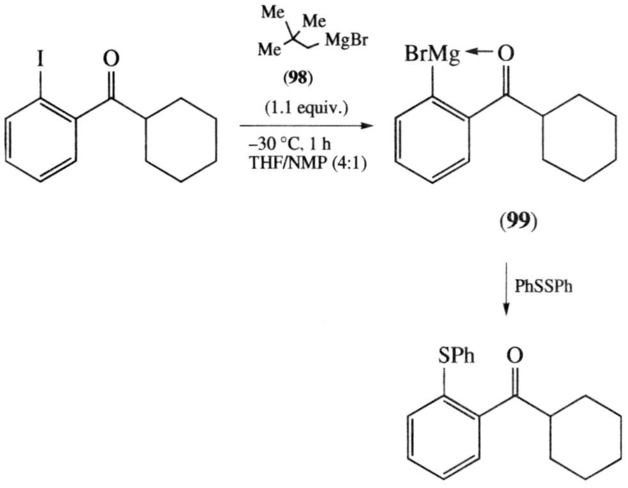

12. Functionalized organomagnesium compounds: Synthesis and reactivity 535

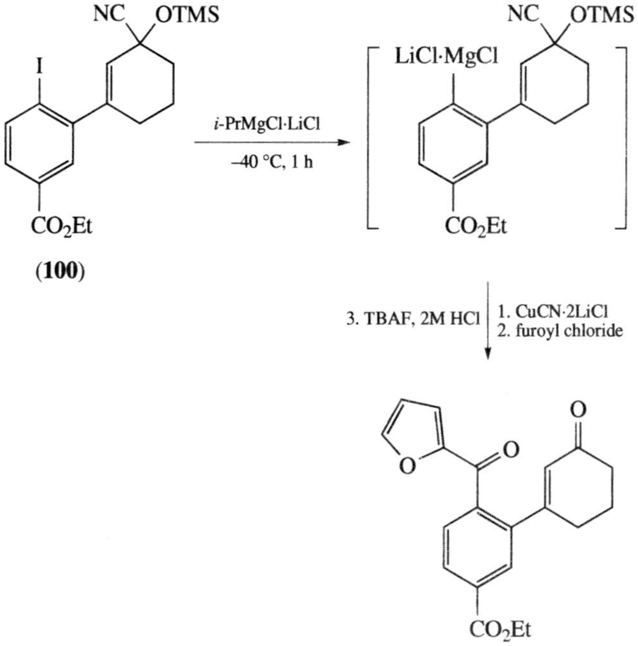

(50)

(51)

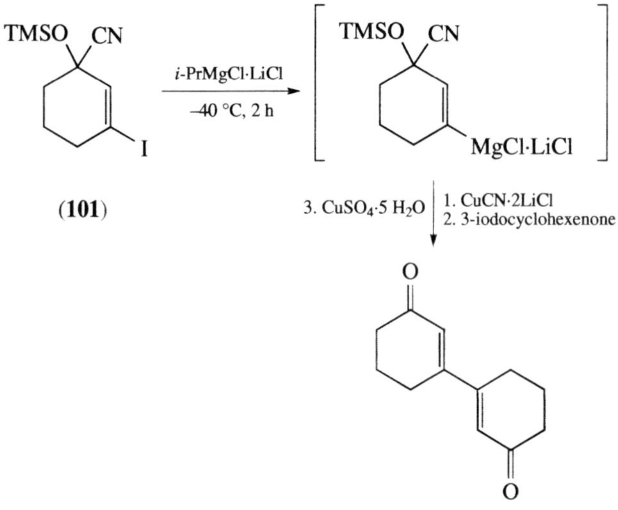

(**101**) (**103**) 76% (52)

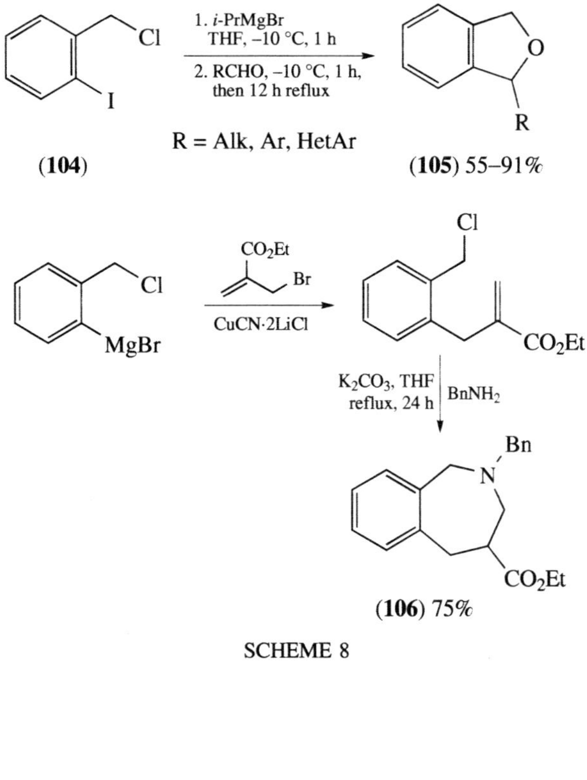

(**104**) (**105**) 55–91%

(**106**) 75%

SCHEME 8

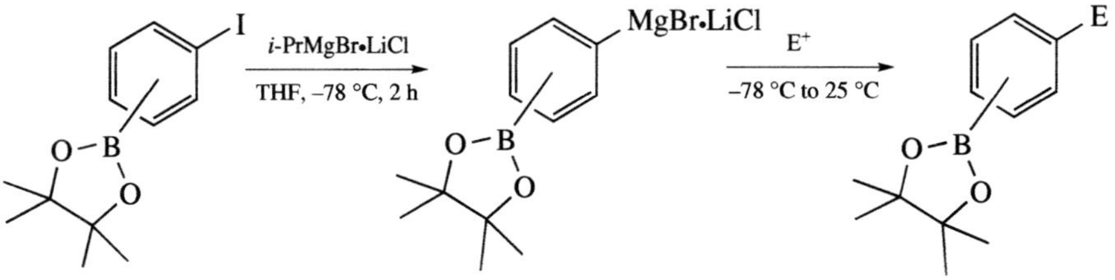

C. Metalation Reactions with Magnesium Amide Bases

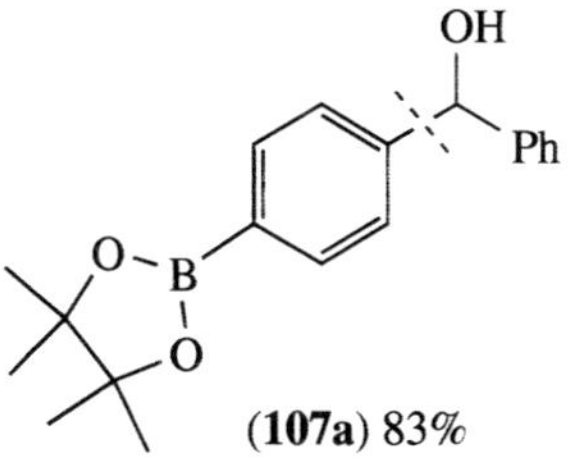

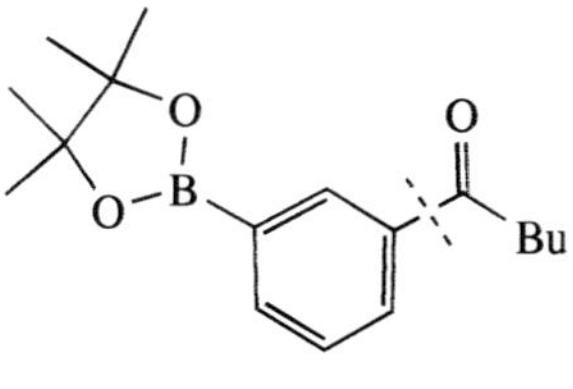

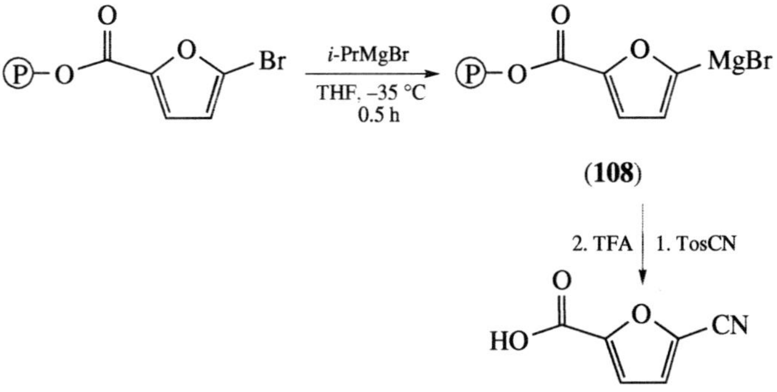

Alkyllithium reagents (RLi) or lithium dialkylamides (R_2NLi) have been widely used for the *ortho*-metalation reactions of aromatic and heteroaromatic compounds[102], although their use is usually complicated by the presence of undesired side reactions as a result of their reactivity and strong nucleophilicity and by a low compatibility of functional groups. To overcome low reaction temperatures (−78 to −90 °C) and ensure the presence of various functional groups, magnesium amides and to a smaller extent Grignard reagents have been developed as metalating agents. Alkylmagnesium reagents are strongly basic. Nevertheless, their low kinetic basicity allows only successful magnesiations in a few cases such as the pyridyl amide **109**. The activating group both directs the Grignard reagent and breaks magnesium aggregates. Unfortunately, the attempted metalation often competes with addition reactions (equation 55)[103].

3-Substituted pyridines undergo exclusively 1,4-addition, while 4-substituted pyridines give a mixture of products[103]. The metalation of alkynes by *n*-BuMgCl is often used[104].

Reaction of alkylmagnesium reagents with sterically hindered amines leads to the formation of magnesium amides **110–112**[105], reacting much faster than the parent alkylmagnesium derivatives with C–H acidic substrates (Scheme 9).

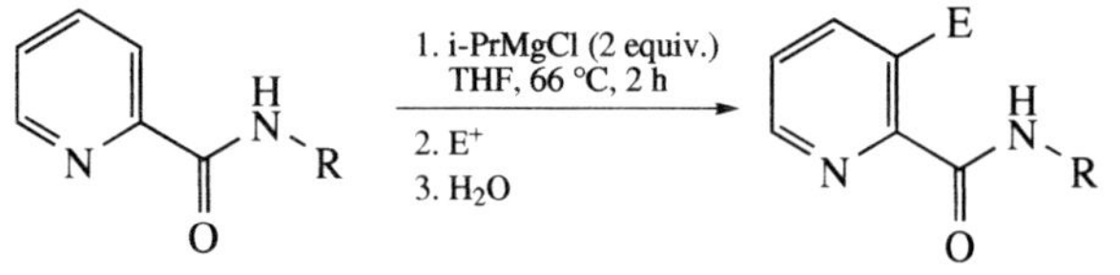

$$
(109) \xrightarrow[\substack{2.\ E^+ \\ 3.\ H_2O}]{\substack{1.\ i\text{-PrMgCl (2 equiv.)} \\ \text{THF, 66 °C, 2 h}}} 25\text{–}97\%
\tag{55}
$$

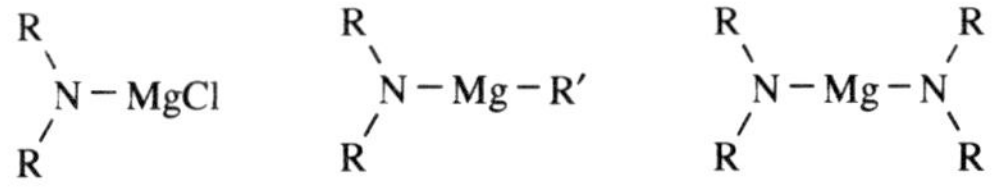

(**110**) R$_2$N–MgCl   (**111**) R$_2$N–Mg–R′   (**112**) R$_2$N–Mg–NR$_2$

SCHEME 9

The low solubility of the amides R$_2$NMgCl (**110**), R$_2$NMgR′ (**111**) or (R$_2$N)$_2$Mg (**112**) has hampered a general application of these bases. Usually, a large excess (up to 5 equiv.) is necessary to ensure a complete magnesiation. This is disadvantageous, since the range of electrophiles added to quench the newly generated magnesium reagent may be limited due to side reactions of some electrophiles with the excess of magnesium base. Nevertheless, cyclopropyl amides such as **113** can be functionalized (equation 56)[106]. Also, the magnesiation of indoles like **114** can be realized (equation 57)[107, 108]. Finally, the catalytic generation of the magnesium base may be advantageous in the case of the magnesiation of pyrrole **115** (equation 58)[109].

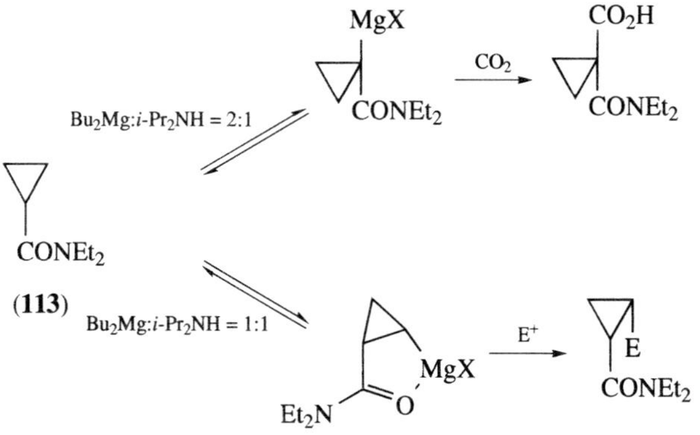

(56)

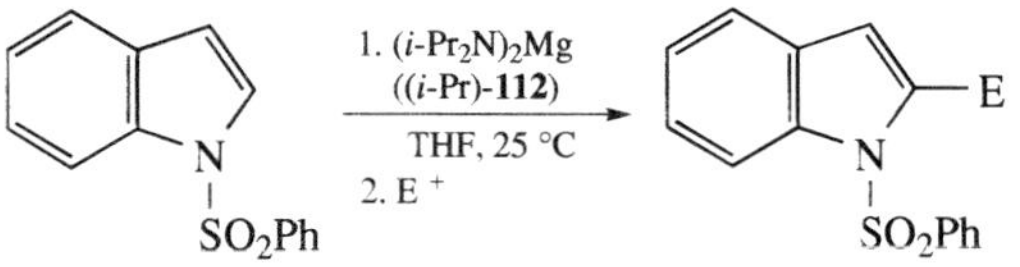

(57) 60–93%

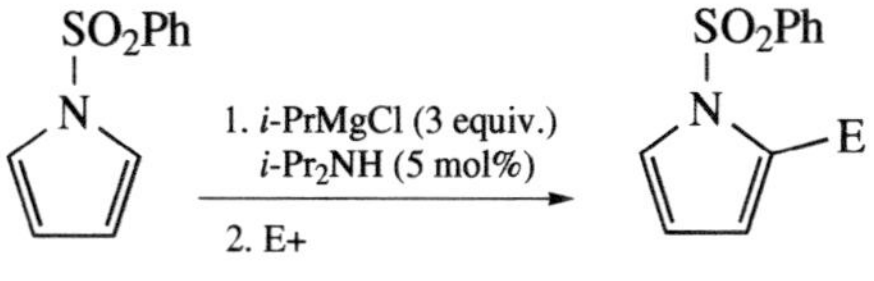

12. Functionalized organomagnesium compounds: Synthesis and reactivity 539

$$\text{(115)} \xrightarrow[\text{2. E+}]{\text{1. }i\text{-PrMgCl (3 equiv.)} \atop i\text{-Pr}_2\text{NH (5 mol\%)}} \text{24–57\%} \qquad (58)$$

E = I, Alk, TMS, Ar, HetAr

The chemistry of magnesium bisamides has been reviewed[110]. They can be used for the regio- and stereoselective formation of enolates[111], while chiral magnesium amides are applied in asymmetric synthesis for enantioselective enolisations[112].

A large excess of the magnesium amides can be avoided by using highly soluble mixed Mg/Li amides R_2NMgCl•LiCl 116[113]. Reaction of i-PrMgCl•LiCl with sterically hindered secondary amines affords Mg/Li reagents 116 (e.g. 116a,b) which display high kinetic activity, excellent solubility combined with a better stability upon storage as THF solutions (equation 59).

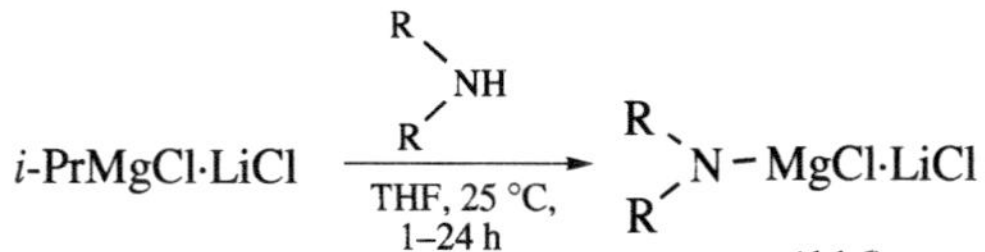

$$i\text{-PrMgCl·LiCl} \xrightarrow[\text{THF, 25 °C,} \atop 1–24\text{ h}]{R_2\text{NH}} \text{(116)} \qquad (59)$$

(116a) R = i-Pr
(116b) R_2N = 2,2,6,6-tetramethylpiperidyl
(TMPMgCl·LiCl)

TMPMgCl·LiCl (116b) is an especially efficient base for the regioselective magnesiation of various heteroaromatics (117 and 118) and aromatics (119 and 120) species (Scheme 10)[113, 114].

Even a multiple functionalization of a bisubstituted aromatic compound such as 121 can be achieved by successive magnesiation with TMPMgCl·LiCl (116b) and quench-

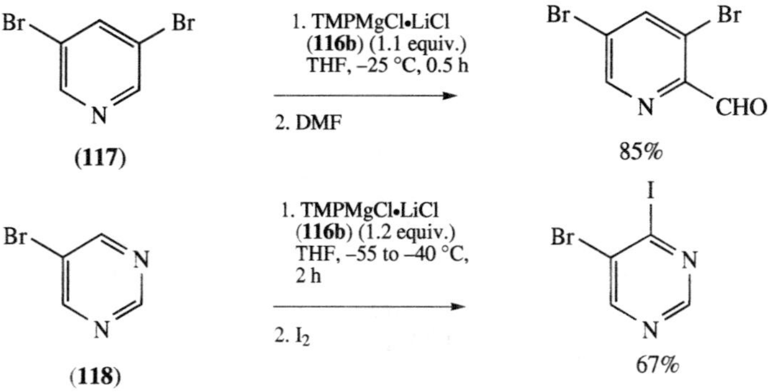

SCHEME 10

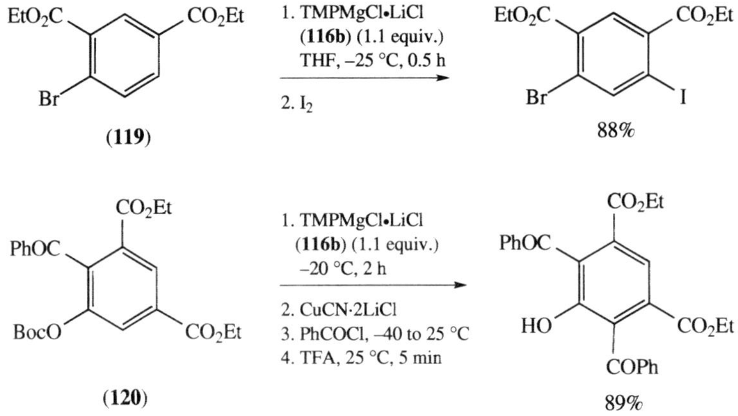

SCHEME 10

ing with various electrophiles leading to hexasubstituted benzene derivatives **122a–d** (Scheme 11)[114].

D. Miscellaneous Methods

Arynes are valuable intermediates in synthetic organic chemistry[115]. The triple bond of an aryne is highly reactive toward reactions with nucleophiles. For example, functionalized arynes **123**, prepared from *ortho*-iodoaryl sulfonates **124** by an iodine–magnesium exchange followed by the elimination reaction of *ortho*-magnesio-arylsulfonates, react with a number of heteroatomic nucleophiles, like R_2NMgX, RSMgX, RSeMgX and partially with R_2PMgX, generating novel Grignard species of type **125**. These reagents can be trapped by electrophiles leading to functionalized aromatics **126**. Alternatively, arynes **123** undergo cyclization reactions with furan leading to cycloadducts of type **127** (Scheme 12)[116]. Following this protocol, various functionalized arynes **126a, b** and **127a, c** and heteroaryne **127d** have been prepared.

The nature of the sulfonate leaving group proved to be crucial for an efficient elimination reaction leading to arynes of type **123** (Scheme 12). Best results are obtained with 4-chlorobenzenesulfonate as leaving group. Similarly, the reaction of arylmagnesium reagents with arynes, prepared *in situ* from 2-fluorophenyllithium, gave sterically encumbered substituted 2-iodobiphenyls after iodolysis[117]. The addition of MgH_2 to olefins is leading to Grignard reagents[118]. Also, the reaction of Mg metal with 1,4-diarylbutadienes gives magnesium derivatives behaving as Grignard reagents (equation 60)[119].

Hydromagnesiation of acetylenes, catalyzed by titano- and zirconocenes, can be readily achieved. A regioselective reaction occurs only if one of the substituents on the triple bond is silicon or an aryl group **128** (equation 61). Mechanistic studies on this hydromagnesiation have been reported[120, 121]. This reaction has been applied to the synthesis of polysubstituted alkenes of type **129**[122]. The non-catalyzed hydromagnesiation of 1,3-alkadienes with i-PrMgCl gives only mixtures of magnesiated alkenes[123]. Reaction of 2-alkyl-1,3-butadienes with n-PrMgX in the presence of Cp_2TiCl_2 affords allylmagnesium

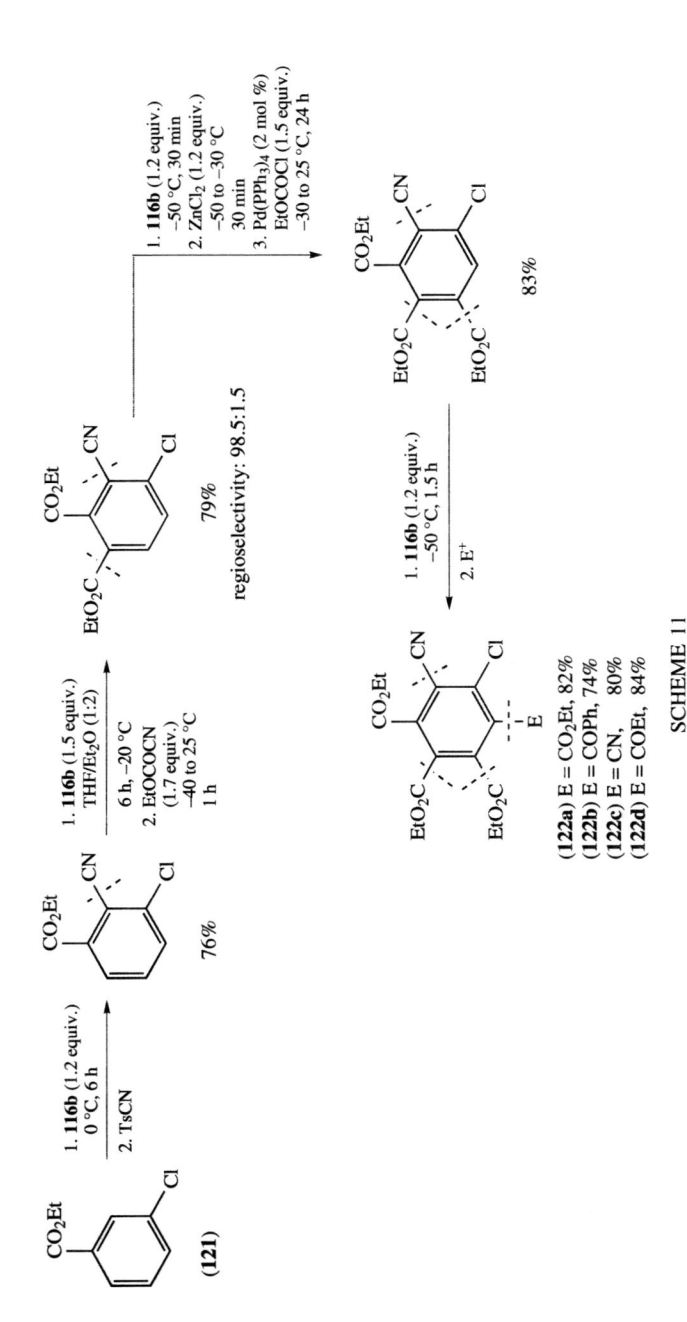

(122a) E = CO₂Et, 82%
(122b) E = COPh, 74%
(122c) E = CN, 80%
(122d) E = COEt, 84%

SCHEME 11

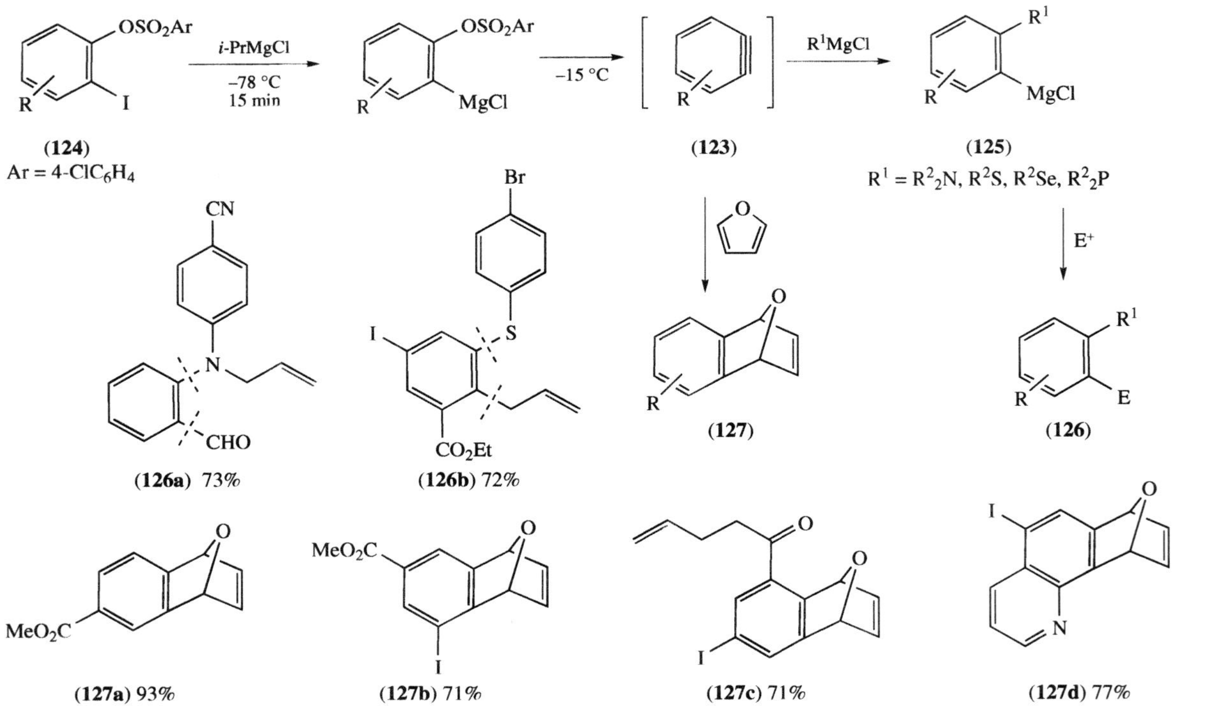

SCHEME 12

reagents as single regioisomers[124].

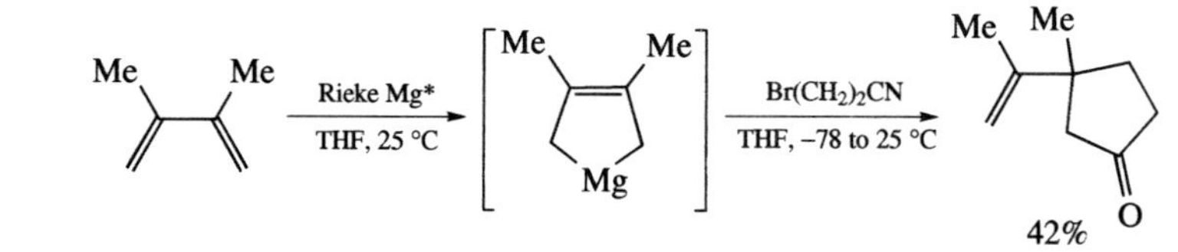

$$\text{Me}\diagdown\diagdown\text{Me} \xrightarrow[\text{THF, 25 °C}]{\text{Rieke Mg*}} \left[\begin{array}{c}\text{Me}\diagdown\diagdown\text{Me}\\ \diagup\diagdown\\ \text{Mg}\end{array}\right] \xrightarrow[\text{THF, -78 to 25 °C}]{\text{Br(CH}_2)_2\text{CN}} \underset{42\%}{\overset{\text{Me Me}}{\diagup}\diagdown\diagdown\text{O}} \quad (60)$$

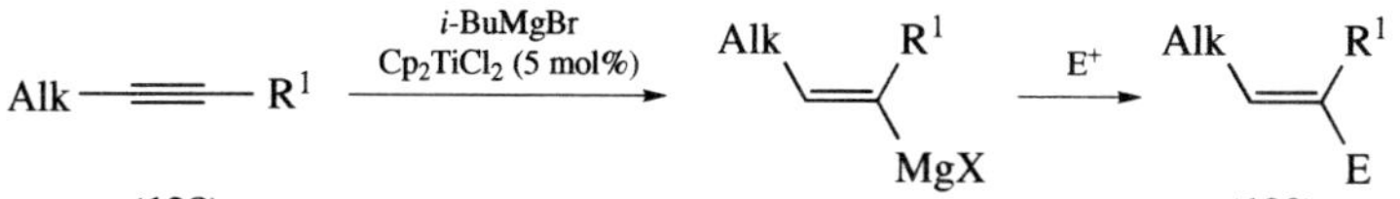

$$\text{Alk}\!\!=\!\!\text{R}^1 \xrightarrow[\text{Cp}_2\text{TiCl}_2\text{ (5 mol\%)}]{i\text{-BuMgBr}} \underset{\text{MgX}}{\overset{\text{Alk}\quad\text{R}^1}{\diagup\!\!=\!\!\diagdown}} \xrightarrow{\text{E}^+} \underset{\text{E}}{\overset{\text{Alk}\quad\text{R}^1}{\diagup\!\!=\!\!\diagdown}} \quad (61)$$

(128) \hspace{4cm} (129)
$R^1 = \text{SiMe}_3$, Ar \hspace{2cm} E = PhS, PhTe, Alk$_3$Sn, vinyl iodide (Pd cat.)

III. REACTIVITY OF GRIGNARD REAGENTS

Organomagnesium compounds are versatile reagents for organic synthesis[125]. They undergo a multitude of reactions, which will be divided in this section into two major groups: substitution and addition reactions. Cross-coupling reactions[126], allylic substitutions and ring-opening of small cyclic molecules will be considered as substitution reactions. Carbomagnesiation and 1,4-addition reactions will be discussed in the addition part. Cases of mechanistically complex reactions will be classified according to the structure of the final product, i.e. addition–elimination reactions will be included in the substitution part. The substitution reactions are further subdivided into three groups according to the degree of unsaturation at the electrophilic center of the substrate.

A. Substitutions at an sp³-center

1. Transition-metal-catalyzed cross-coupling reactions

Pd- and Ni-catalyzed coupling reactions follow a mechanism which is described in Scheme 13.

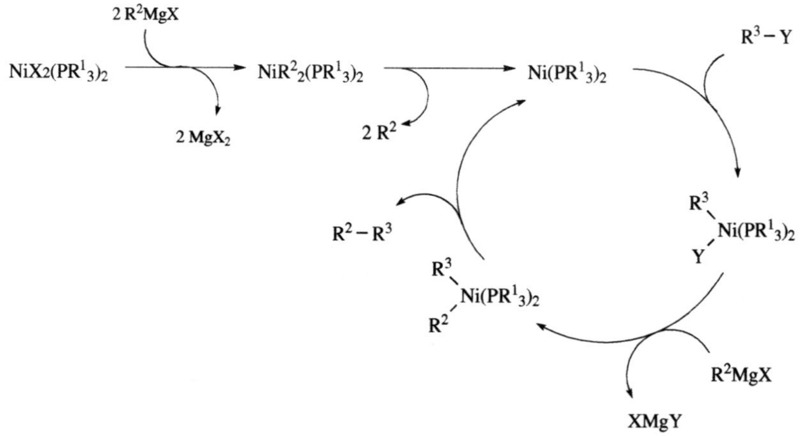

SCHEME 13

Compared to the formation of Csp^2-Csp^2 bonds, Csp^3-Csp^2 couplings are difficult to perform. Unactivated alkyl electrophiles only reluctantly undergo the oxidative addition to a metal center, while β-hydride elimination is generally fast. *n*-Alkyl bromides and tosylates like **130** can be coupled successively with aryl- and alkylmagnesium reagents in the presence of 1,3-butadiene, using 3% $Pd(acac)_2$ as a catalyst (equation 62)[127]. Using 1 mol% $NiCl_2$ and 10 mol% butadiene, the reactions can even be performed at 0 °C (equation 63)[128]. With 1,3,8,10-tetraenes like **131** as additives, the nickel-catalyzed cross-coupling of alkyl halides with organozinc or Grignard reagents proceeds readily (equation 64). In the presence of strong electron-donating ligands (PCy_3, IMes), arylmagnesium bromides can be coupled with primary alkyl chlorides like **132** using Pd catalysis (equation 65)[129, 130].

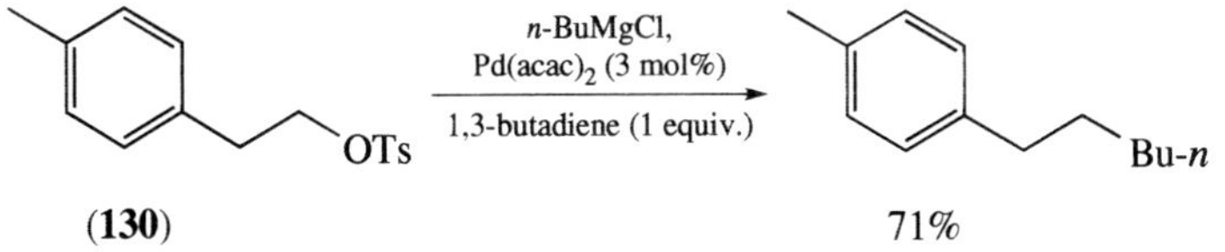

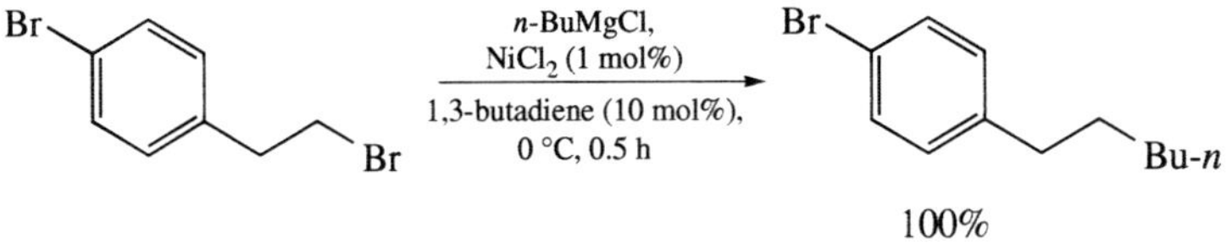

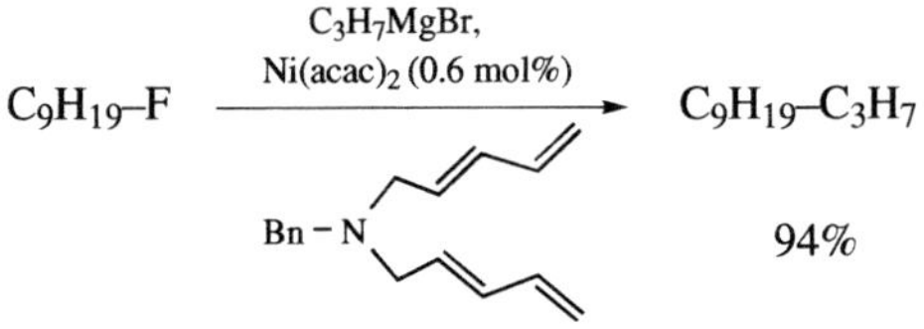

Enol phosphates can be coupled with alkyl or aryl magnesium reagents using $NiCl_2$(dppe)[131] or $PdCl_2(PPh_3)_2$[132]. Copper-catalyzed cross-coupling of alkyl halides or sulfonates with Grignard reagents has become a popular method for constructing alkyl chains. For example, Li_2CuCl_4 in THF-NMP efficiently catalyses the reaction of alkyl- and vinylmagnesium reagents with primary alkyl halides. Functional groups (ketone, ester, nitrile, sulfonate) are tolerated in this reaction (equation 66)[133].

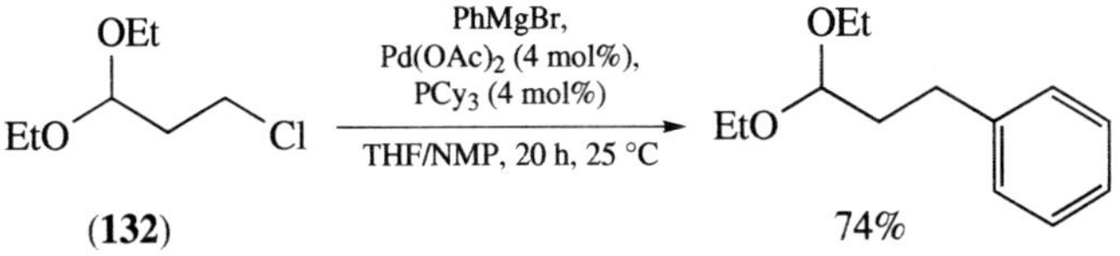

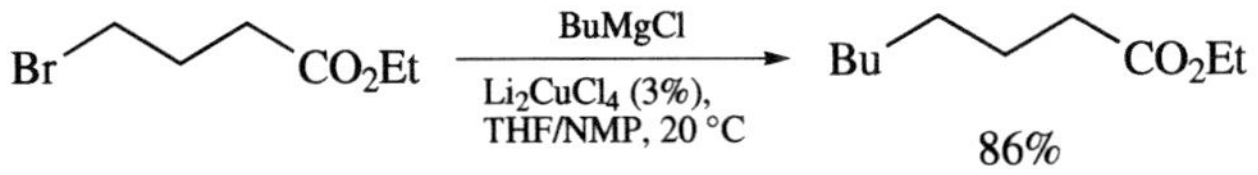

If an organomagnesium reagent bears a remote leaving group, cyclizations can be achieved. Starting from the tosylate **133**, a stereoselective substitution using $CuCN \cdot 2LiCl$

afforded the benzotetrahydrofuran **134** without loss of optical purity (equation 67).

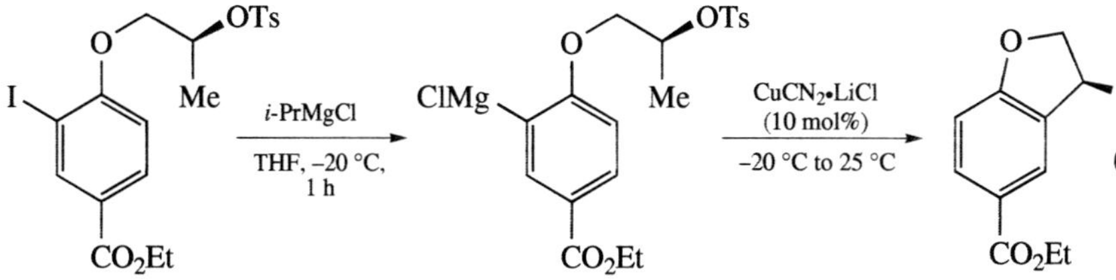

$$\text{(133) 60\% ee} \xrightarrow[\text{THF, }-20\,^\circ\text{C, 1 h}]{i\text{-PrMgCl}} \text{ClMg intermediate} \xrightarrow[-20\,^\circ\text{C to 25\,^\circ\text{C}}]{\text{CuCN}_2\cdot\text{LiCl (10 mol\%)}} \text{(134) 83\%; 60\% ee} \quad (67)$$

Copper thiophenolate-LiBr has proven to be a superior catalyst for special cases like the reaction of arylmagnesium compounds with primary alkyl tosylates and n-alkylmagnesium bromides with secondary tosylates[134]. Primary alkyl fluorides react smoothly with tertiary alkylmagnesium halides in the presence of $CuCl_2$, while primary and secondary alkyl magnesium reagents require the addition of butadiene. Arylmagnesium derivatives only react at elevated temperatures, while alkyl chlorides give poor results[135]. The latter can be efficiently coupled in the presence of 1-phenylpropyne as an additive (equation 68)[136]. Various alkylmagnesium chlorides can be reacted with alkyl bromides in the presence of an amino-organomanganese complex and $CuCl$[137]. A combination of $CuCl$, triethyl phosphate and tetrabutylammonium iodide (TBAI) efficiently couples aryl- or heteroarylmagnesium halides with benzylic phosphates **135** forming polyfunctionalized diarylmethanes **136** (equation 69)[138].

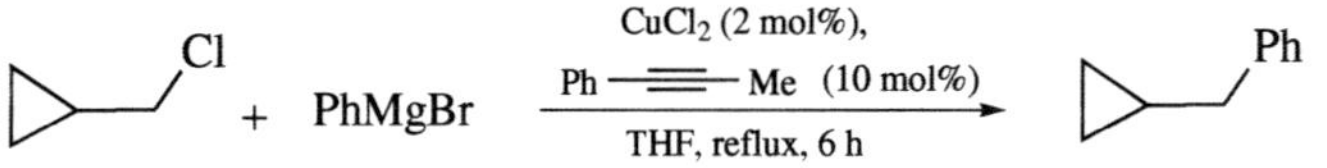

$$\text{cyclopropyl-CH}_2\text{Cl} + \text{PhMgBr} \xrightarrow[\text{THF, reflux, 6 h}]{\substack{\text{CuCl}_2\text{ (2 mol\%),}\\ \text{Ph}\equiv\text{Me (10 mol\%)}}} \text{cyclopropyl-CH}_2\text{Ph} \quad 98\% \quad (68)$$

$$\text{(135)} + \text{FG}^2\text{-Ar-MgX} \xrightarrow[\text{DME, 60\,^\circ\text{C, 1 h}}]{\substack{\text{TBAI (10 mol\%)}\\ \text{CuCl (10 mol\%), P(OEt)}_3\text{ (20 mol\%)}}} \text{(136) up to 88\%} \quad (69)$$

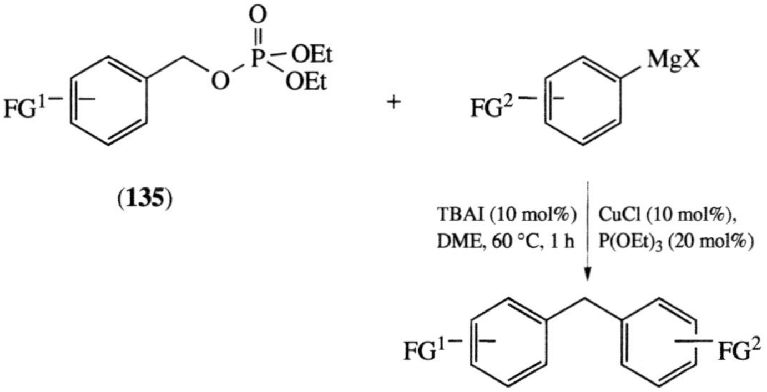

Other possible catalysts for the reaction of various alkyl bromides with allylmagnesium halides in THF are cobaltbis(1,3-diphenylphosphino)propane complex (equation 70)[139] or cobalt chloride in combination with a diamine (equation 71)[140]. Cyclization products can be obtained in good yields in the presence of a suitably placed double bond[141].

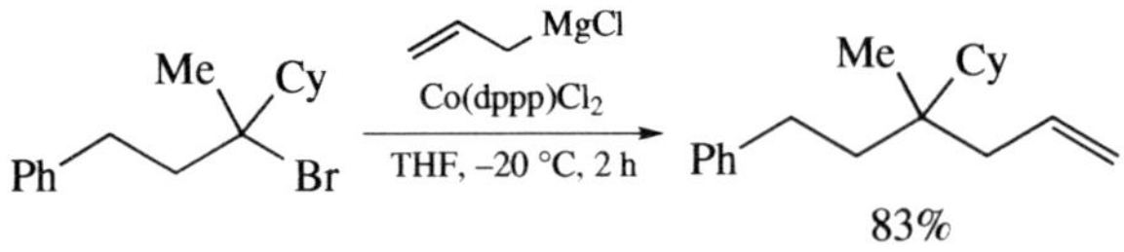

$$\text{Ph}\diagup\diagdown\underset{\text{Br}}{\overset{\text{Me Cy}}{\diagdown}} + \diagup\diagdown\text{MgCl} \xrightarrow[\text{THF, }-20\,°\text{C, 2 h}]{\text{Co(dppp)Cl}_2} \text{Ph}\diagup\diagdown\underset{83\%}{\overset{\text{Me Cy}}{\diagdown}}\diagdown\diagup \quad (70)$$

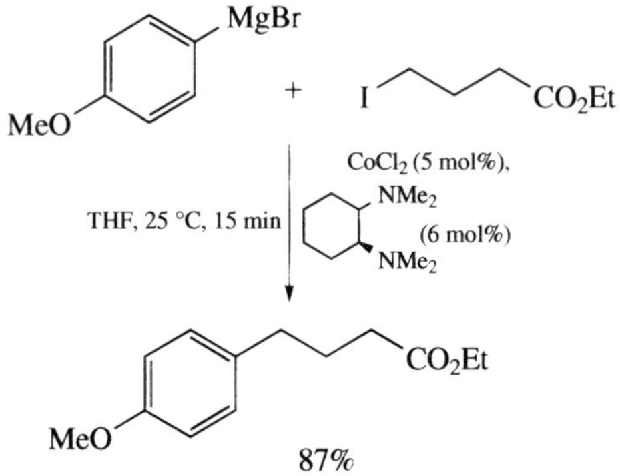

(71)

Fe^{III} salts have been successfully used for coupling reactions of various organomagnesium reagents with alkyl electrophiles. Thus, $FeCl_3$ and TMEDA[142], $Fe(acac)_3$ (equation 72)[143] or Fe^{III} salen-type complexes (**137**, equation 73)[144] catalyze the coupling of arylmagnesium reagents with primary and secondary alkyl bromides. $Fe(MgX)_2$ is believed to be the active catalyst. It is formed in the reaction mixture by the *in situ* reduction of Fe^{III} salts by the Grignard reagent[145]. The Fe^{II} complex $[Li(TMEDA)]_2[Fe(C_2H_4)_4]$ is another efficient catalyst for the cross-coupling reaction between alkyl electrophiles and arylmagnesium compounds[145] affording complete conversion within minutes even at $-20\,°C$.

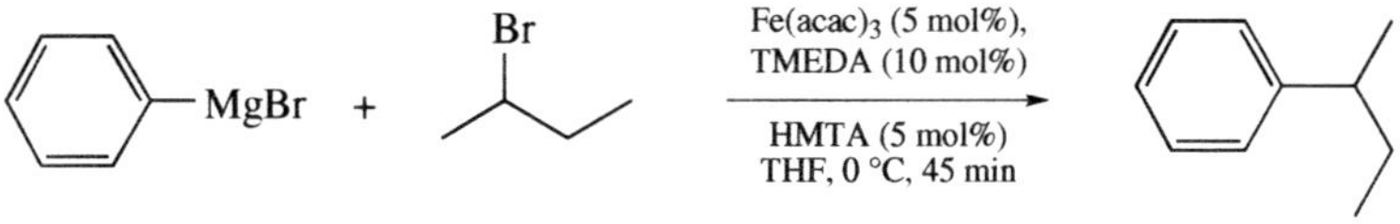

(72)

HMTA = hexamethylenetetramine

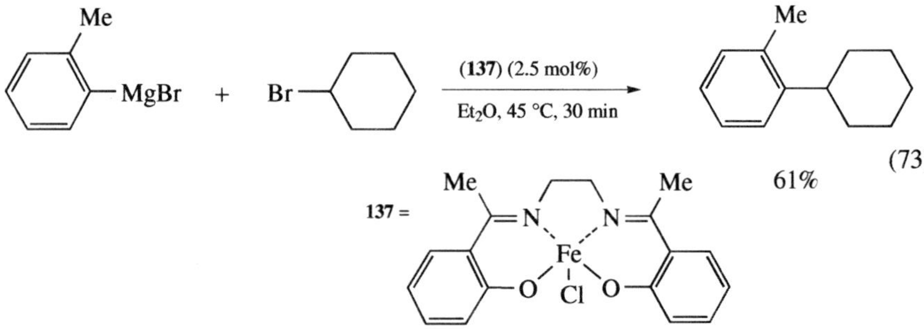

(73)

12. Functionalized organomagnesium compounds: Synthesis and reactivity 547

Oxidative homo-coupling of alkyl magnesium reagents possessing β-hydrogens is achieved in the presence of silver tosylate (AgOTs, 1 mol%) as a catalyst and 1,2-dibromoethane as a reoxidant[146].

2. Ring-opening of small cycles

Organomagnesium reagents can effect ring-opening of aziridines in the presence of a catalytic amount of Cu^I salt[147]. The aziridines must bear a phosphinoyl, sulfonyl or carbamate group on the nitrogen. This method can be used for the synthesis of chiral β-(het)arylalkylamines[148] or α-amino acids like **138** (equation 74)[149]. The easily available chiral β-propiolactones of type **139** undergo similar ring-opening reactions, thus offering an alternative to enantioselective 1,4-addition reactions (equation 75)[150].

$$\underset{\text{CO}_2\text{Bu-}t}{\underset{N}{\overset{\text{Boc}}{\triangle}}} \xrightarrow[\text{THF or PhMe}]{\text{RMgX (1.5–3 equiv.)} \atop \text{CuBr·SMe}_2 \text{ (0.1–0.3 equiv.)}} \underset{\text{CO}_2\text{Bu-}t}{\underset{(\textbf{138})}{R \diagup\!\!\!\diagdown \text{NHBoc}}} \quad (74)$$

R = Alk, alkenyl, Ar

$$\underset{(\textbf{139})}{O\!\!=\!\!\overset{O}{\square}\!\!-\!\!R^1} \xrightarrow[\text{THF, –30 °C}]{\text{R}^2\text{MgBr, TMSCl} \atop \text{CuBr·SMe}_2} \text{HO}_2\text{C}\diagup\!\!\!\underset{R^1}{\overset{R^2}{\diagdown}} \quad (75)$$

Terminal epoxides **140** are deprotonated at 0 °C using lithium tetramethylpiperidide (LiTMP). The resulting anion reacts with alkyl- and alkenylmagesium reagents under ring-opening. Li$_2$O is eliminated and alkenes **141** are formed (equation 76)[151].

$$\text{H}_{21}\text{C}_{10}\!\!-\!\!\overset{O}{\triangle} \xrightarrow[0\,°\text{C to 25 }°\text{C, 3 h}]{\text{LiTMP (2.0 equiv.)} \atop \text{RMgCl (1.4 equiv.)}} \text{H}_{21}\text{C}_{10}\diagup\!\!\!\diagdown\!\!R \quad (76)$$

(**140**) R = Alk, Alkenyl (**141**)

Reaction of ketone dithioacetals with Grignard reagents opens synthetic routes to a variety of substituted alkenes (Scheme 14)[152]. For the reactions of simple aliphatic dithioacetals the presence of a Ni-trialkylphosphine catalyst is needed[153].

Chiral acetals undergo diastereoselective ring-opening with Grignard reagents in toluene[154]. Ketals, derived from ω-bromoketones, react with Mg/MgBr$_2$, giving cycloalkanol ethers after Lewis-acid-assisted ring-opening and intramolecular quench of the alkylmagnesium species. Substituted cyclopropanes and cyclobutanes are obtained by this method[155].

3. Diverse reactions

Benzotriazole (Bt) may serve as a leaving group in reactions with organometallic species[156]. Thus, polysubtituted α-aminobenzotriazoles **142** react with Grignard reagents as imine equivalents and the use of alkenyl- or propargylmagnesium reagents allows the synthesis of allyl- or propargylamines **143** and **144** in good yields (Scheme 15)[157].

This reaction can furthermore be applied on chiral aminals, affording a straightforward route to optically pure *trans*-2,5-pyrrolidines[158] or chiral alkyl-substituted 1,3-oxazolidines. This method was used for the enantioselective synthesis of substituted piperidines[159].

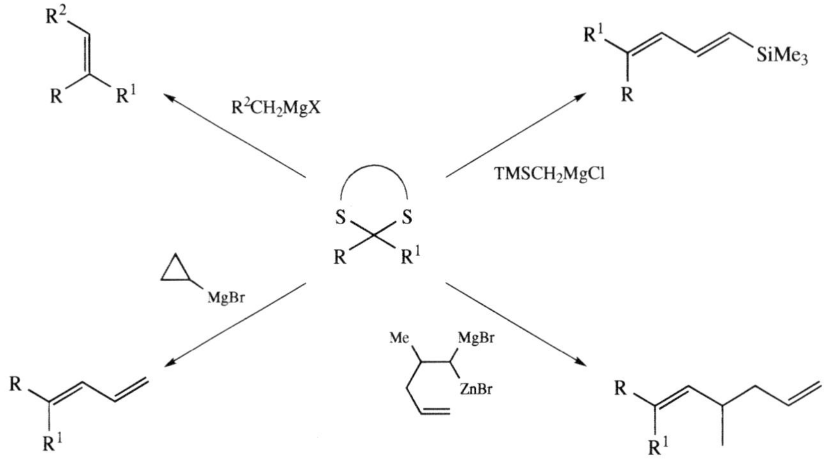

SCHEME 14

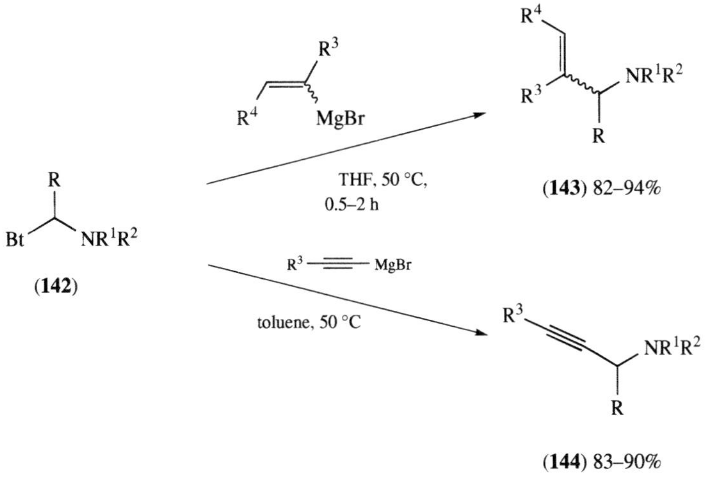

SCHEME 15

An alternative route to tertiary allyl- and propargylamines **145** and **146** is the reaction of Grignard compounds with iminium triflates **147** and **148** (equations 77 and 78). The intermediate iminium triflates **147** and **148** are obtained from the corresponding aminals by reaction with Tf$_2$O. Primary propargylamines can be prepared from tetraallylated aminals[160].

Other functionalities than halides, sulfonates or benzotriazoles (Scheme 15) might serve as leaving groups in substitution reactions with organomagnesium reagents. For example, benzylic α-azidoethers react with a substitution of the azide group[161]. Primary and benzylic alcohols **149** can be converted into good leaving groups by their transformation into

12. Functionalized organomagnesium compounds: Synthesis and reactivity   549

diphenylphosphinites followed by quaternization with MeI. Reaction with alkyl-, benzyl- and arylmagnesium compounds gives the corresponding coupling products of type **150** (equation 79)[162].

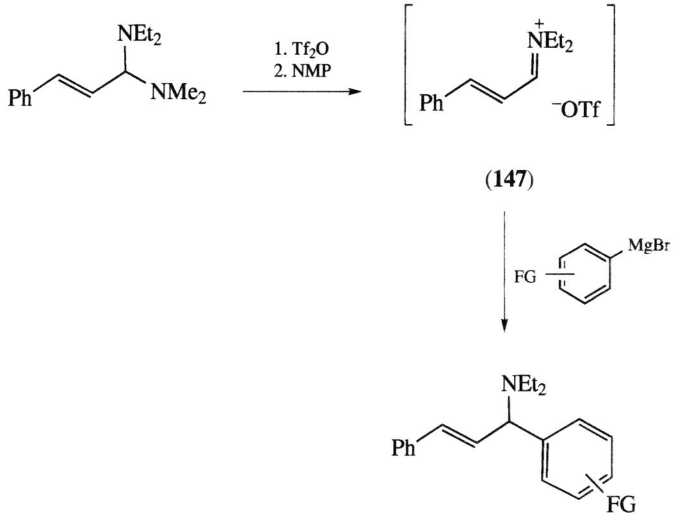

(**147**)

(77)

(**145**) 50–81%

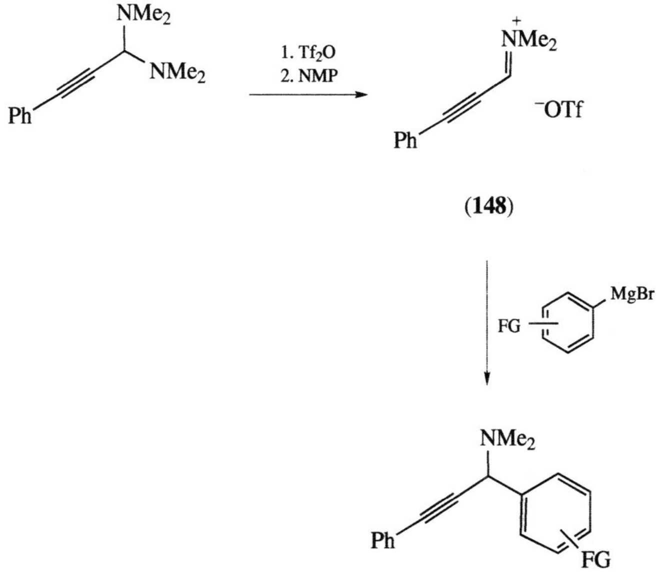

(**148**)

(78)

(**146**) 55–80%

ROH (**149**) $\xrightarrow[\text{2. MeI}]{\text{1. BuLi-Ph}_2\text{PCl}}$ $\underset{\text{Me}}{\text{Ph}_2\overset{+}{\text{P}}-\text{OR}}$ $\xrightarrow[\text{THF, 25 °C, 1 h}]{R^1\text{MgX}}$ R−R^1 (79)

R^1 = Ar, Bn, n-Alk (**150**) 76–99%

The transmetalation of alkylmagnesium reagents to Cu, Mn and Zn was performed by using an optically enriched Grignard reagent. Transmetalation to zinc proceeds with complete retention of configuration (concerted mechanism), while the transformations to copper and manganese organometallics[163] as well as trialkyltin halides[164] are rather complicated.

B. Substitutions at an sp^2-center

1. Transition-metal-catalyzed cross-coupling reactions

The reaction of aryl electrophiles with organomagnesium compounds is known as Kumada or Kumada–Tamao–Corriu reaction. The most common leaving groups in the electrophile are halogen atoms and, among them, chlorine is the most wanted due to the good availability and the low price of aryl or heteroaryl chlorides. Unfortunately, the oxidative addition of a metal center to an aryl chloride is a difficult reaction and many efforts have been made to overcome existing limitations.

More efficient ligands were recently developed. In the case of Pd and Ni, electron-rich ligands like bulky trialkylphosphines (equation 80)[165] or stable carbenes like N-heterocyclic carbenes (NHC, **151**, equation 81)[166] and PEPPSI (equation 82)[167] can be used with good success. The combination of Ni with a NHC ligand allows coupling at room temperature and permits the coupling with aryl fluorides[168]. These fluorides can be coupled as well using triarylphosphine ligand **152** bearing a hydroxyl group in close vicinity (equation 83)[169]. Using biaryl ligand **153**, the coupling reactions can be carried out at temperatures below 0 °C, thus allowing the use of functionalized Grignard reagents like **154** (equation 84)[170].

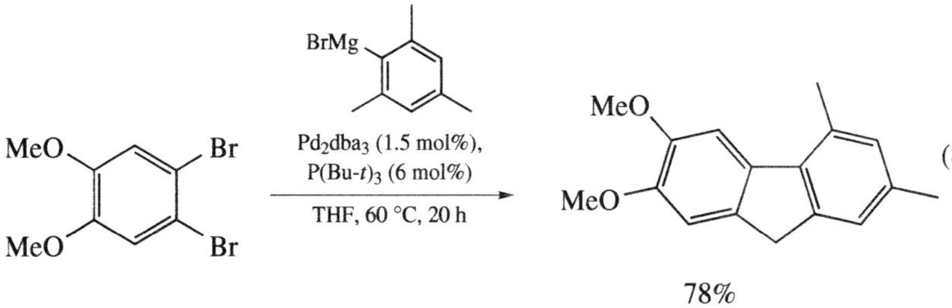

(80)

78%

Dialkylphosphine oxides are another class of highly efficient ligands for coupling reactions using Pd[171] or Ni[172] (equation 85). Electron-poor fluoroazines and -diazines react already in the presence of NiCl$_2$(dppp)[173], while hindered and electron-rich P(Bu-t)$_3$[126] and (t-Bu)$_2$P(S)H[174] can be used for the coupling of unactivated aryl chlorides in the presence of Ni.

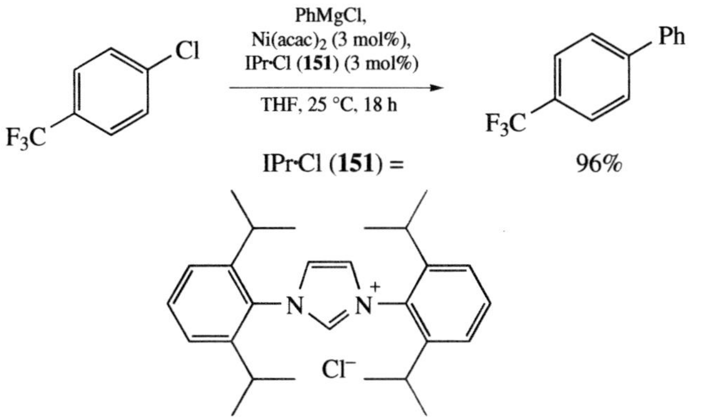

12. Functionalized organomagnesium compounds: Synthesis and reactivity 551

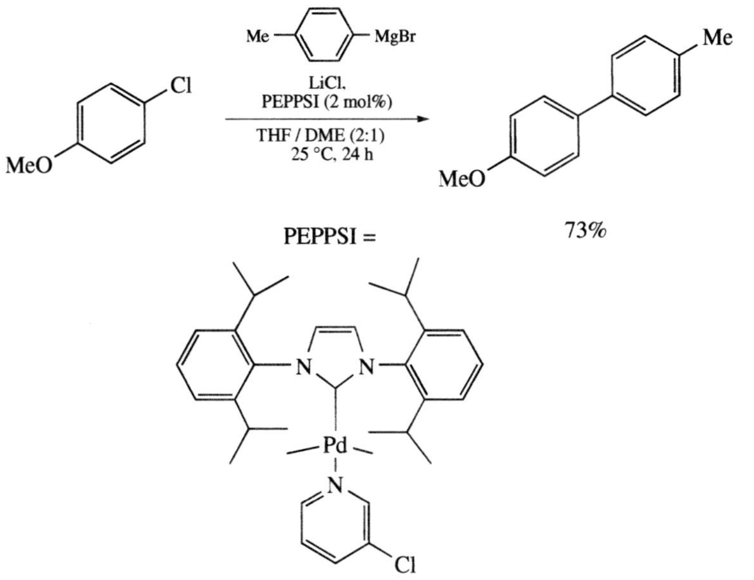

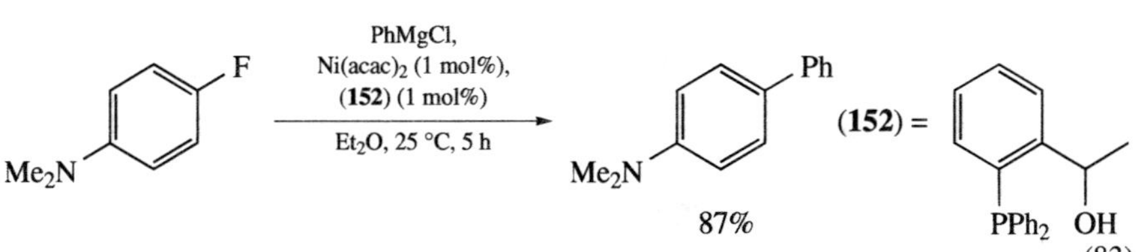

(81)

IPr·Cl (**151**) =

(82)

PEPPSI =

(83)

(**152**) =

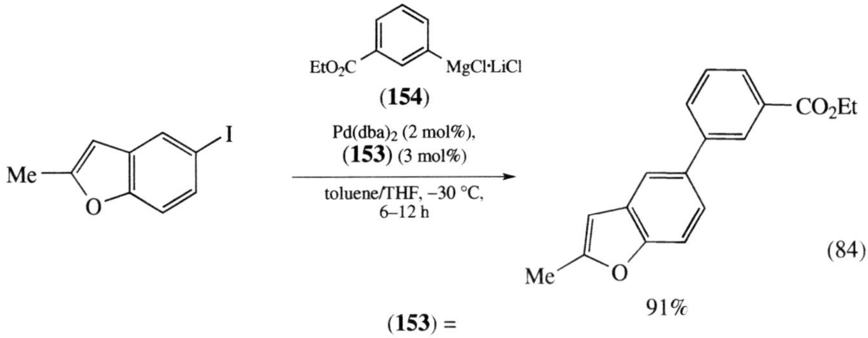

$$\text{(84)}$$

(153) =

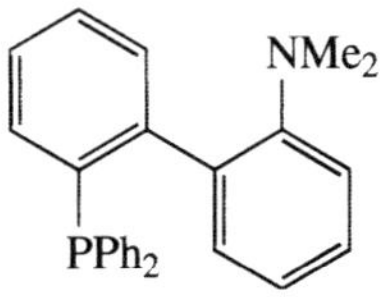

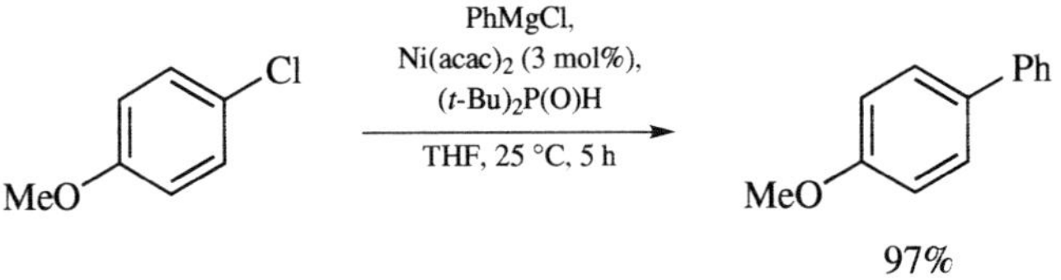

$$\text{(85)}$$

Transmetalation of organomagnesium compounds to zinc reagents opens new pathways for their coupling. It allows the use of conventional ligands like PPh_3 for nickel-catalyzed coupling reactions[175]. Manganese salts in NMP are effective catalysts for the coupling of aryl chlorides bearing an *ortho*-coordinating group with alkyl, alkenyl or aryl Grignard reagents (equation 86)[176]. Using this method, other halogens or even a methoxy group can serve as a leaving group (equation 87). Heteroaromatic chlorides can be also coupled with aryl- and alkylmagnesium halides using manganese chloride as catalyst[177].

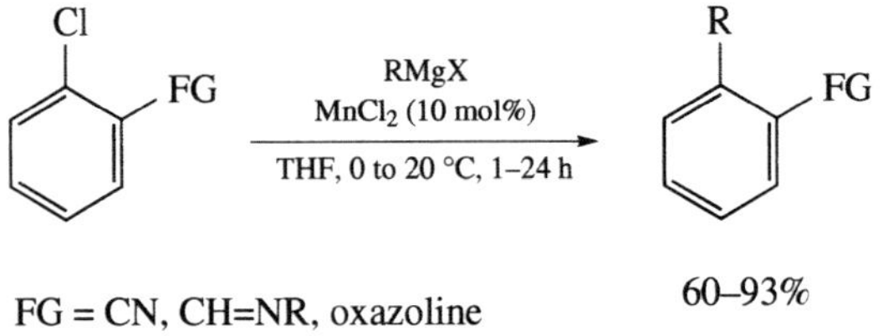

FG = CN, CH=NR, oxazoline

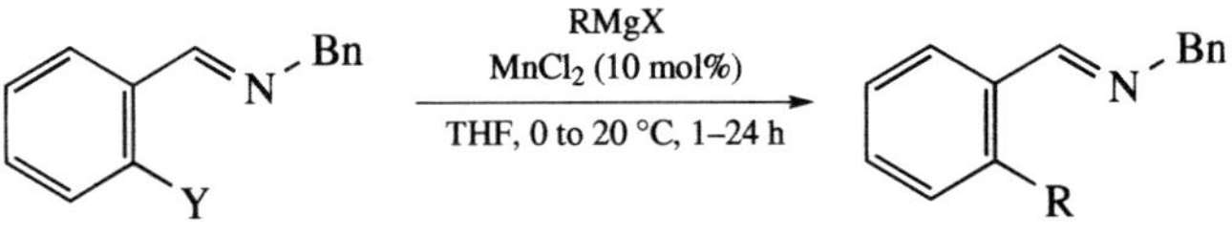

$$\text{(86)}$$

Y = Cl, F, OMe, Br

$$\text{(87)}$$

12. Functionalized organomagnesium compounds: Synthesis and reactivity 553

Besides developing new ligand systems, much effort has been made toward other leaving groups than halogens. Sulfonates[178], sulfones[179], tosylates[180], nitriles[181], alkyl ethers[182] or sulfonamides[183] can serve as leaving groups. The N,N-dialkyl sulfamate group can further function as *ortho*-directing group, thus allowing sequential functionalization of aromatic compounds like **155** (equation 88)[184]. Of special interest is the selectivity of coupling reactions of an electrophile bearing various leaving groups. Although iodine is almost always the most active leaving group, the choice between bromine or triflate can be made by selecting the appropriate catalyst (equations 89 and 90)[185].

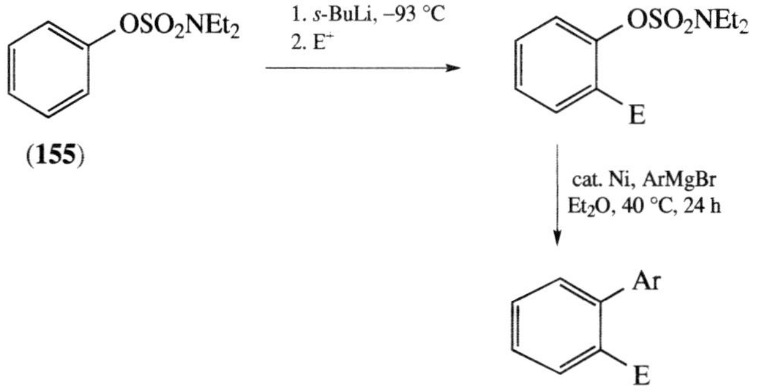

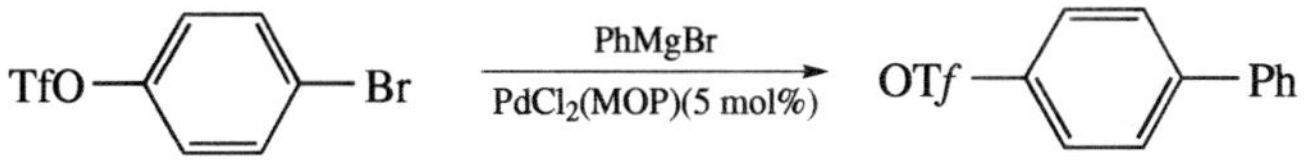

(88)

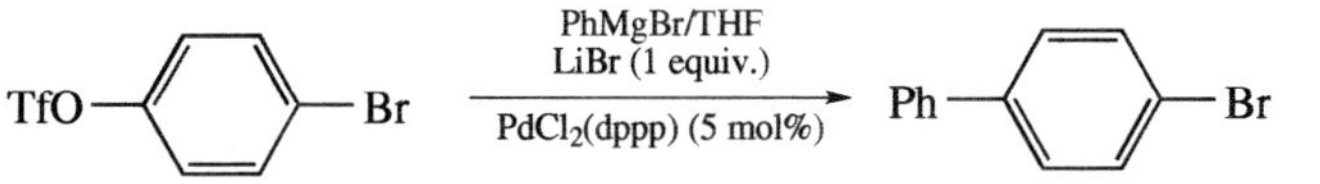

(89) 97%

(90) 68%

MOP= 2-diphenylphosphino-2'-methoxy-1,1'-binaphthyl

Substituted pyridines, quinolines and diazines react with polyfunctionalized arylmagnesium reagents under very mild conditions in the presence of PdCl$_2$(dppf) as a catalyst (equation 91)[186].

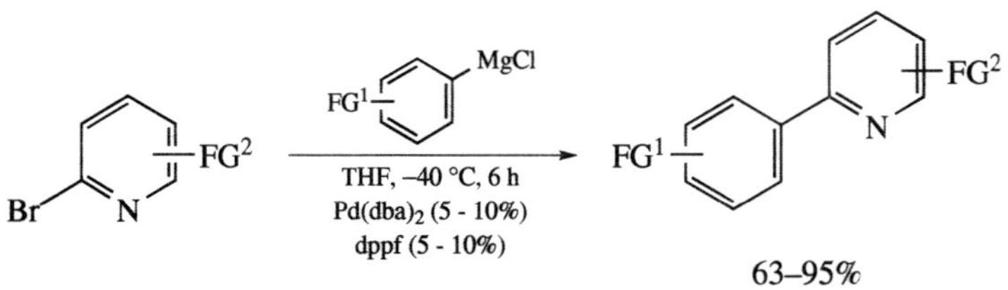

(91) 63–95%

Bromo- and iodoanilines, -phenols and -benzoic acids are first deprotonated *in situ* by an excess of the organomagnesium reagent and then coupled (PdCl$_2$(dppf), 1 mol%, THF, 25 °C, 3 h) with organomagnesium halides, thus avoiding tedious protection–deprotection steps[187]. Iron(III) salts can also serve as appropriate catalysts for various cross-coupling

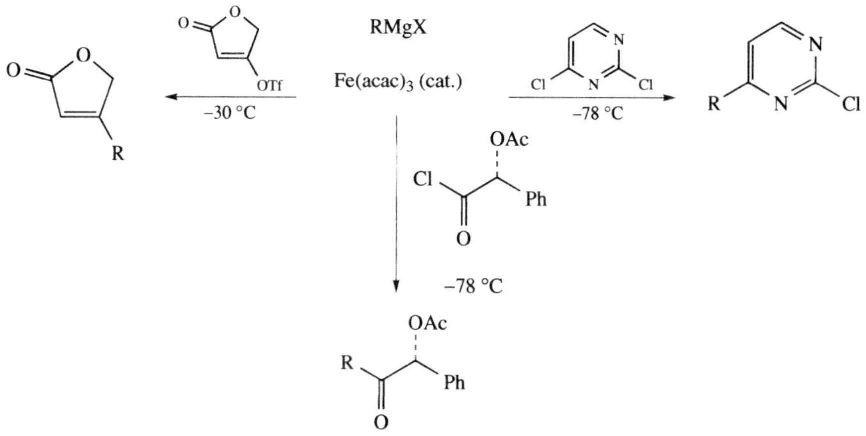

SCHEME 16

reactions[188]. They are cheap, non-toxic, environmentally friendly and effective for Csp^3–Csp^2 couplings, allowing the reactions of alkylmagnesium reagents with aryl chlorides, triflates and tosylates as well as heteroaryl chlorides (Scheme 16)[189].

However, arylmagnesium organometallics can only be used in a few cases[190] like an iron-catalyzed homocoupling (equation 92)[191]. Iron catalysis is further used for the dechlorination of electron-rich aryl chlorides by Grignard reagents[192].

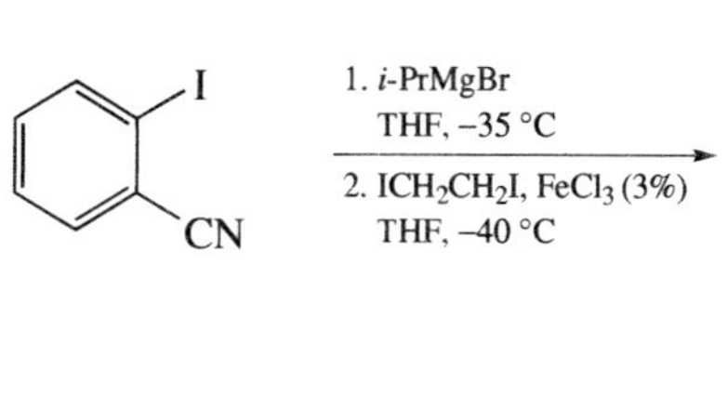

Another way toward more cost-efficient and environmentally friendly catalysts is their immobilization on a solid phase as could be realized with nickel on charcoal[193].

Substituted dibenzothiophenes **156** react with Grignard reagents to give products of the thiophene ring cleavage. With a chiral Ni catalyst, axially chiral biaryl compounds **157** can be obtained in high enantioselectivities (equation 93)[194].

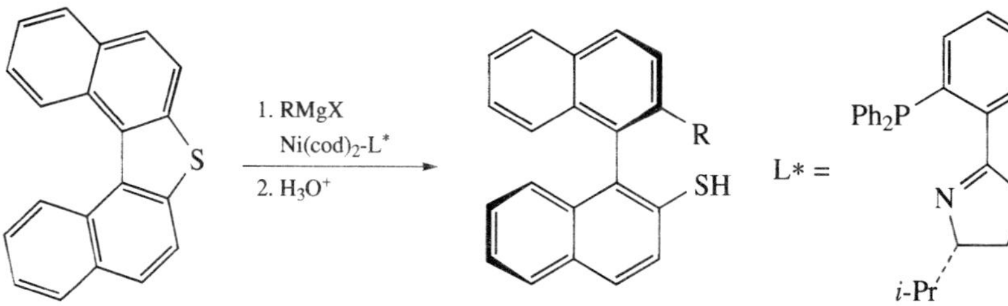

12. Functionalized organomagnesium compounds: Synthesis and reactivity 555

Substituted alkenes can serve as electrophiles in cross-coupling reactions. For example, chloroenynes **158** and chlorodienes react with Grignard reagents with retention of the stereochemistry using Pd[195], Mn[196] or Fe[197] (equation 94). Iron catalysis usually proceeds under mild reaction conditions[198] as shown in the coupling of alkenyl sulfides **159** (equation 95)[199]. Alkenyl chlorides like **160** undergo smooth cross-couplings with arylmagnesiums using nickel catalysis in the presence of hydroxyphosphine ligand **152** (equation 96)[200].

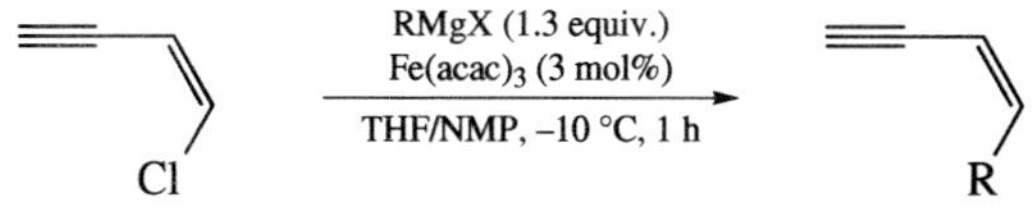

$$\text{(158)} \quad \xrightarrow[\text{THF/NMP, }-10\,°\text{C, 1 h}]{\text{RMgX (1.3 equiv.)} \atop \text{Fe(acac)}_3 \text{ (3 mol\%)}} \quad \text{65-85\%} \tag{94}$$

$$\text{(159) PhS} \xrightarrow[\text{THF, 25 °C, 22 h}]{\text{MeO-C}_6\text{H}_4\text{-MgBr} \atop \text{Fe(acac)}_3 \text{ (5\%)}} \quad 74\% \tag{95}$$

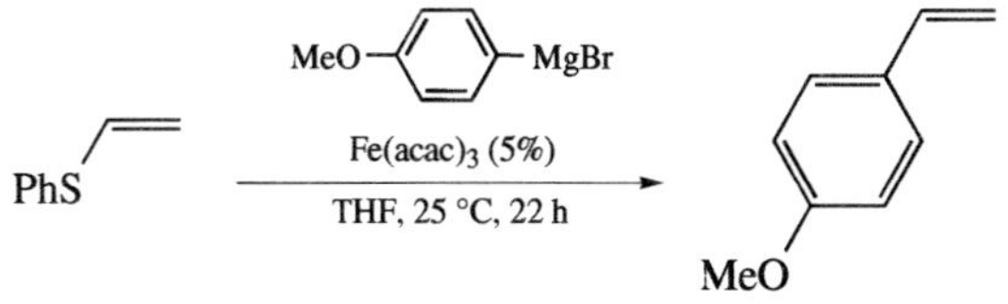

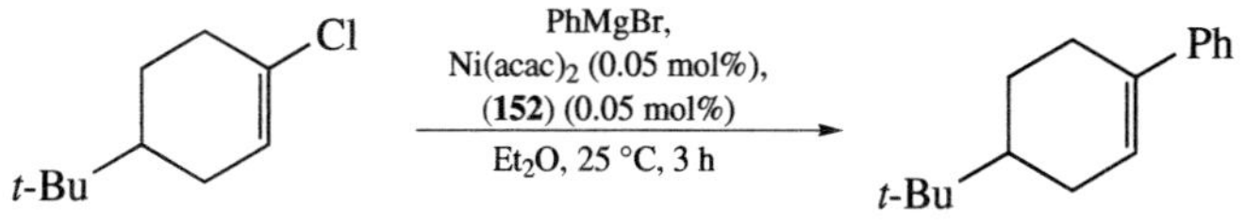

$$\text{(160)} \xrightarrow[\text{Et}_2\text{O, 25 °C, 3 h}]{\text{PhMgBr,} \atop \text{Ni(acac)}_2 \text{ (0.05 mol\%),} \atop \text{(152) (0.05 mol\%)}} \quad 95\% \tag{96}$$

(152) = 2-(1-hydroxyethyl)phenyl-PPh₂ (OH, PPh₂)

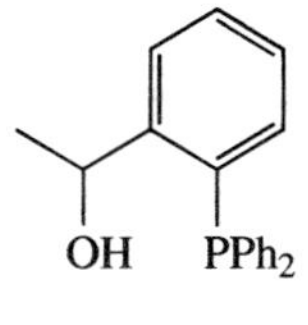

1,3-Dienyl triflates[201] and enol triflates derived from β-ketoesters[202] can be coupled with Grignard reagents in the presence of Cu^I species. Enol triflates **161** have been successfully coupled with NiCl$_2$(dppp) as a catalyst (equation 97)[203]. High yields in the coupling of dienyl phosphates can only be achieved in the presence of nickel salts[204], whereas enol phosphates, which can be derived *in situ* from ketones, can be coupled with an arylmagnesium species using a palladium catalyst[205].

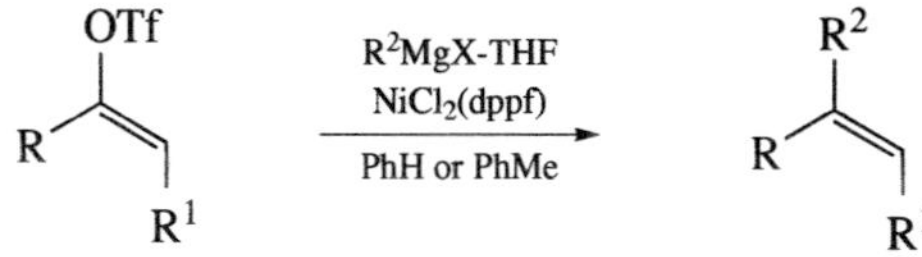

$$\text{(161)} \quad \underset{R^1}{\overset{\text{OTf}}{R}} \xrightarrow[\text{PhH or PhMe}]{R^2\text{MgX-THF} \atop \text{NiCl}_2\text{(dppf)}} \underset{R^1}{\overset{R^2}{R}} \quad R^2 = \text{Alk, Bn, Ph} \quad 75-95\% \tag{97}$$

2-Bromostyrenes **162** react with 1-ethylphenylmagnesium chloride under Ni catalysis providing 1,3-diaryl-1-butenes **163**. When chiral nickel complexes are used, the products

163 are obtained with up to 78% ee (equation 98)[206, 207].

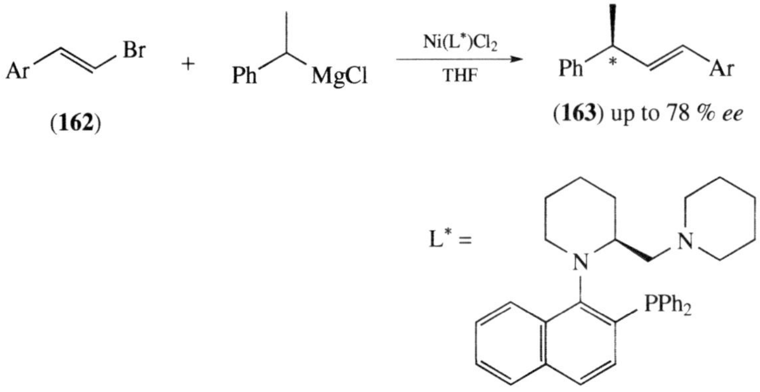

$$\text{Ar} \diagup\!\!\!\diagup \text{Br} \;+\; \text{Ph-CH(CH}_3\text{)-MgCl} \xrightarrow[\text{THF}]{\text{Ni(L*)Cl}_2} \text{Ph-*CH(CH}_3\text{)-CH=CH-Ar}$$

(**162**) (**163**) up to 78 % ee

L* = (chiral bis-piperidine naphthyl PPh$_2$ ligand)

(98)

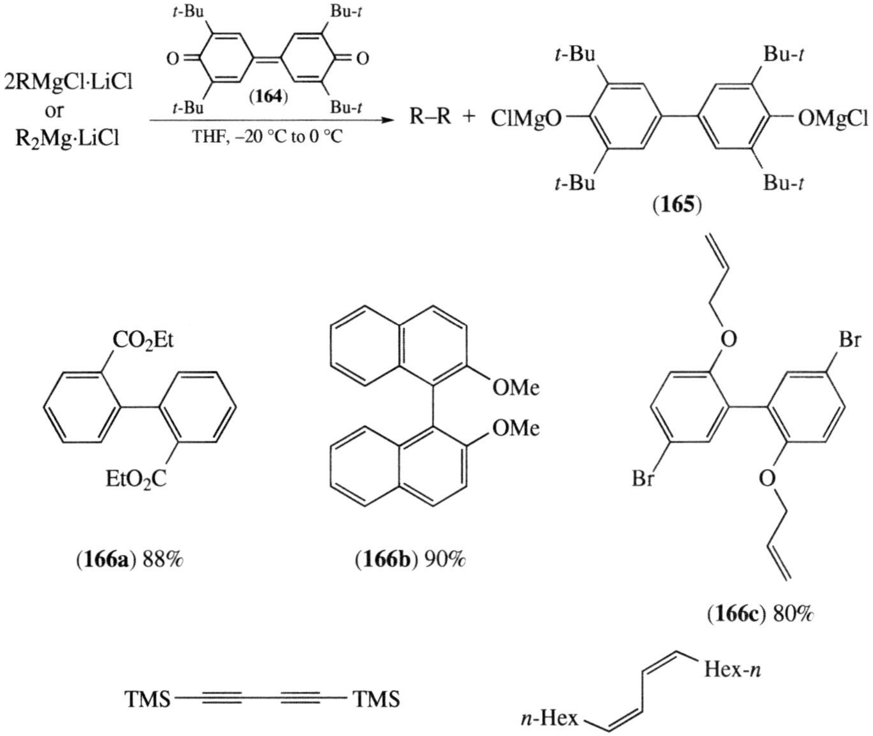

$$2\text{RMgCl·LiCl} \text{ or } \text{R}_2\text{Mg·LiCl} \; \xrightarrow[\text{THF, }-20\,°\text{C to }0\,°\text{C}]{(\mathbf{164})} \; \text{R–R} \;+\; \text{ClMgO–Ar–Ar–OMgCl}$$

(**164**) = 3,3′,5,5′-tetra-t-Bu-diphenoquinone
(**165**) = 3,3′,5,5′-tetra-t-Bu-biphenol bis(MgCl) salt

(**166a**) 88% — diethyl biphenyl-2,2′-dicarboxylate
(**166b**) 90% — 2,2′-dimethoxy-1,1′-binaphthyl
(**166c**) 80% — 2,2′-bis(allyloxy)-5,5′-dibromobiphenyl

TMS—≡—≡—TMS

(**167**) 90%

n-Hex—CH=CH—CH=CH—Hex-n

(**168**) 88%, Z,Z/E,Z > 99:1

(99)

2. Transition-metal-free cross-coupling reactions

Besides all achievements in the area of transition-metal catalysis, there is a need for finding new catalytic systems which meet the requirements of high efficiency, low

12. Functionalized organomagnesium compounds: Synthesis and reactivity 557

price and environmental friendliness. Such conditions are realized in transition-metal-free homocoupling of organomagnesium compounds, where mono- or diorganomagnesium compounds, that are complexed with lithium chloride, are oxidatively coupled using the readily available 3,3′,5,5′-tetra-*tert*-butyldiphenoquinone (**164**) (equation 99)[208]. The resulting biphenyldiolate **165** can be easily separated from the reaction mixture and reoxidized to **164** with air. The method allows the synthesis of various biaryls **166a–c**, diynes **167** and dienes **168** with retention of the double bond configuration.

3. Allylic substitution reactions

Reactions of organomagnesium compounds with allylic electrophiles usually require transition metal catalysis in order to achieve good regio- and stereoselectivities[209]. If Cu^I salts are used as catalysts, increased reaction temperatures and amounts of catalyst as well as slow addition of Grignard reagent favor the formation of the γ-adduct **169**[210]. The transmetalation of Grignard reagents to zinc organometallics prior to the addition of the Cu^I catalyst strongly favor S_N2'-substitutions. In contrast, Fe^{III} catalysis for the reaction of Grignard reagents with allyl diphenylphosphates leads almost exclusively to the α-substitution products **170** (Scheme 17)[211].

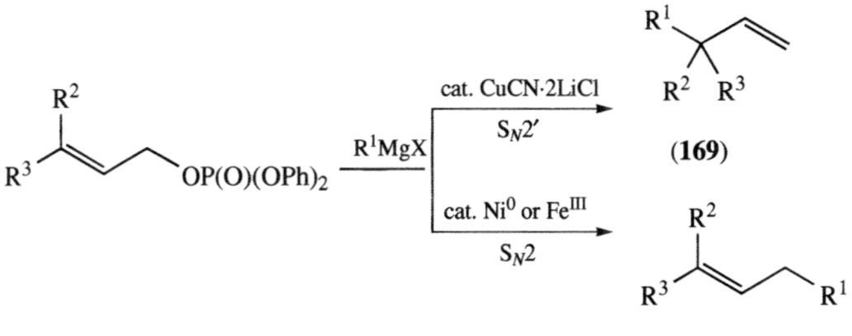

SCHEME 17

The presence of electron-donor sites in the substrates like a diphenylphosphinyl moiety[212] or *o*-diphenylphosphinobenzoate as leaving group[213] allows high levels of stereoselectivity (*syn* S_N2'). High regio- and stereoselectivities can also be obtained using allylic carbamates **171**[214] or allylic cyclic carbonates[215] (equation 100). Allylic ethers are coupled using cobalt or rhodium catalysis[216].

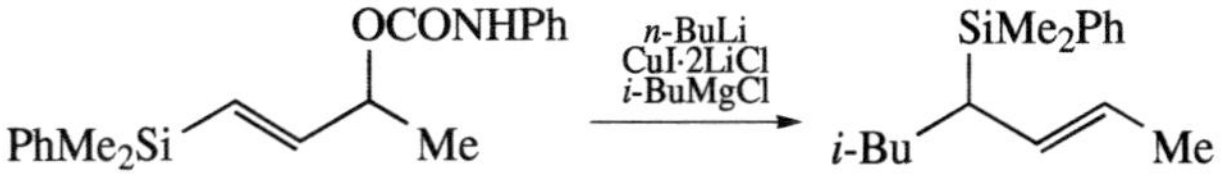

(100)

E/Z (reactant)	Yield	E/Z (product)
98:2	90	6:94
2:98	93	97:3

Allylic substitution reactions are valuable for the asymmetric synthesis of complex organic molecules, since they allow enantioselective C−C bond formations. One approach is to use readily available chiral allylic alcohols or derivatives. The resulting substitution products **172** and **173**, which contain tertiary or quaternary chiral centers, are obtained in high diastereomeric purities (equations 101 and 102)[217, 218].

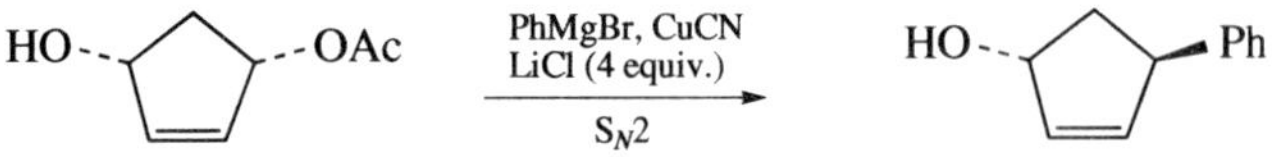

(**172**) 94%, d.r. = 97:3 (101)

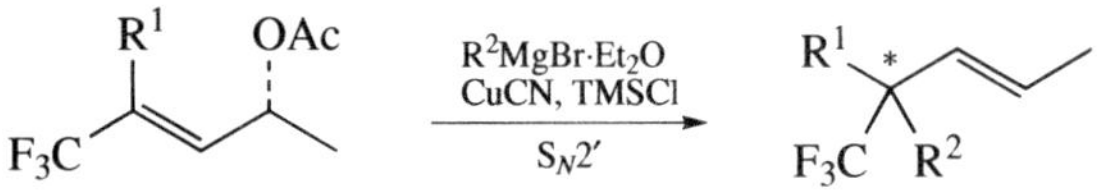

R^1, R^2 = Alk, Ar (**173**) (102)

The second approach starts from achiral allylic substrates and uses a chiral metal catalyst. Among all ligands which have been tested[218, 219], binaphtol-derived phosphoramidites like **174** have led to excellent regio- and enantioselectivities (equation 103)[220]. Highly stereoselective substitutions are also obtained using ligand **175** (*Taniaphos*, equation 104)[221].

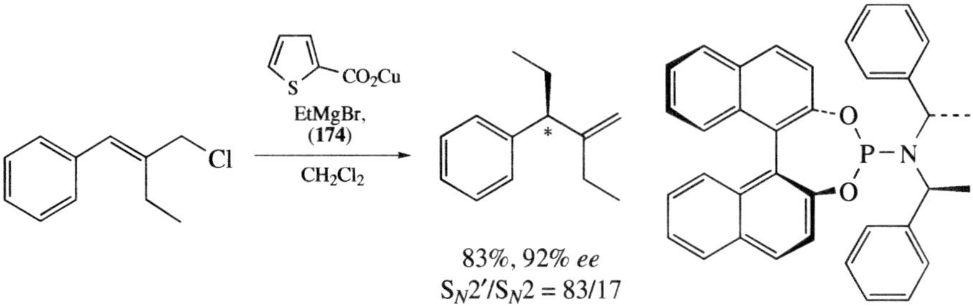

83%, 92% ee
$S_N 2'/S_N 2 = 83/17$

(**174**) (103)

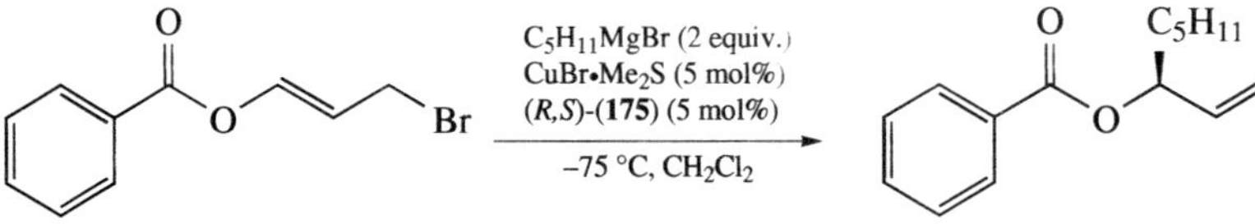

99%, 97% ee

(104)

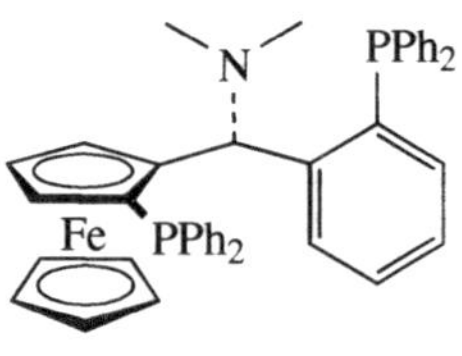

(R,S)-(**175**) (Taniaphos)

4. Synthesis of carbonyl compounds

The acylation of organometallic reagents with acyl chlorides has been reviewed[222]. The presence of catalytic amounts of Fe(acac)$_3$ allows these reactions to proceed at $-78\,°C$ and thus undesired side reactions, like a subsequent attack on the resulting ketone, are completely suppressed[223]. Even aroyl cyanides can be coupled using this method to form benzophenones. Besides N-*tert*-butoxy-N-methyl amides, Weinreb amides are efficient acylation reagents which suppress the formation of side products[224]. Alternatively, acyl halides first react with tri-*n*-butylphosphine to form acylphosphonium salts. These salts react smoothly with Grignard reagents giving ketones in good yields[225]. Various carboxylic acids are converted quantitatively with 2-chloro-4,6-dimethoxy-1,3,5-triazine to activated amides, which react *in situ* with Grignard reagents in the presence of copper iodide. Quantitative yields of ketones have been achieved[226]. The conversion of α-amino acids into α-aminoketones without epimerisation of the chiral center has been achieved[227,228]. Thus, protected pseudoephedrine glycinamide **176** is first diastereoselectively alkylated in the presence of a base. Treatment of the resulting alkylated amino acid derivative **177** with organomagnesium compounds gives the protected α-aminoketones **178** in good yields with complete retention of configuration at the α-carbon (equation 105)[229].

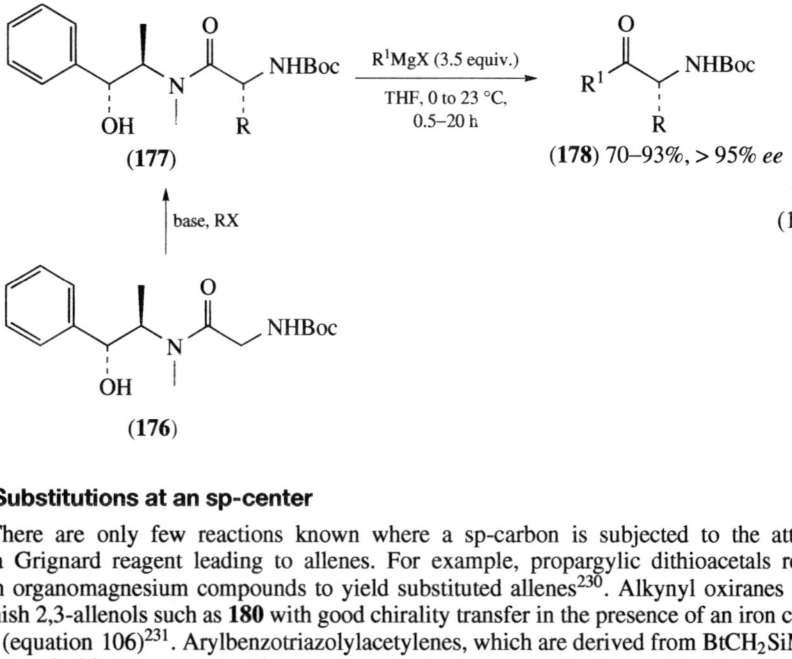

(105)

C. Substitutions at an sp-center

There are only few reactions known where a sp-carbon is subjected to the attack of a Grignard reagent leading to allenes. For example, propargylic dithioacetals react with organomagnesium compounds to yield substituted allenes[230]. Alkynyl oxiranes **179** furnish 2,3-allenols such as **180** with good chirality transfer in the presence of an iron catalyst (equation 106)[231]. Arylbenzotriazolylacetylenes, which are derived from BtCH$_2$SiMe$_3$ and aroyl chlorides, react with organomagnesium compounds to provide disubstituted acetylenes[232].

D. Addition of Organomagnesium Reagents to Multiple Bonds

1. Addition to carbon–carbon bonds

a. Catalyzed addition to non-activated C=C bonds. The uncatalyzed reaction of Grignard reagents with a non-activated double bond is generally difficult, with the exception of allylmagnesium and/or conjugated dienes as substrates. Another exception is the addition to cyclopropenes, occurring in a highly stereoselective fashion[233]. Recently, a number of synthetically useful addition methods involving transition metal catalysis were developed. Stereoselective addition of Grignard reagents to alkenes, mostly catalyzed by nickel and zirconium, has been reviewed[234]. The nickel complex $NiCl_2(dppf)$ efficiently promotes the three-component coupling of alkyl halides, butadienes and arylmagnesium halides (equation 107)[235].

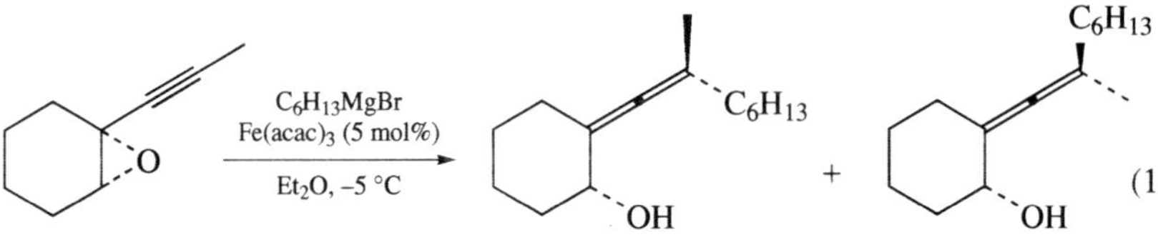

(179) 93% ee   syn-(180) 90% ee   anti-(180) 85% ee

syn:anti = 6:1

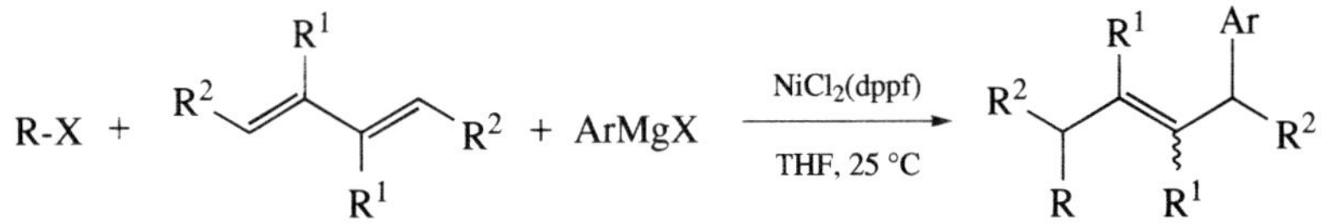

(106)

R-X + R²–(R¹)=–(R¹)–R² + ArMgX $\xrightarrow{NiCl_2(dppf), THF, 25\,°C}$ product 60–91%   (107)

A combination of $CoCl_2$ and 1,6-bis(diphenylphosphino)hexane catalyzes a similar reaction of an alkyl bromide with a 1,3-diene and trimethylsilylmethylmagnesium chloride, giving homoallylic silanes (equation 108)[236].

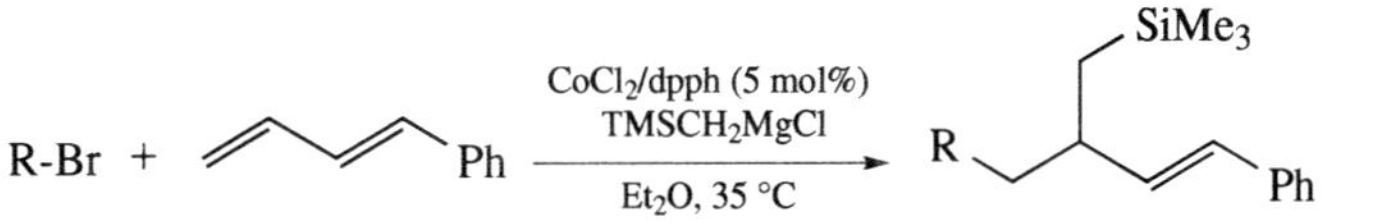

R-Br + diene-Ph $\xrightarrow{CoCl_2/dpph\ (5\ mol\%),\ TMSCH_2MgCl,\ Et_2O, 35\,°C}$ product   (108)

A similar Co-catalyzed reaction with arylmagnesium halides and alkenes **181**, bearing a halogen in a suitable position, occurs via a radical intermediate, leading to cyclic acetals **182** (equation 109)[237].

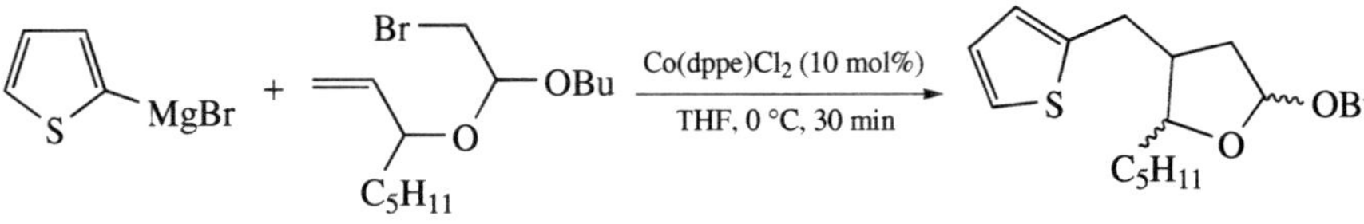

(181)   (182) 63%

(109)

6-Halo-1-hexene and heteroatom-substituted analogues **183** react with allyldimethylsilylmethylmagnesium chloride, giving 5-membered cyclic products **184** (equation 110)[238]. This method has been used for the synthesis of substituted pyrrolidines, otherwise difficult to prepare.

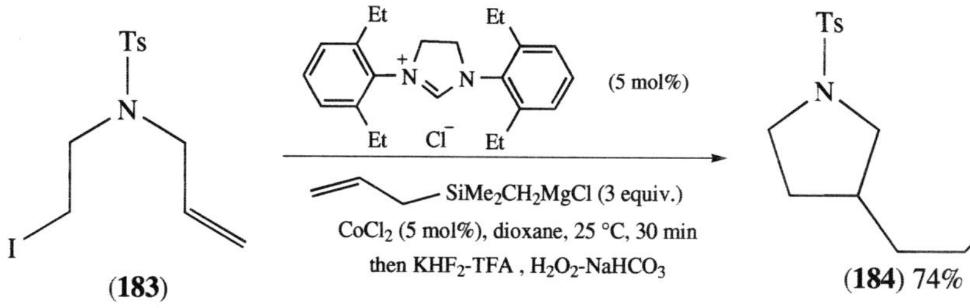

[Equation (110) scheme: compound (183) — N-Ts amine with iodoalkyl and allyl groups — treated with an imidazolinium chloride (5 mol%) bearing 2,6-diethylphenyl groups, SiMe$_2$CH$_2$MgCl (3 equiv.), CoCl$_2$ (5 mol%), dioxane, 25 °C, 30 min; then KHF$_2$-TFA, H$_2$O$_2$-NaHCO$_3$ → N-Ts pyrrolidine bearing a CH$_2$CH$_2$OH side chain (184) 74%]

Allenes **185** react with arylmagnesium chlorides in the presence of trialkylsilyl chlorides and a Pd^0 catalyst, furnishing substituted allylsilanes **186** with high (Z)-stereoselectivity (equation 111). Alkyl halides afford in this reaction mixtures of regioisomeric trisubstituted alkenes (equation 112)[239].

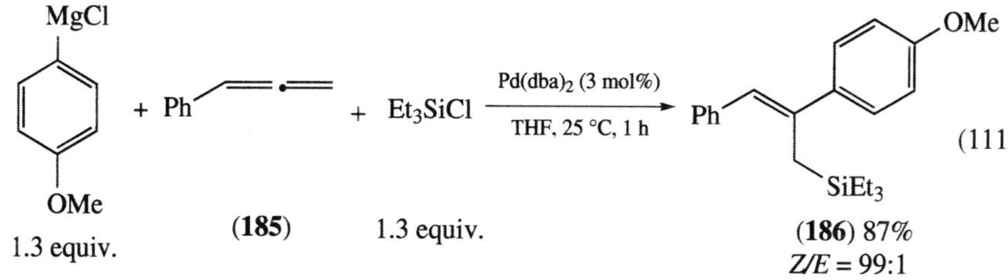

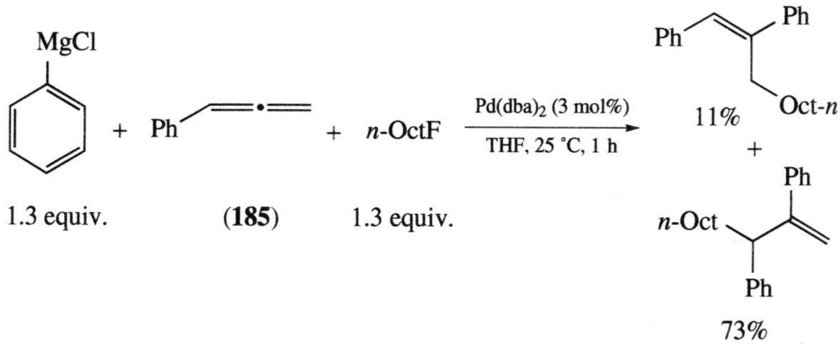

[Equation (111): 4-MeO-C$_6$H$_4$-MgCl (1.3 equiv.) + Ph-CH=C=CH$_2$ (**185**, 1.3 equiv.) + Et$_3$SiCl, Pd(dba)$_2$ (3 mol%), THF, 25 °C, 1 h → allylsilane (**186**) 87%, Z/E = 99:1]

[Equation (112): PhMgCl (1.3 equiv.) + **185** + n-OctF (1.3 equiv.), Pd(dba)$_2$ (3 mol%), THF, 25 °C, 1 h → two regioisomeric trisubstituted alkenes, 11% and 73%]

Titanocene and zirconocene dichlorides efficiently catalyze the addition of Grignard reagents to unactivated alkenes. In the presence of Cp$_2$ZrCl$_2$, alkylmagnesium bromides react with monosubstituted alkenes and alkyl tosylates leading potentially to three different types of products (equation 113). In many cases the reaction is highly selective, providing the formal carbomagnesiation product which reacts with oxygen or electrophiles like NBS yielding the corresponding alcohols or bromides (equation 114)[240].

Titanocene dichloride also catalyzes a regioselective carbomagnesiation of alkenes **187** (equation 115) and dienes **188** (equation 116). The reaction proceeds at 0 °C in THF in the presence of Cp$_2$TiCl$_2$, an organic halide and n-BuMgCl which leads to the catalytic species, affording benzyl, allyl or α-silyl alkylmagnesium halides, which are trapped with electrophiles (equation 117)[241].

A dimerization reaction of alkenylmagnesium reagents in the presence of chlorosilanes, catalyzed by Cp$_2$TiCl$_2$, furnishing 1,4-disilyl-2-butenes has been reported[242]. Transition-metal-catalyzed carbon–carbon bond formation, promoted by Mn, Cr, Fe and Co, has been reviewed[243].

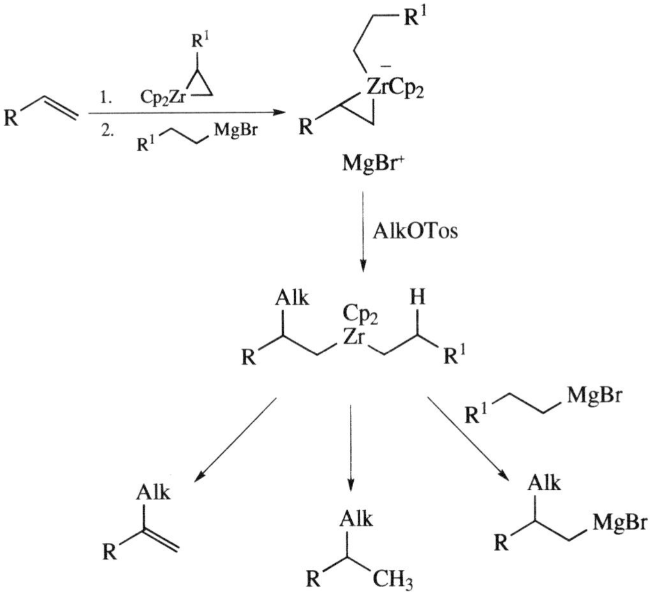

(113)

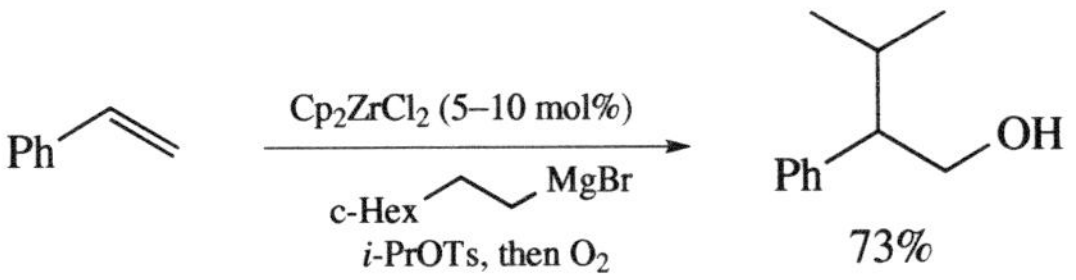

(114)

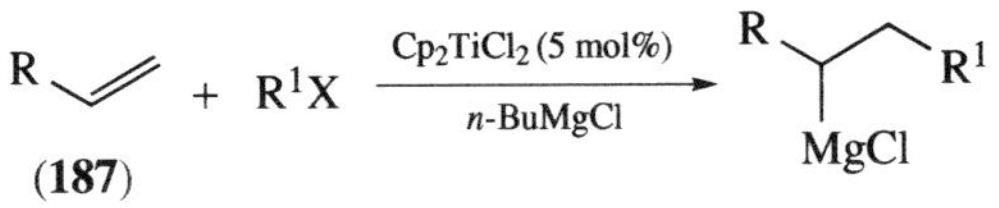

(115)

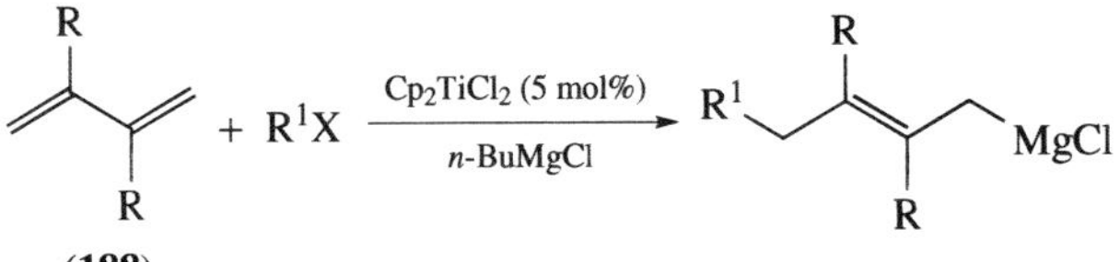

(116)

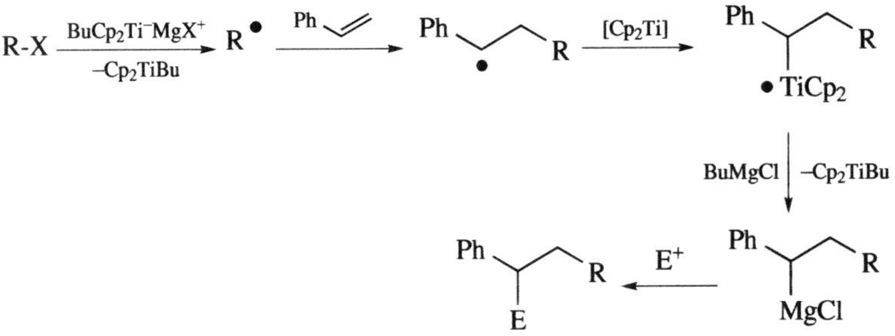

(117)

12. Functionalized organomagnesium compounds: Synthesis and reactivity   563

b. 1,4-Addition to Michael acceptors. The Michael addition of Grignard reagents to various unsaturated carbonyl compounds has been extensively studied. New developments in this field were recently reviewed[244]. Functionalized organomagnesium reagents, obtained by a low-temperature iodine–magnesium exchange reaction, add in 1,4-fashion to α,β-enones in the presence of CuCN·2LiCl (5–10 mol%)[245]. Treatment of aromatic, aliphatic or α-aminomethyl carboxylates **189** with an excess of an alkenylmagnesium bromide in the presence of a catalytic amount of copper salts provides homoallylic ketones **190** in 26–77% yield. The reaction proceeds as a sequence of a Grignard acylation followed by a 1,4-addition (equation 118)[246]. 3-Substituted glutarate diesters are easily obtained in good yields by the reaction of various Grignard reagents with dimethyl 1,3-propenedicarboxylate[247].

$$\underset{\textbf{(189)}}{\overset{\text{NHBoc}}{\underset{\text{CO}_2\text{Me}}{\bigwedge}}} \quad \xrightarrow[\substack{\text{CuCN-LiCl (50 mol\%)} \\ \text{THF, }-45\,°\text{C}}]{\text{MgCl (4 equiv)}} \quad \underset{\textbf{(190)} \quad 77\%}{\overset{\text{NHBoc}}{\underset{\text{O}}{\bigwedge\!\!\!\bigvee}}} \quad (118)$$

Substituted acrylic acids **191** and amides **192** usually do not react with organomagnesium reagents, but MeLi allows a smooth addition even at −15 °C, if an excess of Grignard reagent is used (equations 119 and 120)[248].

$$\underset{\textbf{(191)}}{\text{Ph}\!\!\overset{\text{O}}{\underset{}{\diagup\!\!\!\diagdown}}\!\!\text{OH}} \quad \xrightarrow[\substack{\text{MeLi (12 equiv.)} \\ \text{THF/Et}_2\text{O, }-15\,°\text{C, 1.5 h}}]{\textit{s}\text{-BuMgCl (4 equiv.),}} \quad \underset{60\%}{\text{Ph}\!\!\overset{\textit{s}\text{-Bu}\;\;\;\text{O}}{\underset{}{\bigwedge\!\!\!\bigvee}}\!\!\text{OH}} \quad (119)$$

$$\underset{\textbf{(192)}}{\overset{\text{O}}{\underset{\text{H}}{\diagup\!\!\!\diagdown\!\!\!\underset{}{\text{N}}\!\!\text{-Ph}}}} \quad \xrightarrow[\substack{\text{MeLi (12 equiv.)} \\ \text{THF/Et}_2\text{O, }-10\,°\text{C, 1 h}}]{\textit{t}\text{-BuMgCl (4 equiv.),}} \quad \underset{66\%}{\overset{\textit{t}\text{-Bu}\;\;\;\;\text{O}}{\underset{\text{H}}{\bigwedge\!\!\!\bigvee\!\!\!\underset{}{\text{N}}\!\!\text{-Ph}}}} \quad (120)$$

Conjugated allenyl ketones **193** react smoothly with Grignard reagents in ether at −78 °C without a catalyst, yielding α,β-enones in excellent yields and with complete (*E*)-selectivity (equation 121)[249].

$$\underset{\textbf{(193)}}{\overset{\text{O}}{\underset{}{=\!\!=\!\!\bigwedge\!\!\text{Bu}}}} \quad \xrightarrow[\text{Et}_2\text{O, }-78\,°\text{C}]{\textit{t}\text{-BuMgCl}} \quad \underset{95\%}{\textit{t}\text{-Bu}\!\!\overset{\text{O}}{\underset{}{\diagdown\!\!\!=\!\!\!\diagup}}\!\!\text{Bu}} \quad (121)$$

Cyclic γ-oxonitriles **194** react sequentially with two different Grignard reagents, affording enamides **195** with high diastereoselectivity (equation 122)[250]. A number of chiral auxiliaries have been developed for performing enantioselective 1,4-addition to α,β-unsaturated carbonyl compounds. Chiral oxazolidinones are often highly effective, affording the Michael adducts with up to 99% ee. Optically active amidoacrylates, prepared from acryloyl oxazolidinone in four steps, react with Grignard reagents in the presence of a catalytic amount of Cu$^\text{I}$. This reaction was applied for the synthesis of α-amino acids with up to 97% ee[251]. Asymmetric conjugate addition of an arylmagnesium reagent to cinnamoyloxazolidinone **196** is the key step in the synthesis of (+)-tolterodine[252], a pharmacologically important muscarinic receptor agonist (equation 123).

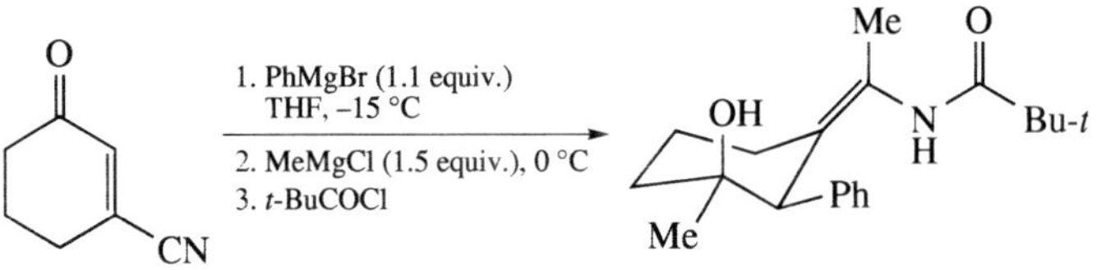

(**194**) (**195**) 73% (122)

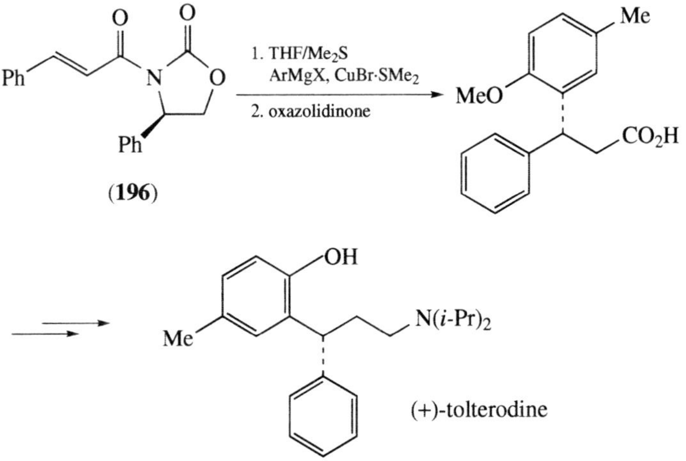

(**196**) (+)-tolterodine (123)

Similar acryloyl imidazolidinones have been used for the asymmetric 1,4-addition of organomagnesium compounds in the presence of a Lewis acid. The diastereoselectivity is variable and was found to be highly depending on the nature of all substrates[253].

The proline-derived carbamoylphosphine **197** catalyzes the asymmetric 1,4-addition of Grignard reagents to 2-cyclohexenone although a high catalyst loading is required in this case (equation 124). Changing the solvent to diethyl ether and lowering the catalyst loading to 3 mol% leads to the 1,4-adduct with 67% *ee*[254].

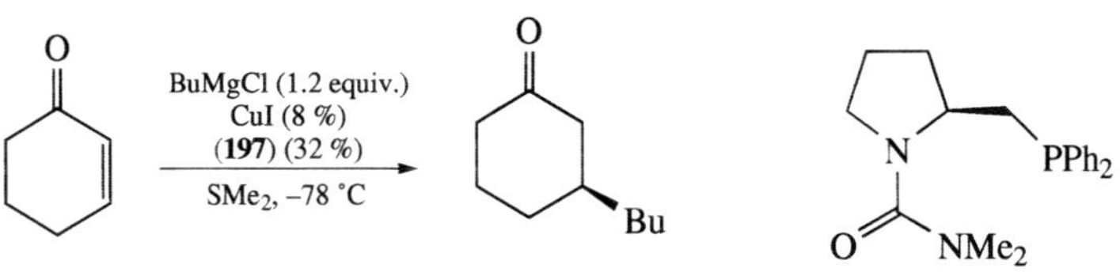

82% yield, 77% *ee* (**197**) (124)

Ferrocene-derived ligand (*R*,*S*)-Josiphos, which is widely used for catalytic asymmetric hydrogenation reactions, is also a good catalyst for the asymmetric copper-catalyzed 1,4-addition. Reaction in *t*-BuOMe in the presence of 6 mol% of this ligand gives products with up to 98% *ee*[255].

Unsaturated esters like **198** react with Grignard reagents under similar conditions, giving excellent yields and enantioselectivities. The catalyst is a Cu^I–Tol-BINAP complex

12. Functionalized organomagnesium compounds: Synthesis and reactivity 565

(equation 125)[256].

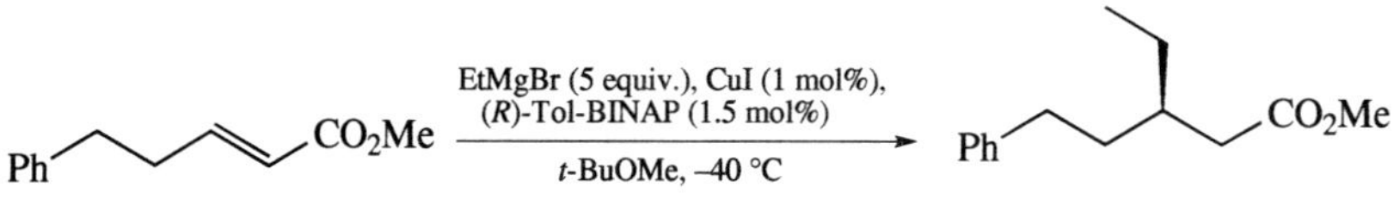

$$\text{(198)} \xrightarrow[\text{\textit{t}-BuOMe, }-40\,°\text{C}]{\text{EtMgBr (5 equiv.), CuI (1 mol\%)},\ (R)\text{-Tol-BINAP (1.5 mol\%)}} \text{Ph-CH}_2\text{CH}_2\text{CH(Et)CH}_2\text{CO}_2\text{Me}\quad 88\%,\ 93\%\ ee \tag{125}$$

The creation of all-carbon quaternary chiral centers by asymmetric conjugate addition is a challenging task. A chiral heterocyclic carbene **199** has been used as a ligand for this reaction. Chiral 3,3-disubstituted cyclohexanones **200** were obtained by this method with up to 85% ee (equation 126)[257].

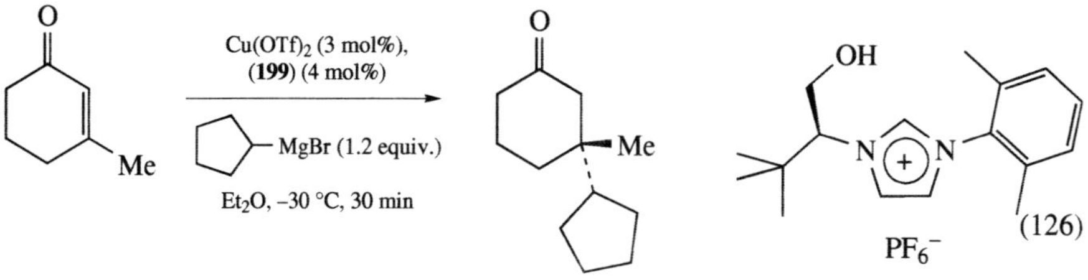

$$\tag{126}$$

(**200**) 100%, 85% ee   (**199**)

c. Addition to other activated alkenes. Addition to α,β-unsaturated nitriles has been reviewed[258]. Addition to cyclic oxo-nitriles **201** can be directed either by using steric effects or by chelation (equation 127). γ-Hydroxy- α,β-unsaturated nitriles like **202** are subjected to a one-pot addition–alkylation sequence, leading to polysubstituted nitriles. In the case of open-chain systems, diastereoselectivities up to 6.6:1 are achieved, while cyclic systems may lead to highly diastereoselective reactions (equation 128)[259]. ω-Chloroalkyl-alkenylmagnesium halides **203** give bicyclic hydroxylnitriles **204** (equation 129)[260, 261].

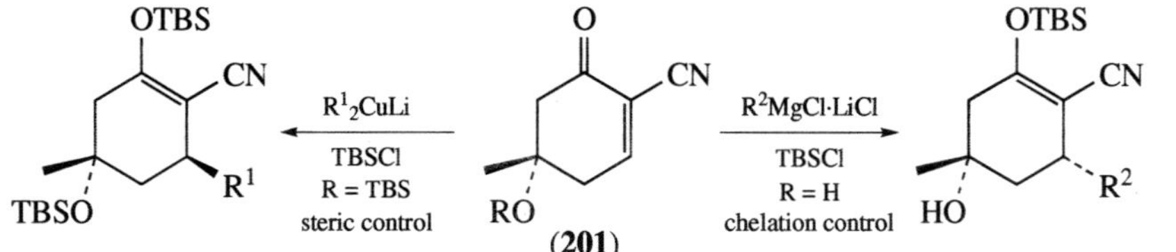

$$\tag{127}$$

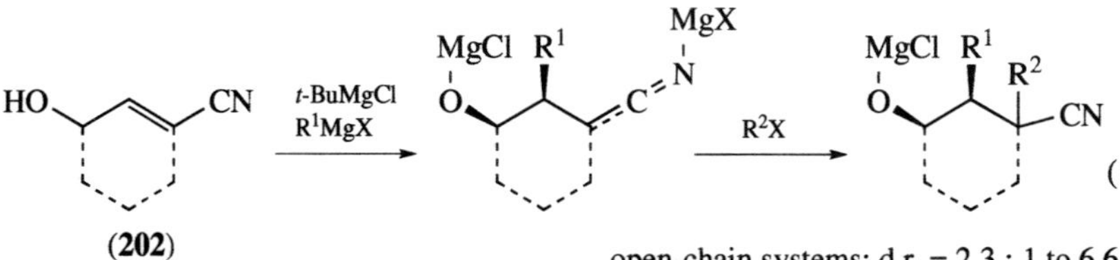

$$\tag{128}$$

(**202**)
R^1 = Me, Ar
R^2X = PhCHO, BnBr

open-chain systems: d.r. = 2.3 : 1 to 6.6 : 1
cyclic systems: completely stereoselective

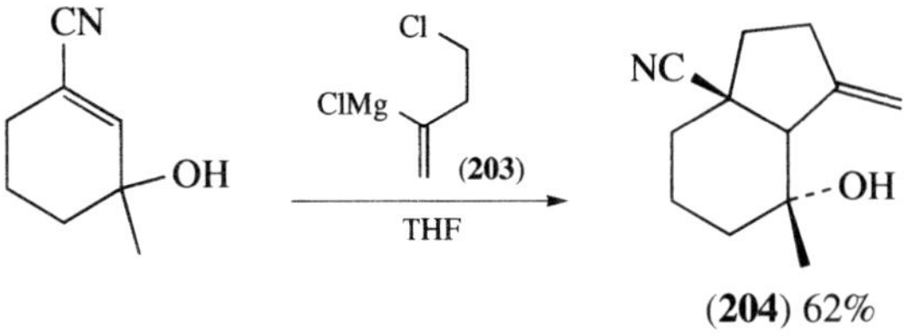

(129)

The addition of Grignard reagents to nitroalkenes like **205** gives *aci*-salts **206**, which can be further transformed into nitroalkanes, hydroxymoyl halides or carboxylic acids (equation 130)[262]. Reaction of RMgX with nitroalkenes in the presence of $CeCl_3$, followed by treatment with 100% acetic acid, was developed as efficient synthesis of complex nitroalkanes[263].

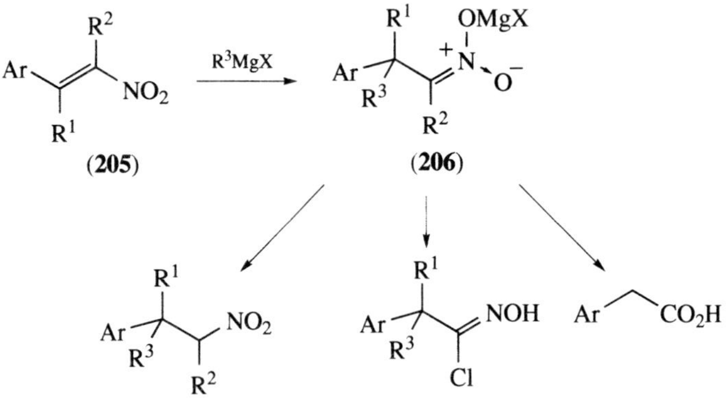

(130)

Organomagnesium compounds react with imines, prepared from 3-methoxy-2-naphthaldehydes by a 1,4-addition mechanism. This reaction can be performed with high diastereoselectivity. The method was applied for the synthesis of optically pure β-tetralones[264]. Vinylmagnesium bromide reacts as an acceptor with a ketone dimethyl hydrazone zincate **207**, yielding a 1,1-bimetallic species, which can be reacted sequentially with two different electrophiles (equations 131 and 132)[265]. The reaction proceeds via a metalla-aza-Claisen rearrangement, where the dimethylhydrazone anion behaves as an 'aza-allylic' system[266].

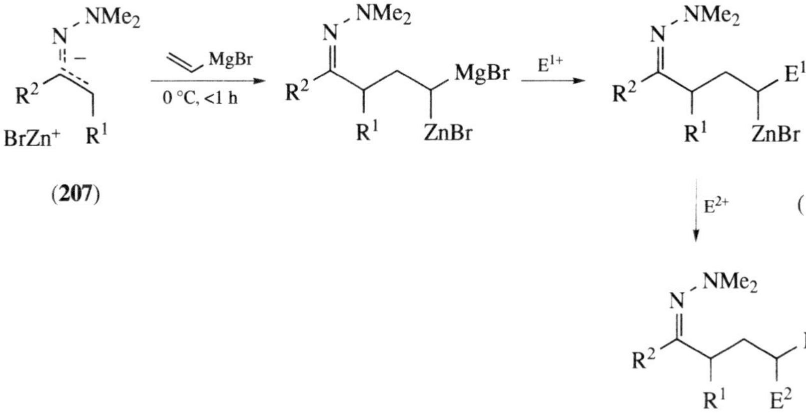

(131)

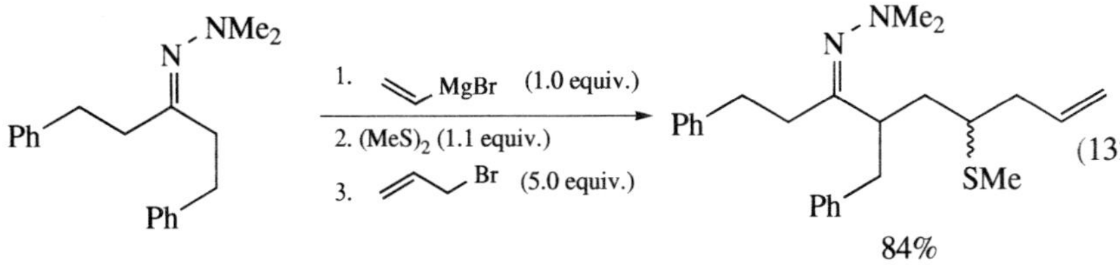

(132) reaction scheme shown, 84%

Vinylphosphonium bromide **208** reacts with Grignard reagents, forming alkylphosphonium ylides. These ylides react with aldehydes, giving alkenes in a one-pot sequence. In this reaction, catalytic amounts of both copper and silver salts are necessary (equation 133)[267].

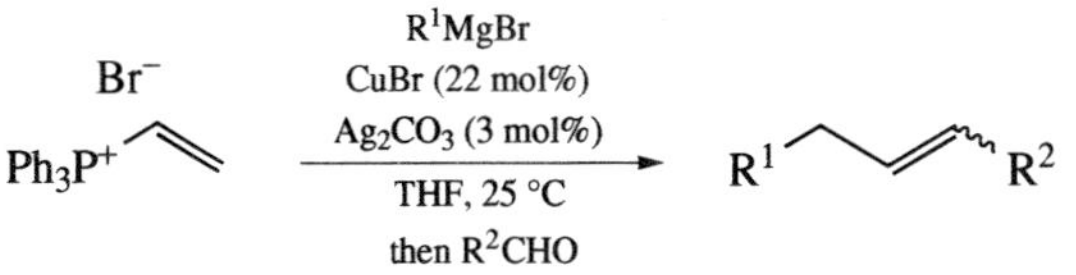

(**208**) → 68–94% (133)

d. Addition to carbon–carbon triple bonds. Addition of Grignard reagents to an unactivated alkyne usually requires a coordinating group and/or a transition metal catalyst. Propargyl alcohols react with RMgX regioselectively without catalysis. This reaction allows a number of synthetically useful transformations (Scheme 18)[268].

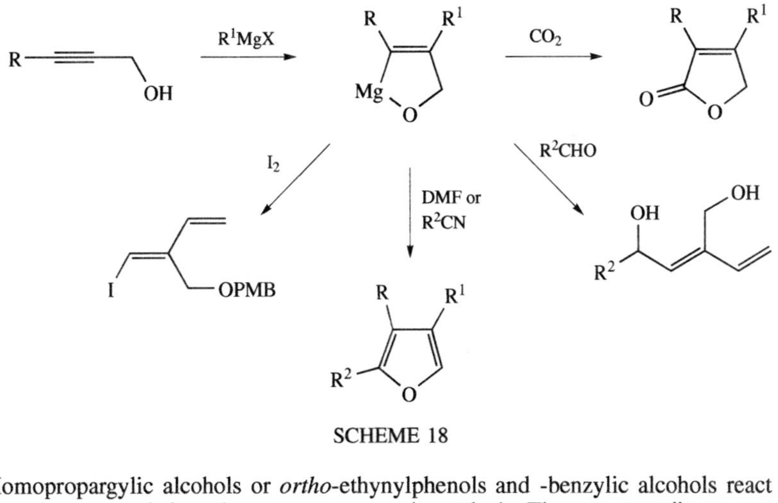

SCHEME 18

Homopropargylic alcohols or *ortho*-ethynylphenols and -benzylic alcohols react with Grignard compounds by using a manganese salt catalysis. The corresponding propargylic alcohols give under these conditions substituted allenes (Scheme 19)[269, 270]. Secondary and tertiary propargylic alcohols react with primary alkylmagnesium reagents with a high selectivity under Cu^I catalysis[271].

Unactivated alkynes **209** undergo the addition of arylmagnesium reagents under cooperative iron and copper catalysis, yielding trisubstituted alkenylmagnesium species. They can be trapped with electrophiles, giving tetrasubstituted alkenes such as **210**. In some

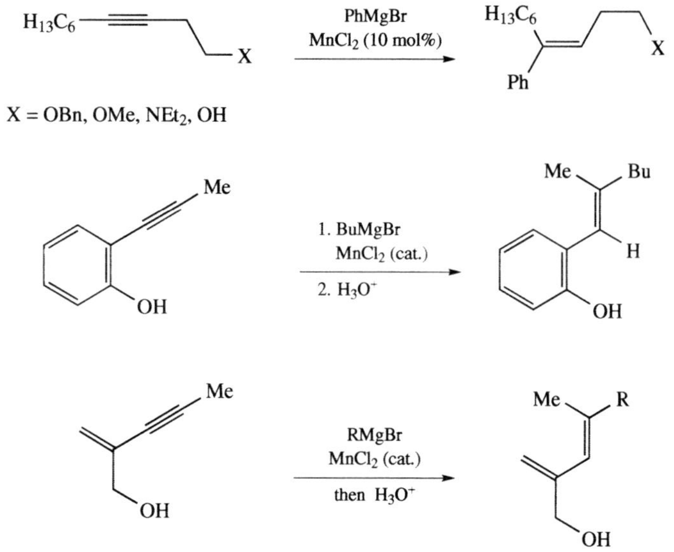

SCHEME 19

cases, unsymmetrical alkynes react completely regioselectively (equation 134)[272]. A similar process has been developed using chromium catalysis and toluene as the solvent[273]. In this case, the reactions with aldehydes provide highly substituted allylic alcohols (equation 135).

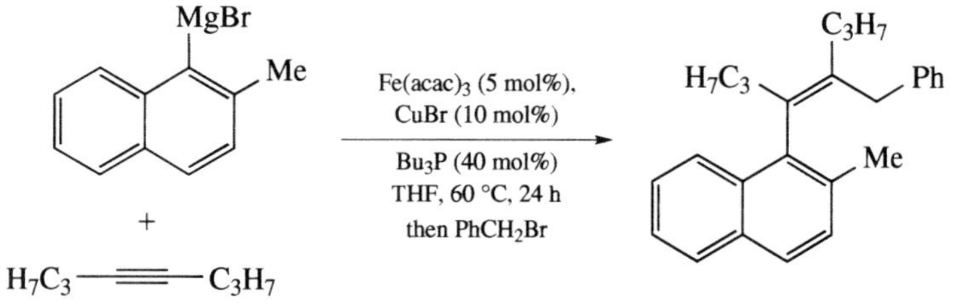

(**209**) (**210**) 49%

(134)

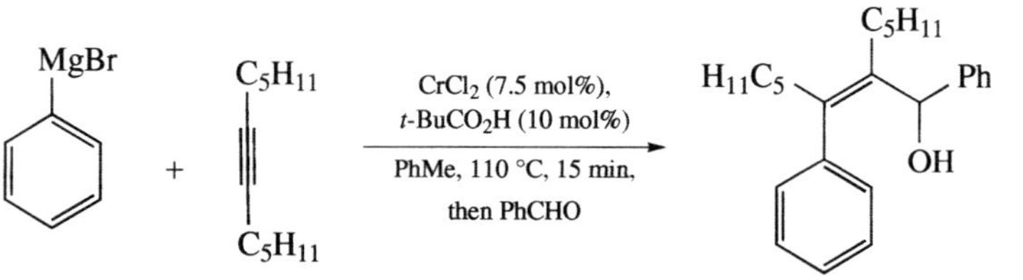

74%

E/Z 7:93

(135)

2. Addition to carbon–oxygen bonds

Recent advances in the selective addition of organometallic reagents to carbonyl compounds have been reviewed[274]. A computational study of the mechanism of 1,2-addition reactions of Grignard compounds to carbonyl compounds was also reported[275]. The addition of organomagnesium species to carbonyl compounds is usually complicated by side processes like reduction or enolization. Often, these processes are even dominating. Many efforts were dedicated to the development of more selective and practical processes. The use of rare-earth metal salts like cerium(III) chloride, giving after transmetalation reaction highly nucleophilic, but less basic species, is one of the most widespread methods[276]. However, the heterogenicity of the reaction mixture due to the low solubility of $CeCl_3$ in THF causes significant problems; besides, this method is usually not suitable for reactions with highly functionalized Grignard reagents. The addition of LiCl increases the solubility of metal salts. This allowed the development of an excellent and very general method for the addition of various Grignard reagents to carbonyl compounds, making use of soluble complexes of lanthanide salts with LiCl in THF (equation 136)[277]. This protocol is particularly useful in the case of polyfunctionalized arylmagnesium reagents.

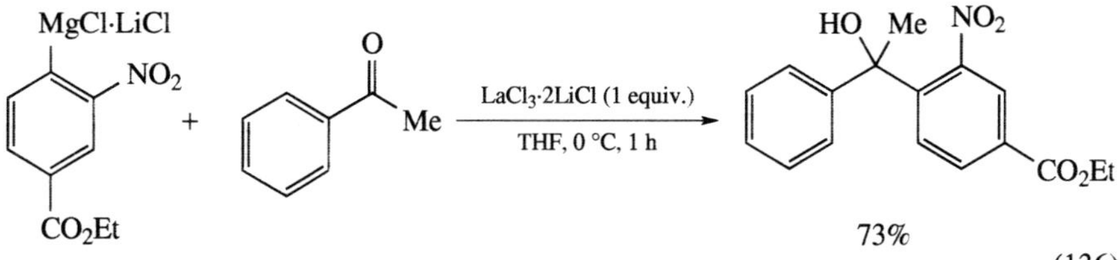

No product was formed in this reaction in the absence of the soluble lanthanide salt or even in the presence of $CeCl_3$. Heteroaryl Grignard reagents react smoothly in the presence of $LaCl_3 \cdot 2LiCl$ even with highly sterically hindered ketones like **211** (equation 137).

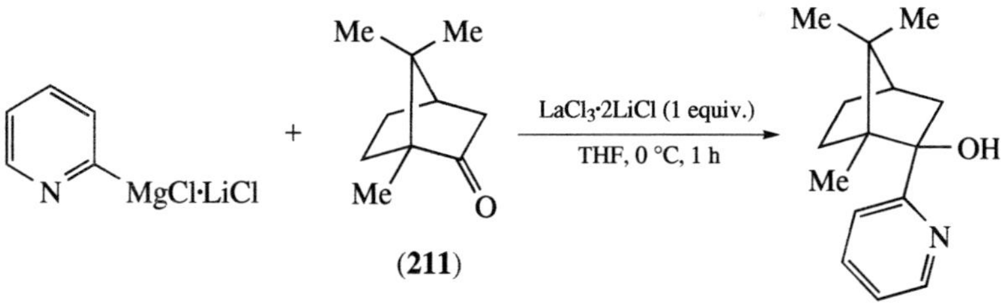

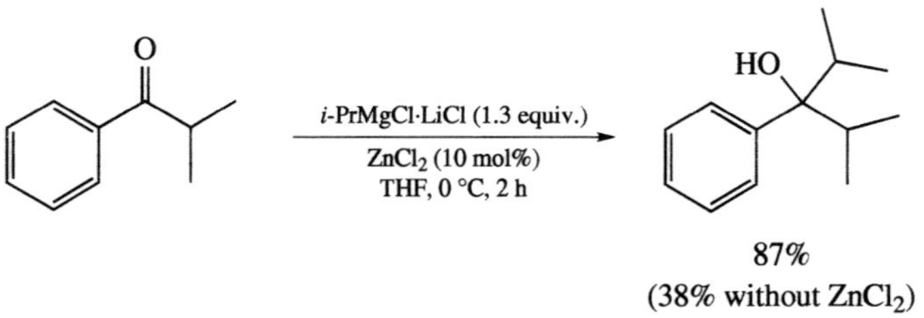

In the absence of additives the yield of the alcohol **212** is only 17%, and 53% in the presence of CeCl$_3$. Catalytic amounts of zinc chloride in the presence of LiCl were found to have a similar positive effect on the outcome of the 1,2-addition reaction of alkylmagnesium halides to enolizable ketones (equations 138 and 139)[278]. Due to the low price of ZnCl$_2$, this method seems promising for large-scale applications.

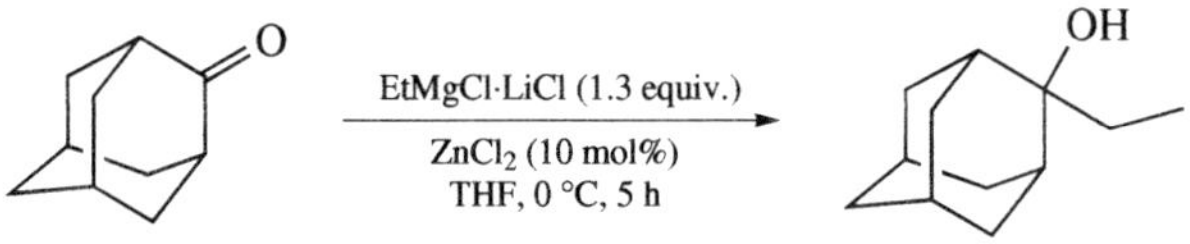

$$\text{adamantanone} \xrightarrow[\substack{\text{ZnCl}_2 \text{ (10 mol\%)} \\ \text{THF, 0 °C, 5 h}}]{\text{EtMgCl·LiCl (1.3 equiv.)}} \text{2-ethyl-2-adamantanol}$$

81%
(29% without ZnCl$_2$)

(139)

Addition of a primary alkyl group to enolizable ketones can be performed using magnesium–ate complexes[279]. The additional presence of 2,2′-bipyridyl (1 equiv.) in the reaction mixture improves the yields. The ate complexes are prepared *in situ* from the corresponding Grignard reagents and alkyllithium compounds (equations 140 and 141).

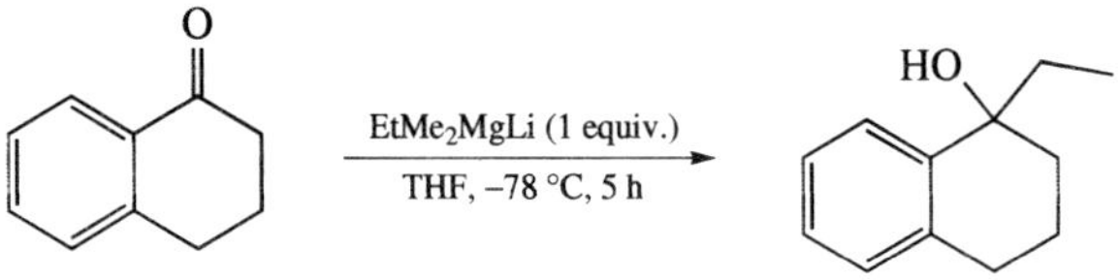

$$\text{1-tetralone} \xrightarrow[\text{THF, }-78\text{ °C, 5 h}]{\text{EtMe}_2\text{MgLi (1 equiv.)}} \text{1-ethyl-1-tetralol}$$

89%
(36% with EtMgBr)

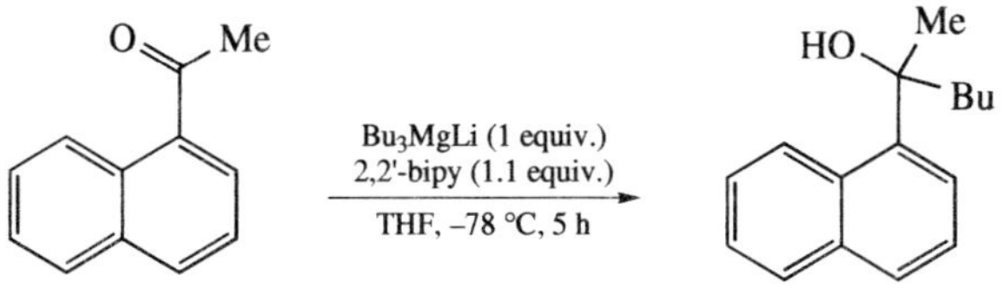

(140)

$$\text{1-acetylnaphthalene} \xrightarrow[\substack{\text{2,2'-bipy (1.1 equiv.)} \\ \text{THF, }-78\text{ °C, 5 h}}]{\text{Bu}_3\text{MgLi (1 equiv.)}} \text{product}$$

71%
(45% with BuLi)

(141)

An interesting selectivity in the transfer of alkyl groups (*n*-Alk>Me) is observed in these addition reactions. The regioselectivity of the reaction of crotylmagnesium chloride (**213**) with benzaldehyde strongly depends on the presence of various rare-earth metal chlorides. The α- to γ ratio of products can be switched to the opposite by using only another metal salt. Yttrium trichloride gives exclusively γ-product, while neodymium trichloride leads to 89% of the α-attack (with 92% of (*E*)-isomer) (equation 142)[280].

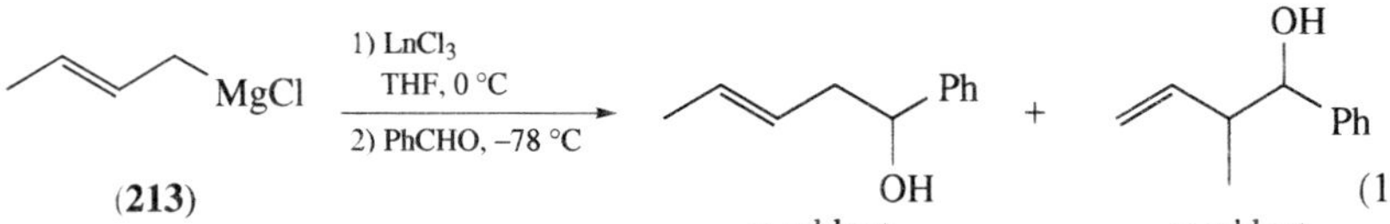

$$\text{crotyl-MgCl} \xrightarrow[\text{2) PhCHO, }-78\text{ °C}]{\substack{\text{1) LnCl}_3 \\ \text{THF, 0 °C}}} \text{α-adduct} + \text{γ-adduct}$$

(**213**)

98%, M = Nd, α:γ = 89:11
88%, M = Y, α:γ = 1:99

(142)

12. Functionalized organomagnesium compounds: Synthesis and reactivity 571

The enantioselective addition of organomagnesium compounds to ketones can be most conveniently performed by using a chiral auxiliary in the substrate molecule. Primary alkylmagnesium reagents react with aryl and heteroaryl ketones in the presence of magnesium TADDOLate at $-100\,°C$, yielding products with up to 98% *ee* (equation 143)[281]. Chiral α-ketoacetals **214**, prepared in two steps from α-substituted cinnamic aldehydes, add organomagnesium species with up to 98% diastereoselectivity (equation 144).

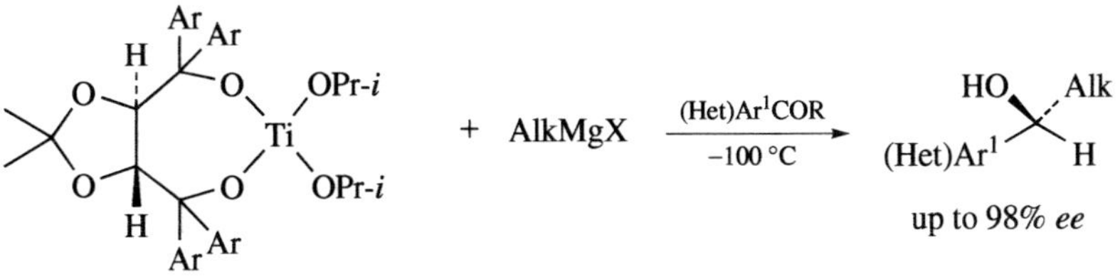

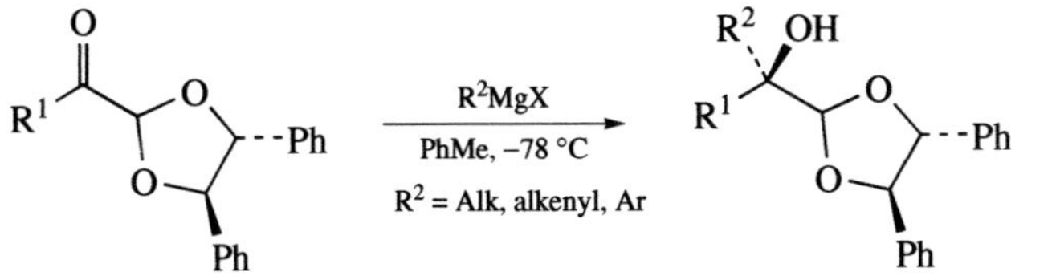

N-Boc-leucinal may react with allyl- and alkenylmagnesium halides giving *syn*- and *anti*-products in ca 9:1 ratio. This method was used for the asymmetric synthesis of important amino acids like statine and norstatine[282]. An enantioselective desymmetrization of anhydrides was reported. Arylmagnesium chlorides react in toluene in the presence of (−)-sparteine (1 equiv.) with 3-substituted glutaric anhydrides **215**, giving aryl ketones with 87–92% *ee* (equation 145)[283].

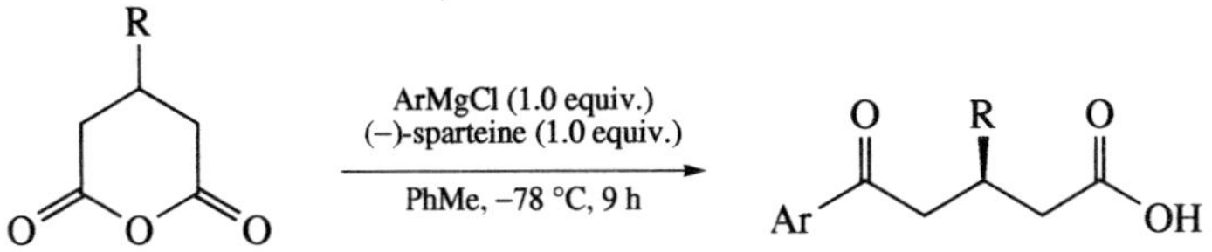

3. Addition to carbon–nitrogen bonds

The reaction of Grignard reagents with imines and nitriles is an important method for the preparation of primary amines. In comparison with organolithium reagents, the addition of RMgX to imines shows a complex behavior which depends on the nature of the reagents. A directing group on the aldimine facilitates the reaction. A stoichiometric amount of a Lewis acid can be added to enhance the rate and 1,2-selectivity[284]. The addition reaction of organometallic reagents to imines has been reviewed[285, 286].

Grignard reagents add with difficulty to imines derived from enolizable carbonyl compounds. The activation of the C=N bond can be achieved either by attachment of an electron-withdrawing group or N-coordination with a Lewis acid[285]. The use of a catalytic amount of the soluble rare-earth metal complex $LnCl_3·2LiCl$ allows the addition of various Grignard reagents to imines (equations 146 and 147)[276].

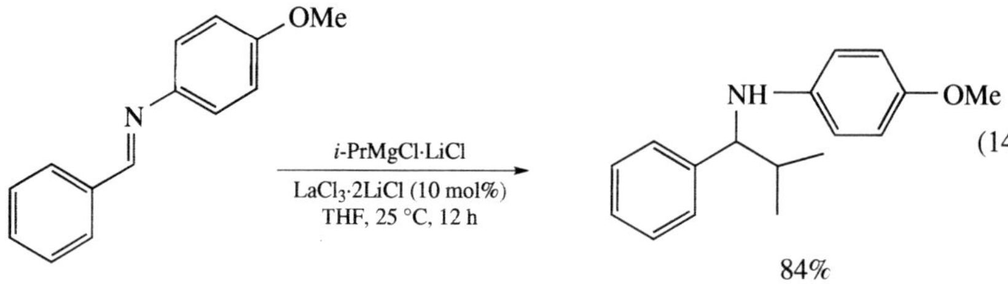

[Equation (146): PhCH=N(4-MeOC₆H₄) + i-PrMgCl·LiCl, LaCl₃·2LiCl (10 mol%), THF, 25 °C, 12 h → PhCH(i-Pr)NH(4-MeOC₆H₄), 84%]

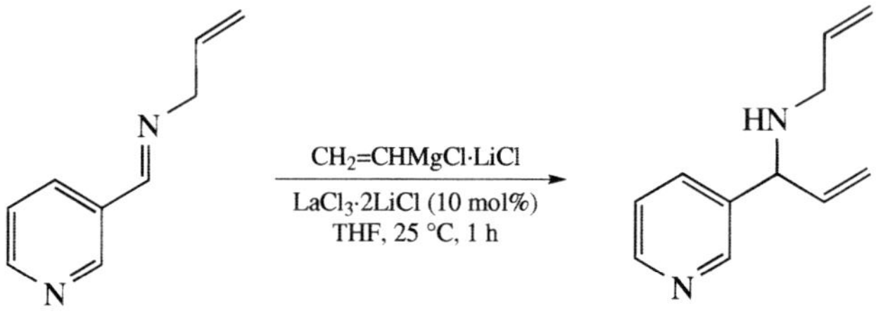

[Equation (147): 3-pyridyl-CH=N-CH₂CH=CH₂ + CH₂=CHMgCl·LiCl, LaCl₃·2LiCl (10 mol%), THF, 25 °C, 1 h → 3-pyridyl-CH(CH=CH₂)-NH-CH₂CH=CH₂, 87%]

The addition of allylmagnesium halides (2 equiv.) to 1,3-azadienes affords after *in situ* alkylation dihomoallylamines, which are useful intermediates in the synthesis of azepines or related heterocycles[287]. Activation of the C=N moiety of aldimines by 1-benzotriazolyltrimethylsilane minimizes side reactions. The mechanism involves reversible addition of BtTMS (**216**) to the imine **217** followed by displacement of the benzotriazolyl group by a Grignard reagent (equation 148)[288].

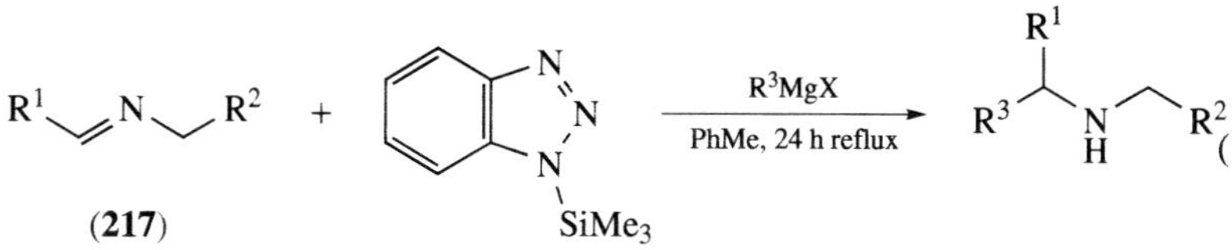

[Equation (148): R¹CH=N-CH₂R² (**217**) + Bt-SiMe₃ (**216**) → R³MgX, PhMe, 24 h reflux → R³CH(R¹)-N(H)-CH₂R²]

Similarly, reaction of enamines such as **218** with benzotriazole affords α-aminoalkyl-benzotriazoles **219**, which react smoothly with organomagnesium compounds, giving tertiary alkyl carbylamines **220** (equation 149)[289]. The whole sequence can therefore be considered as the addition of Grignard reagents to imines.

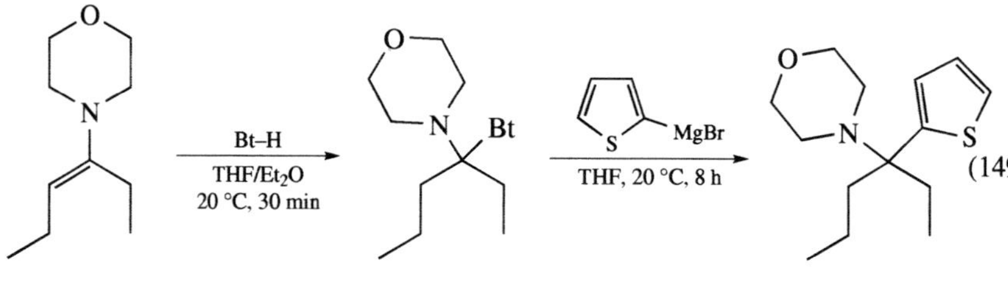

[Equation (149): morpholine-enamine **218** + Bt–H, THF/Et₂O, 20 °C, 30 min → **219** (N-Bt aminoalkyl) + 2-thienylMgBr, THF, 20 °C, 8 h → **220**, 66%]

12. Functionalized organomagnesium compounds: Synthesis and reactivity 573

Imines, derived from O-benzyllactaldehyde and benzylamine, react with non-stabilized Grignard reagents in ether yielding products with excellent diastereoselectivities (equation 150)[290].

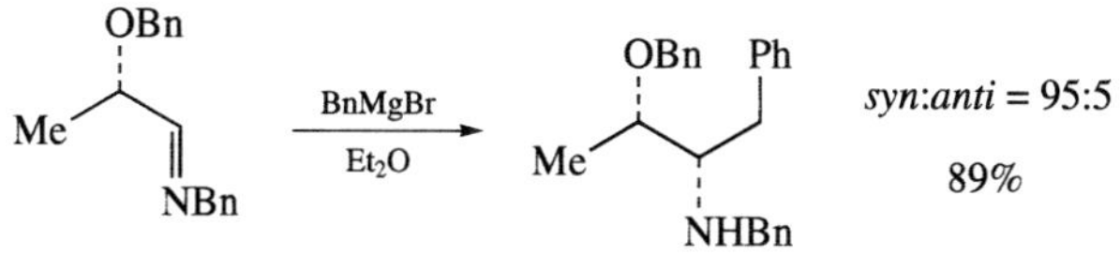

$$\text{Me-CH(OBn)-CH=NBn} \xrightarrow[\text{Et}_2\text{O}]{\text{BnMgBr}} \text{Me-CH(OBn)-CH(Ph)-NHBn} \quad syn:anti = 95:5 \quad 89\% \tag{150}$$

High yields and diastereoselectivities have also been observed for the addition of Grignard reagents to imines like **221** derived from phenylglycinol (**222**), which are existing in equilibria with 1,3-oxazolidines. Also, the imine derived from methoxyacetone affords amino-ethers with excellent diastereoselectivities. The addition of a Lewis acid (MgBr$_2$) has a strong effect on both the yield and the selectivity (equation 151)[291].

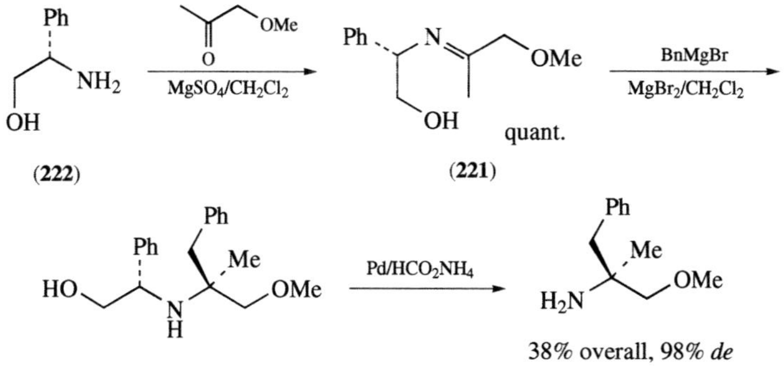

(151)

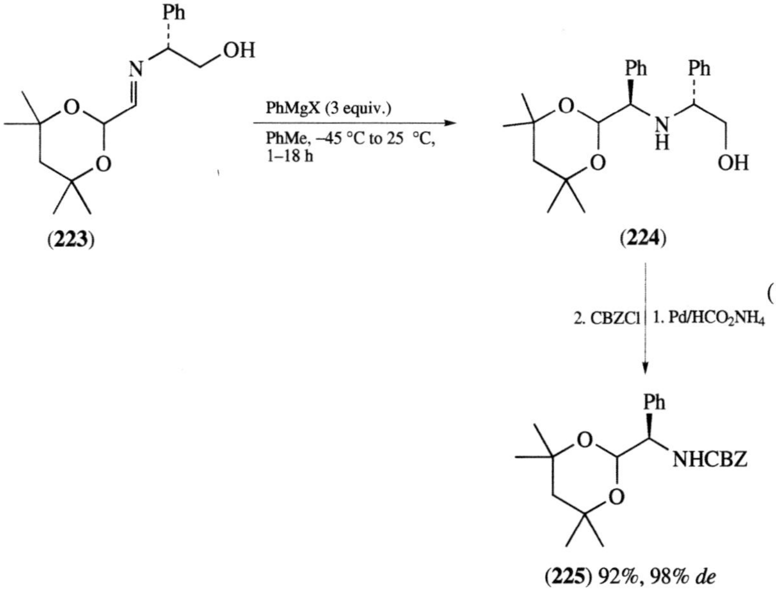

(222) → (221) quant. → [BnMgBr / MgBr$_2$/CH$_2$Cl$_2$] → intermediate → Pd/HCO$_2$NH$_4$ → product, 38% overall, 98% de

(152)

(223) → [PhMgX (3 equiv.), PhMe, −45 °C to 25 °C, 1–18 h] → (224) → [1. Pd/HCO$_2$NH$_4$; 2. CBZCl] → (225) 92%, 98% de

Imines **223** derived from glyoxal acetals react with various organomagnesium compounds with high diastereoselectivity (equation 152)[292]. The 1,2-aminoalcohols **224** can be converted into the protected enantiopure aminoaldehydes **225**. For these reactions toluene was found to be a superior solvent.

The addition reaction of 1,2-bisimine **226**, prepared from glyoxal and chiral α-phenylethylamine, gives diastereomerically pure product **227**, which was converted to the chiral 1,2-diamine **228** (equation 153)[293]. Decreasing the temperature below 50 °C leads to a sharp drop of the stereoselectivity.

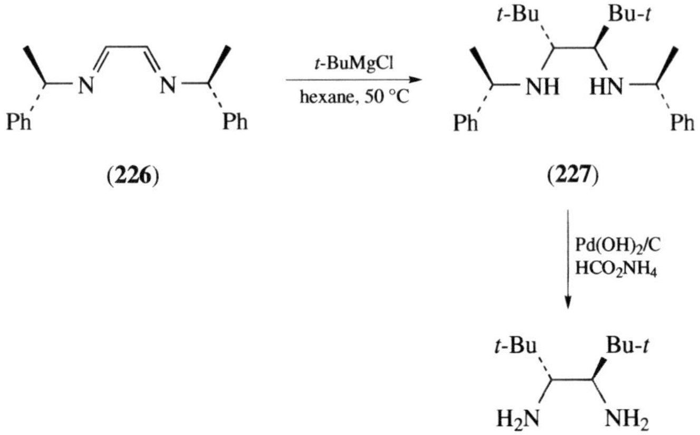

$$\text{(226)} \xrightarrow[\text{hexane, 50 °C}]{t\text{-BuMgCl}} \text{(227)} \xrightarrow[\text{HCO}_2\text{NH}_4]{\text{Pd(OH)}_2/\text{C}} \text{(228)} > 95\% \text{ de} \quad (153)$$

Chiral oxime ethers **229** of (R)- and (S)-O-(1-phenylbutyl)-hydroxylamine (ROPHy/SOPHy) react with Grignard reagents in the presence of BF$_3$·OEt$_2$ in toluene at −78 °C yielding addition products with high diastereoselectivities (equation 154)[294]. The resulting chiral hydroxylamine derivatives have been converted enantioselectively to primary amines, or (when R = allyl) to β-amino acids.

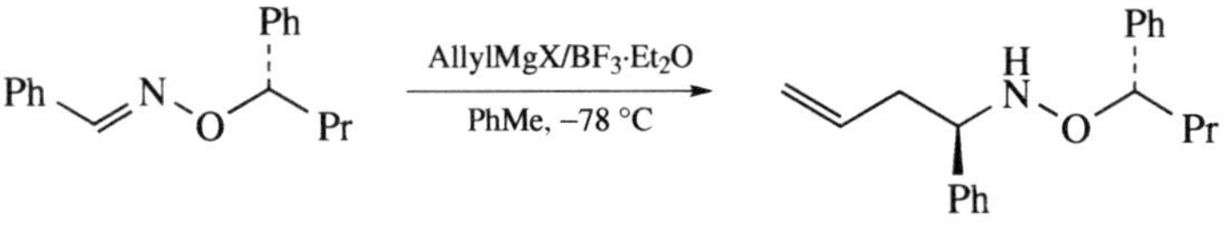

$$\text{(229)} \xrightarrow[\text{PhMe, −78 °C}]{\text{AllylMgX/BF}_3\cdot\text{Et}_2\text{O}} \quad 100\%, 92\% \text{ de} \quad (154)$$

O-Benzyllactaldehyde dimethylhydrazone **230** allows a substrate control in the addition reaction of organomagnesium halides, leading almost exclusively to the syn-isomer **231** (equation 155)[295]. The resulting hydrazide can be reduced on Raney Ni to the corresponding syn-aminoalcohol **232**. The stereoselective Grignard addition to a similar N-formyl hydrazone **233** proceeds with 92% diastereoselectivity (equation 156). The silylation of the amide nitrogen by TMSCl provides the pure syn-adduct[296].

A convenient synthesis of aryl glycines **234** is performed by the addition of arylmagnesium chlorides to N-Boc-iminomalonate (**235**) (prepared from diethyl mesoxalate and BocN=PPh$_3$). This reaction proceeds smoothly at low temperatures and is tolerating many functional groups in the Grignard reagent leading to amino acids of type **234** (equation 157)[297].

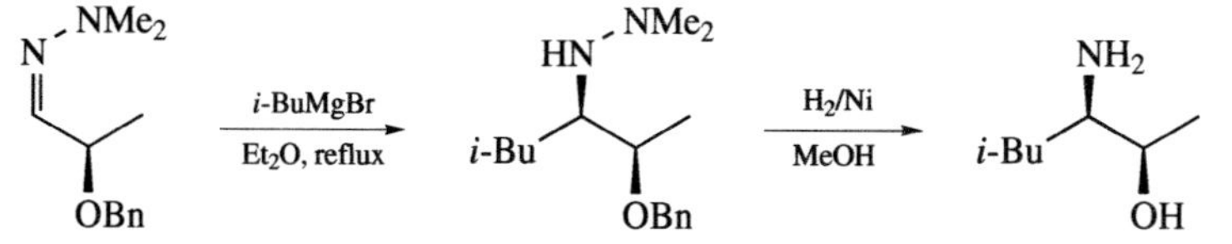

$$\text{(230)} \xrightarrow[\text{Et}_2\text{O, reflux}]{i\text{-BuMgBr}} \text{(231)} \xrightarrow[\text{MeOH}]{\text{H}_2/\text{Ni}} \text{(232)} > 96\% \ de \quad (155)$$

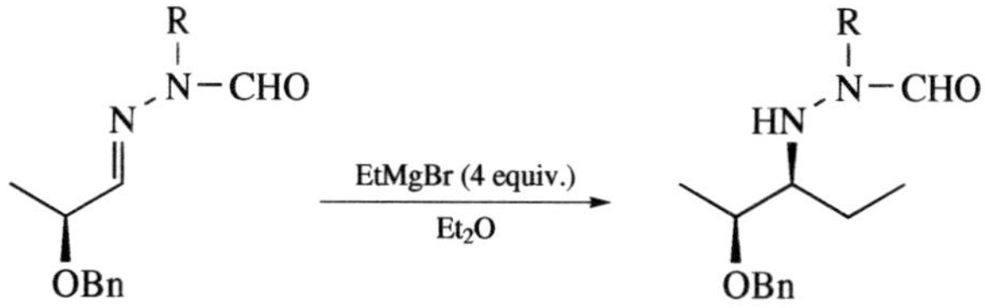

$$\text{(233)} \xrightarrow[\text{Et}_2\text{O}]{\text{EtMgBr (4 equiv.)}} \text{55\%} \quad (156)$$

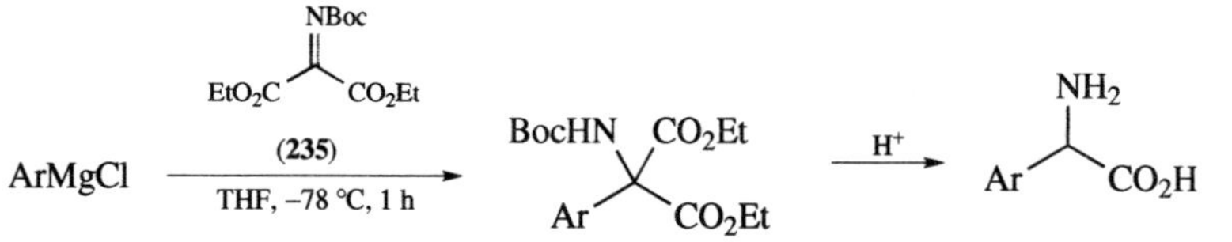

R = H: 92% de
R = SiMe$_3$: 99.6% de

$$\text{ArMgCl} + \text{(235)} \xrightarrow{\text{THF, }-78\ °\text{C, 1 h}} \text{BocHN-C(Ar)(CO}_2\text{Et)}_2 \xrightarrow{\text{H}^+} \text{(234) 80–85\%} \quad (157)$$

Chiral sulfinimines **236** are very useful intermediates for the preparation of enantiomerically pure primary amines **237** (equation 158)[298]. This reaction has been applied to the synthesis of α-amino acids[299]. For sulfinimines obtained from simple ketones, lithium reagents are preferable for the addition[298b], while for cyclic ketones organomagnesium compounds gave the best results. Addition of alkyl and aryl Grignard compounds to sulfinimines, derived from 3- and 4-substituted cyclohexanones, proceeds with excellent diastereoselectivity, depending on the stereochemistry of the ring substituents rather than the sulfinyl group[300].

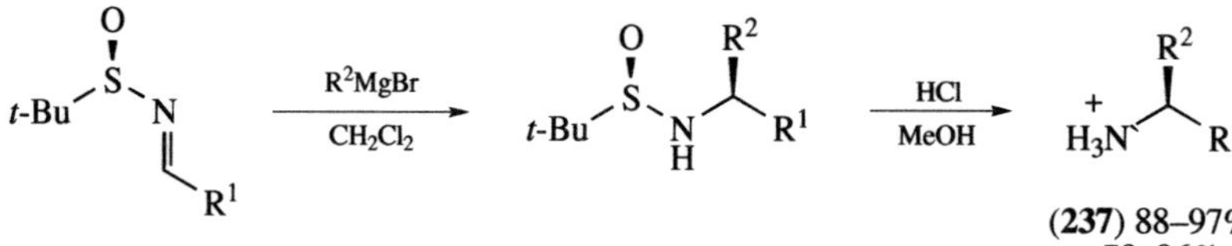

$$\text{(236)} \xrightarrow[\text{CH}_2\text{Cl}_2]{\text{R}^2\text{MgBr}} \xrightarrow[\text{MeOH}]{\text{HCl}} \text{(237) 88–97\%, 78–96\% ee} \quad (158)$$

E. Diverse Reactions

1. Amination reactions

a. Electrophilic amination reactions. The reaction of nucleophilic Grignard reagents with electrophilic aromatic nitrogen compounds in a higher oxidation state is a versatile method for the synthesis of amines[301]. The reaction of functionalized arylmagnesium halides such as **238** with nitroarenes **239** followed by a reductive workup[302] provides polyfunctionalized diarylamines **240a, b** and heterocyclic amines like **241** in high yields (equation 159). The use of two equivalents of the Grignard reagent is crucial to obtain complete conversion as explained by considering the reaction mechanism (equation 160)[303]. The intermediate arylnitroso derivative **242** is a proposed reactive species. Thus, nitrosoarenes like **243** can be directly used as starting materials, allowing the synthesis of various functionalized diarylamines like **244** (equation 161)[304].

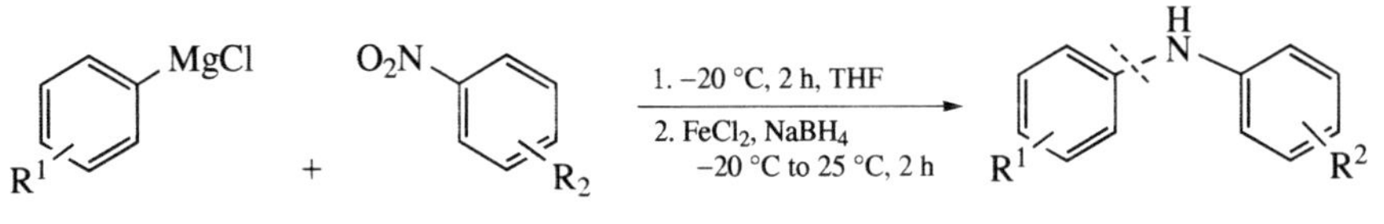

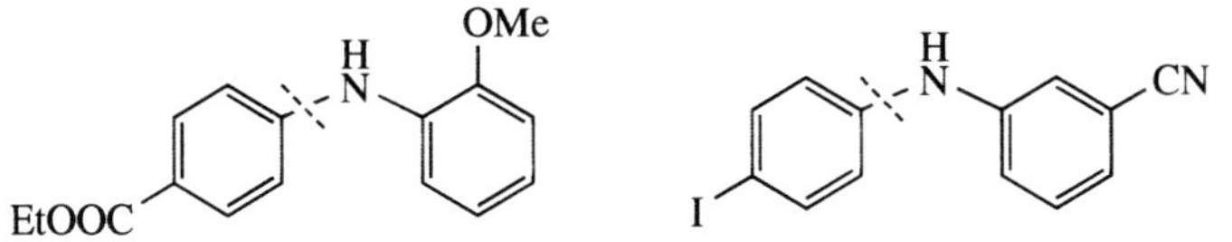

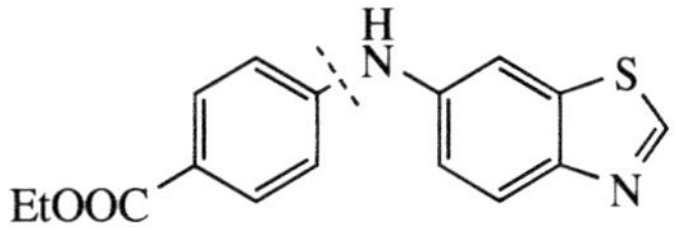

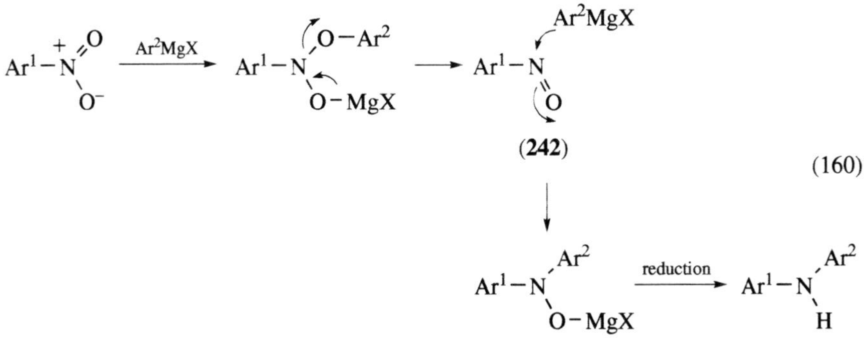

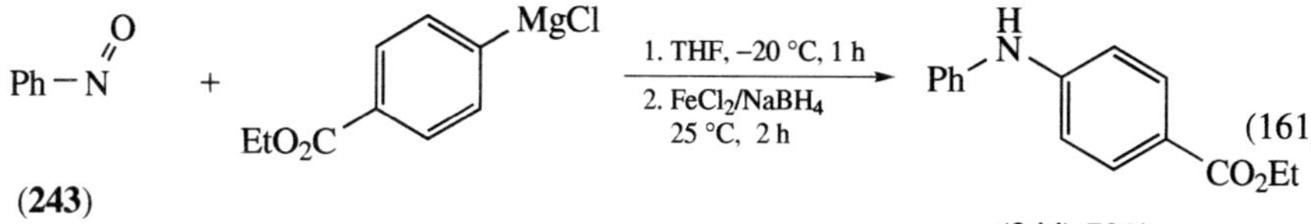

Nitroarenes, bearing bulky substituents next to the nitro function, are not reduced to the corresponding diarylamines[305]. Although the formation of the intermediate nitrosoarene **242** is still observed, due to steric hindrance the second equivalent of the Grignard reagent adds to the oxygen atom resulting in the formation of the nitrene **245**. This reactive intermediate **245** can be used for the mild synthesis of benzimidazoles like **246** or indoles like **247** bearing a broad range of functional groups (Scheme 20).

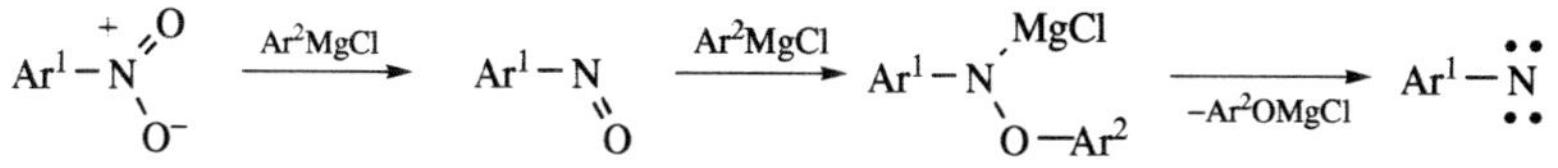

SCHEME 20

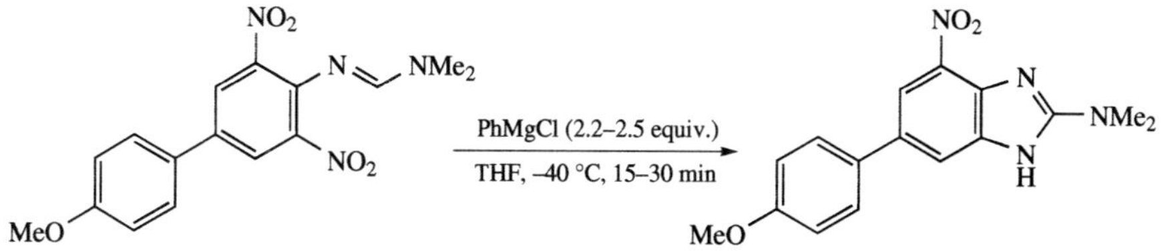

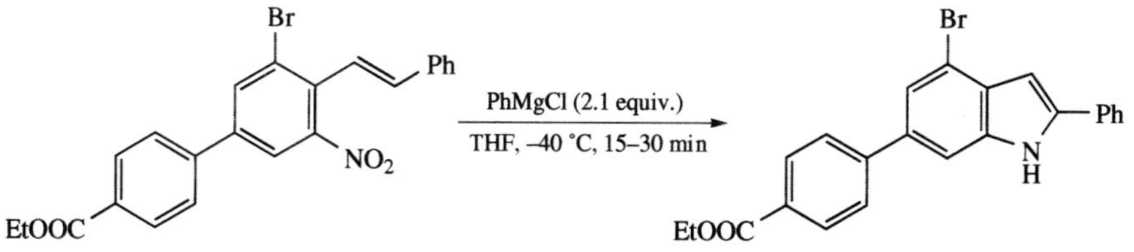

Arylazo tosylates of type **248**, which are readily obtained from aniline derivatives **249** in a two-step procedure (equation 162), can be alternatively used as starting materials[306]. This electrophilic nitrogen equivalent **248** reacts with a broad range of functionalized Grignard reagents under mild conditions. Subsequent allylation of the addition products with allyl iodide, followed by reductive cleavage of the resulting hydrazine derivatives

250, furnishes polyfunctionalized amines **251a–e** (equation 163).

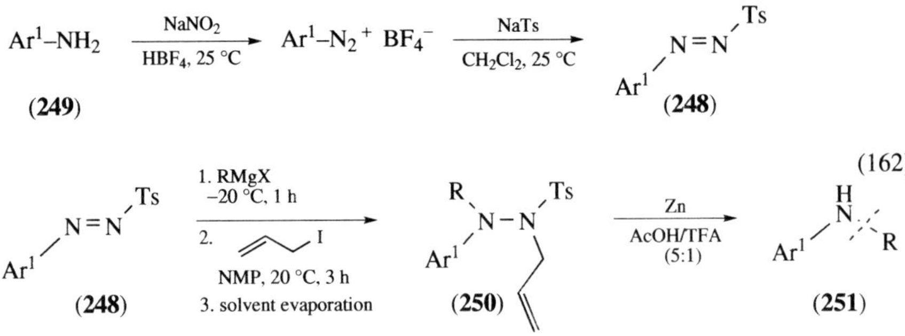

$$Ar^1-NH_2 \xrightarrow[\text{HBF}_4, 25\,°C]{\text{NaNO}_2} Ar^1-N_2^+ \, BF_4^- \xrightarrow[\text{CH}_2\text{Cl}_2, 25\,°C]{\text{NaTs}} Ar^1-N=N-Ts$$

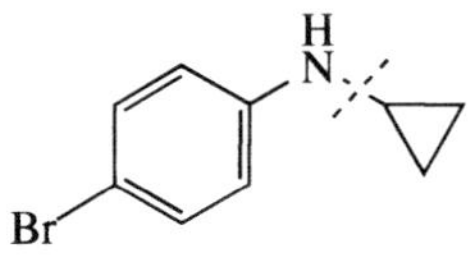

(249) → (248)

(162)

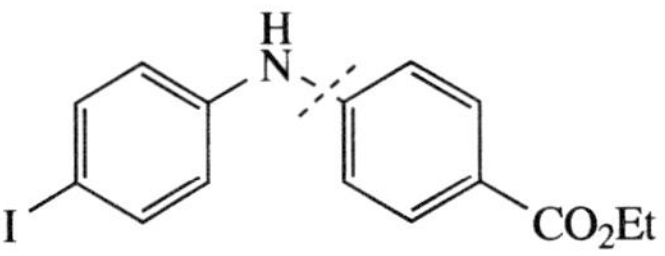

$$Ar^1-N=N-Ts \xrightarrow[\substack{\text{1. RMgX}\\-20\,°C,\,1\,h\\ \text{2. allyl–I}\\ \text{NMP, 20\,°C, 3\,h}\\ \text{3. solvent evaporation}}]{} Ar^1-N(R)-N(Ts)(\text{allyl}) \xrightarrow[\text{(5:1)}]{\text{Zn}\atop \text{AcOH/TFA}} Ar^1-NH-R$$

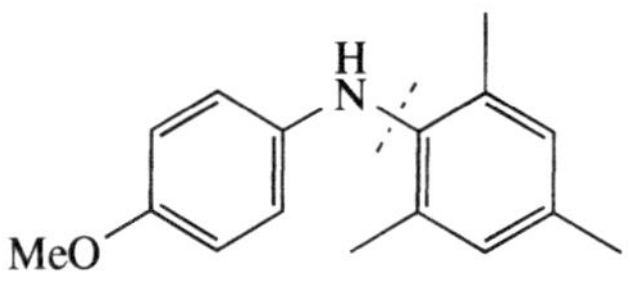

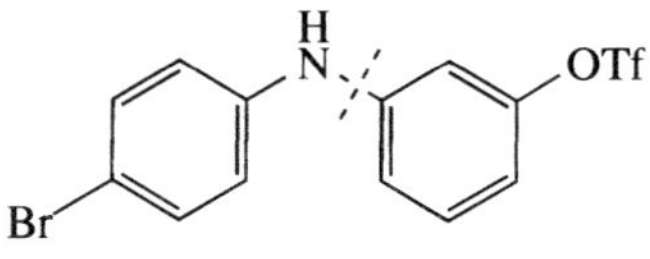

(248) → (250) → (251)

(**251a**) 71% — 4-I-C6H4-NH-C6H4-CO2Et-4

(**251b**) 67% — 4-Br-C6H4-NH-cyclopropyl

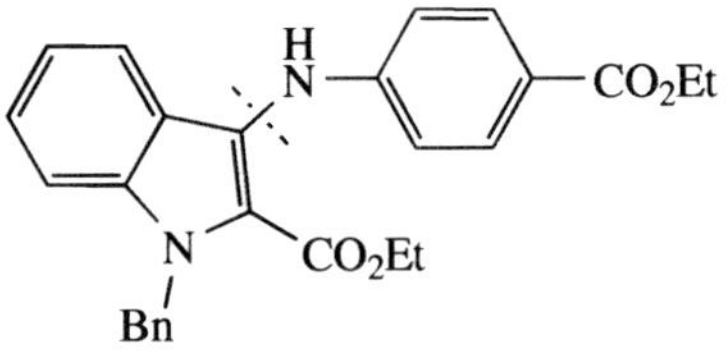

(163)

(**251c**) 71% — N-benzyl-2-(CO2Et)-3-(NH-C6H4-CO2Et)-indole

(**251d**) 75% — 4-MeO-C6H4-NH-(2,4,6-trimethylphenyl)

(**251e**) 70% — 4-Br-C6H4-NH-C6H4-OTf-3

The electrophilic amination reaction of organometallic species using mono-, di- and trihaloamines has attracted a lot of attention for the synthesis of amines. Only a few cases have been reported using alkylchloroamines as precursors for the synthesis of tertiary amines[307]. One example is the reaction of functionalized arylmagnesium compounds with benzyl-N-chloroamines **252** providing polyfunctional tertiary amines **253** (equation 164)[308]. The procedure was also applied for the preparation of chiral N-chloroamines with retention of chirality at the α-carbon. However, the amination process is limited to benzyl-N-chloroamines.

(164)

(252) (253) 67%

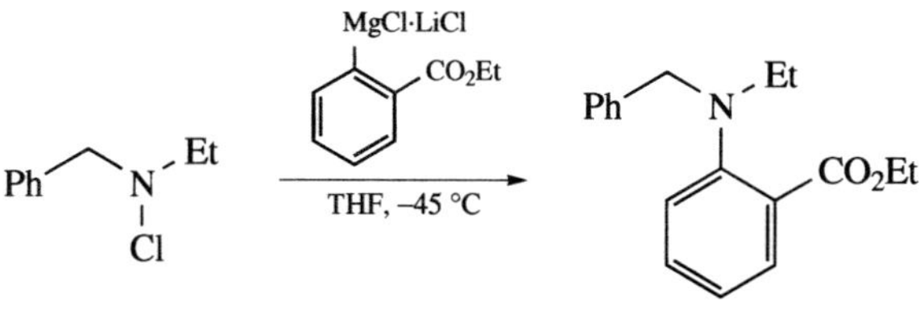

Another electrophilic amination method uses 4,4,5,5-tetramethyl-1,3-dioxolan-2-one *O*-phenylsulfoxime **254** as an electrophilic nitrogen equivalent[309]. It proved suitable for the amination of alkyl- and arylmagnesium reagents affording the respective primary alkyl or aryl amines **255** (equation 165). Electrophilic amination of Grignard reagents can further be achieved using *O*-benzoyl-*N*,*N*-dialkylhydroxylamines **256** and a catalytic amount of $CuCl_2$ (equation 166)[310]. A three-component coupling reaction of thioformamides **257** with organolithium and Grignard reagents allows the formation of tertiary amines like **258** (equation 167)[311].

(254) (1.1 equiv.)

(165)

(255)

overall yield: 96%

(166)

(256) 89%

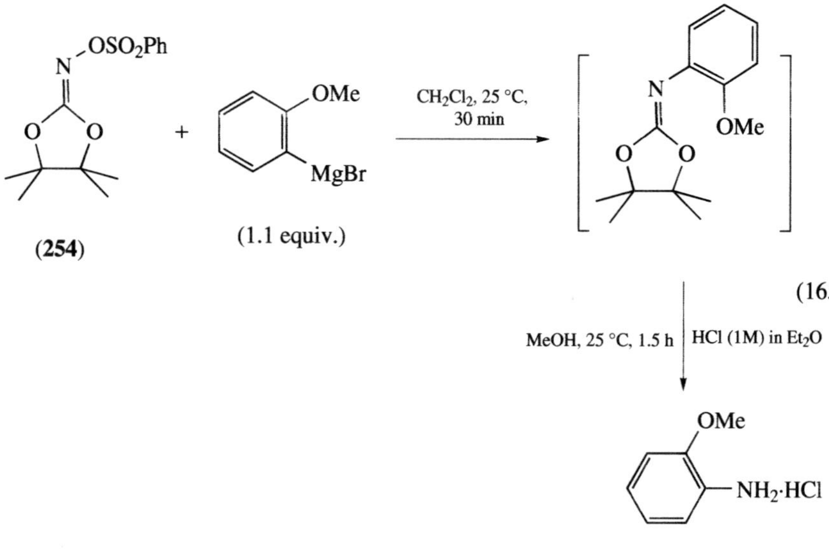

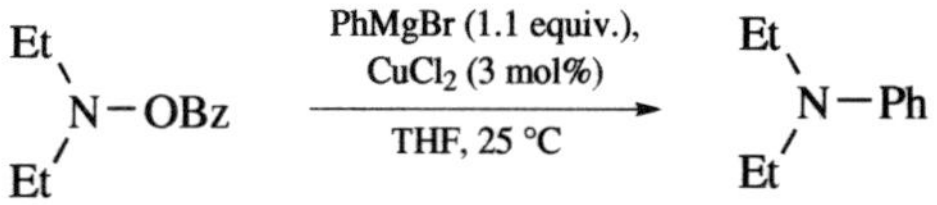

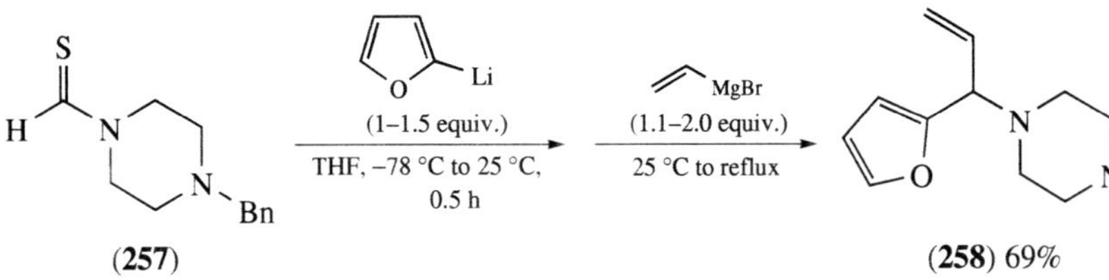

b. Oxidative coupling of polyfunctional aryl and heteroaryl amidocuprates. The oxidative amination of amidocuprates is a complement to the electrophilic amination reaction. Previous work[312, 313] focused on the use of oxygen as oxidant for converting amidocuprates to various amines.

In a new synthetic protocol for the preparation of polyfunctional primary, secondary and tertiary aryl and heteroaryl amines **259**, chloranil (**260**) proves to be an efficient oxidant[314]. The required functionalized amidocuprates **261** are prepared starting from organomagnesium reagents **262** by transmetalation with CuCl·2LiCl followed by treatment with a lithium amide **263**. Oxidation of **261** with chloranil (**260**) affords the amines **259** in 70–80% yield (equation 168). This sequence can be performed with arylmagnesium reagents **262** bearing various functional groups such as a methoxy, an iodide and an amide group, as well as with various lithium amides **263** bearing functional groups like a bromide, a nitrile and an ester group, allowing the synthesis of the polyfunctionalized amines **259**. Steric hindrance of either the copper reagent or the lithium amide **263** is well tolerated (equation 169).

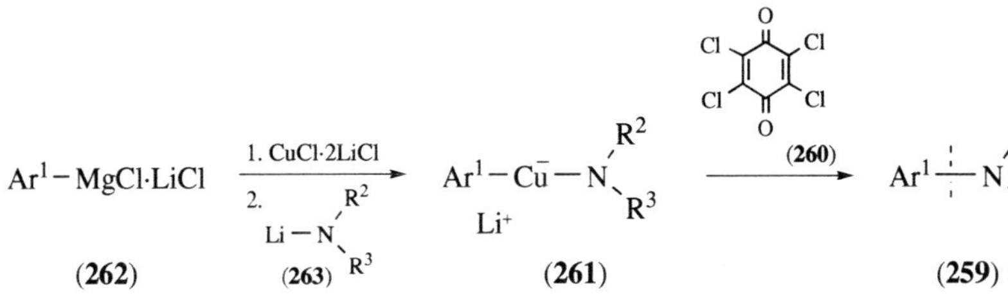

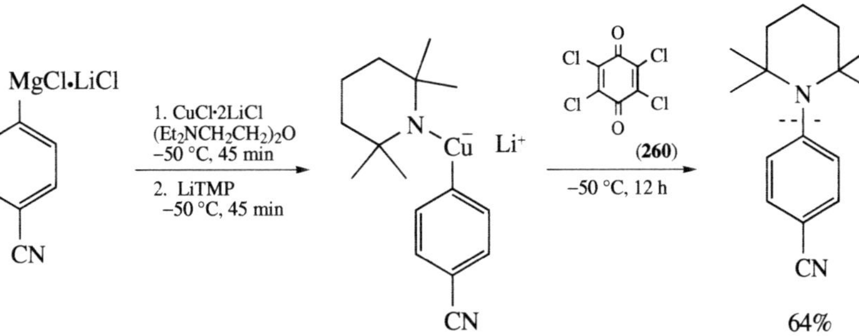

For the synthesis of primary amines, the diester **264** is magnesiated with TMPMgCl·LiCl and transmetalated with CuCl·2LiCl, affording the corresponding arylcopper derivative **265**. Addition of LiHMDS furnishes the corresponding amidocuprate, which is reacted with chloranil (**260**) leading to the *N,N*-bis(trimethylsilyl)amine derivative **266**. Desilylation with TBAF results in the formation of the arylamine **267** in 72% yield (equation 170).

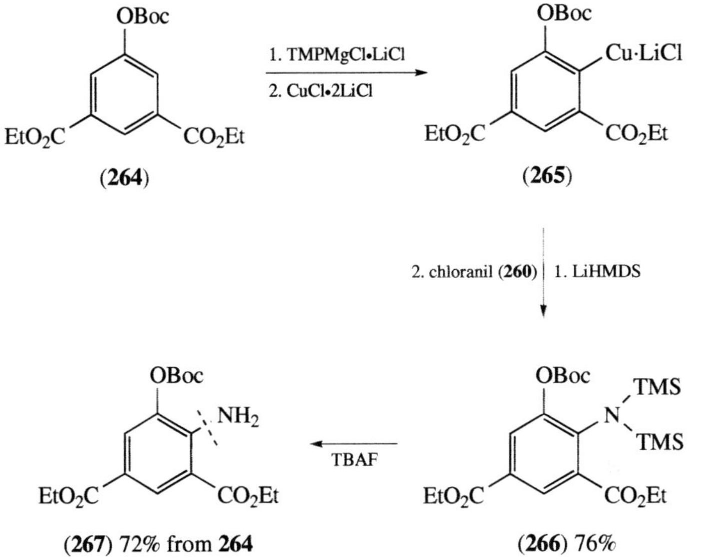

(170)

Secondary amines **268** are prepared using TBS-protected lithium amides **269** (Scheme 21), while the preparation of polyfunctional triarylamines applies lithium amides which are derived from secondary amines.

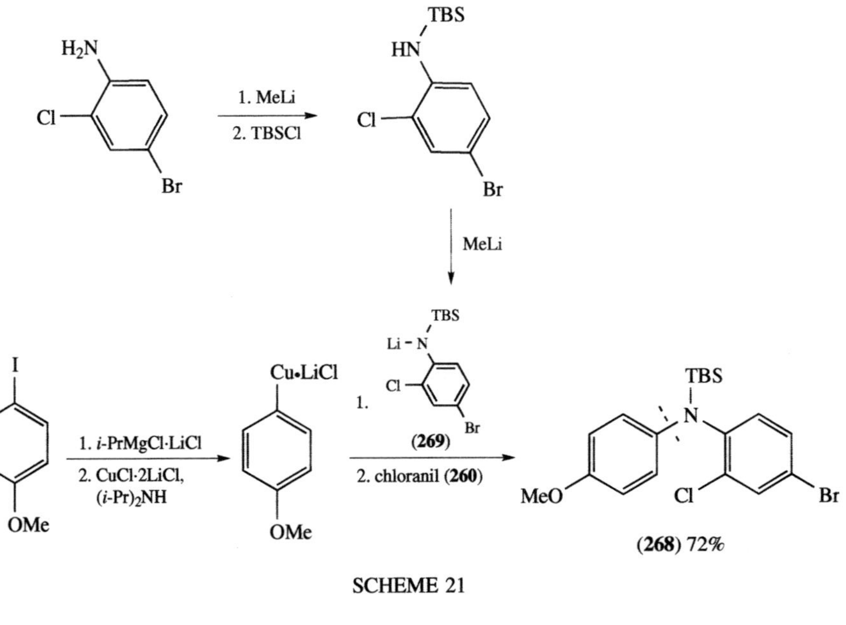

SCHEME 21

2. Synthesis of cyclopropanes

Cyclopropylamines and cyclopropanols can be prepared from alkylmagnesium halides[315]. The reaction is catalyzed by titanium alcoholates and its mechanism includes the formation of a dialkoxytitanacyclopropane **270**, which reacts with a carbonyl compound or nitrile (Scheme 22). The use of chiral titanium alcoholates allows the reaction to be performed with up to 78% *ee* (equation 171)[316].

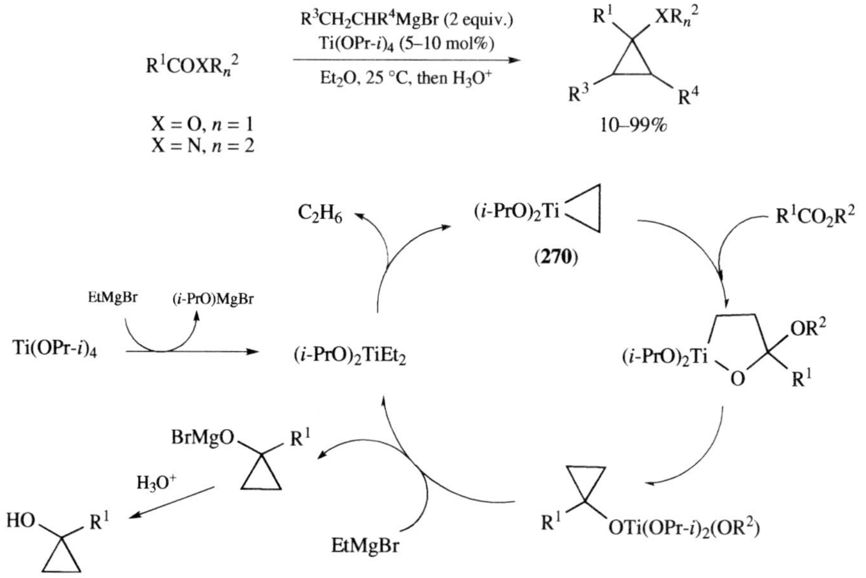

SCHEME 22

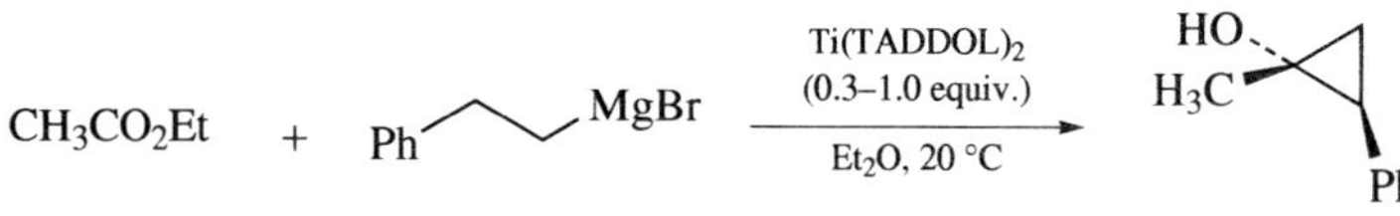

65–72%, up to 78% *ee* (171)

TADDOL =

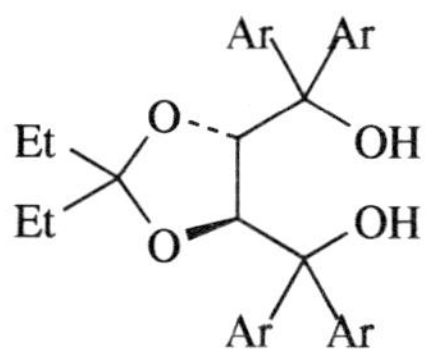

Ar = 3,5-bis(trifluoromethyl)phenyl

3. Synthesis of chiral sulfoxides

Chiral sulfoxides are useful intermediates in asymmetric synthesis. A number of methods can be used for their preparation. For example, enantiomerically pure *p*-tolylsulfoxides can be obtained by displacing a dimethylphosphonylmethyl moiety, a carbon leaving group, from sulfur by Grignard reagents[317]. Optically pure menthyl 4-bromophenyl sulfinate can be sequentionally displaced, yielding unsymmetrical dialkyl sulfoxides in 60–97% yield and>98% ee[318]. A simple one-pot synthesis of chiral sulfoxides **271** in high optical purity was developed recently. It starts from norephedrine-derived sulfamidites **272**[319] (equation 172).

$$\underset{\underset{Ph\quad Me}{(\mathbf{272})}}{\overset{O}{\underset{O}{S}}\underset{N}{\diagdown}-CO_2Bn} \quad \xrightarrow[\text{one-pot}]{\begin{array}{l}1.\ RMgX\\2.\ HBF_4\\3.\ R^1MgX\end{array}} \quad \underset{(\mathbf{271})\ 50-78\%,\ >93\%\ ee}{R\overset{O}{\underset{}{\overset{\shortparallel}{S}}}R^1} \quad (172)$$

IV. REFERENCES

1. V. Grignard, *Compt. Rend. Acad. Sci. Paris*, **130**, 1322 (1900).
2. B. H. Lipshutz and S. Sengupta, *Org. React.*, **41**, 135 (1992).
3. (a) H. G. Richey, Jr., *Grignard Reagents*, Wiley, New York, 2000.
   (b) M. S. Kharasch and O. Reinmuth, *Grignard Reactions of Nonmetallic Substances*, Prentice-Hall, New York, 1954.
   (c) G. S. Silverman and P. E. Rakita, *Handbook of Grignard Reagents*, Marcel Dekker, New York, 1996.
   (d) B. J. Wakefield, *Organomagnesium Methods in Organic Synthesis*, Academic Press, London, 1995.
4. W. E. Lindsell, in *Comprehensive Organometallic Chemistry II* (Eds. E. W. Abel, F. G. A. Stone and G. Wilkinson), Vol. 1 (Vol. Ed. C. E. Housecroft), Chap. 3, Pergamon Press, Oxford, 1995, pp. 72–78 and references therein.
5. (a) C. Hamdouchi and H. M. Walborsky, in *Handbook of Grignard Reagents* (Eds. G. S. Silverman and P. E. Rakita), Marcel Dekker, New York, 1996, p. 145.
   (b) J. F. Garst and F. Ungvary, in *Grignard Reagents: New Developments* (Ed. H. G. Richey, Jr.), Wiley, Chichester, 2000, p. 185.
   (c) J. F. Garst, F., Ungváry and J. T. Baxter, *J. Am. Chem. Soc.*, **119**, 253 (1997).
   (d) J. F. Garst and M. P. Soriaga, *Coord. Chem. Rev.*, **248**, 623 (2004).
6. (a) F. Bickelhaupt, in *Grignard Reagents: New Developments* (Ed. H. G. Richey, Jr.), Wiley, Chichester, 2000, p. 299.
   (b) H. L. Uhm, in *Handbook of Grignard Reagents* (Eds. G. S. Silverman and P. E. Rakita), Marcel Dekker, New York, 1996, p. 117.
7. P. E. Rakita, J. F. Aultman and L. Stapleton, *Chem. Eng.*, **97**, 110 (1990).
8. (a) A. Tuulmets, M. Marvi and D. Panov, *J. Organomet. Chem.*, **523**, 133 (1996).
   (b) M. Sassian, D. Panov and A. Tuulmets, *Appl. Organomet. Chem.*, **16**, 525 (2002).
9. H. Simuste, D. Panov, A. Tuulmets and B. T. Nguyen, *J. Organomet. Chem.*, **690**, 3061 (2005).
10. J. Wiss and G. Ermini, *Org. Process Res. Dev.*, **10**, 1282 (2006).
11. J. F. Garst and M. P. Seriaga, *Coord. Chem. Rev.*, **248**, 623 (2004).
12. (a) H. M. Walborsky and J. Topolski, *J. Am. Chem. Soc.*, **114**, 3455 (1992).
    (b) C. E. Teerlinck and W. J. Bowyer, *J. Org. Chem.*, **61**, 1059 (1996).
13. For factors controlling Grignard reagent formation, see: W. E. Lindsell, in *Comprehensive Organometallic Chemistry I* (Eds. G. Wilkinson, F. G. Stone and G. E. Ebel), Vol. 1, Chap. 3, Pergamon Press, Oxford, 1982, pp. 155–252 and references therein.
14. (a) H. M. Walborsky, *Acc. Chem. Res.*, **23**, 286 (1990).
    (b) J. F. Garst, *Acc. Chem. Res.*, **24**, 95 (1991).
    (c) C. Walling, *Acc. Chem. Res.*, **24**, 255 (1991).
    (d) H. R. Rogers, C. L. Hill, Y. Fujiwara, R. J. Rogers, H. L. Mitchell and G. M. Whitesides, *J. Am. Chem. Soc.*, **102**, 217 (1980).

(e) K. S. Root, C. L. Hill, L. M. Lawrence and G. M. Whitesides, *J. Am. Chem. Soc.*, **111**, 5404 (1989).
(f) E. C. Ashby and J. Oswald, *J. Org. Chem.*, **53**, 6068 (1988).
15. F. Kanoufi, C. Combellas, H. Hazimeh, J.-M. Mattalia, C. Marchi-Delapierre and M. Chanon, *J. Phys. Org. Chem.*, **19**, 847 (2006).
16. U. Tilstam and H. Weinmann, *Org. Process Res. Dev.*, **6**, 906 (2002).
17. B. Bogdanović and M. Schwickardi, *Angew. Chem., Int. Ed.*, **39**, 4610 (2000).
18. H. Gold, M. Larhed and P. Nilsson, *Synlett*, 1596 (2005).
19. (a) R. D. Rieke and M. V. Hanson, *Tetrahedron*, **53**, 1925 (1997).
(b) R. D. Rieke, M. S. Sell, W. R. Klein, T.-A. Chen, J. D. Brown and M. U. Hansen, in *Active Metals. Preparation, Characterization, Application* (Ed. A. Fürstner), Wiley-VCH, Weinheim, 1996, p. 1.
20. J. Lee, R. Velarde-Ortiz, A. Guijarro, J. R. Wurst and R. D. Rieke, *J. Org. Chem.*, **65**, 5428 (2000).
21. (a) O. Sugimoto, S. Yamada and K. Tanji, *Tetrahedron Lett.*, **43**, 3355 (2002).
(b) O. Sugimoto, S. Yamada and K. Shigeru, *J. Org. Chem.*, **68**, 2054 (2003).
22. B. Bogdanović, *Acc. Chem. Res.*, **21**, 261 (1988).
23. D. Cheng, S. Zhu, Z. Yu and T. Cohen, *J. Am. Chem. Soc.*, **123**, 30 (2001).
24. J. Beckmann, D. Dakternieks, M. Draeger and A. Duthie, *Angew. Chem., Int. Ed.*, **45**, 6509 (2006).
25. S. T. Handy, *J. Org. Chem.*, **71**, 4659 (2006).
26. M. C. Law, K.-Y. Wong and T. H. Chan, *Chem. Commun.*, 2457 (2006).
27. C. Prévost, *Bull. Soc. Chim. Fr.*, 1372 (1931).
28. (a) W. F. Bailey and J. J. Patricia, *J. Organomet. Chem.*, **352**, 1 (1988).
(b) H. J. Reich, N. H. Phillips and I. L. Reich, *J. Am. Chem. Soc.*, **107**, 4101 (1985).
(c) W. B. Farnham and J. C. Calabrese, *J. Am. Chem. Soc.*, **108**, 2449 (1986).
(d) A. Krasovskiy, B. F. Straub and P. Knochel, *Angew. Chem., Int. Ed.*, **45**, 159 (2006).
29. (a) O. R. Pierce, A. F. Meiners and E. T. McBee, *J. Am. Chem. Soc.*, **75**, 2516 (1953).
(b) E. T. McBee, C. W. Roberts and A. F. Meiners, *J. Am. Chem. Soc.*, **79**, 335 (1957).
(c) P. Moreau, R. Albachi and A. Commeyras, *Nouv. J. Chim.*, **1**, 497 (1977).
30. (a) R. D. Chambers, W. K. R. Musgrave and J. Savory, *J. Chem. Soc.*, 1993 (1962).
(b) For a review on fluorinated organometallics, see: D. J. Burton and Z. Y. Yang, *Tetrahedron*, **48**, 189 (1992).
31. (a) J. Villiéras, *Bull. Soc. Chim. Fr.*, 1520 (1967).
(b) J. Villiéras, B. Kirschleger, R. Tarhouni and M. Rambaud, *Bull. Soc. Chim. Fr.*, 470 (1986).
32. For recent examples, see:
(a) A. Müller, M. Marsch, K. Harms, J. C. W. Lohrenz and G. Boche, *Angew. Chem., Int. Ed.*, **35**, 1518 (1996).
(b) R. W. Hoffmann, M. Julius, F. Chemla, T. Ruhland and G. Frenzen, *Tetrahedron*, **50**, 6049 (1994).
33. C. Tamborski and G. J. Moore, *J. Organomet. Chem.*, **26**, 153 (1971).
34. (a) N. Furukawa, T. Shibutani and H. Fujihara, *Tetrahedron Lett.*, **28**, 5845 (1987).
(b) D. J. Burton and Z. Y. Yang, *Tetrahedron*, **48**, 189 (1992).
(c) R. D. Chambers, W. K. R. Musgrave and J. Savory, *J. Chem. Soc.*, 1993 (1962).
(d) H. H. Paradies and M. Görbing, *Angew. Chem., Int. Ed. Engl.*, **8**, 279 (1969).
(e) G. Cahiez, D. Bernard and J. F. Normant, *J. Organomet. Chem.*, **113**, 107 (1976).
(f) D. Seyferth and R. L. Lambert, *J. Organomet. Chem.*, **54**, 123 (1973).
(g) H. Nishiyama, K. Isaka, K. Itoh, K. Ohno, H. Nagase, K. Matsumoto and H. Yoshiwara, *J. Org. Chem.*, **57**, 407 (1992).
(h) C. Bolm and D. Pupowicz, *Tetrahedron Lett.*, **38**, 7349 (1997).
35. L. Boymond, M. Rottländer, G. Cahiez and P. Knochel, *Angew. Chem., Int. Ed.*, **37**, 1701 (1998).
36. A. E. Jensen, W. Dohle, I. Sapountzis, D. M. Lindsay, V. A. Vu and P. Knochel, *Synthesis*, 565 (2002).
37. J. T. Reeves, M. Sarvestani, J. J. Song, Z. Tan, L. Nummy, H. Lee, N. K. Yee and C. H. Senanayake, *Org. Process Res. Dev.*, **10**, 1258 (2006).
38. P. Cali and M. Begtrup, *Synthesis*, 6 (2002).

12. Functionalized organomagnesium compounds: Synthesis and reactivity 585

39. (a) R. Kober, W. Hammes and W. Steglich, *Angew. Chem., Int. Ed.*, **21**, 203 (1982).
    (b) D. von der Brück, R. Bühler and H. Plieninger, *Tetrahedron*, **28**, 791 (1972).
40. T. Graening, W. Friedrichsen, J. Lex and H.-G. Schmalz, *Angew. Chem., Int. Ed.*, **41**, 1524 (2002).
41. (a) M. Abarbri, J. Thibonnet, L. Bérillon, F. Dehmel, M. Rottländer and P. Knochel, *J. Org. Chem.*, **65**, 4618 (2000).
    (b) P. Knochel, E. Hupe, W. Dohle, D. M. Lindsay, V. Bonnet, G. Quéguiner, A. Boudier, F. Kopp, S. Demay, N. Seidel, M. I. Calaza, V. A. Vu, I. Sapountzis and T. Bunlaksananusorn, *Pure Appl. Chem.*, **74**, 11 (2002).
42. P. Knochel, W. Dohle, N. Gommermann, F. F. Kneisel, T. Korn, I. Sapountzis and V. A. Vu, *Angew. Chem., Int. Ed.*, **42**, 4302 (2003).
43. L. Bérillon, A. Leprêtre, A. Turck, N. Ple, G. Quéguiner, G. Cahiez and P. Knochel, *Synlett*, 1359 (1998).
44. M. Abarbri and P. Knochel, *Synlett*, 1577 (1999).
45. T. Tobrman and D. Dvorak, *Org. Lett.*, **5**, 4289 (2003).
46. F. Dehmel, M. Abarbri and P. Knochel, *Synlett*, 345 (2000).
47. A. Staubitz, W. Dohle and P. Knochel, *Synthesis*, 233 (2003).
48. C. Jaramillo, J. C. Carretero, J. E. de Diego, M. del Prado, C. Hamdouchi, J. L. Roldán and C. Sánchez-Martínez, *Tetrahedron Lett.*, **43**, 9051 (2002).
49. (a) M. Bergauer and P. Gmeiner, *Synthesis*, 2281 (2001).
    (b) J. Felding, J. Kristensen, T. Bjerregaard, L. Sander, P. Vedsø and M. Begtrup, *J. Org. Chem.*, **64**, 4196 (1999).
50. H. Kromann, F. A. Slok, T. N. Johansen and P. Krogsgaard-Larsen, *Tetrahedron*, **57**, 2195 (2001).
51. (a) M. Abarbri, F. Dehmel and P. Knochel, *Tetrahedron Lett.*, **40**, 7449 (1999).
    (b) G. Varchi, A. E. Jensen, W. Dohle, A. Ricci, G. Cahiez and P. Knochel, *Synlett*, 477 (2001).
52. J. L. Leazer Jr., R. Cvetovich, F.-R. Tsay, U. Dolling, T. Vickery and D. Bachert, *J. Org. Chem.*, **68**, 3695 (2003).
53. (a) E. C. Ashby and D. M. Al-Ferki, *J. Organomet. Chem.*, **390**, 275 (1990).
    (b) M. C. Jones, *Plant Oper. Prog.*, **8**, 200 (1989).
    (c) J. Broeke, B.-J. Deelman and G. van Koten, *Tetrahedron Lett.*, **42**, 8085 (2001).
    (d) P. Pinho, D. Guijarro and P. Andersson, *Tetrahedron*, **54**, 7897 (1998).
    (e) F. A. R. Kaul, G. T. Puchta, H. Schneider, M. Grosche, D. Mihalios and W. A. Herrmann, *J. Organomet. Chem.*, **621**, 184 (2001).
54. H. Nishiyama, K. Isaka, K. Itoh, K. Ohno, H. Nagase, K. Matsumoto and H. Yoshiwara, *J. Org. Chem.*, **57**, 407 (1992).
55. A. Krasovskiy and P. Knochel, *Angew. Chem., Int. Ed.*, **43**, 3333 (2004).
56. D. Hauk, S. Lang and A. Murso, *Org. Process Res. Dev.*, **10**, 733 (2006).
57. F. Kopp, A. Krasovskiy and P. Knochel, *Chem. Commun.*, 2288 (2004).
58. X.-J. Wang, X. Sun, L. Zhang, Y. Xu, D. Krishnamurthy and C. H. Senanayake, *Org. Lett.*, **8**, 305 (2006).
59. K. Kitagawa, A. Inoue, H. Shinokubo and K. Oshima, *Angew. Chem., Int. Ed.*, **39**, 2481 (2000).
60. F. Trécourt, G. Breton, V. Bonnet, F. Mongin, F. Marsais and G. Quéguiner, *Tetrahedron Lett.*, **40**, 4339 (1999).
61. H. Ren and P. Knochel, *Chem. Commun.*, 726 (2006).
62. N. Boudet and P. Knochel, *Org. Lett.*, **8**, 3737 (2006).
63. S. Kii, A. Akao, T. Iida, T. Mase and N. Yasuda, *Tetrahedron Lett.*, **47**, 1877 (2006).
64. C. Christophersen, M. Begtrup, S. Ebdrup, H. Petersen and P. Vedsø, *J. Org. Chem.*, **68**, 9513 (2003).
65. A. Spiess, G. Heckmann and T. Bach, *Synlett*, 131 (2004).
66. (a) H. Ila, O. Baron, A. J. Wagner and P. Knochel, *Chem. Commun.*, 583 (2006).
    (b) H. Ila, O. Baron, A. J. Wagner and P. Knochel, *Chem. Lett.*, **35**, 2 (2006).
67. J. Thibonnet, V. A. Vu, L. Bérillon and P. Knochel, *Tetrahedron*, **58**, 4787 (2000).
68. I. Sapountzis, W. Dohle and P. Knochel, *Chem. Commun.*, 2068 (2001).
69. V. A. Vu, I. Marek and P. Knochel, *Synthesis*, 1797 (2003).

70. T. Satoh, K. Takano, H. Ota, H. Someya, K. Matsuda and M. Koyama, *Tetrahedron*, **54**, 5557 (1998).
71. M. Watanabe, M. Nakamura and T. Satoh, *Tetrahedron*, **61**, 4409 (2005).
72. (a) H. Ren, A. Krasovskiy and P. Knochel, *Org. Lett.*, **6**, 4215 (2004).
    (b) H. Ren, A. Krasovskiy and P. Knochel, *Chem. Commun.*, 543 (2005).
73. H. Shinokubo, H. Miki, T. Yokoo, K. Oshima and K. Utimoto, *Tetrahedron*, **51**, 11681 (1995).
74. T. Satoh, A. Kondo and J. Musashi, *Tetrahedron*, **60**, 5453 (2004).
75. R. W. Hoffmann, B. Holzer, O. Knopff and K. Harms, *Angew. Chem., Int. Ed.*, **39**, 3072 (2000).
76. R. W. Hoffmann and P. G. Nell, *Angew. Chem., Int. Ed.*, **38**, 338 (1999).
77. A. Inoue, J. Kondo, H. Shinokubo and K. Oshima, *Chem. Eur. J.*, **8**, 1730 (2002).
78. R. W. Hoffmann, O. Knopff and A. Kusche, *Angew. Chem., Int. Ed.*, **39**, 1462 (2000).
79. O. Knopff, H. C. Stiasny and R. W. Hoffmann, *Organometallics*, **23**, 705 (2004).
80. R. W. Hoffmann, *Chem. Soc. Rev.*, **32**, 225 (2003).
81. (a) R. W. Hoffmann and A. Kusche, *Chem. Ber.*, **127**, 1311 (1994).
    (b) M. Rottländer, L. Boymond, L. Bérillon, A. Leprêtre, G. Varchi, S. Avolio, H. Laaziri, G. Quéguiner, A. Ricci, G. Cahiez and P. Knochel, *Chem. Eur. J.*, **6**, 767 (2000).
    (c) S. Avolio, C. Malan, I. Marek and P. Knochel, *Synlett*, 1820 (1999).
82. M. S. Baird, A. V. Nizovtsev and I. G. Bolesov, *Tetrahedron*, **58**, 1581 (2002).
83. (a) V. A. Vu, I. Marek and P. Knochel, *Angew. Chem., Int. Ed.*, **41**, 351 (2002).
    (b) J. Kondo, A. Inoue, H. Shinokubo and K. Oshima, *Angew. Chem., Int. Ed.*, **40**, 2085 (2001).
84. F. Kopp, G. Sklute, K. Polborn, I. Marek and P. Knochel, *Org. Lett.*, **7**, 3789 (2005).
85. (a) F. F. Fleming and B. C. Shook, *Tetrahedron*, **58**, 1 (2002).
    (b) S. Arseniyadis, K. S. Kyler and D. S. Watt, *Org. React.*, **31**, 1 (1984).
86. A. J. Fatiadi, in *The Chemistry of Triple-Bonded Functional Groups. Supplement C* (Eds. S. Patai and Z. Rappoport), Wiley, New York, 1983, p. 1157.
87. F. F. Fleming, Z. Zhang, W. Liu and P. Knochel, *J. Org. Chem.*, **70**, 2200 (2005).
88. A. H. Stoll, A. Krasovskiy and P. Knochel, *Angew. Chem., Int. Ed.*, **45**, 606 (2006).
89. G. Varchi, C. Kofink, D. M. Lindsay, A. Ricci and P. Knochel, *Chem. Commun.*, 396 (2003).
90. S. Kato, N. Nonoyama, K. Tomimoto and T. Mase, *Tetrahedron Lett.*, **43**, 7315 (2002).
91. N. Ono, *The Nitro Group in Organic Synthesis*, Wiley-VCH, New York, 2001.
92. (a) I. Sapountzis and P. Knochel, *Angew. Chem., Int. Ed.*, **41**, 1610 (2002).
    (b) I. Sapountzis, H. Dube, R. Lewis and P. Knochel, *J. Org. Chem.*, **70**, 2445 (2005).
93. (a) S. Bräse, *Acc. Chem. Res.*, **37**, 804 (2004).
    (b) D. B. Kimball and M. M. Haley, *Angew. Chem., Int. Ed.*, **41**, 3338 (2002).
    (c) A. de Meijere, P. von Zezschwitz, H. Nuske and B. Stulgies, *J. Organomet. Chem.*, **653**, 129 (2002).
94. W. B. Wan, R. C. Chiechi, T. J. R. Weakley and M. M. Haley, *Eur. J. Org. Chem.*, **18**, 3485 (2001).
95. C.-Y. Liu and P. Knochel, *Org. Lett.*, **7**, 2543 (2005).
96. For the utility of neopentyl organometallics in zinc and copper chemistry, see:
    (a) P. Jones, C. K. Reddy and P. Knochel, *Tetrahedron*, **54**, 1471 (1998).
    (b) P. Jones and P. Knochel, *J. Chem. Soc., Perkin Trans. 1*, 3117 (1997).
    (c) F. F. Kneisel and P. Knochel, *Synlett*, 1799 (2002).
97. C.-Y. Liu, H. Ren and P. Knochel, *Org. Lett.*, **8**, 617 (2006).
98. T. Delacroix, L. Bérillon, G. Cahiez and P. Knochel, *J. Org. Chem.*, **65**, 8108 (2000).
99. O. Baron and P. Knochel, *Angew. Chem., Int. Ed.*, **44**, 3133 (2005).
100. L. Boymond, M. Rottlander, G. Cahiez and P. Knochel, *Angew. Chem., Int. Ed.*, **37**, 1701 (1998).
101. R. G. Franzen, *Tetrahedron*, **56**, 685 (2000).
102. (a) V. Snieckus, *Chem. Rev.*, **90**, 879 (1990).
    (b) E. J.-G. Anctil and V. Snieckus, *J. Organomet. Chem.*, **653**, 150 (2002).
103. V. Bonnet, F. Mongin, F. Trécourt and G. Quéguinér, *J. Chem. Soc., Perkin Trans. 1*, 4245 (2000).
104. A. B. Holmes and C. N. Sporikou, *Org. Synth.*, **65**, 61 (1987).
    (b) H. J. Bestman, T. Brosche, K. H. Koschatzky, K. Michaelis, H. Platz, K. Roth, J. Suess, O. Vostrowsky and W. Knauf, *Tetrahedron Lett.*, **23**, 4007 (1982).

(c) L. Poncini, *Bull. Soc. Chim. Belg.*, **92**, 215 (1983).
105. (a) M.-X. Zhang and P. E. Eaton, *Angew. Chem., Int. Ed.*, **41**, 2169 (2002).
(b) Y. Kondo, Y. Akihiro and T. Sakamoto, *J. Chem. Soc., Perkin Trans. 1*, 2331 (1996).
(c) P. E. Eaton, C. H. Lee and Y. Xiong, *J. Am. Chem. Soc.*, **111**, 8016 (1989).
(d) P. E. Eaton, M.-X. Zhang, N. Komiya, C.-G., Yang, I. Steele, and R. Gilardi, *Synlett*, 1275 (2003).
(e) P. E. Eaton and R. M. Martin, *J. Org. Chem.*, **53**, 2728 (1988).
(f) M. Shilai, Y. Kondo and T. Sakamoto, *J. Chem. Soc., Perkin Trans. 1*, 442 (2001).
106. P. E. Eaton and M.-X. Zhang, *Angew. Chem., Int. Ed.*, **41**, 2169 (2002).
107. W. Schlecker, A. Huth, E. Ottow and J. Mulzer, *J. Org. Chem.*, **60**, 8414 (1995).
108. Y. Kondo, A. Yoshida and T. Sakamoto, *J. Chem. Soc., Perkin Trans. 1*, 2331 (1996).
109. A. Dinsmore, D. G. Billing, K. Mandy, J. P. Michael, D. Mogano and S. Patil, *Org. Lett.*, **6**, 293 (2004).
110. K. W. Henderson and W. J. Kerr, *Chem. Eur. J.*, **7**, 3431 (2001).
111. (a) D. Benafoux, M. Bordeau, C. Biran and J. Dunogues, *J. Organomet. Chem.*, **993**, 27 (1995).
(b) G. Lesseue, R. Tripoli, P. Cazeau, C. Brian and M. Bordeau, *Tetrahedron Lett.*, **40**, 4037 (1999).
112. (a) D. A. Evans and S. G. Nelson, *J. Am. Chem. Soc.*, **119**, 6452 (1997).
(b) K. W. Hendersson, W. J. Kerr and J. H. Mair, *Tetrahedron*, **58**, 4573 (2002).
113. A. Krasovskiy, V. Krasovskaya and P. Knochel, *Angew. Chem., Int. Ed.*, **45**, 2958 (2006).
114. W. Lin, O. Baron and P. Knochel, *Org. Lett.*, **8**, 5673 (2006).
115. For reviews, see:
(a) A. Sauer and R. Huisgen, *Angew. Chem.*, **72**, 294 (1960).
(b) R. W. Hoffmann, *Dehydrobenzene and Cycloalkynes*, Academic Press, New York, 1967.
(c) L. Castedo and E. Guitian, in *Studies in Natural Products Chemistry*, Vol. 3, Part B (Ed. A. U. Rahman), Elsevier, Amsterdam, 1989, p. 417.
(d) S. V. Kessar, in *Comprehensive Organic Synthesis*, Vol. 4 (Eds. B. M. Trost, I. Fleming and H. F. Semmelhack), Pergamon, Oxford, 1991, p. 483.
(e) R. F. C. Brown, *Synlett*, 9 (1993).
(f) W. Sander, *Acc. Chem. Res.*, **32**, 669 (1999).
(g) H. Pelissier and M. Santelli, *Tetrahedron*, **59**, 701 (2003).
116. (a) I. Sapountzis, W. Lin, M. Fischer and P. Knochel, *Angew. Chem., Int. Ed.*, **43**, 4364 (2004).
(b) W. Lin, I. Sapountzis and P. Knochel, *Angew. Chem., Int. Ed.*, **44**, 4258 (2005).
(c) W. Lin, F. Ilgen and P. Knochel, *Tetrahedron Lett.*, **47**, 1941 (2006).
117. N. Hartmann and M. Niemeyer, *Synth. Commun.*, **31**, 3839 (2001).
118. B. Bogdanović, P. Bons, S. Konstantinović, M. Schwickardi and U. Westeppe, *Chem. Ber.*, **126**, 1371 (1993).
119. (a) Y. Kai, N. Kanegisa, K. Miki, N. Kasai, K. Mashima, H. Yasuda and A. Nakamura, *Chem. Lett.*, 3976 (1982).
(b) R. D. Rieke and H. Xiong *J. Org. Chem.*, **56**, 3109 (1991).
120. Y. Gao and F. Sato, *J. Chem. Soc., Chem. Commun.*, 659 (1995).
121. M. Cai, J. Xia and G. Chen, *J. Organomet. Chem.*, **689**, 2531 (2004).
122. (a) M. Cai, J. Xia, and W. Hao, *Heteroat. Chem.*, **16**, 65 (2005).
(b) M. Cai, W. Hao, H. Zhao and J. Xia, *J. Organomet. Chem.*, **689**, 3593 (2004).
(c) M. Cai, J. Xia and G. Chen, *J. Organomet. Chem.*, **689**, 1714 (2004).
(d) H. Shinokubo, H. Miki, T. Yokoo, K. Oshima and K. Utimoto, *Tetrahedron*, **51**, 11681 (1995).
123. N. B. Victorov and L. M. Zubritskii, *Russ. J. Org. Chem.*, **33**, 1706 (1997).
124. F. Sato and H. Vrade, 'Hydromagnesiation of alkenes and alkynes', in *Grignard Reagents— New Developments* (Ed. H. G. Richey, Jr.), Wiley, New York, 2000, pp. 65–105.
125. P. Knochel, A. Krasovskiy and I. Sapountzis, 'Polyfunctional magnesium organometallics for organic synthesis', Chapter 4, in *Handbook of Functionalized Organometallics* (Ed. P. Knochel.), Wiley-VCH, 2005.
126. P. Knochel, I. Sapountzis and N. Gommermann, 'Carbon–carbon bond-forming reactions mediated by organomagnesium reagents', in *Metal-Catalyzed Cross-Coupling Reactions* (Eds. A. de Meijere and F. Diederich), 2nd edn., Wiley-VCH, 2004, pp. 619–670.

127. J. Terao, Y. Naitoh, H. Kuniyasu and N. Kambe, *Chem. Lett.*, **32**, 890 (2003).
128. J. Terao, H. Hideyuki, A. Ikumi, H. Kuniyasu and N. Kambe, *J. Am. Chem. Soc.*, **124**, 4222 (2002).
129. A. C. Frisch, N. Shaikh, A. Zapf and M. Beller, *Angew. Chem., Int. Ed.*, **41**, 4056 (2002).
130. A. C. Frisch, F. Rataboul, A. Zapf and M. Beller, *J. Organomet. Chem.*, **687**, 403 (2003).
131. A. Sofia, E. Karlström, K. Itami and J.-E. Bäckvall, *J. Org. Chem.*, **64**, 1745 (1999).
132. J. A. Miller, *Tetrahedron Lett.*, **43**, 7111 (2002).
133. G. Cahiez, C. Chaboche and M. Jezequel, *Tetrahedron*, **56**, 2733 (2000).
134. D. H. Burns, J. D. Miller, H.-K. Chan and M. O. Delaney, *J. Am. Chem. Soc.*, **119**, 2125 (1997).
135. J. Terao, A. Ikumi, H. Kuniyasu and N. Kambe, *J. Am. Chem. Soc.*, **125**, 5646 (2003).
136. J. Terao, H. Todo, S. A. Begum, H. Kuniyasu and N. Kambe, *Angew. Chem., Int. Ed.*, **46**, 2086 (2007).
137. J. G. Donkervoort, J. L. Vicario, J. T. B. H. Jastrzebski, R. A. Gossage, G. Cahiez and G. van Koten, *J. Organomet. Chem.*, **558**, 61 (1998).
138. C. C. Kofink and P. Knochel, *Org. Lett.*, **8**, 4121 (2006).
139. (a) T. Tsuji, H. Yorimitsu and K. Oshima, *Angew. Chem., Int. Ed.*, **41**, 4137 (2002).
    (b) H. Ohmiya, T. Tsuji, Y. Hideki and K. Oshima, *Chem. Eur. J.*, **10**, 5640 (2004).
140. H. Ohmiya, H. Yorimitsu and K. Oshima, *J. Am. Chem. Soc.*, **128**, 1886 (2006).
141. H. Ohmiya, T. Tsuji, H. Yorimitsu and K. Oshima, *Chem. Eur. J.*, **10**, 5640 (2004).
142. M. Nakamura, K. Matsuo, S. Ito and E. Nakamura, *J. Am. Chem. Soc.*, **126**, 3686 (2004).
143. (a) T. Nagano and T. Hayashi, *Org. Lett.*, **6**, 1297, (2004).
    (b) G. Cahiez, V. Habiak, C. Duplais and A. Moyeux, *Angew. Chem., Int. Ed.*, **46**, 4364 (2007).
144. R. B. Bedford, D. W. Bruce, R. M. Frost, J. W. Goodby and M. Hird, *Chem. Commun.*, 2822 (2004).
145. M. Ruben and A. Fürstner, *Angew. Chem., Int. Ed.*, **43**, 3955 (2004).
146. T. Nagano and T. Hayashi, *Chem. Lett.*, **34**, 1152 (2005).
147. (a) H. M. I. Osborn, J. B. Sweeney and W. Howson, *Tetrahedron Lett.*, **35**, 2739 (1994).
    (b) T. Gajda, A. Napieraj, K. Osowska-Pacewicka, S. Zawadzki and A. Zwierzak, *Tetrahedron*, **53**, 4935 (1997).
    (c) K. Osowska-Pacewicka and A. Zwierzak, *Synthesis*, 333 (1996).
148. V. G. Nenajdenko, A. S. Karpov and E. S. Balenkova, *Tetrahedron: Asymmetry*, **12**, 2517 (2001).
149. J. E. Baldwin, C. N. Farthing, A. T. Russell, C. J. Schofield and A. C. Spivey, *Tetrahedron Lett.*, **37**, 3761 (1996).
150. S. G. Nelson, Z. Wan and M. A. Stan, *J. Org. Chem.*, **67**, 4680 (2002).
151. D. M. Hodgson, M. J. Fleming and S. J. Stanway, *J. Am. Chem. Soc.*, **126**, 12250 (2004).
152. T.-Y. Luh, *J. Organomet. Chem.*, **653**, 209 (2002) and references therein.
153. L.-F. Huang, C.-H. Huang, B. Stulgies, A. de Meijere and T.-Y. Luh, *Org. Lett.*, **5**, 4489 (2003).
154. (a) T.-M. Yuan, S.-M. Yeh, Y.-T. Hsieh and T.-Y. Luh, *J. Org. Chem.*, **59**, 8192 (1994).
    (b) W.-L. Cheng, Y.-J. Shaw, S.-M. Yeh, P. P. Kanakamma, Y.-H. Chen, C. Chen, J.-C. Shieu, S.-J. Yiin, G.-H. Lee, Y. Wang and T.-Y. Luh, *J. Org. Chem.*, **64**, 532 (1999).
155. J.-W. Huang, C.-D. Chen and M.-K. Leung, *Tetrahedron Lett.*, **40**, 8647 (1999).
156. A. R. Katritzky, X. Lan, J. Z. Yang and O. V. Denisko, *Chem. Rev.*, **98**, 409 (1998).
157. (a) A. R. Katritzky, S. K. Nair and G. Qiu, *Synthesis*, 199 (2002).
158. A. R. Katritzky, X.-L. Cui, B. Yang and P. J. Steel, *Tetrahedron Lett.*, **39**, 1698 (1998).
159. H. Poerwono, K. Higashiyama and H. Takahashi, *J. Org. Chem.*, **63**, 2711 (1998).
160. N. Gommermann, C. Koradin and P. Knochel, *Synthesis*, 2143 (2002).
161. M. Baruah and M. Bols, *J. Chem. Soc., Perkin Trans. 1*, 509 (2002).
162. T. Shintou, W. Kikuchi and T. Mukaiyama, *Chem. Lett.*, **32**, 676 (2003).
163. R. W. Hoffmann and B. Hoelzer, *J. Am. Chem. Soc.*, **124**, 4204 (2002).
164. D. Dakternieks, K. Dunn, D. J. Henry, C. H. Schiesser and E. R. Tiekink, *Organometallics*, **18**, 3342 (1999).
165. C.-G. Dong and Q.-S. Hu, *Angew. Chem., Int. Ed.*, **45**, 2289 (2006).
166. (a) J. Huang and S. P. Nolan, *J. Am. Chem. Soc.*, **121**, 9889 (1999).

(b) V. P. W. Böhm, T. Weskamp, C. W. K. Gstöttmayr and W. A. Herrmann, *Angew. Chem., Int. Ed.*, **39**, 1602 (2000).
167. M. G. Organ, M. Abdel-Hadi, S. Avola, N. Hadei, J. Nasielski, C. O'Brien and C. Valente, *Chem. Eur. J.*, **13**, 150 (2007).
168. W. A. Herrmann, V. P. W. Böhm, C. W. K. Gstöttmayr, M. Grosche, C.-P. Reisinger and T. Weskamp, *J. Organomet. Chem.*, **617–618**, 616 (2001).
169. N. Yoshikai, H. Mashima and E. Nakamura, *J. Am. Chem. Soc.*, **127**, 17978 (2005).
170. R. Martin and S. L. Buchwald, *J. Am. Chem. Soc.*, **129**, 3844 (2007).
171. (a) G. Y. Li, *J. Organomet. Chem.*, **653**, 63 (2002).
    (b) L. Ackermann and A. Althammer, *Org. Lett.*, **8**, 3457 (2006).
172. L. Ackermann, R. Born, J. H. Spatz and D. Meyer, *Angew. Chem., Int. Ed.*, **44**, 7216 (2005).
173. F. Mongin, L. Mojovic, B. Guillamet, F. Trécourt and G. Quéguiner, *J. Org. Chem.*, **67**, 8991 (2002).
174. G. Y. Li and W. J. Marshall, *Organometallics*, **21**, 590 (2002).
175. J. Miller and R. P. Farrell, *Tetrahedron Lett.*, **39**, 7275 (1998).
176. G. Cahiez, F. Lepifre and P. Ramiandrasoa, *Synthesis*, 2138 (1999).
177. M. Rueping and W. Ieawsuwan, *Synlett*, 247 (2007).
178. (a) A. H. Roy and J. F. Hartwig, *J. Am. Chem. Soc.*, **125**, 8704 (2003).
    (b) C.-H. Cho, H.-S. Yun and K. Park, *J. Org. Chem.*, **68**, 3017 (2003).
    (c) C.-H. Cho, M. Sun, Y.-S. Seo, C.-B. Kim and K. Park, *J. Org. Chem.*, **70**, 1482 (2005).
179. J. Clayden, J. J. A. Cooney and M. Julia, *J. Chem. Soc., Perkin Trans. 1*, 7 (1995).
180. A. M. Roy and J. F. Hartwig, *J. Am. Chem. Soc.*, **125**, 8704 (2003).
181. J. A. Miller, *Tetrahedron Lett.*, **42**, 6991 (2001).
182. J. W. Dankwardt, *Angew. Chem., Int. Ed.*, **43**, 2428 (2004).
183. R. R. Milburn and V. Snieckus, *Angew. Chem., Int. Ed.*, **43**, 888 (2004).
184. T. K. Macklin and V. Snieckus, *Org. Lett.*, **7**, 2519 (2005).
185. T. Kamikawa and T. Hayashi, *Tetrahedron Lett.*, **38**, 7087 (1997).
186. (a) V. Bonnet, F. Mongin, F. Trécourt, G. Quéguiner and P. Knochel, *Tetrahedron Lett.*, **42**, 5717 (2001).
    (b) V. Bonnet, F. Mongin, F. Trécourt, G. Quéguiner and P. Knochel, *Tetrahedron*, **58**, 4429 (2002).
187. N. A. Bumagin and E. V. Luzikova, *J. Organomet. Chem.*, **532**, 271 (1997).
188. Review on advances in iron-catalyzed cross-coupling reactions: A. Fürstner and R. Martin, *Chem. Lett.*, **34**, 624 (2005).
189. (a) A. Fürstner, A. Leitner, M. Mendez and H. Krause, *J. Am. Chem. Soc.*, **124**, 13856 (2002).
    (b) B. Scheiper, M. Bonnekessel, H. Krause and A. Fürstner, *J. Org. Chem.*, **69**, 3943 (2004).
    (c) A. Fürstner and A. Leitner, *Angew. Chem., Int. Ed.*, **41**, 609 (2002).
    (d) G. Seidel, D. Laurich and A. Fürstner, *J. Org. Chem.*, **69**, 3950 (2004).
190. J. Quintin, X. Franck, R. Hocquemiller and B. Figadère, *Tetrahedron Lett.*, **43**, 3547 (2002).
191. (a) T. Nagano and T. Hayashi, *Org. Lett.*, **7**, 491 (2005).
    (b) G. Cahiez, C. Chaboche, F. Mahuteau-Betzer and M. Ahr, *Org. Lett.*, **7**, 1943 (2005).
192. H. Guo, K.-I. Kanno and T. Takahashi, *Chem. Lett.*, **33**, 1356 (2004).
193. (a) S. Tasler and B. H. Lipshutz, *J. Org. Chem.*, **68**, 1190 (2003).
    (b) B. H. Lipshutz, S. Tasler, W. Chrisman, B. Spliethoff and B. Tesche, *J. Org. Chem.*, **68**, 1177 (2003).
194. Y.-H. Cho, A. Kina, T. Shimada and T. Hayashi, *J. Org. Chem.*, **69**, 3811 (2004).
195. P. Ramiandrasoa, B. Brehon, A. Thivet, M. Alami and G. Cahiez, *Tetrahedron Lett.*, **38**, 2447 (1997).
196. M. Alami, P. Ramiandrasoa and G. Cahiez, *Synlett*, 325 (1998).
197. M. Seck, X. Franck, R. Hocquemiller, B. Figadere, J.-F. Peyrat, O. Provot, J.-D. Brion and M. Alami, *Tetrahedron Lett.*, **45**, 1881 (2004).
198. (a) W. Dohle, F. Kopp, G. Cahiez and P. Knochel, *Synlett*, 1901 (2001).
    (b) G. Cahiez and H. Avedissian, *Synthesis*, 1199 (1998).
    (c) M. Dos Santos, X. Franck, R. Hocquemiller, B. Figadere, J.-F. Peyrat, O. Provot, J.-D. Brion and M. Alami, *Synlett*, 2697 (2004).
199. K. Itami, S. Higashi, M. Mineno and J.-I. Yoshida, *Org. Lett.*, **7**, 1219 (2005).
200. N. Yoshikai, H. Mashima and E. Nakamura, *J. Am. Chem. Soc.*, **127**, 17978 (2005).

201. A. S. E. Karlstroem, M. Roenn, A. Thorarensen and J.-E. Bäckvall, *J. Org. Chem.*, **63**, 2517 (1998).
202. M. Die and M. Nakata, *Synlett*, 1511 (2001).
203. C. A. Busacca, M. C. Erikkson and R. Fiaschisi, *Tetrahedron Lett.*, **40**, 3101 (1999).
204. A. S. E. Karlstroem, K. Itami and J.-E. Bäckvall, *J. Org. Chem.*, **64**, 1745 (1999).
205. J. A. Miller, *Tetrahedron Lett.*, **43**, 7111 (2002).
206. G. C. Lloyd-Jones and C. P. Butts, *Tetrahedron*, **54**, 901 (1998).
207. H. Horibe, K. Kazuta, M. Kotoku, K. Kondo, H. Okuno, Y. Murakami and T. Aoyama, *Synlett*, 2047 (2003).
208. A. Krasovskiy, A. Tishkov, V. del Amo, H. Mayr and P. Knochel, *Angew. Chem., Int. Ed.*, **45**, 5010 (2006).
209. A. Kar and N. P. Argade, *Synthesis*, 2995 (2005).
210. J.-E. Bäckvall, E. S. M. Persson and A. Brombrun, *J. Org. Chem.*, **59**, 4126 (1994).
211. A. Yanagisawa, N. Nomura and H. Yamamoto, *Tetrahedron*, **50**, 6017 (1994).
212. M. T. Didiuk, J. P. Morken and A. H. Hoveyda, *J. Am. Chem. Soc.*, **117**, 7273 (1995).
213. B. Breit, P. Demel and C. Studte, *Angew. Chem., Int. Ed.*, **43**, 3786 (2004).
214. (a) J. H. Smitrovich and K. A. Woerpel, *J. Am. Chem. Soc.*, **120**, 12998 (1998).
    (b) J. H. Smitrovich and K. A. Woerpel, *J. Org. Chem.*, **65**, 1601 (2000).
215. S.-K. Kang, D.-G. Cho, C.-H. Park, E.-Y. Namkoong and J.-S. Shin, *Synth. Commun.*, **25**, 1659 (1995).
216. H. Yasui, K. Mizutani, H. Yorimitsu and K. Oshima, *Tetrahedron*, **62**, 1410 (2006).
217. (a) Y. Kobayashi, K. Nakata and T. Ainai, *Org. Lett.*, **7**, 183 (2005).
    (b) M. Ito, M. Matsuumi, M. G. Murugesh and Y. Kobayashi, *J. Org. Chem.*, **66**, 5881 (2001).
218. (a) M. Kimura, T. Yamazaki, T. Kitazume and T. Kubota, *Org. Lett.*, **6**, 4651 (2004).
    (b) G. J. Meuzelaar, A. S. E. Karlstrom, M. Van Klaveren, E. Persson, A. Del Villar, G. Van Koten and J.-E. Bäckvall, *Tetrahedron*, **56**, 2895 (2000).
    (c) B. Heckmann, C. Mioskowski, R. K. Bhatt and J. R. Falck, *Tetrahedron Lett.*, **37**, 1421 (1996).
219. (a) A. Alexakis, C. Malan, L. Lea, C. Benhaim and X. Fournioux, *Synlett*, 927 (2001).
    (b) K.-G. Chung, Y. Miyake and S. Uemura, *J. Chem. Soc., Perkin Trans. 1*, 2725 (2000).
220. (a) C. A. Falciola, K. Tissot-Croset and A. Alexakis, *Angew. Chem., Int. Ed.*, **45**, 5995 (2006).
    (b) K. Tissot-Croset, D. Polet and A. Alexakis, *Angew. Chem., Int. Ed.*, **43**, 2426 (2004).
    (c) K. Tissot-Croset and A. Alexakis, *Tetrahedron Lett.*, **45**, 7375 (2004).
    (d) B. L. Feringa, *Acc. Chem. Res.*, **33**, 346 (2000).
221. K. Geurts, S. P. Fletcher and B. L. Feringa, *J. Am. Chem. Soc.*, **128**, 15572 (2006).
222. R. K. Dieter, *Tetrahedron*, **55**, 4177 (1999).
223. C. Duplais, F. Bures, I. Sapountzis, T. J. Korn, G. Cahiez and P. Knochel, *Angew. Chem., Int. Ed.*, **43**, 2968 (2004).
224. O. Labeeuw, P. Phansavath and J.-P. Genet, *Tetrahedron Lett.*, **45**, 7107 (2004).
225. (a) H. Maeda, K. Takahashi and H. Ohmori, *Tetrahedron*, **54**, 12233 (1998).
    (b) H. Maeda, N. Hino, Y. Yamauchi and H. Ohmori, *Chem. Pharm. Bull.*, **48**, 1196 (2000).
226. L. De Luca, G. Giacomelli and A. Porcheddu, *Org. Lett.*, **3**, 1519 (2001).
227. R. C. Klix, S. A. Chamberlin, A. V. Bhatia, D. A. Davis, T. K. Hayes, F. G. Rojas and R. W. Koops, *Tetrahedron Lett.*, **36**, 1791 (1995).
228. B. F. Bonini, M. Comes-Franchini, M. Fochi, G. Mozzanti, A. Ricci and G. Varchi, *Synlett*, 1013 (1998).
229. A. G. Myers and T. Yoon, *Tetrahedron Lett.*, **36**, 9429 (1995).
230. (a) H.-R. Tseng and T.-Y. Luh, *J. Org. Chem.*, **61**, 8685 (1996).
    (h) H.-R. Tseng, C.-F. Lee, L.-M. Yang and T.-Y. Luh, *J. Org. Chem.*, **64**, 8582 (1999).
231. A. Fürstner and M. Mendez, *Angew. Chem., Int. Ed.*, **42**, 5355 (2003).
232. A. R. Katritzky, A. A. A. Abdel-Fattah and M. Wang, *J. Org. Chem.*, **67**, 7526 (2002).
233. (a) X. Liu and J. M. Fox, *J. Am. Chem. Soc.*, **128**, 5600 (2006).
    (b) S. Simaan and I. Marek, *Org. Lett.*, **9**, 2569 (2007).
234. A. H. Hoveyda, N. M. Heron and J. A. Adams, *Grignard Reagents*, Wiley, Chichester, 2000, pp. 107–137.
235. J. Terao, S. Nii, F. A. Chowdhury, A. Nakamura and N. Kambe, *Adv. Synth. Catal.*, **346**, 905 (2004).
236. K. Mizutani, H. Shinokubo and K. Oshima, *Org. Lett.*, **5**, 3959 (2003).

237. H. Ohmiya, K. Wakabayashi, H. Yorimitsu and K. Oshima, *Tetrahedron*, **62**, 2207 (2006).
238. H. Someya, H. Ohmiya, H. Yorimitsu and K. Oshima, *Org. Lett.*, **9**, 1565 (2007).
239. Y. Fujii, J. Terao, H. Kuniyasu and N. Kambe, *J. Organomet. Chem.*, **692**, 375 (2007).
240. J. de Armas and A. H. Hoveyda, *Org. Lett.*, **3**, 2097 (2001).
241. S. Nii, J. Terao and N. Kambe, *J. Org. Chem.*, **69**, 573 (2004).
242. H. Watabe, J. Terao and N. Kambe, *Org. Lett.*, **3**, 1733 (2001).
243. H. Shinokubo and K. Oshima, *Eur. J. Org. Chem.*, **10**, 2081 (2004).
244. F. Lopez, A. J. Minnaard and B. L. Feringa, *Acc. Chem. Res.*, **40**, 179 (2007).
245. G. Varchi, A. Ricci, G. Cahiez and P. Knochel, *Tetrahedron*, **56**, 2727 (2000).
246. K. A. Hansford, J. E. Dettwiler and W. D. Lubell, *Org. Lett.*, **5**, 4887 (2003).
247. G. J. Leotta, L. E. Overman and G. S. Welmaker, *J. Org. Chem.*, **59**, 1946 (1994).
248. M. Kikuchi, S. Niikura, N. Chiba, N. Terauchi and M. Asaoko, *Chem. Lett.*, **36**, 736 (2007).
249. N. Chinkov, N. Morlender-Vais and I. Marek, *Tetrahedron Lett.*, **43**, 6009 (2002).
250. F. F. Fleming, G. Wei, Z. Zhang and O. W. Steward, *Org. Lett.*, **8**, 4903 (2006).
251. P. A. Lander and L. S. Hegedus, *J. Am. Chem. Soc.*, **116**, 8126 (1994).
252. P. G. Andersson, H. E. Schink and K. Oesterlund, *J. Org. Chem.*, **63**, 8067 (1998).
253. A. Bongini, G. Cardillo, A. Mingardi and C. Tomasini, *Tetrahedron: Asymmetry*, **7**, 1457 (1996).
254. M. Kanai and K. Tomioka, *Tetrahedron Lett.*, **36**, 4275 (1995).
255. (a) F. Lopez, S. R. Harutyunyan, A. J. Minaard and B. L. Feringa, *J. Am. Chem. Soc.*, **126**, 12784 (2004).
     (b) A. Alexakis and C. Benhaim, *Eur. J. Org. Chem.*, **19**, 3221 (2002).
256. S. Wang, S. Ji and T. Loh, *J. Am. Chem. Soc.*, **129**, 276 (2007).
257. D. Martin, S. Kehrli, M. D'Augustin, H. Clavier, M. Mauduit and A. Alexakis, *J. Am. Chem. Soc.*, **128**, 8416 (2006).
258. F. F. Fleming and Q. Wang, *Chem. Rev.*, **103**, 2035 (2003).
259. (a) F. F. Fleming, V. Gudipati and O. W. Steward, *Org. Lett.*, **4**, 659 (2002).
     (b) F. F. Fleming, V. Gudipati and O. W. Steward, *Tetrahedron*, **59**, 5585 (2003).
     (c) F. F. Fleming, Q. Wang and O. W. Steward, *J. Org. Chem.*, **68**, 4235 (2003).
     (d) F. F. Fleming, Q. Wang and O. W. Steward, *Org. Lett.*, **2**, 1477 (2000).
260. F. F. Fleming, Z. Zhang, Q. Wang and O. W. Steward, *Angew. Chem., Int. Ed.*, **43**, 1126 (2004).
261. F. F. Fleming, Z. Zhang, Q. Wang and O. W. Steward, *J. Org. Chem.*, **68**, 7646 (2003).
262. C.-F. Yao, K.-H. Kao, T.-J. Liu, C.-M. Chu, Y. Wang, W.-C. Chen, Y.-M. Lin, W.-W. Lin, M.-C. Yan, J.-Y. Liu, M.-C. Chuang and J.-L. Shiue, *Tetrahedron*, **54**, 791 (1998).
263. G. Bartoli, M. Bosco, L. Sembri and E. Marcantoni, *Tetrahedron Lett.*, **35**, 8651 (1994).
264. S. V. Kolotuchin and A. I. Meyers, *J. Org. Chem.*, **65**, 3018 (2000).
265. E. Nakamura, K. Kubota and G. Sakata, *J. Am. Chem. Soc.*, **119**, 5457 (1997); mechanism: I. Marek and J.-F. Normant, *Chem. Rev.*, **96**, 3241 (1996).
266. I. Sapountzis, W. Lin, M. Fischer and P. Knochel, *Angew. Chem., Int. Ed.*, **43**, 4364 (2004).
267. Y. Shen and J. Yao, *J. Org. Chem.*, **61**, 8659 (1996).
268. (a) P. Forgione and A. G. Fallis, *Tetrahedron Lett.*, **41**, 11 (2000).
     (b) P. E. Tessier, A. J. Penwell, F. E. S. Souza and A. G. Fallis, *Org. Lett.*, **5**, 2989 (2003).
     (c) A. G. Fallis, *Acc. Chem. Res.*, **32**, 464 (1999).
269. H. Yorimitsu, J. Tang, K. Okada, H. Shinokubo and K. Oshima, *Chem. Lett.*, 11 (1998).
270. S. Nishimae, R. Inoue, H. Shimokubo and K. Oshima, *Chem. Lett.*, 785 (1998).
271. S. Ma and Z. Lu, *Adv. Synth. Catal.*, **348**, 1894 (2006).
272. E. Shirakawa. T. Yamagami, K. Takafumi, S. Yamaguchi and T. Hayashi *J. Am. Chem. Soc.*, **127**, 17164 (2005).
273. K. Murakami, H. Ohmiya, H. Yorimitsu and K. Oshima, *Org. Lett.*, **9**, 1569 (2007).
274. M. Hatano, T. Miyamoto and K. Ishihara, *Curr. Org. Chem.*, **11**, 127 (2007).
275. S. Yamazaki and S. Yamabe, *J. Org. Chem.*, **67**, 9346 (2002).
276. H.-J. Liu, K.-S. Shia, X. Shang and B.-Y. Zhu, *Tetrahedron*, **55**, 3803 (1999).
277. A. Krasovskiy, F. Kopp and P. Knochel, *Angew. Chem., Int. Ed.*, **45**, 497 (2006).
278. M. Hatano, S. Suzuki and K. Ishihara, *J. Am. Chem. Soc.*, **128**, 9998 (2006).
279. M. Hatano, T. Matsumura and K. Ishihara, *Org. Lett.*, **7**, 573 (2005).
280. S. Matsukawa, Y. Funabashi and T. Imamoto, *Tetrahedron Lett.*, **44**, 1007 (2003).
281. B. Weber and D. Seebach, *Tetrahedron*, **50**, 6117 (1994).

282. G. Verresha and A. Datta, *Tetrahedron Lett.*, **38**, 5223 (1997).
283. R. Shintani and G. C. Fu, *Angew. Chem., Int. Ed.*, **41**, 1057 (2002).
284. S. E. Denmark, N. Nakajama and O. J.-C. Nicaise, *J. Am. Chem. Soc.*, **116**, 8797 (1994).
285. R. Bloch, *Chem. Rev.*, **98**, 1407 (1998).
286. S. E. Denmark and O. J.-C. Nicaise, *Chem. Commun.*, 999 (1996).
287. H. Urabe, D. Shikanai, K. Arayama, T. Sato and R. Tanaka, *Chem. Lett.*, **36**, 556 (2007).
288. A. R. Katritzky, Q. Hong and Z. Yang, *J. Org. Chem.*, **60**, 3405 (1995).
289. A. R. Katritzky, H. Yang and S. K. Singh, *J. Org. Chem.*, **70**, 286 (2005).
290. T. Franz, M. Hein, U. Veith, V. Jäger, E.-M. Peters, K. Peters and H. G. von Schnering, *Angew. Chem., Int. Ed.*, **33**, 1298 (1994).
291. A. G. Steinig and D. M. Spero, *J. Org. Chem.*, **64**, 2406 (1999).
292. K. R. Muralidharan, M. K. Mokhallalati and L. N. Pridgen, *Tetrahedron Lett.*, **35**, 7489 (1994).
293. S. Roland and P. Mangeney, *Eur. J. Org. Chem.*, 611 (2000).
294. J. C. A. Hunt, C. Lloyd, C. J. Moody, A. M. Z. Slawin and A. K. Takle, *J. Chem. Soc., Perkin Trans. 1*, 3443 (1999).
295. A. Solladié-Cavallo and F. Bonne, *Tetrahedron: Asymmetry*, **7**, 171 (1996).
296. A. K. Saksena, V. M. Girijavallabhan, H. Wang, R. G. Lovey, F. Guenter, I. Mergelsberg and M. S. Puar, *Tetrahedron Lett.*, **45**, 8249 (2004).
297. P. Cali and M. Begtrup, *Synthesis*, 63 (2002).
298. (a) J. A. Ellman, T. D. Owens and T. P. Tang, *Acc. Chem. Res.*, **35**, 984 (2002).
    (b) J. A. Ellman, *Pure Appl. Chem.*, **75**, 39 (2003).
    (c) G. Liu, D. A. Cogan and J. A. Ellman, *J. Am. Chem. Soc.*, **119**, 9913 (1997).
299. (a) F. A. Davis and W. McCoull, *J. Org. Chem.*, **64**, 3396 (1999).
    (b) D. A. Cogan and J. A. Ellman, *J. Am. Chem. Soc.*, **121**, 268 (1999).
    (c) D. A. Cogan, G. Liu and J. A. Ellman, *Tetrahedron*, **55**, 8883 (1999).
300. J. P. McMahon and J. A. Ellman, *Org. Lett.*, **6**, 1645 (2004).
301. (a) I. Sapountzis and P. Knochel, *J. Am. Chem. Soc.*, **124**, 9390 (2002).
    (b) Review on the reaction between nitroarenes and Grignard reagents: A. Ricci and M. Fochi, *Angew. Chem., Int. Ed.*, **42**, 1444 (2003).
    (c) Review on modern amination methods: M. Kienle, S. R. Dubbaka, K. Brade and P. Knochel, *Eur. J. Org. Chem.*, **20**, 4166 (2007).
302. A. Ono, H. Sasaki and F. Yaginuma, *Chem. Ind. (London)*, 480 (1983).
303. (a) A. K. Feldman, B. Colasson and V. V. Fokin, *Org. Lett.*, **6**, 3897 (2004).
    (b) C. W. Tornøe, C. Christensen and M. Meldal, *J. Org. Chem.*, **67**, 3057 (2002).
    (c) V. V. Rostovtsev, L. G. Green, V. V. Fokin and K. B. Sharpless, *Angew. Chem., Int. Ed.*, **41**, 2596 (2002).
    (d) J. S. Yadav, B. V. Subba Reddy and V. Geetha, *Synlett*, 513 (2002).
    (e) G. Koebrich and P. Buck, *Chem. Ber.*, **103**, 1412 (1970).
304. F. Kopp, I. Sapountzis and P. Knochel, *Synlett*, 885 (2003).
305. W. Dohle, A. Staubitz and P. Knochel, *Chem. Eur. J.*, **9**, 5323 (2003) and references cited therein.
306. (a) I. Sapountzis and P. Knochel, *Angew. Chem., Int. Ed.*, **43**, 897 (2004).
    (b) P. Sinha and P. Knochel, *Synlett*, 3304 (2006).
307. (a) G. H. Coleman, *J. Am. Chem. Soc.*, **55**, 3001 (1933).
    (b) F. Klages, G. Neber and F. Kirchner, *Liebigs Ann. Chem.*, **547**, 25 (1941).
308. P. Sinha and P. Knochel, *Synlett*, 3304 (2006).
309. (a) M. Kitamura, T. Suga, S. Chiba and K. Narasaka, *Org. Lett.*, **6**, 4619 (2004).
    (b) M. Kitamura, S. Chiba and K. Narasaka, *Bull. Chem. Soc. Jpn.*, **76**, 1063 (2003).
    (c) H. Tsutsui, T. Ichikawa and K. Narasaka, *Bull. Chem. Soc. Jpn.*, **72**, 1869 (1999).
    (d) E. Erdik and T. Daskapan, *J. Chem. Soc., Perkin Trans. 1*, 3139 (1999).
    (e) E.-U. Würthwein and R. Weigmann, *Angew. Chem., Int. Ed.*, **26**, 923 (1987).
    (f) R. A. Hagopian, M. J. Therien and J. R. Murdoch, *J. Am. Chem. Soc.*, **106**, 5753 (1984).
310. M. J. Campbell and J. S. Johnson, *Org. Lett.*, **9**, 1521 (2007).
311. T. Murai and F. Asai, *J. Am. Chem. Soc.*, **129**, 780 (2007).
312. H. Yamamoto and K. Maruoka, *J. Org. Chem.*, **45**, 2739 (1980).
313. (a) A. Casarini, P. Dembech, D. Lazzari, E. Marini, G. Reginato, A. Ricci and G. Seconi, *J. Org. Chem.*, **58**, 5620 (1993).

12. Functionalized organomagnesium compounds: Synthesis and reactivity   593

(b) A. Alberti, F. Canè, P. Dembech, D. Lazzari, A. Ricci and G. Seconi, *J. Org. Chem.*, **61**, 1677 (1996).
(c) F. Canè, D. Brancaleoni, P. Dembech, A. Ricci and G. Seconi, *Synthesis*, 545 (1997).
(d) P. Bernardi, P. Dembech, G. Fabbri, A. Ricci and G. Seconi, *J. Org. Chem.*, **64**, 641 (1999).

314. (a) V. del Amo, S. R. Dubbaka, A. Krasovskiy and P. Knochel, *Angew. Chem., Int. Ed.*, **45**, 7838 (2006).
(b) M. Kienle, S. R. Dubbaka, V. del Amo and P. Knochel, *Synthesis*, 1272 (2007).

315. (a) O. Kulinkovich, S. V. Sviridov and D. A. Vasilevski, *Synthesis*, 234 (1991).
(b) Review on the Kulinkovich reaction: O. G. Kulinkovich and A. de Meijere, *Chem. Rev.*, **100**, 2789 (2000).
(c) Review on the titanium-mediated synthesis of cyclopropylamines from Grignard reagents and nitriles: A. de Meijere, S. I. Kozhushkov and A. I. Savchenko, *J. Organomet. Chem.*, **689**, 2033 (2004).

316. E. J. Corey, S. A. Rao and M. C. Noe, *J. Am. Chem. Soc.*, **116**, 9345 (1994).

317. C. Cardellicchio, A. Iacuone, F. Naso and P. Tortorella, *Tetrahedron Lett.*, **37**, 6017 (1996).

318. (a) M. A. M. Capozzi, C. Cardellicchio, F. Naso, G. Spina and P. Tortorella, *J. Org. Chem.*, **66**, 5933 (2001).
(b) M. A. M. Capozzi, C. Cardellicchio, F. Naso and P. Tortorella, *J. Org. Chem.*, **65**, 2843 (2000).
(c) M. A. M. Capozzi, C. Cardellicchio, F. Naso and V. Rosito, *J. Org. Chem.*, **67**, 7289 (2002).

319. (a) J. L. Garcia Ruano, C. Alemparte, M. Teresa Aranda and M. M. Zarzuelo, *Org. Lett.*, **5**, 75 (2003).
(b) Z. Han, D. Krishnamurthy, P. Grover, H. S. Wilkinson, Q. K. Fang, X. Su, Z.-H. Lu, D. Magiera and C. H. Senanayake, *Angew. Chem., Int. Ed.*, **42**, 2032 (2003).

CHAPTER 13

Iron-Catalyzed Reactions of Grignard Reagents

GÉRARD CAHIEZ and CHRISTOPHE DUPLAIS

Laboratoire de Synthèse Organique Sélective et de Chimie Organométallique (SOSCO), UMR 8123 CNRS-ESCOM-UCP, 5 Mail Gay Lussac, Neuville ^s/Oise, F-95092 Cergy-Pontoise, France
Fax: +33 1 34 25 73 83; e-mail: g.cahiez@escom.fr

I. INTRODUCTION	595
II. DEPROTONATION OF KETONES	596
III. IRON-CATALYZED ACYLATION OF GRIGNARD REAGENTS	599
IV. IRON-CATALYZED CROSS-COUPLING REACTIONS	604
A. Iron-catalyzed Alkenylation of Grignard Reagents	604
1. From alkenyl halides	604
2. From other alkenyl derivatives	608
B. Iron-catalyzed Arylation and Heteroarylation of Alkyl Grignard Reagents	610
C. Iron-catalyzed Heteroarylation of Aryl Grignard Reagents	614
D. Iron-catalyzed Alkylation of Aryl Grignard Reagents	615
V. IRON-CATALYZED HOMOCOUPLING OF AROMATIC GRIGNARD REAGENTS	619
VI. OTHER REACTIONS	621
A. Iron-catalyzed Substitution	621
B. Carbometallation	622
C. Radical Cyclization	623
D. Addition to Conjugated Unsaturated Carbonyl Compounds	624
VII. APPLICATIONS OF IRON-CATALYZED REACTIONS OF GRIGNARD REAGENTS IN ORGANIC SYNTHESIS	625
VIII. REFERENCES	628

I. INTRODUCTION

The first attempts to modify the course of the reaction of Grignard reagents with various substrates by using iron salts as a catalyst were reported by Kharasch and coworkers[1-8]

The chemistry of organomagnesium compounds
Edited by Z. Rappoport and I. Marek © 2008 John Wiley & Sons, Ltd

from 1941 on. The reports mainly treat of the reactions with organic halides and unsaturated ketones. Undoubtedly, Kharasch is the pioneer of the iron-catalyzed reactions of Grignard reagents. Some time afterwards, in 1971, Kochi and coworkers[9-17] published interesting results about the mechanism of the reaction between alkenyl halides and Grignard reagents in the presence of iron salts. Surprisingly, the chemistry of iron-catalyzed Grignard reagents had never been significantly developed until recent years. Indeed, for about 25 years (1975–2000), most of the studies concerning transition-metal-catalyzed Grignard reactions involved copper[18-21] or palladium[22,23] and nickel complexes[24], and the impressive number of results obtained with these metals[25-28] put the development of the iron chemistry on the back burner.

As a major tool resulting from this period, palladium- and nickel-catalyzed cross-coupling reactions are now increasingly used for industrial applications. However, sustainable development is currently an essential part in the strategy of chemical industries and the search for more economic and more eco-friendly synthetic methods is of vital concern. As a consequence, palladium and nickel complexes, which are toxic and/or expensive, have to be replaced by other more convenient catalysts. Iron salts are good candidates since iron is a very cheap metal having no significant toxicological properties. Because of this, in the last decade, the iron-catalyzed cross-coupling reactions of Grignard reagents have been extensively studied. Since the first preparative procedure reported by Cahiez and Avedissian[29] in 1998, the number of publications[30-32] has clearly increased.

II. DEPROTONATION OF KETONES

During their investigations, Kharasch and Tawney[2,6] studied the reaction of methylmagnesium bromide with isophorone in the presence of various transition metal salts in diethyl ether. In a first report[2], they showed that with iron(III) chloride a major product is formed, but they failed to characterize it. A few years later[6], they established that this product is the deconjugated ketone (Scheme 1).

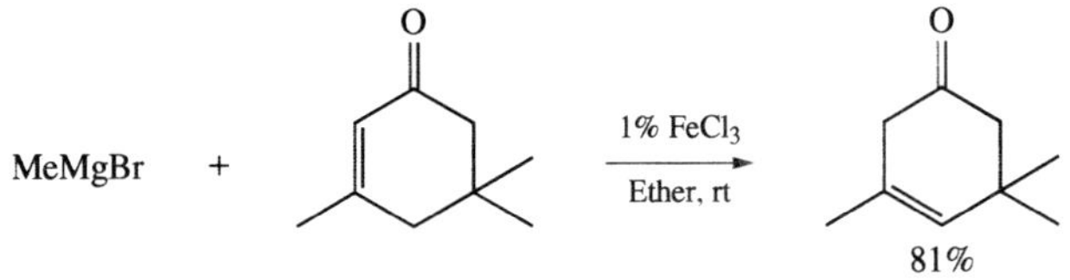

MeMgBr + [isophorone] $\xrightarrow{1\% \text{ FeCl}_3}_{\text{Ether, rt}}$ [deconjugated ketone] 81%

SCHEME 1

This result received no attention until 1984 when Krafft and Holton[33,34] decided to reinvestigate this reaction. The expected enolate was trapped with trimethylchlorosilane to form the corresponding silyl enol ether. Their first attempts showed that unsaturated and saturated ketones react with Grignard reagents in the presence of iron salts to produce substantial amount of a tertiary alcohol resulting from a 1,2-addition. They discovered that this side reaction can be suppressed by using a reagent ('Fe') (see Section IV.B) prepared by addition of three equivalents of methylmagnesium bromide to one equivalent of iron(III) chloride (Scheme 2).

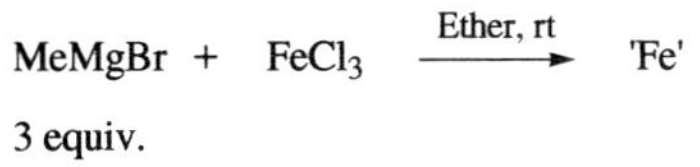

MeMgBr + FeCl$_3$ $\xrightarrow{\text{Ether, rt}}$ 'Fe'
3 equiv.

SCHEME 2

TABLE 1. Regioselective preparation of trimethylsilyl enol ether

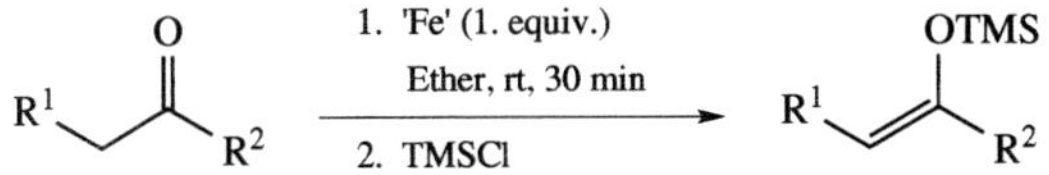

Ketone	Trimethylsilyl enol ethers (ratio)	Yield (%)
4-tert-butylcyclohexanone	4-tert-butyl-1-(trimethylsilyloxy)cyclohexene	98
acetophenone	1-phenyl-1-(trimethylsilyloxy)ethylene	92
2-methylcycloalkanone, n = 1; n = 2	(97:3); (93:7)	99; 90
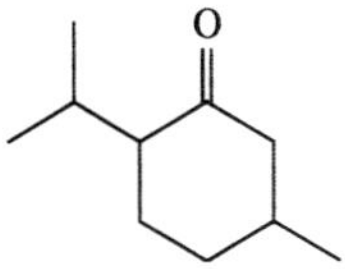 menthone-type	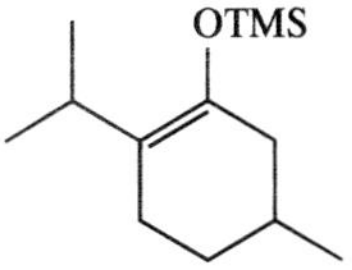 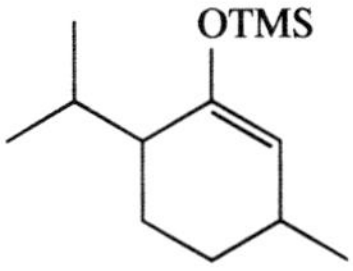 (98:2)	97
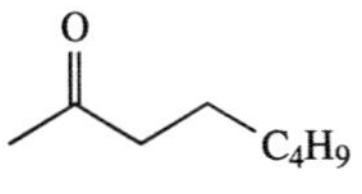 2-octanone (with C$_4$H$_9$)	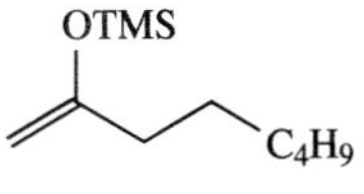 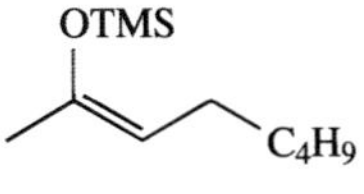 (80:20)	95

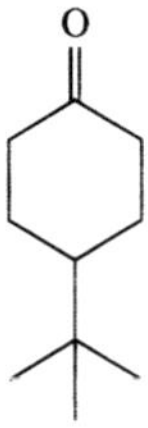

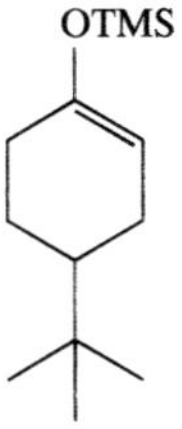

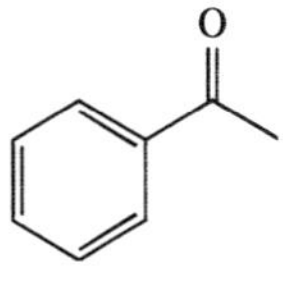

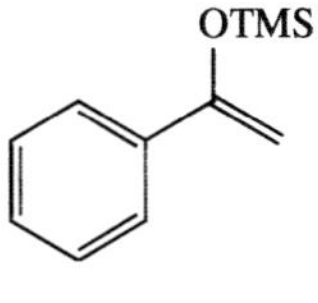

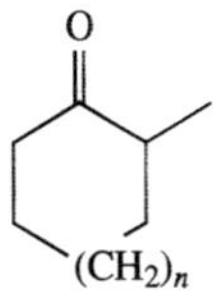

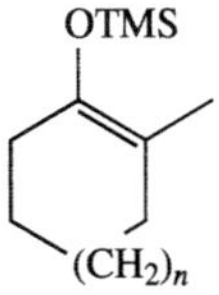

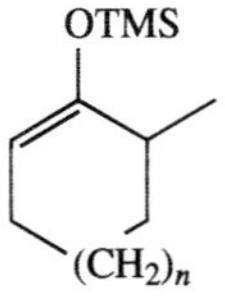

Under these conditions[33], various cyclic and acyclic ketones were successfully deprotonated. After addition of trimethylchlorosilane, the thermodynamically more stable trimethylsilyl enol ether was obtained as a major product in excellent yield (Table 1).

It is important to note that, under the same conditions, isophorone or other β-alkyl cyclic enones predominantly afford the exocyclic conjugated silyl dienol ether[34] (Scheme 3).

On the contrary, the endocyclic conjugated silyl dienol ether is almost exclusively formed under the Kharasch conditions[33] (Scheme 4).

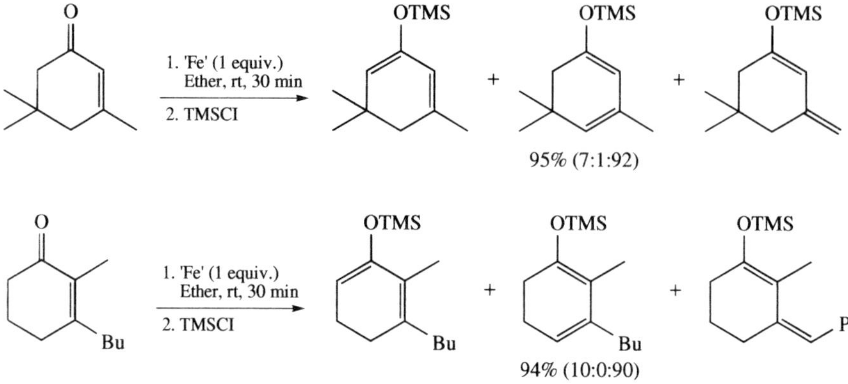

SCHEME 3

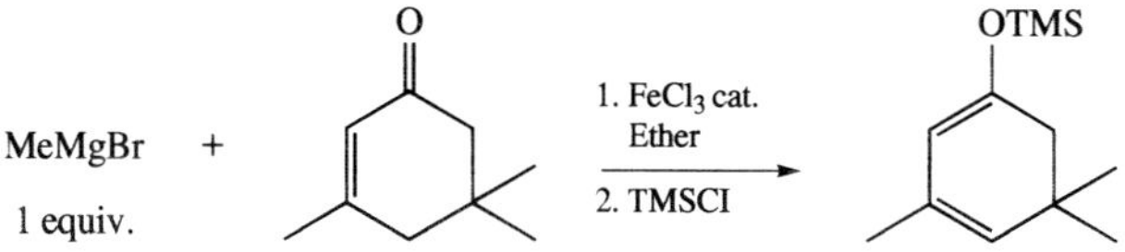

SCHEME 4

The endocyclic silyl dienol ether[34] is also obtained as the major product when the conjugated cyclic enone is successively treated with the 'Fe' complex prepared according to Scheme 2, then with one equivalent of methylmagnesium bromide (Table 2). This is interesting since these products are difficult to obtain otherwise. The mechanism of this unusual reaction remains obscure.

Recently, Fürstner and coworkers[35] have prepared a 'super-ate' complex of iron(II) as shown in Scheme 5. The structure was fully characterized by X-ray crystallography. They have shown that methylmagnesium bromide reacts with pulegone in the presence of this complex to give the corresponding endocyclic silyl dienol ether. Consequently, they have proposed that a similar 'ate-complex' is probably involved when the reaction is performed under the Kharasch conditions.

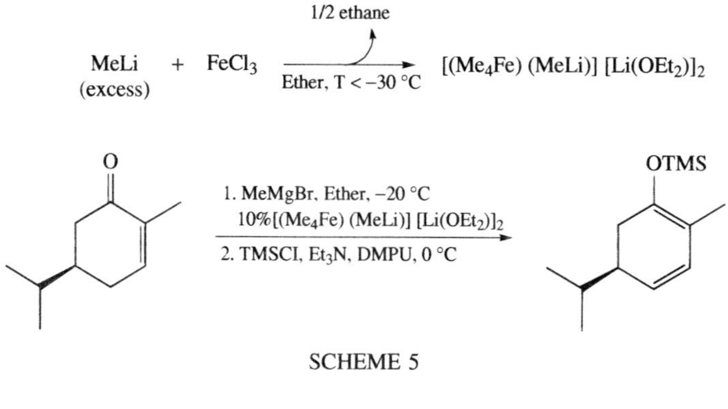

SCHEME 5

TABLE 2. Regioselective preparation of trimethylsilyl dienol ethers

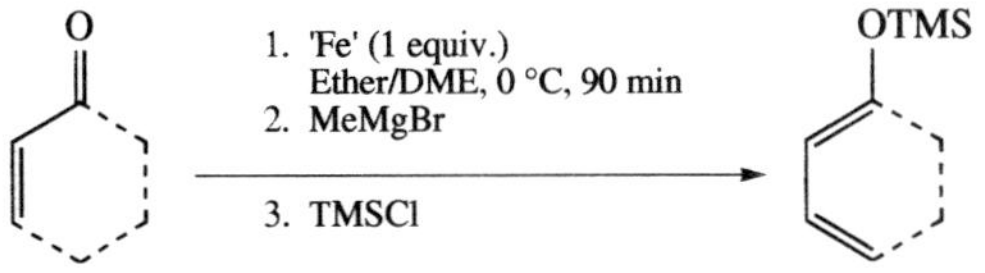

Enone	Dienol ethers (ratio)			Yield (%)
[enone 1]	[OTMS isomer]	[OTMS isomer]	[OTMS isomer]	99
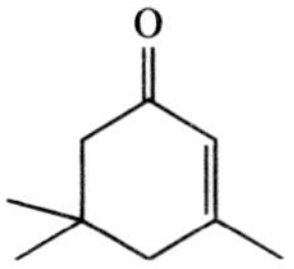	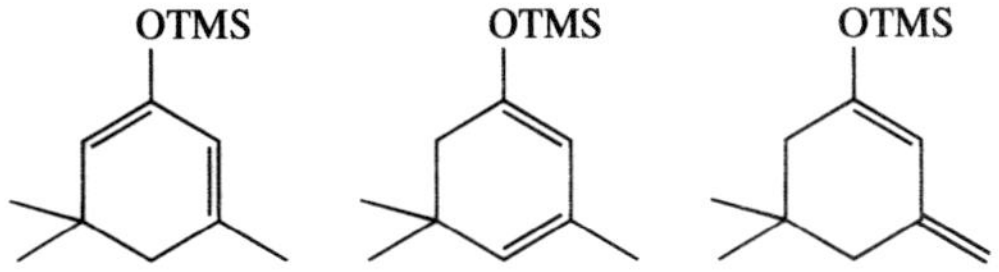 (2:96:2)			
[enone 2, Bu]	[OTMS isomer, Bu]	[OTMS isomer, Bu]	[OTMS isomer, Pr]	98
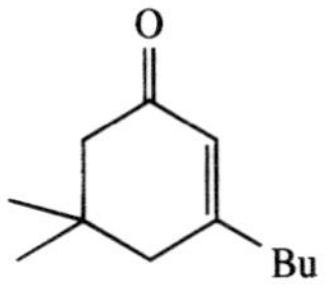	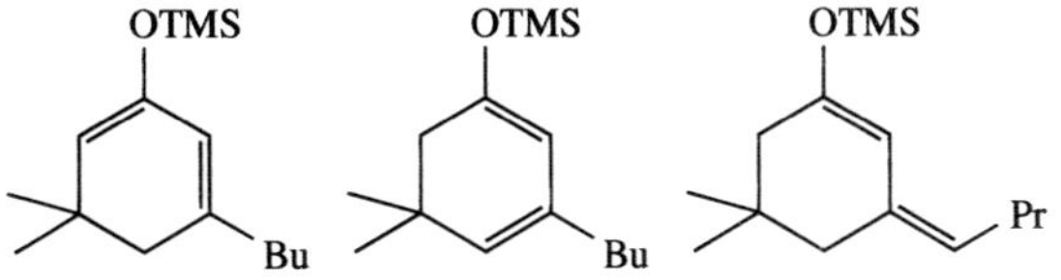 (1:99:0)			
[enone 3]	[OTMS isomer]	[OTMS isomer]	[OTMS isomer]	90
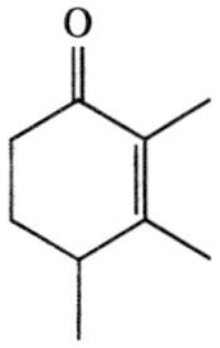	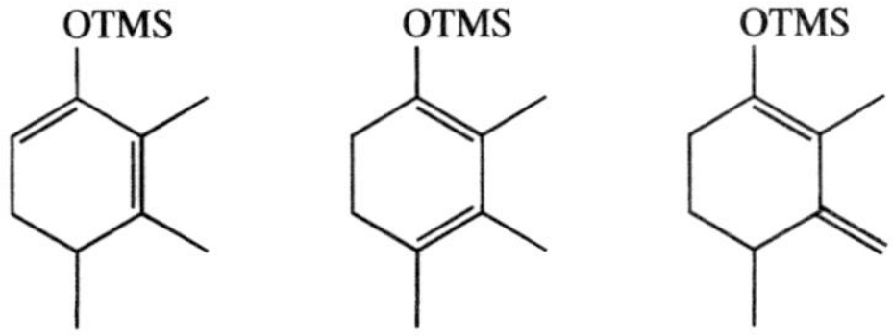 (4:92:4)			
[enone 4]	[OTMS isomer]	[OTMS isomer]		93
	(5:95)			

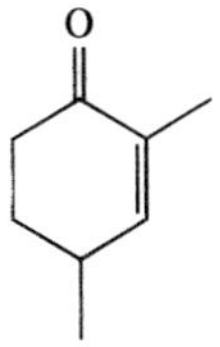

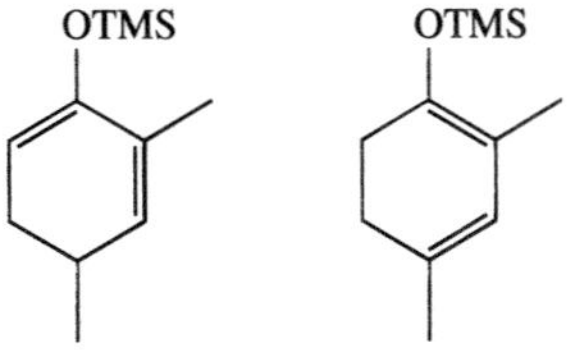

III. IRON-CATALYZED ACYLATION OF GRIGNARD REAGENTS

The first iron-catalyzed acylation of Grignard reagents was described by Kharasch and coworkers[5] in 1944 (Scheme 6). They reported only one example of reaction between methylmagnesium iodide and mesitoyl chloride in the presence of iron(III) chloride in diethyl ether.

In 1953, Percival and coworkers[36] showed that aliphatic ketones can be prepared under these conditions. Unfortunately, in the case of linear aliphatic ketones the reaction only gives moderate yields, since the Grignard reagent adds to the ketone or behaves as a base

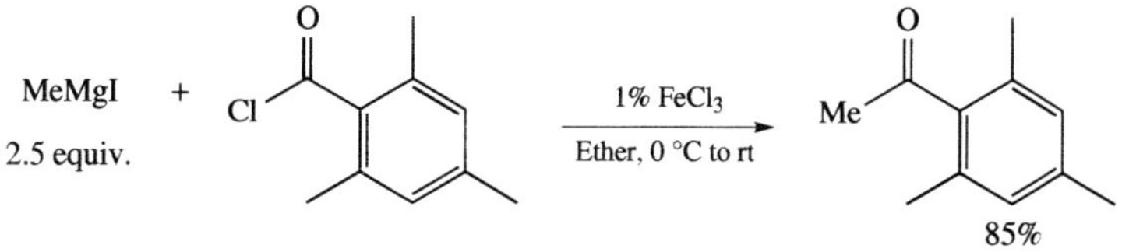

SCHEME 6

MeMgI (2.5 equiv.) + 2,4,6-trimethylbenzoyl chloride $\xrightarrow{\text{1% FeCl}_3, \text{ Ether, 0 °C to rt}}$ 2',4',6'-trimethylacetophenone (85%)

to deprotonate the starting carboxylic acid chloride (dehydrochlorination). However, in the case of hindered ketones both side reactions are considerably limited. As an example, 2,5-dimethylhexan-3-one was obtained in 88% yield by reacting isobutyryl chloride with i-butylmagnesium bromide, in diethyl ether at 5 °C, in the presence of 2% iron(III) chloride (Scheme 7).

i-BuMgBr (1 equiv.) + i-PrCOCl $\xrightarrow{\text{2% FeCl}_3, \text{ Ether, 5 °C, 3 h}}$ i-BuCOPr-i 88% (2 mol scale)

SCHEME 7

It is necessary to operate under reflux to obtain good yields when more hindered Grignard reagents are used (Scheme 8).

t-BuMgCl + t-BuCOCl $\xrightarrow{\text{2% FeCl}_3, \text{ Ether, Reflux, 3 h}}$ t-BuCOBu-t 84% (2 mol scale)

SCHEME 8

With linear aliphatic carboxylic acid chlorides, such as acetyl chloride, Percival and coworkers pointed out that it is possible to prevent the side reactions previously described by performing the reaction at low temperature (Scheme 9).

BuMgCl (1.5 equiv.) + MeCOCl $\xrightarrow{\text{2% FeCl}_3, \text{ Ether}}$ BuCOMe   15 °C: 47%   −65 °C: 72%

SCHEME 9

In 1961, to improve these results, Cason and Kraus[37] used a diethyl ether/toluene mixture as a solvent. Unfortunately, yields of ketone are never better than 80% (Scheme 10).

BuMgBr (1.35 equiv.) + PentCOCl $\xrightarrow{\text{2% FeCl}_3, \text{ Ether/Toluene, −60 °C}}$ BuCOPent   76%

SCHEME 10

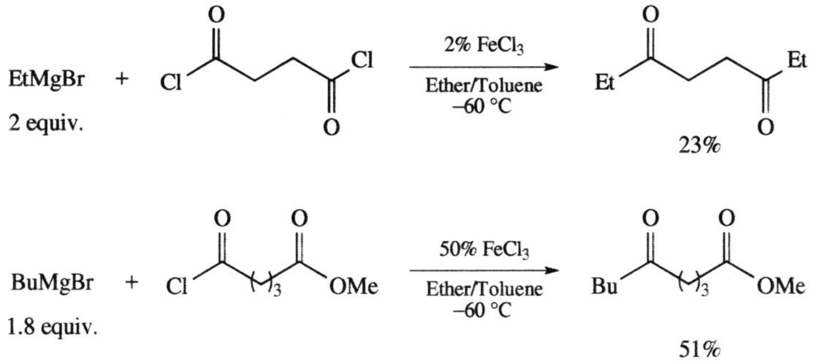

SCHEME 11

EtMgBr (2 equiv.) + ClCOCH$_2$CH$_2$COCl →(2% FeCl$_3$, Ether/Toluene, −60 °C) EtCOCH$_2$CH$_2$COEt (23%)

BuMgBr (1.8 equiv.) + Cl(CO)(CH$_2$)$_3$CO(OMe) →(50% FeCl$_3$, Ether/Toluene, −60 °C) Bu(CO)(CH$_2$)$_3$CO(OMe) (51%)

In addition, the reaction is not chemoselective, thus carboxylic diacid chlorides and carboxylic acid chlorides bearing an ester group give unsatisfactory yields of ketone (Scheme 11)[38,39].

Twenty-five years later, a dramatic improvement was reported by Fiandanese, Marchese and coworkers[40–42]. They discovered that excellent yields of ketone were obtained when diethyl ether is replaced by THF. Moreover, iron acetylacetonate is used as a catalyst instead of iron(III) chloride because it is not hygroscopic and easier to handle. The scope of the procedure is very large and the reaction occurs highly chemoselectively under mild conditions (0 °C). It should be noted that excellent yields are obtained from stoichiometric amounts of Grignard reagents (Table 3).

Unfortunately, the yield depends highly on the concentration of the reaction mixture (Scheme 12). As shown below[43], in the case of the reaction of 6-bromohexanoyl chloride with propylmagnesium chloride, the yield jumps from 38% to 91% when the concentration decreases from 1.2 M to 0.4 M. It is a drawback for large-scale applications since the concentration cannot be higher than 0.5 M.

PrMgCl + Br(CH$_2$)$_5$COCl →(3% Fe(acac)$_3$, THF, 0 °C to 10 °C, 30 min) PrCO(CH$_2$)$_5$Br

$C = 0.4$ M 91%
$C = 1.2$ M 38%

SCHEME 12

The main limitation of this procedure is the preparation of aromatic ketones from aryl Grignard reagents and aromatic carboxylic acid chlorides that are clearly less reactive than their aliphatic analogues. In this case, the ketones are only obtained in moderate yields because of the formation of a large amount of homocoupling product (diaryl). It is possible to improve the yield of diaryl ketone by using an excess of aryl Grignard reagent, but the purification of the product is then very delicate due to the presence of a huge amount of homocoupling product. Cahiez, Knochel and coworkers[44] showed that the side reaction is completely avoided when the starting aromatic carboxylic acid chloride is replaced by the corresponding cyanide (Scheme 13).

Functionalized aryl Grignard compounds, prepared by iodine–magnesium exchange according to the procedure reported by Cahiez, Knochel and coworkers[45] in 1998, can be efficiently acylated. It is thus possible to prepare various polyfunctionalized diaryl ketones (Scheme 14)[44].

TABLE 3. Iron-catalyzed acylation of Grignard reagents in THF

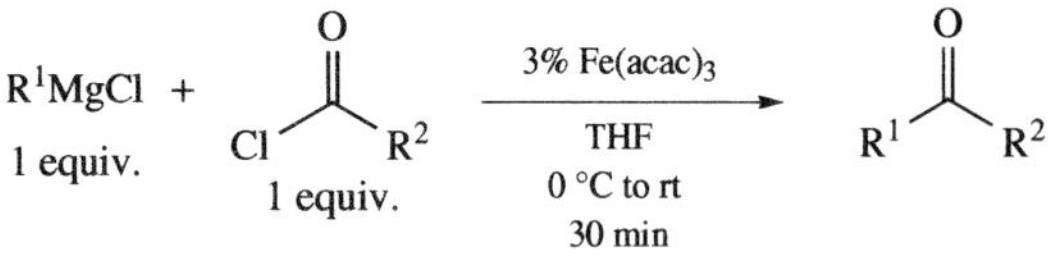

$$R^1MgCl\ (1\ equiv.) + \underset{1\ equiv.}{Cl-C(=O)-R^2} \xrightarrow[\substack{THF \\ 0\ °C\ to\ rt \\ 30\ min}]{3\%\ Fe(acac)_3} R^1-C(=O)-R^2$$

R^1	R^2COCl	Product	Yield (%)
Me	DecCOCl	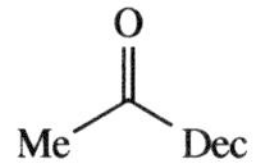 Me-C(=O)-Dec	84
Pent	PentCOCl	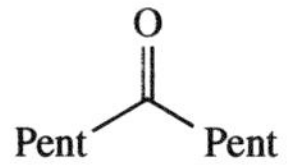 Pent-C(=O)-Pent	82
Dec	MeCOCl	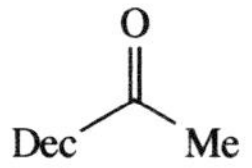 Dec-C(=O)-Me	80
i-Pr	BuCOCl	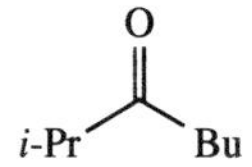 i-Pr-C(=O)-Bu	80
t-Bu	BuCOCl	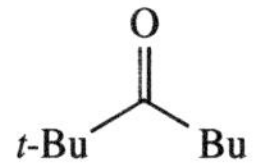 t-Bu-C(=O)-Bu	70
Ph	BuCOCl	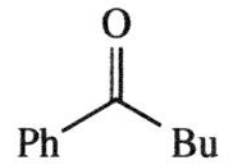 Ph-C(=O)-Bu	83
Ph	i-PrCOCl	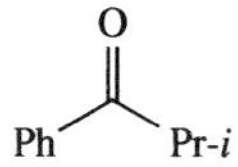 Ph-C(=O)-Pr-i	92
Bu	PhCOCl	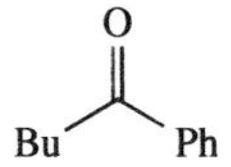 Bu-C(=O)-Ph	90
Et[a]	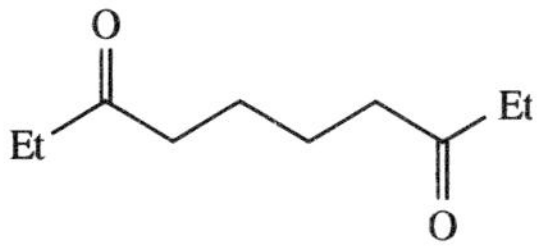 Et-C(=O)-(CH₂)₄-C(=O)-Et	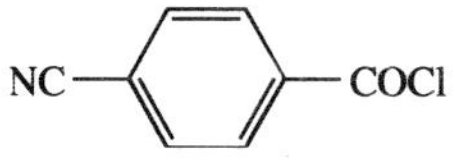 NC-C₆H₄-COCl	90
Me	NCC₆H₄COCl	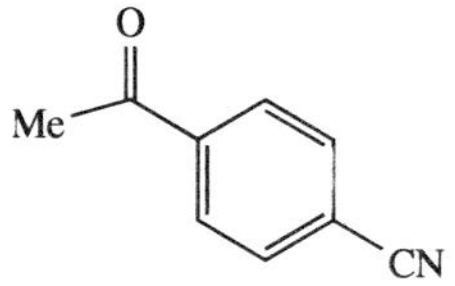 Me-C(=O)-C₆H₄-CN	75

TABLE 3. (*continued*)

R¹	R²COCl	Product	Yield (%)
Bu	MeOOCC$_6$H$_4$COCl	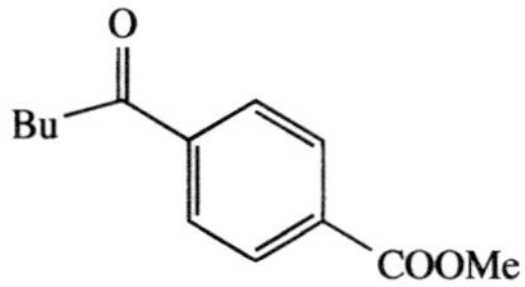 Bu-C(O)-C$_6$H$_4$-COOMe	78

a 2 equivalents of EtMgBr were used.

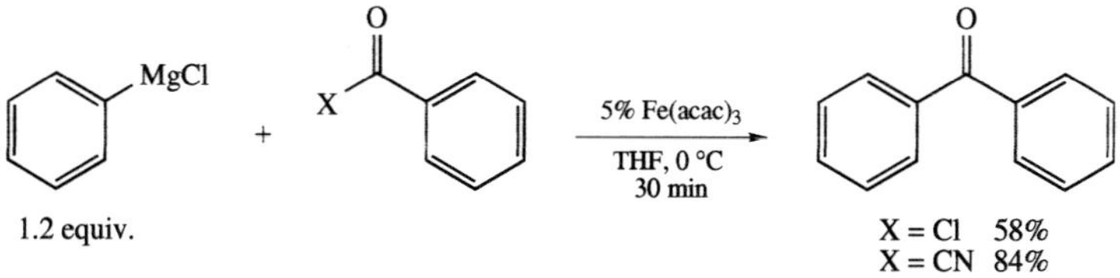

SCHEME 13

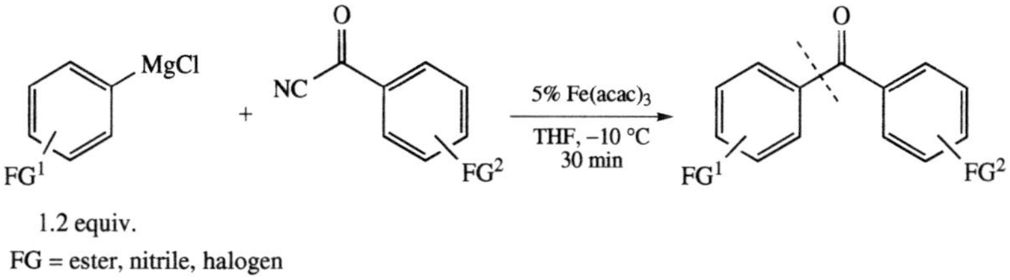

PhMgCl (1.2 equiv.) + PhCOX → benzophenone
5% Fe(acac)$_3$, THF, 0 °C, 30 min
X = Cl 58%
X = CN 84%

SCHEME 14

ArMgCl (1.2 equiv., FG¹ = ester, nitrile, halogen) + Ar'COCN (FG²) → diaryl ketone
5% Fe(acac)$_3$, THF, −10 °C, 30 min

The following diaryl ketones were prepared according to this procedure.

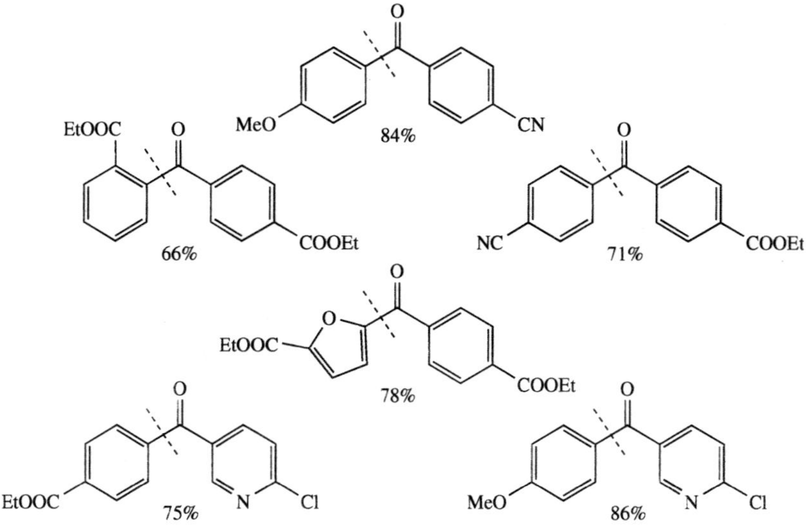

- MeO-C$_6$H$_4$-CO-C$_6$H$_4$-CN 84%
- 2-EtOOC-C$_6$H$_4$-CO-C$_6$H$_4$-COOEt 66%
- NC-C$_6$H$_4$-CO-C$_6$H$_4$-COOEt 71%
- 5-EtOOC-furan-2-yl-CO-C$_6$H$_4$-COOEt 78%
- EtOOC-C$_6$H$_4$-CO-(6-Cl-pyridin-3-yl) 75%
- MeO-C$_6$H$_4$-CO-(6-Cl-pyridin-3-yl) 86%

IV. IRON-CATALYZED CROSS-COUPLING REACTIONS
A. Iron-catalyzed Alkenylation of Grignard Reagents
1. From alkenyl halides

The first example of iron-catalyzed cross-coupling reaction between Grignard reagents and alkenyl bromides was reported by Kharasch and Fuchs[7] in 1945 (Scheme 15).

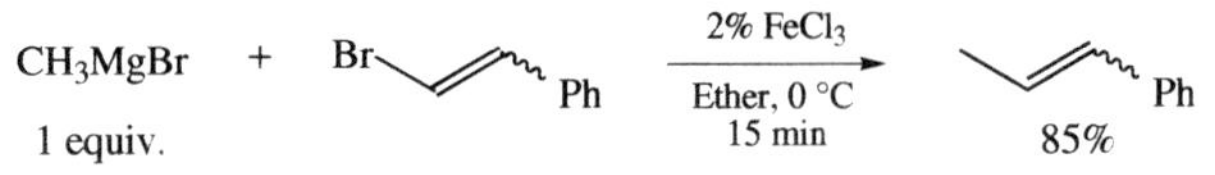

$$CH_3MgBr + Br\text{-CH=CH-Ph} \xrightarrow[\text{Ether, 0 °C, 15 min}]{2\% \text{ FeCl}_3} \text{CH}_3\text{-CH=CH-Ph} \quad 85\%$$

1 equiv.

SCHEME 15

In 1971, Tamura and Kochi[9, 10] described the reaction of alkyl Grignard reagents with alkenyl bromides in the presence of iron(III) chloride in THF. Only very reactive substrates such as vinyl and propenyl bromide were used (Scheme 16). Yields of coupling product are moderate to good but, unfortunately, a large excess of alkenyl bromide is required (3 to 9 equivalents). (E)-Bromopropene reacts 15 times faster than the (Z)-isomer. It should be noted that the reaction is stereoselective.

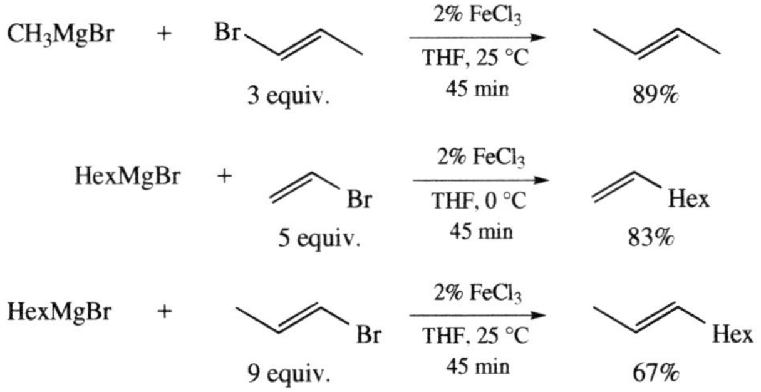

$$CH_3MgBr + \text{Br-CH=CH-CH}_3 \xrightarrow[\text{THF, 25 °C, 45 min}]{2\% \text{ FeCl}_3} \text{CH}_3\text{-CH=CH-CH}_3 \quad 89\%$$
3 equiv.

$$HexMgBr + \text{CH}_2\text{=CH-Br} \xrightarrow[\text{THF, 0 °C, 45 min}]{2\% \text{ FeCl}_3} \text{Hex-CH=CH}_2 \quad 83\%$$
5 equiv.

$$HexMgBr + \text{CH}_3\text{-CH=CH-Br} \xrightarrow[\text{THF, 25 °C, 45 min}]{2\% \text{ FeCl}_3} \text{CH}_3\text{-CH=CH-Hex} \quad 67\%$$
9 equiv.

SCHEME 16

The results published thereafter by Kochi's group are especially interesting from a mechanistic point of view[13–15]. Indeed, for preparative chemistry the yields are not satisfactory and the reaction is limited to reactive alkenyl bromides such as propenyl and styryl bromides (Table 4). Neumann and Kochi[13] were the first to replace iron(III) chloride by iron(III) acetylacetonate or related complexes such as Fe(dbm)$_3$ (iron tris-dibenzoylmethanato) that are less hygroscopic and easier to handle.

In 1983, Molander and coworkers[46] studied the coupling of aromatic Grignard reagents with 2-bromostyrene. They showed that the use of dimethoxyethane (DME) instead of tetrahydrofuran as a solvent significantly increases the yields (from ca 30% to 60–70%). However, the reaction is always limited to reactive alkenyl bromides. Moreover, while (E)-alkenyl bromides always lead to satisfactory yields, the corresponding (Z)-stereomers give either low yields of coupling product or mixtures of stereomers (Scheme 17). Similar results were observed by Smith and Kochi[14] in other cases when a hindered Grignard reagent was used.

In 1998, Cahiez and Avedissian[29] discovered that the addition of N-methylpyrrolidinone (NMP, 4 to 9 equivalents) to the reaction mixture allows one to obtain excellent yields (Scheme 18). In addition, only a stoichiometric amount of alkenyl halide is then required. Under these conditions, the reaction takes place almost instantaneously even from low

TABLE 4. Iron-catalyzed coupling of Grignard reagents with alkenyl bromides

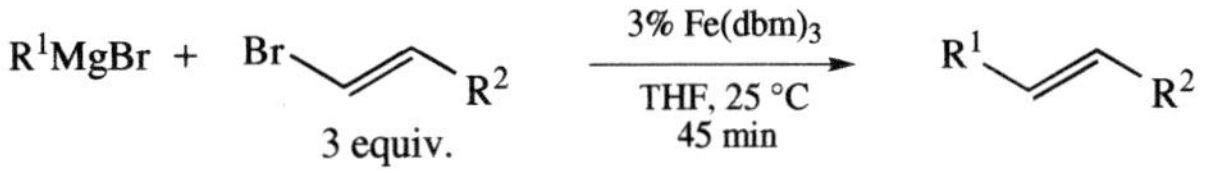

$$R^1MgBr + Br\diagup\!\!\!\diagdown R^2 \xrightarrow[\text{THF, 25 °C}]{3\% \text{ Fe(dbm)}_3} R^1\diagup\!\!\!\diagdown R^2$$
3 equiv.   45 min

R^1	R^2	Yield (%)
Et	Me	58
Ph	Ph	32
Et	Ph	59
i-Pr	Me	60
c-Hex	Me	54
t-Bu	Me	27

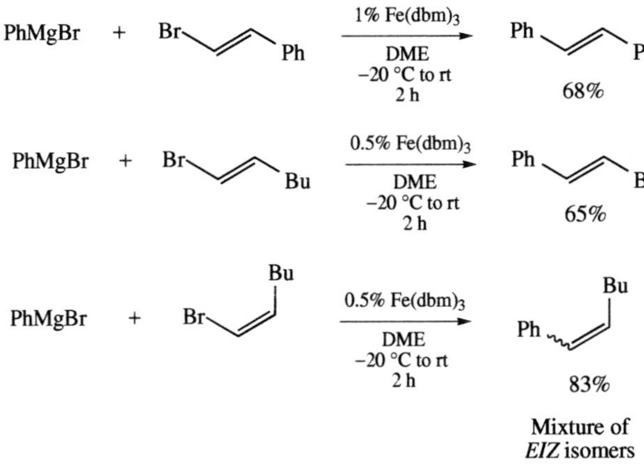

PhMgBr + Br—CH=CH—Ph $\xrightarrow[\text{DME, }-20\text{ °C to rt, 2 h}]{1\% \text{ Fe(dbm)}_3}$ Ph—CH=CH—Ph 68%

PhMgBr + Br—CH=CH—Bu $\xrightarrow[\text{DME, }-20\text{ °C to rt, 2 h}]{0.5\% \text{ Fe(dbm)}_3}$ Ph—CH=CH—Bu 65%

PhMgBr + Br—CH=C(Bu) $\xrightarrow[\text{DME, }-20\text{ °C to rt, 2 h}]{0.5\% \text{ Fe(dbm)}_3}$ Ph—CH=C(Bu) 83%

Mixture of E/Z isomers

SCHEME 17

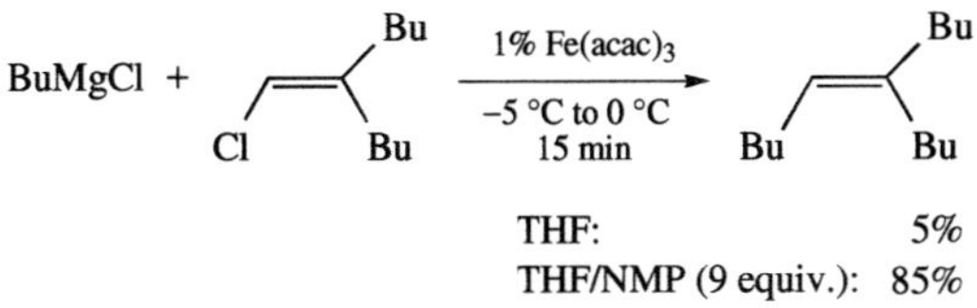

BuMgCl + (Cl)(Bu)C=C(Bu) $\xrightarrow[-5\text{ °C to 0 °C, 15 min}]{1\% \text{ Fe(acac)}_3}$ (Bu)(Bu)C=C(Bu)

THF: 5%
THF/NMP (9 equiv.): 85%

SCHEME 18

reactive substrates such as β,β-disubstituted alkenyl chlorides. It should be noted that NMP is a very cheap additive that was not frequently used before with Grignard reagents.

Only 1 to 3% iron(III) acetylacetonate are required and the scope of the reaction is very large. A vast array of alkenyl iodides, bromides and even chlorides can be successfully used and the stereoselectivity is excellent even in the case of the Z-alkenyl halides (Scheme 19).

In addition, the reaction is very chemoselective and even a keto group is tolerated (Table 5). It should be underlined that this procedure compares advantageously to the corresponding palladium- or nickel-catalyzed coupling reactions.

TABLE 5. Iron-catalyzed coupling of Grignard reagents with alkenyl halides

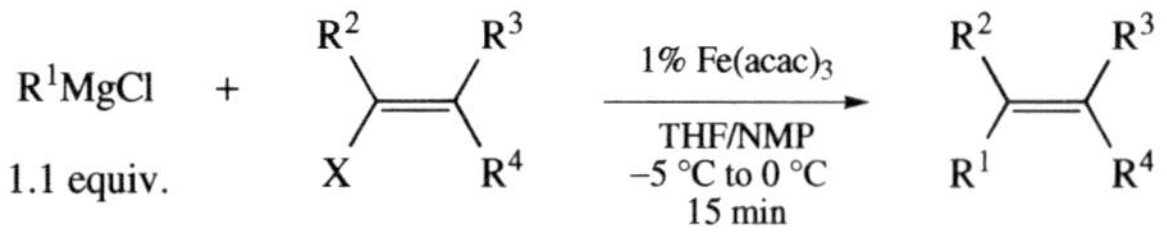

$$R^1MgCl \; (1.1 \text{ equiv.}) + \underset{X \quad R^4}{\overset{R^2 \quad R^3}{\diagup\!\!=\!\!\diagdown}} \xrightarrow[\substack{\text{THF/NMP} \\ -5\,°C \text{ to } 0\,°C \\ 15 \text{ min}}]{1\% \text{ Fe(acac)}_3} \underset{R^1 \quad R^4}{\overset{R^2 \quad R^3}{\diagup\!\!=\!\!\diagdown}}$$

R^1	Alkenyl Halides	Yield (%)
s-Bu	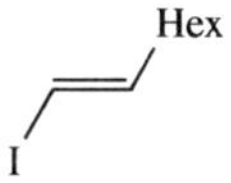 (I)CH=CH-Hex	80
c-Hex	 (Br)CH=CH-Bu (cis)	89
t-Bu	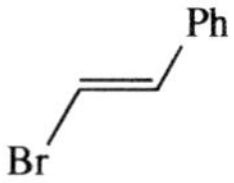 (Br)CH=CH-Ph	64
Bu	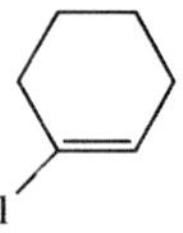 1-chlorocyclohexene	75
i-Pr	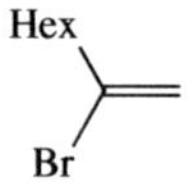 CH$_2$=C(Br)-Hex	72
Oct	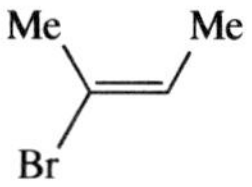 (Me)(Br)C=CH-Me	84
Bu	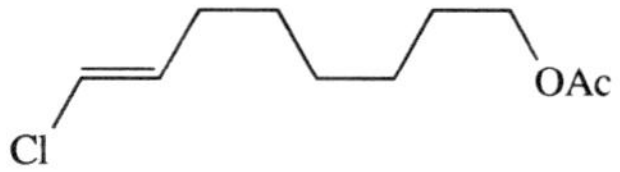 Cl-CH=CH-(CH$_2$)$_4$-CH$_2$OAc	80
Bu	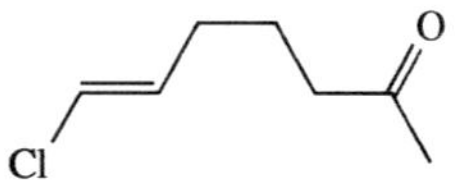 Cl-CH=CH-CH$_2$CH$_2$-C(=O)-CH$_3$	80
Bu	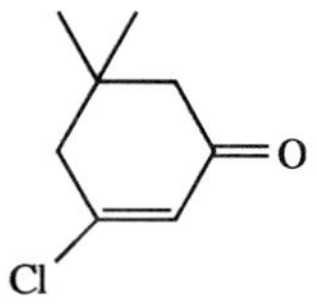 3-chloro-5,5-dimethylcyclohex-2-enone	79
Bu	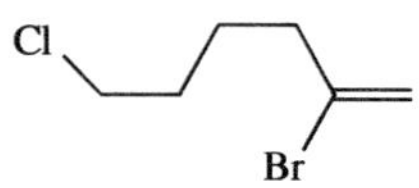 Cl-CH$_2$CH$_2$CH$_2$CH$_2$-C(Br)=CH$_2$	79

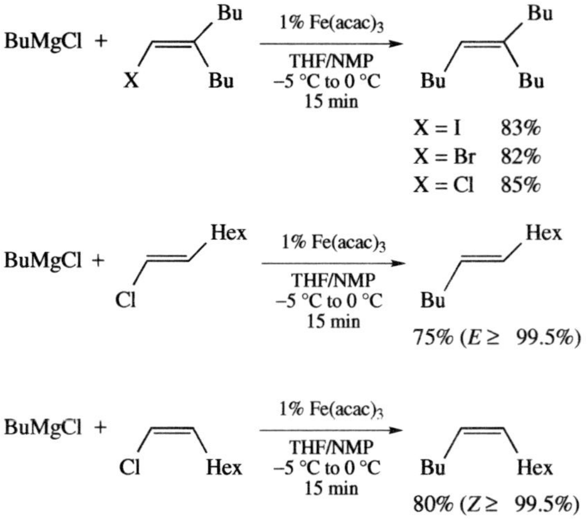

13. Iron-Catalyzed Reactions of Grignard Reagents 607

SCHEME 19

In 2001, Cahiez, Knochel and coworkers[47] reported an extension of this work to the coupling of functionalized arylmagnesium compounds with alkenyl bromides or iodides (Scheme 20). It should be noted that with aryl Grignard reagents the use of NMP is not necessary.

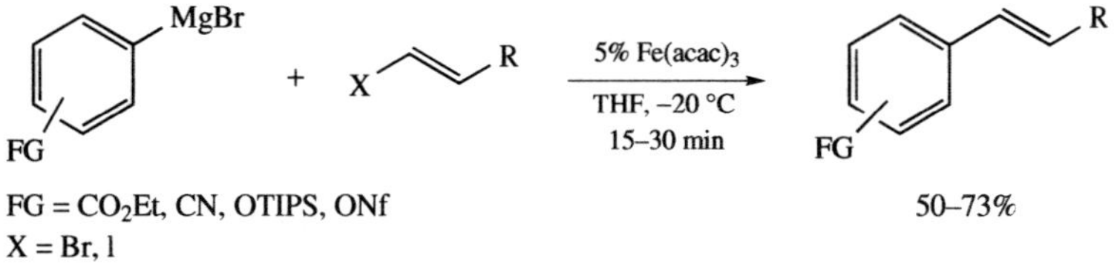

SCHEME 20

The conditions described above were used by Figadère, Alami and coworkers[48] to prepare analogues of meglumine antimonate (Glucantime®), a product used in chemotherapy, from chloroenynes or chlorodienes (Scheme 21). The reaction was further extended to various chloroenynes (Scheme 22).

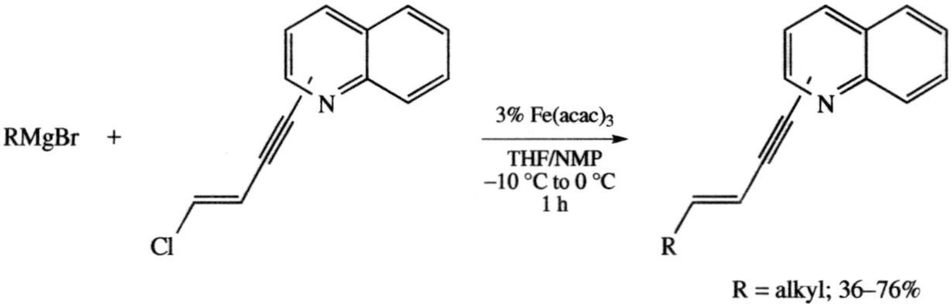

SCHEME 21

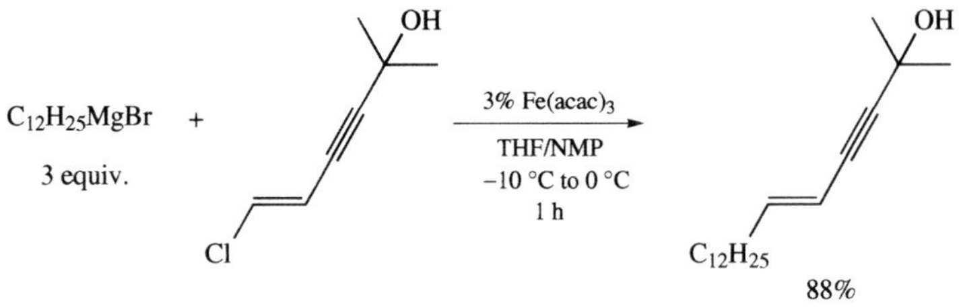

608   Gérard Cahiez and Christophe Duplais

SCHEME 22

In 2002, Figadère and coworkers[49, 50] reported the mono-reduction of 2-aryl (or heteroaryl)-1,1-dibromo-1-alkenes (Scheme 23). The reaction is achieved with one equivalent of isopropylmagnesium chloride in the presence of iron(III) acetylacetonate. Pure (E)-alkenyl bromides are obtained. With two equivalents of alkyl Grignard reagent, the monosubstituted product is obtained in moderate yield.

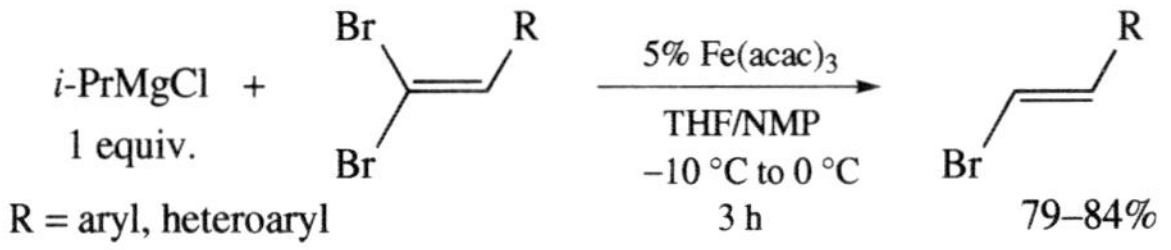

SCHEME 23

It is noteworthy that 1,1-dichloro-1-alkenes behave differently; the reduction is not observed and both chlorine atoms are substituted[51]. Satisfactory yields are obtained by using an excess of Grignard reagent (Scheme 24).

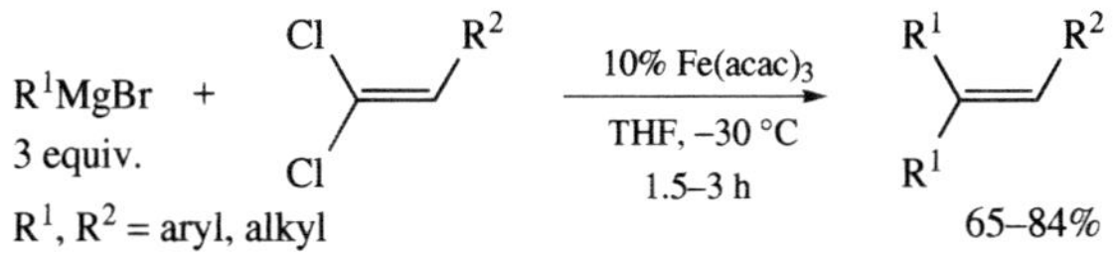

SCHEME 24

2. From other alkenyl derivatives

In 1982, Julia's group[52-55] showed that vinyl sulfones react with Grignard reagents in the presence of iron salts to afford moderate yields of coupling product (Scheme 25). The reaction is stereoselective but its scope is limited. Thus, with secondary alkylmagnesium halides, only the reduction product is formed.

The iron-catalyzed cross-coupling between Grignard reagents and alkenyl sulfides was also studied (Scheme 26)[56]. Unfortunately, this reaction is very sensitive to steric and electronic effects and only two products were synthesized in modest yields. In fact, the scope of the reaction is very limited since only vinyl sulfides can be used.

In 1998, Cahiez and Avedissian[29] reported that enol phosphates can be used instead of alkenyl halides (Scheme 27).

As expected, enol triflates also couple under these conditions (Scheme 28). Thus, Fürstner and coworkers[57] have recently shown that various cyclic and acyclic enol triflates can be used successfully.

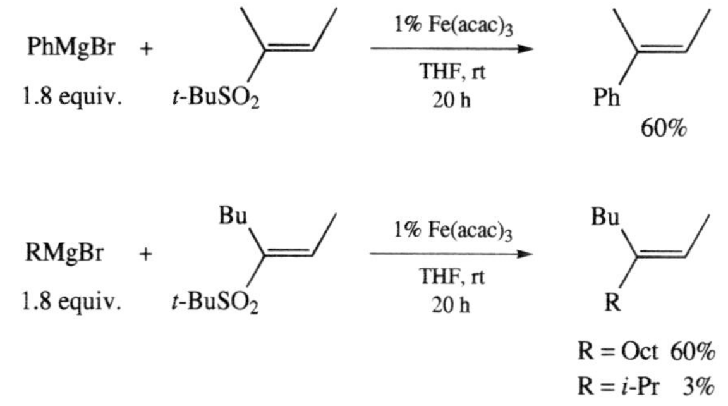

13. Iron-Catalyzed Reactions of Grignard Reagents

PhMgBr + [CH₂=C(t-BuSO₂)CH₂CH₃], 1.8 equiv. → (1% Fe(acac)₃, THF, rt, 20 h) → Ph-substituted alkene, 60%

RMgBr + [Bu-C(t-BuSO₂)=CH–CH₃], 1.8 equiv. → (1% Fe(acac)₃, THF, rt, 20 h) → Bu/R alkene
R = Oct 60%
R = i-Pr 3%

SCHEME 25

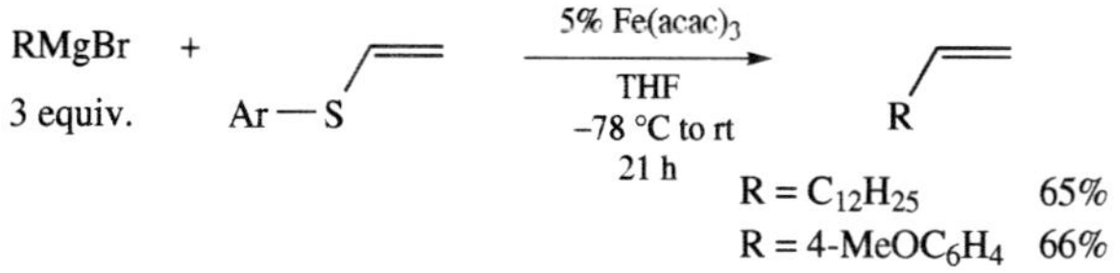

RMgBr + Ar–S–CH=CH₂, 3 equiv. → (5% Fe(acac)₃, THF, −78 °C to rt, 21 h) → R–CH=CH₂
R = $C_{12}H_{25}$ 65%
R = 4-MeOC₆H₄ 66%

SCHEME 26

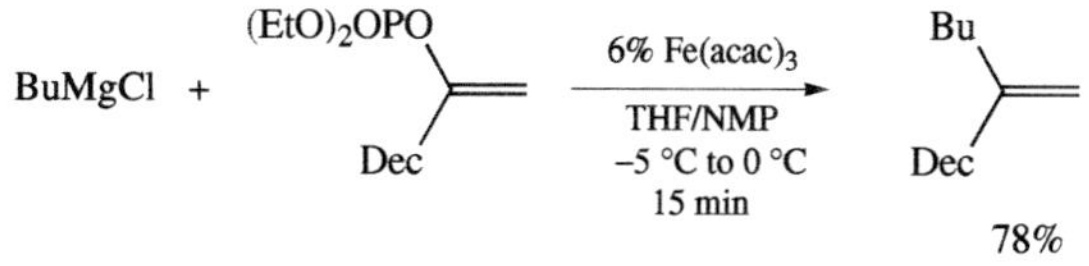

BuMgCl + [(EtO)₂OPO–C(Dec)=CH₂] → (6% Fe(acac)₃, THF/NMP, −5 °C to 0 °C, 15 min) → Bu–C(Dec)=CH₂, 78%

SCHEME 27

R^1MgCl + [R^2R^3C=C(R^4)OTf], 1.1 to 2.4 equiv. → (5% Fe(acac)₃, THF/NMP, −30 °C, 15 min) → R^2R^3C=C(R^1)(R^4)

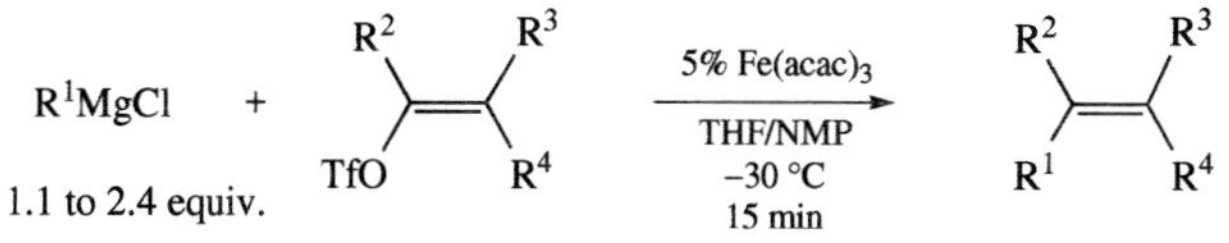

The following coupling products were prepared according to this procedure

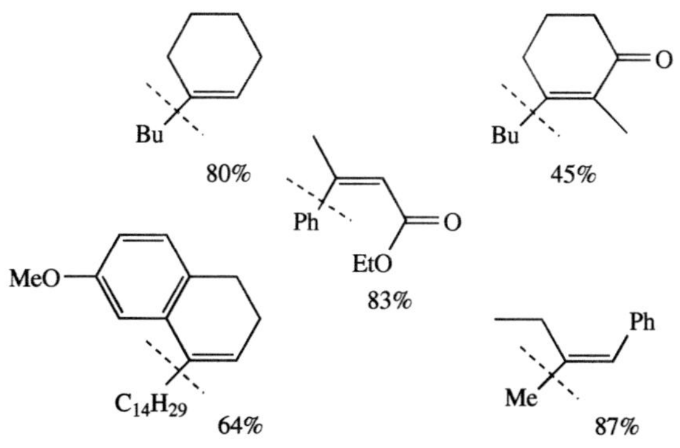

- cyclohexenyl–Bu, 80%
- 2-methyl-3-Bu-cyclohex-2-enone, 45%
- Ph–C(Me)=CH–C(O)OEt, 83%
- 6-MeO-1-($C_{14}H_{29}$)-3,4-dihydronaphthalene, 64%
- Et–C(Me)=CH–Ph, 87%

SCHEME 28

SCHEME 29

It is interesting to note that the 'super' iron(II) ate complex prepared by Fürstner and coworkers[35] leads to good yields of methylated product (Scheme 29).

B. Iron-catalyzed Arylation and Heteroarylation of Alkyl Grignard Reagents

In 2002, Fürstner and coworkers[58, 59] reported that aryl halides react with Grignard reagents under the conditions previously used for the coupling of alkenyl halides. They proposed that the active iron species is $Fe(MgX)_2$, a complex described by Bogdanovic and coworkers[60] a few years earlier (Scheme 30). This iron(-II) species is formed by addition of four equivalents of the Grignard compounds to $FeCl_2$.

$$FeCl_2 + 4\ RCH_2CH_2MgX \longrightarrow [Fe(MgX)_2] + 2\ MgX_2$$

$$RCH_2CH_3 + RCH{=}CH_2 + RCH_2CH_2CH_2R$$

SCHEME 30

The tentative catalytic cycle is presented in Figure 1. The scope of this reaction is unusual since excellent yields are obtained from aryl chlorides, tosylates and triflates

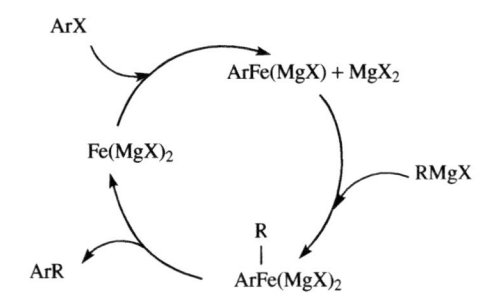

FIGURE 1

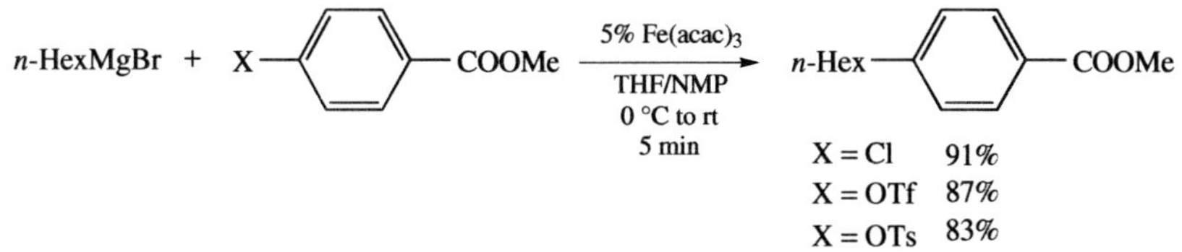

SCHEME 31

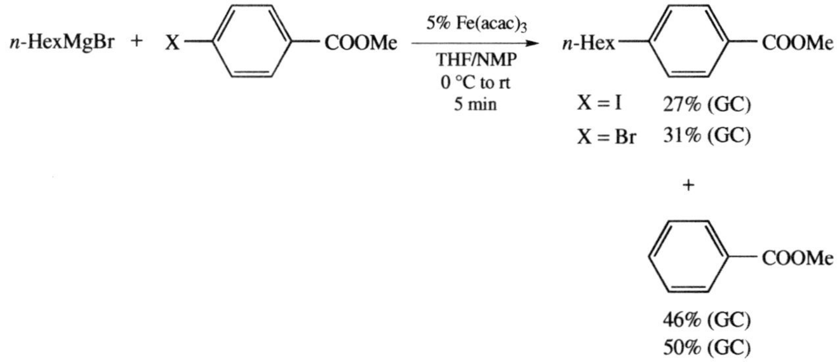

n-HexMgBr + X–C₆H₄–COOMe → (5% Fe(acac)₃, THF/NMP, 0 °C to rt, 5 min) → n-Hex–C₆H₄–COOMe

X = Cl   91%
X = OTf 87%
X = OTs 83%

SCHEME 32

n-HexMgBr + X–C₆H₄–COOMe → (5% Fe(acac)₃, THF/NMP, 0 °C to rt, 5 min) → n-Hex–C₆H₄–COOMe

X = I   27% (GC)
X = Br 31% (GC)

+

C₆H₅–COOMe

46% (GC)
50% (GC)

(Scheme 31), whereas the corresponding bromides and iodides lead to a mixture of coupling and reduction products (Scheme 32).

All electron-poor aryl chlorides, tosylates, triflates and heteroaryl chlorides react to give good to excellent yields. However, in the case of electron-rich aryl groups, only aryl triflates are reactive enough to give satisfactory yields of coupling product. It should be noted that a vast array of heteroaryl chlorides were used successfully (Table 6). The scope of the reaction is more limited regarding the nature of the Grignard reagents. Thus, only primary alkyl Grignard reagents afford good yields. With secondary alkylmagnesium halides, the yield of coupling product never rises above 50%.

Hocek and coworkers[61–63] studied the regioselectivity of the cross-coupling reaction between methylmagnesium bromide and various dichloropurines. With 2,6- and 2,8-dichloropurines, it is possible to obtain the monomethylated product with an excellent regioselectivity (Scheme 33). It is important to notice that such a regioselectivity is not observed under palladium or nickel catalysis.

However, the selectivity depends closely on the difference in reactivity between the two chlorine atoms. As an example, 6,8-dichloropurine gives a mixture of monosubstituted products (Scheme 34).

Sometimes, by using an excess of Grignard reagents, both chlorine atoms can be substituted (Scheme 35).

Recently, Olsson and coworkers[64] have reported a cross-coupling reaction between seven-membered cyclic imidoyl chlorides and alkyl Grignard reagents (Scheme 36). The corresponding substituted imines, which are very difficult to prepare otherwise, are synthesized in good to excellent yields.

Later, Olsson and coworkers[64] proposed a general method to prepare simple acyclic imines from amides according to a two-step procedure (Scheme 37).

TABLE 6. Cross-coupling between alkyl Grignard reagents and aryl or heteroaryl chlorides, triflates, and tosylates

$$n\text{-HexMgBr} + \text{ArX} \xrightarrow[\substack{\text{THF/NMP} \\ 0\,°\text{C to rt} \\ 20\,\text{min}}]{5\%\ \text{Fe(acac)}_3} n\text{-Hex}-\text{Ar}$$

1.2 equiv.

X = Cl, OTs, OTf

ArX	Yield (%)
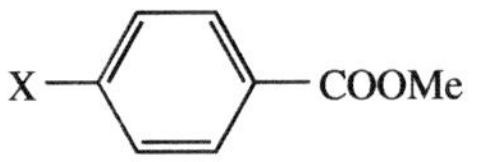 X—C₆H₄—COOMe (para)	91 (X = Cl) 87 (X = OTf) 83 (X = OTs)
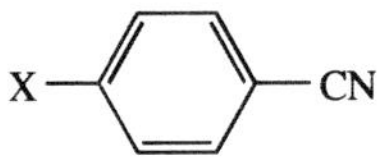 X—C₆H₄—CN (para)	91 (X = Cl) 87 (X = OTf) 83 (X = OTs)
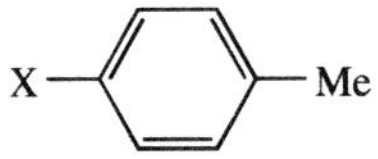 X—C₆H₄—Me (para)	0 (X = Cl) 81 (X = OTf)
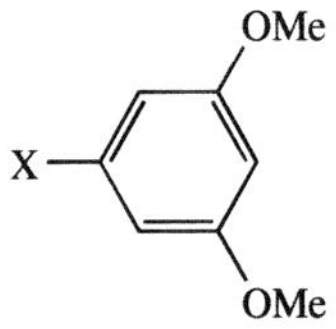 X—C₆H₃(OMe)₂ (3,5-dimethoxy)	0 (X = Cl) 90 (X = OTf)
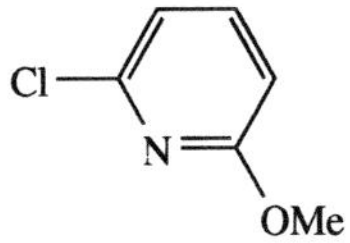 2-Cl-6-OMe-pyridine	95
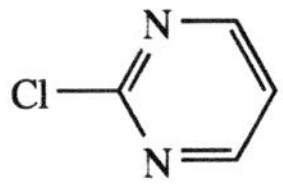 2-Cl-pyrimidine	93
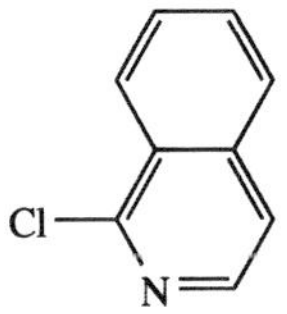 1-Cl-isoquinoline	95
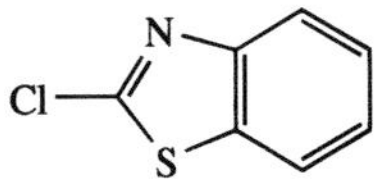 2-Cl-benzothiazole	68

TABLE 6. (continued)

ArX	Yield (%)
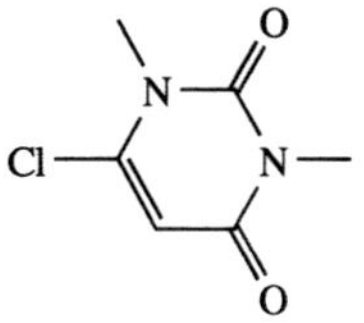 6-chloro-1,3-dimethyluracil	60

SCHEME 33

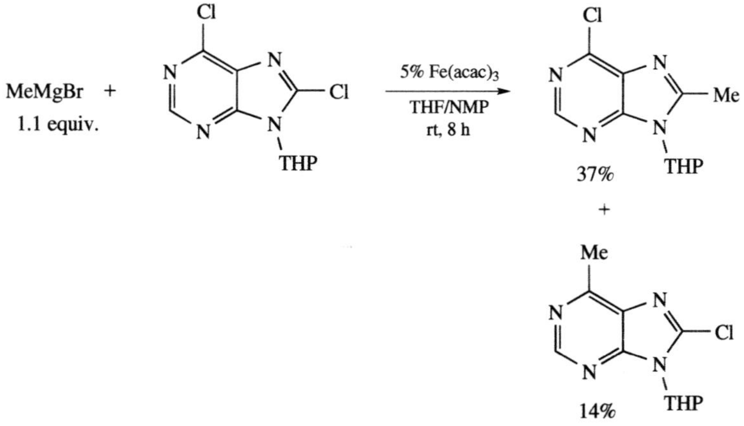

MeMgBr + 2,6-dichloro-9-benzylpurine (1 equiv.) →[5% Fe(acac)$_3$, THF/NMP, rt, 8 h] 2-chloro-6-methyl-9-benzylpurine (72%)

MeMgBr + 2-chloro-6-phenyl-8-chloro-9-benzylpurine (3 equiv.) →[5% Fe(acac)$_3$, THF/NMP, rt, 8 h] 2-chloro-6-phenyl-8-methyl-9-benzylpurine (79%)

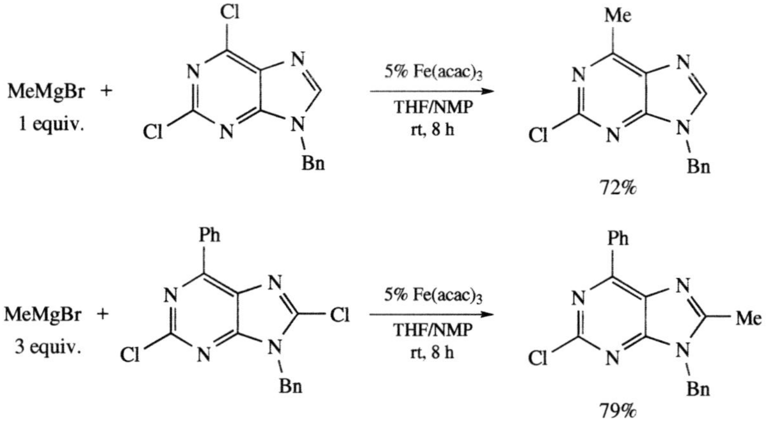

SCHEME 34

MeMgBr + 6,8-dichloro-9-THP-purine (1.1 equiv.) →[5% Fe(acac)$_3$, THF/NMP, rt, 8 h] 6-chloro-8-methyl-9-THP-purine (37%) + 6-methyl-8-chloro-9-THP-purine (14%)

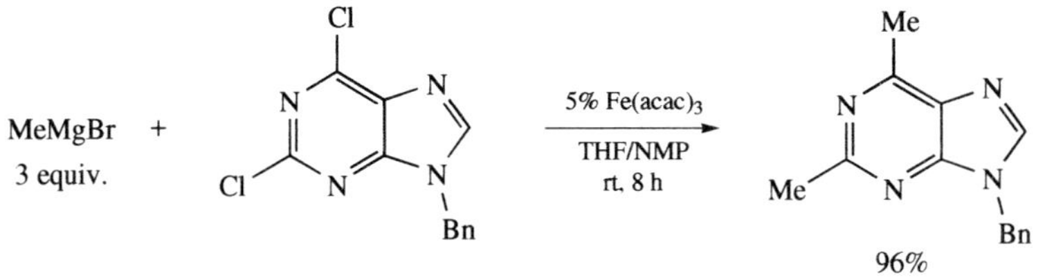

SCHEME 35

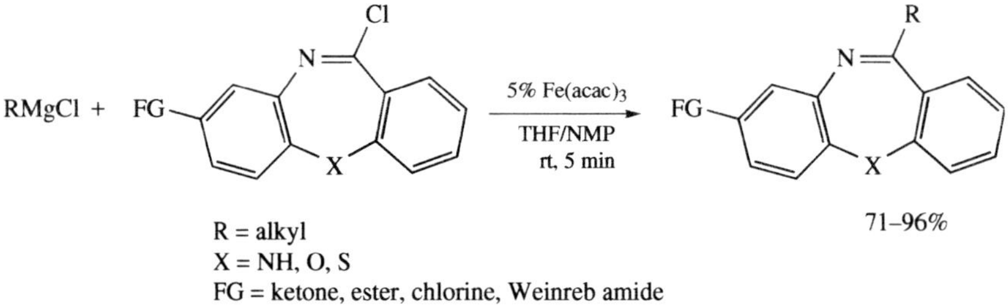

SCHEME 36

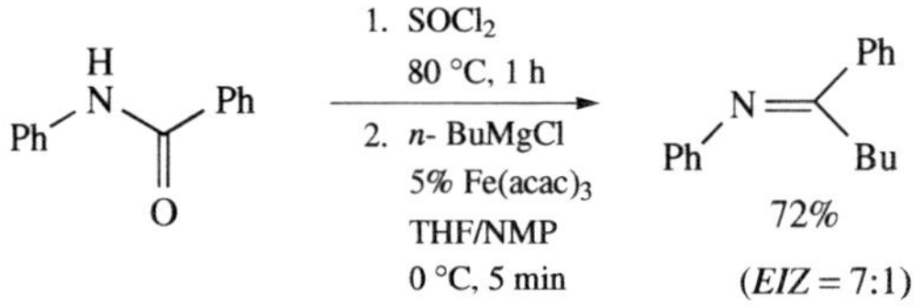

R = alkyl
X = NH, O, S
FG = ketone, ester, chlorine, Weinreb amide

SCHEME 37

C. Iron-catalyzed Heteroarylation of Aryl Grignard Reagents

Iron-catalyzed aryl–aryl coupling reactions generally lead to poor yields of cross-coupling product since the homocoupling of the starting Grignard reagent mainly occurs (see above). However, in the case of heteroaryl halides, moderate yields of cross-coupling product can be obtained by using an excess of aryl Grignard reagent. Thus, Figadère and coworkers[65] synthesized 2-phenylquinoline in 65% yield (Scheme 38).

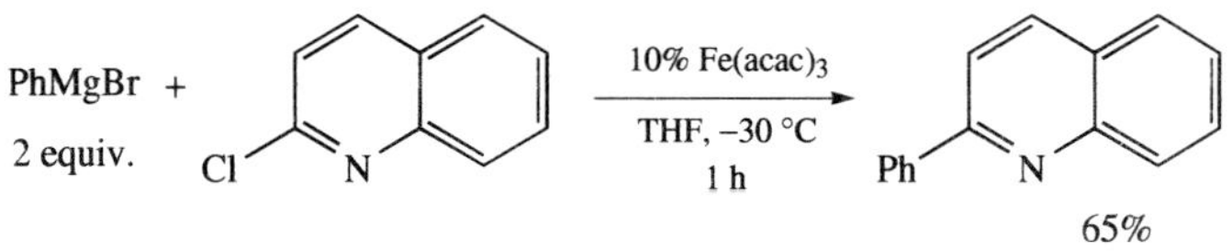

SCHEME 38

Fürstner and coworkers[58, 59] also reported various examples of coupling with nitrogen heteroaryl chlorides (Table 7). In all cases, at least two equivalents of phenylmagnesium bromide are required and the yield never rises above 71%.

TABLE 7. Cross-coupling between phenylmagnesium bromide and heteroaryl chlorides

PhMgBr + RCl $\xrightarrow[\text{THF, }-30\,°C]{5\%\text{ Fe(acac)}_3}$ Ar—R

2.3 equiv.

20 min

R = heteroaryl

RCl	Yield (%)
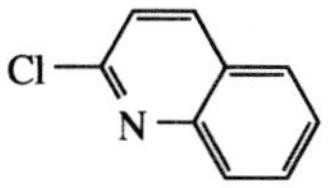 2-chloroquinoline	71
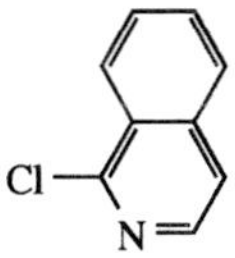 1-chloroisoquinoline	57
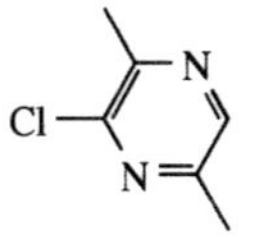 2-chloro-4,6-dimethylpyrimidine	64
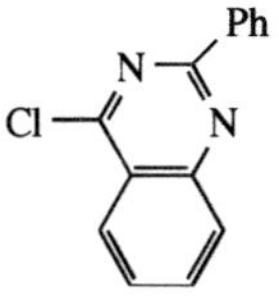 4-chloro-2-phenylquinazoline	66
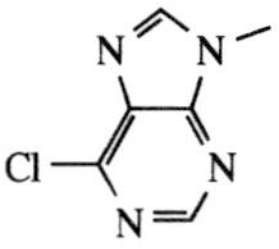 6-chloro-9-methylpurine	60

D. Iron-catalyzed Alkylation of Aryl Grignard Reagents

Recently, this reaction has been extensively studied since it is currently the only method to couple aryl Grignard reagents with secondary alkyl halides[12, 16]. Indeed, secondary alkyl halides do not react under palladium or nickel catalysis[66]. On the other hand, let us recall that the coupling of secondary alkyl Grignard reagents with aryl halides leads to poor results (see above).

The reaction can be performed in diethyl ether in the presence of iron(III) acetylacetonate, as reported by Nagano and Hayashi[67] in 2004, or in the presence of FeCl(salen) or iron(III) chloride/triethylamine, as described by Bedford and coworkers[68] (Scheme 39). The latter compared several ligands (amines[69], phosphines[70]) and the best results were obtained with triethylamine, TMEDA or DABCO. In all cases, the reactions have to be performed in refluxing diethyl ether and the Grignard reagent has to be added at once! Unfortunately, these reaction conditions are only useable on a very small scale (1 mmol) but they cannot be used for large-scale applications.

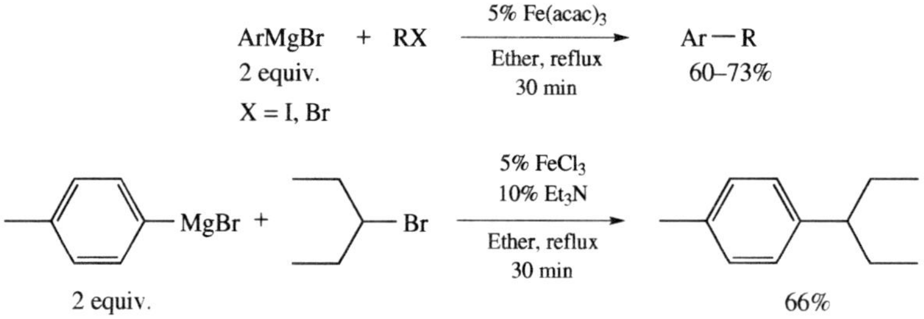

ArMgBr + RX →[5% Fe(acac)₃][Ether, reflux, 30 min] Ar—R
2 equiv.
X = I, Br
60–73%

[p-MeC₆H₄-MgBr] (2 equiv.) + [sec-BuBr] →[5% FeCl₃, 10% Et₃N][Ether, reflux, 30 min] [p-MeC₆H₄-sec-Bu] 66%

SCHEME 39

As shown by Nakamura and coworkers[71] in 2004, the reaction also takes place in THF in the presence of iron(III) chloride (Scheme 40). However, it is necessary to add a huge amount of TMEDA (120% for 5% FeCl₃). Primary and secondary alkyl halides were coupled successfully. It is important to note that the results clearly depend on the origin of the iron(III) chloride[72].

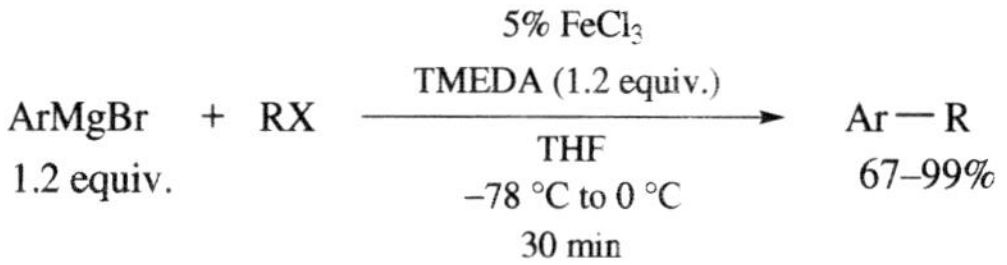

ArMgBr + RX →[5% FeCl₃, TMEDA (1.2 equiv.)][THF, −78 °C to 0 °C, 30 min] Ar—R
1.2 equiv.
67–99%

SCHEME 40

Martin and Fürstner[73] showed that an iron(-II) complex can be used as a precatalyst. This result is very interesting for mechanistic considerations. Unfortunately, this sophisticated complex is not very attractive for large-scale applications. The reaction is remarkably chemoselective, thus the presence of ketone, ester, isocyanide, chloride and nitrile is tolerated (Scheme 41).

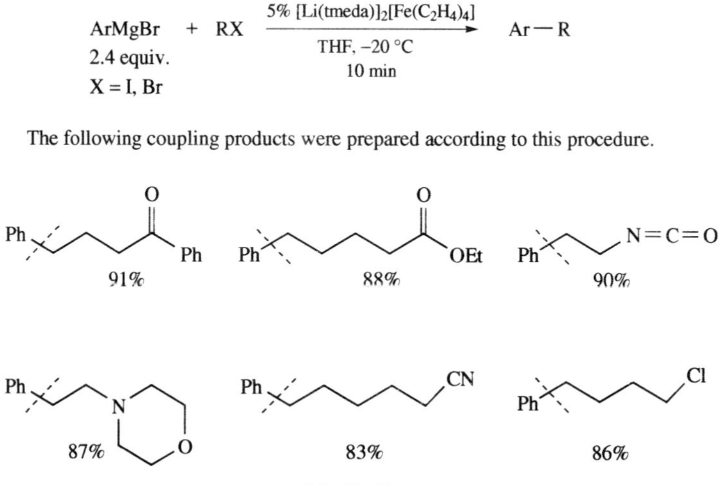

ArMgBr + RX →[5% [Li(tmeda)]₂[Fe(C₂H₄)₄]][THF, −20 °C, 10 min] Ar—R
2.4 equiv.
X = I, Br

The following coupling products were prepared according to this procedure.

Ph~~~C(O)~~~Ph 91%
Ph~~~~C(O)OEt 88%
Ph~~N=C=O 90%

Ph~~N-morpholine 87%
Ph~~~~CN 83%
Ph~~~~Cl 86%

SCHEME 41

13. Iron-Catalyzed Reactions of Grignard Reagents

In 2007, Cahiez and coworkers[72] disclosed two new catalytic systems to improve the reaction, especially for large-scale syntheses. In the first catalytic system, iron(III) acetylacetonate is used instead of iron(III) chloride. In addition, the use of hexamethylenetetramine (HMTA, 5%), a very cheap new ligand, allows one to significantly lower the amount of TMEDA (10% instead of 120%) (Scheme 42). This catalytic system is based on a synergy between TMEDA and HMTA.

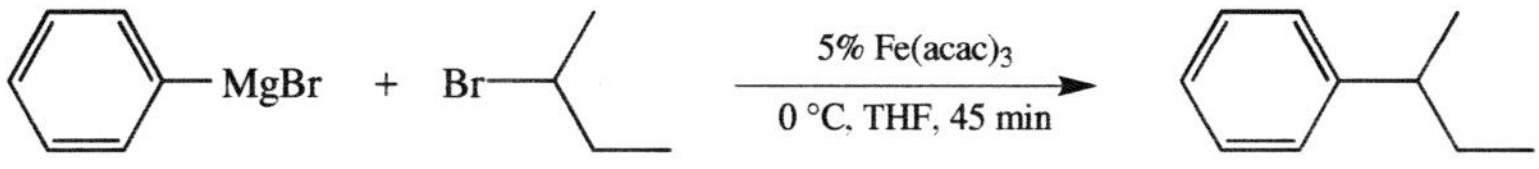

PhMgBr + iPrBr → (5% Fe(acac)$_3$, 0 °C, THF, 45 min) → PhCH(CH$_3$)CH$_2$CH$_3$

1.3 equiv.

no ligand : 45%
5% HMTA : 80%
50% TMEDA : 90%
10% TMEDA/ 5% HMTA : 92%

SCHEME 42

The reaction can be applied to various secondary and primary alkyl bromides (Table 8). The second procedure uses 1.5% [(FeCl$_3$)$_2$(tmeda)$_3$] as a catalyst (Scheme 43). The complex, which is not hygroscopic, is very easily obtained by adding 1.5 equivalents of TMEDA to a solution of iron(III) chloride in THF (Scheme 44). It is quantitatively isolated by filtration.

TABLE 8. Cross-coupling between aryl Grignard reagents and alkyl bromides

ArMgX + RBr → (5% Fe(acac)$_3$, 10% TMEDA, 5% HMTA, THF, 0 °C, 45 min) → Ar—R

1.3 equiv.

ArMgBr	RBr	Yield (%)
PhMgBr	 Br-CH$_2$CH$_2$CH$_2$CH$_3$	75
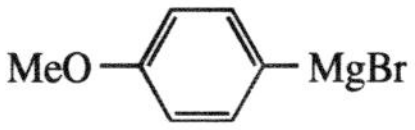 MeO-C$_6$H$_4$-MgBr	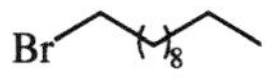 Br-(CH$_2$)$_8$-CH$_3$ (branched)	72
PhMgBr	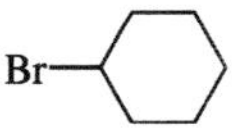 Br-cyclohexyl	94
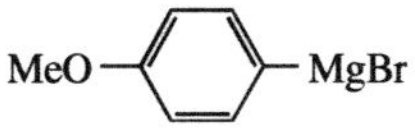 MeO-C$_6$H$_4$-MgBr	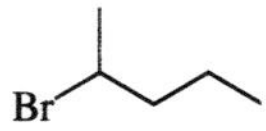 Br-CH(CH$_3$)CH$_2$CH$_3$	93
 MeO-C$_6$H$_4$-MgBr	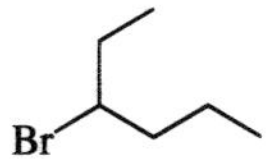 sec-Bu-Br	88
 Me$_2$N-C$_6$H$_4$-MgBr	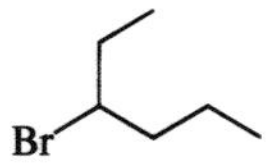 sec-Bu-Br	85

618   Gérard Cahiez and Christophe Duplais

C₆H₅—MgBr + Br—CH(CH₃)(C₂H₅) $\xrightarrow[\text{THF, 20 °C, 90 min}]{1.5\% \,[(FeCl_3)_2(tmeda)_3]}$ PhCH(CH₃)(C₂H₅)

1.3 equiv.   78%
                                                                       0.1 mol scale

SCHEME 43

$FeCl_3$ + TMEDA $\xrightarrow{\text{THF, rt}}$ $(FeCl_3)_2(tmeda)_3$
         1.5 equiv.

SCHEME 44

Recently, Fu and coworkers[66] have shown that secondary alkyl halides do not react under palladium catalysis since the oxidative addition is too slow. They have demonstrated that this lack of reactivity is mainly due to steric effects. Under iron catalysis, the coupling reaction is clearly less sensitive to such steric influences since cyclic and acyclic secondary alkyl bromides were used successfully. Such a difference could be explained by the mechanism proposed by Cahiez and coworkers[72] (Figure 2). Contrary to Pd°, which reacts with alkyl halides according to a concerted oxidative addition mechanism, the iron-catalyzed reaction could involve a two-step monoelectronic transfer.

Finally, Bica and Gaertner[74] have recently shown that an ionic liquid can be used as a solvent to perform the reaction.

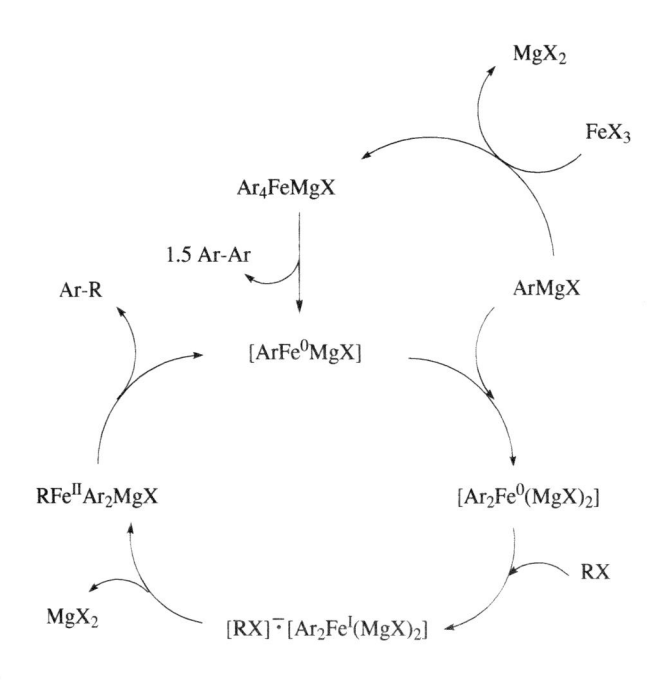

FIGURE 2

V. IRON-CATALYZED HOMOCOUPLING OF AROMATIC GRIGNARD REAGENTS

The first iron-mediated homocoupling reaction was described in 1939 by Gilman and Lichtenwalter[75]. They showed that phenylmagnesium iodide reacts with a sub-stoichiometric amount of iron(II) chloride, in diethyl ether, to give biphenyl in 98% yield (Scheme 45).

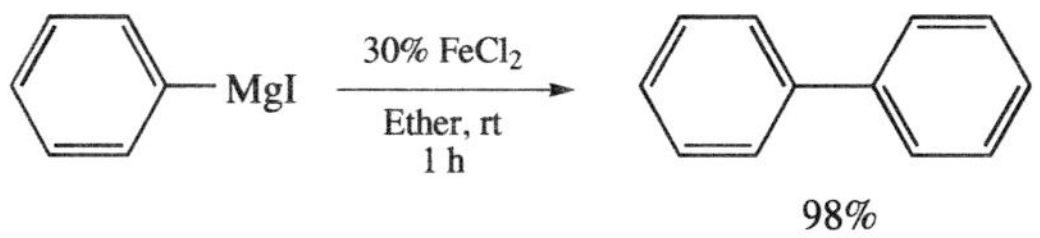

SCHEME 45

In 2005, Cahiez and coworkers[76] and Nagano and Hayashi[77] showed that a catalytic amount of iron(III) chloride is efficient when a suitable oxidant is added to the reaction mixture (Figure 3). Nagano and Hayashi used an excess of 1,2-dichloroethane (the stoichiometric amount is 0.5 equivalent) in refluxing diethyl ether to couple various aryl Grignard reagents in high yield (Scheme 46).

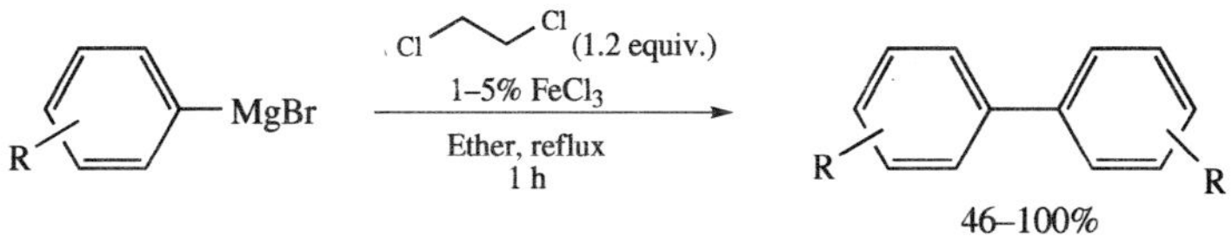

SCHEME 46

Cahiez decided to develop the reaction in THF. Indeed, diethyl ether is not convenient for large-scale applications (especially at reflux). Moreover, many aromatic Grignard reagents can only be obtained in THF e.g. by preparation of arylmagnesium chlorides from aryl chlorides or preparation of functionalized arylmagnesium halides from the corresponding aryl iodides by iodide–magnesium exchange[45].

The coupling procedure requires only 0.6 equivalent of 1,2-dihalogenoethane as an oxidant and can be applied to various simple aryl and heteroaryl Grignard reagents. It should be noted that in the case of hindered and functionalized aryl Grignard reagents, a dramatic improvement was observed by using 1,2-dibromo- or 1,2-diiodoethane instead of 1,2-dichloroethane (Table 9).

Pei and coworkers[78] reported that the homocoupling product can be obtained by adding an aryl bromide to a mixture of magnesium and 2% iron salts in THF (Scheme 47). All reactions were performed on a 1 mmol scale.

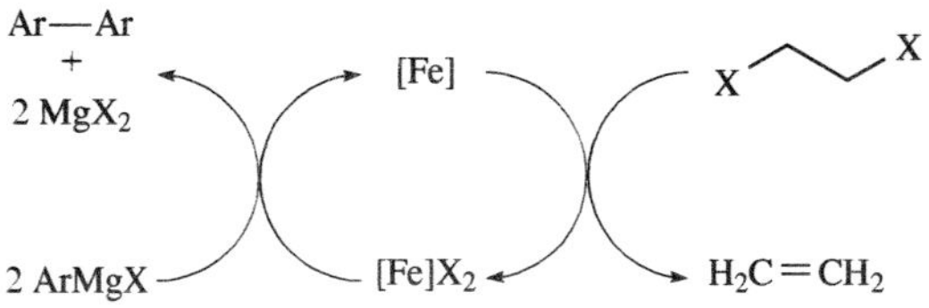

FIGURE 3

TABLE 9. Homocoupling of aryl and heteroaryl Grignard reagents

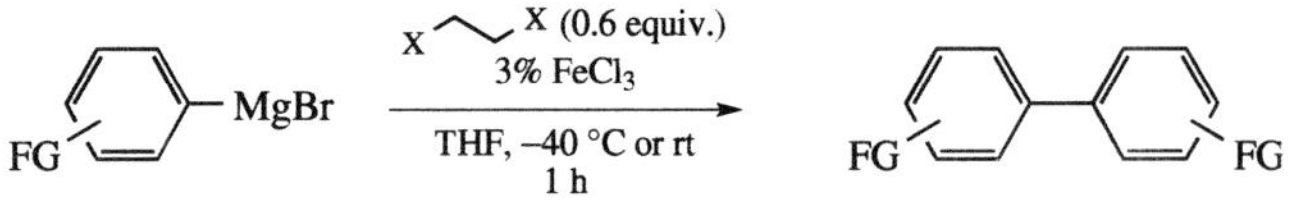

ArMgBr	X	Yield (%)
 MeO–C₆H₄–MgBr	Cl	90
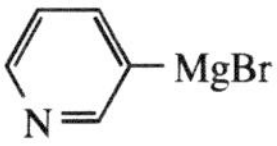 3-pyridyl-MgBr	Cl	82
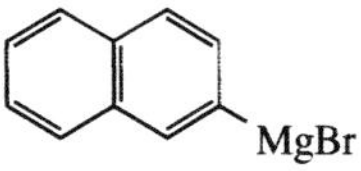 2-naphthyl-MgBr	Cl	81
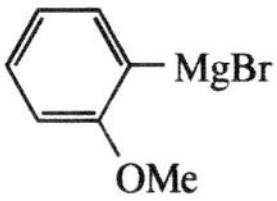 2-(OMe)C₆H₄–MgBr	Cl	21
	Br	73
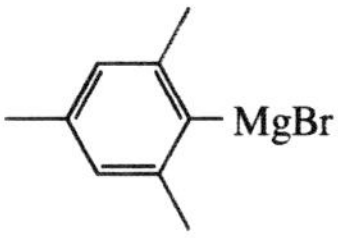 mesityl-MgBr	Cl	5
	Br	60
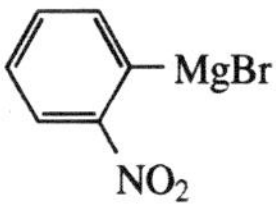 2-(NO₂)C₆H₄–MgBr	Br	41
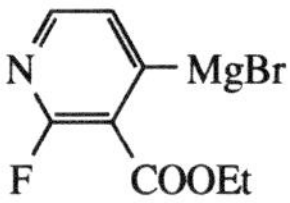 2-F-3-(COOEt)-pyridyl-MgBr	I	33
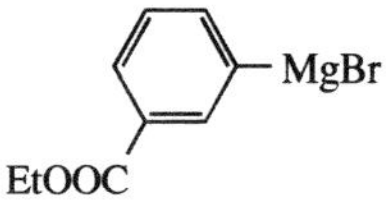 3-(EtOOC)C₆H₄–MgBr	I	67
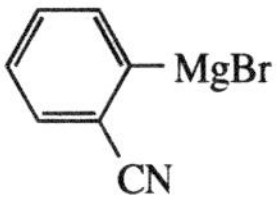 2-(CN)C₆H₄–MgBr	I	75
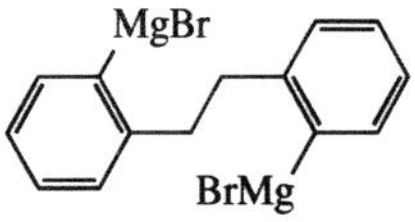 1,2-bis(2-BrMg-phenyl)ethane	Br	76

13. Iron-Catalyzed Reactions of Grignard Reagents

$$\text{ArBr} \xrightarrow[\substack{\text{THF, rt} \\ \text{Mg (2 equiv.)}}]{2\% \text{ Fe(acac)}_3 \text{ or Fe(dbm)}_3} \text{Ar–Ar}$$

1 mmol   59–92%

SCHEME 47

According to our experience, these results are not reproducible on a larger scale since, in all cases, the main product is the Grignard reagent and not the homocoupling product. In fact, it is not surprising since iron salts have been successfully used by Bogdanovic and Schwickardi[79] to catalyze the formation of aryl Grignard reagents from aryl halides and magnesium.

VI. OTHER REACTIONS

A. Iron-catalyzed Substitution

In 1976, Pasto and coworkers[80] described the S_N2' reaction of primary and secondary alkyl Grignard reagents with terminal and non-terminal propargylic chlorides (Scheme 48). Only 0.1% iron(III) chloride is necessary to obtain various allenes in good yields.

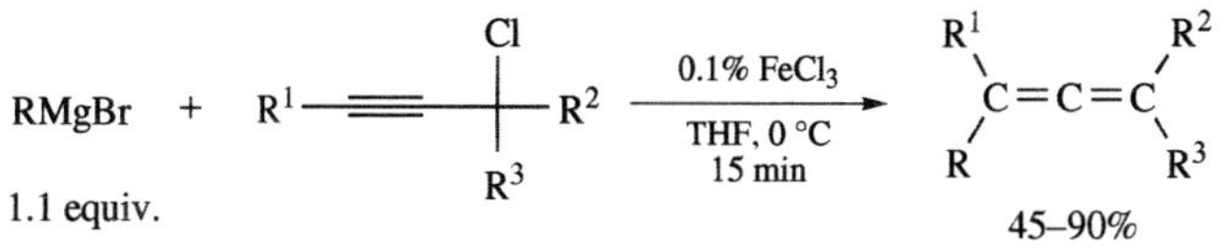

RMgBr + R¹–≡–C(Cl)(R³)–R² $\xrightarrow[\text{15 min}]{0.1\% \text{ FeCl}_3, \text{ THF, 0 °C}}$ R¹R(C=C=C)R²R³

1.1 equiv.   45–90%

SCHEME 48

Yamamoto and coworkers[81,82] studied the substitution of allylic phosphates by Grignard reagents in the presence of copper or iron salts. Only the S_N2' product is formed under copper catalysis whereas, in the presence of iron(III) acetylacetonate, the S_N2 product is generally obtained with an excellent selectivity (Scheme 49). It should be noted that aryl-, alkenyl-, alkynyl- and alkylmagnesium halides can be used successfully.

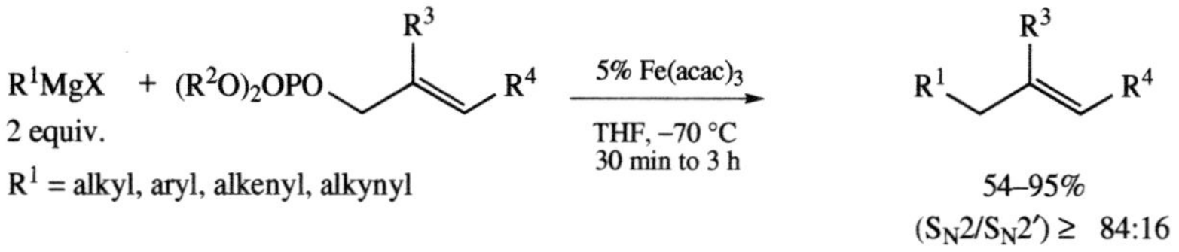

R¹MgX + (R²O)₂OPO–CH(R³)–CH=CH–R⁴ $\xrightarrow[\substack{\text{THF, −70 °C} \\ \text{30 min to 3 h}}]{5\% \text{ Fe(acac)}_3}$ R¹–CH(R³)–CH=CH–R⁴

2 equiv.

R¹ = alkyl, aryl, alkenyl, alkynyl

54–95%

$(S_N2/S_N2') \geq 84{:}16$

SCHEME 49

In 2003, Fürstner and Méndez[83] reported an elegant reaction between optically active propargylic epoxides and organomagnesium compounds in the presence of iron(III) acetylacetonate (Scheme 50). Interestingly, the chirality of the starting product is transferred to the final 2,3-allenols which are obtained with a good enantiomeric purity.

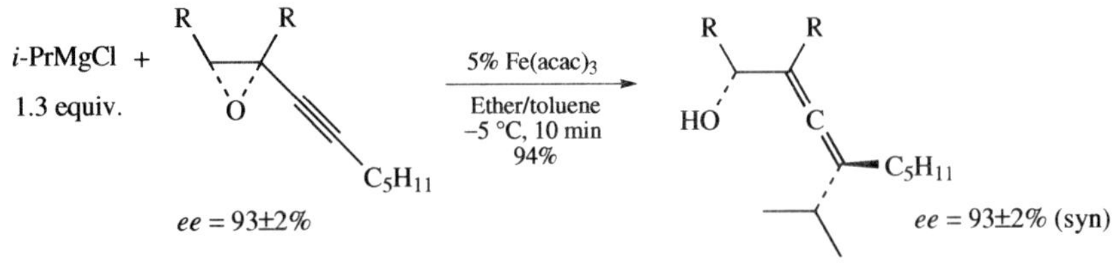

622   Gérard Cahiez and Christophe Duplais

SCHEME 50

In the same year, Nakamura and coworkers[84] published an iron-catalyzed ring opening reaction of oxabicyclic alkenes by Grignard reagents (Scheme 51). This reaction is highly regio- and stereoselective.

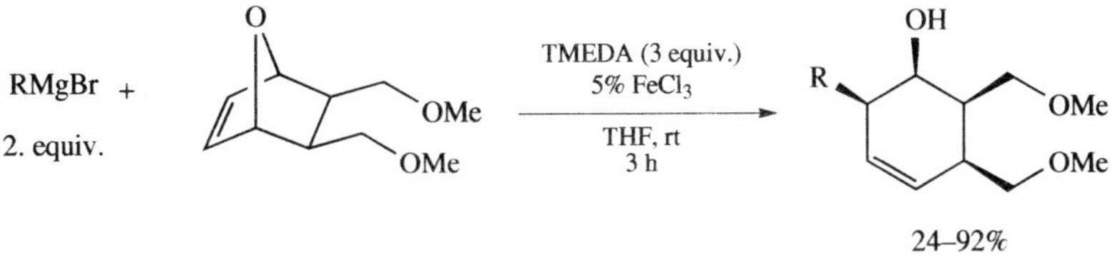

SCHEME 51

Finally, Tanabe and coworkers[85] reported that treatment of *gem*-dichlorocyclopropanes with methylmagnesium bromide in the presence of 5% Fe(dbm)$_3$ and 4-methoxytoluene (1 equiv.) affords the dimethylated product (Scheme 52).

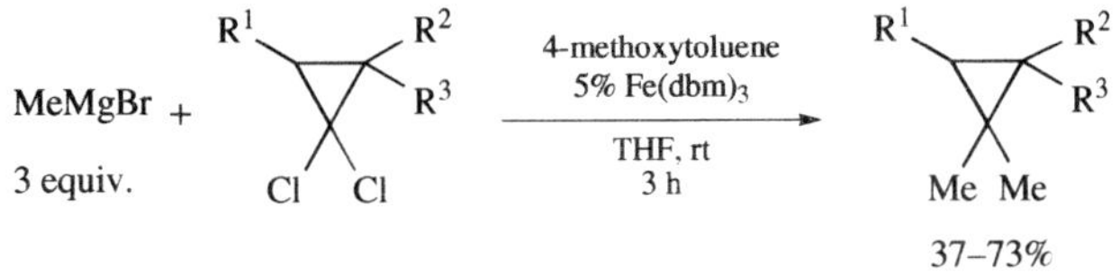

SCHEME 52

B. Carbometallation

In 2000, Nakamura and coworkers[86] described the iron-catalyzed addition of Grignard reagents to the cyclopropenone derivatives depicted in Scheme 53. Phenyl, vinyl and alkyl Grignard reagents lead to good to excellent yields. The carbometallation is highly stereoselective and the resulting cyclopropylmagnesium halide reacts with various electrophiles with retention of configuration.

Recently, Zhang and Ready[87] published an iron-catalyzed carbometallation of propargylic and homopropargylic alcohols. This reaction generates regioselectively tri- and tetrasubstituted olefins in satisfactory yields (Scheme 54). Unfortunately, a large excess of the Grignard reagents is required and sub-stoichiometric amounts of iron salts are necessary.

In 2005, Hayashi and coworkers[88] reported the iron/copper co-catalyzed arylmagnesiation of alkynes. This method gives only moderate yields but the stereoselectivity is generally excellent (Scheme 55).

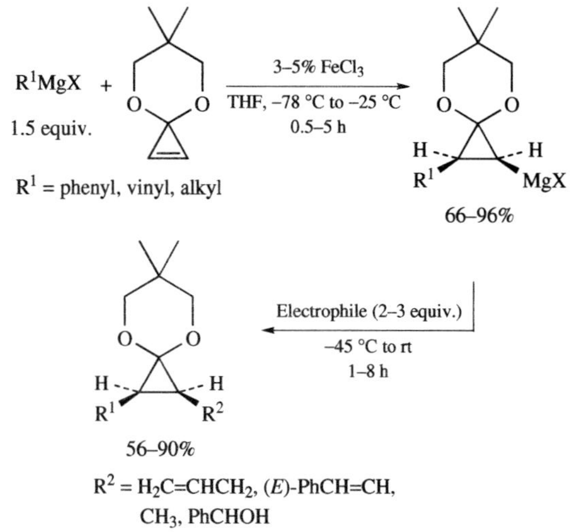

13. Iron-Catalyzed Reactions of Grignard Reagents

SCHEME 53

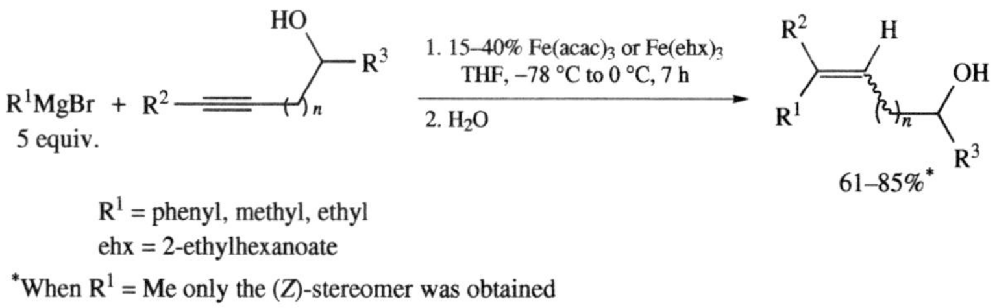

SCHEME 54

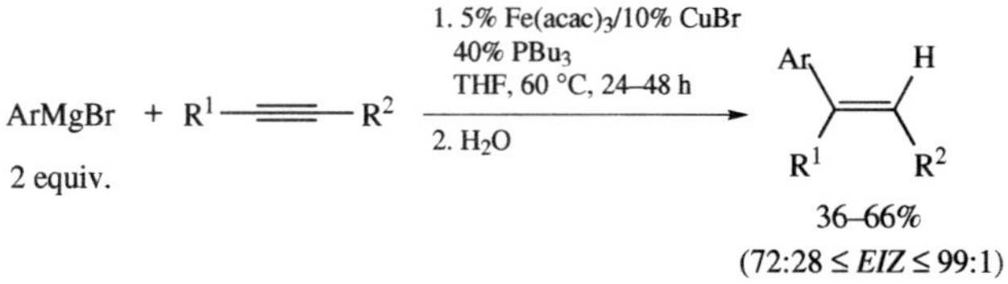

R^1 = phenyl, methyl, ethyl
ehx = 2-ethylhexanoate

*When R^1 = Me only the (Z)-stereomer was obtained

SCHEME 55

Two years later, Hayashi and coworkers[89] showed that the use of an N-heterocyclic carbene ligand (IPr) allows one to perform the iron-catalyzed addition of arylmagnesium bromides to conjugated arylalkynes in high yields and with a good stereoselectivity (Scheme 56).

C. Radical Cyclization

In 1998, Oshima and coworkers[90] reported few examples of radical cyclization mediated by Grignard reagents in the presence of iron(II) chloride (Scheme 57). However, this reaction often gives moderate yields.

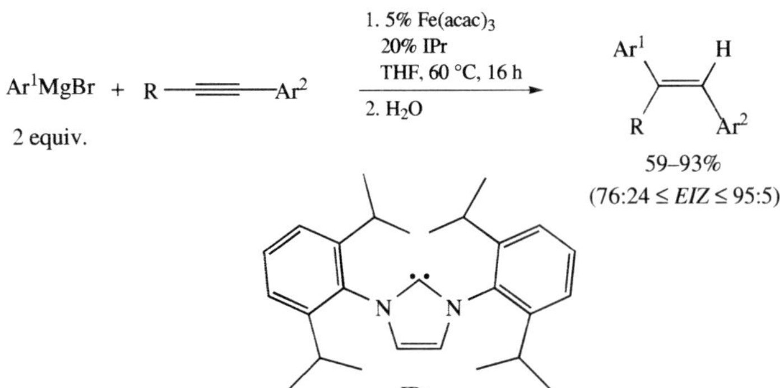

SCHEME 56

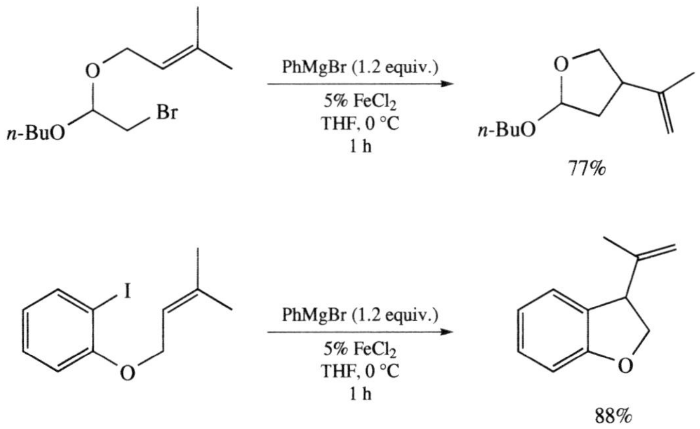

SCHEME 57

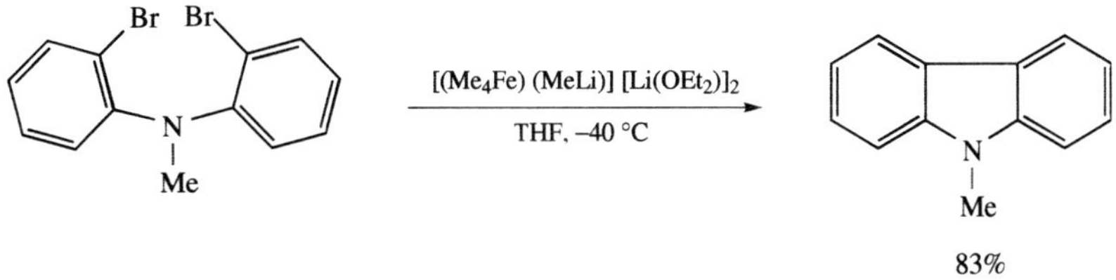

SCHEME 58

It is noteworthy that the 'super' iron(II) ate complex described by Fürstner and co-workers[35] is also able to promote such a radical cyclization (Scheme 58).

D. Addition to Conjugated Unsaturated Carbonyl Compounds

In 1941, Kharasch and Sayles[4] reported the 1,4-addition of methylmagnesium bromide to chalcone in the presence of iron salts (Scheme 59).

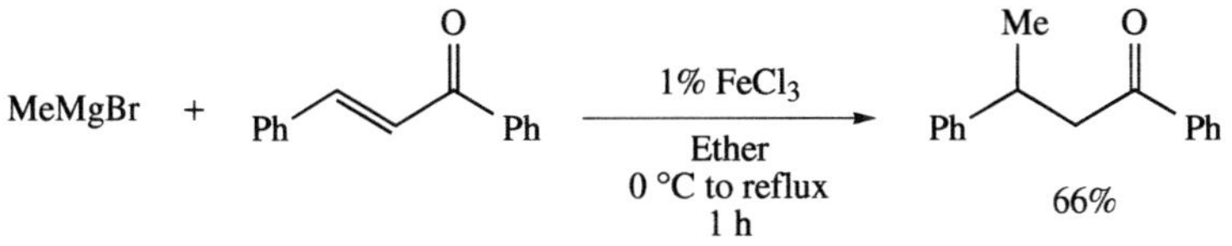

13. Iron-Catalyzed Reactions of Grignard Reagents

MeMgBr + Ph—CH=CH—C(O)—Ph → (1% FeCl₃, Ether, 0 °C to reflux, 1 h) → Ph—CH(Me)—CH₂—C(O)—Ph 66%

SCHEME 59

Many years later, Fukuhara and Urabe[91] showed that aryl Grignard reagents react with ethyl 2,4-dienoates or 2,4-dienamides, in the presence of iron salts, to give the 1,6-addition products (Scheme 60). This reaction is stereoselective and leads to the (Z)-trisubstituted olefins.

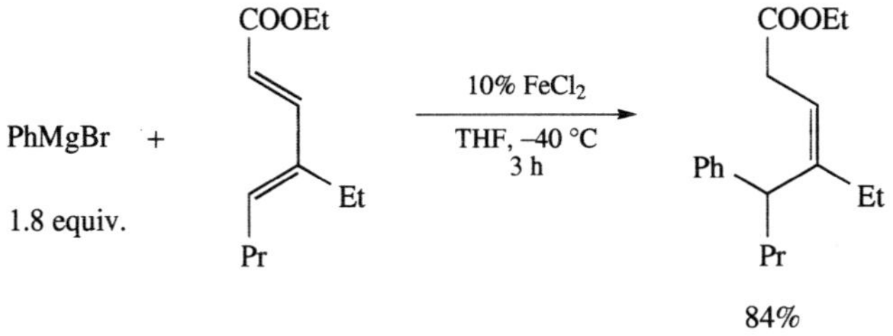

PhMgBr (1.8 equiv.) + dienoate with COOEt, Et, Pr substituents → (10% FeCl₂, THF, −40 °C, 3 h) → 1,6-addition product 84%

SCHEME 60

VII. APPLICATIONS OF IRON-CATALYZED REACTIONS OF GRIGNARD REAGENTS IN ORGANIC SYNTHESIS

These last years, several syntheses of natural products using an iron-catalyzed reaction of Grignard reagents have been published. In 1969, Meinwald and Hendry[92] used the reaction discovered by Kharasch and Tawney[2,6] to prepare an allenic sesquiterpenoid isolated from the grasshopper *Romalea Microptera* (Scheme 61).

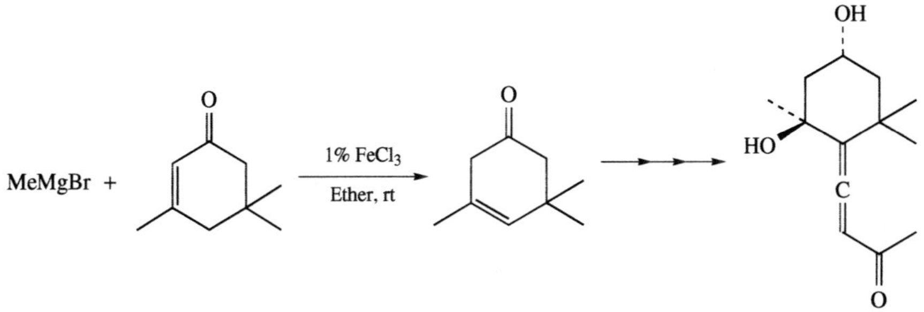

MeMgBr + isophorone → (1% FeCl₃, Ether, rt) → 1,4-addition product → allenic sesquiterpenoid (with OH, HO, C=C=CH–C(O)–)

SCHEME 61

The Kharasch reaction has also been employed by Gennari and coworkers[93] to synthesize, from (L)-carvone, a potential functionalized precursor of sarcodictyins and eleutherobin (Scheme 62).

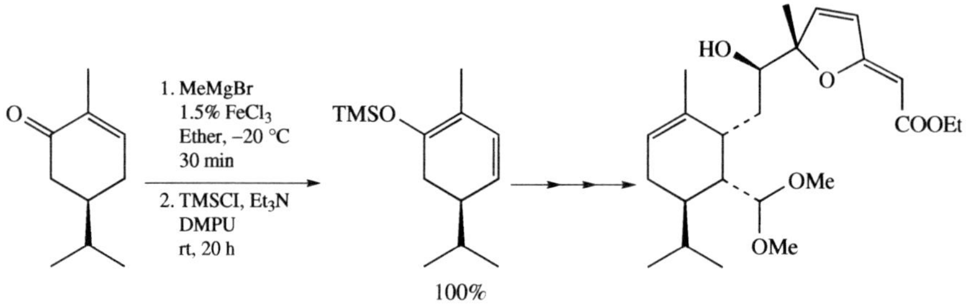

SCHEME 62

Iron-catalyzed alkenylation of Grignard reagents was used by Cahiez and Avedissian[29] to prepare the pheromone of *Argyroplace Leucotetra* in three steps from 1,2-(*E*)-dichloroethene (Scheme 63). Two successive alkenylation reactions, the first involving a cobalt catalysis, the second an iron catalysis, allow one to obtain the desired product in 45% overall yield.

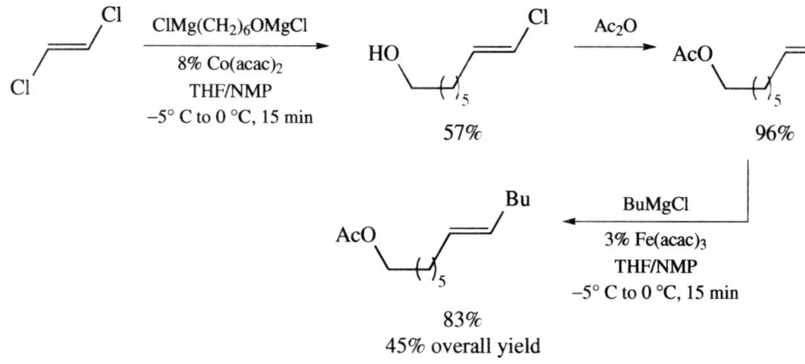

SCHEME 63

The iron-catalyzed homocoupling of Grignard reagents was successfully applied to the synthesis of *N*-methylcrinasiadine by Cahiez and coworkers[76] (Scheme 64).

Fürstner and Leitner[94] have recently described the synthesis of (*R*)-(+)-muscopyridine. This elegant strategy is based on two regioselective iron-catalyzed heteroaryl–alkyl coupling reactions (Scheme 65).

A similar cross-coupling reaction was used for the synthesis of the immunosuppressive agent FTY720 (Scheme 66)[95].

The iron-catalyzed addition of Grignard reagents to propargylic epoxides developed by Fürstner and Méndez[83] allows one to prepare a *syn*-allenol, which is an important intermediate for the synthesis of a precursor of the amphidinolide X[96] (Scheme 67).

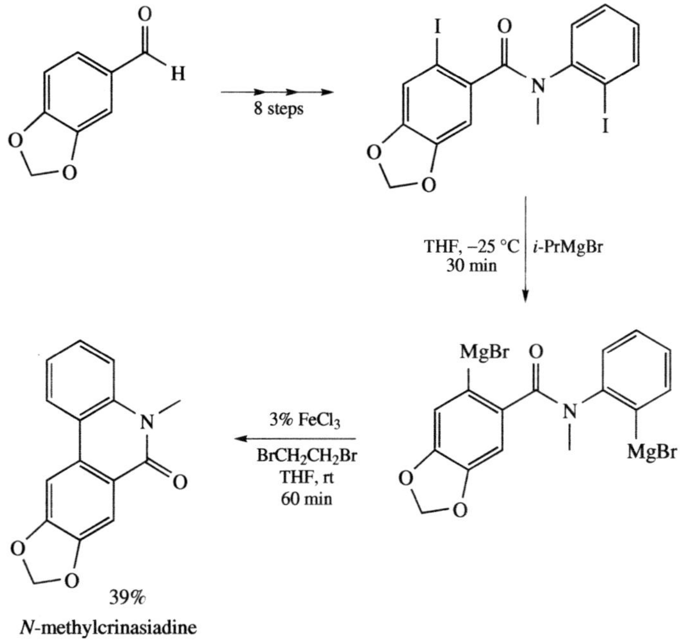

SCHEME 64

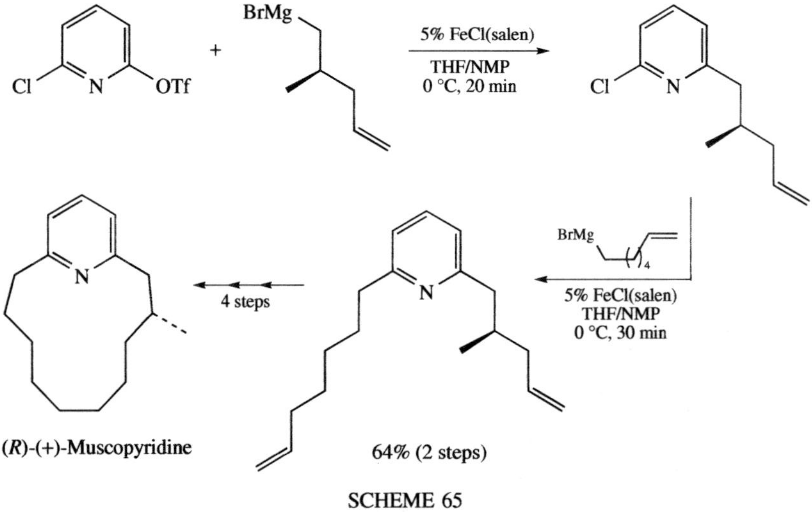

SCHEME 65

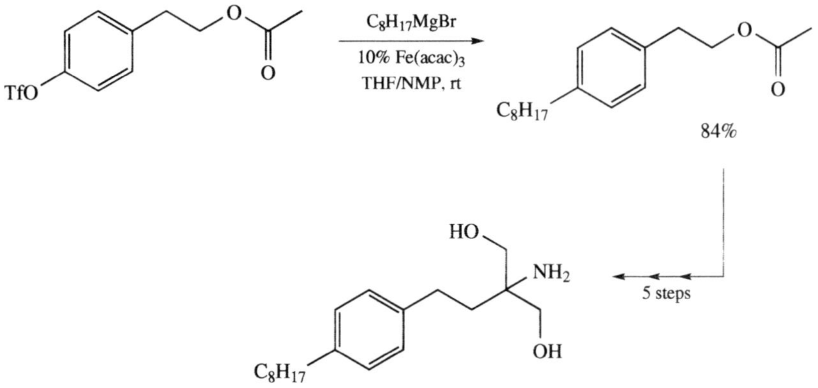

SCHEME 66

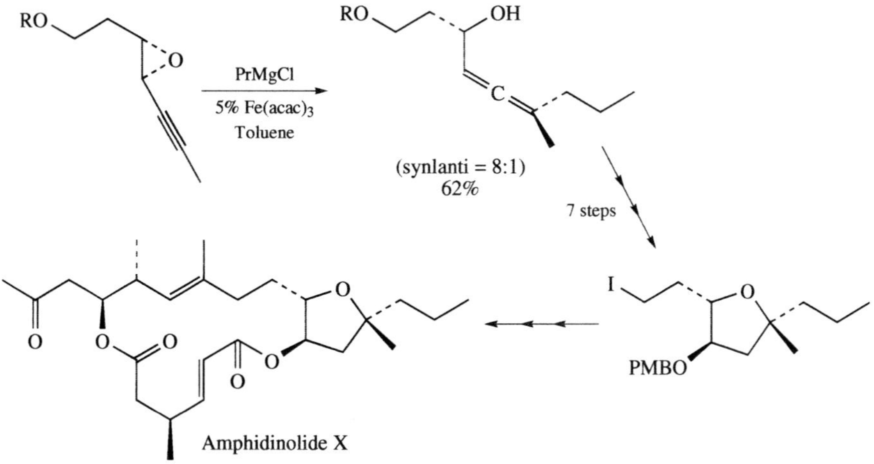

SCHEME 67

VIII. REFERENCES

1. M. S. Kharasch, S. C. Kleiger, J. A. Martin and F. R. Mayo, *J. Am. Chem. Soc.*, **63**, 2305 (1941).
2. M. S. Kharasch and P. O. Tawney, *J. Am. Chem. Soc.*, **63**, 2308 (1941).
3. M. S. Kharasch and E. K. Fields, *J. Am. Chem. Soc.*, **63**, 2316 (1941).
4. M. S. Kharasch and D. C. Sayles, *J. Am. Chem. Soc.*, **64**, 2972 (1942).
5. M. S. Kharasch, R. T. Morrison and W. H. Hurry, *J. Am. Chem. Soc.*, **66**, 368 (1944).
6. M. S. Kharasch and P. O. Tawney, *J. Am. Chem. Soc.*, **67**, 128 (1945).
7. M. S. Kharasch and C. F. Fuchs, *J. Org. Chem.*, **10**, 292 (1945).
8. M. S. Kharasch, M. Weiner, W. Nudenberg, A. Bhattacharya, T. Wang and N. C. Yang, *J. Am. Chem. Soc.*, **83**, 3232 (1961).
9. M. Tamura and J. K. Kochi, *J. Am. Chem. Soc.*, **93**, 1487 (1971).
10. M. Tamura and J. K. Kochi, *Synthesis*, 303 (1971).
11. S. M. Neumann and J. K. Kochi, *J. Organomet. Chem.*, **31**, 289 (1971).

13. Iron-Catalyzed Reactions of Grignard Reagents 629

12. J. K. Kochi, *Acc. Chem. Res.*, **7**, 351 (1974).
13. S. M. Neumann and J. K. Kochi, *J. Org. Chem.*, **40**, 599 (1975).
14. R. S. Smith and J. K. Kochi, *J. Org. Chem.*, **41**, 502 (1976).
15. C. L. Kwan and J. K. Kochi, *J. Am. Chem. Soc.*, **98**, 4903 (1976).
16. K. L. Rollick, W. A. Nugent and J. K. Kochi, *J. Organomet. Chem.*, **225**, 279 (1982).
17. J. K. Kochi, *J. Organomet. Chem.*, **653**, 11 (2002).
18. G. H. Posner, *Org. React.*, **19**, 1 (1972).
19. G. H. Posner, *Org. React.*, **22**, 253 (1975).
20. B. H. Lipshutz and S. Sengupta, *Org. React.*, **41**, 135 (1992).
21. N. Krause (Ed.), *Modern Organocopper Chemistry*, Wiley-VCH, Weinheim, 2002.
22. N. Miyaura and A. Suzuki, *Chem. Rev.*, **95**, 2457 (1995).
23. J. P. Corbet and G. Mignani, *Chem. Rev.*, **106**, 2651 (2006).
24. Y. Tamaru (Ed.), *Modern Organonickel Chemistry*, Wiley-VCH, Weinheim, 2005.
25. S. Stanford, *Tetrahedron*, **54**, 263 (1998).
26. J. Hassan, M. Sévignon, C. Gozzi, E. Schultz and M. Lemaire, *Chem. Rev.*, **102**, 1359 (2002).
27. A. C. Frisch and M. Beller, *Angew. Chem., Int. Ed.*, **44**, 674 (2005).
28. A. de Meijere and P. J. Stang (Eds.), *Metal-Catalyzed Cross-Coupling Reactions*, 2nd edn., Wiley-VCH, Weinheim, 2004.
29. G. Cahiez and H. Avedissian, *Synthesis*, 1199 (1998).
30. C. Bolm, J. Legros, J. Le Paith and L. Zani, *Chem. Rev.*, **104**, 6217 (2004).
31. H. Shinokubo and K. Oshima, *Eur. J. Org. Chem.*, 2081 (2004).
32. A. Fürstner and R. Martin, *Chem. Lett.*, **34**, 624 (2005).
33. M. E. Krafft and R. A. Holton, *J. Org. Chem.*, **49**, 3669 (1984).
34. M. E. Krafft and R. A. Holton, *J. Am. Chem. Soc.*, **106**, 7619 (1984).
35. A. Fürstner, H. Krause and C. W. Lehmann, *Angew. Chem., Int. Ed.*, **45**, 440 (2006).
36. W. C. Percival, R. C. Wagner and N. C. Cook, *J. Am. Chem. Soc.*, **75**, 3731 (1953).
37. J. Cason and K. W. Kraus, *J. Org. Chem.*, **26**, 1768 (1961).
38. J. Cason and E. J. Reist, *J. Org. Chem.*, **23**, 1668 (1958).
39. J. Cason and K. W. Kraus, *J. Org. Chem.*, **26**, 1772 (1961).
40. V. Fiandanese, G. Marchese, V. Martina and L. Ronzini, *Tetrahedron Lett.*, **25**, 4805 (1984).
41. C. Cardellicchio, V. Fiandanese, G. Marchese and L. Ronzini, *Tetrahedron Lett.*, **26**, 3595 (1985).
42. C. Cardellicchio, V. Fiandanese, G. Marchese and L. Ronzini, *Tetrahedron Lett.*, **28**, 2053 (1987).
43. G. Cahiez, Unpublished results.
44. C. Duplais, F. Bures, I. Sapountzis, T. J. Korn, G. Cahiez and P. Knochel, *Angew. Chem., Int. Ed.*, **43**, 2968 (2004).
45. L. Boymond, M. Röttländer, G. Cahiez and P. Knochel, *Angew. Chem., Int. Ed.*, **37**, 1701 (1998).
46. G. A. Molander, B. J. Rahn, D. C. Shubert and S. E. Bonde, *Tetrahedron Lett.*, **24**, 5449 (1983).
47. W. Dohle, F. Kopp, G. Cahiez and P. Knochel, *Synlett*, **12**, 1901 (2001).
48. M. Seck, X. Franck, R. Hocquemiller, B. Figadère, J. F. Peyrat, O. Provot, J. D. Brion and M. Alami, *Tetrahedron Lett.*, **45**, 1881 (2004).
49. M. A. Fakhfakh, X. Franck, R. Hocquemiller and B. Figadère, *J. Organomet. Chem.*, **624**, 131 (2001).
50. M. A. Fakhfakh, X. Franck, A. Fournier, R. Hocquemiller and B. Figadère, *Synth. Commun.*, **32**, 2863 (2002).
51. M. Dos Santos, X. Franck, R. Hocquemiller, B. Figadère, J. F. Peyrat, O. Provot, J. D. Brion and M. Alami, *Synlett*, 2697 (2004).
52. J. L. Fabre, M. Julia and J. N. Verpeaux, *Tetrahedron Lett.*, **23**, 2469 (1982).
53. J. L. Fabre, M. Julia and J. N. Verpeaux, *Bull. Soc. Chim. Fr.*, **5**, 772 (1985).
54. E. Alvarez, T. Cuvigny, C. Hervé du Penhoat and M. Julia, *Tetrahedron*, **44**, 111 (1988).
55. E. Alvarez, T. Cuvigny, C. Hervé du Penhoat and M. Julia, *Tetrahedron*, **44**, 119 (1988).
56. K. Itami, S. Higashi, M. Mineno and J. Yoshida, *Org. Lett.*, **7**, 1219 (2005).
57. B. Scheiper, M. Bonnekessel, H. Krause and A. Fürstner, *J. Org. Chem.*, **69**, 3943 (2004).
58. A. Fürstner and A. Leitner, *Angew. Chem., Int. Ed.*, **41**, 609 (2002).
59. A. Fürstner, A. Leitner, M. Méndez and H. Krause, *J. Am. Chem. Soc.*, **124**, 13856 (2002).

60. L. E. Aleandri, B. Bogdanovic, P. Bons, C. Dürr, A. Gaidies, T. Hartwig, S. C. Huckett, M. Lagarden, U. Wilczok and R. A. Brand, *Chem. Mater.*, **7**, 1153 (1995).
61. M. Hocek and H. Dvoráková, *J. Org. Chem.*, **68**, 5773 (2003).
62. M. Hocek, D. Hocková and H. Dvoráková, *Synthesis*, 889 (2004).
63. M. Hocek and R. Pohl, *Synthesis*, 2869 (2004).
64. L. K. Ottesen, F. Ek and R. Olsson, *Org. Lett.*, **8**, 1771 (2006).
65. J. Quintin, X. Franck, R. Hocquemiller and B. Figadère, *Tetrahedron Lett.*, **43**, 3547 (2002).
66. I. D. Hills, N. R. Netherton and G. C. Fu, *Angew. Chem., Int. Ed.*, **42**, 5749 (2003).
67. T. Nagano and T. Hayashi, *Org. Lett.*, **6**, 1297 (2004).
68. R. B. Bedford, D. W. Bruce, R. M. Frost, J. W. Goodby and M. Hird, *Chem. Commun.*, 2822 (2004).
69. R. B. Bedford, D. W. Bruce, R. M. Frost and M. Hird, *Chem. Commun.*, 4161 (2005).
70. R. B. Bedford, M. Betham, D. W. Bruce, A. A. Danopoulos, R. M. Frost and M. Hird, *J. Org. Chem.*, **71**, 1104 (2006).
71. M. Nakamura, K. Matsuo, S. Ito and E. Nakamura, *J. Am. Chem. Soc.*, **126**, 3686 (2004).
72. G. Cahiez, V. Habiak, C. Duplais and A. Moyeux, *Angew. Chem., Int. Ed.*, **46**, 4364 (2007).
73. R. Martin and A. Fürstner, *Angew. Chem., Int. Ed.*, **43**, 3955 (2004).
74. K. Bica and P. Gaertner, *Org. Lett.*, **8**, 733 (2006).
75. H. Gilman and M. Lichtenwalter, *J. Am. Chem. Soc.*, **61**, 957 (1939).
76. G. Cahiez, C. Chaboche, F. Mahuteau-Betzer and M. Ahr, *Org. Lett.*, **7**, 1943 (2005).
77. T. Nagano and T. Hayashi, *Org. Lett.*, **7**, 491 (2005).
78. X. Xu, D. Cheng and W. Pei, *J. Org. Chem.*, **71**, 6637 (2006).
79. B. Bogdanovic and M. Schwickardi, *Angew. Chem., Int. Ed.*, **39**, 4610 (2000).
80. D. J. Pasto, G. F. Hennion, R. H. Shults, A. Waterhouse and S. K. Chou, *J. Org. Chem.*, **41**, 3496 (1976).
81. A. Yanagisawa, N. Nomura and H. Yamamoto, *Synlett*, 513 (1991).
82. A. Yanagisawa, N. Nomura and H. Yamamoto, *Tetrahedron*, **50**, 6017 (1994).
83. A. Fürstner and M. Méndez, *Angew. Chem., Int. Ed.*, **42**, 5355 (2003).
84. M. Nakamura, K. Matsuo, T. Inoue and E. Nakamura, *Org. Lett.*, **5**, 1373 (2003).
85. Y. Nishii, K. Wakasugi and Y. Tanabe, *Synlett*, 67 (1998).
86. M. Nakamura, A. Hirai and E. Nakamura, *J. Am. Chem. Soc.*, **122**, 978 (2000).
87. D. Zhang and J. M. Ready, *J. Am. Chem. Soc.*, **128**, 15050 (2006).
88. E. Shirakawa, T. Yamagami, T. Kimura, S. Yamaguchi and T. Hayashi, *J. Am. Chem. Soc.*, **127**, 17164 (2005).
89. T. Yamagami, R. Shintani, E. Shirakawa and T. Hayashi, *Org. Lett.*, **9**, 1045 (2007).
90. Y. Hayashi, H. Shinokubo and K. Oshima, *Tetrahedron Lett.*, **39**, 63 (1998).
91. K. Fukuhara and H. Urabe, *Tetrahedron Lett.*, **46**, 603 (2005).
92. J. Meinwald and L. Hendry, *Tetrahedron Lett.*, 1657 (1969).
93. S. M. Ceccarelli, U. Piarulli and C. Gennari, *Tetrahedron.*, **57**, 8531 (2001).
94. A. Fürstner and A. Leitner, *Angew. Chem., Int. Ed.*, **42**, 308 (2003).
95. G. Seidel, D. Laurich and A. Fürstner, *J. Org. Chem.*, **69**, 3950 (2004).
96. O. Lepage, E. Kattnig and A. Fürstner, *J. Am. Chem. Soc.*, **126**, 15970 (2004).

CHAPTER 14

Carbomagnesiation reactions

KENICHIRO ITAMI

Department of Chemistry and Research Center for Materials Science, Nagoya University, Chikusa-ku, Nagoya 464-8602, Japan
Fax: +81-52-788-6098; e-mail: itami@chem.nagoya-u.ac.jp

and

JUN-ICHI YOSHIDA

Department of Synthetic Chemistry and Biological Chemistry, Graduate School of Engineering, Kyoto University, Nishikyo-ku, Kyoto 615-8510, Japan
Fax: +81-75-383-2727; e-mail: yoshida@sbchem.kyoto-u.ac.jp

I. INTRODUCTION	632
II. CARBOMAGNESIATION REACTIONS OF ALKYNES	633
A. Intramolecular Addition to Simple Alkynes	633
B. Intermolecular Addition to Simple Alkynes	635
C. Addition to Alkynylsilanes	639
D. Addition to Nitrogen-, Oxygen- and Sulfur-Attached Alkynes	641
E. Addition to Propargyl Alcohols	645
F. Addition to Propargyl Amines	652
III. CARBOMAGNESIATION REACTIONS OF ALKENES	653
A. Addition to Simple Alkenes	653
B. Allylmagnesiation of Alkenes	655
C. Addition to Strained Alkenes	657
D. Addition to Vinylsilanes	661
E. Addition to Allyl and Homoallyl Alcohols	664
F. Zirconium-catalyzed Ethylmagnesiation of Alkenes	671
IV. CARBOMAGNESIATION REACTIONS OF AROMATICS	674
V. SUMMARY	675
VI. REFERENCES	676

The chemistry of organomagnesium compounds
Edited by Z. Rappoport and I. Marek © 2008 John Wiley & Sons, Ltd

I. INTRODUCTION

Among various reactions of organomagnesium compounds (Grignard reagents), the addition reaction of a carbon–magnesium bond of an organomagnesium compound across a carbon–carbon multiple bond (a so-called carbomagnesiation reaction) has found many uses in organic synthesis (Scheme 1)[1]. It not only forms a carbon–carbon bond but also produces a new organomagnesium compound, which can react with other reactants.

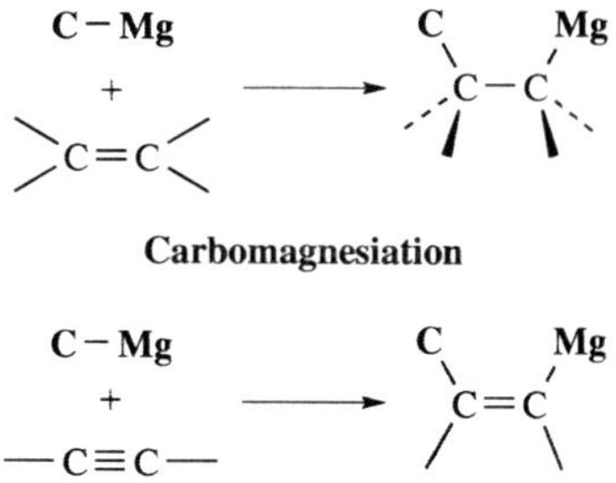

SCHEME 1

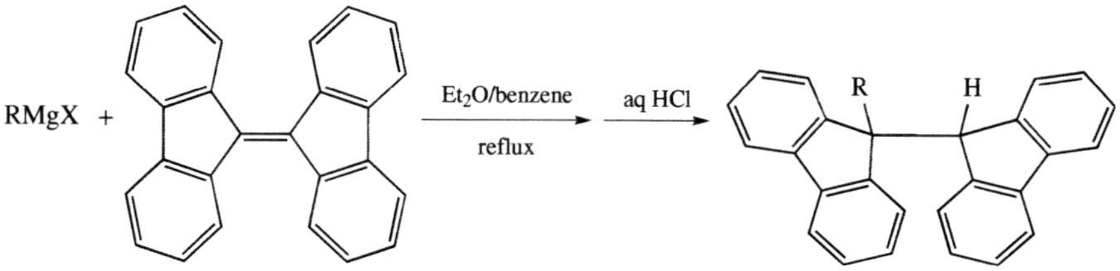

RMgX	Yield (%)
t-BuMgCl	68
PhCH$_2$MgCl	58
MeMgI	0
PhMgBr	0

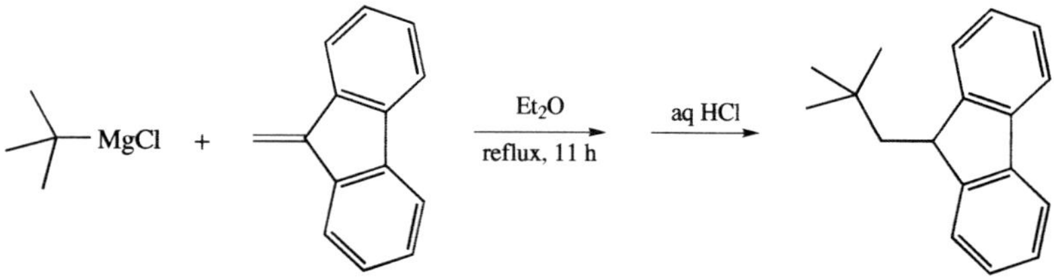

SCHEME 2

14. Carbomagnesiation reactions

One of the earliest examples of the addition of Grignard reagents to a carbon–carbon multiple bond may be the following results reported by Fuson and Porter in 1948[2]. They demonstrated that reactive Grignard reagents such as t-BuMgCl and PhCH$_2$MgCl could add across the C=C bond of fulvene derivatives **1** and **2** (Scheme 2)[2–5]. The addition does not take place with MeMgI or PhMgBr. Although it is obvious that the carbomagnesiation is made possible by a judicious combination of highly reactive Grignard reagents and C=C bonds, this represents the first example of carbomagnesiation of alkenes.

However, it has been known that the addition of a simple organomagnesium compound across an isolated (non-activated) carbon–carbon multiple bond is a sluggish process (Scheme 3)[6–8]. Over the last half-century of extensive worldwide research, synthetic chemists have devised a number of solutions to overcome the low reactivity of organomagnesium compounds and carbon–carbon multiple bonds[1]. Those include (i) the use of transition metal catalysts, (ii) the use of electronically activated alkenes and alkynes, (iii) the use of alkenes and alkynes bearing metal-directing functionalities and (iv) the use of functionalized organomagnesium compounds. This chapter summarizes the state of the art as well as synthetic utilities of carbomagnesiation reactions in organic synthesis. The Michael addition (1,4-addition) of Grignard reagents to simple α,β-unsaturated carbonyl compounds, which perhaps is one of the most studied organic reactions, will not be treated as a major topic in this chapter[9], but only some selected examples for this type of carbomagnesiation will be discussed.

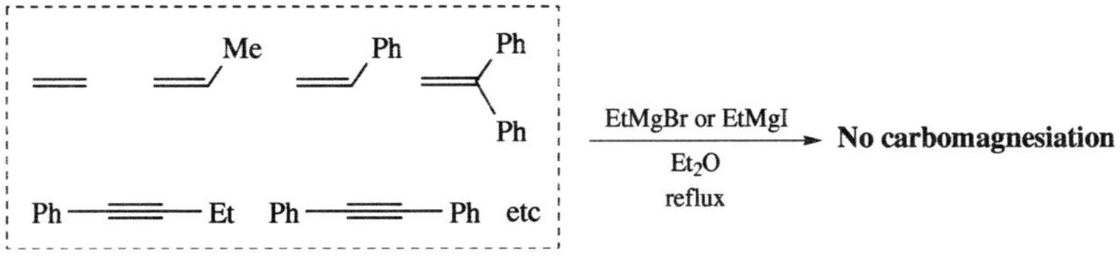

SCHEME 3

II. CARBOMAGNESIATION REACTIONS OF ALKYNES
A. Intramolecular Addition to Simple Alkynes

As stated in the introduction, the addition of a simple organomagnesium compound (Grignard reagent) across an isolated carbon–carbon triple bond is very hard to achieve[7]. However, intramolecular carbomagnesiation across a carbon–carbon triple bond proceeds more easily. For example, treatment of iodobenzene **3** bearing a tolane unit in *ortho* position undergoes cyclization to produce the ring-closed alkenylmagnesium compound, which yields benzylidenefluorene **4** upon hydrolysis (Scheme 4)[10]. Although this might be one of the oldest examples of intramolecular carbomagnesiation reactions of alkynes, there is no evidence for the presence of uncyclized organomagnesium compound.

Intramolecular carbomagnesiation across an unconjugated (unactivated) carbon–carbon triple bonds in aliphatic systems is mechanistically interesting. Heating a THF solution of 5-heptynylmagnesium chloride (**5**) for 6 days at 100 °C followed by hydrolysis furnishes ethylidenecyclopentane (**6**) in 90% yield (Scheme 5)[11]. The carbomagnesiation proceeds exclusively with 5-*exo-dig* fashion and the corresponding 6-*endo-dig* cyclization product (1-methylcyclohexene) is not observed. Kinetic experiments indicates that $t_{1/2}$ of carbomagnesiation is *ca* 50 hours at 100 °C. In contrast, hydrolysis immediately after preparation of organomagnesium compound **5** in refluxing THF furnishes only a few percent of ethylidenecyclopentane (**6**), which indicates that the cyclization occurs after the formation of the Grignard reagent.

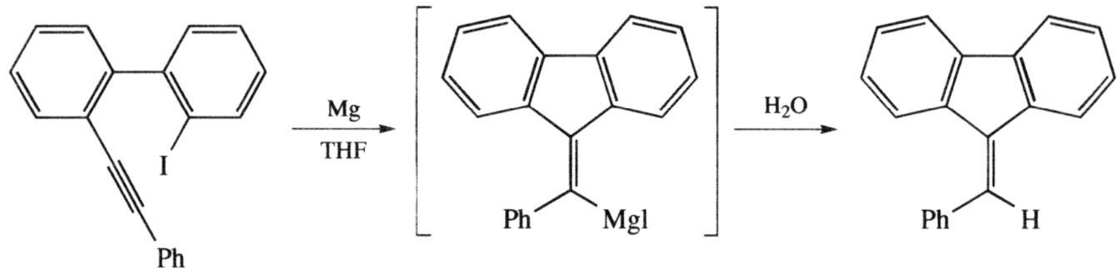

SCHEME 4

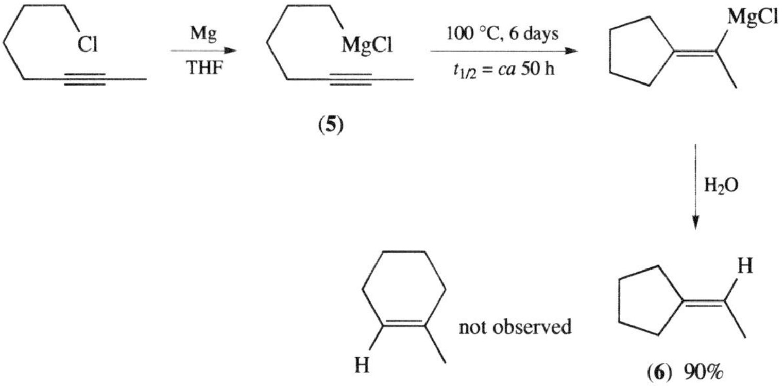

SCHEME 5

Similar experiments using substrates having a methyl group α to magnesium **7** furnishes a mixture of stereoisomers **8** and **9** (Scheme 6)[11]. Whether the mixture of *cis* and *trans* isomers is formed in the carbomagnesiation step, or by isomerization of a single isomer after addition, has not been discussed. Kinetic experiments indicate that $t_{1/2}$ of cyclization using bromide is *ca* 15 hours at 100 °C.

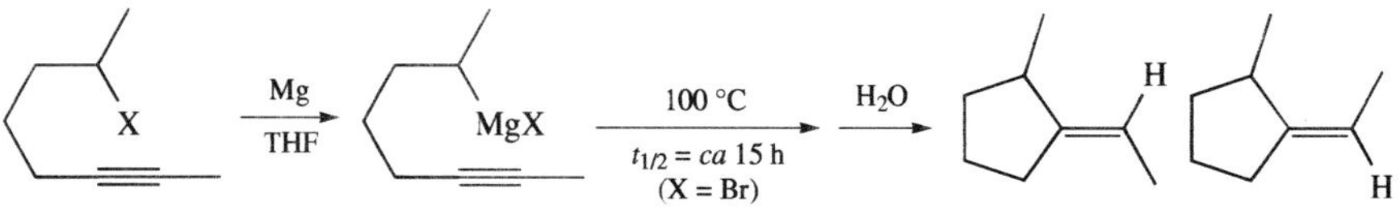

SCHEME 6

Copper salts serve as good catalysts for intramolecular carbomagnesiation. For instance, Crandall and coworkers reported that the treatment of acetylenic iodide **10** with an excess of *n*-BuMgBr in the presence of a catalytic amount of CuI affords the cyclic product **11** after hydrolytic workup (Scheme 7)[12]. Quenching with D_2O leads to 85% incorporation of deuterium at vinylic position in the cyclized product. The reaction probably proceeds by metal–halogen exchange leading to an organomagnesium species which is subsequently

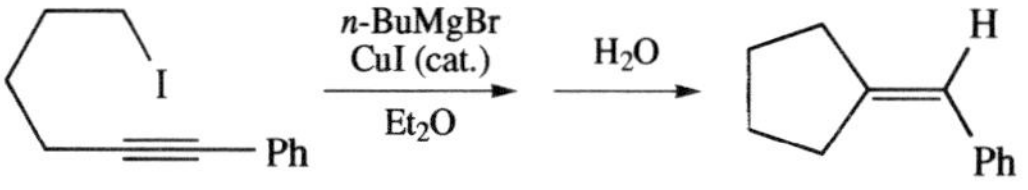

SCHEME 7

transmetalated into an organocopper and cyclizes into the corresponding alkenylmagnesium compound. Overall, this reaction represents a catalytic carbomagnesiation reaction.

B. Intermolecular Addition to Simple Alkynes

The development of a catalytic system leading to an intermolecular carbomagnesiation of simple alkynes has been a challenge. In 1972, Duboudin and Jousseaume reported the Ni-catalyzed carbomagnesiation of alkynes (Scheme 8)[13–15]. Although this may be one of the early successes in the intermolecular carbomagnesiation of alkynes, the scope is somewhat limited: low yielding with dialkylacetylenes, and inapplicable to alkyl Grignard reagents other than MeMgBr.

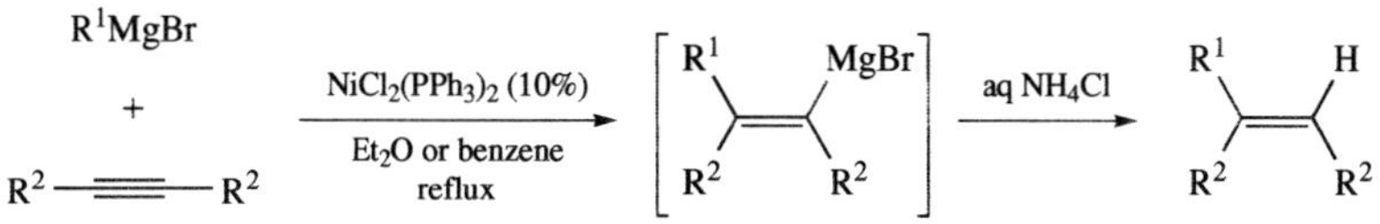

R^1	R^2	Yield (%)
Me	Ph	65
Ph	Ph	50
Ph	Et	30
Ph	n-Pr	31

SCHEME 8

Normant and coworkers demonstrated that the intermolecular carbomagnesiation across acetylene (HC≡CH) could be catalyzed by a copper salt[16]. The treatment of n-heptylmagnesium bromide with acetylene in the presence of a catalytic amount of CuBr (5 mol%) in Et_2O at $-20\,°C$ followed by the reaction with C_2H_5CHO and quenching with H_2O results in the formation of allylic alcohol **12** in 31% yield (Scheme 9)[16]. The carbomagnesiation takes place in a *syn*-addition manner.

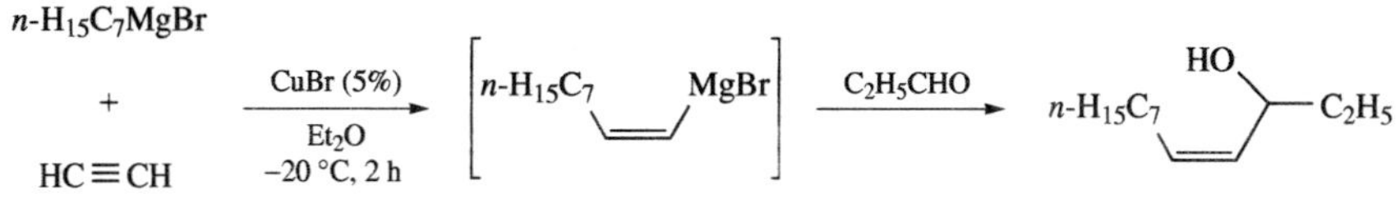

SCHEME 9

Since this report on the copper catalysis in carbomagnesiation across acetylene, a variety of Cu-catalyzed carbomagnesiation reactions of alkynes have been reported. However, the applicable alkynes are somewhat limited to electronically biased (activated) or heteroatom-containing alkynes, which will be discussed later.

Oshima and coworkers reported that $MnCl_2$ could catalyze the phenylmagnesiation of arylalkynes (Scheme 10)[17]. Although phenylacetylene itself cannot be used, the phenylmagnesiation takes place with a range of arylalkynes at 100 °C in toluene. The methoxy and dimethylamino groups in the *ortho* position of benzene ring of arylalkynes dramatically facilitate the reaction most likely through chelation assistance. They further extended such a chelation-assisted carbomagnesiation to 2-alkynylphenols[18].

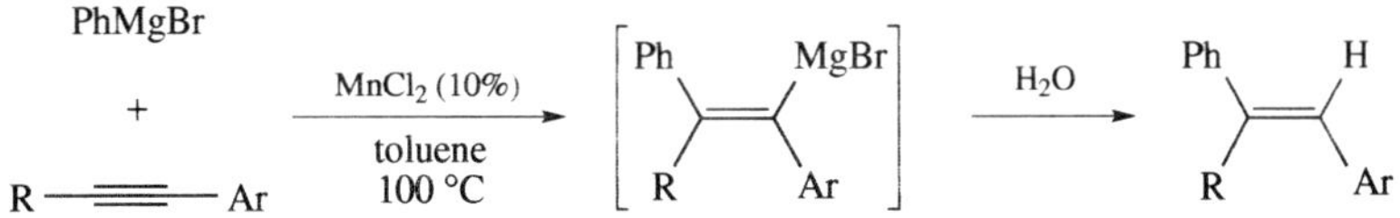

R	Ar	Yield (%)
n-C_6H_{13}	Ph	66
Ph	Ph	60
n-C_6H_{13}	2-$MeOC_6H_4$	80
n-C_6H_{13}	3-$MeOC_6H_4$	63
n-C_6H_{13}	4-$MeOC_6H_4$	38
n-C_6H_{13}	2-$Me_2NC_6H_4$	94

SCHEME 10

As in the previous examples, the presence of polar functionality near the alkyne affects the rate, regioselectivity and stereoselectivity of carbometalation reaction. Oshima, Utimoto and coworkers reported such a substituent effect in the Mn-catalyzed allylmagnesiation of alkynes[19]. The treatment of homopropargylic alcohol methyl ether **13** with allylmagnesium bromide in the presence of MnI_2 catalyst (3 mol%) provides the allylated alkenylmagnesium species **14** in a regio- and stereoselective fashion (Scheme 11). The Mn-catalyzed allylmagnesiation proceeds well with the corresponding benzyl and tetrahydropyranyl ethers also. The thus-generated alkenylmagnesium species **14** is allowed to react with electrophiles such as aldehydes and allyl bromides to give tetrasubstituted olefins **15** in good yields (Scheme 11)[19]. The *syn* addition of an allylmetal component has been also confirmed. The reaction is clearly oxygen-assisted since 6-dodecyne is completely recovered unchanged even at elevated temperature.

Chromium salts can catalyze the carbomagnesiation reactions of 1,6-enynes with cyclization (Scheme 12)[20, 21]. This reaction is probably initiated by carbometalation of the alkyne unit. The resultant organomagnesium species **16** undergoes further functionalization upon treatment with various electrophiles (Scheme 12)[21].

Yorimitsu, Oshima and coworkers reported the Cr-catalyzed arylmagnesiation of simple alkynes[22]. In early experiments, they found that the phenylmagnesiation of 6-dodecyne could be catalyzed by $CrCl_2$ in toluene at 100 °C (Scheme 13). They further found that some alcohols as additives have an accelerating effect on the reaction. The investigation

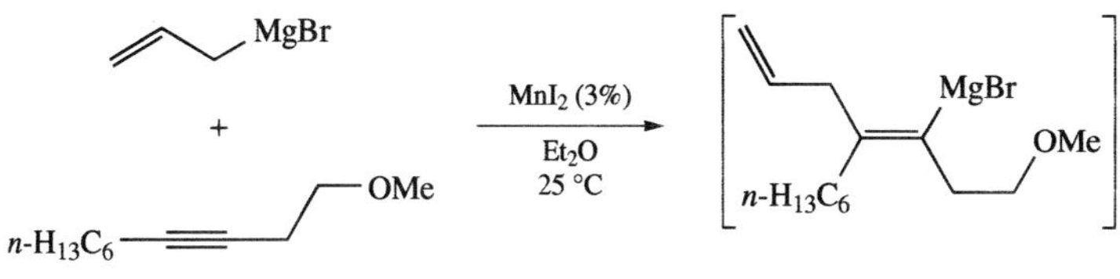

SCHEME 11

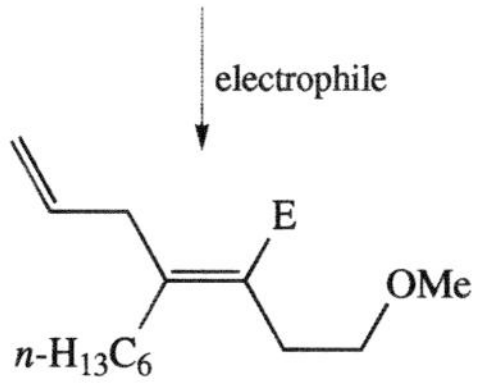

Electrophile	E	Yield (%)
H$_2$O	H	83
PhCHO	PhCH(OH)-	92
n-C$_6$H$_{13}$CHO	n-C$_6$H$_{13}$CH(OH)-	87
CH$_2$=CHCH$_2$Br	CH$_2$=CHCH$_2$-	77

SCHEME 12

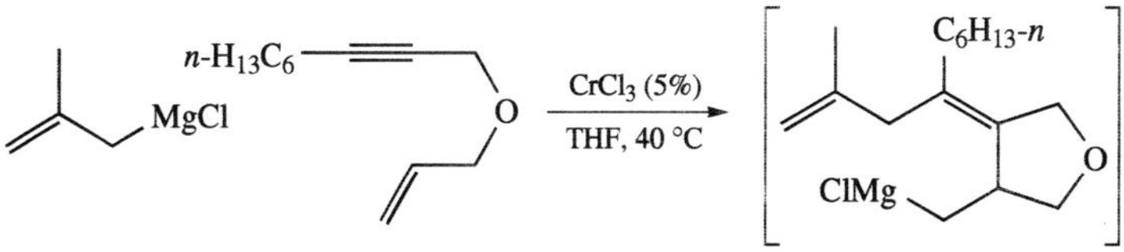

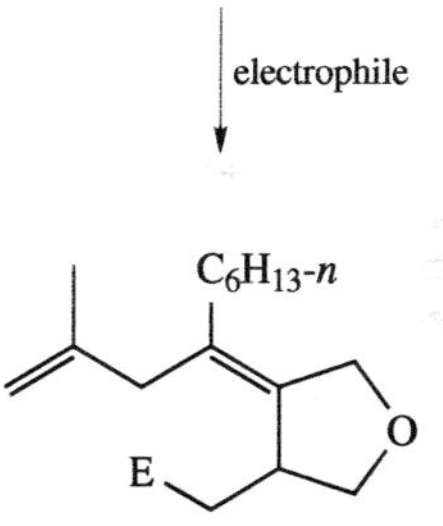

Electrophile	E	Yield (%)
PhCHO	PhCH(OH)-	84
I$_2$	I	82
CH$_3$COCl	CH$_3$C(=O)-	87[a]
CH$_2$=CHCH$_2$Br	CH$_2$=CHCH$_2$-	82[a]

[a]With 20 mol% CuCN·2LiCl.

into the additive effect in this transformation finally led to the discovery of pivalic acid (t-BuCOOH) as an optimal promoter, providing the addition product in high yield with virtually complete stereoselectivity (>99% E). Although the reason for the dramatic effect of additives is not clear, the regio- and stereoselective arylmagnesiation takes place with a range of dialkylacetylenes and aryl(alkyl)acetylenes. The use of silyl-substituted alkynes results in low stereoselectivity.

Shirakawa, Hayashi and coworkers disclosed that the arylmagnesiation of simple alkynes could be effectively catalyzed by a catalytic system consisting of iron and copper (Scheme 14)[23]. For example, the treatment of 3,5-dimethylphenylmagnesium bromide (2 equiv) with 4-octyne (1 equiv) under the catalytic influence of Fe(acac)$_3$ (5 mol%), CuBr

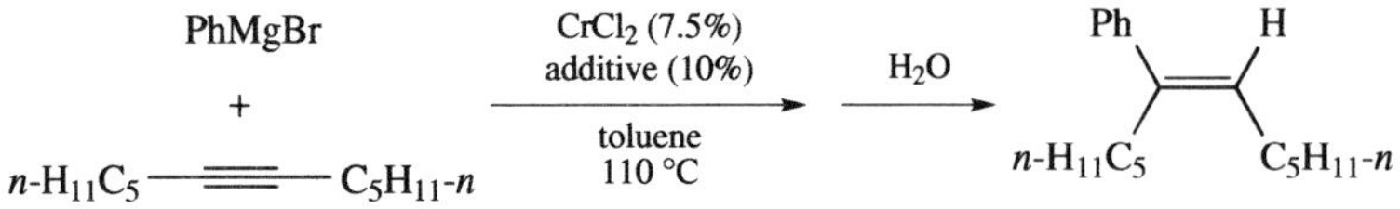

PhMgBr + n-H₁₁C₅—≡—C₅H₁₁-n → [CrCl₂ (7.5%), additive (10%), toluene, 110 °C] → H₂O → Ph\(n-H₁₁C₅)=C(H)(C₅H₁₁-n)

Additive	Time (h)	Yield (%)	E/Z
none	18	81	91:9
MeOH	2	77	95:5
PhOH	2	77	95:5
PhCOOH	0.25	81	>99:1
t-BuCOOH	**0.25**	**87**	**>99:1**

SCHEME 13

(10 mol%) and PBu$_3$ (40 mol%) in THF at 60 °C for 24 h followed by hydrolysis furnishes 74% of hydroarylation product **17** with 95% *E* selectivity. Quenching the reaction with D$_2$O results in the formation of arylalkene with exclusive incorporation of deuterium at its olefinic methyne, indicating the carbomagnesiation process. Omission of either Fe(acac)$_3$ or CuBr results in dramatic decrease in carbomagnesiation efficiency (26% and 0% yield, respectively). These control experiments clearly indicate a unique iron/copper cooperative catalysis in this reaction. This catalytic system can be applied to various aryl Grignard reagents. On the basis of several control experiments, a possible mechanism, given in Scheme 15, has been proposed.

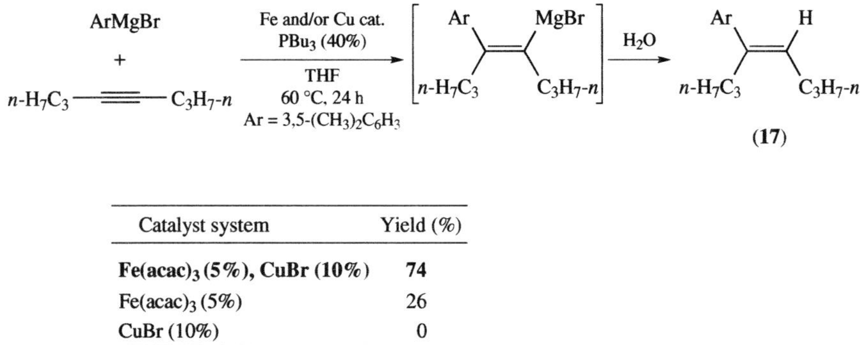

ArMgBr + n-H₇C₃—≡—C₃H₇-n → [Fe and/or Cu cat., PBu$_3$ (40%), THF, 60 °C, 24 h, Ar = 3,5-(CH$_3$)$_2$C$_6$H$_3$] → [Ar\(n-H₇C₃)=C(MgBr)(C₃H₇-n)] → H₂O → Ar\(n-H₇C₃)=C(H)(C₃H₇-n) (**17**)

Catalyst system	Yield (%)
Fe(acac)$_3$ (5%), CuBr (10%)	**74**
Fe(acac)$_3$ (5%)	26
CuBr (10%)	0

SCHEME 14

In connection with the aforementioned Fe/Cu co-catalyzed arylmagnesiation, Shirakawa, Hayashi and coworkers further investigated the iron catalysis focusing on aryl(alkyl)acetylenes as substrates[24]. As already stated in their previous work, a combination of iron salt with a phosphine does not have high activity (Scheme 16). However, the Fe-catalyzed arylmagnesiation of aryl(alkyl)acetylenes is greatly improved by addition of a catalytic amount of 1,3-bis-(2,6-diisopropylphenyl)imidazol-2-ylidene (IPr) (Scheme 16)[24]. With this improved catalyst system, a range of arylmagnesium bromides can be added across various aryl(alkyl)acetylenes.

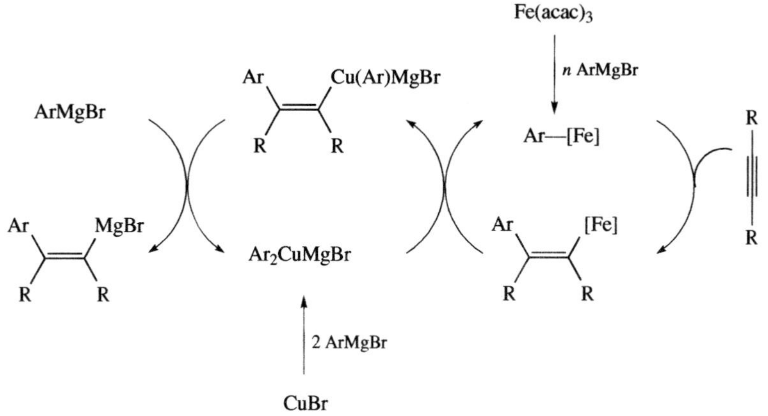

SCHEME 15

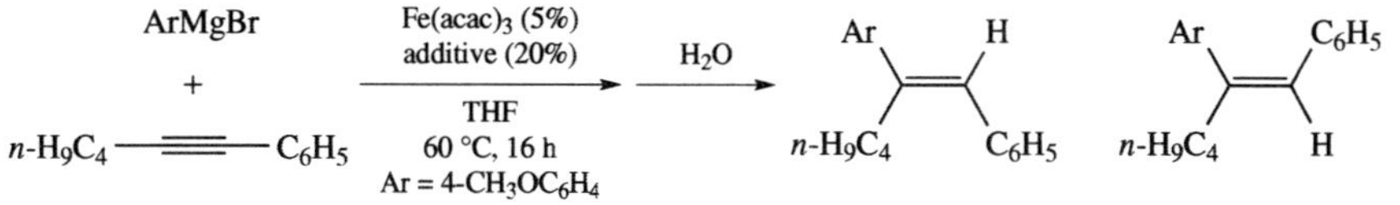

(18) **(19)**

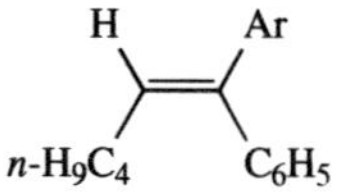

(20)

Additive	Yield	**18**:**19**:**20**
None	23%	85:13:2
$P(Bu\text{-}n)_3$	16%	72:21:7
$P(Bu\text{-}t)_3$	23%	82:17:1
PPh_3	33%	86:12:2
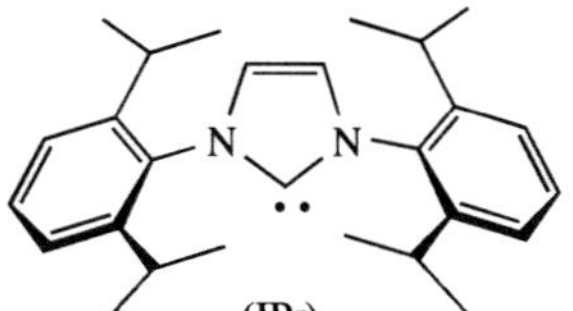 (IPr)	**91%**	**88:11:1**

SCHEME 16

The alkenylmagnesium species generated by the Fe-catalyzed arylmagnesiation can be trapped by electrophiles. For example, the cross-coupling reaction of the alkenylmagnesium species with an aryl halide is achieved with a nickel catalyst, giving a tetrasubstituted olefin **21** in good overall yield (Scheme 17)[24].

C. Addition to Alkynylsilanes

A carbomagnesiation of alkynylsilanes is an attractive method generating synthetically versatile silyl-substituted alkenylmagnesium species. Snider and coworkers reported the

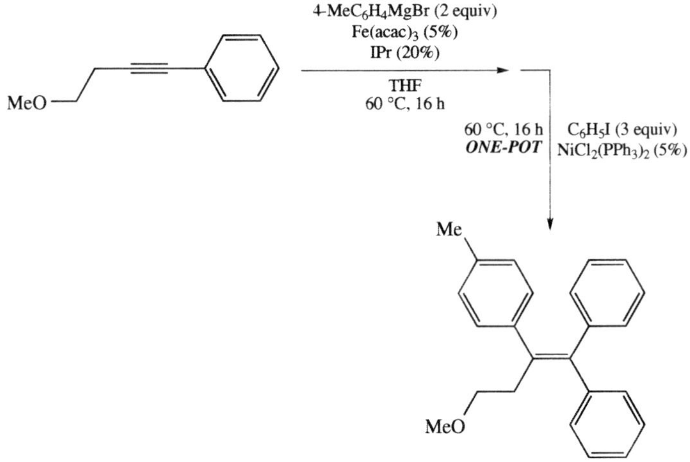

(21) 72%

SCHEME 17

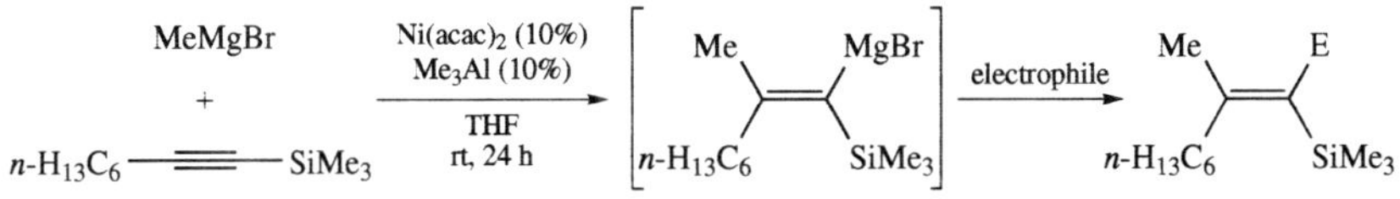

(22)

Electrophile	E	Yield (%)	Z/E
H_2O	H	80	90:10
D_2O	D	75	90:10
CH_3CHO	$CH_3CH(OH)$-	75	85:15
CH_2=CHBr	CH_2=CH-	48	85:15

SCHEME 18

Ni-catalyzed addition of MeMgBr to alkynylsilanes (Scheme 18)[25,26]. A combination of $Ni(acac)_2$ and Me_3Al seems to be crucial; $NiBr_2$ and $Ni(acac)_2$ show lower activity, and $NiCl_2$(phosphine) complexes are totally inactive. Some of the methyl groups transferred may arise from Me_3Al since the use of $Ni(acac)_2/(i\text{-}Bu)_2AlH$ as catalyst leads to small amounts of hydrometalation. Although this catalytic system is somewhat limited to methylmagnesiation, the resultant alkenylmagnesium species **22** can be trapped by various electrophiles (Scheme 18)[25]. Quenching the reaction with H_2O and D_2O leads to the incorporation of H and D, respectively. The reaction with acetaldehyde gives an allylic alcohol. The reaction with vinyl bromide affords a 1,3-diene that is most likely nickel-catalyzed (Tamao–Kumada–Corriu-type cross-coupling). The presence of a coordinating heteroatom on the side chain of the alkyne influences the stereoselectivity of methylmagnesiation reaction of alkynylsilanes[26].

Itami, Yoshida and coworkers reported an efficient protocol for the carbomagnesiation of alkynylsilanes bearing a metal-coordinating 2-pyridyl group on silicon

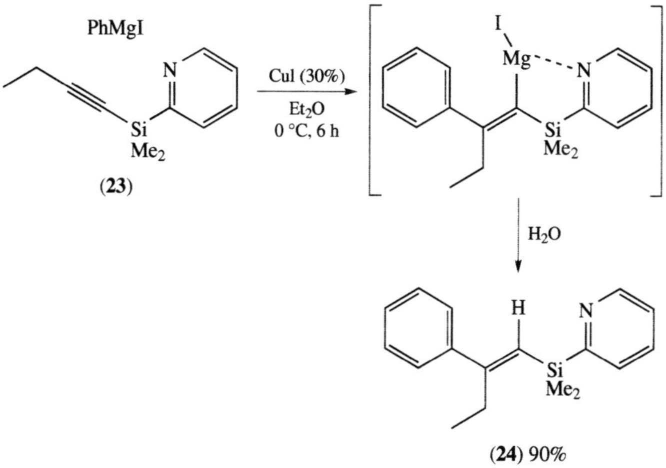

(23)

(24) 90%

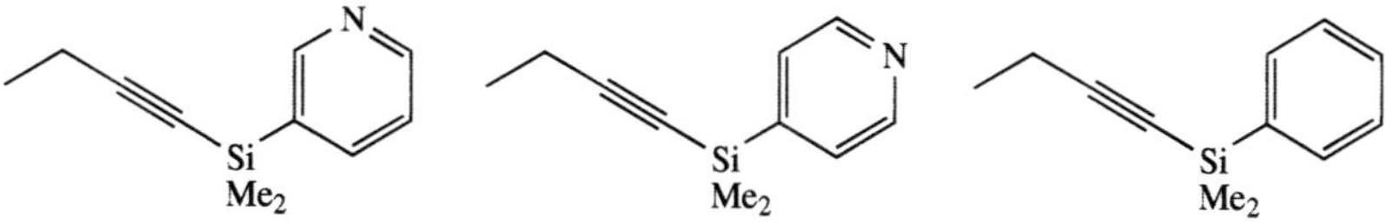

(25) *No addition* (26) *No addition* (27) *No addition*

SCHEME 19

(Scheme 19)[27,28]. The regio- and stereoselective addition of arylmagnesium iodides (ArMgI) to 2-pyridylsilylalkyne **23** proceeds in the presence of CuI catalyst, furnishing alkenylsilanes **24** in high yields after quenching the reaction with water (Scheme 19). The use of ArMgBr, ArMgCl or Ar_2Mg in place of ArMgI results in lower addition efficiency. The Cu-catalyzed addition does not occur at all with the corresponding 3-pyridyl, 4-pyridyl and phenylsilanes (**25**–**27**), which clearly implicates the strong directing effect of the 2-pyridyl group on silicon.

By using this reaction as a key step, a programmed and diversity-oriented synthesis of tamoxifen-type tetrasubstituted olefins can be accomplished (Scheme 20)[27,28]. Thus, by adding a Pd catalyst and aryl iodides (Ar^2I) to the solution of alkenylmagnesium species generated by the Cu-catalyzed carbomagnesiation of 1-butynyldimethyl(2-pyridyl)silane **23**, a regio- and stereoselective introduction of two aryl groups onto the alkynylsilane is achieved in one-pot. Although direct C–Si bond arylation of the resultant alkenylsilanes **28** is not feasible, a borodesilylation/cross-coupling sequence of the alkenylsilanes allows the installation of the third aryl group at C–Si bonds yielding the tamoxifen-type tetrasubstituted olefins **29**.

D. Addition to Nitrogen-, Oxygen- and Sulfur-Attached Alkynes

Marek and coworkers have developed the regio- and stereocontrolled carbomagnesiation of alkynyl amine derivatives. For example, PhMgBr adds across the triple bond of ynamide **30** in the presence of a catalytic amount of $CuBr{\bullet}Me_2S$ (10 mol%). The following

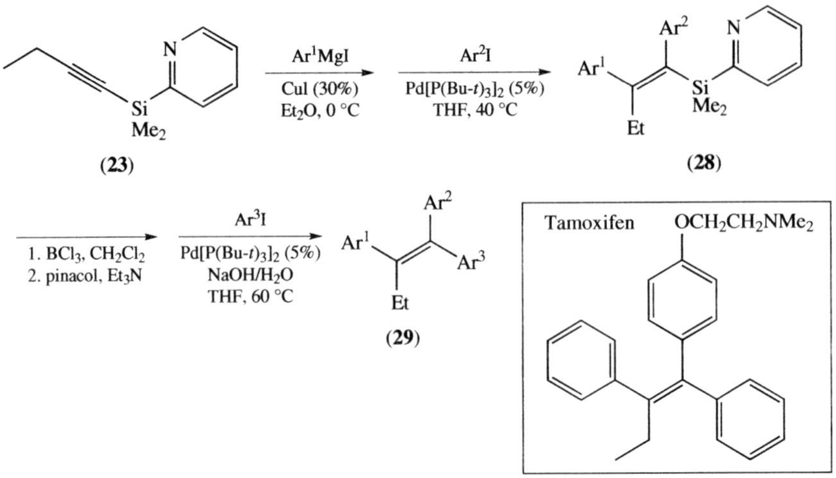

SCHEME 20

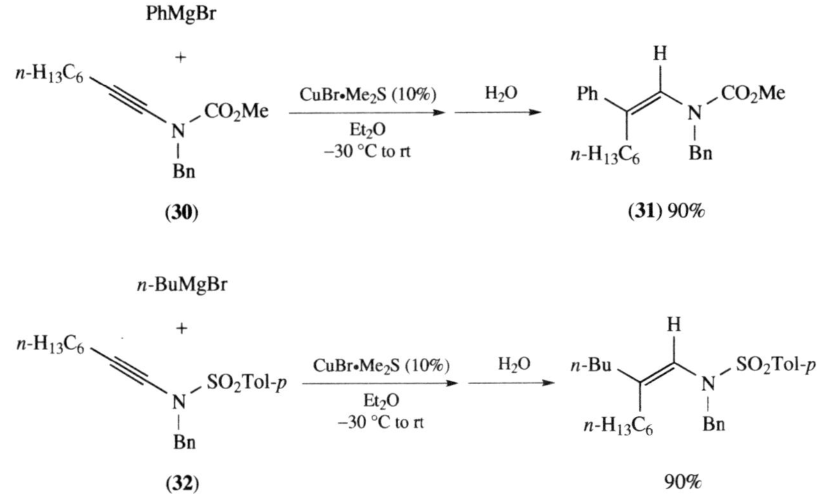

SCHEME 21

hydrolysis affords the enamide **31** in 90% yield (Scheme 21)[29]. The stereochemical assignment of the product by NOE experiments indicates that the carbometalation proceeds in a *syn*-addition fashion. As for catalyst precursor, CuBr•Me$_2$S is optimal. The Cu-catalyzed carbomagnesiation also takes place with sulfonyl-substituted ynamide **32** with high regio- and stereoselectivity (Scheme 21)[29]. The chelation of the carbamoyl and sulfonyl moieties to the organometallic species has been proposed to be of primary importance for controlling the regiochemistry of reaction. In line with such an assumption, the use of coordinating solvent such as THF erodes the regioselectivity of the carbomagnesiation.

14. Carbomagnesiation reactions

The Cu-catalyzed carbomagnesiation of alkynylsulfonamide **33** having an allyl group on the nitrogen atom also takes place. Interestingly, the organomagnesium species **34** thus generated undergoes aza-Claisen rearrangement to yield pentenenitrile **35** (Scheme 22)[30].

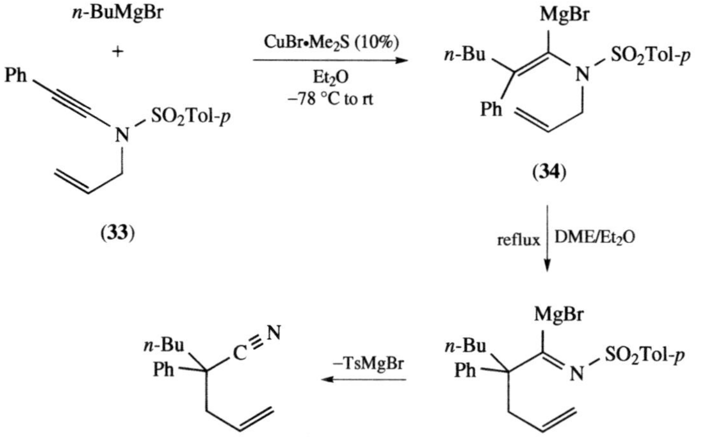

SCHEME 22

Marek and Chechik-Lankin have demonstrated that the stereoselective Cu-catalyzed carbomagnesiation reactions of alkynyl carbamates is a straightforward means for the preparation of synthetically versatile alkenyl enol carbamates. When ethynylcarbamate **36** is added to a stoichiometric amount of *n*-BuMgBr in Et$_2$O in the presence of CuI (10 mol%), the addition takes place smoothly at $-40\,°$C giving (*E*)-alkenyl carbamate **37** in 70% yield (Scheme 23)[31]. The addition of benzaldehyde as an electrophile gives **38** in 69% yield. An intramolecular coordination of the sp^2 organometallic species (Cu or Mg) by the carbamate moiety has been proposed.

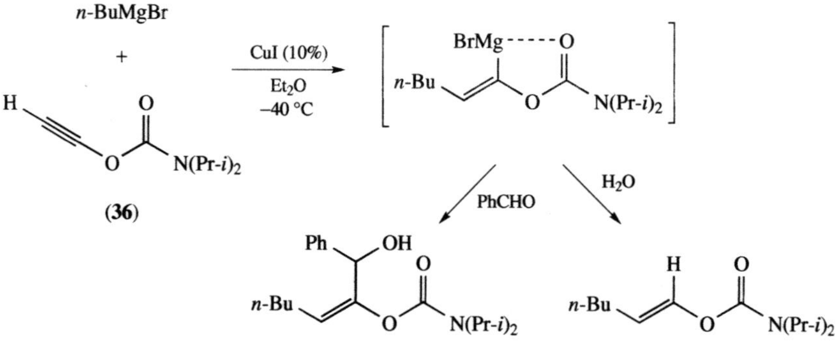

SCHEME 23

The carbomagnesiation of alkynyl sulfones takes place in the presence of copper catalyst. For example, the addition of PhMgBr to *p*-tolylsulfonylheptyne **39** in the presence of CuCN (10 mol%) occurs at $-20\,°C$ in THF/CH$_2$Cl$_2$ yielding the corresponding alkenylmagnesium species **40**, which then is allowed to react with allyl bromide to afford trisubstituted vinyl sulfone **41** in 73% yield (Scheme 24)[32]. The stereochemistry of the product has been verified by NMR experiment (NOESY) and X-ray crystal structure analysis. These results demonstrate that the addition of Grignard reagent occurs in a *syn* fashion. The advantage of this catalytic protocol is obvious since it has been known that (i) the reaction of Grignard reagent with alkynyl sulfones typically yields the products of overall substitution of sulfone moiety, and (ii) organocopper reagents add to alkynyl sulfones with low stereoselectivity.

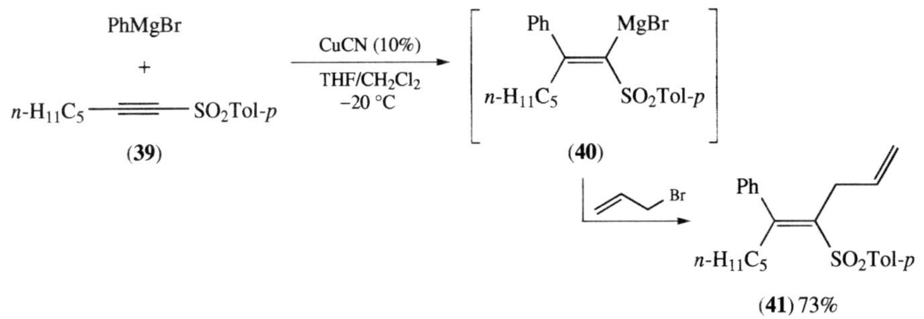

SCHEME 24

When the phenyl-substituted alkynyl sulfone **42** is used as a substrate for the Cu-catalyzed carbomagnesiation, interesting nucleophile-dependent stereoselectivity is observed. While the use of allyl Grignard reagent results in a *syn* addition, the use of aryl Grignard reagent results in an *anti*-carbomagnesiation (Scheme 25)[33].

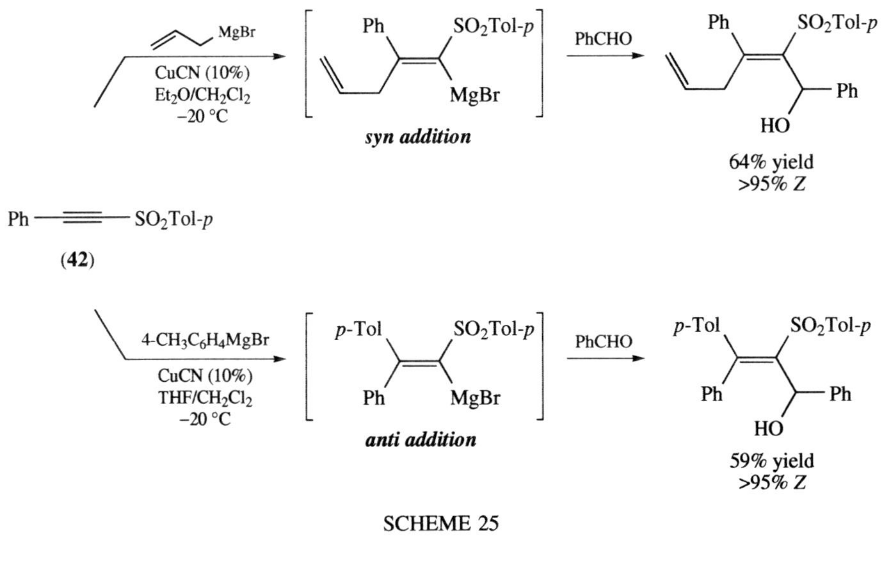

SCHEME 25

E. Addition to Propargyl Alcohols

Although the typical carbomagnesiation to alkynes proceeds in a *syn*-addition fashion, propargyl alcohols react with organomagnesium compounds in an *anti*-addition manner (Scheme 26). The reactions are believed to proceed through the formation of a magnesium–oxygen bond, making a five-membered cyclic organomagnesium compound **43**, which then reacts with electrophiles to give substituted allylic alcohols **44** stereoselectively (Scheme 26). The unique *anti*-carbomagnesiation process may be closely related to the *anti*-hydroalumination of propargyl alcohols with LiAlH$_4$ or Red-Al[34].

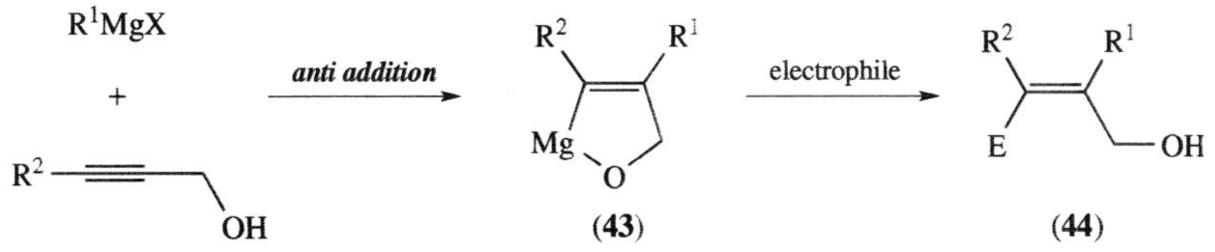

SCHEME 26

For example, the treatment of 2-butyn-1-ol (**45**) with allylmagnesium chloride in refluxing ether followed by hydrolysis affords **46** in 85% yield (Scheme 27)[35, 36]. Although the use of vinylmagnesium chloride also provides the corresponding *anti*-addition product in

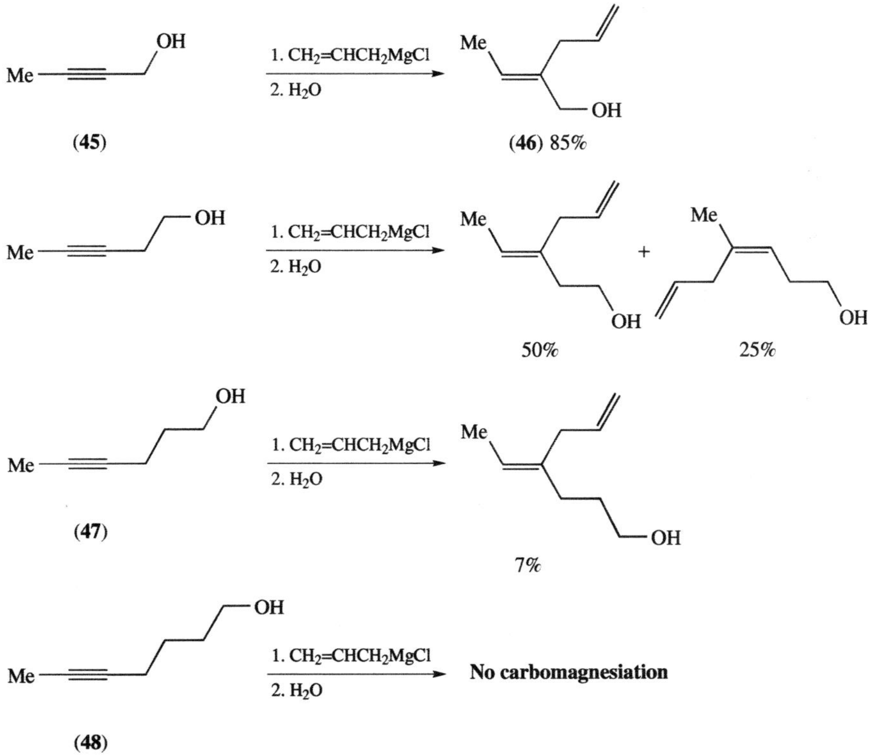

SCHEME 27

60% yield, no addition takes place with PhMgBr, MeMgCl and *t*-BuMgCl. The reaction is clearly oxygen-assisted since the yields of addition product substantially decrease in the case of **47** (7%) and **48** (0%), where the oxygen atom is too far remote for efficient assistance. In the homopropargyl series, both regioisomers are obtained.

Useful 1,3-dienylmagnesium species can be generated through an *anti*-vinylmagnesiation of 2-butyn-1-ol (Scheme 28)[37]. The resultant organomagnesium species **49** is allowed to react with electrophiles, such as iodine, aldehydes and ketones, to furnish a range of substituted 1,3-dienes **50** in reasonable overall yields (Scheme 28)[37, 38].

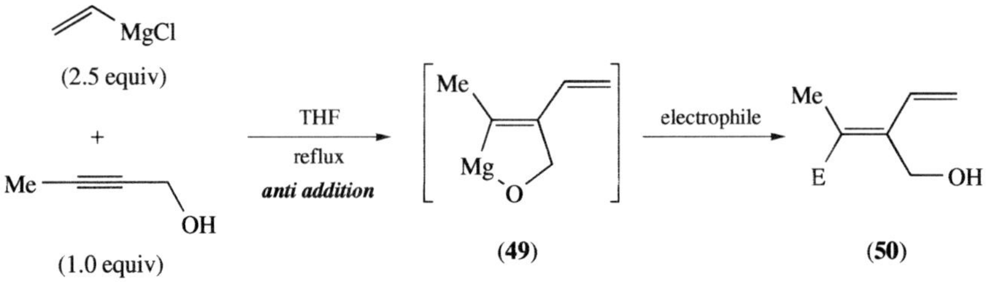

Electrophile	E	Yield (%)
aq NH$_4$Cl	H	75
I$_2$	I	60
PhCHO	PhCH(OH)-	80
PhCOMe	PhCMe(OH)-	44

SCHEME 28

Propargylic alcohols having a cyano group **51** also undergo *anti*-carbomagnesiation (Scheme 29)[39]. Mechanistically, the reaction most likely proceeds through initial formation of halomagnesium alkoxide **52** followed by halogen–alkyl exchange. Alkyl transfer from the resulting alkoxide **53** leads to an intermediate chelate **54**. Fleming and coworkers found that *t*-BuMgCl serves as an excellent sacrificial base for initial deprotonation[39].

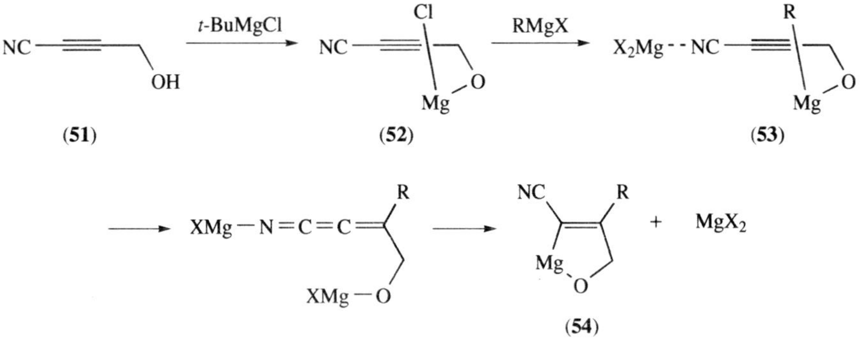

SCHEME 29

The carbomagnesiation reaction proceeds with a range of organomagnesium compounds (Scheme 30)[39,40]. Not only aryl, alkenyl and alkynyl groups, but also alkyl groups were found to add across a triple bond. The enhanced reactivity of cyano-substituted alkynes is worthwhile, and this may be due to accelerated alkyl transfer from **53** with activation of the cyano group by MgX_2 (Scheme 29).

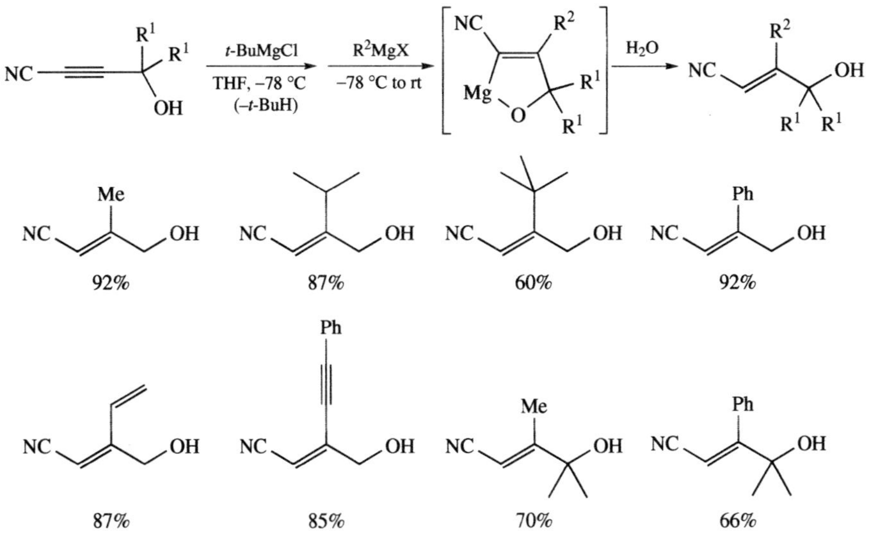

SCHEME 30

Chelation is essential for this carbomagnesiation. A control experiment, in which the THP-protected cyanoalkyne **55** is exposed to *n*-BuMgCl, leads to 90% recovery of unchanged alkyne (Scheme 31). In addition, homopropargylic alcohol **56** does not undergo carbomagnesiation[40].

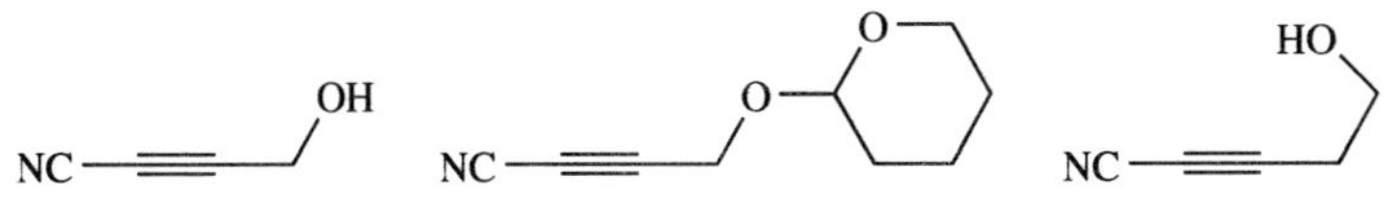

(51) **(55)** **(56)**

efficient addition ***no addition*** ***no addition***

SCHEME 31

When enyne alcohol **57** is subjected to *anti*-carbomagnesiation using vinylmagnesium chloride, the resultant triene **58** undergoes further electrocyclization (Scheme 32)[41].

Fallis and coworkers reported the synthesis of a tricyclic ABC ring-system of paclitaxel using the three-component assembly of Grignard reagent, propargylic alcohol and aldehyde as a key step (Scheme 33)[42,43].

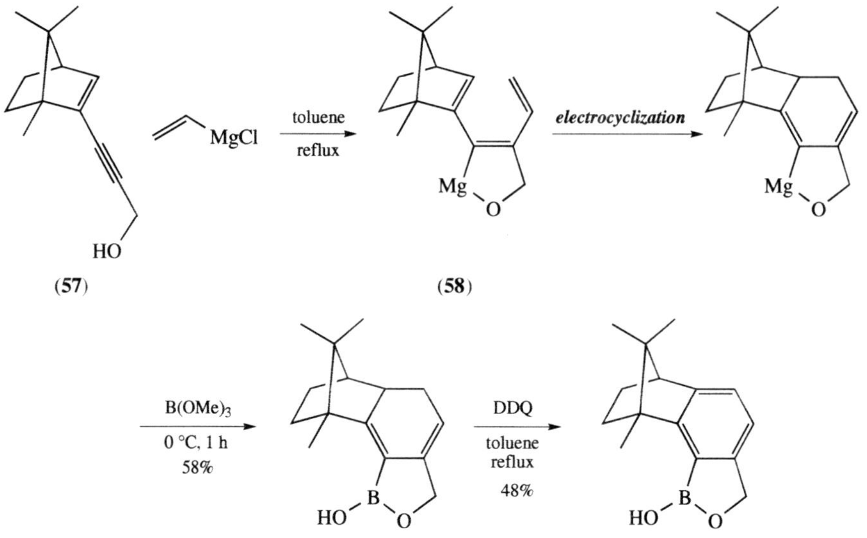

SCHEME 32

By using dimethylformamide or benzonitrile as an electrophile after the *anti*-carbomagnesiation of propargylic alcohols, substituted furans **59** are obtained by treatment with *p*-TsOH (Scheme 34)[44].

The use of CO_2 and $SOCl_2$ as electrophiles furnishes furanones and sultines, respectively (Scheme 35)[45–47]. By using this method, Fallis and coworkers demonstrated a facile synthesis of Vioxx, Merck's anti-inflammatory drug, as well as 'thio-Vioxx' which has been revealed to be a selective COX-2 (cyclooxygenase-2) inhibitor[45].

As electrophiles for post-functionalization of *anti*-carbomagnesiation, aryl and alkenyl halides can also be used when Pd catalyst is employed. By merging such Pd-catalyzed arylation, tamoxifen can be synthesized in a stereoselective manner (Scheme 36)[48].

Several catalytic procedures have been also developed for the carbomagnesiation of propargylic alcohols. For example, Duboudin and Jousseaume reported that the Cu-catalyzed carbomagnesiation of propargylic alcohols **60** leads to the selective formation of 2,3-disubstituted propyn-2-ols **61** via the hydroxyl-controlled *anti*-carbomagnesiation (Scheme 37)[49, 50]. The copper catalyst not only allows the addition under much milder conditions (0 °C), but also allows the use of a wide range of organomagnesium compounds (Me, Et, *i*-Pr, *t*-Bu, Ph, $PhCH_2$, $CH_2=CHCH_2$).

Quite interestingly, propargyl alcohol itself can be applied in the copper-catalyzed protocol. In the absence of copper catalyst, only the metalation (deprotonation) of terminal hydrogen occurs (Scheme 38)[49, 50].

The use of secondary propargylic alcohols affords a mixture of two regioisomeric products, while the reaction of tertiary propargylic alcohols gives the linear products selectively. Ma and Lu reported a highly regioselective Cu-catalyzed *anti*-carbomagnesiation of secondary terminal propargylic alcohols affording 2-substituted allylic alcohols[51,52]. For example, when 3-butyn-2-ol (**62**) is treated with a THF solution of *n*-pentylmagnesium bromide in the presence of CuI (0.5 equiv), *anti*-carbomagnesiation occurs (Scheme 39). When iodine is added as a quenching agent, alkenyl iodide **63** is obtained in high

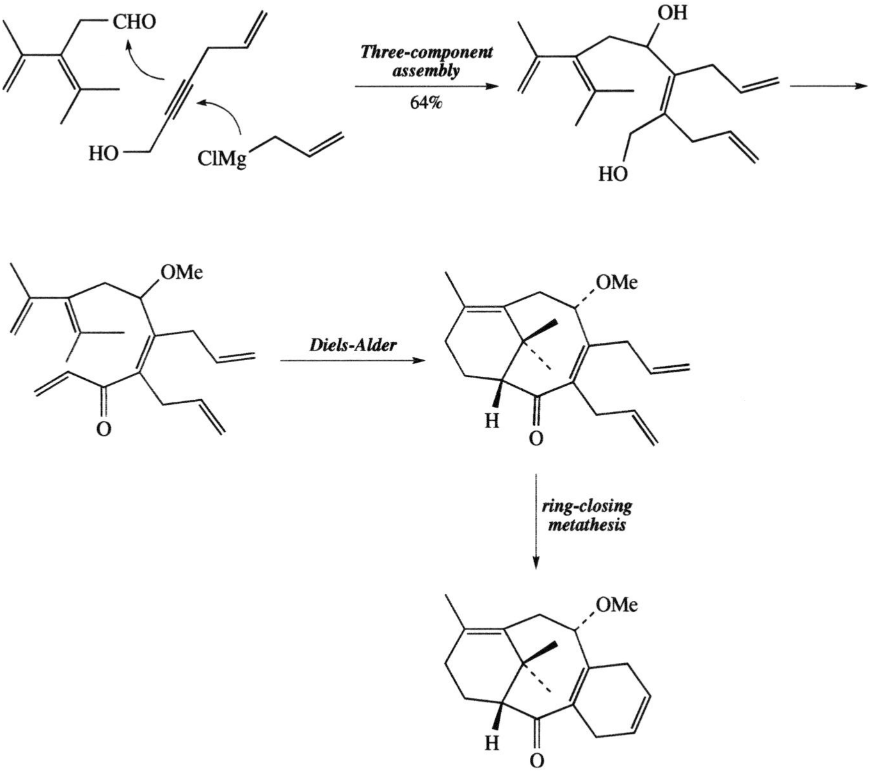

14. Carbomagnesiation reactions

SCHEME 33

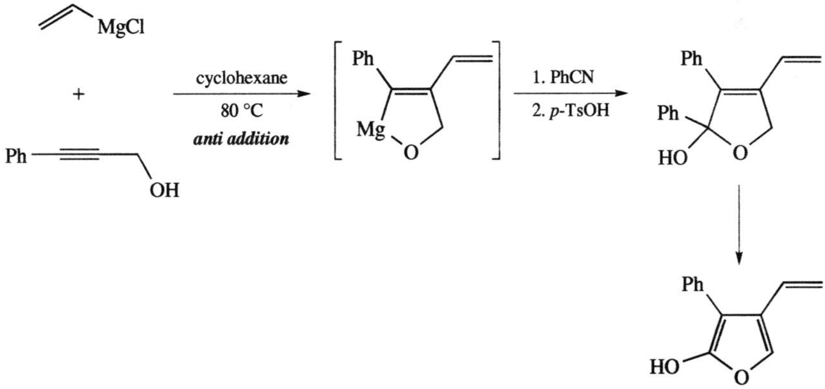

SCHEME 34

(59) 62%

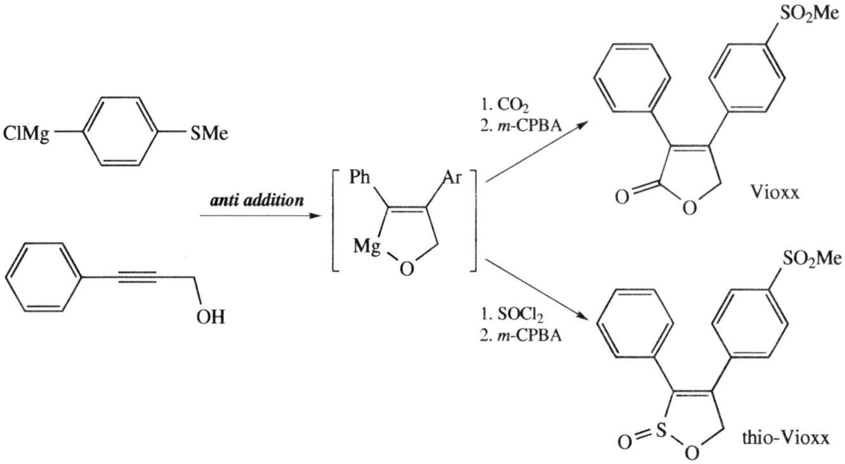

SCHEME 35

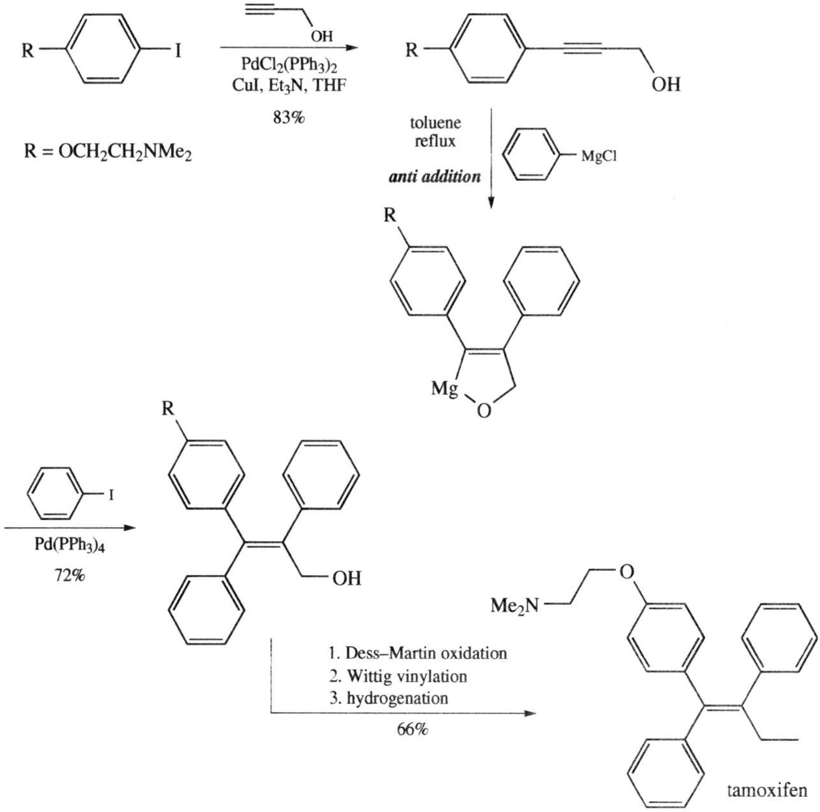

SCHEME 36

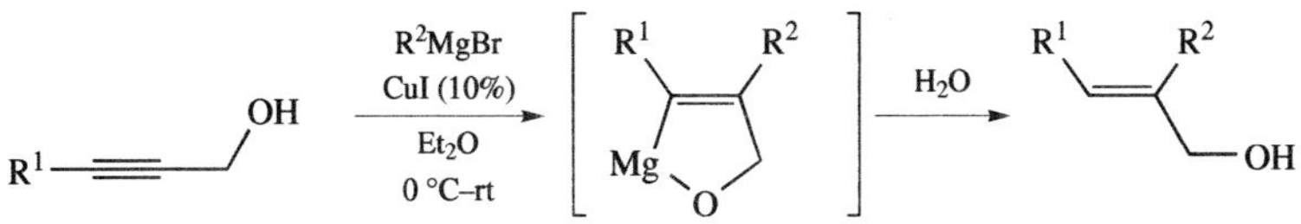

14. Carbomagnesiation reactions

R^1	R^2	Yield (%)	R^1	R^2	Yield (%)
Me	allyl	80	Ph	Ph	55
Me	Ph	40	Ph	Me	58
Me	Et	55	Ph	Et	70
Me	i-Pr	45	Ph	i-Pr	60
Me	t-Bu	20	Ph	t-Bu	55

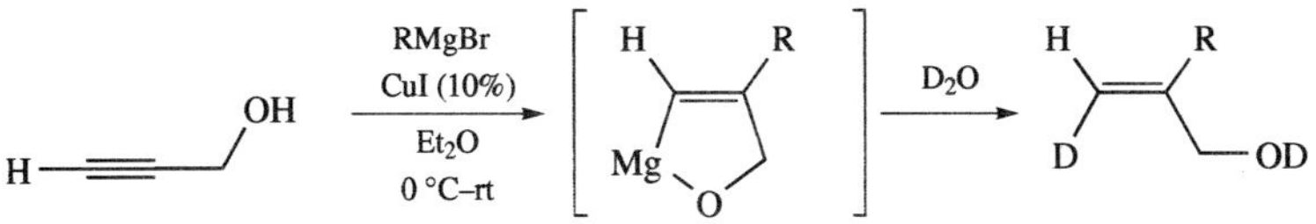

SCHEME 37

R	Yield (%)	R	Yield (%)
Allyl	60	Me	53
PhCH$_2$	80	Et	60
Ph	73	t-Bu	45

SCHEME 38

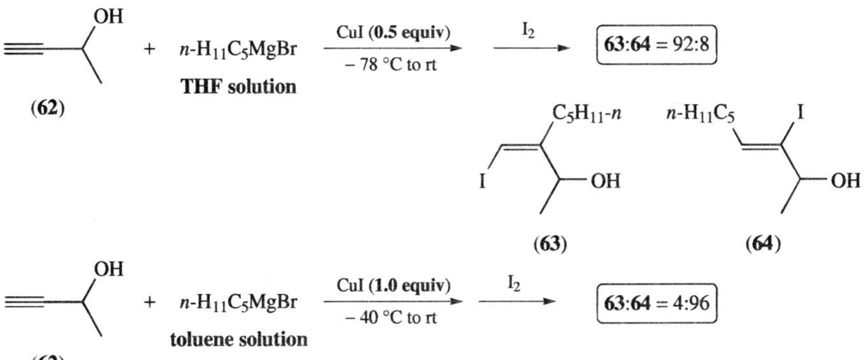

SCHEME 39

selectivity. The reaction also takes place with other copper salts such as CuCl, albeit with lower efficiency. Quite interestingly, however, the use of a stoichiometric amount of CuI results in reverse selectivity in product distribution (**63** vs **64**). Changing the solvent of Grignard reagent from THF to toluene results in exclusive formation of **64** (Scheme 39). This highly stereoselective *syn*-carbometalation can also be applied to tertiary propargylic alcohols.

When an iron salt is used as a catalyst in the addition of Grignard reagents to propargylic alcohols, *syn*-carbomagnesiation takes place (Scheme 40)[53]. For example, when a catalytic amount of Fe(acac)$_2$ and Ph$_2$PCH$_2$CH$_2$PPh$_2$ (dppe) are subjected to the reaction of MeMgBr and propargylic alcohol **65**, the carbomagnesiation takes place to give allylic alcohol **66** selectively. Important findings are not only that MeMgBr does not add to **65** in the absence of iron catalyst, but also that the present catalysis provides access to a different isomer of the addition product. Other metal salts such as Co(OAc)$_2$ and Ni(acac)$_2$ also show catalytic activity, but the efficiency is not as high as iron salts. Homopropargylic alcohol **67** also provides the corresponding homoallylic alcohol **68**. The directing effect of oxygen atom has been proposed as a possible scenario for highly regio- and stereoselective carbometalation[53].

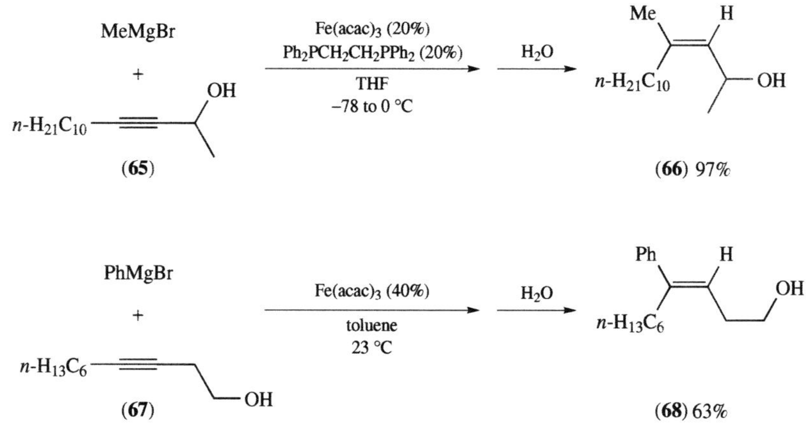

SCHEME 40

F. Addition to Propargyl Amines

Similarly to propargylic alcohols, propargylic amines also undergo regio- and stereoselective carbomagnesiation (Scheme 41)[54, 55]. For example, when propargylic amine **69** is treated with allylmagnesium chloride in refluxing THF, allylated product **70** is obtained in 51% yield[54].

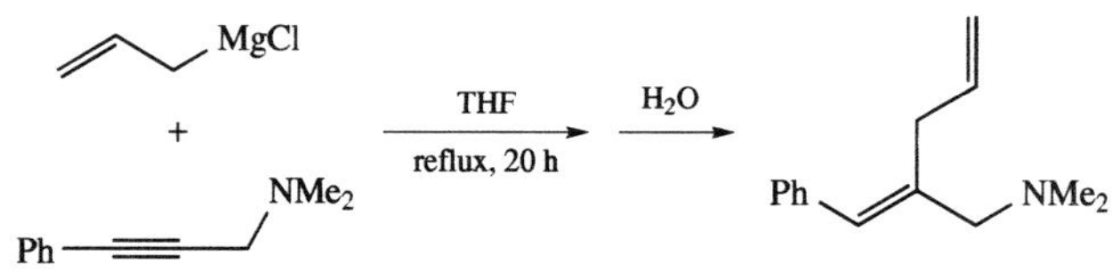

SCHEME 41

14. Carbomagnesiation reactions

When an alkyne bears both a hydroxyl and an amino group at different propargylic positions, there is a regioselectivity issue with regard to the addition of organomagnesium compounds. When **71** is subjected to the reaction with organomagnesium compounds, organic groups are introduced at the carbon proximal to hydroxyl group (Scheme 42)[56]. These results clearly indicate the superior directing power of the hydroxyl group. By using such hydroxyl-assisted carbomagnesiation, a short synthesis of *trans*-zeatin has been accomplished[56].

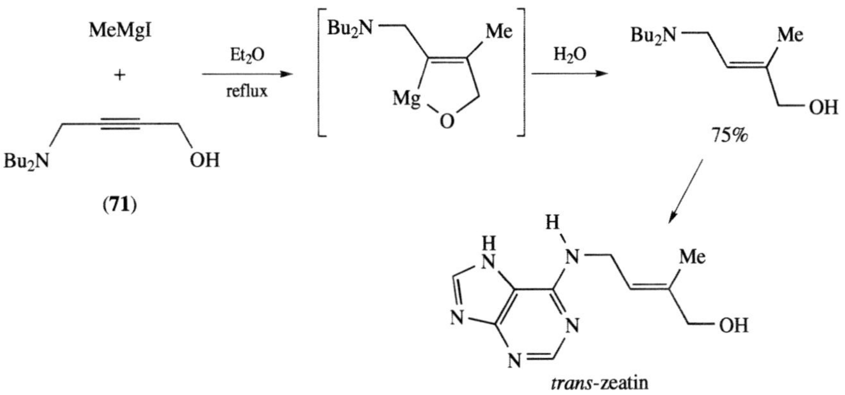

SCHEME 42

III. CARBOMAGNESIATION REACTIONS OF ALKENES
A. Addition to Simple Alkenes

As already stated in the introductory part of this chapter, the careful investigations of Gilman and coworkers revealed that ethylmagnesium halides do not add across simple alkenes in refluxing ether[6,8]. However, the carbomagnesiation of simple alkenes does take place under forcing conditions. In 1958, Podall and Foster reported that diethylmagnesium in ether reacts with ethylene at 50 atm and 100 °C to yield dibutylmagnesium (Scheme 43)[57]. Later, Shepherd[58] and Lehmkuhl and coworkers[59-66] obtained a 1:1 ratio of ethylene and other 1-alkenes by the addition of *sec*-alkyl, *tert*-alkyl and allylic Grignard reagents under high pressure and temperature (30–70 atm and 50–175 °C).

$$H_5C_2-Mg-C_2H_5 + H_2C=CH_2 \xrightarrow[100\,°C,\,43\,h]{Et_2O} n\text{-}H_9C_4-Mg-C_4H_9\text{-}n$$
(50 atm)

33% yield (55% conversion)

SCHEME 43

The presence of transition metal assists the insertion of C=C bond into the carbon–magnesium bond of Grignard reagents[67]. For example, Job and Reich as early as 1924 noticed that a mixture of PhMgBr and $NiCl_2$ in a Et_2O solution absorbs ethylene[68]. Later, it was found that $NiCl_2$ does catalyze an insertion of ethylene into the Ph–Mg bond giving $PhCH_2CH_2MgBr$ (Scheme 44)[69-71]. However, it was also found that this catalytic carbomagnesiation is accompanied by subsequent alkyl–olefin exchange reaction, which is also catalyzed by $NiCl_2$ (Scheme 44)[69-71]. Because of the occurrence of the alkyl–olefin

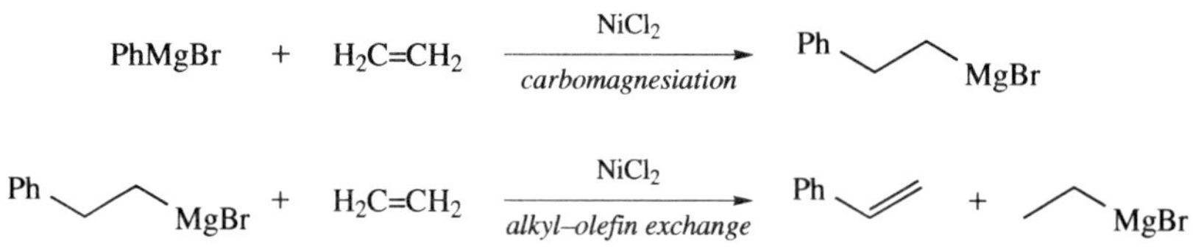

PhMgBr + H$_2$C=CH$_2$ $\xrightarrow[\text{carbomagnesiation}]{\text{NiCl}_2}$ Ph~~~MgBr

Ph~~~MgBr + H$_2$C=CH$_2$ $\xrightarrow[\text{alkyl–olefin exchange}]{\text{NiCl}_2}$ Ph~~= + ~~MgBr

SCHEME 44

exchange, this catalytic carbomagnesiation has not received much attention in organic synthesis.

Although it may be regarded as an example of catalytic conjugate addition, Ni-catalyzed carbomagnesiation is possible for 4-vinylpyridines (Scheme 45)[72]. Under the influence of nickel catalyst, the addition of phenyl, vinyl and benzyl Grignard reagents takes place to give the addition products in good yields. As nickel catalysts, NiCl$_2$(dppp), NiCl$_2$(PPh$_3$)$_2$ and Ni(acac)$_2$ have sufficient activity. Unfortunately, Grignard reagents bearing β-hydrogen atom(s) cannot be applied. By using this method, a rapid synthesis of CDP840, a phosphodiesterase IV inhibitor, has been accomplished (Scheme 45)[72].

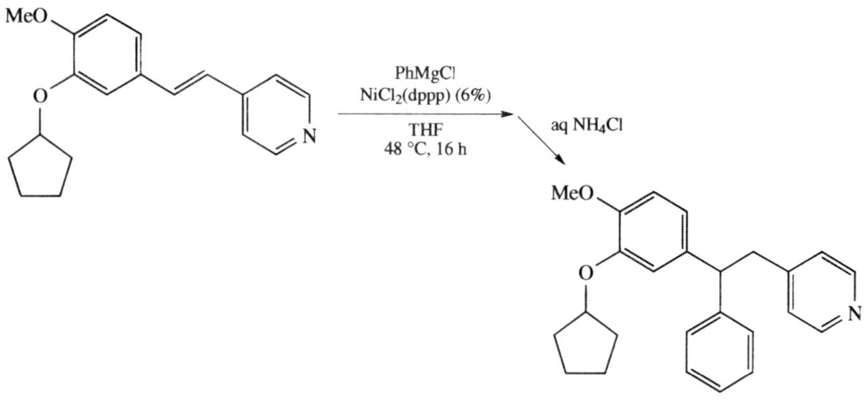

SCHEME 45

Similar to the carbomagnesiation of alkynes, the low reactivity of simple alkenes can be alleviated by performing intramolecular carbomagnesiation[73]. For example, it has been reported that intramolecular carbomagnesiation is involved in the rearrangements of butenyl- and pentenylmagnesium compounds (Scheme 46)[74–78]. However, cyclic products are often not observed because they are usually much less stable than the ring-opening products.

A facile intramolecular carbomagnesiation becomes possible by inserting one more carbon between the reactive magnesium center and the double bond. For example, when a 6-chloro-1-heptene was refluxed with magnesium, 1,2-dimethylcyclopentane (*cis/trans* = *ca* 1/4) was obtained in 88% yield after hydrolysis (Scheme 47)[79, 80]. The cyclization shows a 5-*exo-trig* selectivity and the product derived from 6-*endo-trig* cyclization (methylcyclohexane) is not observed.

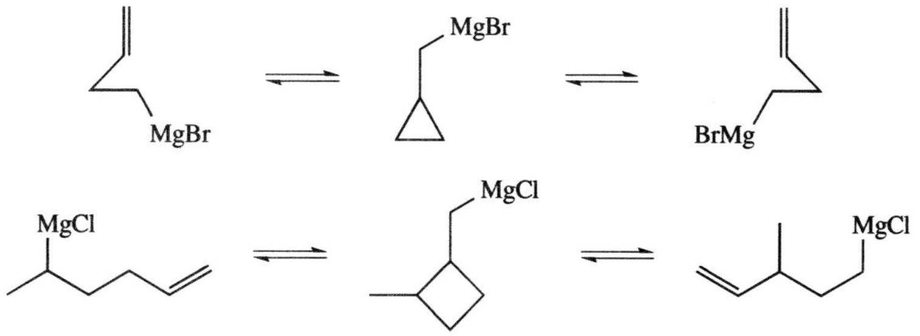

SCHEME 46

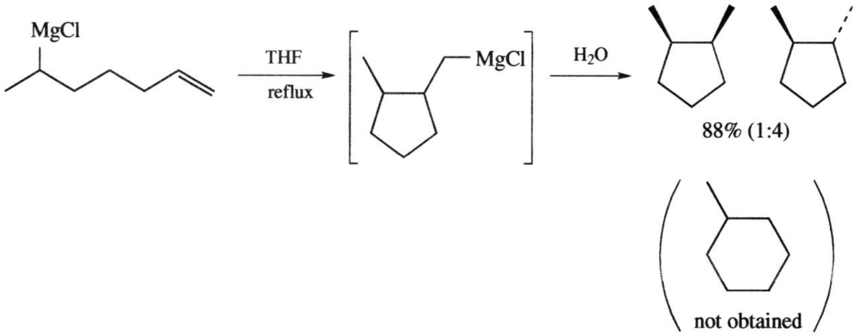

SCHEME 47

B. Allylmagnesiation of Alkenes

Allylic Grignard reagents are known to possess exceptionally high reactivity toward alkenes in comparison with other Grignard reagents (Scheme 48)[59–61,63–65,81]. The process has been now recognized as a metallo-ene reaction. Even with enhanced reactivity, however, intermolecular carbomagnesiation to simple alkenes is still low-yielding, and thus has received virtually no attention as a strategic tool in organic synthesis.

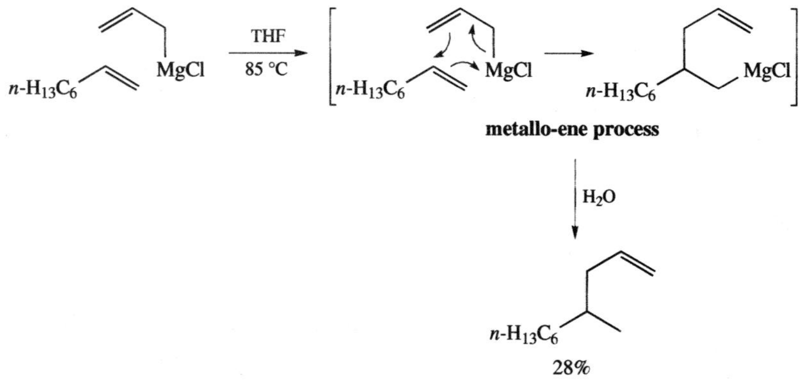

metallo-ene process

SCHEME 48

In contrast to intermolecular versions, intramolecular allylmagnesiation of alkenes (metallo-ene reaction) is entropically favored, and thus more efficient and selective[82]. Felkin and coworkers have demonstrated that 2,7-octadienylmagnesium bromide **73** prepared from **72** undergoes intramolecular allylmagnesiation in refluxing ether to give **74** stereoselectively (Scheme 49)[83].

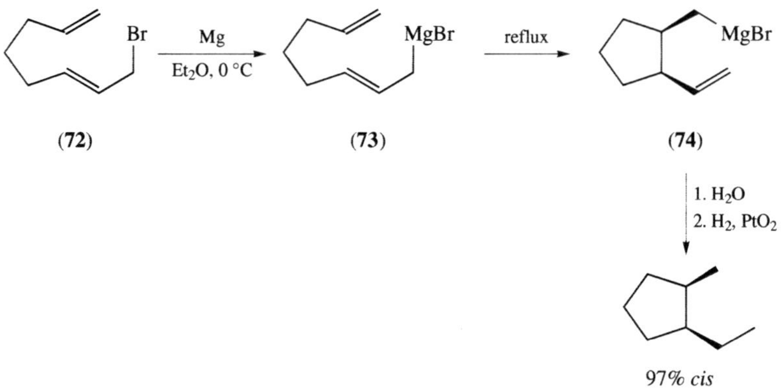

SCHEME 49

Oppolzer and coworkers extensively utilized such intramolecular 'magnesium-ene' reaction in the synthesis of complex natural products[82]. For example, $\Delta^{9(12)}$-capnellene has been synthesized by using this intramolecular allylmagnesiations iteratively (Scheme 50)[84].

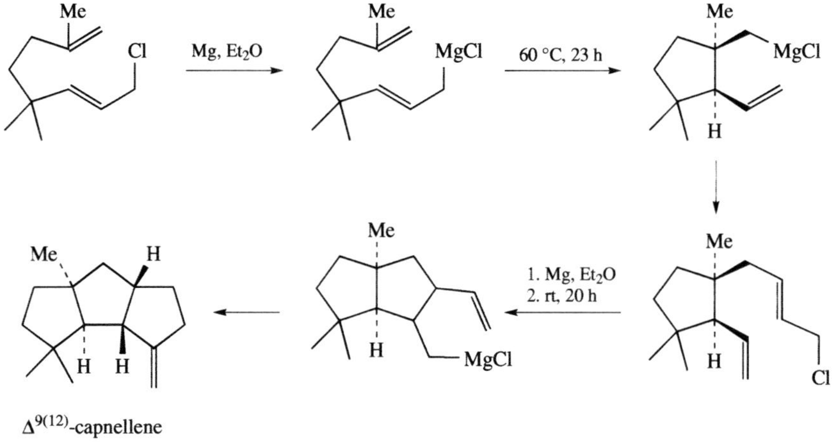

SCHEME 50

Oppolzer and coworkers further demonstrated that an alternative mode of cyclization[85] led to the total synthesis of khusimone (Scheme 51)[86].

Allylic Grignard reagents also react with dienes such as butadiene or isoprene[87]. However, the reaction tends to produce oligomeric products. Otsuka and Akutagawa have

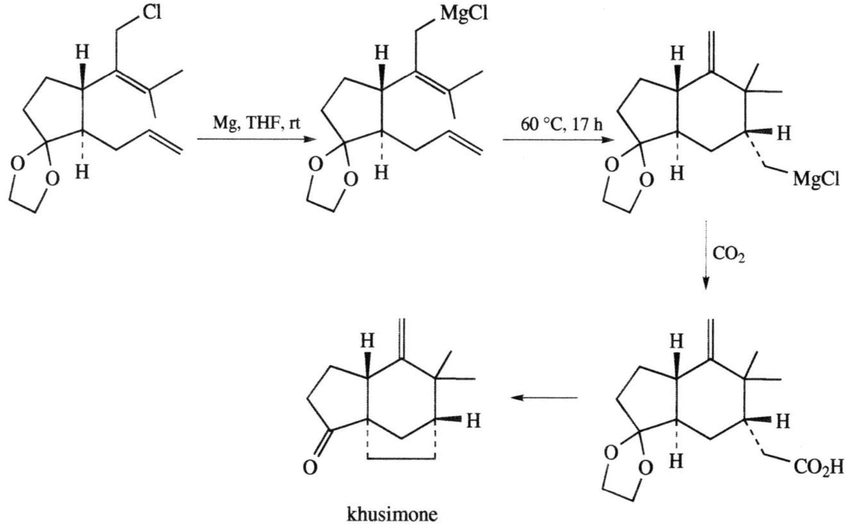

SCHEME 51

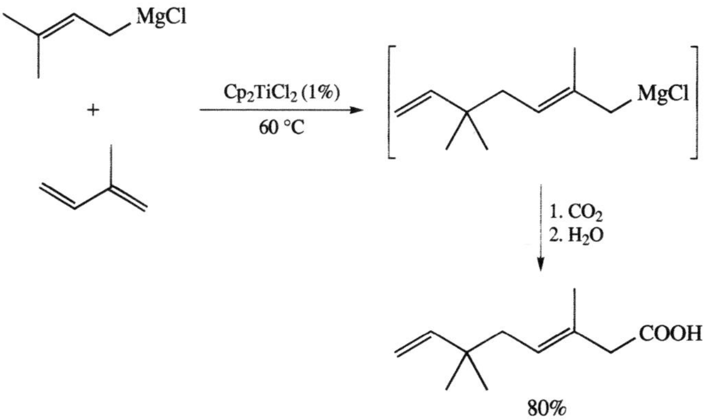

khusimone

SCHEME 52

demonstrated that the use of Cp_2TiCl_2 or $TiCl_2(OEt)_2$ as a catalyst promotes the selective formation of 1:1 adduct (Scheme 52)[88, 89]. By using this method, a range of natural terpenes, such as lanceol and lavandurol, have been synthesized[88].

C. Addition to Strained Alkenes

Cyclopropenes possess appreciable high reactivity toward carbomagnesiation[90]. For example, dialkylmagnesium compounds react with spiro[2.4]hept-1-ene (**75**) to give the carbometalated products **76** (Scheme 53)[91]. The *syn*-addition has been confirmed by D_2O quenching of the reaction.

Nakamura and coworkers reported that the carbomagnesiation of cyclopropenone acetal **77** is significantly promoted by the addition of a catalytic amount of $FeCl_3$ (Scheme 54)[92].

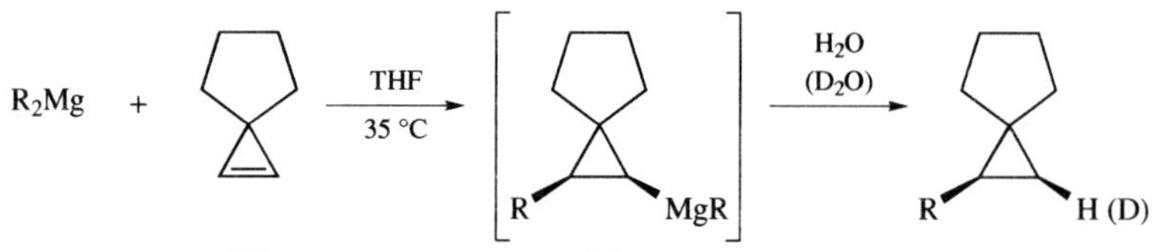

658   Kenichiro Itami and Jun-ichi Yoshida

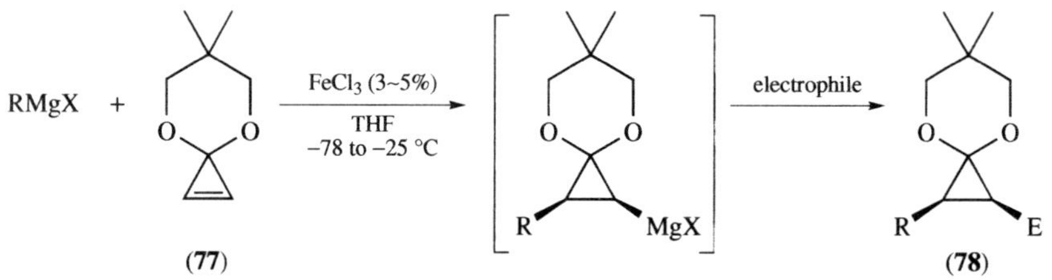

SCHEME 53

R = Me, Et, i-Pr, t-Bu

RMgX	Electrophile	E	Yield (%)
C_6H_5MgBr	H_2O	H	96
C_6H_5MgBr	$CH_2=CHCH_2Br$	$CH_2=CHCH_2-$	85
C_6H_5MgBr	CH_3I	CH_3	90
C_6H_5MgBr	C_6H_5CHO	$C_6H_5CH(OH)-$	56
$CH_2=CHMgBr$	H_2O	H	75
CH_3MgBr	H_2O	H	66
$C_6H_5CH_2CH_2MgCl$	H_2O	H	85

SCHEME 54

The reactions using phenyl, vinyl and alkyl Grignard reagents afford the substituted cyclopropanone acetals **78** in good to excellent yields after hydrolysis. Notably, the reaction of the Grignard reagent possessing β-hydrogen atoms takes place in good yield. The *syn*-carbometalation has been confirmed by employing carbon electrophiles.

This Fe-catalyzed carbomagnesiation reaction can be extended to other strained alkenes as well. For example, the treatment of aryl Grignard reagents with 7-oxabicyclo[2.2.1] heptane derivative **79** induces the stereoselective arylative ring-opening reaction in the presence of FeCl$_3$ catalyst to give the densely substituted cyclohexene derivative **80** in good yields (Scheme 55)[92,93]. Addition of N,N,N',N'-tetramethylethylenediamine (TMEDA) to the reaction mixture facilitates the ring-opening reaction[93]. The reaction takes place in such a manner that the aryl group attacks the carbon–carbon double bond from the *exo*-face of the substrate to give all-*cis*-substituted cyclohexenol product **80** after subsequent β-eliminative ring opening of the oxygen bridge. The reaction can be extended to alkenyl Grignard reagents but the use of EtMgBr and *i*-PrMgBr results in the production of vinyl- and hydrogen-transferred products, respectively ($R^1 \neq R^2$)[93]. Lautens and coworkers reported that the nickel-based catalytic system is also effective in the ring-opening reactions of oxabicyclic alkenes[94].

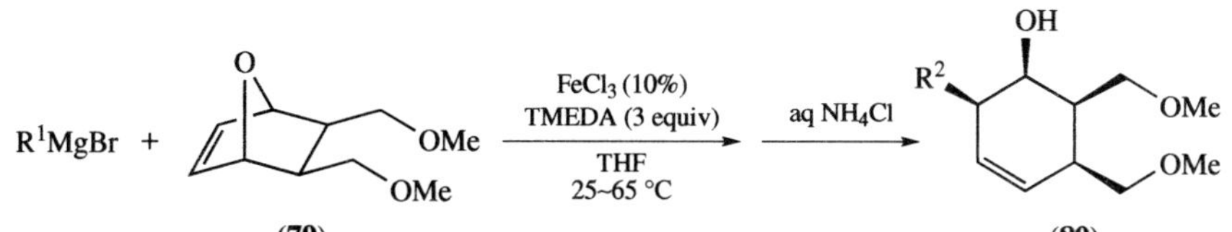

14. Carbomagnesiation reactions

SCHEME 55

R¹	R²	Yield (%)
C_6H_5	C_6H_5	74
4-MeC$_6$H$_4$	4-MeC$_6$H$_4$	72
2-MeC$_6$H$_4$	2-MeC$_6$H$_4$	75
CH$_2$=CH	CH$_2$=CH	41
CH$_3$CH$_2$	CH$_2$=CH	24
(CH$_3$)$_2$CH	H	92

Arrayás, Carretero and coworkers have demonstrated that the CuCl/PPh$_3$ system also induces the ring opening of oxabicylic alkenes but with completely opposite (*anti*) stereoselectivity (Scheme 56)[95]. The introduction of methyl and alkyl groups, which was not possible with the iron-based system (Scheme 55), is also feasible with high fidelity. Miller and coworkers reported the Cu-catalyzed regio- and stereoselective ring openings of 3-aza-2-oxabicyclo[2.2.1]hept-5-ene systems with Grignard reagents[96].

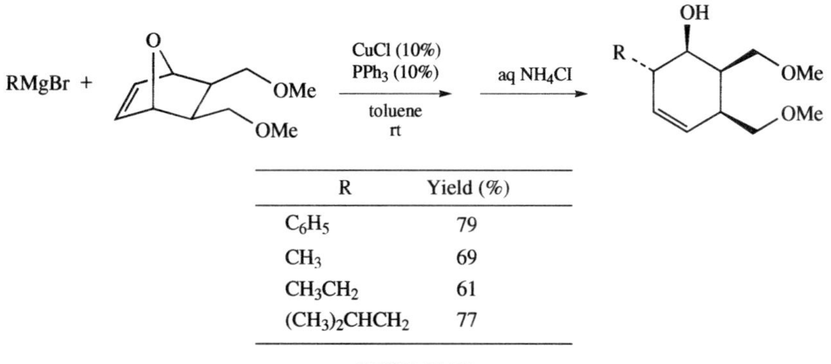

SCHEME 56

R	Yield (%)
C_6H_5	79
CH$_3$	69
CH$_3$CH$_2$	61
(CH$_3$)$_2$CHCH$_2$	77

Hydroxymethylated cyclopropenes, which can be readily prepared by Rh-catalyzed reaction of diazoesters and alkynes, are good substrates for uncatalyzed[97] and Cu-catalyzed[98] carbomagnesiation. For example, a range of substituted cyclopropanes **82** can be synthesized in a regio- and stereoselective fashion by the Cu-catalyzed addition of Grignard reagents to (3-hydroxymethyl)cyclopropenes **81** (Scheme 57)[98,99].

Fox and Lui have demonstrated that the addition of *N*-methylprolinol can induce high enantioselectivity in the methylmagnesiation of (3-hydroxymethyl)cyclopropene **83** (Scheme 58)[100]. The fact that the reaction produces only one diastereomer, where the methyl and hydroxymethyl groups on cyclopropane ring are *syn*, may be rationalized by

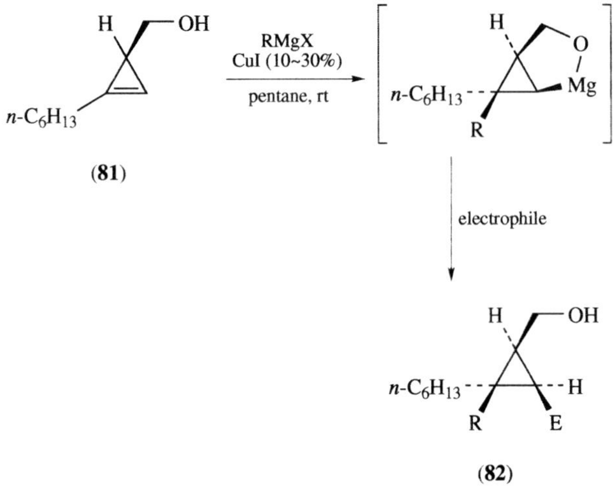

R	Electrophile	E	Yield (%)
Methyl	H_2O	H	83
Methyl	I_2	I	81
Benzyl	H_2O	H	78
Vinyl	H_2O	H	81
Vinyl	I_2	I	83
Vinyl	Bu_3SnCl	Bu_3Sn	71
Phenylethynyl	H_2O	H	75

SCHEME 57

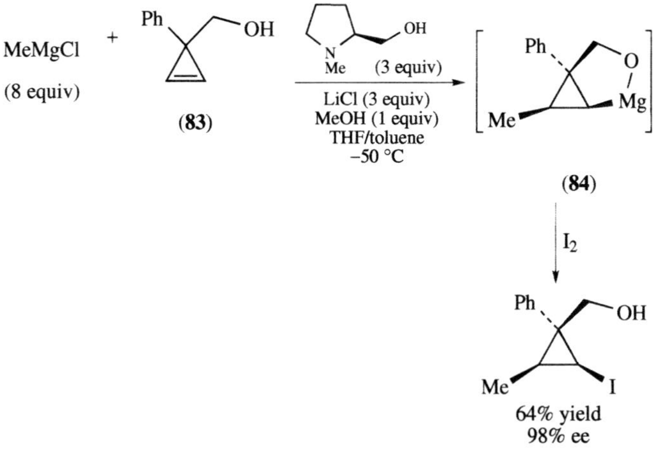

SCHEME 58

the intermediacy of magnesium chelate **84**. The application to other organomagnesium compounds results in much lower enantioselectivity.

D. Addition to Vinylsilanes

The carbomagnesiation of vinylsilanes is a powerful method for the generation of synthetically useful α-silyl carbanions. However, simple vinylsilanes such as trimethyl(vinyl) silane do not undergo carbometalation with Grignard reagents (Scheme 59)[101]. In early days, only limited success had been achieved by using perfluorovinylsilanes[102].

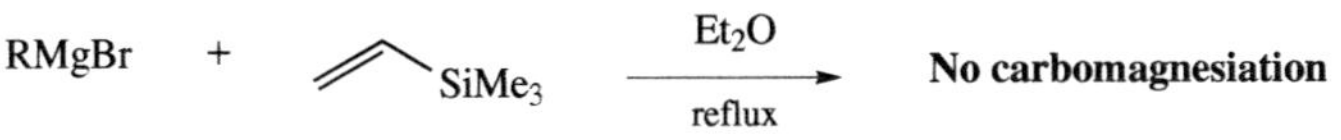

RMgBr + CH$_2$=CH-SiMe$_3$ $\xrightarrow[\text{reflux}]{\text{Et}_2\text{O}}$ **No carbomagnesiation**

SCHEME 59

One of the classical solutions to overcome the low reactivity is to render the carbomagnesiation intramolecular. For example, Utimoto and coworkers reported that the reaction of (*E*)-6-bromo-3-methyl-1-trimethylsilyl-1-hexene (**85**) with magnesium produces the corresponding Grignard reagent **86**, which intramolecularly adds to the vinylsilane moiety from the less hindered side affording a single stereoisomer of cyclized product **87** (Scheme 60)[103].

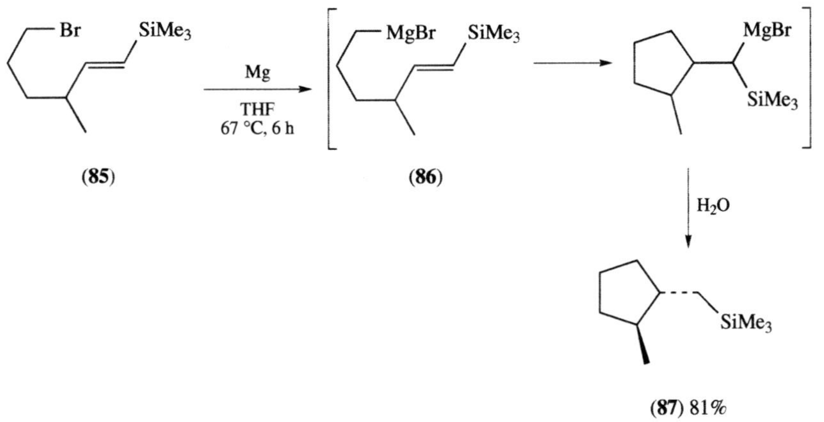

SCHEME 60

Hoffmann and coworkers have further carefully examined the intramolecular carbomagnesiation of a vinylsilane (Scheme 61)[104]. Both (*E*)- and (*Z*)-isomers of **88** undergo an intramolecular carbomagnesiation and a stereospecific (>95%) *syn*-addition of the carbon–magnesium bond to the double bond takes place. The resulting α-silylalkylmagnesium compounds **89** are not configurationally stable under the reaction conditions. They epimerize with a half-life of 2.7 days at room temperature.

Electron-withdrawing groups such as alkoxy and chloro groups on silicon activate vinylsilanes toward the addition of secondary and tertiary alkylmagnesium halides (Scheme 62)[101, 105].

Subsequent reactions of thus-generated organomagnesium species are also possible[106, 107]. For example, α-silylorganomagnesium compound **91** is allowed to react with

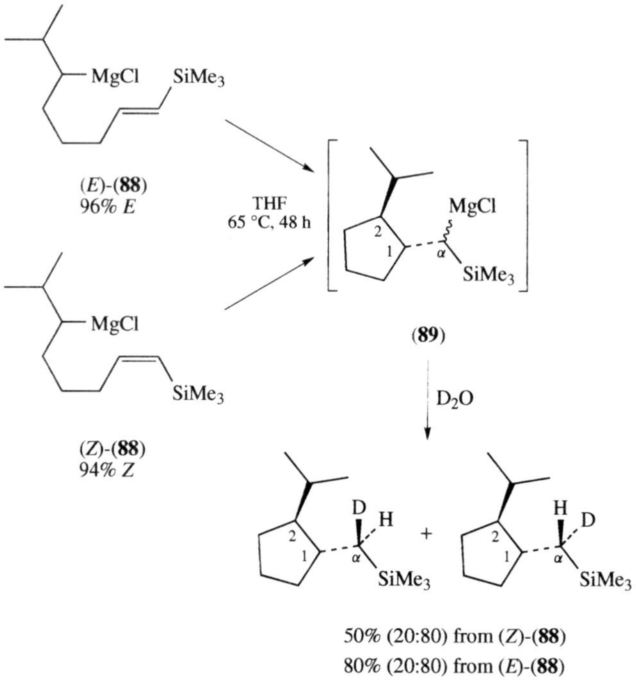

662   Kenichiro Itami and Jun-ichi Yoshida

SCHEME 61

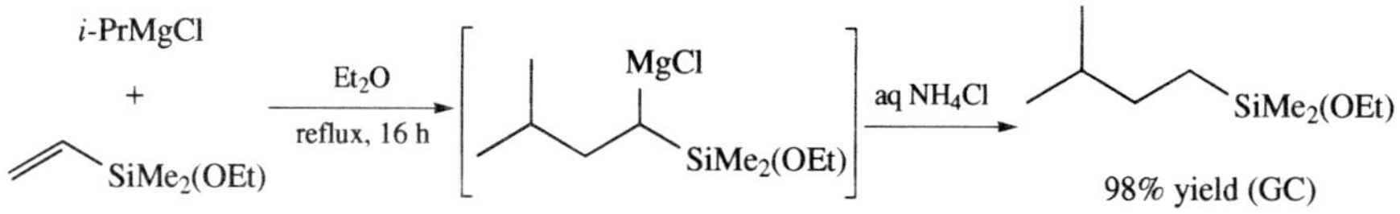

SCHEME 62

allyl bromide in the presence of CuI catalyst to give **92** in 69% yield (Scheme 63)[106]. The subsequent oxidative cleavage of the carbon–silicon bond then affords the secondary alcohol **93** in 69% yield (Scheme 63).

Although Grignard reagents become viable for the addition reaction to vinylsilanes, serious limitations still exist. For example, substitution reactions at the silicon atom are often observed as unavoidable side reactions when activating groups (e.g. chloro, alkoxy and amino groups) are used. In addition, primary alkyl Grignard reagents are not applicable. Itami, Yoshida and coworkers have developed a novel strategy for intermolecular carbomagnesiation of vinylsilanes by exploiting the 2-pyridylsilyl group as a removable directing group (Scheme 64)[108, 109]. The reaction presumably involves a pre-equilibrium complex of **94** and a Grignard reagent, making the subsequent carbomagnesiation intramolecular in nature. The importance of this pre-equilibrium complex was further supported by the observation of a dramatic solvent effect: weakly coordinating solvents such as Et_2O favor this reaction, whereas strongly coordinating solvents such as THF disfavor it. In addition to this kinetic preference, stabilization of the resultant organomagnesium species **95** by intramolecular coordination of the pyridyl group might

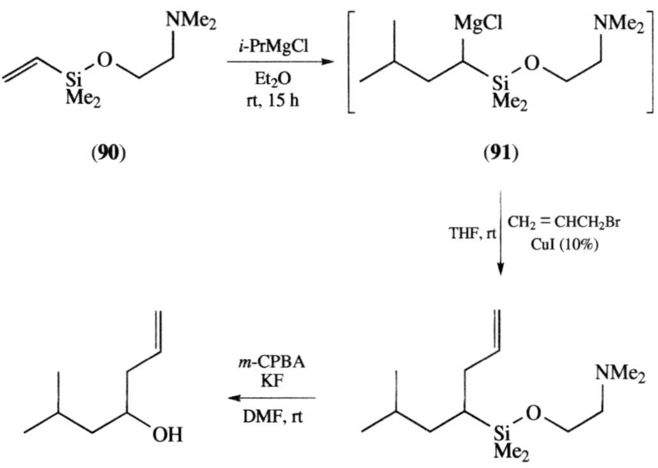

(**90**) (**91**)

(**93**) 69% (**92**) 69% from **90**

SCHEME 63

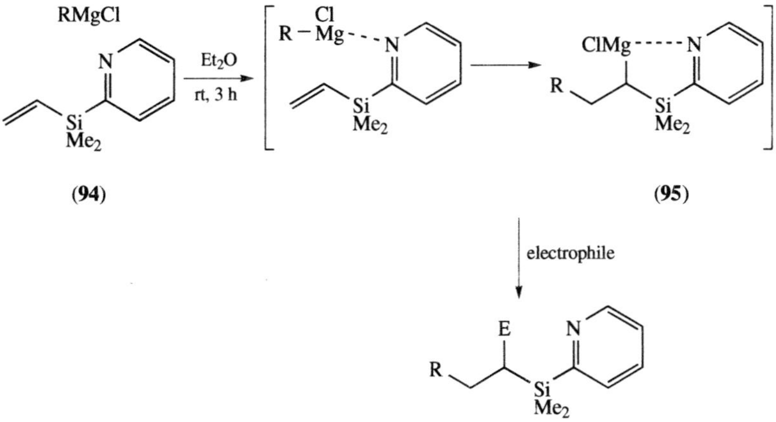

(**94**) (**95**)

R	Electrophile	E	Yield (%)
n-Butyl	D_2O	D	83
n-Butyl	$CH_2=CHCH_2Br$	$CH_2=CHCH_2$-	91
n-Butyl	$CH_2=N^+Me_2I^-$	Me_2NCH_2-	93
n-Butyl	C_6H_5I [5% $Pd(PPh_3)_4$]	C_6H_5	77
i-Propyl	$CH_2=CHCH_2Br$	$CH_2=CHCH_2$-	91
Allyl	4-ClC_6H_4Br [2.5% $Pd(PPh_3)_4$]	4-ClC_6H_4	93
Phenyl	H_2O	H	73

SCHEME 64

also be responsible for the efficiency of this carbomagnesiation process. Nevertheless, by using this protocol, facile addition of primary alkyl Grignard reagents to vinylsilanes has been realized for the first time. Secondary alkyl, allyl and phenyl Grignard reagents also add across **94**. The addition of various electrophiles in the subsequent reactions is also possible (Scheme 64).

The carbomagnesiation also takes place with β-substituted vinylsilane **96**, which represents a more difficult class of substrate (Scheme 65)[108]. The three-component assembled product **97** is converted to the corresponding secondary alcohol **98** by fluoride-mediated oxidative cleavage of the carbon–silicon bond (Scheme 65).

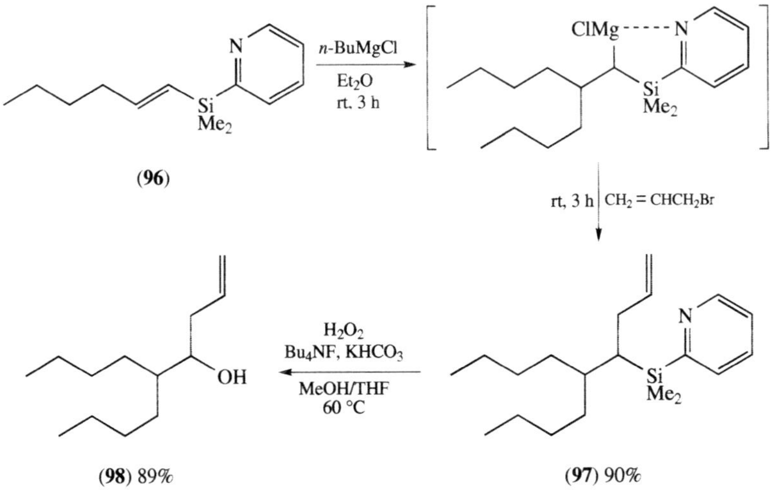

SCHEME 65

E. Addition to Allyl and Homoallyl Alcohols

The presence of a neighboring hydroxyl group facilitates the carbomagnesiation of alkenes. For example, Eisch and Husk unexpectedly found that homoallylic alcohol **99** reacts with an excess amount of allylmagnesium bromide at room temperature to give the carbomagnesiation product **102** in 56% yield (Scheme 66)[110]. At this time, it was one of the few examples of carbomagnesiation to unconjugated ethylenic linkage taking place under mild conditions. However, the reaction is somewhat limited to allyl Grignard reagent: t-BuMgBr and PhCH$_2$MgBr exhibit extremely low reactivity, and no reaction occurs with PhMgBr[111]. An analysis of this facile carbomagnesiation in terms of substrate, magnesium reagent and medium has led to the suggestion that the reaction occurs via the intramolecular rearrangement of **100** to **101**[111, 112]. In line with such a hydroxyl-assisted proximity effect, an insertion of methylene spacer(s) between the hydroxyl group and the double bond leads to substantial decrease in reaction efficiency[112]. In some instances when the carbomagnesiation is slow, the dehydration of the initial product often occurs[112].

When allylmagnesium bromide is allowed to react with 3-cyclopentenol, the allylated product (3-allylcyclopentanol) is obtained as an 80:20 mixture of *cis*- and *trans*-isomers (Scheme 67)[113]. The reaction of 3-cyclopentenol with diallylmagnesium in refluxing benzene proceeds slowly but cleanly to yield only the *cis*-isomer.

Ni(acac)$_2$ acts as a catalyst for this carbomagnesiation[112]. The remarkable effect of nickel salts on this allylation reaction has three interconnected advantages: more rapid

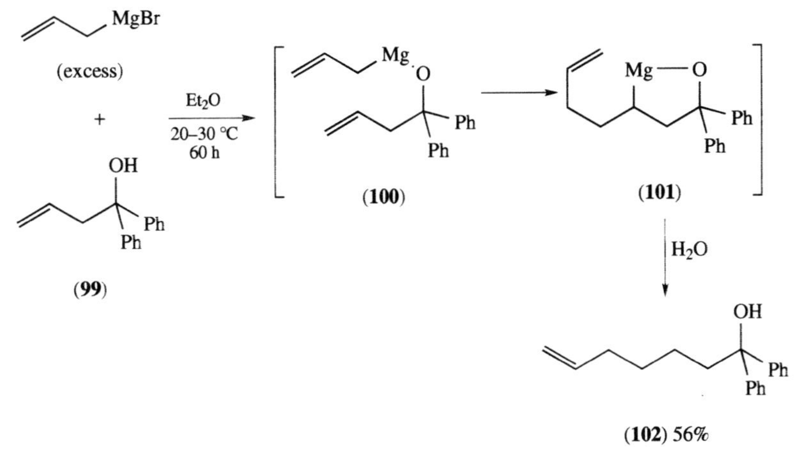

SCHEME 66

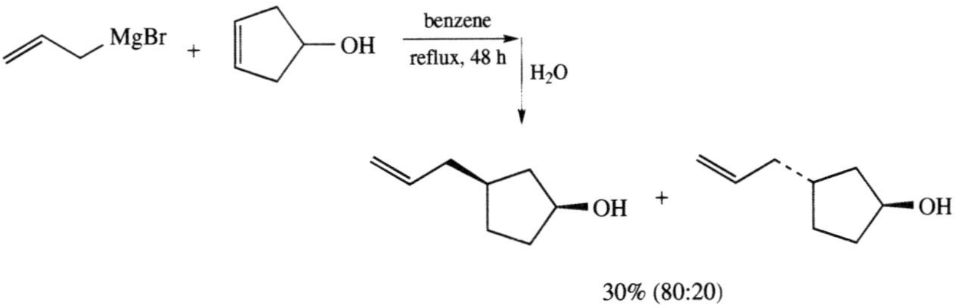

SCHEME 67

reaction, higher conversion to product and less dehydration of the initial alcohol. Although (allyl)nickel species seems to be involved, the mechanism of nickel catalysis is still unknown.

At about the same time that Eisch and coworkers discovered the hydroxyl-assisted carbomagnesiation of homoallylic alcohol, Felkin and coworkers reported a different carbomagnesiation that was also assisted by hydroxyl group. Allylmagnesium bromide adds regioselectively to allyl alcohol, and the alcohol **103** was obtained in 50% yield after hydrolysis (Scheme 68)[114]. The use of PhCH$_2$MgCl affords a 10% yield of 2-benzylpropanol (benzene, reflux, 170 h), but the reaction does not take place with t-BuMgCl, i-BuMgCl, i-PrMgCl and EtMgBr. Although cinnamyl alcohol is an excellent substrate for the reaction with allylmagnesium bromide (99% yield, Et$_2$O, room temperature)[115], crotyl alcohol and β-methallyl alcohol are poor substrates (<5%, Et$_2$O, reflux, 170 h)[114]. The protection of the hydroxyl group completely suppresses the reaction, which clearly indicates that the reaction is assisted by the hydroxyl group. It has been proposed that it follows the magnesium-ene mechanism, which is promoted by coordination of alkene π-bond to a covalently bound magnesium atom (Scheme 68). This is in sharp contrast to the mechanism proposed for the allylmagnesiation of homoallylic alcohol, where the allyl group is transferred intramolecularly (Scheme 66).

A concerted addition of this kind, involving simultaneous intramolecular electrophilic assistance by the magnesium bound to oxygen, and intermolecular nucleophilic attack by

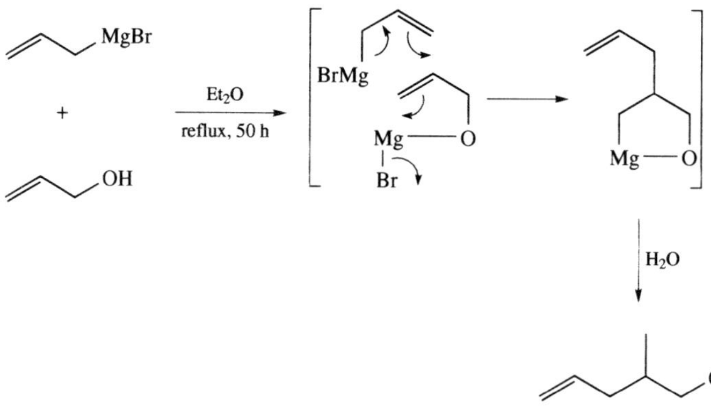

(**103**) 50%

SCHEME 68

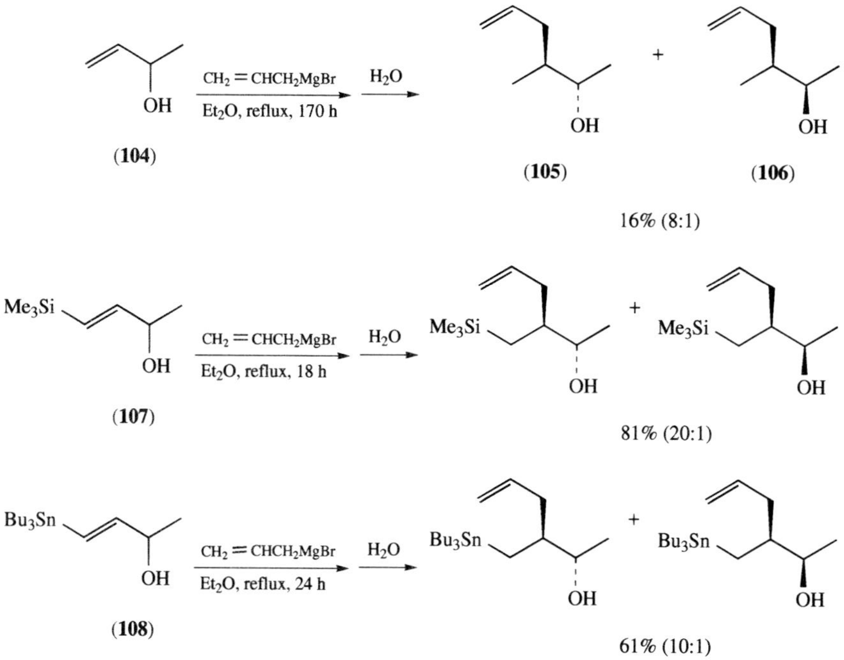

(**104**) (**105**) (**106**) (**107**) (**108**)

SCHEME 69

allylmagnesium bromide, is consistent not only with the kinetic data, but also with the stereochemical course of the reaction. For example, the reaction of α-methallyl alcohol (**104**) with allylmagnesium bromide gives the product alcohol **105** and **106** (16% yield) in a ratio of 8:1 (Scheme 69)[114]. The reactions of silylated and stannylated α-methallyl alcohols (**107** and **108**) give carbometalated products in much higher yields with preference for the *erythro* isomers[116].

14. Carbomagnesiation reactions   667

Similarly to the case of the hydroxyl group, the amino group also exhibits an acceleration effect in carbomagnesiation reaction. For example, allylic amines such as **109** and **111** undergo regioselective carbomagnesiation with allylmagnesium chloride giving the adducts **110** and **112** in 23% and 44% yield, respectively (Scheme 70)[54]. The failure to observe addition to the related unsaturated hydrocarbon (1-phenyl-1-propene), even when amine function are present in other molecules in the solution, suggests that the amine functions promotes the additions to **109** and **111**. Hence, an amine-assisted carbomagnesiation mechanism similar to that proposed for allylic alcohol (Scheme 68) has been suggested.

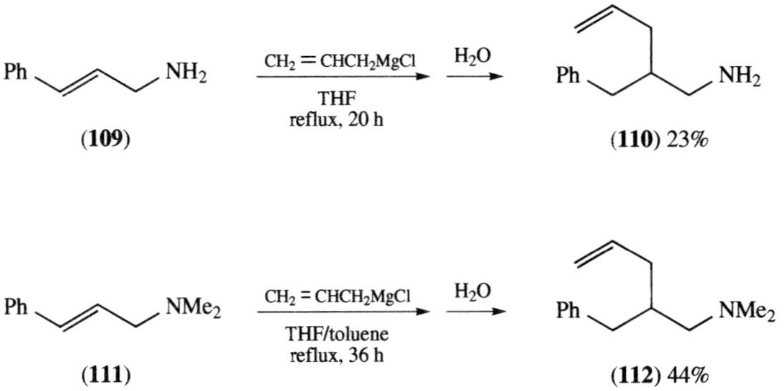

Ph⎯⎯⎯NH₂ $\xrightarrow[\text{THF}]{\text{CH}_2=\text{CHCH}_2\text{MgCl}}$ $\xrightarrow{\text{H}_2\text{O}}$ Ph⎯⎯⎯NH₂
(**109**)   reflux, 20 h   (**110**) 23%

Ph⎯⎯⎯NMe₂ $\xrightarrow[\text{THF/toluene}]{\text{CH}_2=\text{CHCH}_2\text{MgCl}}$ $\xrightarrow{\text{H}_2\text{O}}$ Ph⎯⎯⎯NMe₂
(**111**)   reflux, 36 h   (**112**) 44%

SCHEME 70

As in the cases of many types of carbomagnesiation reactions mentioned before, a chelating auxiliary on olefinic substrates has a strong impact on the reactivity and selectivity in conjugate addition. Chelation-controlled conjugate additions have been investigated with quinol alkoxide **113** that reacts with Grignard reagents to afford conjugate addition products **114** (Scheme 71)[117–119]. Chelation between the alkoxide and the Grignard reagent results in a transient complex that delivers the nucleophile from the same face and establishes the *syn* stereochemistry between the installed organic group and the hydroxyl group.

Such an alkoxide-assisted (hydroxyl-assisted) conjugate addition methodology has been exploited in the key step toward (±)-euonyminol (Scheme 72)[120].

Addition of Grignard reagents to hydroxyl-containing unsaturated nitrile **115** provides conjugate addition products with virtually complete stereocontrol (Scheme 73)[121]. Mechanistic evidence supports a chelation-controlled conjugate addition via alkylmagnesium alkoxide intermediates. Diverse Grignard reagents having sp³-, sp²- and sp-hybridized carbons react with high efficiency (−78 °C, 5 min).

Fleming and coworkers further extended the chelation-controlled conjugate addition strategy to γ-hydroxy unsaturated nitriles (Scheme 74)[122,123]. A treatment of a Grignard reagent with γ-hydroxy unsaturated nitrile **116** results in clean and regioselective carbomagnesiation. Mechanistically, deprotonation of the hydroxyl group generates the alkoxymagnesium halide **117** that rapidly engages in halogen–alkyl exchange with the Grignard reagent. The resulting alkylmagnesium alkoxide initiates a smooth conjugate addition, ultimately generating the conjugate adduct **118** after treatment with acetic acid. The substantial rate difference between deprotonation and conjugate addition allows *t*-BuMgCl to be employed as a sacrificial base. The hydroxyl-assisted mechanism resembles that suggested for the carbomagnesiation of allylic alcohol. In line with such a chelation-controlled mechanism, control experiments involving addition of *n*-BuMgCl to

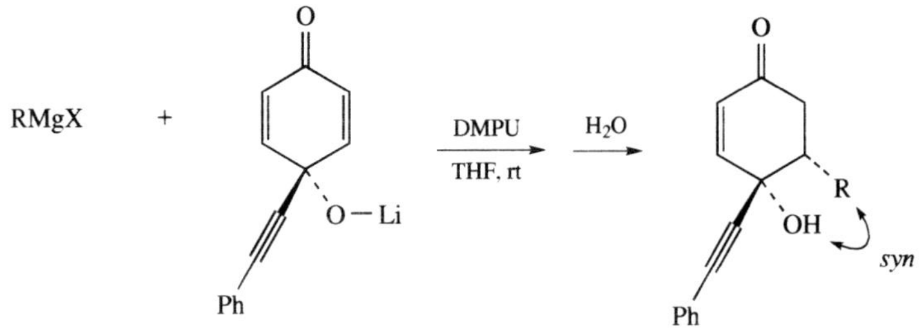

RMgX	Yield (%)
MeMgCl	85
EtMgBr	86
n-BuMgCl	83
CH$_2$=CHMgCl	68
PhMgBr	81

SCHEME 71

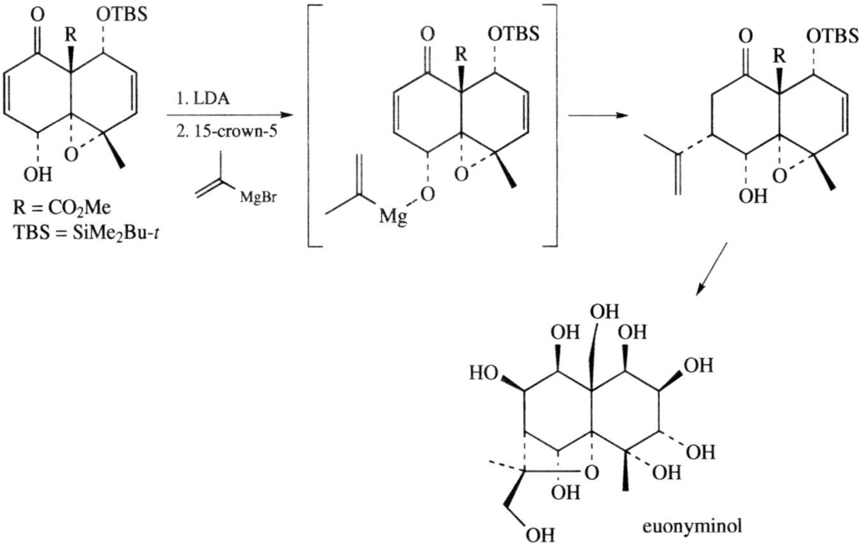

SCHEME 72

the substrates lacking the hydroxyl group (**119**) afford only recovered starting materials. Chelation apparently requires the hydroxyl group adjacent to the double bond since relocating the hydroxyl group three or four carbons away, as in **120** and **121**, precludes conjugate addition.

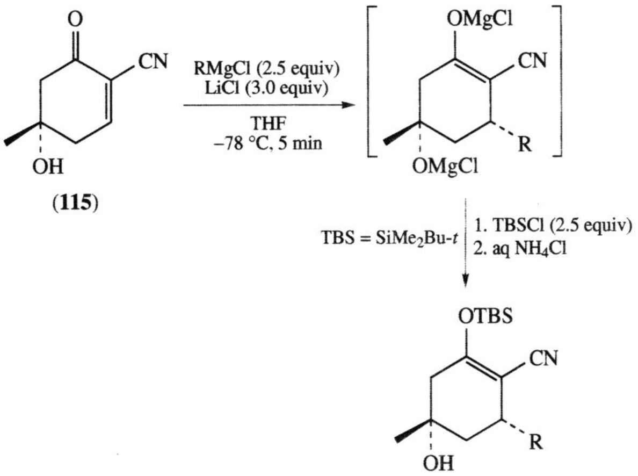

14. Carbomagnesiation reactions

R	Yield (%)
Methyl	58
t-Butyl	43
Phenyl	58
Vinyl	49
Ethynyl	47

SCHEME 73

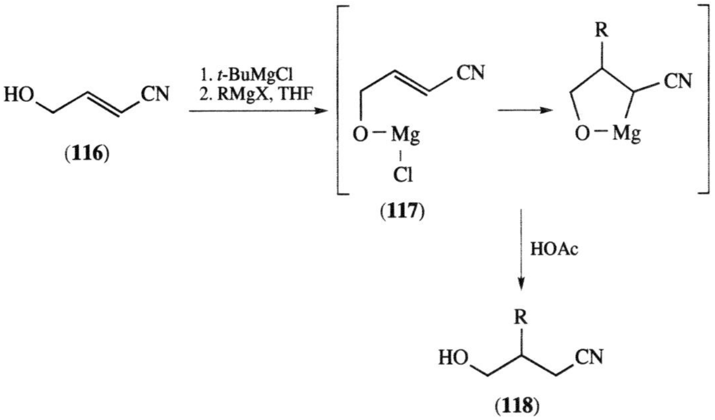

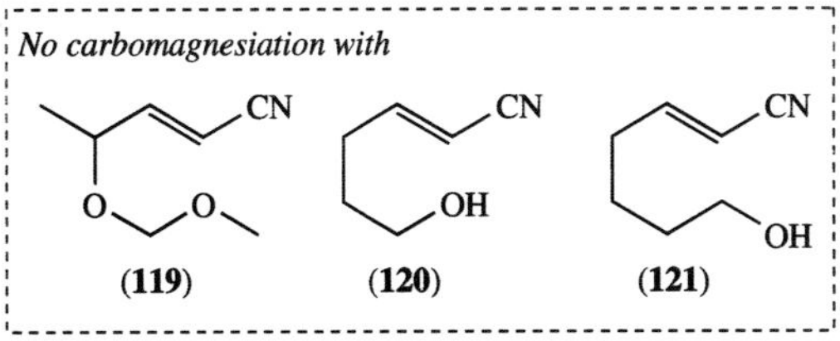

R	Yield (%)
Methyl	74
n-Butyl	80
Phenyl	76
Vinyl	63
Phenylethynyl	62

No carbomagnesiation with (119), (120), (121)

SCHEME 74

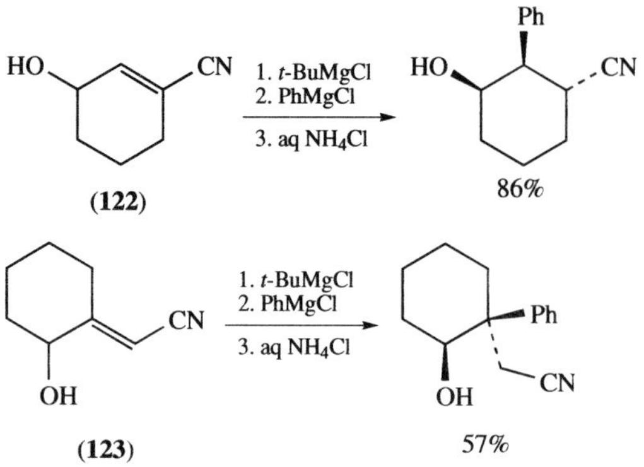

SCHEME 75

Conjugate additions to cyclic unsaturated nitriles such as **122** and **123** also proceed smoothly with virtually complete stereocontrol (Scheme 75)[123]. Given the long-standing difficulty of performing conjugate additions to α,β-unsaturated nitrile compounds[124], the present chelation-assisted methodology is extremely powerful and might provide a practical solution.

The clean and facile chelation-controlled carbomagnesiation of hydroxylated unsaturated nitriles offers an opportunity for further multicomponent assembly through the reactions of the resultant organomagnesium species with electrophiles. The following example accessing terpenoid structures may illustrate the power of a multicomponent assembling reaction (Scheme 76)[125]. The sequential addition of MeMgCl, second Grignard reagent **125** and MeI to cyanocyclohexenone **124** gave **126** in 54% yield. Obviously MeMgCl adds to carbonyl group forming the alkoxide, which triggers the chelation-controlled conjugate addition of **125** to the C=C bond. The follow-up methylation of the resultant organomagnesium species with MeI then furnishes the entire abietane carbon

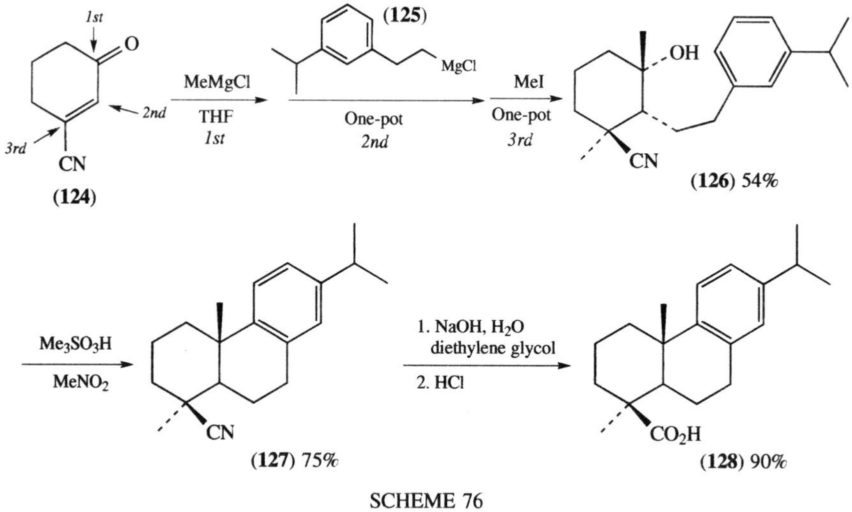

SCHEME 76

skeleton in one-pot. Intramolecular Friedel–Crafts alkylation of **126** affords predominantly the *cis*-abietane **127**. Finally, the nitrile hydrolysis completes the synthesis of *epi*-dehydroabietic acid **128**.

Fleming and coworkers have developed a number of such multicomponent assembling reactions using alkyl halides, aldehydes, ketones and acid chlorides as electrophiles for post-functionalization of carbomagnesiation[126–130]. The stereoselectivity of the reaction with electrophiles is not only generally high but also highly dependent on the nature of electrophiles.

F. Zirconium-catalyzed Ethylmagnesiation of Alkenes

Despite impressive development in the addition reaction of Grignard reagents to carbon–carbon multiple bonds, a facile intermolecular carbomagnesiation reaction of simple alkenes has been a long-standing goal. Although somewhat limited to ethylmagnesium compounds, the zirconium-catalyzed ethylmagnesiation of terminal alkenes, originally discovered by Dzhemilev and extensively investigated by Hoveyda and others, has attracted great attention in the synthetic community.

In 1983, Dzhemilev and coworkers reported that Cp_2ZrCl_2 catalyzes the addition of ethylmagnesium compounds, such as ethylmagnesium halides and diethylmagnesium, to simple alkenes. For example, under the catalytic influence of Cp_2ZrCl_2 (1 mol%), Et_2Mg reacts with a range of terminal alkenes at room temperature in Et_2O furnishing the ethylmagnesiation product in good to excellent yields (Scheme 77)[131–133]. Although the addition to internal alkenes such as 2-hexene and cyclohexene does not take place, strained cyclic alkenes such as norbornene and norbornadiene undergo ethylmagnesiation. The use of other alkylmagnesium compounds results in the formation of many unwanted side products[133–135].

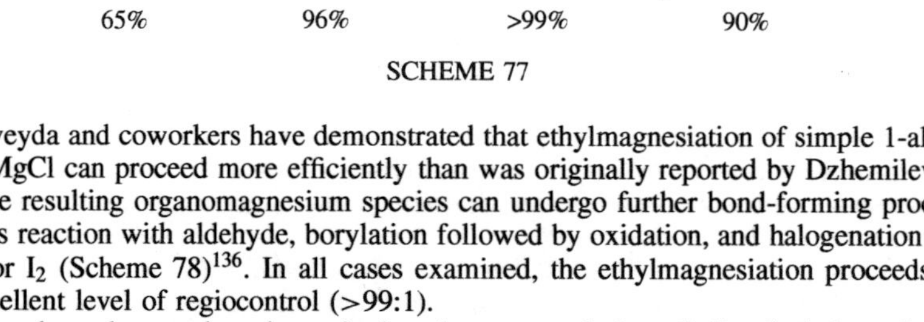

SCHEME 77

Hoveyda and coworkers have demonstrated that ethylmagnesiation of simple 1-alkenes by EtMgCl can proceed more efficiently than was originally reported by Dzhemilev, and that the resulting organomagnesium species can undergo further bond-forming processes such as reaction with aldehyde, borylation followed by oxidation, and halogenation using NBS or I_2 (Scheme 78)[136]. In all cases examined, the ethylmagnesiation proceeds with an excellent level of regiocontrol (>99:1).

Hoveyda and coworkers have further demonstrated that allylic alcohols and ethers are good substrates exhibiting notable diastereocontrol (Scheme 79)[136]. The ethylmagnesiation of allylic alcohol **129** affords the *syn* diol with 95:5 diastereoselectivity (70% yield). On the other hand, the reaction of the corresponding methyl ether **130** affords the monoprotected alcohol with opposite sense of diastereoselectivity (*syn:anti* = 11:89). The corresponding *t*-butyldimethylsilyl ether **131** is recovered unchanged, and oxygen-free substrate **132** provides an equal mixture of diastereomers. The chelation between the

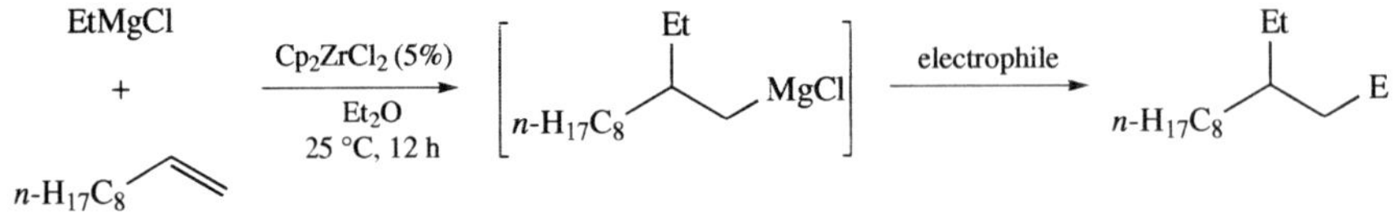

SCHEME 78

Electrophile	E	Yield (%)
CH$_3$CHO	CH$_3$CH(OH)-	55
NBS	Br	60
I$_2$	I	65
B(OMe)$_3$ then H$_2$O$_2$	OH	60

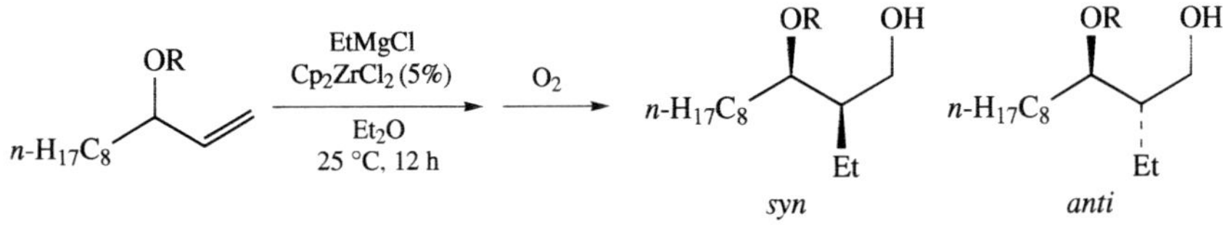

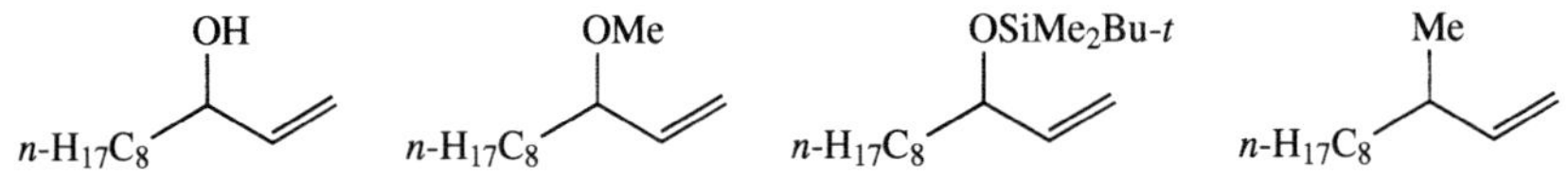

SCHEME 79

(129) 70% yield, 95% *syn*

(130) 80% yield, 89% *anti*

(131) No reaction

(132) No selectivity

Lewis basic heteroatom and a metal center (Zr or Mg) in a transition state organization seems to be responsible for high diastereocontrol[137].

In analogy to other zirconium-catalyzed reactions, a possible pathway has been thought to involve oxidative cyclization of Cp$_2$Zr(CH$_2$=CH$_2$) and 1-alkene giving zirconacyclopentane species, which further reacts with EtMgCl leading to the formal ethylmagnesiation product. On the basis of extensive mechanistic studies, the mechanistic picture that includes a biszirconocene complex as its centerpiece has been proposed by Hoveyda and coworkers (Scheme 80)[137]. The proposed mechanistic hypothesis rationalizes (i) the requirement for excess EtMgCl, (ii) the necessity for the presence of a Lewis basic heteroatom for diastereocontrol and (iii) the reaction for the highly regioselective rupture of the intermediate zirconacyclopentane.

The dramatic influence of internal Lewis bases on reactivity/selectivity can be seen in other systems such as homoallylic alcohols[138] and substituted norbornenes[139]. For example, when *endo*-5-norbornen-2-ol (133) is used as a substrate, the Zr-catalyzed carbomagnesiation occurs with a range of alkyl Grignard reagents in an *anti*-addition manner (Scheme 81)[139]. It is reasonable to suggest that with the Mg salt of 133, the heteroatom binds and delivers the magnesium ion to initiate a complete reversal in the regioselective zirconacyclopentane cleavage.

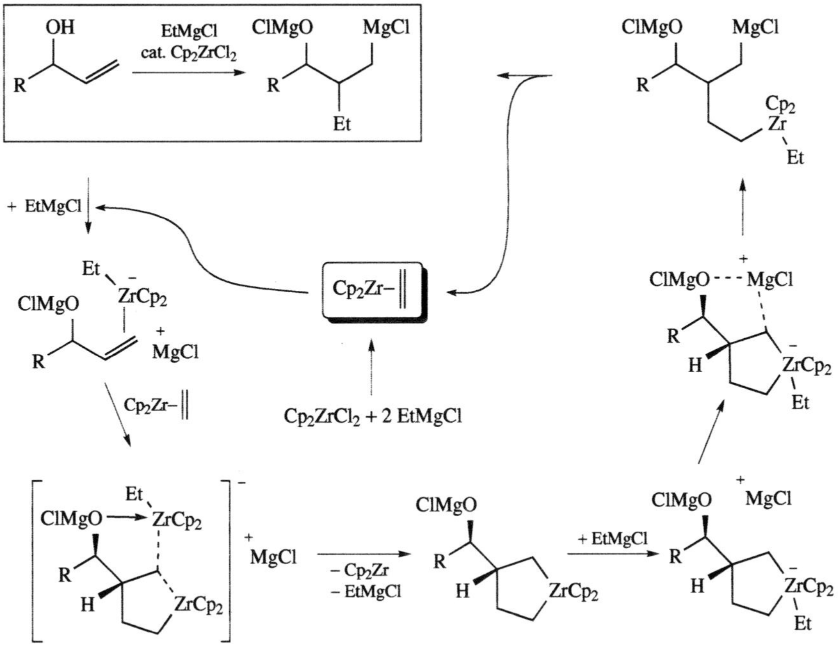

SCHEME 80

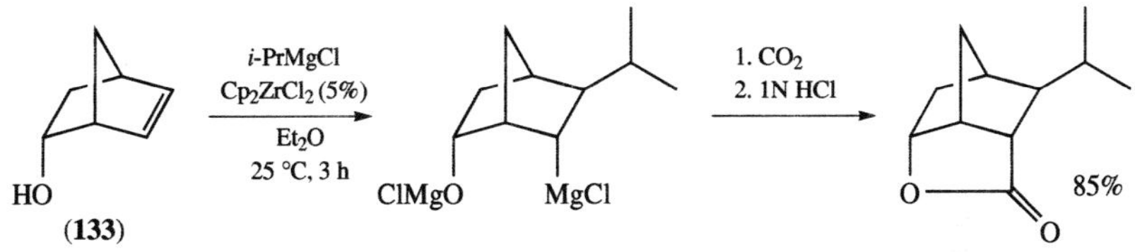

SCHEME 81

Enantioselective ethylmagnesiation can be achieved by using a chiral Zr catalyst[140–149]. Whitby and coworkers reported that the C_1-symmetric zirconocene dichloride **134** serves as a cheap, active and selective catalyst precursor for the enantioselective ethylmagnesiation of unactivated terminal alkenes (Scheme 82)[140, 141]. For example, the reaction of N-allylaniline (**135**) and EtMgCl in Et$_2$O followed by quenching with MeSSMe gives **136** in 81% ee (90% yield). Dramatic decrease in enantioselectivity is observed when the NH functionality of **135** is protected as NMe moiety (**137**). The reaction using allyl benzyl ether gives high enantioselectivity (83% ee) whereas the use of simple 4-phenyl-1-butene as a substrate results in very poor yield and enantioselectivity (Scheme 82)[141]. Attractive heteroatom-coordination to the metal (Zr or Mg) may be involved in the stereo-determining transition structure of this reaction.

Hoveyda and coworkers have also developed a number of useful asymmetric transformations by judicious modification of Zr-catalyzed ethylmagnesiation of alkenes into other reactions[142–149].

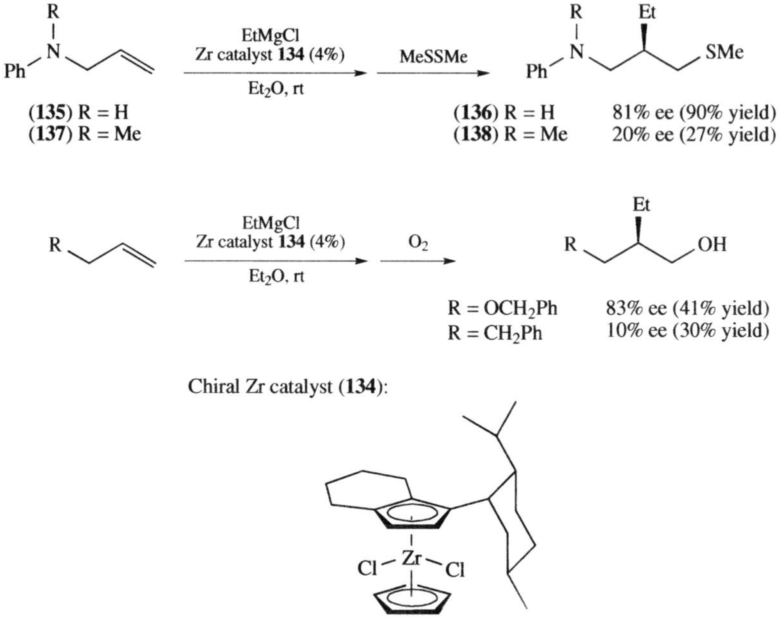

SCHEME 82

IV. CARBOMAGNESIATION REACTIONS OF AROMATICS

Under forcing conditions, Grignard reagents add to certain aromatic compounds. For example, the addition of PhMgBr to naphthalene takes place at 200 °C (Scheme 83)[150]. After treatment of the crude product with chloranil (for dehydrogenation), 1-phenylnaphthalene is obtained in 34% yield.

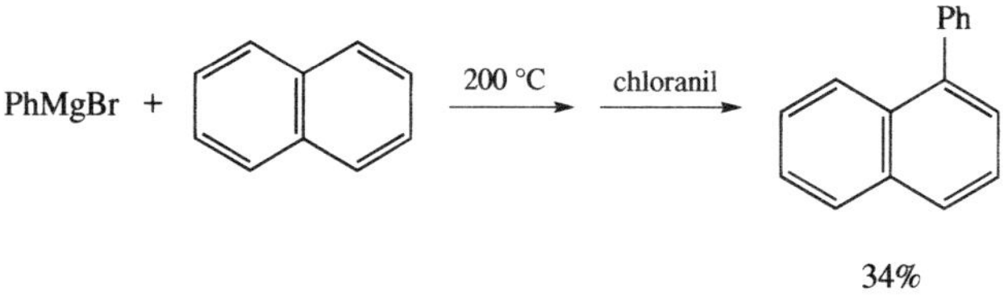

SCHEME 83

On the other hand, aromatic compounds substituted by electron-withdrawing groups such as nitro group are surprisingly susceptible to attack by Grignard reagents[151]. Such a reaction may be regarded as a variant of conjugate addition. It was also reported that the addition of Grignard reagent to naphthyloxazolidine **139** takes place in much milder conditions (Scheme 84)[152,153]. It has been proposed that the Grignard reagents add to the naphthylimine **140**, which is formed by the ring opening of oxazolidine ring with the action of Grignard reagent, in a 1,4-addition manner.

It is well-known that [60]fullerene is a fairly electronegative system having a low-lying LUMO. Therefore, [60]fullerene behaves like an electron-poor conjugated polyolefin

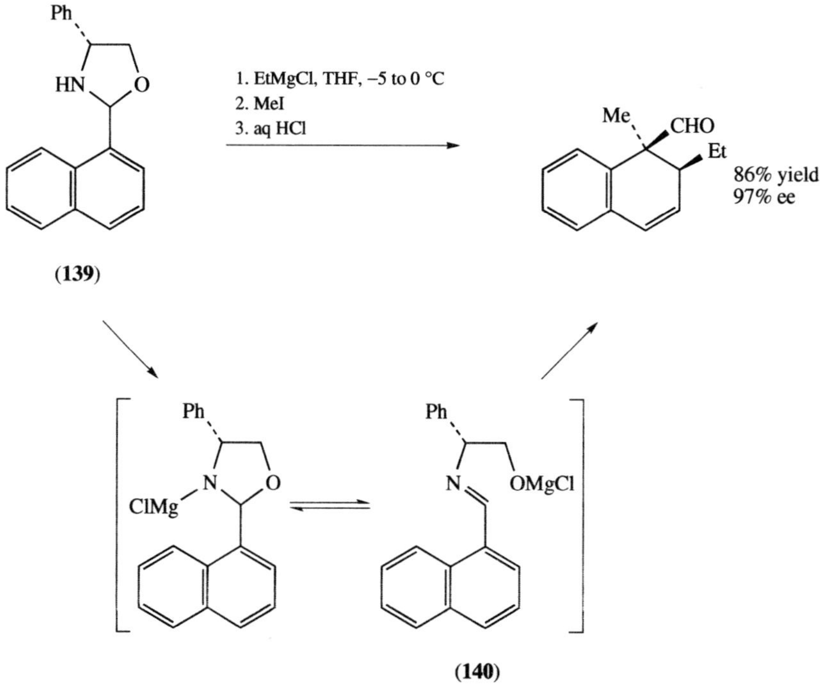

SCHEME 84

rather than a 'superarene'. Consequently, [60]fullerene undergoes nucleophilic additions with various nucleophiles[154]. In 1992, Hirsch and coworkers demonstrated that Grignard reagents can be used as nucleophiles for [60]fullerene (Scheme 85)[155, 156]. Phenyl and alkyl Grignard reagents add across [60]fullerene, but it is typically necessary to monitor the reaction by titration in order to maximize the yield of mono-adducts $RC_{60}H$[156]. The addition takes place across the C=C bond between two six-membered rings of [60]fullerene. An interesting solvent effect in the addition of silylmethyl Grignard reagents to [60]fullerene has been demonstrated by Nagashima and coworkers. Although the use of THF results in the production of mono-adduct $RC_{60}H$ as usual, the use of toluene provides bis-adduct $RC_{60}R$[157, 158]. Hirsch and coworkers have also demonstrated that [70]fullerene also reacts with PhMgBr in a regioselective fashion giving $PhC_{70}H$ as an isometrically pure compound[156].

V. SUMMARY

Carbomagnesiation reactions became an important and powerful tool in organic synthesis. Although a truly universal reaction system that allows the regio- and stereoselective carbomagnesiation of unactivated alkenes and alkynes has not yet been described, the last half-century of extensive worldwide research has resulted in impressive progress for making useful organic frameworks. We believe that the development of new catalysts, reagents and strategies in carbomagnesiation will be a topic of unparalleled importance in all aspects of pure and applied chemistry.

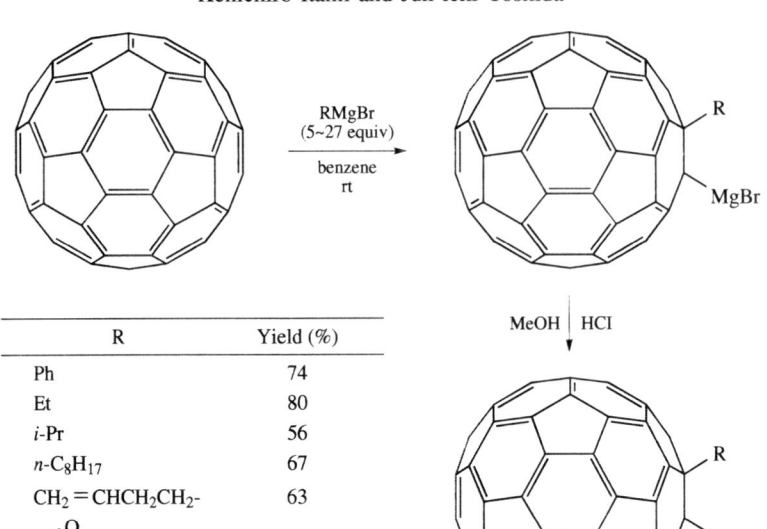

R	Yield (%)
Ph	74
Et	80
i-Pr	56
n-C_8H_{17}	67
CH_2=$CHCH_2CH_2$-	63
(1,3-dioxolan-2-yl)-CH_2CH_2-	52

SCHEME 85

VI. REFERENCES

1. I. Marek, N. Chinkov and D. Banon-Tenne, in *Metal-Catalyzed Cross-Coupling Reactions*, 2nd edn. (Eds. A. de Meijere and F. Diederich), Chap. 7, Wiley-VCH, Weinheim, 2004; I. Marek and J. Normant, in *Metal-Catalyzed Cross-Coupling Reactions* (Eds. F. Deiderich and P. J. Stang), Chap. 7, Wiley-VCH, New York, 1998; B. J. Wakefield, *Organomagnesium Methods in Organic Synthesis*, Academic Press, London, 1995; A. Inoue and K. Oshima, in *Main Group Metals in Organic Synthesis* (Eds. H. Yamamoto and K. Oshima), Vol. 1, Chap. 3, Wiley-VCH, Weinheim, 2004; P. Knochel, A. Gavryushin, A. Krasovskiy and H. Leuser, in *Comprehensive Organometallic Chemistry III* (Eds. D. M. P. Mingos, R. H. Crabtree and P. Knochel), Vol. 9, Chap. 9.03, Elsevier, Oxford, 2007; J. F. Normant and A. Alexakis, *Synthesis*, 841 (1981); A. G. Fallis and P. Forgione, *Tetrahedron*, **57**, 5899 (2001).
2. R. C. Fuson and H. D. Porter, *J. Am. Chem. Soc.*, **70**, 895 (1948).
3. R. C. Fuson, H. A. DeWald and R. Gaertner, *J. Org. Chem.*, **16**, 21 (1951).
4. R. C. Fuson and F. E. Mumford, *J. Org. Chem.*, **17**, 255 (1952).
5. R. C. Fuson and O. York, Jr., *J. Org. Chem.*, **18**, 570 (1953).
6. H. Gilman and H. M. Crawford, *J. Am. Chem. Soc.*, **45**, 554 (1923).
7. H. Gilman and J. B. Shumaker, *J. Am. Chem. Soc.*, **47**, 514 (1925).
8. H. Gilman and J. M. Peterson, *J. Am. Chem. Soc.*, **48**, 423 (1926).
9. G. H. Posner, *Org. React.*, **19**, 1 (1972); N. Krause (Ed.), *Modern Organocopper Chemistry*, Wiley-VCH, Weinheim, 2002.
10. S. A. Kandil and R. E. Dessy, *J. Am. Chem. Soc.*, **88**, 3027 (1966).
11. H. G. Richey, Jr. and A. M. Rothman, *Tetrahedron Lett.*, 1457 (1968).
12. J. K. Crandall, P. Battioni, J. T. Wehlacz and R. Bindra, *J. Am. Chem. Soc.*, **97**, 7171 (1975).
13. J.-G. Duboudin and B. Jousseaume, *J. Organomet. Chem.*, **44**, C1 (1972).
14. J.-G. Duboudin and B. Jousseaume, *J. Organomet. Chem.*, **96**, C47 (1975).
15. J.-G. Duboudin and B. Jousseaume, *J. Organomet. Chem.*, **162**, 209 (1978).
16. A. Alexakis, G. Cahiez and J. F. Normant, *J. Organomet. Chem.*, **177**, 293 (1979).
17. H. Yorimitsu, J. Tang, K. Okada, H. Shinokubo and K. Oshima, *Chem. Lett.*, 11 (1998).
18. S. Nishimae, R. Inoue, H. Shinokubo and K. Oshima, *Chem. Lett.*, 785 (1998).
19. K. Okada, K. Oshima and K. Utimoto, *J. Am. Chem. Soc.*, **118**, 6076 (1996).

20. T. Nishikawa, H. Shinokubo and K. Oshima, *J. Am. Chem. Soc.*, **123**, 4629 (2001).
21. T. Nishikawa, H. Shinokubo and K. Oshima, *Org. Lett.*, **4**, 2795 (2002).
22. K. Murakami, H. Ohmiya, H. Yorimitsu and K. Oshima, *Org. Lett.*, **9**, 1569 (2007).
23. E. Shirakawa, T. Yamagami, T. Kimura, S. Yamaguchi and T. Hayashi, *J. Am. Chem. Soc.*, **127**, 17164 (2005).
24. T. Yamagami, R. Shintani, E. Shirakawa and T. Hayashi, *Org. Lett.*, **9**, 1045 (2007).
25. B. B. Snider, M. Karras and R. S. E. Conn, *J. Am. Chem. Soc.*, **100**, 4624 (1978).
26. B. B. Snider, R. S. E. Conn and M. Karras, *Tetrahedron Lett.*, 1679 (1979).
27. K. Itami, T. Kamei and J. Yoshida, *J. Am. Chem. Soc.*, **125**, 14670 (2003).
28. T. Kamei, K. Itami and J. Yoshida, *Adv. Synth. Catal.*, **346**, 1824 (2004).
29. H. Chechik-Lankin, S. Livshin and I. Marek, *Synlett*, 2098 (2005).
30. H. Yasui, H. Yorimitsu and K. Oshima, *Chem. Lett.*, **36**, 32 (2007).
31. H. Chechik-Lankin and I. Marek, *Org. Lett.*, **5**, 5087 (2003).
32. M. Xie and X. Huang, *Synlett*, 477 (2004).
33. M. Xie, L. Liu, J. Wang and S. Wang, *J. Organomet. Chem.*, **690**, 4058 (2005).
34. J. A. Marshal and B. S. DeHoff, *J. Org. Chem.*, **51**, 863 (1986).
35. H. G. Richey, Jr. and F. W. Von Rein, *J. Organomet. Chem.*, **20**, P32 (1969).
36. F. W. Von Rein and H. G. Richey, Jr., *Tetrahedron Lett.*, 3777 (1971).
37. T. Wong, M. W. Tjepkema, H. Audrain, P. D. Wilson and A. G. Fallis, *Tetrahedron Lett.*, **37**, 755 (1996).
38. P. Forgione and A. G. Fallis, *Tetrahedron Lett.*, **41**, 11 (2000).
39. F. F. Fleming, V. Gudipati and O. W. Steward, *Org. Lett.*, **4**, 659 (2002).
40. F. F. Fleming, V. Gudipati and O. W. Steward, *Tetrahedron*, **59**, 5585 (2003).
41. P. E. Tessier, N. Nguyen, M. D. Clay and A. G. Fallis, *Org. Lett.*, **7**, 767 (2005).
42. N. P. Villalva-Servín, A. Laurent, G. P. A. Yap and A. G. Fallis, *Synlett*, 1263 (2003).
43. P. Forgione, P. D. Wilson, G. P. A. Yap and A. G. Fallis, *Synlett*, 921 (2000).
44. P. Forgione, P. D. Wilson and A. G. Fallis, *Tetrahedron Lett.*, **41**, 17 (2000).
45. D. V. Smil, F. E. S. Souza and A. G. Fallis, *Bioorg. Med. Chem. Lett.*, **15**, 2057 (2005).
46. M. A. Franks, E. A. Schrader, E. C. Pietsch, D. R. Pennella, S. V. Torti and M. E. Welker, *Bioorg. Med. Chem.*, **13**, 2221 (2005).
47. F. C. Engelhardt, Y.-J. Shi, C. J. Cowden, D. A. Conlon, B. Pipik, G. Zhou, J. M. McNamara and U.-H. Dolling, *J. Org. Chem.*, **71**, 480 (2006).
48. P. E. Tessier, A. J. Penwell, F. E. S. Souza and A. G. Fallis, *Org. Lett.*, **5**, 2989 (2003).
49. B. Jousseaume and J.-G. Duboudin, *J. Organomet. Chem.*, **91**, C1 (1975).
50. J.-G. Duboudin and B. Jousseaume, *J. Organomet. Chem.*, **168**, 1 (1979).
51. Z. Lu and S. Ma, *J. Org. Chem.*, **71**, 2655 (2006).
52. S. Ma and Z. Lu, *Adv. Synth. Catal.*, **348**, 1894 (2006).
53. D. Zhang and J. M. Ready, *J. Am. Chem. Soc.*, **128**, 15050 (2006).
54. H. G. Richey, W. E. Erickson and A. S. Heyn, *Tetrahedron Lett.*, 2183 (1971).
55. R. Mornet and L. Gouin, *Bull. Soc. Chim. Fr.*, 737 (1977).
56. R. Mornet and L. Gouin, *Tetrahedron Lett.*, 167 (1977).
57. H. E. Podall and W. E. Foster, *J. Org. Chem.*, **23**, 1848 (1958).
58. L. H. Shepherd, Jr., U.S. Patent 3 597 488; *Chem. Abstr.*, **75**, 88751 (1971).
59. H. Lehmkuhl and D. Reinehr, *J. Organomet. Chem.*, **25**, C47 (1970).
60. H. Lehmkuhl and D. Reinehr, *J. Organomet. Chem.*, **34**, 1 (1972).
61. H. Lehmkuhl and D. Reinehr, *J. Organomet. Chem.*, **57**, 29 (1973).
62. H. Lehmkuhl, D. Reinehr, J. Frandt and G. Schroth, *J. Organomet. Chem.*, **57**, 39 (1973).
63. H. Lehmkuhl, D. Reinehr, D. Henneberg and G. Schroth, *J. Organomet. Chem.*, **57**, 49 (1973).
64. H. Lehmkuhl, D. Reinehr, G. Schomburg, D. Henneberg, H. Damen and G. Schroth, *Justus Liebigs Ann. Chem.*, 103 (1975).
65. H. Lehmkuhl, D. Reinehr, D. Henneberg, G. Schomburg and G. Schroth, *Justus Liebigs Ann. Chem.*, 119 (1975).
66. H. Lehmkuhl, O. Olbrysch, D. Reinehr, G. Schomburg and D. Henneberg, *Justus Liebigs Ann. Chem.*, 145 (1975).
67. M. S. Kharasch, *J. Org. Chem.*, **19**, 1600 (1954).
68. A. Job and R. Reich, *Compt. Rend.*, **179**, 330 (1924).
69. L. Farády, L. Bencze and L. Markó, *J. Organomet. Chem.*, **10**, 505 (1967).
70. L. Farády, L. Bencze and L. Markó, *J. Organomet. Chem.*, **17**, 107 (1969).

71. L. Farády and L. Markó, *J. Organomet. Chem.*, **28**, 159 (1971).
72. I. N. Houpis, J. Lee, I. Dorziotis, A. Molina, B. Reamer, R. P. Volante and P. J. Reider, *Tetrahedron*, **54**, 1185 (1998).
73. E. A. Hill, *J. Organomet. Chem.*, **91**, 123 (1975).
74. M. S. Silver, P. R. Shafer, J. E. Nordlander, C. Rüchardt and J. D. Roberts, *J. Am. Chem. Soc.*, **82**, 2646 (1960).
75. E. A. Hill, H. G. Richey, Jr. and T. C. Rees, *J. Org. Chem.*, **28**, 2161 (1963).
76. D. J. Patel, C. L. Hamilton and J. D. Roberts, *J. Am. Chem. Soc.*, **87**, 5144 (1965).
77. M. E. H. Howden, A. Maercker, J. Burdon and J. D. Roberts, *J. Am. Chem. Soc.*, **88**, 1732 (1966).
78. A. Maercker and J. D. Roberts, *J. Am. Chem. Soc.*, **88**, 1742 (1966).
79. H. G. Richey, Jr. and T. C. Rees, *Tetrahedron Lett.*, 4297 (1966).
80. W. C. Kossa, Jr., T. C. Rees and H. G. Richey, Jr., *Tetrahedron Lett.*, 3455 (1971).
81. H. Lehmkuhl, *Bull. Soc. Chim. Fr. Part II*, 87 (1981).
82. W. Oppolzer, *Angew. Chem., Int. Ed. Engl.*, **28**, 38 (1989).
83. H. Felkin, J. D. Umpleby, E. Hagaman and E. Wenkert, *Tetrahedron Lett.*, 2285 (1972).
84. W. Oppolzer and K. Bätting, *Tetrahedron Lett.*, **23**, 4669 (1982).
85. W. Oppolzer, R. Pitteloud and H. F. Strauss, *J. Am. Chem. Soc.*, **104**, 6476 (1982).
86. W. Oppolzer and R. Pitteloud, *J. Am. Chem. Soc.*, **104**, 6478 (1982).
87. H. Lehmkuhl and D. Reinehr, *J. Organomet. Chem.*, **34**, 1 (1972).
88. S. Akutagawa and S. Otsuka, *J. Am. Chem. Soc.*, **97**, 6870 (1975).
89. F. Barbot and P. Miginiac, *J. Organomet. Chem.*, **145**, 269 (1978).
90. B. J. Wakefield, in *Comprehensive Organometallic Chemistry* (Eds. G. Wilkinson, F. G. A. Stone and E. W. Abel), Chap. 44, Pergamon, Oxford,1982.
91. E. K. Watkins and H. G. Richey, Jr., *Organometallics*, **11**, 3785 (1992).
92. M. Nakamura, A. Hirai and E. Nakamura, *J. Am. Chem. Soc.*, **122**, 978 (2000).
93. M. Nakamura, K. Matsuo, T. Inoue and E. Nakamura, *Org. Lett.*, **5**, 1373 (2003).
94. M. Lautens and S. Ma, *J. Org. Chem.*, **61**, 7246 (1996).
95. R. G. Arrayás, S. Carrera and J. C. Carretero, *Org. Lett.*, **5**, 1333 (2003).
96. M. D. Surman, M. J. Mulvihill and M. J. Miller, *J. Org. Chem.*, **67**, 4115 (2002).
97. H. G. Richey, Jr. and R. M. Bension, *J. Org. Chem.*, **45**, 5036 (1980).
98. L.-a. Liao and J. M. Fox, *J. Am. Chem. Soc.*, **124**, 14322 (2002).
99. L.-a. Liao, F. Zhang, N. Yan, J. A. Golen and J. M. Fox, *Tetrahedron*, **60**, 1803 (2004).
100. X. Lui and J. M. Fox, *J. Am. Chem. Soc.*, **128**, 5600 (2006).
101. K. Tamao, R. Kanatani and M. Kumada, *Tetrahedron Lett.*, **25**, 1905 (1984).
102. D. Seyferth and T. Wada, *Inorg. Chem.*, **1**, 78 (1962).
103. K. Utimoto, K. Imi, H. Shiragami, S. Fujikura and H. Nozaki, *Tetrahedron Lett.*, **26**, 2101 (1985).
104. R. W. Hoffmann, O. Knopff and T. Faber, *J. Chem. Soc., Perkin Trans. 2*, 1785 (2001).
105. G. R. Buell, R. Corriu, C. Guerin and L. Spialter, *J. Am. Chem. Soc.*, **92**, 7424 (1970).
106. K. Tamao, T. Iwahara, R. Kanatani and M. Kumada, *Tetrahedron Lett.*, **25**, 1909 (1984).
107. K. Tamao, R. Kanatani and M. Kumada, *Tetrahedron Lett.*, **25**, 1913 (1984).
108. K. Itami, K. Mitsudo and J. Yoshida, *Angew. Chem., Int. Ed.*, **40**, 2337 (2001).
109. K. Itami and J. Yoshida, *Synlett*, 157 (2006).
110. J. J. Eisch and G. R. Husk, *J. Am. Chem. Soc.*, **87**, 4194 (1965).
111. J. J. Eisch and J. H. Merkley, *J. Organomet. Chem.*, **20**, P27 (1969).
112. J. J. Eisch and J. H. Merkley, *J. Am. Chem. Soc.*, **101**, 1148 (1979).
113. J. J. Eisch, J. H. Merkley and J. E. Galle, *J. Org. Chem.*, **44**, 587 (1979).
114. M. Chérest, H. Felkin, C. Frajerman, C. Lion, G. Roussi and G. Swierczewski, *Tetrahedron Lett.*, 875 (1966).
115. H. Felkin and C. Kaeseberg, *Tetrahedron Lett.*, 4587 (1970).
116. P. Kocienski, C. Love and D. A. Roberts, *Tetrahedron Lett.*, **30**, 6753 (1989).
117. K. A. Swiss, W. Hinkley, C. A. Maryanoff and D. C. Liotta, *Synthesis*, 127 (1992).
118. M. Solomon, W. C. L. Jamison, M. McCormick, D. C. Liotta, D. A. Cherry, J. E. Mills, R. D. Shah, J. D. Rodgers and C. A. Maryanoff, *J. Am. Chem. Soc.*, **110**, 3702 (1988).
119. K. A. Swiss, D. C. Liotta and C. A. Maryanoff, *J. Am. Chem. Soc.*, **112**, 9393 (1990).
120. J. D. White, H. Shin, T.-S. Kim and N. S. Cutshall, *J. Am. Chem. Soc.*, **119**, 2404 (1997).
121. F. F. Fleming, J. Guo, Q. Wang and D. Weaver, *J. Org. Chem.*, **64**, 8568 (1999).

122. F. F. Fleming, Q. Wang and O. W. Steward, *Org. Lett.*, **2**, 1477 (2000).
123. F. F. Fleming, Q. Wang, Z. Zhang and O. W. Steward, *J. Org. Chem.*, **67**, 5953 (2002).
124. F. F. Fleming and Q. Wang, *Chem. Rev.*, **103**, 2035 (2003).
125. F. F. Fleming, Z. Zhang, Q. Wang and O. W. Steward, *Angew. Chem., Int. Ed.*, **43**, 1126 (2004).
126. F. F. Fleming, Q. Wang and O. W. Steward, *J. Org. Chem.*, **68**, 4235 (2003).
127. F. F. Fleming, Z. Zhang, Q. Wang and O. W. Steward, *J. Org. Chem.*, **68**, 7646 (2003).
128. F. F. Fleming, Z. Zhang, G. Wei and O. W. Steward, *Org. Lett.*, **7**, 447 (2005).
129. F. F. Fleming, Z. Zhang, G. Wei and O. W. Steward, *J. Org. Chem.*, **71**, 1430 (2006).
130. F. F. Fleming, G. Wei, Z. Zhang and O. W. Steward, *Org. Lett.*, **8**, 4903 (2006).
131. U. M. Dzhemilev, O. S. Vostrikova and R. M. Sultanov, *Izv. Akad. Nauk SSSE, Ser. Khim.*, 218 (1983); *Chem. Abstr.*, **98**, 160304 (1983).
132. U. M. Dzhemilev, O. S. Vostrikova, R. M. Sultanov, A. G. Kukovinets and A. M. Khalilov, *Izv. Akad. Nauk SSSE, Ser. Khim.*, 2053 (1984); *Chem. Abstr.*, **102**, 165971 (1985).
133. U. M. Dzhemilev and O. S. Vostrikova, *J. Organomet. Chem.*, **285**, 43 (1985).
134. C. J. Rousset, E. Negishi, N. Suzuki and T. Takahashi, *Tetrahedron Lett.*, **33**, 1965 (1992).
135. U. M. Dzhemilev, *Tetrahedron*, **51**, 4333 (1995).
136. A. H. Hoveyda and Z. Xu, *J. Am. Chem. Soc.*, **113**, 5079 (1991).
137. A. F. Houri, M. T. Didiluk, Z. Xu, N. R. Horan and A. H. Hoveyda, *J. Am. Chem. Soc.*, **115**, 6614 (1993).
138. A. H. Hoveyda, Z. Xu, J. P. Morken and A. F. Houri, *J. Am. Chem. Soc.*, **113**, 8950 (1991).
139. A. H. Hoveyda, J. P. Morken, A. F. Houri and Z. Xu, *J. Am. Chem. Soc.*, **114**, 6692 (1992).
140. L. Bell, R. J. Whitby, R. V. H. Jones and M. C. H. Standen, *Tetrahedron Lett.*, **37**, 7139 (1996).
141. L. Bell, D. C. Brookings, G. J. Dawson, R. J. Whitby, R. V. H. Jones and M. C. H. Standen, *Tetrahedron*, **54**, 14617 (1998).
142. A. H. Hoveyda and J. P. Morken, *J. Org. Chem.*, **58**, 4237 (1993).
143. J. P. Morken, M. T. Didiuk and A. H. Hoveyda, *J. Am. Chem. Soc.*, **115**, 6997 (1993).
144. J. P. Morken, M. T. Didiuk, M. S. Visser and A. H. Hoveyda, *J. Am. Chem. Soc.*, **116**, 3123 (1994).
145. A. F. Houri, Z. Xu, D. A. Cogan and A. H. Hoveyda, *J. Am. Chem. Soc.*, **117**, 2943 (1995).
146. M. S. Visser and A. H. Hoveyda, *Tetrahedron*, **51**, 4383 (1995).
147. M. S. Visser, J. P. A. Harrity and A. H. Hoveyda, *J. Am. Chem. Soc.*, **118**, 3779 (1996).
148. N. M. Heron, J. A. Adams and A. H. Hoveyda, *J. Am. Chem. Soc.*, **119**, 6205 (1997).
149. J. A. Adams, N. M. Heron, A.-M. Koss and A. H. Hoveyda, *J. Org. Chem.*, **64**, 854 (1999).
150. D. Bryce-Smith and B. J. Wakefield, *Tetrahedron Lett.*, 3295 (1964).
151. G. Bartoli, M. Bosco and G. Baccolini, *J. Org. Chem.*, **45**, 522 (1980).
152. L. N. Pridgen, M. K. Mokhallalati and M.-J. Wu, *J. Org. Chem.*, **57**, 1237 (1992).
153. M. K. Mokhallalati, K. R. Muralidharan and L. N. Pridgen, *Tetrahedron Lett.*, **35**, 4267 (1994).
154. A. Hirsch and M. Brettreich, *Fullerenes: Chemistry and Reactions*, Wiley-VCH, Weinheim, 2005.
155. A. Hirsch, A. Soi and H. R. Karfunkel, *Angew. Chem., Int. Ed. Engl.*, **31**, 766 (1992).
156. A. Hirsch, T. Gröser, A. Skiebe and A. Soi, *Chem. Ber.*, **126**, 1061 (1993).
157. H. Nagashima, H. Terasaki, E. Kimura, K. Nakajima and K. Itoh, *J. Org. Chem.*, **59**, 1246 (1994).
158. H. Nakajima, M. Saito, Y. Kato, H. Goto, E. Osawa, M. Haga and K. Itoh, *Tetrahedron*, **52**, 5053 (1996).

CHAPTER 15

The chemistry of organomagnesium ate complexes

HIDEKI YORIMITSU and KOICHIRO OSHIMA

Department of Material Chemistry, Graduate School of Engineering, Kyoto University, Kyoto-daigaku Katsura, Nishikyo, Kyoto 615-8510, Japan
Fax: +81-75-383-2438; e-mail: yori@orgrxn.mbox.media.kyoto-u.ac.jp; oshima@orgrxn.mbox.media.kyoto-u.ac.jp

I. INTRODUCTION	681
II. PREPARATION OF TRIORGANOMAGNESATES	682
III. REACTIONS OF TRIORGANOMAGNESATES	683
A. Nucleophilic Addition Reactions	683
B. Deprotonation Reactions	686
C. Halogen–Magnesium Exchange	690
IV. ADDITIONAL IMPORTANT CHEMISTRY OF ORGANOMAGNESATES	711
A. Alkoxydialkylmagnesate R_2MgOR	711
B. Lithium Tris(2,2,6,6-tetramethylpiperidino)magnesate $(TMP)_3MgLi$	713
C. Combination of Isopropylmagnesium Chloride and Lithium Chloride $i\text{-PrMgCl}\bullet\text{LiCl}$	714
V. SUMMARY	714
VI. REFERENCES	714

I. INTRODUCTION

Ate complexes have formed a structurally intriguing entity for chemists working on organometallics. Organomagnesium ate complexes, organomagnesates, emerged in 1951, when Wittig and coworkers proposed the formation of lithium triphenylmagnesate Ph_3MgLi by mixing phenyllithium and diphenylmagnesium[1]. NMR studies of the reaction of methyllithium and dimethylmagnesium in 1967 strongly supported the formation of lithium trimethylmagnesate[2]. Finally, in 1978, X-ray crystallographic analysis by Thoennes and Weiss unambiguously provided evidence of the existence of

---

The chemistry of organomagnesium compounds
Edited by Z. Rappoport and I. Marek   © 2008 John Wiley & Sons, Ltd

organomagnesates[3]. Since the latter report, a number of chemists have devoted themselves to crystallographic studies on organomagnesium ate complexes. Some reports help to overview structural studies on organomagnesates[4-6].

Ate complexation is a promising method to enhance the reactivity of organometallic reagents. Ate complexes such as organocuprates, -zincates, -aluminates, -borates and -silicates are hence useful in organic synthesis. In contrast, the application of organomagnesates in organic synthesis had been largely unexplored. During the last decade, organomagnesates have been attracting the increasing attention of synthetic organic chemists. The recent progress is largely due to lithium triorganomagnesates R_3MgLi and alkylmagnesium chloride–lithium chloride complex $RMgCl \cdot LiCl$. This chapter mainly summarizes the preparation and reactions of triorganomagnesate reagents, as the chemistry of $RMgCl \cdot LiCl$ is summarized in Chapter 12 by Knochel[7,8]. The $RMgCl \cdot LiCl$ reagents are now used on industrial scales[9].

II. PREPARATION OF TRIORGANOMAGNESATES

The easiest and reliable way to prepare triorganomagnesates is the reaction of diorganomagnesium with one molar equivalent of organolithium (equation 1). Organometallic chemists synthesized triorganomagnesates by this method[1-3], and sodium triorganomagnesates were also synthesized in a similar fashion[10]. Synthetic organic chemists prefer the more convenient route, the reaction of alkylmagnesium halide with two molar equivalents of organolithium (equation 2)[11]. They use the triorganomagnesates in the same pot without isolation. The triorganomagnesates prepared by the latter method can exist as mixtures of R_2Mg, R_3MgLi, R_4MgLi_2 etc. in solution. One can prepare mixed organomagnesates $R^1_2R^2MgLi$ by using a combination of $R^1_2Mg + R^2Li$ or $R^2MgBr + 2 R^1Li$ (equation 3), although the exact structure in solutions is not clear[11,12].

$$R_2Mg + RLi \longrightarrow R_3MgLi \qquad (1)$$

$$RMgX + 2\ RLi \longrightarrow R_3MgLi + LiX \qquad (2)$$

$$\begin{array}{c} R^1_2Mg + R^2Li \\ \text{or} \\ R^2MgBr + 2\ R^1Li \end{array} \longrightarrow R^1_2R^2MgLi + LiBr \qquad (3)$$

Trialkylmagnesate species could be isolated by treatment of diorganomagnesium compounds with cryptands,[13] 15-crown-5[14] or a tetraazamacrocycle[5]. An equilibrium as shown in equation 4 rationalizes the formation of trialkylmagnesate.

$$\begin{array}{c} 2\ (t\text{-}BuCH_2)_2Mg \\ + \\ 2,1,1\text{-cryptand} \end{array} \longrightarrow [(t\text{-}BuCH_2)_3Mg]^- [t\text{-}BuCH_2Mg(2,1,1\text{-cryptand})]^+ \qquad (4)$$

The reduction of dialkylmagnesiums by alkali metals in hydrocarbon solvent yielded the corresponding ate complexes of definite stoichiometries[15]. Lithium reacted with dialkylmagnesium to form both R_5MgLi_3 and R_3MgLi according to the stoichiometry used (equations 5 and 6). Reduction with sodium afforded R_5Mg_2Na as well as R_3MgNa. Potassium, rubidium and cesium reduced dialkylmagnesium to yield trialkylmagnesates (equation 7).

$$5\ s\text{-}Bu_2Mg + 6\ Li \longrightarrow 2\ s\text{-}Bu_5MgLi_3 + 3\ Mg \qquad (5)$$

$$3\ Bu_2Mg + 2\ Li \longrightarrow 2\ Bu_3MgLi + Mg \qquad (6)$$

$$3\ Bu_2Mg + 2\ K \longrightarrow 2\ Bu_3MgK + Mg \qquad (7)$$

III. REACTIONS OF TRIORGANOMAGNESATES
A. Nucleophilic Addition Reactions

Wittig's pioneering work includes the reactions of benzophenone and chalcone with lithium triphenylmagnesate[1]. The reaction of benzophenone with diphenylmagnesium afforded triphenylmethanol in 67% yield (equation 8). Lithium triphenylmagnesate was more reactive than diphenylmagnesium, and as reactive as phenyllithium, resulting in a quantitative formation of triphenylmethanol (equation 9). The magnesate was more reactive than the corresponding beryllium and zinc ate complexes, which provided triphenylmethanol in 45% and 54% yields, respectively. With chalcone, diphenylmagnesium and lithium triphenylmagnesate both yielded mainly the corresponding 1,4-adduct (equation 10), whereas phenyllithium alone yielded the 1,2-adduct.

$$Ph_2Mg + Ph_2CO \xrightarrow{67\%} Ph_3COH \quad (8)$$

$$Ph_3MLi + Ph_2CO \longrightarrow Ph_3COH \quad M = Mg(100\%), Be(45\%), Zn(54\%) \quad (9)$$

$$Ph_2Mg \text{ or } Ph_3MgLi + Ph\text{–CH=CH–C(O)–Ph} \xrightarrow{>94\%} Ph\text{–CH(Ph)–CH}_2\text{–C(O)–Ph} \quad (10)$$

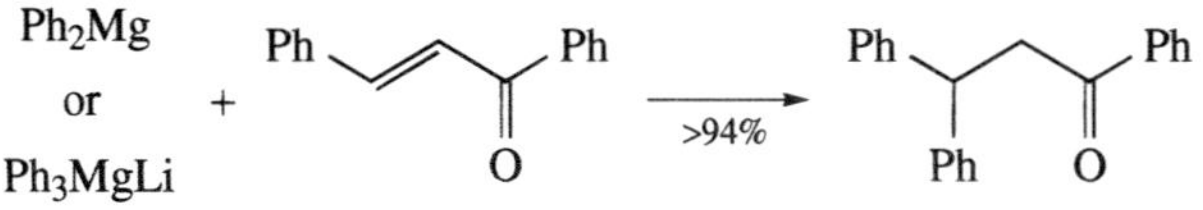

Ashby and coworkers reported in 1974 stereochemical studies on addition of lithium trimethylmagnesate to several ketones[16]. The reactions of 4-*tert*-butylcyclohexanone with methyllithium, with dimethylmagnesium and with lithium trimethylmagnesate in ether afforded mixtures of the corresponding axial and equatorial alcohols in ratios of 65:35, 70:30 and 69:31, respectively (equation 11). Reaction of 3,3,5-trimethylcyclohexanone with lithium trimethylmagnesate yielded exclusively the axial alcohol (equation 12). Reaction of norcamphor provided 95% of the *endo* alcohol and 5% of the *exo* alcohol (equation 13). In contrast, reaction of camphor yielded the *exo* alcohol with high stereoselectivity (equation 14). Among the methylmetals examined, no difference in the stereoselectivity was virtually observed.

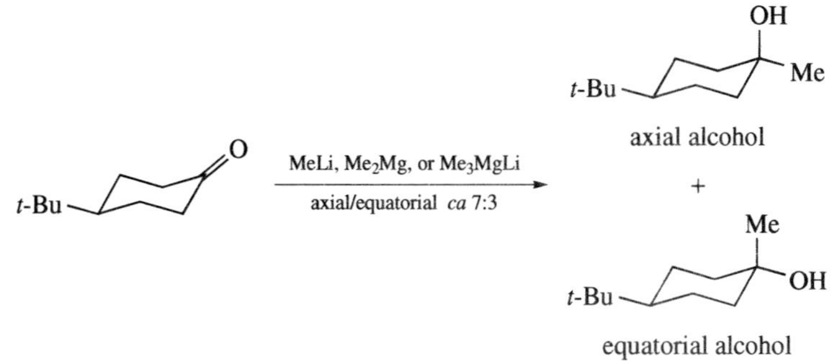

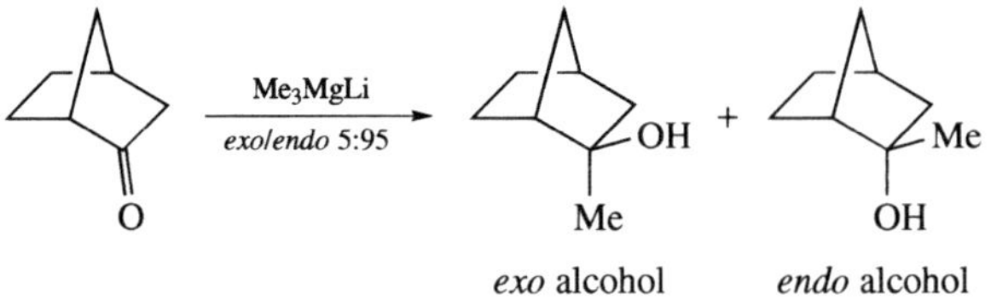

$$\text{[norbornanone]} \xrightarrow[\text{exo/endo 5:95}]{\text{Me}_3\text{MgLi}} \text{[exo alcohol, Me/OH]} + \text{[endo alcohol, Me/OH]} \qquad (13)$$

$$\text{[camphor-like ketone]} \xrightarrow[\text{exo/endo 99:1}]{\text{Me}_3\text{MgLi}} \text{[exo alcohol, Me/OH]} + \text{[endo alcohol, Me/OH]} \qquad (14)$$

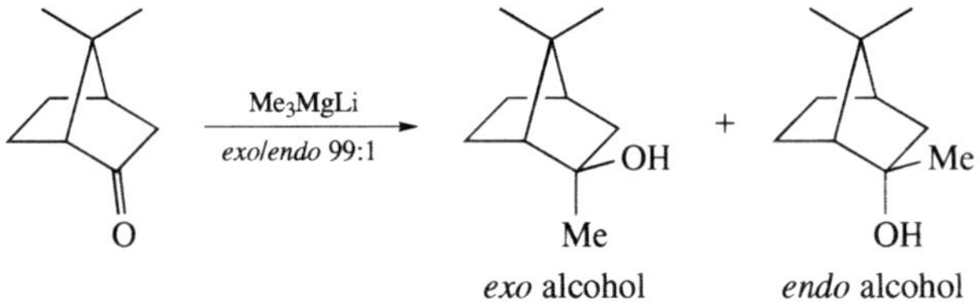

Richey and King found in 1982 that addition of 15-crown-5 to a THF solution of diethylmagnesium and pyridine significantly accelerated the nucleophilic addition of the organomagnesium compound to pyridine (Table 1)[17]. Interestingly, the reaction led to the formation of a significant amount of 4-ethylpyridine. It is noteworthy that diethylmagnesium and pyridine react very slowly in the absence of the crown ether to produce 2-ethylpyridine as a sole product. A similar phenomenon was observed by using cryptand as an additive[13]. Richey conjectured, and finally confirmed, that organomagesate complexes are formed *in situ* (equation 4), and that the complex is responsible for the unusual reactivity. Richey and Farkas showed that reactions of pyridine with solutions prepared by mixing diethylmagnesium and ethyllithium solutions yielded a mixture of 4-ethylpyridine and 2-ethylpyridine[18]. Reaction of 2-cyclohexen-1-one with diethylmagnesium in the presence of 2,1,1-cryptand gave much more of 1,4-addition product, 3-ethylcyclohexanone, than that with either organometallic compound alone (Table 2)[18].

Very recently, Ishihara and coworkers developed a highly efficient nucleophilic addition of magnesate reagents to ketones, which is very useful in organic synthesis[19]. The nucleophilicity of the alkyl group in lithium trialkylmagnesate is markedly enhanced compared to that of the parent alkyllithium or alkylmagnesium halide. Butylation of acetophenone was examined with several organolithium and organomagnesium reagents (Table 3). Reaction with butyllithium gave the corresponding tertiary alcohol in modest yield, along with a small amount of an undesired aldol product (entry 1). The basic nature of butyllithium

TABLE 1. Reactions of pyridine with ethylmetals

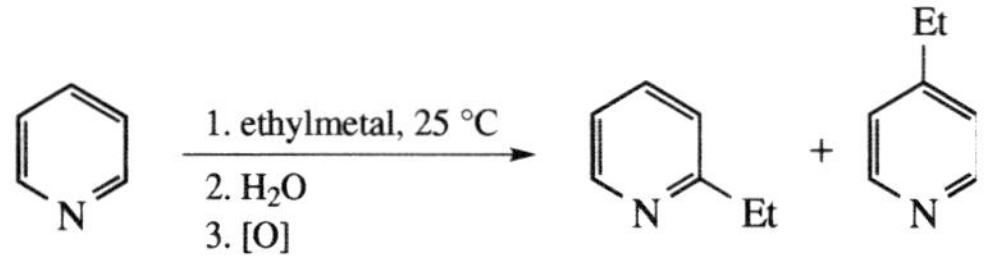

$$\text{pyridine} \xrightarrow[\substack{\text{2. H}_2\text{O} \\ \text{3. [O]}}]{\text{1. ethylmetal, 25 °C}} \text{2-Et-pyridine} + \text{4-Et-pyridine}$$

Ethylmetal	Time	2-Ethylpyridine (%)	4-Ethylpyridine (%)
Et$_2$Mg	22 h	0.3	0
Et$_2$Mg + 15-crown-5	20 h[a]	13	37
Et$_2$Mg + 2,1,1-cryptand	24 h	14	31
Et$_2$Mg + EtLi	22 h	37	6
EtLi	0.5 h	39	0

[a] Performed at 40 °C.

TABLE 2. Reactions of 2-cyclohexen-1-one with ethylmetals

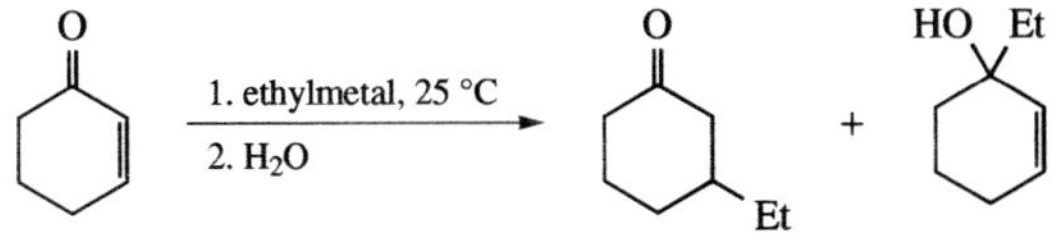

Ethylmetal	1,4-Adduct (%)	1,2-Adduct (%)
Et_2Mg	2	67
Et_2Mg + 15-crown-5	34	57
Et_2Mg + 2,1,1-cryptand	30	40
EtLi	0	73

TABLE 3. Reactions of acetophenone with butylmetals

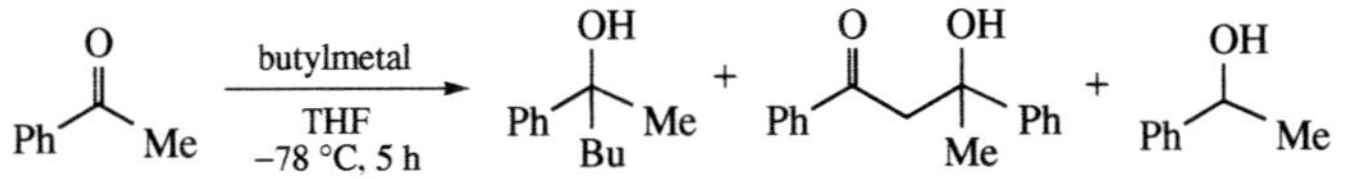

Entry	Butylmetal	1,2-Adduct (%)	Aldol product (%)	Reduced product (%)
1	BuLi	62	7	0
2	BuMgCl	50	9	8
3	Bu_2Mg	48	27	20
4	Bu_3MgLi	99	0	0

would produce the enolate of acetophenone. Use of butylmagnesium chloride also led to a similar result, except for the additional production of a small amount of 1-phenylethanol (entry 2). Although dibutylmagnesium effected butylation, the selectivity of the desired product was considerably diminished (entry 3). When lithium tributylmagnesate, prepared from butylmagnesium chloride and butyllithium in a 1:2 molar ratio, was examined, a highly selective and efficient butylation took place (entry 4).

Interestingly, mixed magnesate reagents, RMe_2MgLi (R ≠ Me), could transfer the R group selectively to ketones (equations 15–18). Surprisingly, readily enolizable ketones such as β-tetralone were ethylated in excellent yields (equation 18).

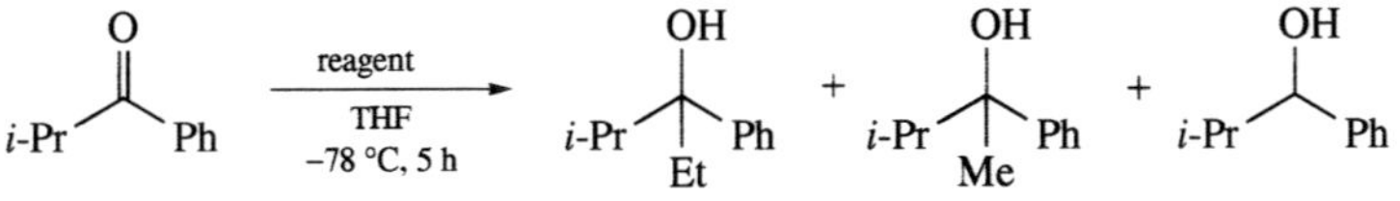

	1,2-Adduct (Et)	1,2-Adduct (Me)	Reduced product
EtMgBr	36%	–	11%
$EtMe_2MgLi$	92%	8%	0%

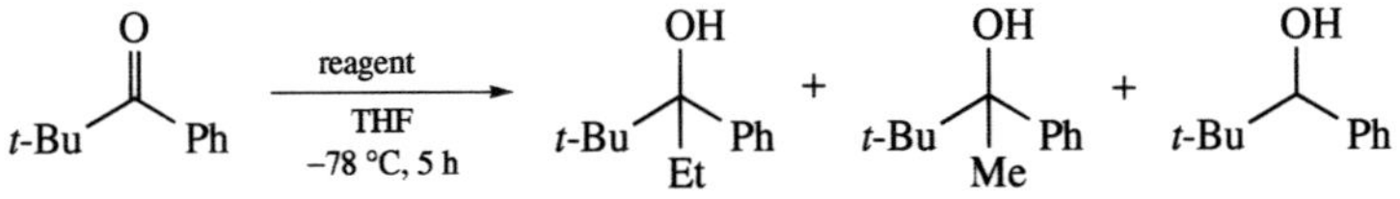

(15)

	1,2-Adduct (Et)	1,2-Adduct (Me)	Reduced product
EtMgBr	0%	–	50%
$EtMe_2MgLi$	89%	11%	0%

(16)

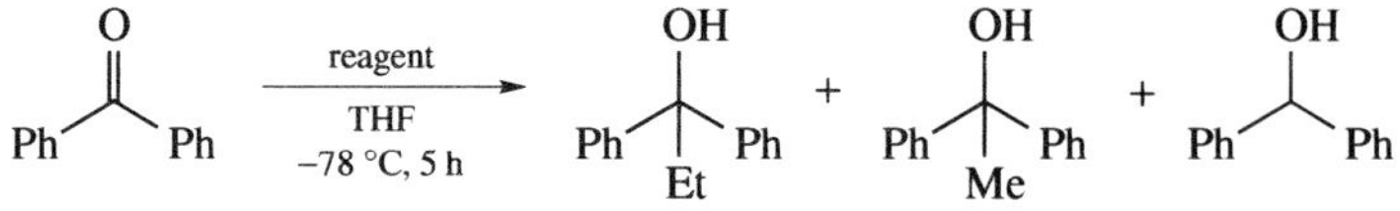

				(17)
	Ph-C(OH)(Et)-Ph	Ph-C(OH)(Me)-Ph	Ph-CH(OH)-Ph	
EtMgBr	14%	–	68%	
EtMe$_2$MgLi	>99%	<1%	0%	

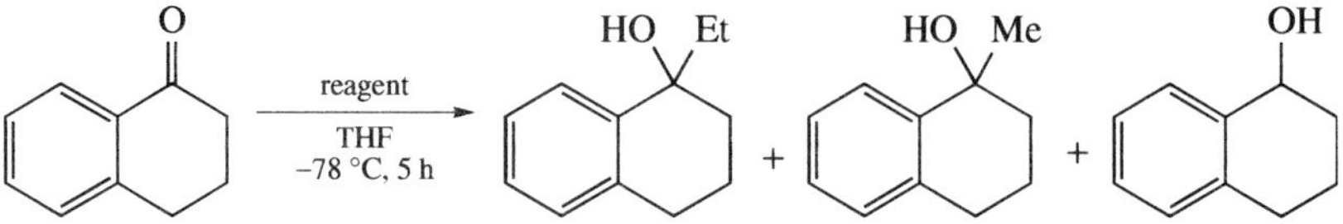

(Equation 17: PhC(O)Ph + reagent, THF, −78 °C, 5 h → products)

(Equation 18: α-tetralone + reagent, THF, −78 °C, 5 h →)

	HO,Et product	HO,Me product	OH product	
EtMgBr	36%	–	9%	(18)
EtMe$_2$MgLi	89%	7%	0%	

Lithium allyldibutylmagnesate can be formed easily by mixing allylmagnesium chloride and two molar equivalents of butyllithium. The allyldibutylmagnesate was nucleophilic enough to allylate pyridine-2-thiones and pyridin-2-ones[20]. Treatment of 1-methylpyridine-2-thione and 1-methylpyridin-2-one with lithium allyldibutylmagnesate gave rise to the allylation at the 6 positions (equation 19). On the other hand, 1-lithiopyridine-2-thione and 1-lithiopyridin-2-one underwent the allylation at the 4 positions (Scheme 1). By careful monitoring of the reaction mixture, it was concluded that the 4-allylpyridine derivatives were obtained via the [3,3] sigmatropic rearrangement of the initially formed 6-allylpyridine intermediates.

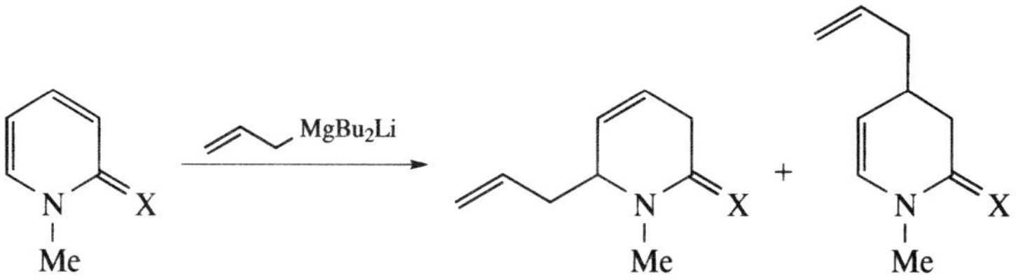

(Equation 19)

X = S 80% combined yield 84 : 16
X = O 61% combined yield 96 : 4

The reaction of a 1-allylpyridin-2-one derivative with lithium allyldibutylmagnesate followed by ring-closing metathesis (RCM) yielded tetrahydroquinolizin-4-one (Scheme 2)[21].

B. Deprotonation Reactions

The efficiency of deprotonation reaction of fluorene was a benchmark of the enhanced reactivity of triorganomagnesates[1,17]. The deprotonation with triorganomagnesates has emerged as a useful tool in modern synthetic organic chemistry since the work by Nakata and coworkers in 1997[22]. As a key step in the asymmetric synthesis of the fourteen-membered unit of methyl sarcophytoate, lithium tributylmagnesate was highly effective to metalate **1** at the allylic sulfide moiety (Scheme 3). The anion generated underwent cyclization to yield **2**, whereas the original protocol using butyllithium/DABCO system or other surrogates suffered from substantial decomposition of **1** as well as poor reproducibility.

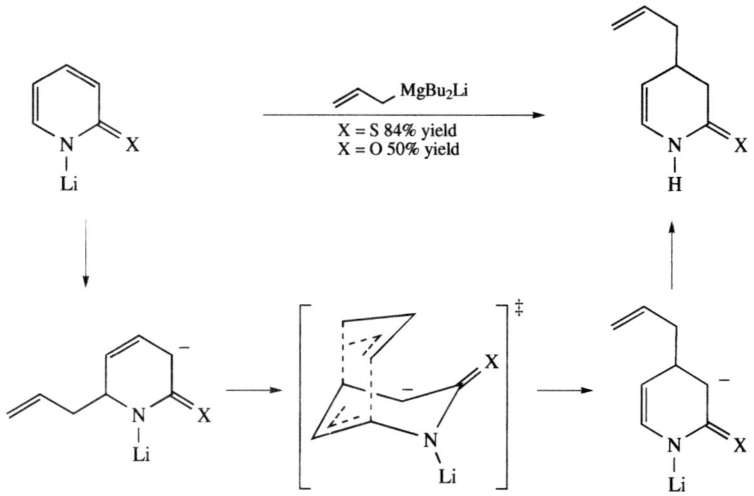

15. The chemistry of organomagnesium ate complexes

SCHEME 1

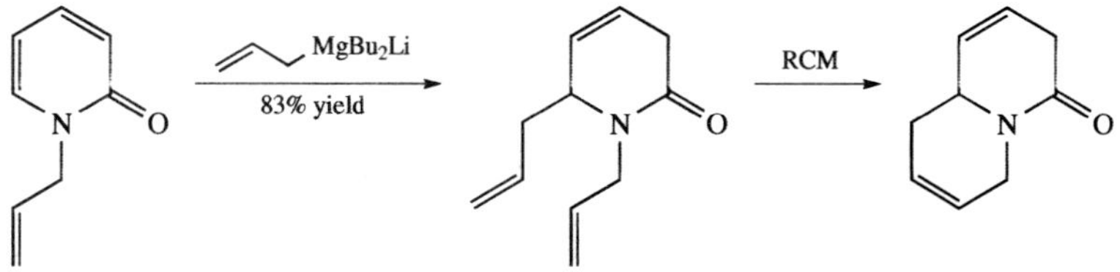

SCHEME 2

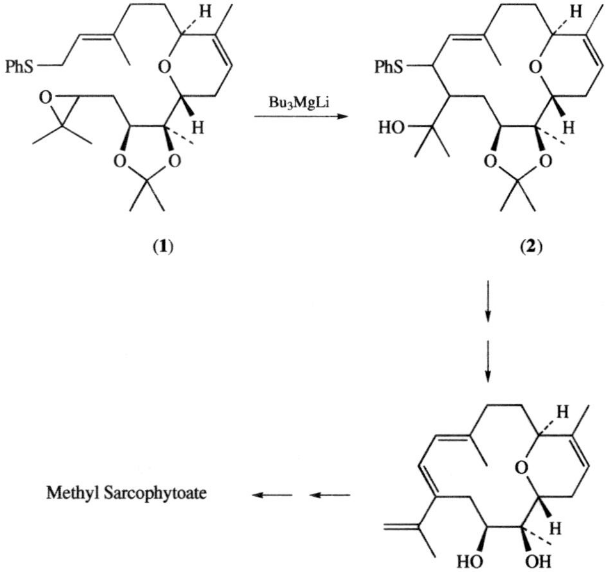

SCHEME 3

Methyl Sarcophytoate

TABLE 4. Generation of sulfur-stabilized anion with organometallic reagents

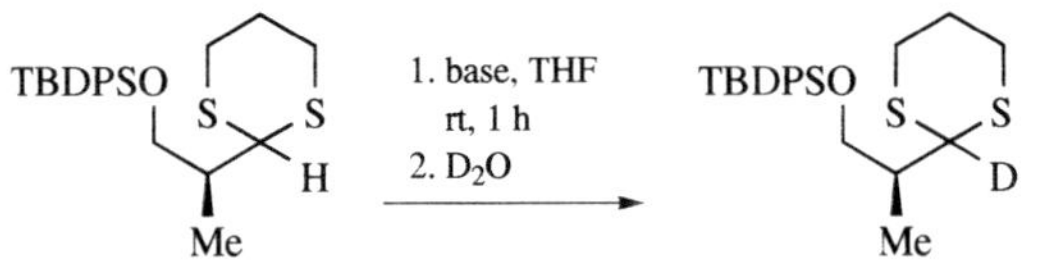

Base	%D
BuLi	86
0.25 BuLi + Bu$_2$Mg	85
1.0 BuLi + Bu$_2$Mg	62
Bu$_2$Mg	0

TABLE 5. Metalation of 3-fluoropyridine

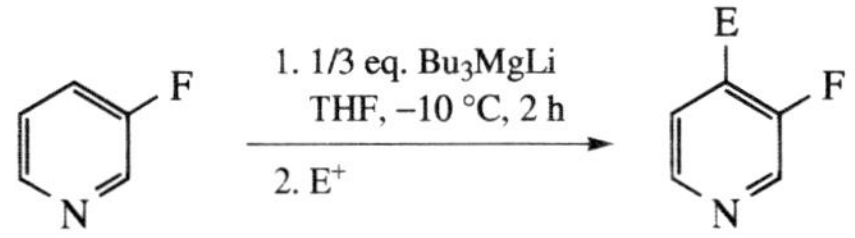

E$^+$	E	Yield (%)
I$_2$	I	64
4-MeOC$_6$H$_4$CHO	4-MeOC$_6$H$_4$CHOH	50
3,4,5-(MeO)$_3$C$_6$H$_2$CHO	3,4,5-(MeO)$_3$C$_6$H$_2$CHOH	55 [a]
CH$_2$=CHCH$_2$Br	CH$_2$=CHCH$_2$	40
2-Bromopyridine	2-Pyridyl	51 [b]

[a] 74% in the presence of 0.33 molar equivalents of TMEDA.
[b] With PdCl$_2$(dppf).

Nakata and coworkers also found that a mixture of butyllithium and dibutylmagnesium generated the anion of a 1,3-dithiane derivative (Table 4)[23]. It is worth noting that their reagent consisted of dibutylmagnesium and 0.25 molar equivalents of butyllithium. The reagent was more effective for the deprotonation than lithium tributylmagnesate. The sulfur-stabilized anion which was generated by the action of their reagent had a longer lifetime than that generated by using butyllithium alone or lithium tributylmagnesate.

Mongin and coworkers have been focusing on deprotonation reactions of aromatic hydrogens with magnesate reagents. Treatment of 3-fluoropyridine with 0.33 molar equivalents of lithium tributylmagnesate in THF at −10 °C resulted in deprotonation at the 4 position (Table 5)[24]. The intermediate lithium tripyridylmagnesate reacted with electrophiles as well as underwent a palladium-catalyzed cross-coupling reaction. Use of TMEDA as an additive proved to enhance the reactivity of the magnesate complexes. Deprotonation also took place at the 2 position of 1,3-difluorobenzene (equation 20). 3,5-Dichloropyridine was involved in a deprotonation reaction in which it was transformed into the tetrachloro-4,4′-bipyridyl via 1,2-migration of the lithium pyridylmagnesate formed by deprotonation (Scheme 4)[25].

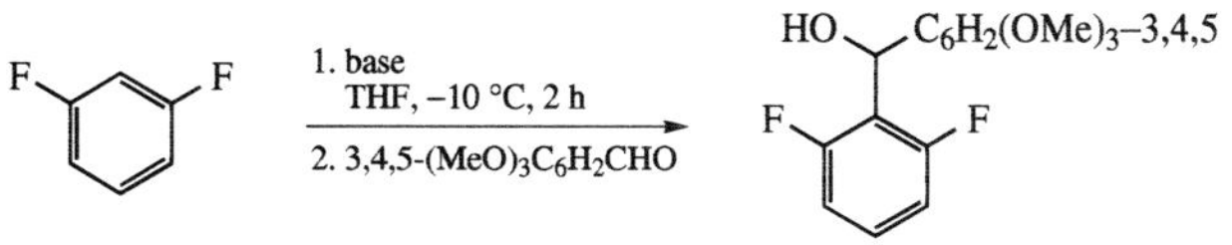

15. The chemistry of organomagnesium ate complexes

$$\begin{array}{c} \text{1. base} \\ \text{THF, }-10\ °\text{C, 2 h} \\ \hline \text{2. 3,4,5-(MeO)}_3\text{C}_6\text{H}_2\text{CHO} \end{array} \quad (20)$$

1/3 eq. Bu$_3$MgLi   74%
1/3 eq. Bu$_3$MgLi + 1/3 eq. TMEDA   86%

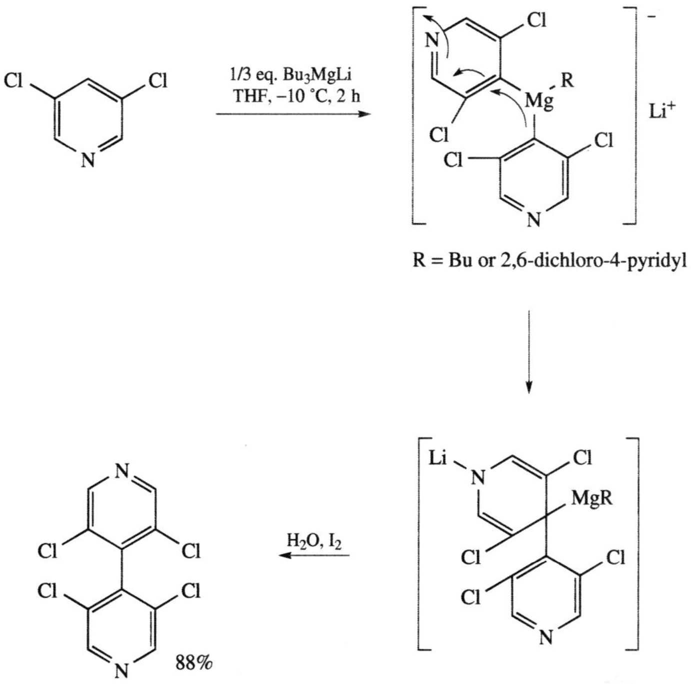

R = Bu or 2,6-dichloro-4-pyridyl

SCHEME 4

Deprotonation reactions of other heteroaromatics were also reported. Thiophene was regioselectively deprotonated at the 2 position by the action of 0.33 molar equivalents of lithium tributylmagnesate in THF at room temperature[26]. Deprotonation took place at the 5 position upon treatment of 2-chloro- and 2-methoxythiophenes (equation 21). The lithium trithienylmagnesate underwent reactions with electrophiles and cross-coupling reactions. Deprotonation reactions of oxazole and benzoxazole proceeded smoothly with lithium tributylmagnesate in THF at room temperature (Scheme 5)[27]. Interestingly, tri(2-oxazolyl)- and tri(2-benzoxazolyl)magnesates isomerized very rapidly and completely to the isocyano-substituted enolate of acetaldehyde and 2-isocyanophenolate, respectively. The reactions of the metalated intermediates with various electrophiles occurred easily with concomitant recovery of the original aromaticity. Furan and benzofuran were deprotonated upon treatment with 0.33 molar equivalents of lithium tributylmagnesate (equation 22)[28]. Dilithium tetrabutylmagnesate, prepared from magnesium dibromide and four molar equivalents of butyllithium, was more reactive than lithium tributylmagnesate.

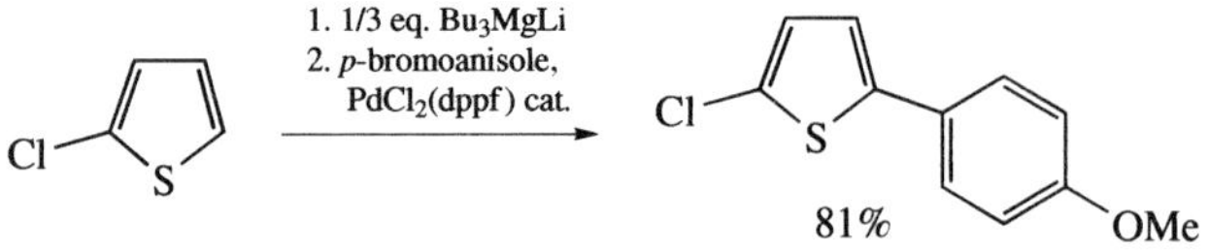

$$\text{Cl}\underset{S}{\diagdown}\quad \xrightarrow[\text{PdCl}_2(\text{dppf}) \text{ cat.}]{\substack{\text{1. 1/3 eq. Bu}_3\text{MgLi} \\ \text{2. }p\text{-bromoanisole,}}} \quad \text{Cl}\underset{S}{\diagdown}\!\!\!-\!\!\!\underset{81\%}{\diagdown}\!\!\!-\text{OMe} \qquad (21)$$

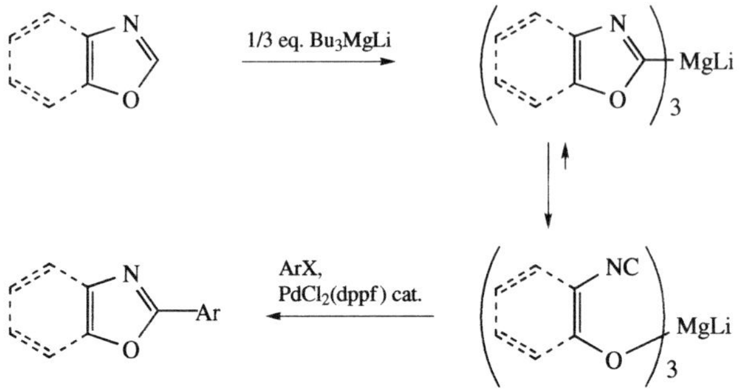

SCHEME 5

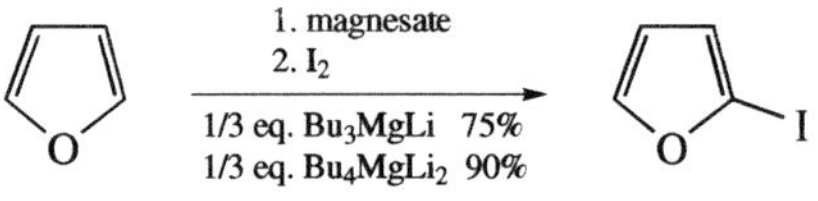

$$\underset{O}{\diagdown}\quad \xrightarrow[\substack{\text{1/3 eq. Bu}_3\text{MgLi 75\%} \\ \text{1/3 eq. Bu}_4\text{MgLi}_2\text{ 90\%}}]{\substack{\text{1. magnesate} \\ \text{2. I}_2}} \quad \underset{O}{\diagdown}\!\!\!-\text{I} \qquad (22)$$

C. Halogen–Magnesium Exchange

Although halogen–lithium exchange reactions are among the most important methods for the preparation of organolithium compounds, the functional group compatibility of the exchange is not satisfactory. Despite their promising functional group compatibility, halogen–magnesium exchange reactions are thought to be slower, to require higher temperature, and thus to be less useful. For the last decade, halogen–magnesium exchange has been recognized as a powerful tool for preparation of organomagnesium reagents[29]. The significant development has been observed since 1998, when Knochel and coworkers reported isopropylmagnesium-induced exchange reactions that are applicable to polyfunctional organomagnesium reagents[30]. Knochel's group also disclosed very recently that an isopropylmagnesium chloride/lithium chloride system is highly efficient for the exchange reactions (Chapter X by Knochel)[7,8].

Halogen–magnesium exchange reactions with triorganomagnesates were reported independently from academia[11] and industry[31] in 2001.

Treatment of various aryl iodides with 1.2 (or 0.5) molar equivalents of lithium tributylmagnesate, prepared from butylmagnesium bromide and butyllithium in a 1:2 ratio in THF, led to iodine–magnesium exchange[11]. The arylmagnesium reagents thus formed were trapped by electrophiles (Table 6). Electron-rich as well as electron-deficient arylmagnesium reagents were prepared. Functional groups such as ester were tolerated during the exchange procedure, as the exchange reactions proceeded even at −78 °C. For instance, treatment of *tert*-butyl *p*-iodobenzoate with 0.5 molar equivalents of lithium tributylmagnesate at −78 °C furnished the corresponding arylmagnesium reagent, which then reacted with allyl bromide under the catalysis of CuCN•2LiCl (Table 6, last entry). Intriguingly,

TABLE 6. Iodine–magnesium exchange of aryl iodides

$$\text{Ar–I} \xrightarrow[\text{THF, }-78\,°\text{C, 30 min}]{\text{Bu}_3\text{MgLi (1.2 eq.)}} \text{Ar–Bu}_2\text{MgLi} \xrightarrow{\text{Electrophile}} \text{Ar–E}$$

Ar–I	Electrophile	Ar–E	Yield (%)
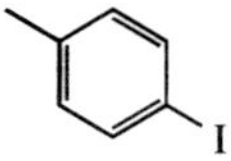 4-MeC₆H₄–I	C_2H_5CHO	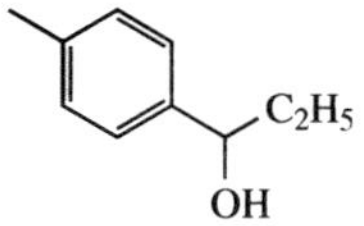 4-MeC₆H₄–CH(OH)C₂H₅	75[a]
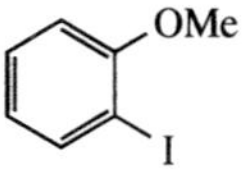 2-MeO-C₆H₄–I	PhCHO	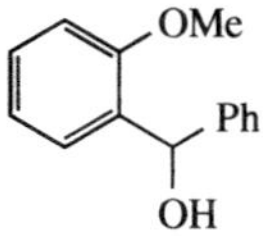 2-MeO-C₆H₄–CH(OH)Ph	92
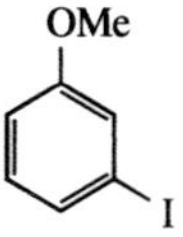 3-MeO-C₆H₄–I	PhCHO	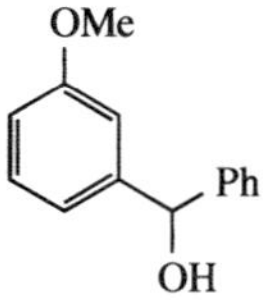 3-MeO-C₆H₄–CH(OH)Ph	93
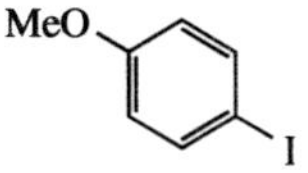 4-MeO-C₆H₄–I	$n\text{-}C_6H_{13}CHO$	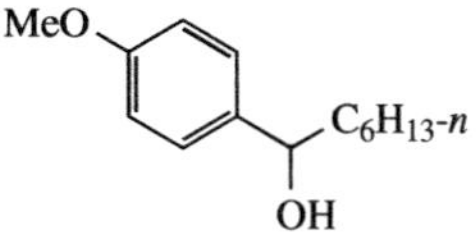 4-MeO-C₆H₄–CH(OH)$C_6H_{13}\text{-}n$	94
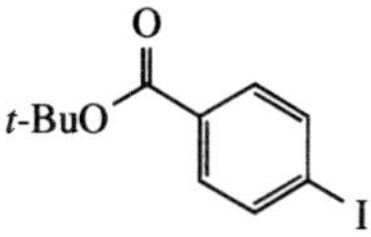 4-(t-BuOC(O))-C₆H₄–I	$CH_2=CHCH_2Br$[b]	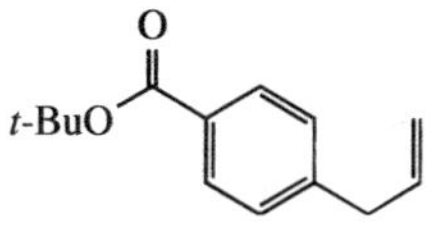 4-(t-BuOC(O))-C₆H₄–CH₂CH=CH₂	88

[a] 0.5 molar equivalents of Bu_3MgLi was used.
[b] A catalytic amount of CuCN•2LiCl was added.

the magnesiation reaction of *t*-butyl *o*-iodobenzoate was faster than the nucleophilic butylation of an aldehyde (equation 23). Addition of lithium tributylmagnesate to a mixture of the iodobenzoate and heptanal afforded the corresponding phthalide in a good yield. Ethyl (2-iodophenoxy)acetate underwent iodine–magnesium exchange to give 3-coumaranone in excellent yield (equation 24).

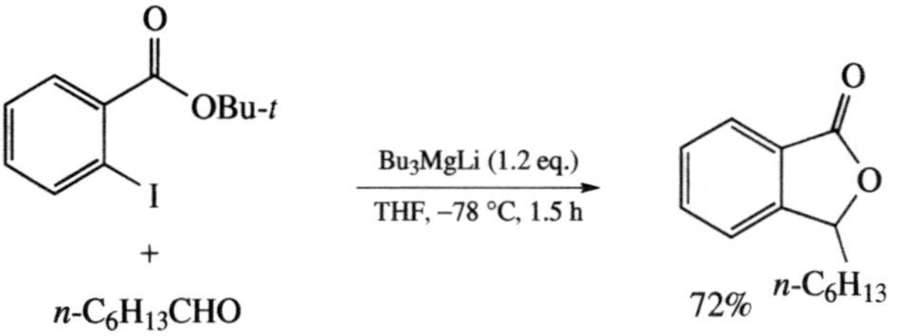

$$\text{2-I-C}_6\text{H}_4\text{-C(O)OBu-}t + n\text{-C}_6\text{H}_{13}\text{CHO} \xrightarrow[\text{THF, }-78\,°\text{C, 1.5 h}]{\text{Bu}_3\text{MgLi (1.2 eq.)}} \text{3-}(n\text{-C}_6\text{H}_{13})\text{phthalide, 72\%} \quad (23)$$

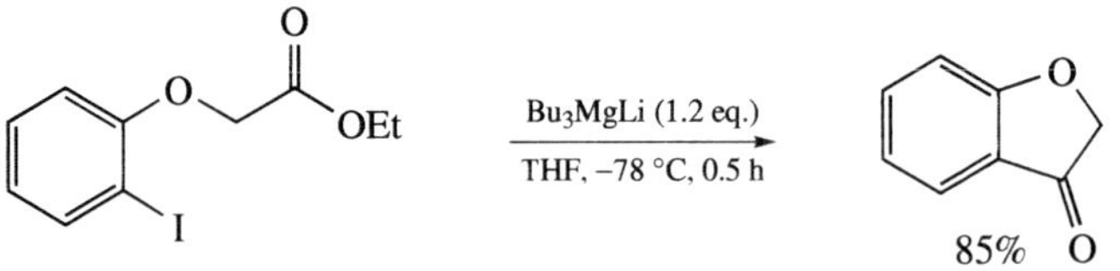

$$\text{(structure: 2-iodophenyl OCH}_2\text{CO}_2\text{Et)} \xrightarrow[\text{THF, }-78\,°\text{C, 0.5 h}]{\text{Bu}_3\text{MgLi (1.2 eq.)}} \text{3-oxo-2,3-dihydrobenzofuran} \quad 85\% \quad (24)$$

Since the exchange reaction of aryl bromides did not proceed to completion at −78 °C in some cases, bromine–magnesium exchange with lithium tributylmagnesate should be performed at 0 °C (Table 7)[11]. Cyano groups survived even at −40 °C (last two entries). Unfortunately, one needs to avoid using aryl bromides with carbonyl groups due to the higher temperature. To attain more efficient bromine–magnesium exchange reactions at −78 °C, lithium dibutylisopropylmagnesate proved to be suitable (Table 8). Functional groups such as carbonyl groups tolerated the magnesate reagent. The exchange reaction of aryl bromide **3** led to the formation of 2-alkoxycoumaran in good yield via intramolecular nucleophilic substitution (equation 25).

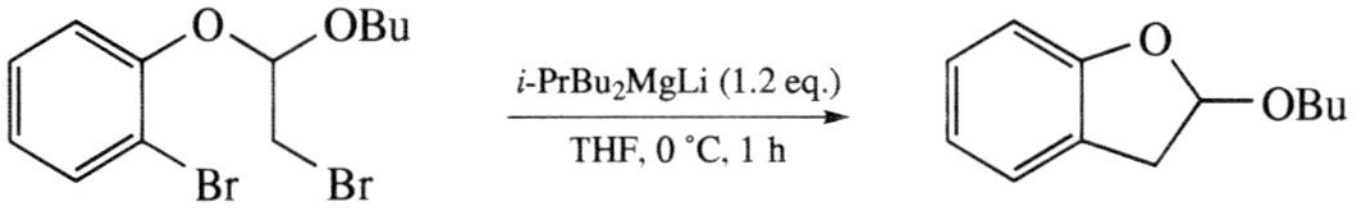

$$\underset{(\mathbf{3})}{\text{(2-bromophenyl OCH(OBu)CH}_2\text{Br)}} \xrightarrow[\text{THF, 0 °C, 1 h}]{i\text{-PrBu}_2\text{MgLi (1.2 eq.)}} \text{2-butoxy-2,3-dihydrobenzofuran} \quad 64\% \quad (25)$$

Controlled metalation of dihaloarenes provides an efficient method to synthesize disubstituted aromatic compounds (Table 9)[11]. In the case of *p*-bromoiodobenzene, no bromine–magnesium exchange took place with a stoichiometric amount of a magnesate reagent (entry 1). Only one of two bromides in *m*- or *p*-dibromobenzene underwent the exchange (entries 2 and 3). On the other hand, *p*-diiodobenzene was converted to dimagnesiated benzene upon treatment with 1.0 molar equivalent of lithium tributylmagnesate (entry 4). Dimetalation of *m*-diiodobenzene needed 2.0 molar equivalents of the reagent (entry 5). Lithium butyldimethylmagnesium allowed for selective monomagnesiation of *p*-diiodobenzene (entry 6). The lower reactivity of the methyl group would decelerate the second exchange reaction. Thus, properly mixed triorganomagnesates can control the reactivity in halogen–magnesium exchange.

Treatment of *o*-dibromobenzene with lithium tributylmagnesate yielded 2-butylphenylmagnesium reagent, which was formed via benzyne (Scheme 6)[11].

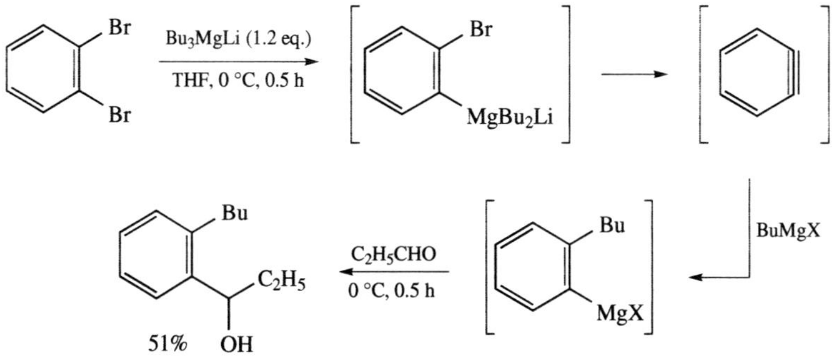

SCHEME 6

TABLE 7. Bromine–magnesium exchange of aryl bromides with Bu_3MgLi

$$Ar\text{–}Br \xrightarrow[\text{THF, 0 °C, 30 min}]{Bu_3MgLi \text{ (1.2 eq.)}} Ar\text{–}Bu_2MgLi \xrightarrow{\text{Electrophile}} Ar\text{–}E$$

Ar–Br	Electrophile	Ar–E	Yield (%)
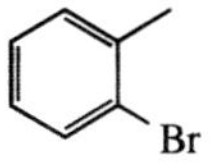 2-methylphenyl bromide	C_2H_5CHO	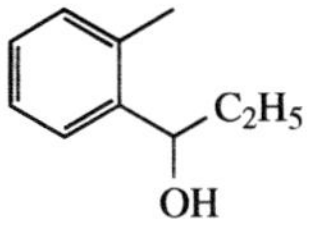 1-(2-methylphenyl)propan-1-ol	88
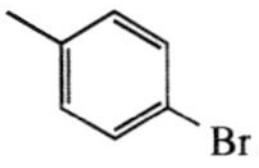 4-methylphenyl bromide	C_2H_5CHO	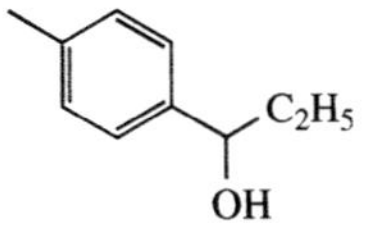 1-(4-methylphenyl)propan-1-ol	85
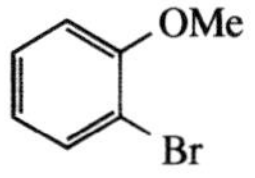 2-bromoanisole	C_2H_5CHO	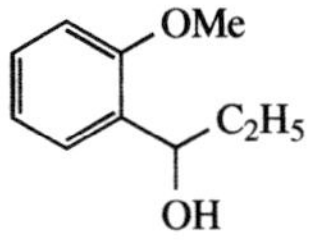 1-(2-methoxyphenyl)propan-1-ol	90 [a]
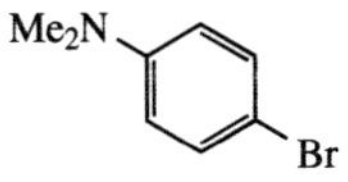 4-bromo-N,N-dimethylaniline	C_2H_5CHO	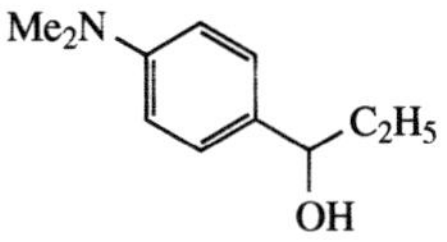 1-(4-dimethylaminophenyl)propan-1-ol	94
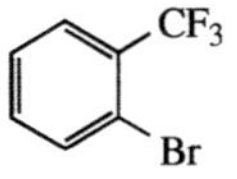 2-bromobenzotrifluoride	$n\text{-}C_6H_{13}CHO$	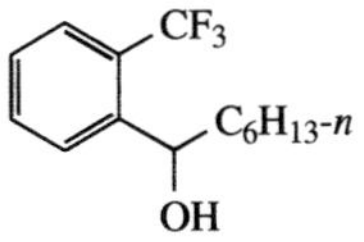 1-(2-trifluoromethylphenyl)heptan-1-ol	76
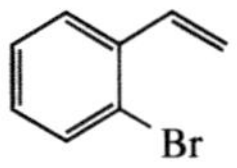 2-bromostyrene	C_2H_5CHO	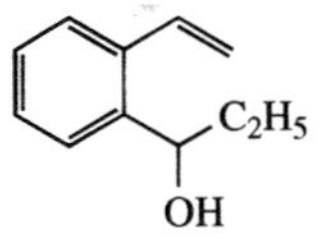 1-(2-vinylphenyl)propan-1-ol	52
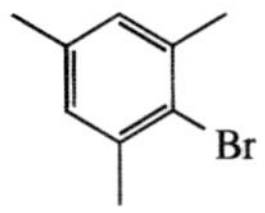 2-bromomesitylene	$CH_2\text{=}CHCH_2Br$ [b]	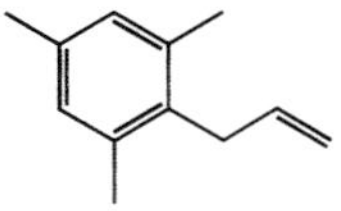 allylmesitylene	93
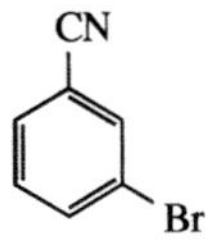 3-bromobenzonitrile	$CH_2\text{=}CHCH_2Br$ [b]	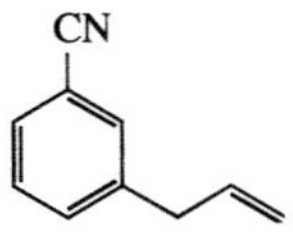 3-allylbenzonitrile	85 [a,c]
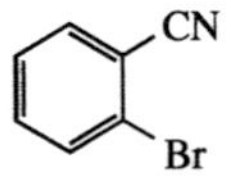 2-bromobenzonitrile	$CH_2\text{=}CHCH_2Br$ [b]	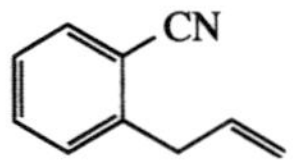 2-allylbenzonitrile	80 [a,c]

[a] 0.5 molar equivalents of Bu_3MgLi was used.
[b] A catalytic amount of CuCN·2LiCl was added.
[c] The reaction was performed at $-40\,°C$.

TABLE 8. Bromine–magnesium exchange of aryl bromides with i-PrBu$_2$MgLi

$$\text{Ar–Br} \xrightarrow[\text{THF, }-78\,°\text{C, 1 h}]{i\text{-PrBu}_2\text{MgLi (1.2 eq.)}} \text{Ar–Bu}_2\text{MgLi} \xrightarrow[-78\,°\text{C, 1 h}]{\text{Electrophile}} \text{Ar–E}$$

Ar–Br	Electrophile	Ar–E	Yield (%)
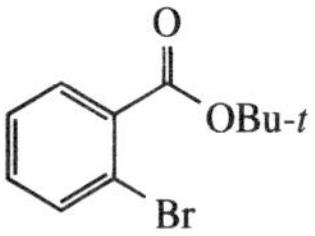 2-Br-C$_6$H$_4$-C(O)OBu-t	CH$_2$=CHCH$_2$Br [a]	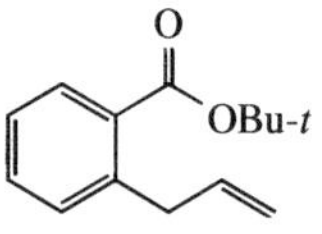 2-(allyl)-C$_6$H$_4$-C(O)OBu-t	99
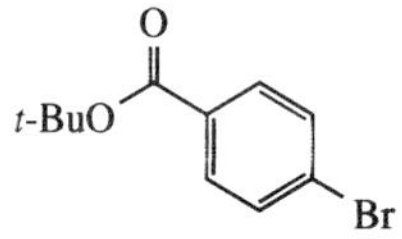 4-Br-C$_6$H$_4$-C(O)O-t-Bu	n-C$_6$H$_{13}$CHO	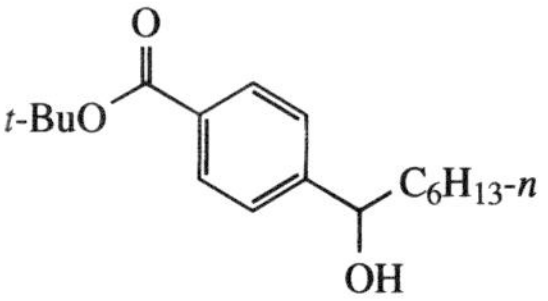 4-[CH(OH)C$_6$H$_{13}$-n]-C$_6$H$_4$-C(O)O-t-Bu	71
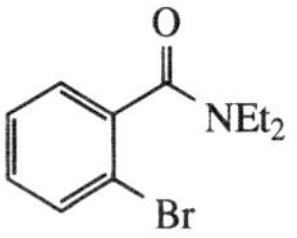 2-Br-C$_6$H$_4$-C(O)NEt$_2$	CH$_2$=CHCH$_2$Br [a]	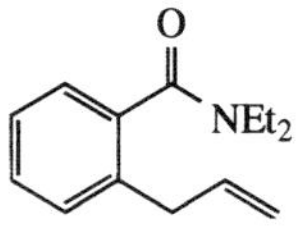 2-(allyl)-C$_6$H$_4$-C(O)NEt$_2$	80
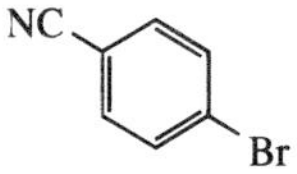 4-Br-C$_6$H$_4$-CN	CH$_2$=CHCH$_2$Br [a]	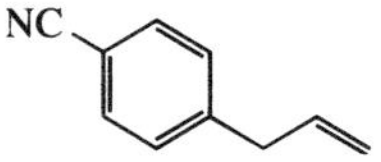 4-(allyl)-C$_6$H$_4$-CN	87

[a] A catalytic amount of CuCN•2LiCl was added.

Lithium butyldimethylmagnesate was preferable to tributylmagnesate in the bromine–magnesium exchange of 3-bromopyridine since the latter afforded a rather complex mixture (Table 10)[11]. 2-Bromopyridine and 2-bromothiophene were also converted to the corresponding magnesium reagents which then reacted with aldehydes.

Stereospecific magnesiation was observed in the exchange reactions of alkenyl iodides with lithium dibutylisopropylmagnesate (Table 11)[11]. The magnesiation could be completed within 1 h at $-78\,°$C, where the ester functionality was compatible with the reaction (last entry). Unfortunately, the bromine–magnesium exchange of alkenyl bromide was disappointing since, due to the slow exchange, sequential dehydrobromination and deprotonation took place to yield the relevant magnesium acetylide as a by-product (equation 26). The exchange reaction of 1-silyl-1-haloalkenes proceeded smoothly, along with considerable isomerization to give mainly the E isomer (Scheme 7). Obviously, these results reflect the strong preference of the bulky silyl group to be in the *trans* position to the alkyl group.

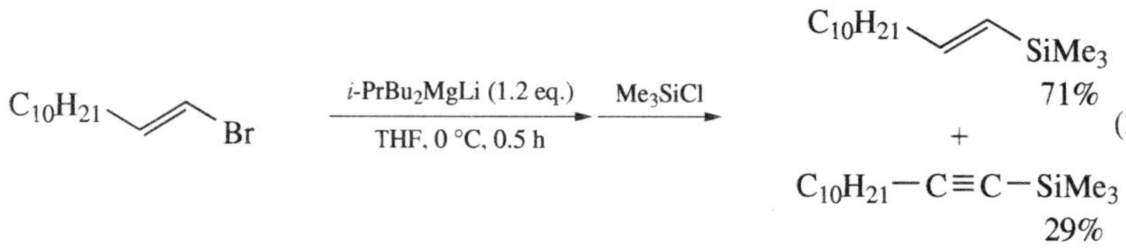

$$\text{C}_{10}\text{H}_{21}\text{CH=CHBr} \xrightarrow[\text{THF, 0 °C, 0.5 h}]{i\text{-PrBu}_2\text{MgLi (1.2 eq.)} \quad \text{Me}_3\text{SiCl}} \begin{array}{l} \text{C}_{10}\text{H}_{21}\text{CH=CHSiMe}_3 \quad 71\% \\ + \\ \text{C}_{10}\text{H}_{21}-\text{C}\equiv\text{C}-\text{SiMe}_3 \quad 29\% \end{array} \quad (26)$$

TABLE 9. Halogen–magnesium exchange of dihaloarenes followed by reaction with propanal

Entry	Ar–I	Conditions	Product	Yield (%)
1	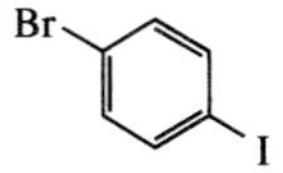 4-Br-C6H4-I	i-PrBu$_2$MgLi (1.0 eq.) −78 °C, 0.5 h	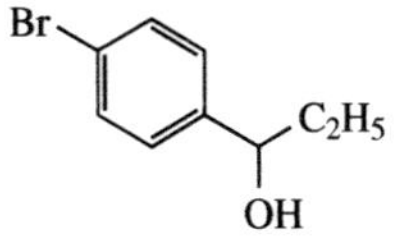 4-Br-C6H4-CH(OH)C$_2$H$_5$	65
2	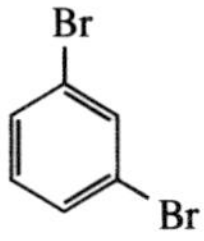 1,3-Br$_2$-C6H4	i-PrBu$_2$MgLi (1.0 eq.) 0 °C, 0.5 h	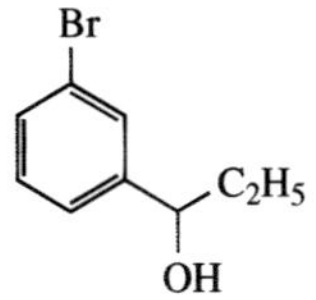 3-Br-C6H4-CH(OH)C$_2$H$_5$	78
3	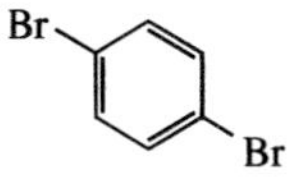 4-Br-C6H4-Br	Bu$_3$MgLi (1.0 eq.) −78 °C, 0.5 h	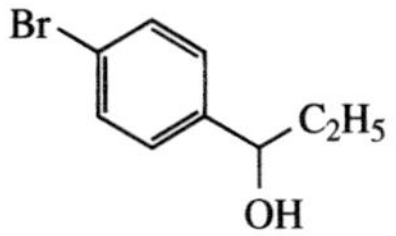 4-Br-C6H4-CH(OH)C$_2$H$_5$	85
4	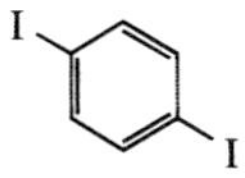 1,4-I$_2$-C6H4	Bu$_3$MgLi (1.0 eq.) −78 °C, 0.5 h	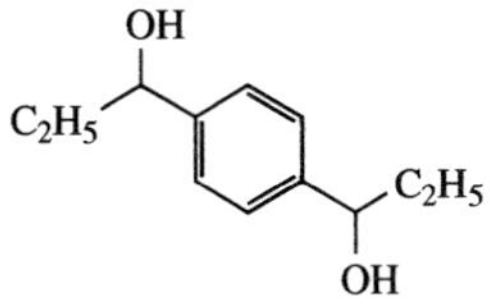 1,4-[C$_2$H$_5$CH(OH)]$_2$-C6H4	80
5	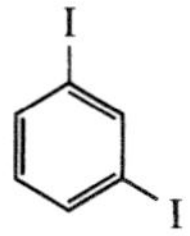 1,3-I$_2$-C6H4	Bu$_3$MgLi (2.0 eq.) −78 °C, 0.5 h	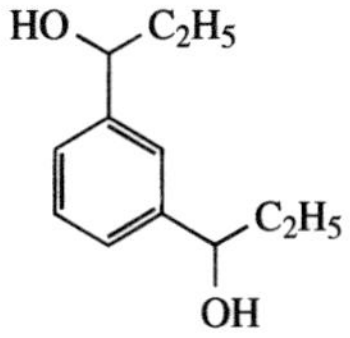 1,3-[C$_2$H$_5$CH(OH)]$_2$-C6H4	48
6	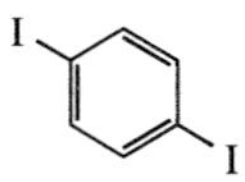 1,4-I$_2$-C6H4	BuMe$_2$MgLi (1.0 eq.) −78 °C, 0.5 h	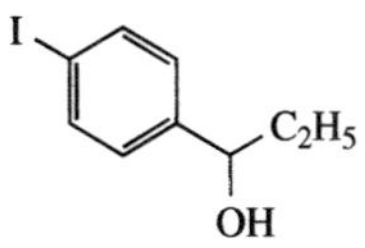 4-I-C6H4-CH(OH)C$_2$H$_5$	64
			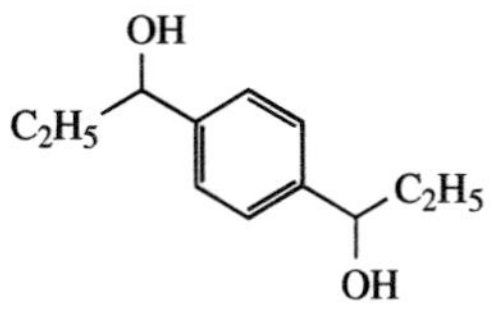 1,4-[C$_2$H$_5$CH(OH)]$_2$-C6H4	9

TABLE 10. Bromine–magnesium exchange of heteroaryl bromides with BuMe$_2$MgLi

$$\text{HeteroAr-Br} \xrightarrow[\text{THF, 0 °C, 0.5 h}]{\text{BuMe}_2\text{MgLi (1.0 eq.)}} \xrightarrow[\text{0 °C, 0.5 h}]{\text{RCHO}} \text{HeteroAr-CH(OH)R}$$

HeteroAr–Br	RCHO	HeteroAr–CH(OH)R	Yield (%)
3-Br-pyridine	C$_2$H$_5$CHO	3-pyridyl-CH(OH)C$_2$H$_5$	73
2-Br-pyridine	C$_2$H$_5$CHO	2-pyridyl-CH(OH)C$_2$H$_5$	67
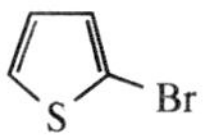 2-Br-thiophene	PhCHO	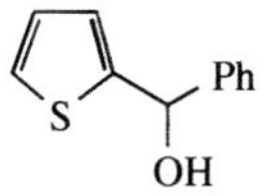 2-thienyl-CH(OH)Ph	78 [a]

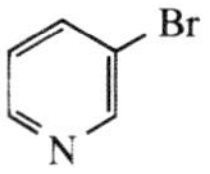

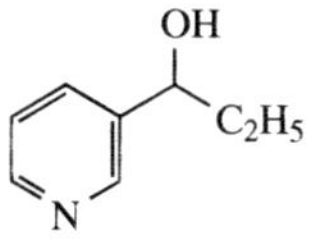

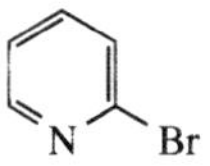

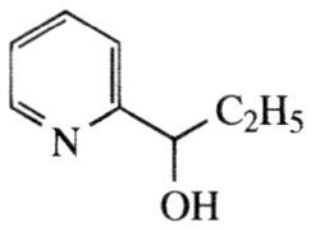

[a] Lithium tributylmagnesate was used.

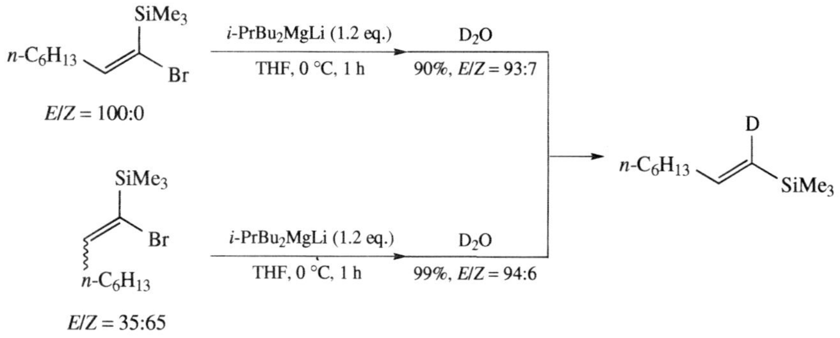

SCHEME 7

Since the carbon–carbon bonds in strained cyclopropane rings have large s-character, cyclopropyl bromides underwent smooth bromine–magnesium exchange by the action of lithium tributylmagnesate (equation 27)[11].

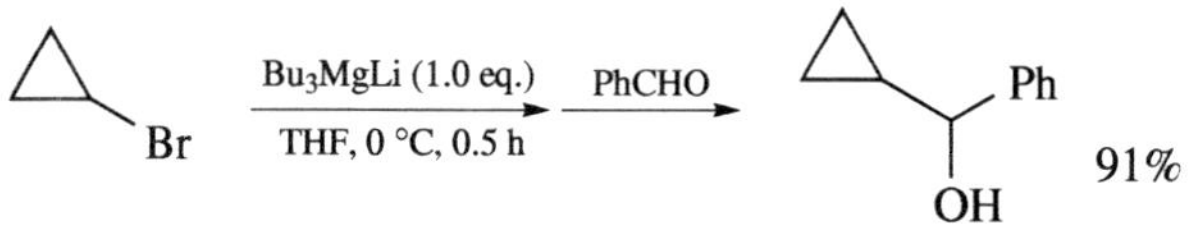

$$\text{cyclopropyl-Br} \xrightarrow[\text{THF, 0 °C, 0.5 h}]{\text{Bu}_3\text{MgLi (1.0 eq.)}} \xrightarrow{\text{PhCHO}} \text{cyclopropyl-CH(OH)Ph} \quad 91\% \qquad (27)$$

Chemists in Banyu Pharmaceutical discovered that lithium tributylmagnesate, prepared from butylmagnesium chloride and two molar equivalents of butyllithium, is quite

TABLE 11. Iodine–magnesium exchange of alkenyl iodides

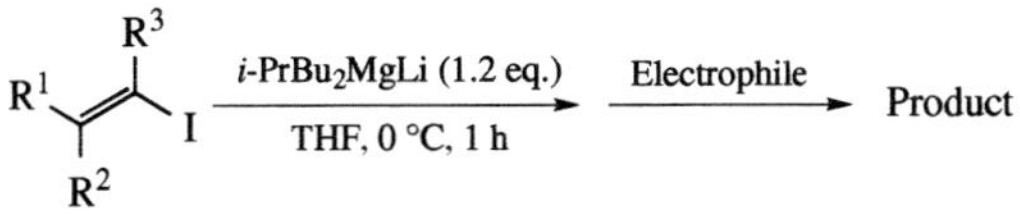

Entry	Alkenyl iodide	Electrophile	Product	Yield (%)
1	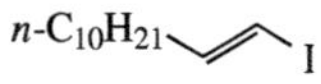 n-$C_{10}H_{21}$–CH=CH–I	Me_3SiCl	 n-$C_{10}H_{21}$–CH=CH–$SiMe_3$	93
2	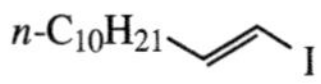 n-$C_{10}H_{21}$–CH=CH–I	CH_2=$CHCH_2Br$ [a]	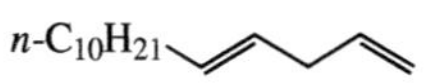 n-$C_{10}H_{21}$–CH=CH–CH_2CH=CH_2	70
3	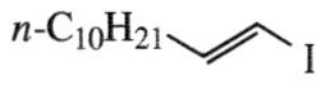 n-$C_{10}H_{21}$–CH=CH–I	Me_2CO	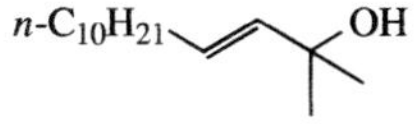 n-$C_{10}H_{21}$–CH=CH–C(Me)$_2$OH	75
4	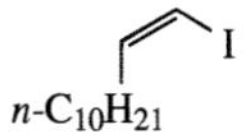 (n-$C_{10}H_{21}$)C=CH–I	Me_3SiCl	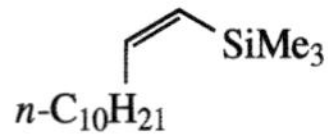 (n-$C_{10}H_{21}$)C=CH–$SiMe_3$	87
5	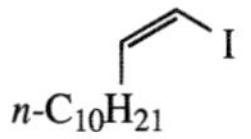 (n-$C_{10}H_{21}$)C=CH–I	CH_2=$CHCH_2Br$ [a]	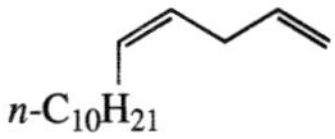 (n-$C_{10}H_{21}$)C=CH–CH_2CH=CH_2	70
6	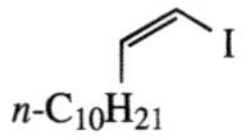 (n-$C_{10}H_{21}$)C=CH–I	Me_2CO	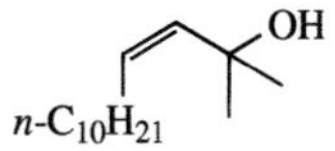 (n-$C_{10}H_{21}$)C=CH–C(Me)$_2$OH	75
7	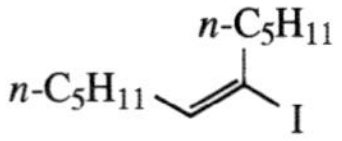 (n-C_5H_{11})(n-C_5H_{11})C=CH–I	PhCHO	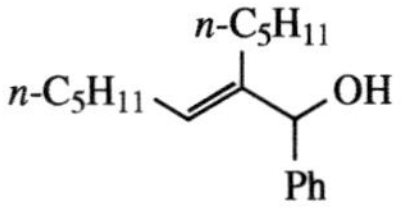 (n-C_5H_{11})(n-C_5H_{11})C=CH–CH(OH)Ph	87
8	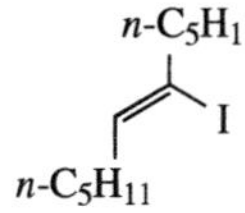 (n-C_5H_{11})(n-C_5H_{11})C=CI	PhCHO	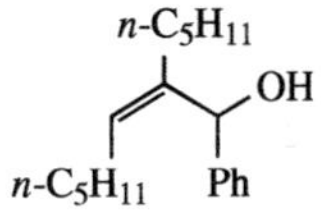 (n-C_5H_{11})(n-C_5H_{11})C=C–CH(OH)Ph	70
9	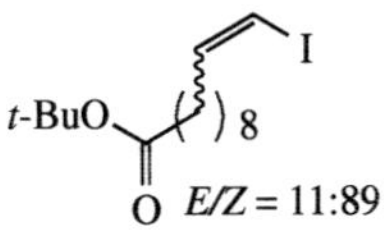 t-BuO-CO-(CH$_2$)$_8$-CH=CH–I, E/Z = 11:89	C_2H_5CHO	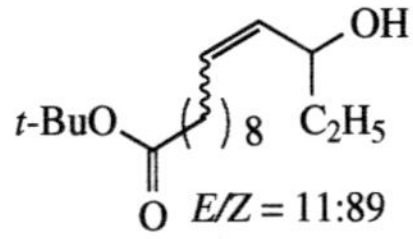 t-BuO-CO-(CH$_2$)$_8$-CH=CH–CH(OH)C_2H_5, E/Z = 11:89	80[b]

[a] A catalytic amount of CuCN·2LiCl was added.
[b] Performed at −78 °C.

efficient for the selective monomagnesiation of 2,6-dibromopyridine (Scheme 8)[31]. The bromine–magnesium exchange reaction of 2,6-dibromopyridine with the magnesate reagent (0.35 molar equivalents) proceeded efficiently under noncryogenic conditions (−10 °C) in toluene, affording a virtually pure mono-magnesiated intermediate. Subsequent treatment with DMF provided the desired aldehyde in 95% yield. In the Banyu protocol, all the three butyl groups in the ate complex participated in the exchange.

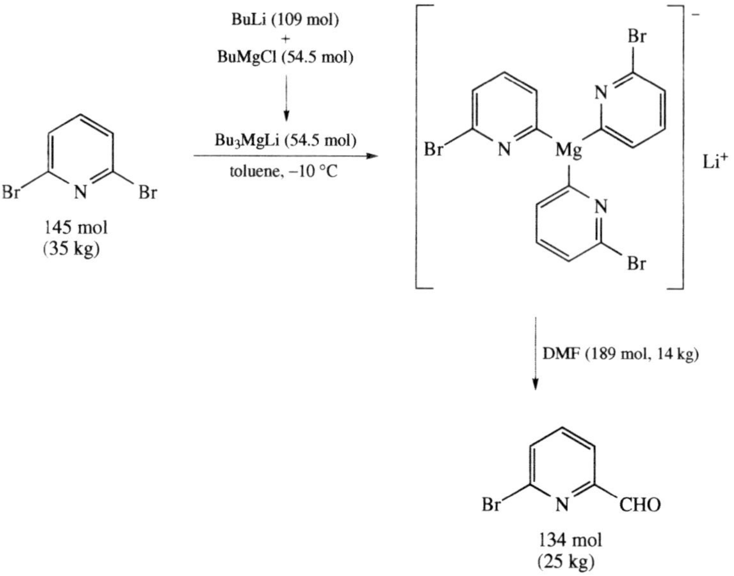

SCHEME 8

TABLE 12. Monoformylation of dibromoarenes by the Banyu protocol [(1) 0.35–0.40 molar equivalents of Bu_3MgLi, (2) 1.3 molar equivalents of DMF, (3) aq. citric acid or aq. acetic acid]

Dibromoarene	Conditions	Magnesiation (mono-/di-)	Product	Yield (%)
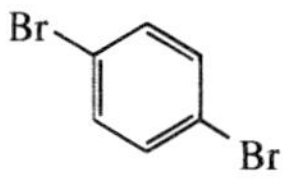 1,4-dibromobenzene	toluene, 0 °C, 5 h	88.5:< 0.1	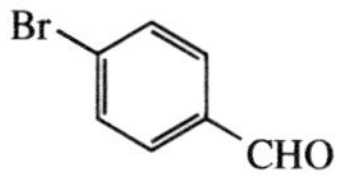 4-bromobenzaldehyde	84
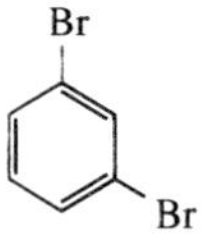 1,3-dibromobenzene	toluene, 0 °C, 1.5 h	92.0:< 0.1	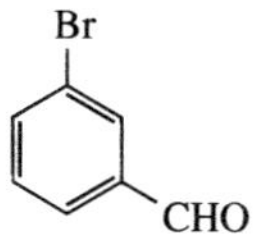 3-bromobenzaldehyde	99
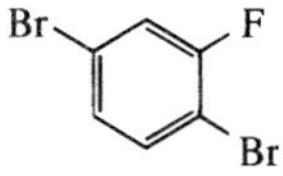 1,4-dibromo-2-fluorobenzene	toluene/THF (5:1), 0 °C, 1 h	91.9:< 2	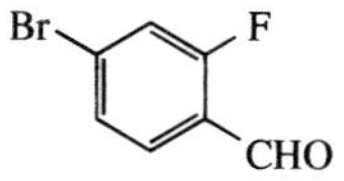 4-bromo-2-fluorobenzaldehyde	92
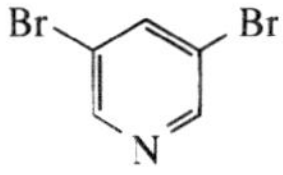 3,5-dibromopyridine	toluene/THF (1:1), −10 °C, 1.5 h	97.7:< 0.1	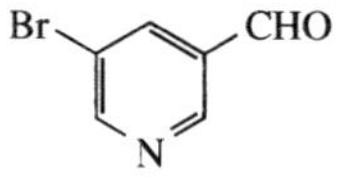 5-bromopyridine-3-carbaldehyde	78
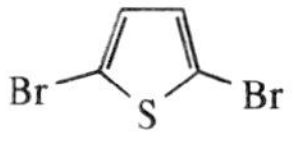 2,5-dibromothiophene	toluene, −10 °C, 3 h	90.7:1.6	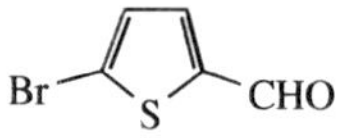 5-bromothiophene-2-carbaldehyde	73

15. The chemistry of organomagnesium ate complexes

The protocol is used for the preparation of 25 kg of the aldehyde in Banyu, and applied to the synthesis of a muscarinic receptor agonist. It is noteworthy that halogen–metal exchange reactions with other metal reagents such as butyllithium or isopropylmagnesium bromide led to more complex mixtures. The Banyu protocol was applicable to similar monoformylation reactions of dibromoheteroarenes (Table 12).

Since the two pioneering reports from academia[11] and industry[31], organomagnesate reagents have been recognized as reliable reagents in organic synthesis. Most of the

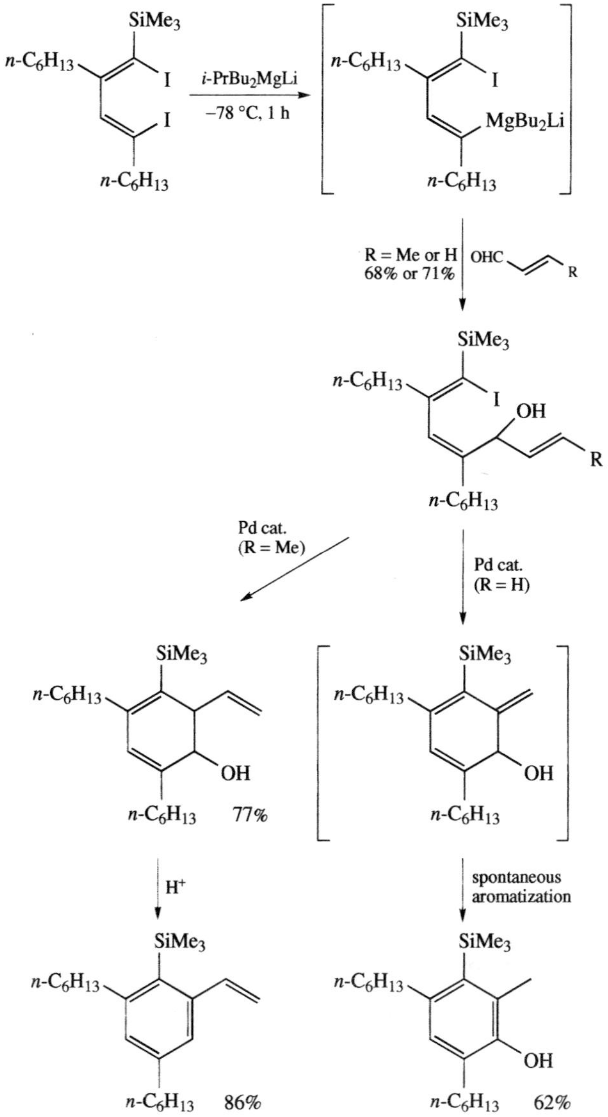

SCHEME 9

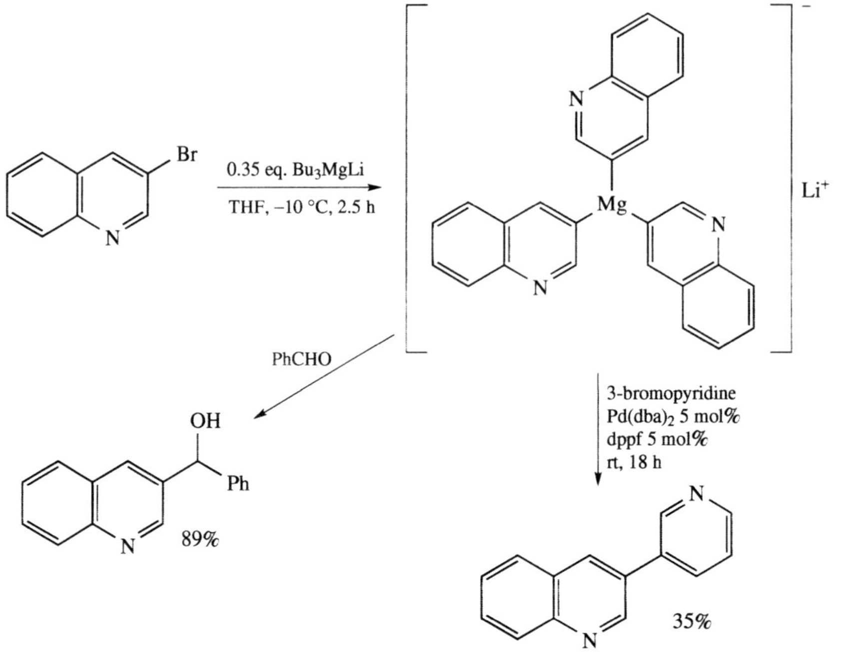

SCHEME 10

examples of the nucleophilic addition reactions and the deprotonation reactions in the previous sections were reported after these two reports. This is also the case for halogen–magnesium exchange reactions.

Sato and coworkers found that site-selective iodine–magnesium exchange reactions of 1,4-diiodo-1,3-alkadienes were attained only by using the organomagnesium ate complex, lithium dibutylisopropylmagnesate (Scheme 9)[32]. The magnesiated iodoalkadienes were transformed into polysubstituted styrenes and phenols.

Although isopropylmagnesium chloride, tert-butylmagnesium chloride and diisopropylmagnesium failed to effect the bromine–magnesium exchange reaction of 3-bromoquinoline, 0.35 molar equivalents of lithium tributylmagnesate smoothly promoted the metalation at $-10\,°C$ in THF (Scheme 10)[33]. Lithium tris(3-quinolyl)magnesate reacted with benzaldehyde to yield the corresponding alcohol. Intriguingly, the yields largely depended on the amount of the magnesate reagent used, i.e. 0% yield with 1 molar equivalent of the reagent, and 5–10% with 0.66 molar equivalents. Butylated products were obtained in these cases through addition of the remaining butylmagnesium species to the quinoline ring. Similar bromine–magnesium exchange occurred in the reactions of 2- or 4-bromoquinolines. The tris(quinolyl)magnesates and related tris(heteroaryl)magnesates are involved in palladium- and nickel-catalyzed cross-coupling reactions, albeit the yields were moderate[34].

Selective halogen–magnesium exchange of *m*- and *p*-halobenzene derivatives having *ortho*-directing groups took place with magnesate reagents (equation 28)[35]. Lithium dibutylisopropylmagnesate proved to be superior in preventing *ortho*-metalation to other metalation agents such as isopropylmagnesium bromide.

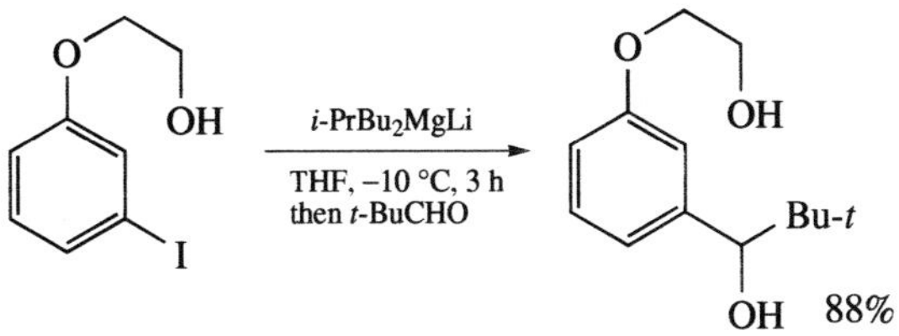

15. The chemistry of organomagnesium ate complexes

(28)

Preparation of some azulenylmagnesium species was achieved by the halogen–magnesium exchange reactions of iodoazulenes with lithium tributylmagnesate at low temperatures (equations 29–33)[36]. The reactions offer access to a variety of functionalized azulenes including azulenylphosphine, -stannane and -boronic ester.

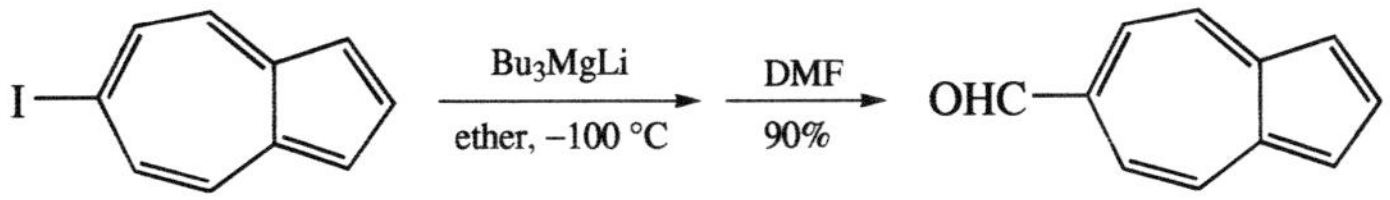

(29)

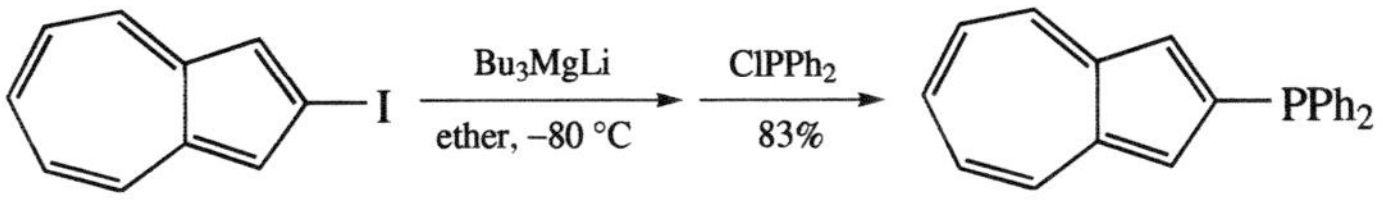

(30)

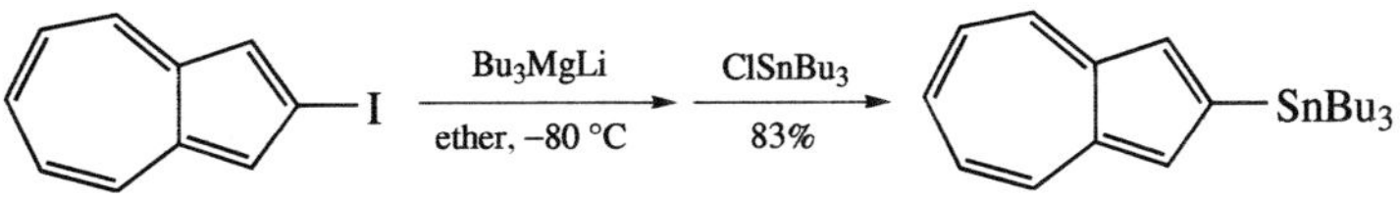

(31)

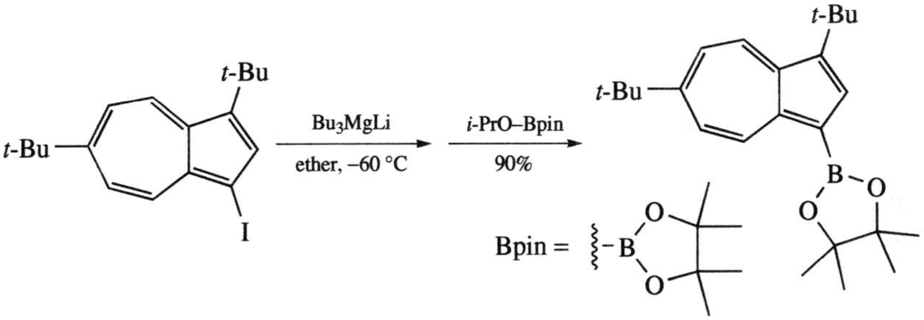

(32)

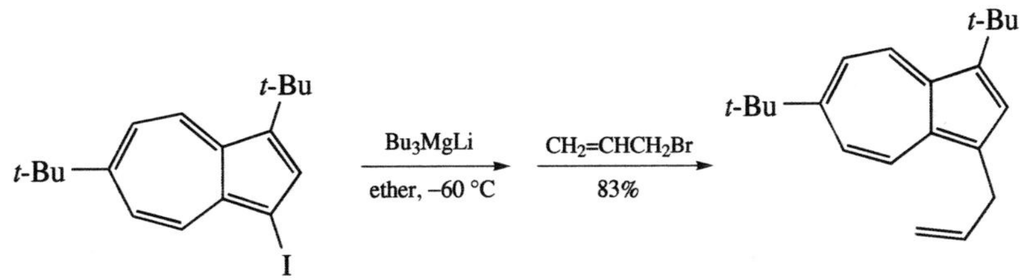

(33)

A halogen–magnesium exchange reaction of 4-iodo-6-phenylthieno[2,3-d]pyrimidine with various magnesate reagents had taken place (Table 13)[37]. Among the magnesate reagents, lithium butyldimethylmagnesium was the most efficient. The magnesate reagent was superior to butyllithium since the reaction with the magnesate under Barbier-type conditions was performed at 0 °C. The reaction with butyllithium at 0 °C afforded none of the desired product.

TABLE 13. Barbier-type reaction of iodopyrimidine derivative with aldehyde

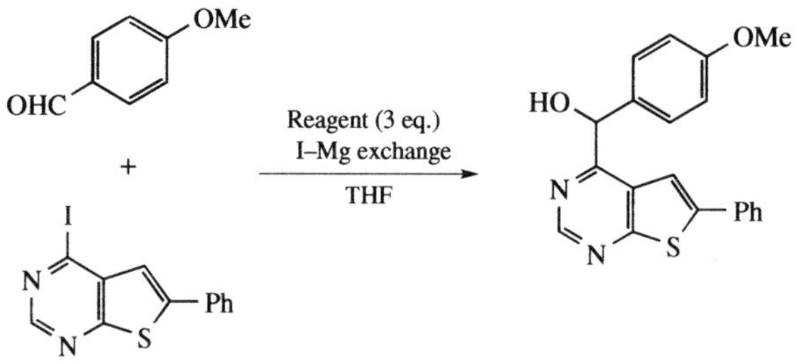

Conditions	Yield (%)
BuMe$_2$MgLi, −76 °C	60
BuMe$_2$MgLi, 0 °C	62
Bu$_3$MgLi, −76 °C	19
PhBu$_2$MgLi, −76 °C	56
BuPh$_2$MgLi, −76 °C	trace
BnPh$_2$MgLi, −76 °C	41
BuLi, −76 °C	57
BuLi, 0 °C	0

Metalation of only the bromine of 2-bromo-3,5-dichloro-6-(trifluoromethyl)-4-trimethylsilylbenzoic acid proceeded smoothly with lithium tributylmagnesate (equation 34)[38].

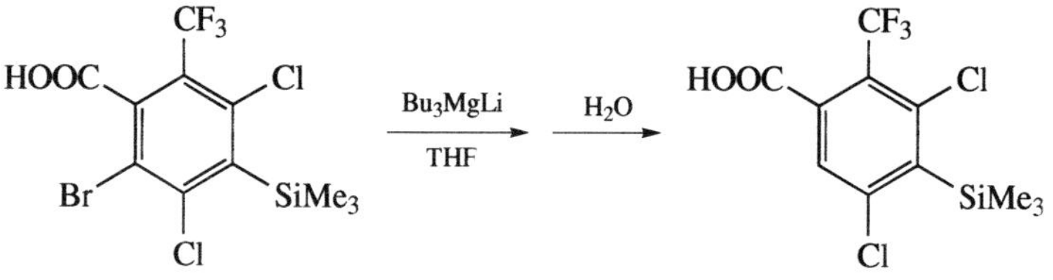

(34)

Lithium dibutylisopropylmagnesate proved to be quite efficient for the bromine–magnesium exchange of 5-bromo-2-picoline at −10 °C (Table 14)[39]. The resulting picolylmagnesium reagent reacted with electrophiles including thiuram disulfide (last entry).

The reactions of 2-iodopyrazine, 2-methylsulfanyl-4-iodopyrimidine and 3-iodo-6-phenylpyridazine with lithium tributylmagnesate resulted in very efficient iodine–magnesium exchange to yield the corresponding heteroarylmagnesium species (equations 35–37)[40]. The reactions with carbonyl compounds and diphenyl disulfide proceeded with good yields. The reactions proceeded smoothly, and neither the starting iodide nor any butylated compounds derived from nucleophilic addition to the heteroaromatic nuclei were observed.

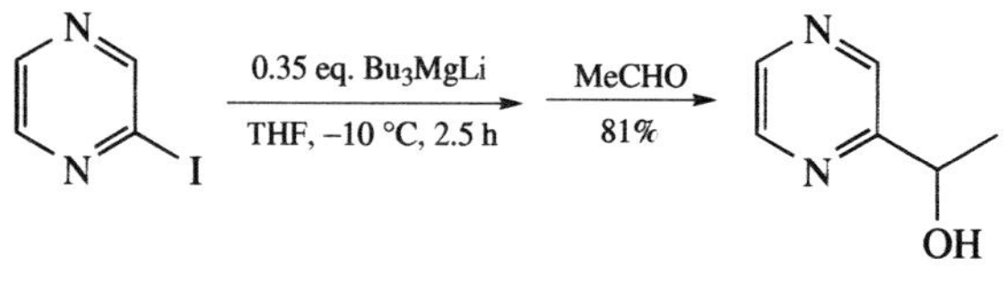

(35)

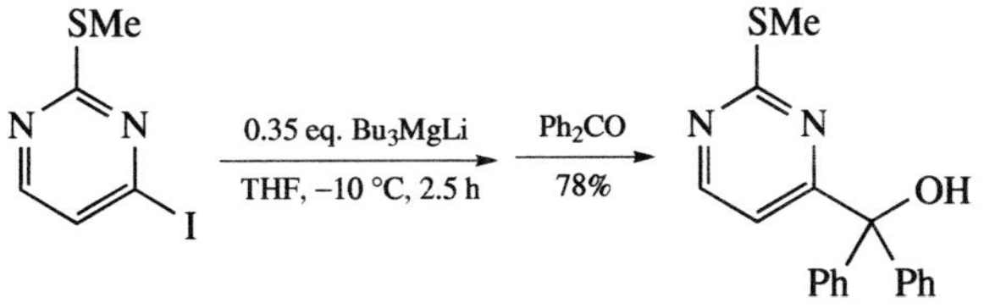

15. The chemistry of organomagnesium ate complexes

$$\underset{I}{\text{2-SMe-pyrimidine}} \xrightarrow[\text{THF, }-10\,°\text{C, 2.5 h}]{0.35 \text{ eq. Bu}_3\text{MgLi}} \xrightarrow[78\%]{\text{Ph}_2\text{CO}} \text{product with C(OH)Ph}_2 \quad (36)$$

$$\underset{I}{\text{3-Ph-pyridazine}} \xrightarrow[\text{THF, }-10\,°\text{C, 2.5 h}]{0.35 \text{ eq. Bu}_3\text{MgLi}} \xrightarrow[62\%]{\text{PhSSPh}} \text{product with SPh} \quad (37)$$

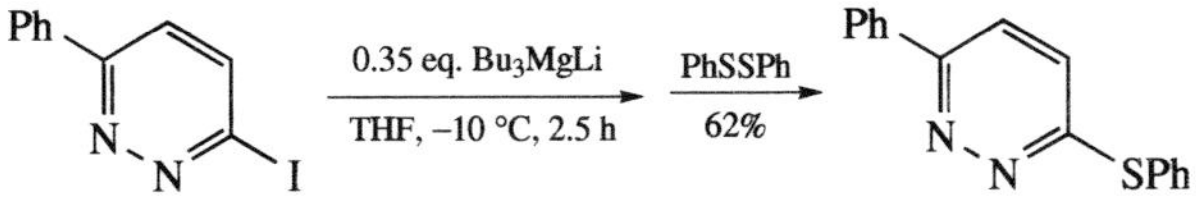

Chemists in Merck sought to develop a cost-efficient and practical synthesis of biphenyl ketone **4**, a precursor to potent cathepsin K Inhibitor **5** (Scheme 11). Instead of an obvious retrosynthetic analysis via a Suzuki–Miyaura cross-coupling between the corresponding arylboronic acid and aryl bromide, they have found the following new route to the ketone **4**[41]. A selective halogen–magnesium exchange/dimethyl disulfide quench protocol on 4,4′-dibromobiphenyl with lithium tributylmagnesate proceeded smoothly under noncryogenic conditions on a large scale. The magnesate showed superb reactivity, whereas other metalation agents such as butyllithium, isopropylmagnesium chloride and Knochel's isopropylmagnesium chloride–lithium chloride system were less selective or completely unreactive.

The reactions of *gem*-dibromocyclopropanes with dialkylcuprates, trialkylzincates and trialkylmanganates were well known, and afforded alkylated cyclopropylmetals. Lithium trialkylmagnesates also participated in similar alkylative metalation (Scheme 12)[42]. Treatment of dibromocyclopropane **6** with lithium tributylmagnesate at low temperatures followed by addition of electrophiles provided the corresponding butylated products as mixtures of diastereomers **7** and **8**. The reactions should be performed at −78 °C to −30 °C. At higher temperatures, formation of 1,2-nonadiene was inevitable. 1,2-Migration of the butyl group was incomplete at −78 °C.

Treatment of dibromomethyl methyldiphenylsilane with lithium tributylmagnesate at −78 °C induced very efficient bromine–magnesium exchange to yield the bromomethylsilane upon protonolysis at −78 °C (Scheme 13)[42,43]. Warming the reaction mixture in the presence of a copper salt before protonolysis led to smooth migration of one of the butyl groups to afford 1-silylpentylmetal.

TABLE 14. Preparation of picolylmagnesium and its reactions with electrophiles

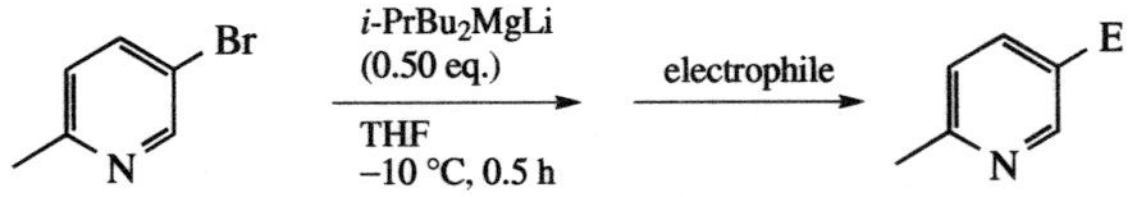

Electrophile	E	Yield (%)
PhCHO	CH(OH)Ph	93
PhCOCl	PhCO	52
t-BuCOCl	*t*-BuCO	57
DMF	CHO	85
CH$_2$=CHCH$_2$Br	CH$_2$=CHCH$_2$	92
i-Pr$_2$NC(S)SSC(S)N(Pr-*i*)$_2$	SC(S)N(Pr-*i*)$_2$	89

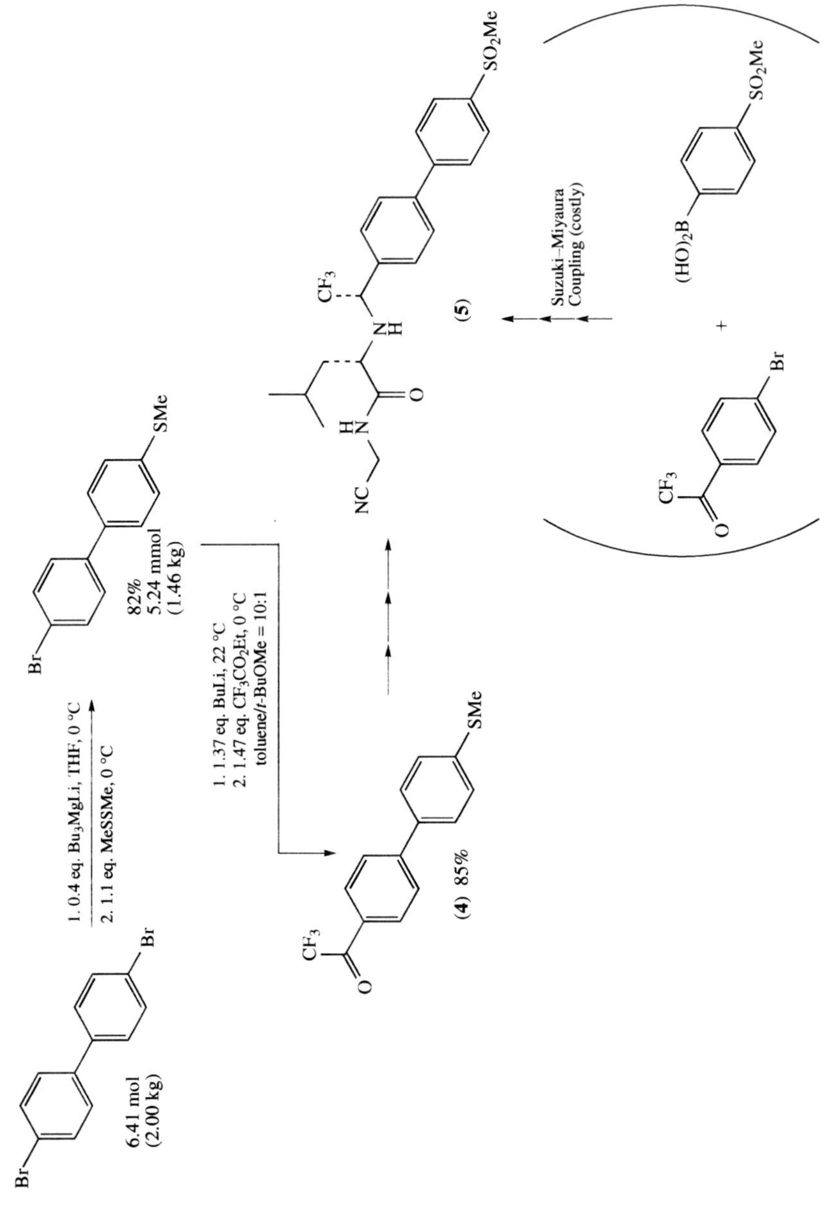
Br
Br
6.41 mol
(2.00 kg)
1. 0.4 eq. Bu$_3$MgLi, THF, 0 °C
2. 1.1 eq. MeSSMe, 0 °C
Br
SMe
82%
5.24 mmol
(1.46 kg)
1. 1.37 eq. BuLi, 22 °C
2. 1.47 eq. CF$_3$CO$_2$Et, 0 °C
toluene/t-BuOMe = 10:1
CF$_3$
O
SMe
(4) 85%
NC
H
N
O
N
H
CF$_3$
SO$_2$Me
(5)
Suzuki–Miyaura
Coupling (costly)
CF$_3$
O
Br
+
(HO)$_2$B
SO$_2$Me

SCHEME 11

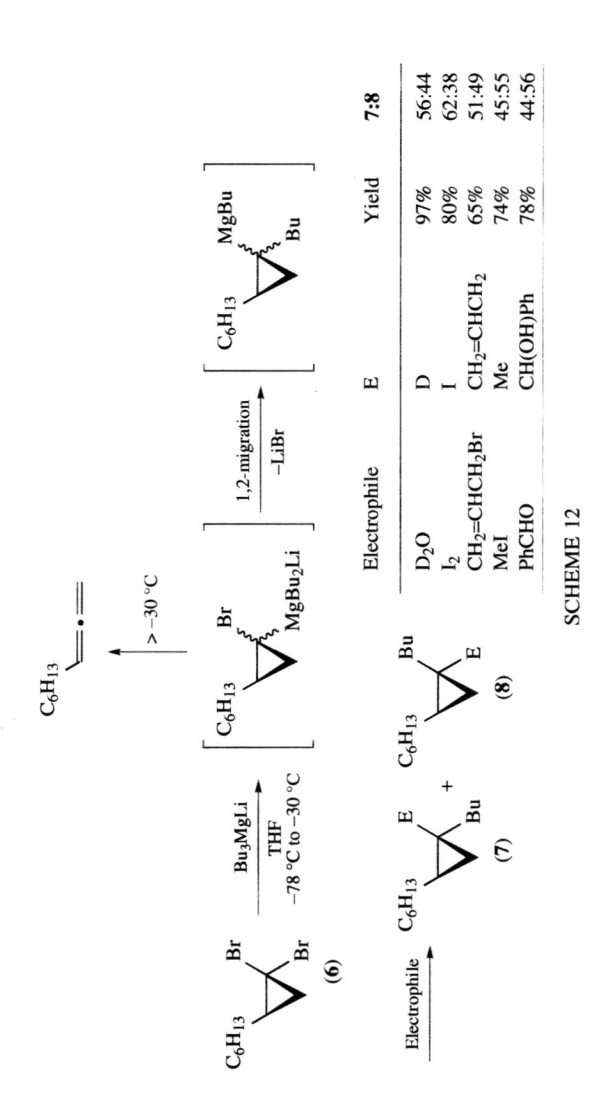

SCHEME 12

Electrophile	E	Yield	**7:8**
D$_2$O	D	97%	56:44
I$_2$	I	80%	62:38
CH$_2$=CHCH$_2$Br	CH$_2$=CHCH$_2$	65%	51:49
MeI	Me	74%	45:55
PhCHO	CH(OH)Ph	78%	44:56

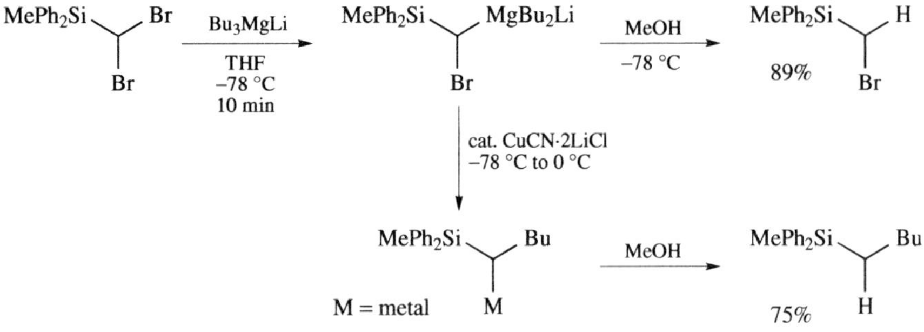

SCHEME 13

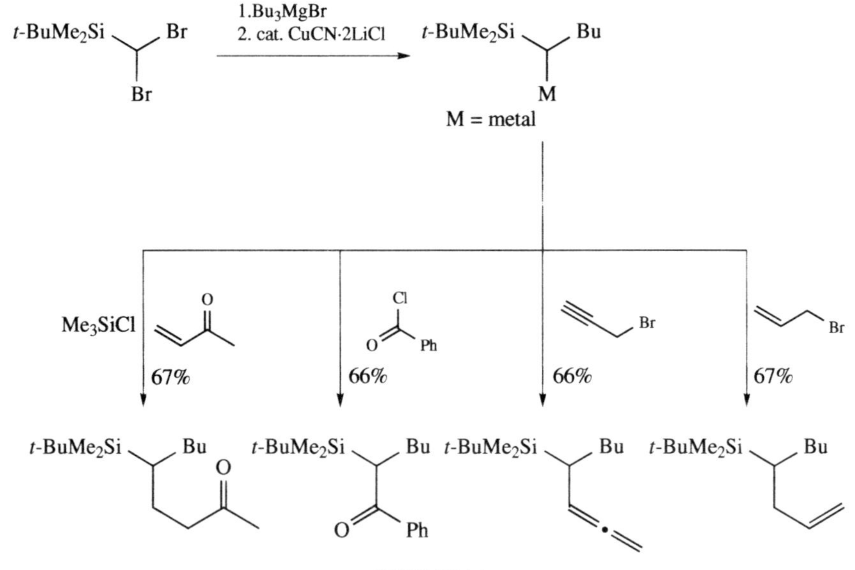

SCHEME 14

The 1-silylpentylmetal species thus formed could react with allyl bromide, propargyl bromide, acid chlorides and α,β-unsaturated ketones (Scheme 14).

Lithium tris(sec-butyl)magnesate underwent bromine–magnesium exchange, which was followed by 1,2-migration in the absence of CuCN•2LiCl at room temperature (Scheme 15). However, CuCN•2LiCl was essential for the acylation and allylation.

The reactions of dibromomethylsilanes with lithium trimethylmagnesate proceeded via a reaction course different from that with lithium tributylmagnesate[42, 44]. One of the two bromine atoms was substituted by the methyl group, and the other bromine atom remained intact (Table 15). Dibromo compounds such as dibromomethylsilane (entry 1), 1,1-dibromoethylsilane (entry 2) and dibromodisilylmethanes (entries 3–8) were transformed into the corresponding monomethylated products in high yields, regardless of the bulkiness of the silyl groups.

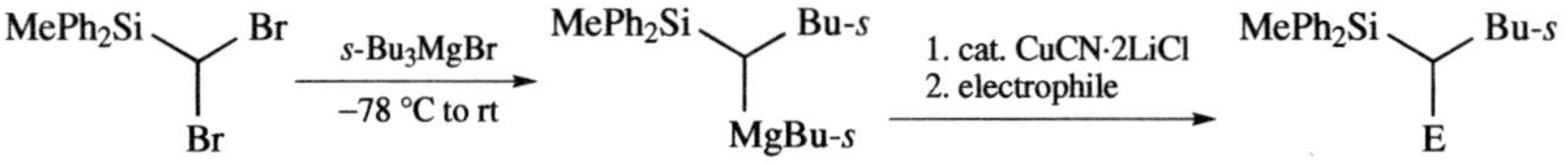

15. The chemistry of organomagnesium ate complexes

MePh₂Si-C(Br)(Br) →[s-Bu₃MgBr, −78 °C to rt] MePh₂Si-C(Bu-s)(MgBu-s) →[1. cat. CuCN·2LiCl; 2. electrophile] MePh₂Si-C(Bu-s)(E)

Electrophile	E	Yield (%)
CH₂=CHCH₂Br	CH₂=CHCH₂	75
n-C₃H₇COCl	n-C₃H₇CO	72
PhCOCl	PhCO	71

SCHEME 15

TABLE 15. Monomethylation with lithium trimethylmagnesate

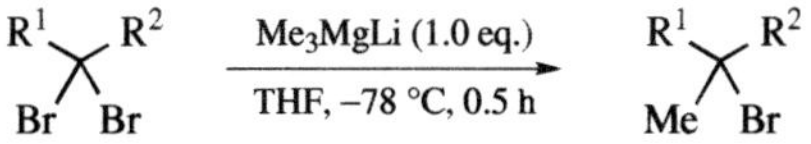

R¹R²C(Br)(Br) →[Me₃MgLi (1.0 eq.), THF, −78 °C, 0.5 h] R¹R²C(Me)(Br)

Entry	R¹	R²	Yield (%)
1	Ph₂MeSi	H	98
2	Ph₂MeSi	Me	89
3	Ph₂MeSi	Me₃Si	93
4	Et₃Si	Et₃Si	90
5	t-BuMe₂Si	Me₃Si	82
6	Me₃Si	Me₃Si	89
7	PhMe₂Si	PhMe₂Si	80
8	Ph₂MeSi	Ph₂MeSi	90

Lithium tributylmagnesate induced iodine–magnesium exchange reaction of 5-alkoxy-3-iodomethyl-1-oxacyclopentanes (Scheme 16)[45]. A following intramolecular nucleophilic substitution led to construction of a cyclopropane with concomitant opening of the oxacyclopentane ring.

Suzuki and coworkers found an application of magnesate reagents in the stereoselective synthesis of (−)-gallocatechin (Scheme 17)[46]. They examined chemoselective iodine–metal exchange of **9** followed by intramolecular cyclization by using various metalation agents. The use of organolithium reagents resulted in limited success. The metalation with isopropylmagnesium chloride resulted in slow halogen–metal exchange at −40 °C, yet afforded no cyclic product **10**. With isopropylmagnesium chloride, uncyclized product **11** was completely deuteriated, which suggests that the corresponding arylmagnesium reagent has a long lifetime. The use of a magnesate reagent, Ph₃MgLi, improved the yield of **10**, compared to isopropylmagnesium chloride. Almost quantitative transformation of **9** to **10** was attained by a combined use of Ph₃MgLi and HMPA.

Trost and coworkers employed a magnesate reagent in the formal synthesis of fostriecin (Scheme 18)[47]. Stereoselective addition of the corresponding alkenylmagnesium to α-alkoxyketone proceeded smoothly in 75% yield with more than 20:1 diastereoselectivity.

Magnesate reagents are reactive enough to enable magnesiations on polymer beads. Schreiber and coworkers reported diversity-oriented synthesis of biaryl-containing medium rings using a one bead/one stock solution platform[48]. For the diversity-oriented synthesis of biaryl-containing medium rings in an atropdiastereoselective fashion, they investigated the development of the oxidation of organocuprates. The reactions were performed on

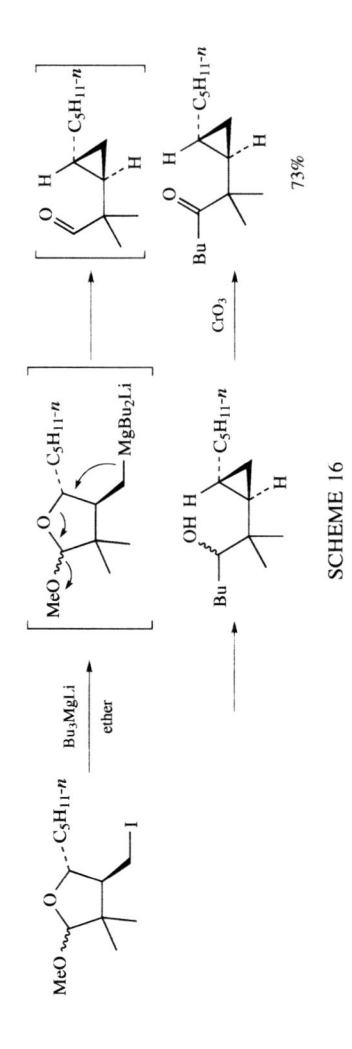

SCHEME 16

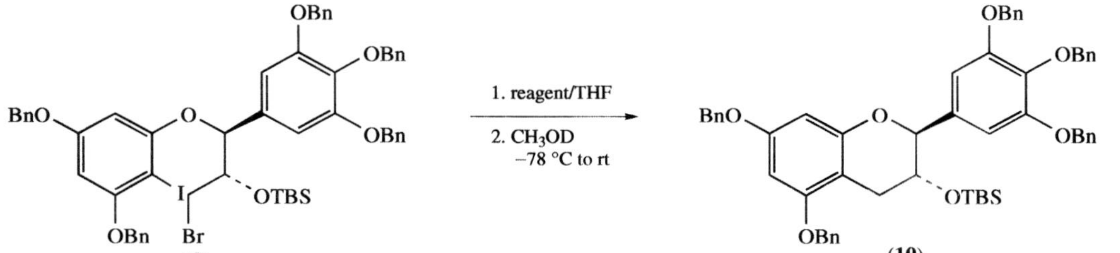

Reagent	**10**	**11** (%D)	**9** (recovery)
BuLi (1.0 eq.)	77%	5% (0)	6%
t-BuLi (2.0 eq.)	53%	–	–
i-PrMgCl (1.2 eq.)	–	28% (100)	72%
Bu_3MgLi (1.2 eq.)	84%	12% (0)	–
Ph_3MgLi (1.2 eq.)	76%	14% (86)	–
Ph_3MgLi (2.5 eq.), HMPA (10 eq.)	96%	–	–

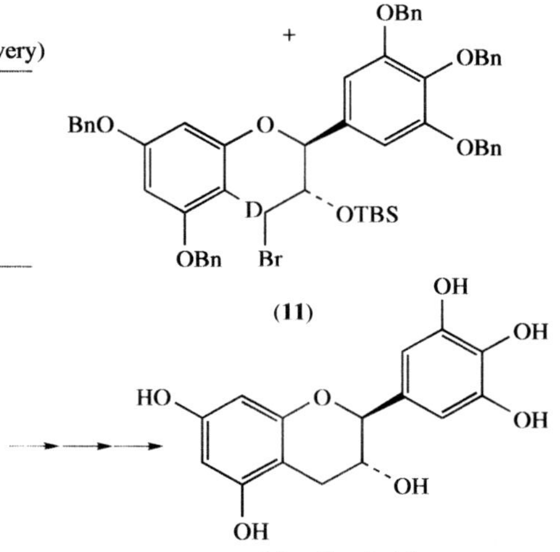

(9) (10) (11)

(–)-gallocatechin

SCHEME 17

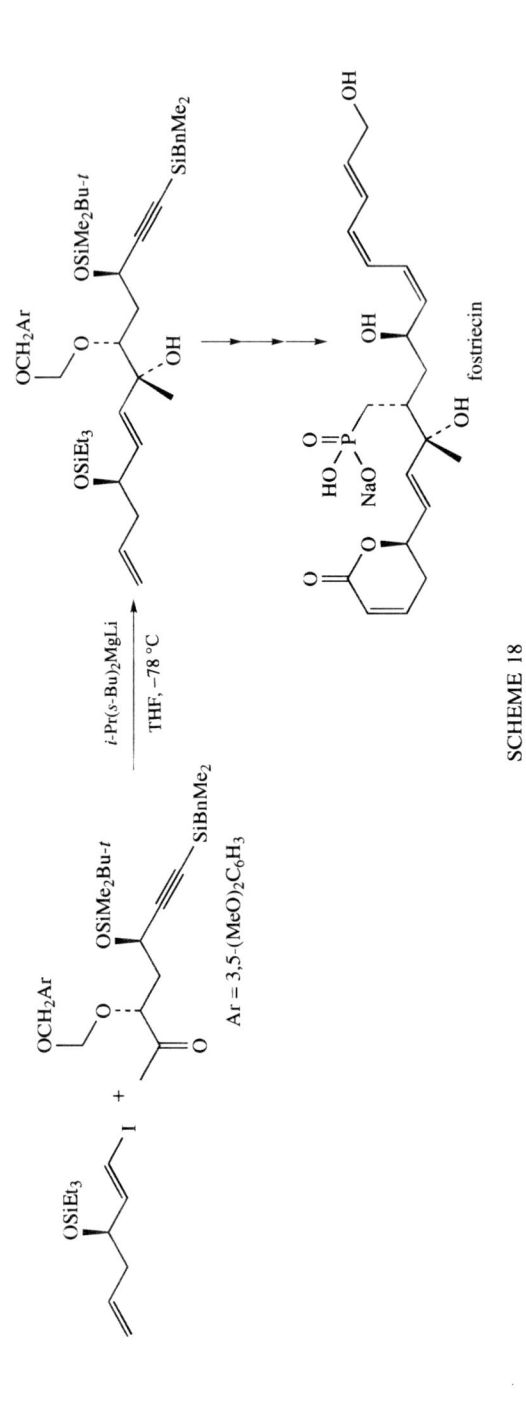

SCHEME 18

polystyrene beads by metalating polymer-supported aryl bromides with lithium dibutylisopropylmagnesate, followed by transmetalation with copper and then oxidation with 1,3-dinitrobenzene (equation 38).

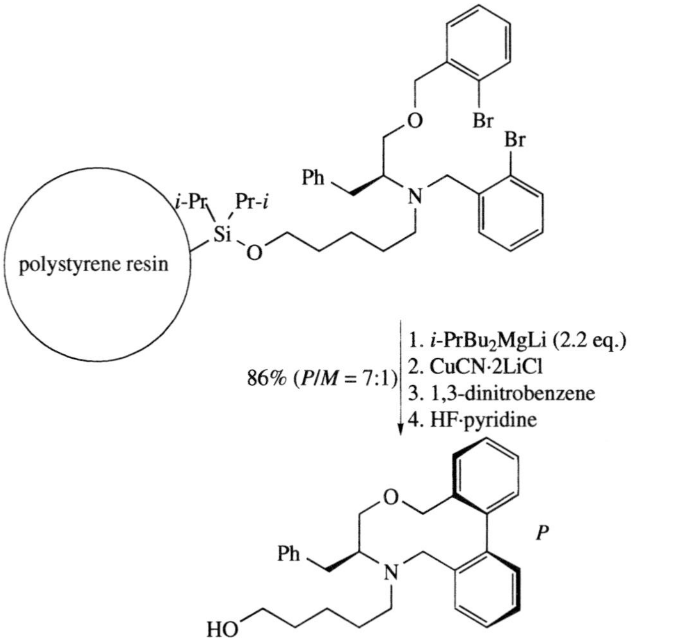

$$86\% \ (P/M = 7:1) \quad \begin{array}{l} 1.\ i\text{-PrBu}_2\text{MgLi (2.2 eq.)} \\ 2.\ \text{CuCN·2LiCl} \\ 3.\ 1,3\text{-dinitrobenzene} \\ 4.\ \text{HF·pyridine} \end{array} \tag{38}$$

Polystyrene beads which are composed of 74% styrene, 25% 4-bromostyrene and 1% divinylbenzene were completely metalated with lithium dibutylisopropylmagnesate (Scheme 19)[49]. The polymagnesiated polystyrene then reacted with a variety of electrophiles to yield high quality solid-supported reagents. For instance, the use of chlorodiphenylphosphine as an electrophile generated polystyrene beads (150–600 μm) having diphenylphosphinophenyl groups. It is noteworthy that treatment with isopropylmagnesium chloride or butyllithium alone resulted in incomplete functionalization of the beads. An additional example is the synthesis of diisopropylsilane-functionalized polystyrene and its use for covalently attaching alcohols onto the polystyrene solid-support. The chemical stability was comparable to that of a triisopropylsilyl protecting group.

Synthesis of copolysiloxanes **14**, a candidate for dental and medical devices and sensors, required the preparation of the substituted allylbenzene **13** (Scheme 20)[50]. However, attempted cross-coupling reaction of **12** with allylzinc bromide resulted in failure. Alternatively, the magnesiation of **12** with lithium tributylmagnesate followed by the addition of allyl bromide provided **13** in excellent yield.

IV. ADDITIONAL IMPORTANT CHEMISTRY OF ORGANOMAGNESATES

A. Alkoxydialkylmagnesate R₂MgOR

Hanawalt and Richey observed that additions of alkali-metal alkoxides to dialkylmagnesium led in some reactions to behavior resembling that of trialkylmagnesates[51]. This includes enhanced reactivities in addition to pyridine leading to 4- or 2-alkyl-substituted

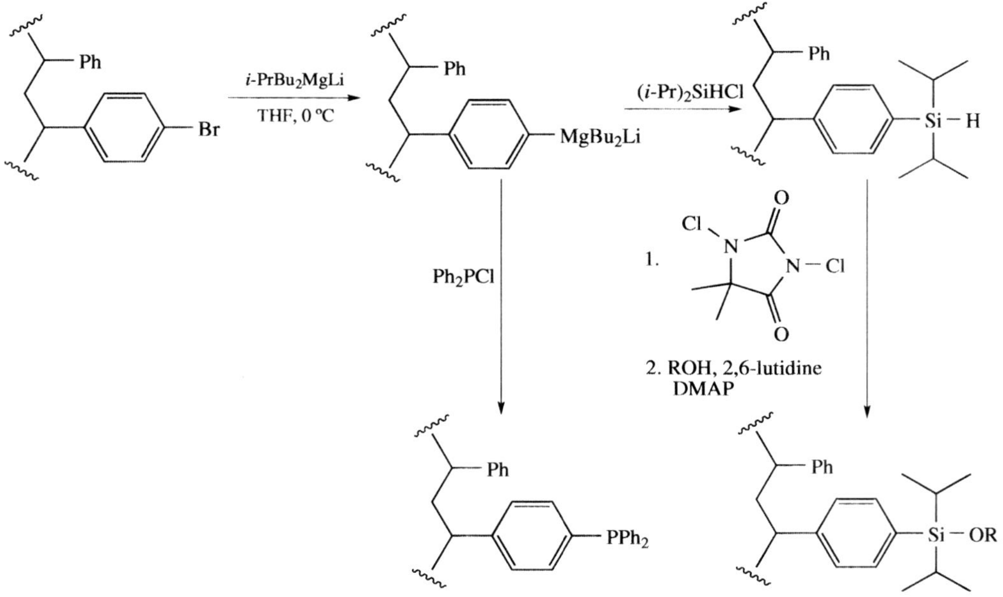

SCHEME 19

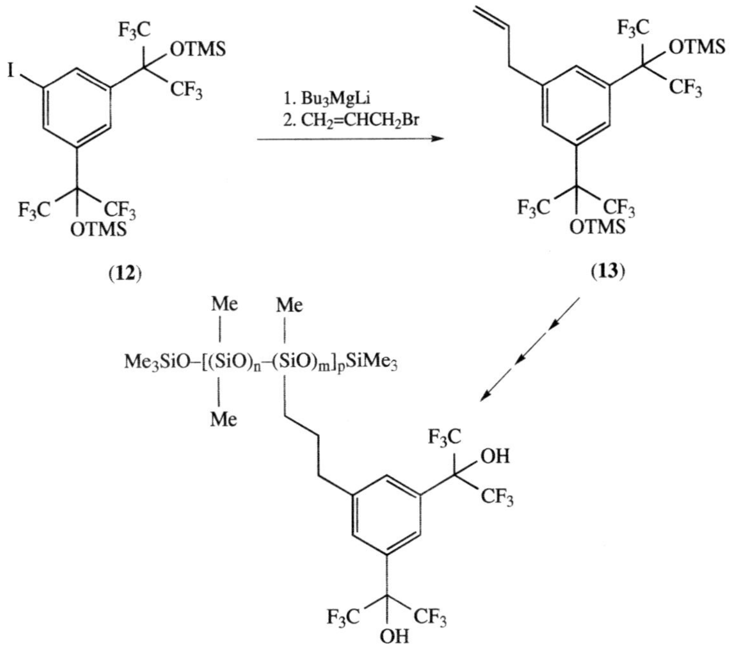

SCHEME 20

pyridines, additions to ketones in which the accompanying reduction of the ketones are suppressed[52] and halogen–metal exchange with aryl halides[53]. NMR studies revealed the formation of the dinuclear magnesate complexes from the dialkylmagnesium and the alkoxides (equation 39).

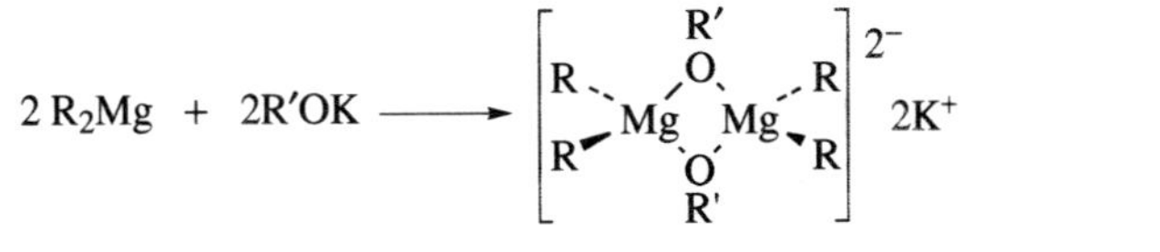

$$2\,R_2Mg + 2R'OK \longrightarrow \left[\begin{array}{c} R'\\ R\cdots\overset{O}{\underset{\underset{R'}{O}}{Mg\ Mg}}\cdots R \\ R \end{array} \right]^{2-} 2K^+ \quad (39)$$

B. Lithium Tris(2,2,6,6-tetramethylpiperidino)magnesate (TMP)$_3$MgLi

The title reagent has no carbon–magnesium bonds. However, the 'inorganic' magnesate reagent is to be noted since it showed excellent reactivity in deprotonation reactions. The reagent is readily available by mixing lithium tributylmagnesate and three molar equivalents of 2,2,6,6-tetramethylpiperidine, and has much less nucleophilicity than triorganomagnesate[24,25]. Triamidomagnesates are structurally intriguing since they are recognized as 'inverse crown ethers.'[54]. Zinc analogs of the reagent are also useful for deprotonation reactions[55].

C. Combination of Isopropylmagnesium Chloride and Lithium Chloride i-PrMgCl•LiCl

Knochel and coworkers showed that i-PrMgCl•LiCl is a useful reagent for the simple and high-yielding preparation of a broad range of functionalized arylmagnesium reagents starting from readily available aryl bromides (Chapter X by Knochel)[7,8]. The exchange reactions proceed under noncryogenic conditions and are scalable. The reactivity is comparable to or a little lower than that of triorganomagnesates, and hence a wider variety of functional groups are tolerant. In order to perform unknown halogen–magnesium exchange reactions, i-PrMgCl•LiCl and lithium trialkylmagnesates are the first choices.

V. SUMMARY

Triorganomagnesates are establishing their positions in modern organic synthesis. Grignard reagents RMgX have a long history and outstanding utility and their ate complexation significantly enhanced their reactivity. We believe that magnesates will find as many applications as Grignard reagents in organic synthesis.

VI. REFERENCES

1. G. Wittig, F. J. Meyer and G. Lange, *Justus Liebigs Ann. Chem.*, **571**, 167 (1951).
2. L. M. Seitz and T. L. Brown, *J. Am. Chem. Soc.*, **88**, 4140 (1966); **89**, 1602 (1967).
3. D. Thoennes and E. Weiss, *Chem. Ber.*, **111**, 3726 (1978).
4. K. M. Waggoner and P. P. Power, *Organometallics*, **11**, 3209 (1992).
5. A. D. Pajerski, D. M. Kushlan, M. Parvez and H. G. Richey, Jr., *Organometallics*, **25**, 1206 (2006).
6. R. E. Mulvey, *Chem. Commun.*, 1049 (2001).
7. A. Krasovskiy and P. Knochel, *Angew. Chem. Int. Ed.*, **43**, 3333 (2004); A. Krasovskiy, B. F. Straub and P. Knochel, *Angew. Chem. Int. Ed.*, **45**, 159 (2006).
8. P. Knochel, A. Krasovskiy and I. Sapountzis, *Polyfunctional Magnesium Organometallics for Organic Synthesis*, in *Handbook of Functionalized Organometallics* (Ed. P. Knochel), Wiley-VCH, Weinheim, 2005.
9. D. Hauk, S. Lang and A. Murso, *Org. Proc. Res. Devel.*, **10**, 733 (2006).
10. M. Geissler, J. Kopf and E. Weiss, *Chem. Ber.*, **122**, 1395 (1989).
11. K. Kitagawa, A. Inoue, H. Shinokubo and K. Oshima, *Angew. Chem. Int. Ed.*, **39**, 2481 (2000); A. Inoue, K. Kitagawa, H. Shinokubo and K. Oshima, *J. Org. Chem.*, **66**, 4333 (2001).
12. G. E. Coates and J. A. Heslop, *J. Chem. Soc., A*, 514 (1968).
13. E. P. Squiller, R. R. Whittle and H. G. Richey, Jr., *J. Am. Chem. Soc.*, **107**, 432 (1985).
14. A. D. Pajerski, M. Parvez and H. G. Richey, Jr., *J. Am. Chem. Soc.*, **110**, 2660 (1988).
15. D. B. Malpass and J. F. Eastham, *J. Org. Chem.*, **38**, 3718 (1973).
16. E. C. Ashby, L.-C. Chao and J. Laemmle, *J. Org. Chem.*, **39**, 3258 (1974).
17. H. G. Richey, Jr. and B. A. King, *J. Am. Chem. Soc.*, **104**, 4672 (1982).
18. H. G. Richey, Jr. and J. Farkas, Jr., *Tetrahedron Lett.*, **26**, 275 (1985); H. G. Richey, Jr. and J. Farkas, Jr., *Organometallics*, **9**, 1778 (1990).
19. M. Hatano, T. Matsumura and K. Ishihara, *Org. Lett.*, **7**, 573 (2005).
20. J. G. Sosnicki, *Tetrahedron Lett.*, **46**, 4295 (2005).
21. J. G. Sosnicki, *Tetrahedron Lett.*, **47**, 6809 (2006).
22. M. Yasuda, M. Ide, Y. Matsumoto and M. Nakata, *Synlett*, 899 (1997); *Bull. Chem. Soc. Jpn.*, **71**, 1417 (1998).
23. M. Ide, M. Yasuda and M. Nakata, *Synlett*, 936 (1998); M. Ide and M. Nakata, *Bull. Chem. Soc. Jpn.*, **72**, 2491 (1999).
24. H. Awad, F. Mongin, F. Trécourt, G. Quéguiner, F. Marsais, F. Blanco, B. Abarca and R. Ballesteros, *Tetrahedron Lett.*, **45**, 6697 (2004).
25. H. Awad, F. Mongin, F. Trécourt, G. Quéguiner and F. Marsais, *Tetrahedron Lett.*, **45**, 7873 (2004).

26. O. Bayh, H. Awad, F. Mongin, C. Hoarau, F. Trécourt, G. Quéguiner, F. Marsais, F. Blanco, B. Abarca and R. Ballesteros, *Tetrahedron*, **61**, 4779 (2005).
27. O. Bayh, H. Awad, F. Mongin, C. Hoarau, L. Bischoff, F. Trécourt, G. Quéguiner, F. Marsais, F. Blanco, B. Abarca and R. Ballesteros, *J. Org. Chem.*, **70**, 5190 (2005).
28. F. Mongin, A. Bucher, J. P. Bazureau, O. Bayh, H. Awad and F. Trécourt, *Tetrahedron Lett.*, **46**, 7989 (2005).
29. A. Inoue and K. Oshima, '*Magnesium in Organic Synthesis*', in *Main Group Metals in Organic Synthesis* (Eds. H. Yamamoto and K. Oshima), Volume 1, Chapter 3.4, Wiley–VCH, Weinheim, 2004; P. Knochel, A. Krasovskiy and I. Sapountzis, '*Functional Magnesium Organometallics for Organic Synthesis*', in *Handbook of Functionalized Organometallics* (Ed. P. Knochel), Volume 1, Chapter 4.2.3, Wiley–VCH, Weinheim, 2005; B. J. Wakefield, *Organomagnesium Methods in Organic Synthesis*, Chapter 3.2.2, Academic Press, London, 1995.
30. L. Boymond, M. Rottländer, G. Cahiez and P. Knochel, *Angew. Chem. Int. Ed.*, **37**, 1701 (1998).
31. T. Iida, T. Wada, K. Tomimoto and T. Mase, *Tetrahedron Lett.*, **42**, 4841 (2001); T. Mase, I. N. Houpis, A. Akao, I. Dorziotis, K. Emerson, T. Hoang, T. Iida, T. Itoh, K. Kamei, S. Kato, Y. Kato, M. Kawasaki, F. Lang, J. Lee, J. Lynch, P. Maligres, A. Molina, T. Nemoto, S. Okada, R. Reamer, J. Z. Song, D. Tschaen, T. Wada, D. Zewge, R. P. Volante, P. J. Reider and K. Tomimoto, *J. Org. Chem.*, **66**, 6775 (2001).
32. K. Fukuhara, Y. Takayama and F. Sato, *J. Am. Chem. Soc.*, **125**, 6884 (2003).
33. S. Dumouchel, F. Mongin, F. Trécourt and G. Quéguiner, *Tetrahedron Lett.*, **44**, 2033 (2003).
34. S. Dumouchel, F. Mongin, F. Trécourt and G. Quéguiner, *Tetrahedron Lett.*, **44**, 3877 (2003).
35. J. Xu, N. Jain and Z. Sui, *Tetrahedron Lett.*, **45**, 6399 (2004).
36. S. Ito, T. Kubo, N. Morita, Y. Matsui, T. Watanabe, A. Ohta, K. Fujimori, T. Murafuji, Y. Sugihara and A. Tajiri, *Tetrahedron Lett.*, **45**, 2891 (2004).
37. F. D. Therkelsen, M. Rottländer, N. Thorup and E. B. Pedersen, *Org. Lett.*, **6**, 1991 (2004).
38. E. Masson, E. Marzi, F. Cottet, C. Bobbio and M. Schlosser, *Eur. J. Org. Chem.*, 4393 (2005).
39. S. Kii, A. Akao, T. Iida, T. Mase and N. Yasuda, *Tetrahedron Lett.*, **47**, 1877 (2006).
40. F. Buron, N. Plé, A. Turck and F. Marsais, *Synlett*, 1586 (2006).
41. S. J. Dolman, F. Gosselin, P. D. O'Shea and I. W. Davis, *Tetrahedron*, **62**, 5092 (2006).
42. A. Inoue, J. Kondo, H. Shinokubo and K. Oshima, *Chem. Eur. J.*, **8**, 1730 (2002).
43. J. Kondo, A. Inoue, H. Shinokubo and K. Oshima, *Angew. Chem. Int. Ed.*, **40**, 2085 (2001).
44. A. Inoue, J. Kondo, H. Shinokubo and K. Oshima, *Chem. Lett.*, **30**, 956 (2001).
45. T. Tsuji, T. Nakamura, H. Yorimitsu, H. Shinokubo and K. Oshima, *Tetrahedron*, **60**, 973 (2004).
46. T. Higuchi, K. Ohmori and K. Suzuki, *Chem. Lett.*, **35**, 1006 (2006).
47. B. M. Trost, M. U. Frederiksen, J. P. N. Papillon, P. E. Harrington, S. Shin and B. T. Shireman, *J. Am. Chem. Soc.*, **127**, 3666 (2005).
48. D. R. Spring, S. Krishnan, H. E. Blackwell and S. L. Schreiber, *J. Am. Chem. Soc.*, **124**, 1354 (2002); S. Krishnan and S. L. Schreiber, *Org. Lett.*, **6**, 4021 (2004).
49. G. L. Thomas, M. Ladlow and D. R. Spring, *Org. Biomol. Chem.*, **2**, 1679 (2004); G. L. Thomas, C. Böhner, M. Ladlow and D. R. Spring, *Tetrahedron*, **61**, 12153 (2005).
50. B. Boutevin, M. Gaboyard, F. Guida-Pietrasanta, L. Hairault, B. Lebret and E. Pasquinet, *J. Polym. Sci., A, Polym. Chem.*, **41**, 1400 (2003).
51. E. M. Hanawalt and H. G. Richey Jr., *J. Am. Chem. Soc.*, **112**, 4983 (1990).
52. J. E. Chubb and H. G. Richey, Jr., *Organometallics*, **21**, 3661 (2002).
53. J. Farkas, Jr., S. J. Stoudt, E. M. Hanawalt, A. D. Pajerski and H. G. Richey, Jr., *Organometallics*, **23**, 423 (2004).
54. R. E. Mulvey, *Organometallics*, **25**, 1060 (2006).
55. Y. Kondo, M. Shirai, M. Uchiyama and T. Sakamoto, *J. Am. Chem. Soc.*, **121**, 3539 (1999); M. Uchiyama and Y. Kondo, *J. Synth. Org. Chem. Jpn.*, **64**, 1180 (2006).

CHAPTER 16

The chemistry of magnesium carbenoids

TSUYOSHI SATOH

Department of Chemistry, Faculty of Science, Tokyo University of Science; Ichigaya-funagawara-machi 12, Shinjuku-ku, Tokyo 162-0826, Japan
Fax: 81-3-5261-4631; e-mail: tsatoh@rs.kagu.tus.ac.jp

I. INTRODUCTION	718
II. GENERATION OF MAGNESIUM CARBENOIDS	718
A. Generation of Magnesium Carbenoids by Halogen–Magnesium Exchange Reaction	718
B. Generation of Magnesium Carbenoids by Sulfoxide–Magnesium Exchange Reaction	720
1. Generation of racemic magnesium carbenoids	720
2. Generation of optically active magnesium carbenoids	721
III. REACTIONS AND SYNTHETIC USES OF MAGNESIUM CARBENOIDS	722
A. Cyclopropanation of Allylic Alcohols with Magnesium Carbenoids . . .	722
B. Electrophilic Reactions of Magnesium Carbenoids	723
1. Reaction of magnesium carbenoids with Grignard reagents	723
2. Synthesis of secondary chiral Grignard reagents	725
3. Reaction of magnesium carbenoids with α-sulfonyl lithium carbanions	727
4. Reaction of magnesium carbenoids with N-lithio arylamines	728
5. 1,3-Carbon–hydrogen (C,H) insertion reaction of magnesium carbenoids	729
IV. MAGNESIUM CYCLOPROPYLIDENES	735
A. Magnesium Cyclopropylidenes as Intermediates in the Doering–LaFlamme Allene Synthesis	735
B. Nucleophilic Reactions of Magnesium Cyclopropylidenes	738
C. Electrophilic Reactions of Magnesium Cyclopropylidenes	739
V. MAGNESIUM ALKYLIDENE CARBENOIDS	742
A. Generation and Nucleophilic Property of Magnesium Alkylidene Carbenoids, the Fritsch–Buttenberg–Wiechell Rearrangement	742

The chemistry of organomagnesium compounds
Edited by Z. Rappoport and I. Marek   © 2008 John Wiley & Sons, Ltd

B. Electrophilic Reactions of Magnesium Alkylidene Carbenoids 746
   1. One-pot synthesis of tetrasubstituted olefins 746
   2. Synthesis of allenes by alkenylation of magnesium alkylidene carbenoids with α-sulfonyl lithium carbanions 748
   3. Reaction of magnesium alkylidene carbenoids with N-lithio arylamines 748
   4. Reaction of magnesium alkylidene carbenoids with N-lithio nitrogen-containing heterocycles, lithium acetylides and lithium thiolates 755
VI. MAGNESIUM β-OXIDO CARBENOIDS 760
VII. OTHER MAGNESIUM CARBENOIDS 767
VIII. CONCLUSIONS AND PERSPECTIVE 767
IX. REFERENCES 767

I. INTRODUCTION

Carbenes and carbenoids[1] have long been recognized as a highly reactive species and are frequently used as intermediates in organic synthesis[2]. From a synthetic perspective, however, most of the carbenes are relatively short-lived and are too reactive to be controlled. Recently, metal–carbene complexes (or metallocarbenes) were found to be easier to control and are nowadays widely used in organic synthesis[2b].

Carbenoids (**2**) have been generated from alkyl halides (**1**; Y = H or halogen) by hydrogen–metal or halogen–metal exchange reactions (equation 1). Especially, lithium carbenoids (**2**; Metal = Li) were generated from alkyl halides with butyllithium; however, they are so reactive that the H–Li or halogen–Li exchange reaction must be usually conducted below −90 °C. On the other hand, from recent cumulative investigations, magnesium carbenoids (**2**; Metal = MgX) could be generated from alkyl iodides (**1**; Y = I) or sulfoxides (**1**; Y = S(O)Ar) by iodine–magnesium or sulfoxide–magnesium exchange reaction and were found to be much more stable as compared to lithium carbenoids. As a result, magnesium carbenoids can be generated around −78 °C, are relatively easy to handle and present interesting reactivities. In this chapter, generation, properties and synthetic uses of magnesium carbenoids will be discussed.

$$R^2 \underset{X}{\overset{R^1}{-}} C - Y \quad \xrightarrow{\text{H – Metal exchange or Halogen – Metal exchange or Sulfoxide – Metal exchange reaction}} \quad R^2 \underset{X}{\overset{R^1}{-}} C - \text{Metal} \qquad (1)$$

(**1**) Y = H, Halogen, S(O)Ar
    X = Halogen

(**2**) X = Halogen
    Metal = Li, MgX

II. GENERATION OF MAGNESIUM CARBENOIDS

A. Generation of Magnesium Carbenoids by Halogen–Magnesium Exchange Reaction

The halogen–metal exchange is a well-known reaction for the generation of alkyl-, alkenyl- and arylmetals from the corresponding halides and alkyl metals. Especially the bromine– or iodine–lithium exchange reaction is widely used for the preparation of lithium carbanions. Lithium carbenoids have also been generated from polyhaloalkyl compounds by halogen–lithium exchange reaction.

16. The chemistry of magnesium carbenoids

On the other hand, a rather limited number of examples were reported for the generation of magnesium carbenoids via the halogen–magnesium exchange reaction before 2000. Seyferth and coworkers reported the synthesis of bromochloromethylmagnesium chloride (**3**) from chlorodibromomethane in THF by bromine–magnesium exchange with isopropylmagnesium chloride at −95 °C (equation 2)[3]. Diiodomethylmagnesium chloride (**4**) was also derived from triiodomethane with isopropylmagnesium chloride at −85 °C by iodine–magnesium exchange reaction (equation 3). The magnesium carbenoids (**3** and **4**) were found to be sufficiently stable at low temperature and could react with electrophiles. For example, treatment of **3** and **4** with chlorotrimethylsilane resulted in the formation of trimethylsilylbromochloromethane and trimethylsilyldiiodomethane in 63 and 77% yield, respectively.

$$HCBr_2Cl \xrightarrow[\text{THF, }-95\,°C]{i\text{-PrMgCl}} \underset{(\mathbf{3})}{HCBr(Cl)MgCl} \xrightarrow[63\%]{Me_3SiCl} Me_3SiCBr(Cl)H \quad (2)$$

$$HCI_3 \xrightarrow[\text{THF, }-85\,°C]{i\text{-PrMgCl}} \underset{(\mathbf{4})}{HCI_2MgCl} \xrightarrow[77\%]{Me_3SiCl} Me_3SiCI_2H \quad (3)$$

Magnesium carbenoids (**5** and **6**) were generated from a geminal diiodoalkane by diastereoselective iodine–magnesium exchange reaction with isopropylmagnesium halide in THF at −78 °C for 45 min (Scheme 1)[4]. Subsequent reaction of the magnesium

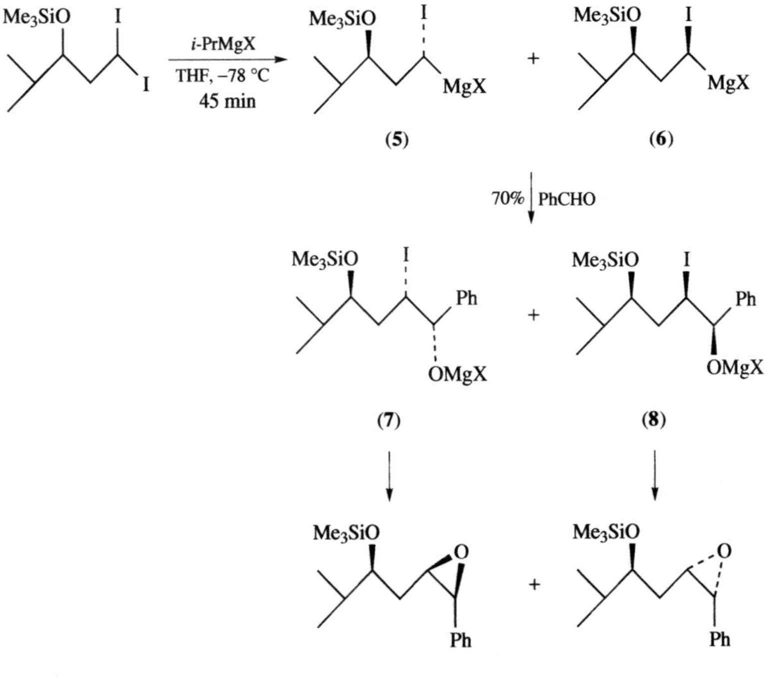

SCHEME 1

carbenoids with benzaldehyde gave cis-disubstituted epoxides (**9** and **10**) via the adducts **7** and **8** in 70% yield as a 4:1 mixture of diastereomers. It was concluded that the configuration of the carbenoid was stable at temperatures up to $-20\,°C$. Further studies on the diastereoselective addition reactions of magnesium carbenoids with benzaldehyde were reported[5]. Mechanistic study of the iodine–magnesium exchange[6] and ^{13}C NMR study of the resulting magnesium carbenoid[7] were reported by Hoffmann and coworkers.

B. Generation of Magnesium Carbenoids by Sulfoxide–Magnesium Exchange Reaction

1. Generation of racemic magnesium carbenoids

Treatment of alkyl aryl sulfoxides with an alkylmetal results in sulfur–aryl (or sulfur–alkyl) bond-cleavage to give a new arylmetal (or alkylmetal) and a new sulfoxide. This reaction is called sulfoxide–metal exchange reaction or ligand exchange reaction of sulfoxides (equation 4)[8]. When the sulfoxide has an alkyl group with a halogen or heteroatom at its α-position, the sulfoxide–metal exchange reaction exclusively takes place between the sulfur–alkyl bond to give carbenoid **11** (equation 5). For example, treatment of α-chlorofluoro sulfoxide (**12**) with phenylmagnesium bromide leads initially to the unstable magnesium carbenoid **13** and then to the alkenylfluoride (equation 6)[9].

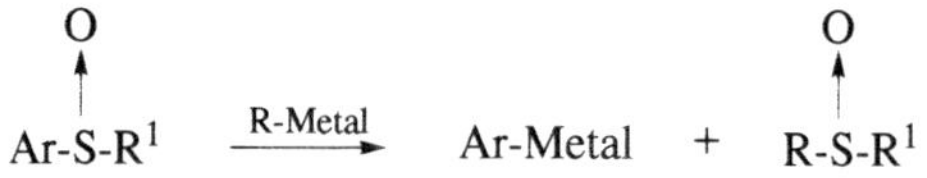

$$Ar\text{-}S\text{-}R^1 \xrightarrow{R\text{-Metal}} Ar\text{-Metal} + R\text{-}S\text{-}R^1 \quad (4)$$

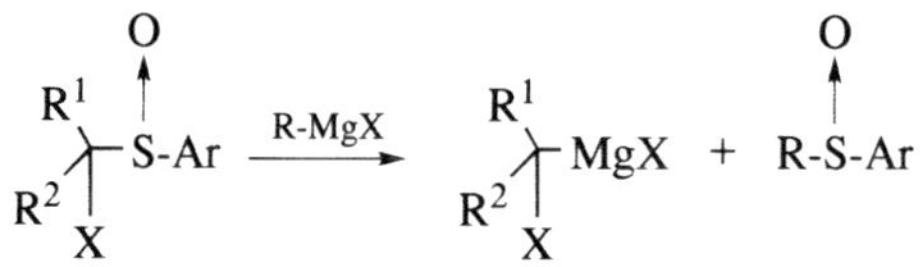

$$\underset{R^2}{\overset{R^1}{\diagdown}}\underset{X}{\overset{}{\diagup}}S\text{-}Ar \xrightarrow{R\text{-MgX}} \underset{R^2}{\overset{R^1}{\diagdown}}\underset{X}{\overset{}{\diagup}}MgX + R\text{-}S\text{-}Ar \quad (5)$$

X= halogen, O, N etc. (**11**)

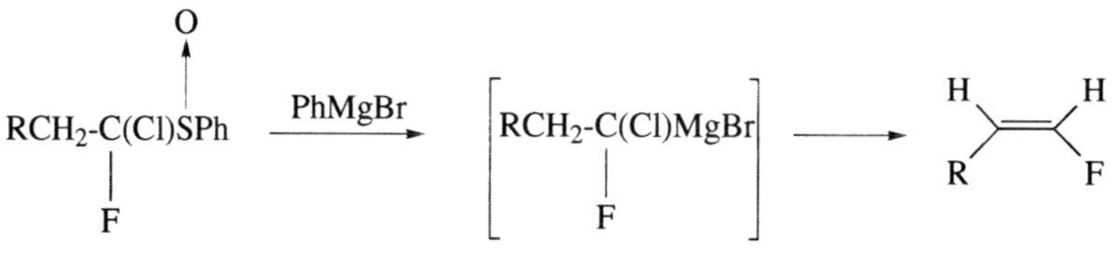

$$RCH_2\text{-}C(Cl)SPh \xrightarrow{PhMgBr} [RCH_2\text{-}C(Cl)MgBr] \longrightarrow \underset{R}{\overset{H}{\diagdown}}C=C\underset{F}{\overset{H}{\diagup}} \quad (6)$$
(with F on first carbon)

(**12**)   (**13**)   Z-only

The sulfoxide–metal exchange reaction using *t*-BuLi and EtMgCl was also investigated (equation 7)[10]. Treatment of α-chloroalkyl *p*-tolyl sulfoxide with *t*-BuLi in THF at $-100\,°C$ for 5 min afforded olefin (**16**) in 86% yield with traces of chloride **15**. This result shows that the intermediate lithium carbenoid **14a** is highly reactive and decomposes rapidly. On the other hand, treatment of the same α-chloroalkyl sulfoxide with EtMgCl at $-78\,°C$ for 5 min gave the chloride **15** in 90% yield with 8% of **16**. Even when the reaction was maintained at $-78\,°C$ for 2 h the chloride **15** was still obtained in 85% yield with 15% of the olefin **16** after hydrolysis. This suggests that magnesium carbenoid **14b** is stable at $-78\,°C$.

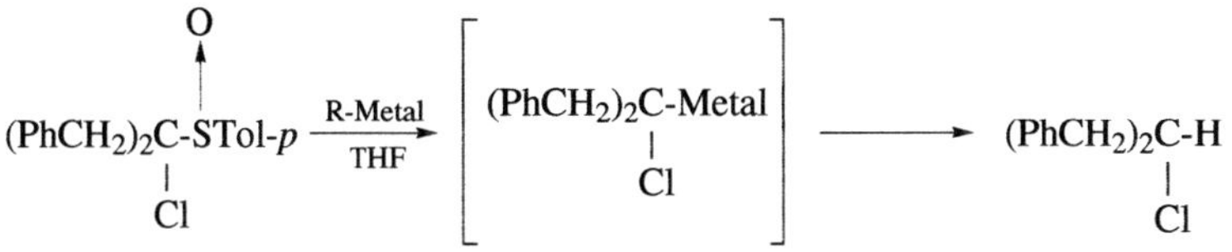

16. The chemistry of magnesium carbenoids

$$(PhCH_2)_2C(O)(STol\text{-}p)(Cl) \xrightarrow[THF]{R\text{-}Metal} [(PhCH_2)_2C(Metal)(Cl)] \longrightarrow (PhCH_2)_2CH(Cl)$$

(14a) Metal = Li
(14b) Metal = MgCl

(15)

+

$PhCH=CHCH_2Ph$

(16)

(7)

Metal	15	16
Li	trace	86%
MgCl	90%	8%

2. Generation of optically active magnesium carbenoids

Optically active α-chloroalkyl aryl sulfoxides can be prepared from optically active alkyl aryl sulfoxides by chlorination with NCS in the presence of K_2CO_3[11]. Hoffmann and Nell reported the preparation of optically active magnesium carbenoid **17**, with over 90% ee from the optically active α-chloroalkyl p-tolyl sulfoxide via a sulfoxide–magnesium exchange reaction (equation 8)[12]. Indeed, 1-chloro-2-phenylethyl p-tolyl sulfoxide of 97% ee was treated with EtMgBr at −78 °C and the resulting magnesium carbenoid **17** was treated with benzaldehyde activated by dimethylaluminum chloride to give an (R,R)-chlorohydrin. The latter was treated with KOH to afford cis-epoxide **18** in 90% yield with 93% ee (equation 8). Three important conclusions were obtained from the above-mentioned investigations. First, the sulfoxide–magnesium exchange reaction can be applied to generate optically enriched Grignard reagent **17**. Second, the sulfoxide–magnesium exchange reaction occurs with retention of configuration at the carbon bearing the chlorine atom. Third, only very slow racemization of magnesium carbenoid **17** took place under the reaction conditions.

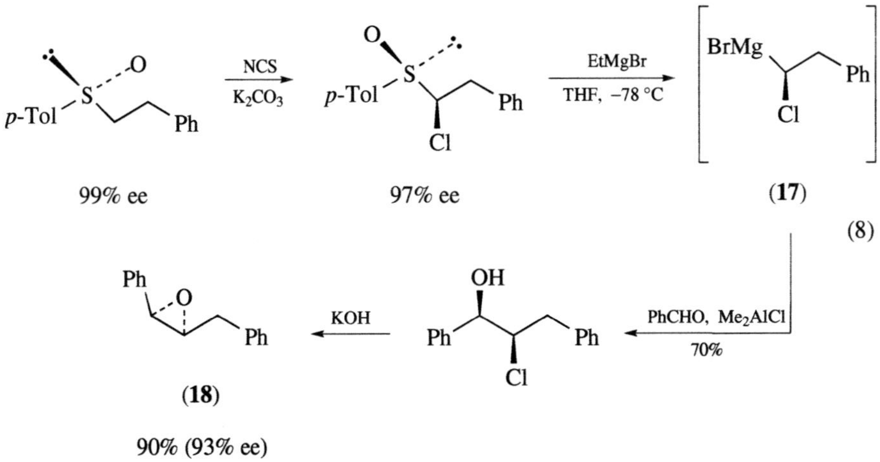

99% ee →(NCS, K_2CO_3)→ 97% ee →(EtMgBr, THF, −78 °C)→ (**17**)

(8)

(**18**) ←(KOH)← chlorohydrin ←(PhCHO, Me_2AlCl, 70%)←

90% (93% ee)

An asymmetric homologation of boronic esters was realized by using chiral magnesium carbenoid **19** (equation 9)[13]. When optically active magnesium carbenoid (**19**), derived from the reaction of α-chloroalkyl p-tolyl sulfoxide with EtMgCl (see equation 8), reacts with boronic ester, the optically active ate-complex **20** was obtained. 1,2-Nucleophilic rearrangement then took place from the complex with inversion of configuration at the migratory terminus to give the boronate intermediate **21**. Then, the optically active secondary alcohol **22** was obtained in 82% ee after oxidation of the boronate **21**.

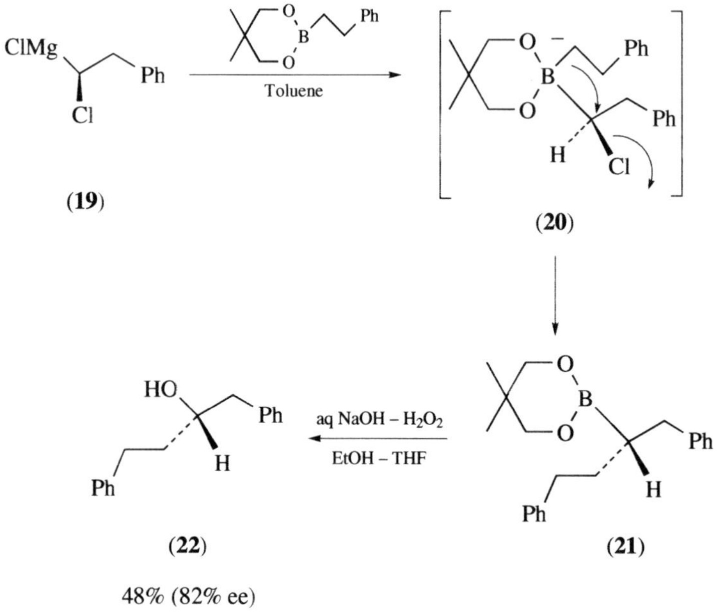

(9)

48% (82% ee)

III. REACTIONS AND SYNTHETIC USES OF MAGNESIUM CARBENOIDS

A. Cyclopropanation of Allylic Alcohols with Magnesium Carbenoids

The Simmons–Smith-type cyclopropanation of olefins[14] is one of the most well-known reactions of carbenes and carbenoids. However, cyclopropanation of simple olefins with magnesium carbenoids is usually very difficult and only cyclopropanation of allylic alcohols was reported[15]. Thus, treatment of allylic alcohols (**23**) in CH_2Cl_2 at $-70\,°C$ with i-PrMgCl and diiodomethane for 48 to 60 h afforded cyclopropanes in up to 82% yield as a mixture of *syn*- and *anti*-isomers. In this reaction, *syn*-isomers were mainly or exclusively obtained (*syn:anti* = 5:1–400:1) (equation 10).

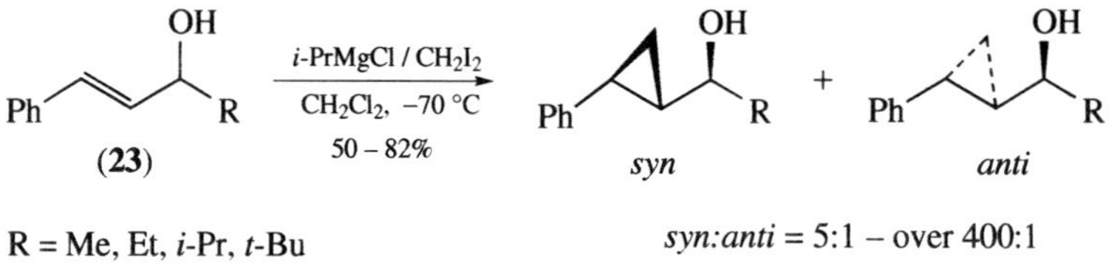

R = Me, Et, i-Pr, t-Bu

syn:anti = 5:1 – over 400:1

(10)

B. Electrophilic Reactions of Magnesium Carbenoids

Carbenoids have both a nucleophilic and an electrophilic nature. This is one of the most striking characteristics of carbenoids. Especially, electrophilic reaction of magnesium carbenoids with carbon and nitrogen nucleophiles has recently received much attention and various new interesting synthetic methods have appeared.

1. Reaction of magnesium carbenoids with Grignard reagents

In 1990, Hahn and Tompkins reported an interesting multi-carbon homologation of alkyl halides by the reaction of magnesium carbenoid with a Grignard reagent (Scheme 2)[16].

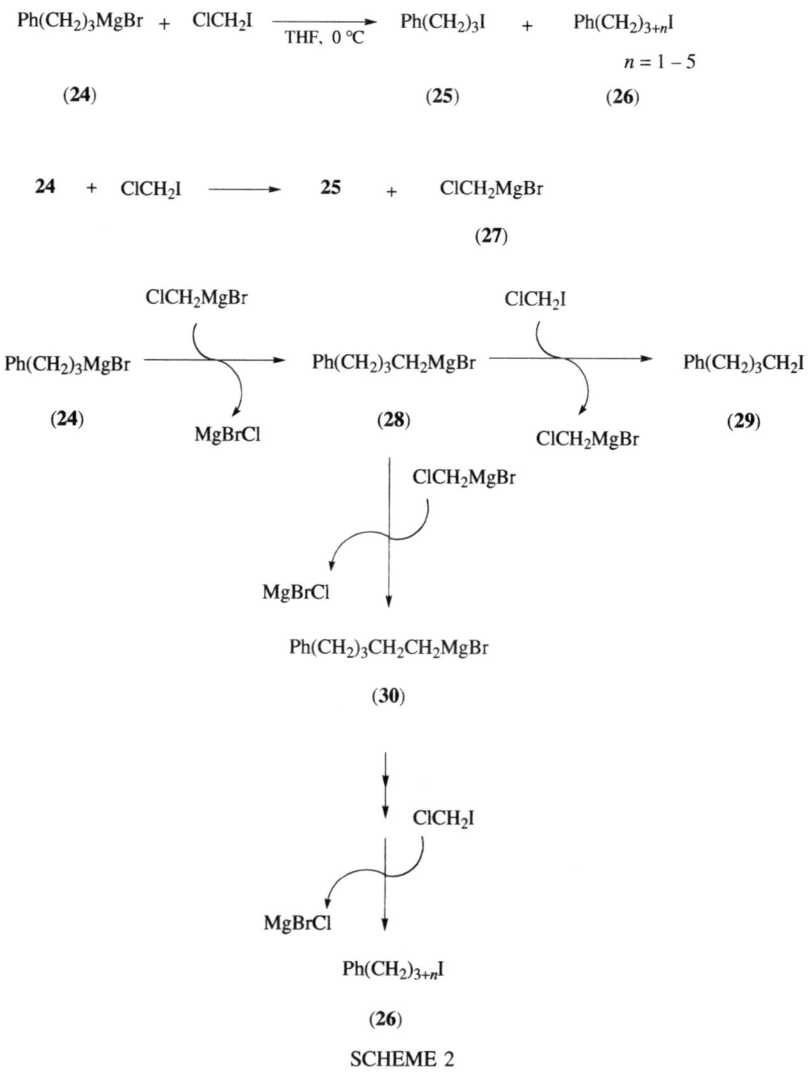

$$Ph(CH_2)_3MgBr + ClCH_2I \xrightarrow{THF,\ 0\ ^\circ C} Ph(CH_2)_3I + Ph(CH_2)_{3+n}I$$

$$n = 1 - 5$$

(24)   (25)   (26)

$$24 + ClCH_2I \longrightarrow 25 + ClCH_2MgBr$$

(27)

$$Ph(CH_2)_3MgBr \xrightarrow[-MgBrCl]{ClCH_2MgBr} Ph(CH_2)_3CH_2MgBr \xrightarrow[-ClCH_2MgBr]{ClCH_2I} Ph(CH_2)_3CH_2I$$

(24)   (28)   (29)

$$\xrightarrow[-MgBrCl]{ClCH_2MgBr} Ph(CH_2)_3CH_2CH_2MgBr$$

(30)

$$\xrightarrow[-MgBrCl]{ClCH_2I} Ph(CH_2)_{3+n}I$$

(26)

SCHEME 2

Thus, reaction of a Grignard reagent (**24**) with chloroiodomethane in THF at 0 °C resulted in the formation of iodide **25** and a mixture of multi-carbon homologated alkyl iodides (**26**). The mechanism of this interesting reaction was thought to be as follows: First, iodine–magnesium exchange reaction between the Grignard reagent **24** and chloroiodomethane occurred to give iodide **25** and magnesium carbenoid **27**. As magnesium carbenoid **27** has an electrophilic nature, the reaction of the Grignard reagent **24** with **27** afforded a one-carbon homologated Grignard reagent **28** and bromochloromagnesium. Reaction of the Grignard reagent **28** with chloroiodomethane gave one-carbon homologated alkyl iodide **29** and magnesium carbenoid **27**. On the other hand, reaction of **28** with magnesium carbenoid **27** resulted in the formation of a two-carbon homologated Grignard reagent **30**.

Formation of homologated reagents by the reaction of a Grignard reagent with magnesium carbenoid was reported (equation 11)[17]. Treatment of a diiodide **31** with i-PrMgCl at −78 °C in THF for 1 h afforded α-iodoalkylmagnesium compound **32**. By warming the reaction mixture to −60 °C, magnesium carbenoid **32** reacted with i-PrMgCl to afford the Grignard reagent **33a**. Hydrolysis gave the corresponding hydrocarbon **34** in 92% yield. Reaction of the Grignard reagent **33a** with ethyl (α-bromomethyl)acrylate gave **35** in 79% yield with some amounts of **36**, which indicated that the intermediate **33a** was contaminated with the 'rearranged' Grignard reagent **33b**. Interestingly, when this reaction was carried out in diisopropyl ether, instead of THF, the 'rearranged' Grignard reagent **33b** was formed as the main product. As Grignard reagents **33a** and **33b** did not interconvert, the Grignard reagents were anticipated to be formed from **32** by reaction with i-PrMgCl in two independent pathways. A plausible mechanism of this very interesting reaction was proposed[17].

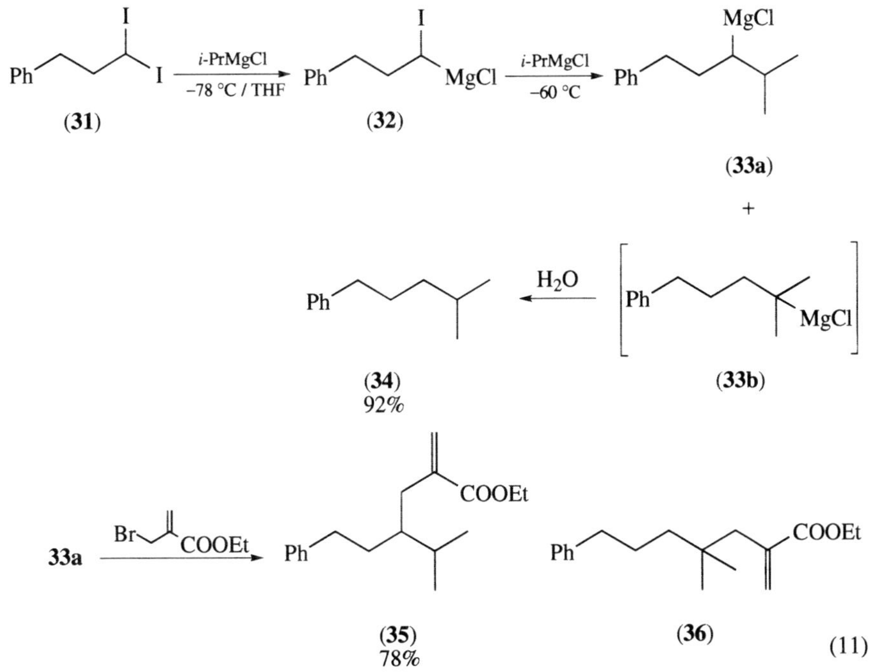

The reaction of a Grignard reagent with magnesium carbenoids, derived from 1-chloroalkyl aryl sulfoxides by sulfoxide–magnesium exchange reaction, was also reported (equation 12)[18]. Thus, treatment of 1-chloroalkyl phenyl sulfoxide (**37**) with 3 eq of EtMgCl at $-80\,°C$ followed by slowly warming the reaction to $-30\,°C$ gave first **39** and then the ethylated product (**40**, $R = CH_2CH_3$) in 80% yield. The sulfoxide–magnesium exchange reaction was found to be completed within 10 min at $-80\,°C$ and the resulting magnesium carbenoid **38** was stable below $-60\,°C$ for a long period of time. By warming the reaction mixture, the magnesium carbenoid **38** reacts with the Grignard reagents in excess to afford first **39** and then the alkylated products **40** in good yields, as indicated in equation 12. Primary and secondary Grignard reagents react well with carbenoid **38**; however, *tert*-BuMgCl did not react with 1-chloroalkyl phenyl sulfoxide (**37**).

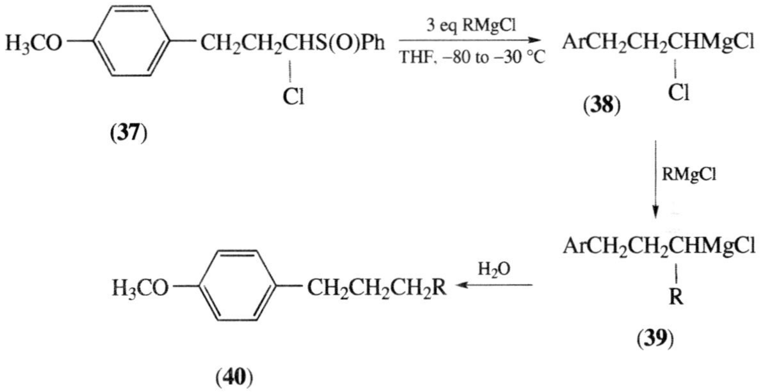

$$H_3CO\text{-Ar-}CH_2CH_2CHS(O)Ph \;\; (Cl) \;\; (\mathbf{37}) \xrightarrow[\text{THF, }-80\text{ to }-30\,°C]{3\text{ eq RMgCl}} ArCH_2CH_2CHMgCl \;\; (Cl) \;\; (\mathbf{38})$$

$$\downarrow RMgCl$$

$$ArCH_2CH_2CHMgCl \;\; (R) \;\; (\mathbf{39}) \xrightarrow{H_2O} H_3CO\text{-Ar-}CH_2CH_2CH_2R \;\; (\mathbf{40}) \quad (12)$$

$R = CH_2CH_3$ (80%), $CH_2(CH_2)_4CH_3$ (87%), $CH(CH_3)_2$ (87%), cyclopentyl (94%), cyclohexyl (75%), *tert*-Bu (0%)

Reaction of phenylthiobromodifluoromethane (**41**) with a Grignard reagent gave the alkyl phenyl sulfide **42** as the main product along with ketenedithioacetal (**43**) (Scheme 3)[19]. The proposed mechanism is as follows: The bromine–magnesium exchange reaction of **41** affords magnesium carbenoid **44**. Then, α-elimination of the dihalomagnesium results in the formation of a carbene-like intermediate, which reacts with RMgCl to give the second magnesium carbenoid **45**. The same reaction takes place to afford the α-sulfur-stabilized Grignard reagent **46**. Acidic work-up of the reaction furnished sulfide **42**. Furthermore, trapping the intermediate **46** with electrophiles, such as iodoalkanes, aldehydes and benzoyl cyanide, were also successfully performed.

2. Synthesis of secondary chiral Grignard reagents

As mentioned above, the configuration of the magnesium carbenoid is rather stable at low temperature; chiral Grignard reagents having over 90% ee could be generated from optically active 1-chloroalkyl aryl sulfoxides (Scheme 4)[20–22].

Treatment of optically pure 1-chloroalkyl aryl sulfoxide **47** with excess EtMgCl gave initially the optically active magnesium carbenoid **19** via a sulfoxide–magnesium exchange reaction. The electrophilic reaction of carbenoid **19** with EtMgCl gave the optically active secondary Grignard reagent **48** with inversion of configuration at the chiral carbon

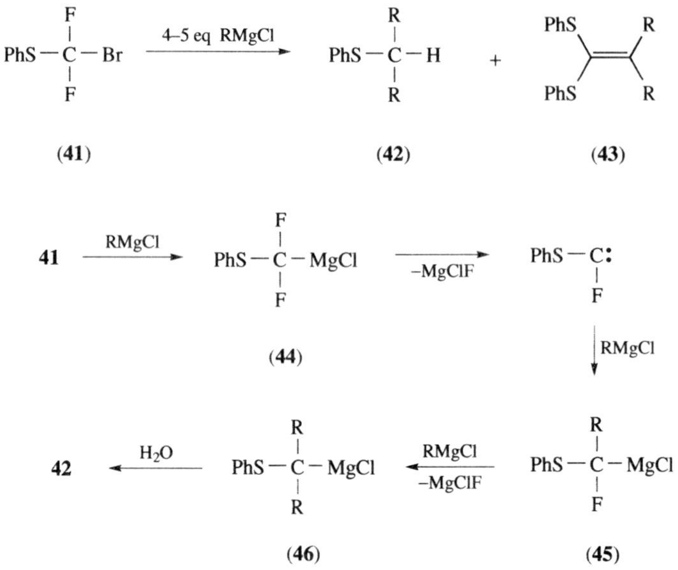

SCHEME 3

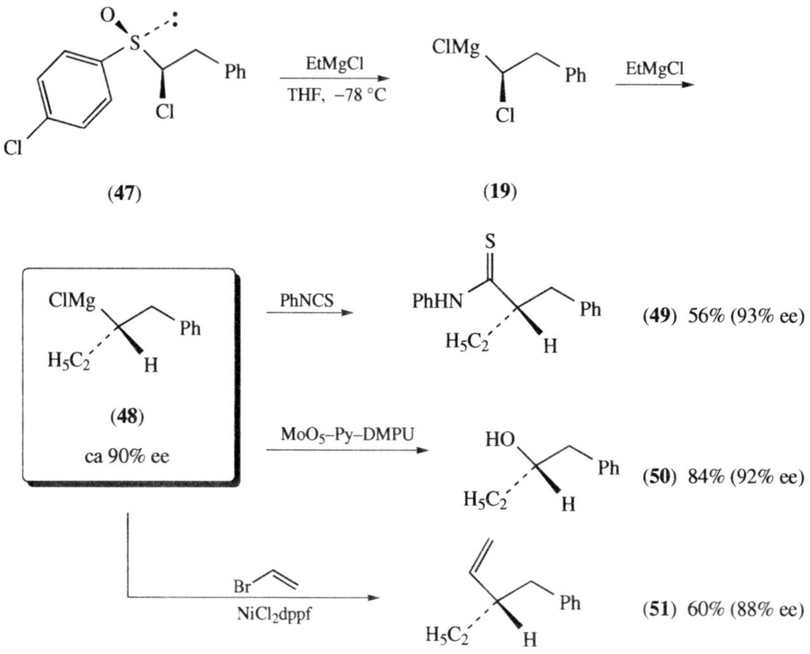

SCHEME 4

center. Quenching this reaction with phenylisothiocyanate gave thioamide **49** in 56% yield with 93% ee. Quite interestingly, from this experiment it appeared that the secondary Grignard reagent **48** is configurationally stable at −78 °C. Oxidation of **48** with molybdenum peroxide gave alcohol **50** with retention of configuration and the enantiomeric purity of **48** was retained. Kumada–Corriu coupling of **48** with vinyl bromide in the presence of Ni-catalyst gave the coupling product **51** with full retention of the configuration[21].

3. Reaction of magnesium carbenoids with α-sulfonyl lithium carbanions

Magnesium carbenoids react not only with Grignard reagents but also with other carbanions such as α-sulfonyl lithium carbanions to afford olefins (equation 13)[23]. Thus, chloromethylmagnesium chloride (**53**), generated in THF at −78 °C from chloroiodomethane and i-PrMgCl, reacts with α-sulfonyl lithium carbanion (**52**) to give the olefin **55**. The mechanism of this reaction was thought to be as described in equation 13. The electrophilic reaction of carbenoid **53** with the carbanion **52** gave a new Grignard reagent having a sulfonyl group at the β-position (**54**). β-Elimination then took place to afford olefin **55**. Moderate to good yields of the olefins **55**, which are summarized in Table 1, were obtained by this reaction.

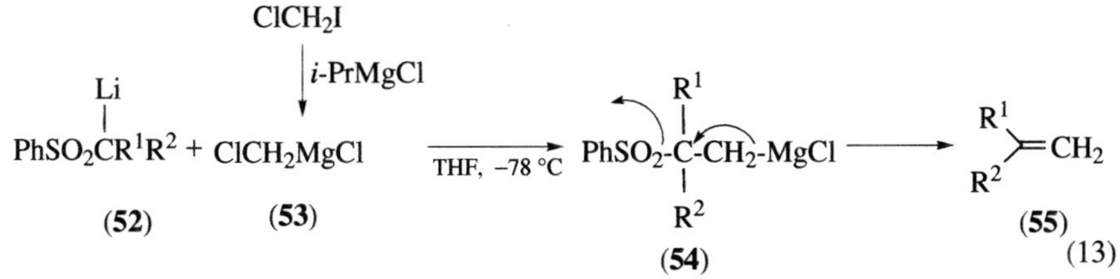

$$\underset{(52)}{PhSO_2CR^1R^2} + \underset{(53)}{ClCH_2MgCl} \xrightarrow[THF,\ -78\ °C]{ClCH_2I \mid i\text{-PrMgCl}} \underset{(54)}{PhSO_2-\underset{R^2}{\overset{R^1}{C}}-CH_2-MgCl} \longrightarrow \underset{(55)}{\overset{R^1}{\underset{R^2}{>}}=CH_2} \quad (13)$$

Similarly, the reaction of 1-chloropentylmagnesium chloride (**56**), derived from 1-chloro-1-iodopentane, with α-sulfonyl lithium carbanions (**52**) affords 1,2-di- or 1,1,2-trisubstituted olefins (**57**) in moderate to good yields (equation 14). This reaction represents an elegant preparation of olefins from sulfones.

Optically pure methylenecyclopropanes **59** were synthesized from cyclopropyl sulfones **58** and the carbenoid, iodomethylmagnesium chloride, in moderate to good yields (equation 15)[24].

The olefin formation described above can be conducted with magnesium carbenoids derived from 1-chloroalkyl phenyl sulfoxides[18]. An example is shown in equation 16.

TABLE 1. Reaction of α-sulfonyl lithium carbanions (**52**) with chloromethylmagnesium chloride (**53**)

α-Sulfonyl lithium carbanion (**52**)		Olefin (**55**)
R^1	R^2	Yield (%)
$CH_2(CH_2)_9CH_3$	H	82
$CH_2(CH_2)_8COOH$	H	60
$CH_2(CH_2)_4CH_3$	CH_3	68
CH_3	Ph	76
$-(CH_2)_6-$		76
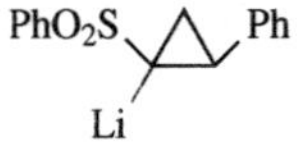 PhO₂S⟨△⟩Ph, Li		 ⟨△⟩=Ph, 65

Thus, magnesium carbenoid **38**, generated from 1-chloroalkyl phenyl sulfoxide (**37**) in THF at $-65\,°C$ with 2.8 eq of i-PrMgCl, reacts with α-sulfonyl lithium carbanion to lead to 1,2-di- and 1,1,2-trisubstituted olefins (**60**). Yields are better in such conditions as compared to the reaction described in equation 14.

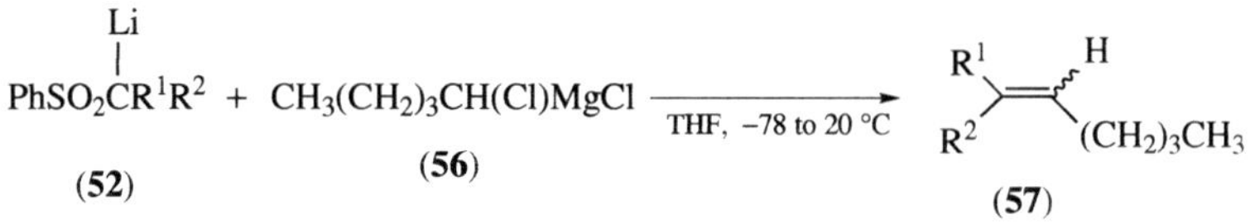

$$\underset{(52)}{PhSO_2\overset{\underset{|}{Li}}{C}R^1R^2} + \underset{(56)}{CH_3(CH_2)_3CH(Cl)MgCl} \xrightarrow{THF,\ -78\ to\ 20\,°C} \underset{(57)}{\underset{R^2}{\overset{R^1}{>}}=\underset{(CH_2)_3CH_3}{\overset{H}{<}}} \quad (14)$$

$R^1 = CH_3(CH_2)_{10}$, $R^2 = H$ (78%; $E/Z = 13/7$)
$R^1 = CH_3$, $R^2 = Ph$ (75%; $E/Z = 2/3$)
$R^1 = R^2 = Ph$ (55%)
$R^1, R^2 = -(CH_2)_6-$ (78%)
$R^1, R^2 = -CH_2\underset{|}{CH}-$ (50%)
$\quad\quad\quad\quad\quad Ph$

$$\underset{(58)}{\text{cyclopropane with } H, SO_2Tol\text{-}p, R^1, R, H, CN}} \xrightarrow[\text{2. ICH}_2\text{MgCl}]{1.\ n\text{-BuLi}} \underset{(59)}{\text{methylenecyclopropane with } R^1, R, H, CN} \quad (15)$$
$\quad\quad -78\,°C\text{ to rt}$

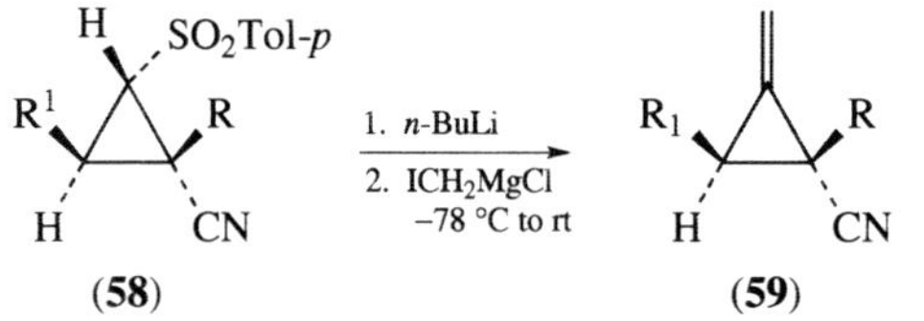

$R = CH_3(CH_2)_3$, $R^1 = H$ (60%)
$R = t\text{-Bu}$, $R^1 = H$ (55%)
$R = CH_3(CH_2)_3$, $R^1 = CH_3$ (70%)
$R = t\text{-Bu}$, $R^1 = CH_3$ (73%)

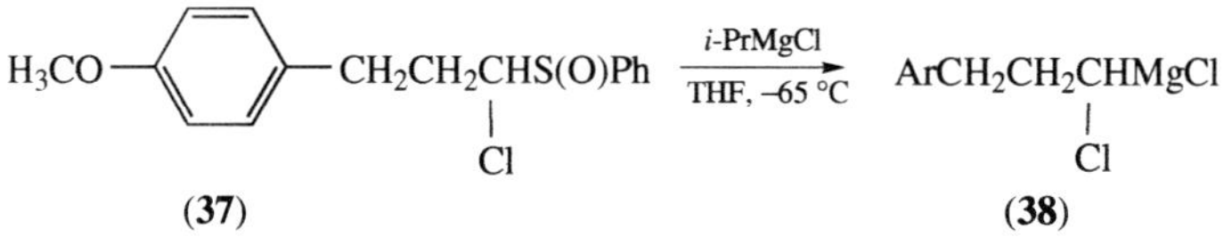

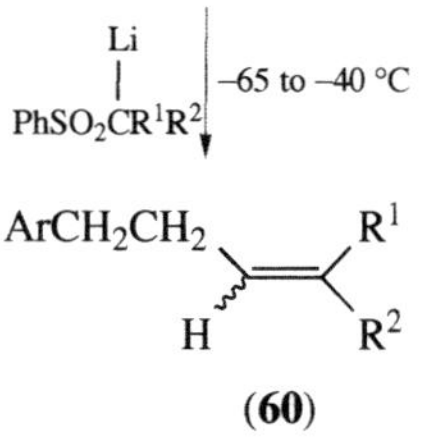

$$\underset{(37)}{H_3CO\text{-}C_6H_4\text{-}CH_2CH_2\underset{\underset{Cl}{|}}{CH}S(O)Ph} \xrightarrow[\text{THF, }-65\,°C]{i\text{-PrMgCl}} \underset{(38)}{ArCH_2CH_2\underset{\underset{Cl}{|}}{CH}MgCl}$$

$R^1 = Ph$, $R^2 = H$ (93%; $E/Z = 1/2$)
$R^1 = 1\text{-Naph}$, $R^2 = H$ (89%; $E/Z = 1/4$)
$R^1 = Ph$, $R^2 = CH_3$ (92%; $E/Z = 1/2$)
$R^1 = R^2 = Ph$ (50%)
$R^1 = CH_3(CH_2)_8$, $R_2 = H$ (86%; $E/Z = 2/3$)
$R^1 = R^2 = CH_3$ (88%)

$$PhSO_2\overset{\underset{|}{Li}}{C}R^1R^2 \xrightarrow{-65\ to\ -40\,°C} \underset{(60)}{\underset{H}{\overset{ArCH_2CH_2}{>}}\sim\underset{R^2}{\overset{R^1}{<}}} \quad (16)$$

4. Reaction of magnesium carbenoids with N-lithio arylamines

The electrophilic reaction of magnesium carbenoids with N-lithio arylamines was found to give non-stabilized α-amino-substituted carbanions (equation 17)[25]. Treatment of magnesium carbenoid **38** with 3.5 eq of N-lithio N-methylaniline at $-70\,°C$ followed by slowly warming the reaction mixture to −40 °C gave the non-stabilized α-amino-substituted carbanion (α-amino-substituted Grignard reagent) **61** in good yield. The presence of the non-stabilized α-amino-substituted carbanion **61** was confirmed by quenching the reaction with deuterated methanol to give *N*-methylaniline having an α-deuteriated alkyl group (**62**) in 77% yield with 91% deuterium incorporation. The preparation of α-amino-substituted carbanions is well recognized to be rather difficult from non-activated amines[26]. Therefore, the results obtained in this study are rewarding and can be considered as an excellent alternative for the preparation of non-stabilized α-amino carbanions.

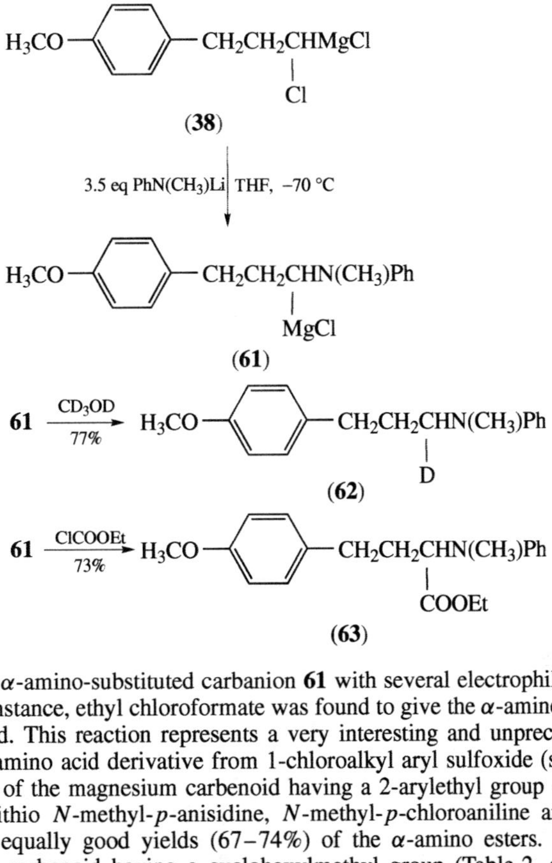

$$H_3CO-\text{C}_6H_4-CH_2CH_2CHMgCl$$
$$|$$
$$Cl$$
(**38**)

↓ 3.5 eq PhN(CH$_3$)Li, THF, −70 °C

$$H_3CO-\text{C}_6H_4-CH_2CH_2CHN(CH_3)Ph$$
$$|$$
$$MgCl$$
(**61**)   (17)

61 $\xrightarrow[77\%]{CD_3OD}$ $H_3CO-\text{C}_6H_4-CH_2CH_2CHN(CH_3)Ph$
                                                               $|$
                                                               D
(**62**)

61 $\xrightarrow[73\%]{ClCOOEt}$ $H_3CO-\text{C}_6H_4-CH_2CH_2CHN(CH_3)Ph$
                                                               $|$
                                                               $COOEt$
(**63**)

Reactions of α-amino-substituted carbanion **61** with several electrophiles were investigated and, for instance, ethyl chloroformate was found to give the α-amino acid derivative **63** in 73% yield. This reaction represents a very interesting and unprecedented one-pot synthesis of α-amino acid derivative from 1-chloroalkyl aryl sulfoxide (see Table 2).

The reaction of the magnesium carbenoid having a 2-arylethyl group (Table 2, entries 1–3) with *N*-lithio *N*-methyl-*p*-anisidine, *N*-methyl-*p*-chloroaniline and *N*-benzyl-*p*-anisidine gave equally good yields (67–74%) of the α-amino esters. The reaction of the magnesium carbenoid having a cyclohexylmethyl group (Table 2, entries 4 and 5) gave a similar yield (68%); however, the reaction of the magnesium carbenoid having a cyclohexyl group showed markedly diminished yield (48%, Table 2, entry 6). Glycine derivatives could be synthesized starting from chloromethyl *p*-tolyl sulfoxide via a magnesium carbenoid, chloromethylmagnesium chloride, in 30–61% yields (Table 2, entries 7–9).

5. *1,3-Carbon–hydrogen (C,H) insertion reaction of magnesium carbenoids*

The carbon–hydrogen insertion (C,H insertion) is one of the most striking reactions of carbenes and carbenoids. The reaction is interesting and very useful for the construction

TABLE 2. Synthesis of α-amino esters from magnesium carbenoids by the reaction with N-lithio N-substituted arylamines followed by ethyl chloroformate

$$\text{R-CHS(O)Ph} \atop \text{Cl} \quad \xrightarrow{i\text{-PrMgCl}} \quad \left[\text{R-CHMgCl} \atop \text{Cl}\right] \quad \xrightarrow[\text{2. ClCOOEt}]{\text{1. ArN(R}^1\text{)Li}} \quad \text{R-CHN(R}^1\text{)Ar} \atop \text{COOEt}$$

R	Ar	R^1	Yield (%)
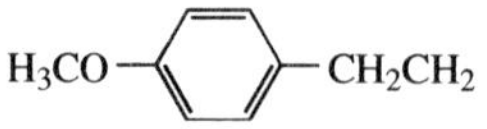 H_3CO-C$_6H_4$-CH_2CH_2	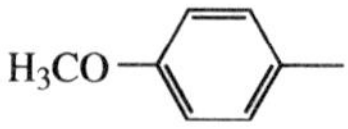 H_3CO-C$_6H_4$-	CH_3	74
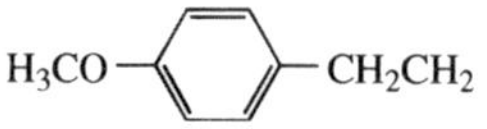 H_3CO-C$_6H_4$-CH_2CH_2	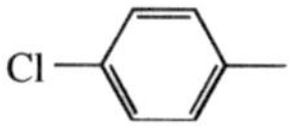 Cl-C$_6H_4$-	CH_3	73
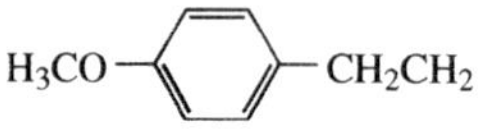 H_3CO-C$_6H_4$-CH_2CH_2	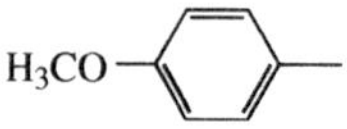 H_3CO-C$_6H_4$-	$PhCH_2$	67
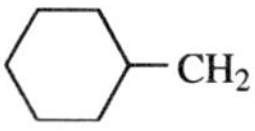 C$_6H_{11}$-CH_2	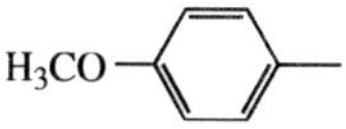 H_3CO-C$_6H_4$-	CH_3	68
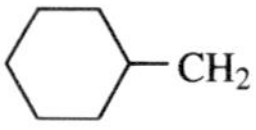 C$_6H_{11}$-CH_2	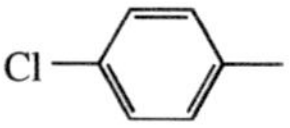 Cl-C$_6H_4$-	CH_3	68
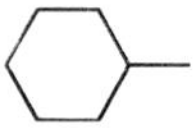 C$_6H_{11}$-	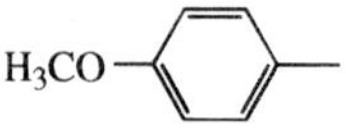 H_3CO-C$_6H_4$-	CH_3	48
H	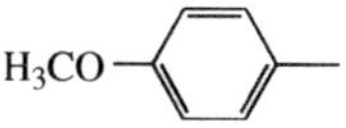 H_3CO-C$_6H_4$-	CH_3	61
H	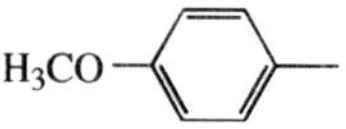 H_3CO-C$_6H_4$-	$PhCH_2$	58
H	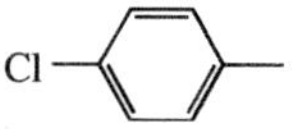 Cl-C$_6H_4$-	CH_3	30

of complex molecules, due to the formation of a new carbon–carbon bond between a carbene (or carbenoid) and an unactivated carbon center. The author studied the C,H insertion of magnesium carbenoids starting from 1-chloroalkyl phenyl sulfoxides (**64**) and a representative example is described in equation 18[27].

When 1-chloroalkyl phenyl sulfoxide (**64**) was treated with 3 eq of i-PrMgCl in THF at $-78\,°C$ and the reaction mixture was slowly allowed to warm to $0\,°C$, the resulting cyclopropanes **66** and **67** were formed in good to high yields. Magnesium carbenoid **65**, first intermediate in this reaction, undergoes a 1,3-C,H insertion reaction with either the methyl or methylene carbon center. Interestingly, when the substituent R has an oxygen functional group, the C,H insertion exclusively takes place with one of the two methyl groups to afford cyclopropane (**66**) after the 1,3-elimination reaction. As recognized from the results in equation 18, the C,H insertion of magnesium carbenoids gives high yields of cyclopropanes under mild conditions.

Clayden and Julia reported the 1,3-C,H insertion reaction of lithium carbenoid (**69**) derived from a primary alkyl chloride (**68**) by H–Li exchange reaction (equation 19)[28]. Treatment of **68** with a mixture of n-BuLi and $tert$-BuOK gave three products. These were silicon-migrated chloride **70** and cyclopropanes **66** and **71**. Obviously, the 1,3-C,H insertion through magnesium carbenoids **65** proceeds more selectively and in better yields.

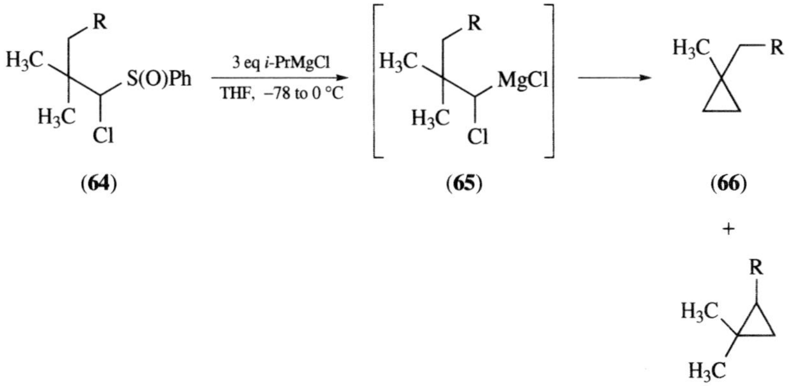

(**64**) (**65**) (**66**) + (**67**) (18)

R	**66** Yield (%)	**67** Yield (%)
1-Naph	65	32
OTHP	97	0
$OSiMe_2Bu$-*t*	85	0
$OCH_2OCH_2CH_2OCH_3$	89	0

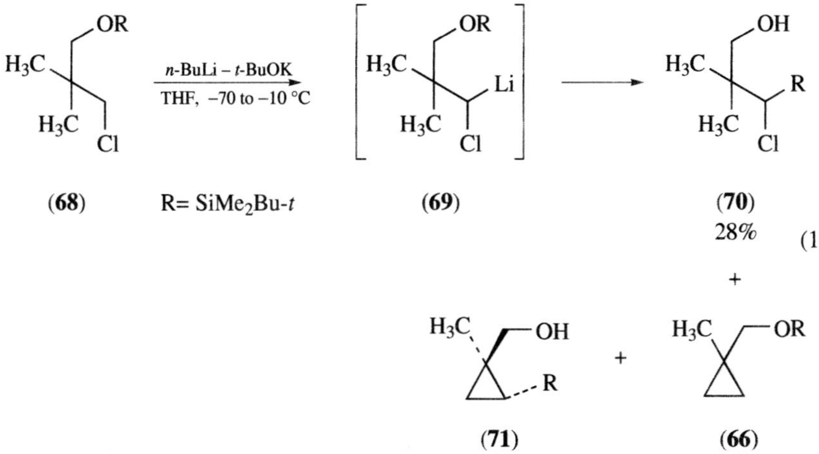

(**68**) R= $SiMe_2Bu$-*t* (**69**) (**70**) 28% (19) + (**71**) 18% + (**66**) 16%

The 1,3-C,H insertion of magnesium carbenoid occurs not only between the carbenoid carbon center and methyl groups but also with the methylene group. For instance, when **72** was treated with *i*-PrMgCl, the cyclopropane (**73**) was obtained in 88% yield as a

5:1 mixture of two diastereomers (equation 20). Interestingly, when 1-chloroalkyl phenyl sulfoxide (**74**) was subjected to the sulfoxide–magnesium exchange, the alkene **75** was isolated in 54% yield suggesting that the 1,2-C,H insertion competes favorably with the 1,3-C,H insertion reaction (equation 21).

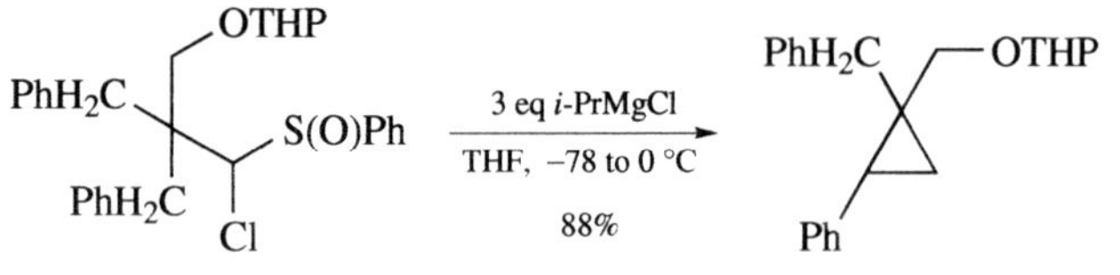

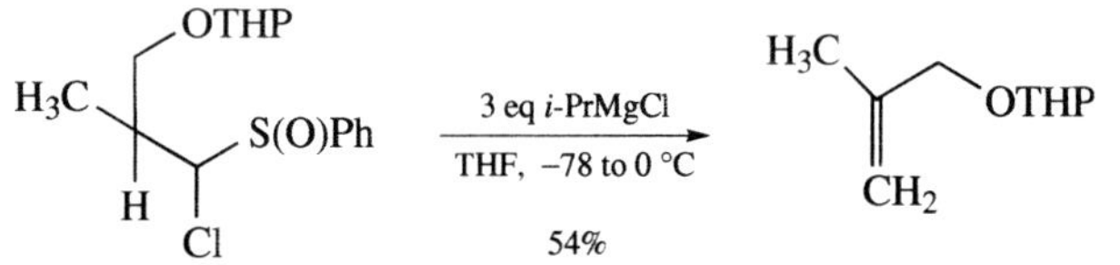

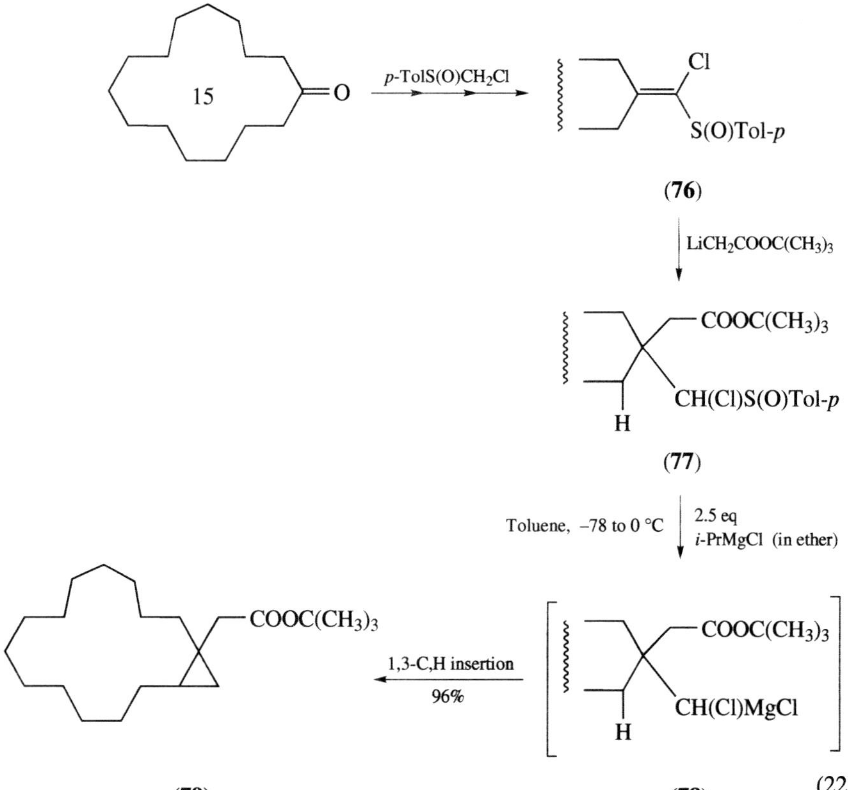

A very interesting synthetic method of bicyclo[n.1.0]alkanes from cyclic ketones via this 1,3-C,H insertion of magnesium carbenoid as a key reaction was reported (equation 22)[29]. 1-Chlorovinyl p-tolyl sulfoxide (**76**) was synthesized from cyclopentadecanone and chloromethyl p-tolyl sulfoxide in three steps in high overall yield. Lithium enolate of *tert*-butyl acetate was added to **76** to give the adduct **77** in quantitative yield. α-Chlorosulfoxide (**77**) in a toluene solution was treated with *i*-PrMgCl in ether at −78 °C and the reaction mixture was slowly warmed to 0 °C to afford the bicyclo[13.1.0]hexadecane derivative **79** in 96% yield through the reaction of the intermediate magnesium carbenoid **78**.

It is worth noting that use of *i*-PrMgCl in ether (not in THF) and toluene as solvent for the reaction is essential for the reaction to proceed. Otherwise, the protonated product of magnesium carbenoid **78**, obtained as by-product, was very difficult to separate from the desired compound (**79**). Interestingly, the 1,3-C,H insertion reaction is highly regioselective.

The synthesis of bicyclo[n.1.0]alkanes (**82**) from various 1-chlorovinyl p-tolyl sulfoxides (**80**) and lithium enolate of *tert*-butyl acetate, propionate and hexanoate through the adducts (**81**) are summarized in Table 3. As shown in Table 3, addition reaction of *tert*-butyl carboxylates to 1-chlorovinyl p-tolyl sulfoxides (**80**) proceeds smoothly to afford the adducts (**81**) in high to quantitative yields. Cyclopropanation of **81** with *i*-PrMgCl

TABLE 3. Synthesis of bicyclo [n.1.0]alkanes (**82**) from 1-chlorovinyl p-tolyl sulfoxides (**80**) with lithium enolate of *tert*-butyl acetate, propionate and hexanoate through the adduct (**81**)

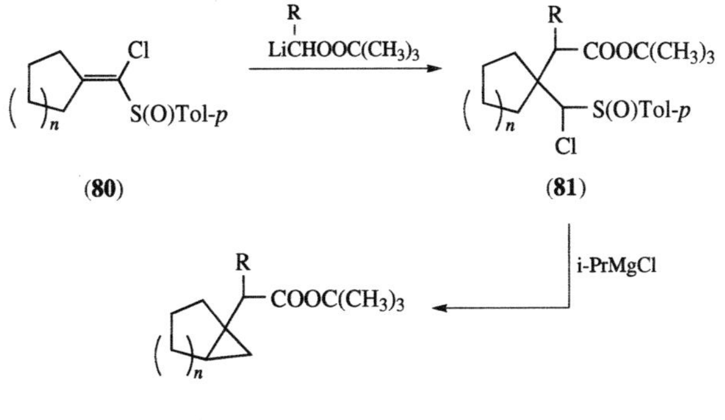

n	R	81 Yield (%)	82 Yield (%)
1	H	98	74
2	H	97	68
4	H	97	90
2	CH_3	99	76[a]
4	CH_3	88	96[a]
11	CH_3	93	95[a]
2	$CH_2CH_2CH_2CH_3$	99	95[a]
4	$CH_2CH_2CH_2CH_3$	78	89[a]
11	$CH_2CH_2CH_2CH_3$	89	99[a]

[a] A single isomer was obtained.

afforded bicyclo[*n*.1.0]alkanes having a *tert*-butyl carboxylate moiety (**82**) in 68–99% yields with very high regioselectivity.

Starting from optically active 1-chlorovinyl *p*-tolyl sulfoxide derived from 2-cyclohexenone, the asymmetric synthesis of cyclopropane derivative (**85**) was realized (equation 23)[29]. Addition reaction of lithium enolate of *tert*-butyl acetate to **83** gave the adduct (**84**) in 96% yield with over 99% ee. Treatment of the latter with *i*-PrMgCl in a similar way as described above afforded optically pure (1*S*,6*R*)-bicyclo[4.1.0]hept-2-ene (**85**) in 90% yield.

(**83**) (**84**) 96% over 99% ee (23)

(1*S*,6*R*)-bicyclo[4.1.0]hept-2-ene

(**85**)

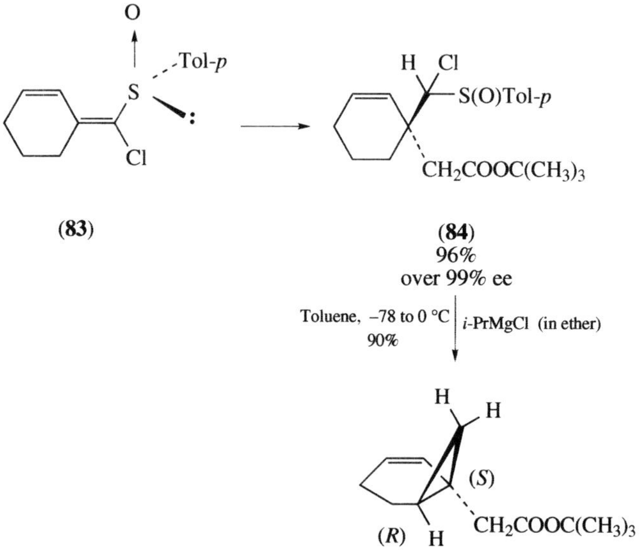

An intramolecular 1,5-C,H insertion reaction was reported from substituted bromoiodoalkane and halogen–magnesium exchange reaction (equation 24)[30]. Indeed, treatment of bromoiodoalkane (**86**) with *i*-PrMgCl in ether at −78 °C to −20 °C resulted in the formation of magnesium carbenoid **87**. The 1,5-C,H insertion of **87** took place to afford a cyclopentane derivative (**88**); however, the yield was not satisfactory.

(24)

(**86**) (**87**) (**88**) *cis*:*trans* = 1:3

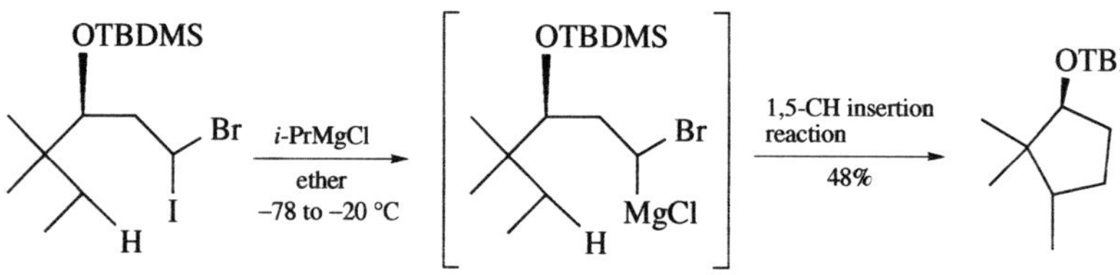

IV. MAGNESIUM CYCLOPROPYLIDENES
A. Magnesium Cyclopropylidenes as Intermediates in the Doering–LaFlamme Allene Synthesis

Cyclopropylidenes (carbenacyclopropanes) are carbenes generated from cyclopropanes and are known to be highly reactive intermediates leading to allenes. They are usually prepared from the reaction of 1,1-dihalocyclopropanes with alkylmetal derivatives. This reaction is now called the Doering–LaFlamme allene synthesis[14]. For example, when 9,9-dibromobicyclo[6.1.0]nonane (**89**), derived from cyclooctene, was treated with magnesium in ether under reflux, 1,2-cyclononadiene (**91**) was obtained in 59% yield. Magnesium cyclopropylidene (**90**) is thought to be the intermediate of this reaction (equation 25)[31].

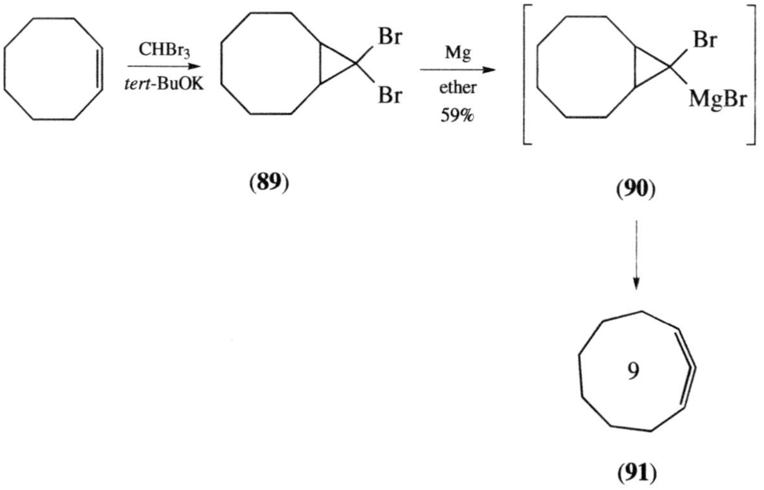

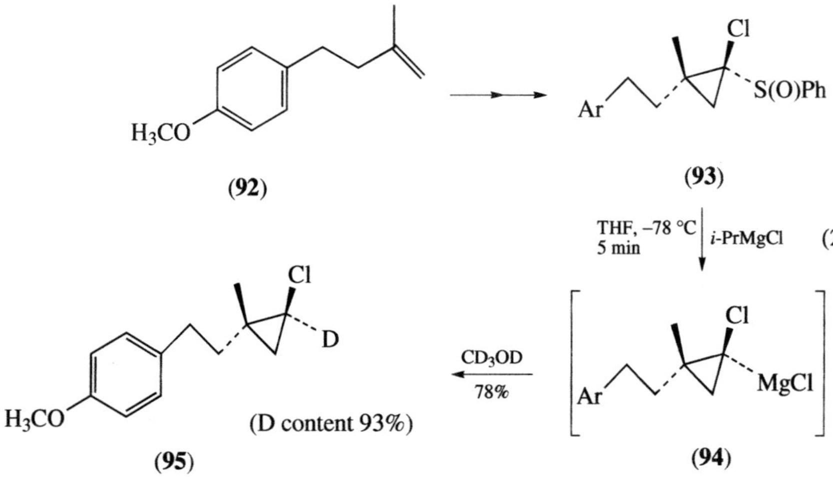

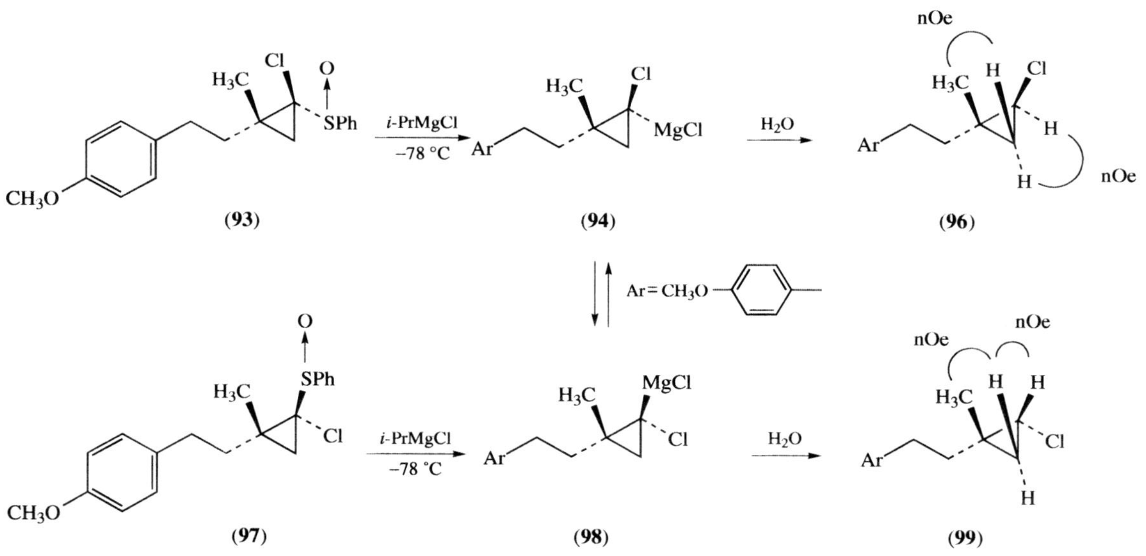

Conditions From	−78 °C, 5 min	−78 to −60 °C, 1 h	−78 to −60 °C, 1 h; then −60 °C, 3 h
93	96 only	96 + trace of 99	96 : 99 = 5 : 2[a]
97	99 only	99 + trace of 96	96 : 99 = 1 : 5[a]

[a] The ratio was determined by ^{1}H NMR.

SCHEME 5

16. The chemistry of magnesium carbenoids   737

Since cyclopropylidenes were generated from 1,1-dihalocyclopropanes with either alkyllithiums or Grignard reagent but as well with lithium, or magnesium metal, at room or even higher temperature, the generated carbenoids quickly afforded allenes (or decomposed). As the stability and chemical nature of magnesium cyclopropylidenes was not investigated, Satoh and coworkers studied, in 2001, the preparation of such magnesium cyclopropylidenes from 1-chlorocyclopropyl phenyl sulfoxides at −78 °C by sulfoxide–magnesium exchange reaction[32].

At first, 1-chlorocyclopropyl phenyl sulfoxide (**93**) was synthesized from the olefin **92** in three steps in good yield. Treatment of sulfoxide **93** with 2.5 eq of *i*-PrMgCl in THF at −78 °C for 5 min followed by quenching with CD$_3$OD afforded the deuterated chlorocyclopropane (**95**) in 78% yield with high deuterium incorporation. From this experiment, it was proved that the intermediate of the reaction is a magnesium cyclopropylidene species **94** (equation 26).

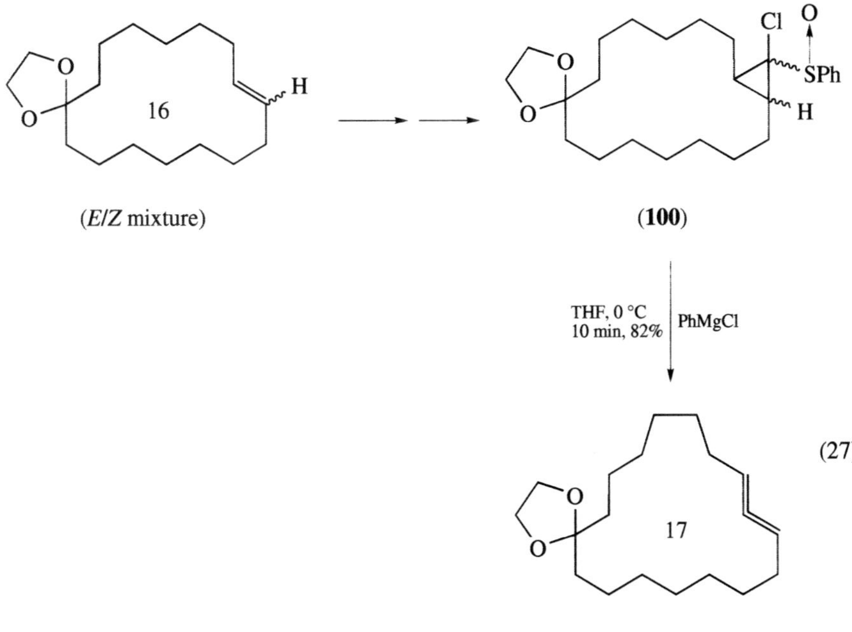

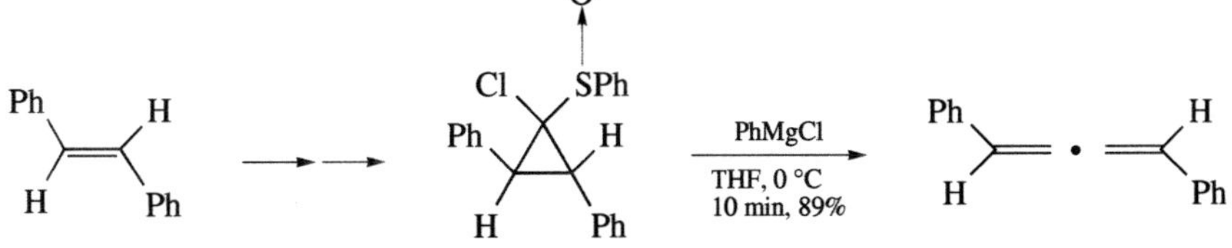

Then, the magnesium cyclopropylidene (**94**) generated at −78 °C was slowly warmed to −60 °C and stirred at this temperature for 3 h. Deuterolysis of the reaction mixture with CD$_3$OD afforded the chlorocyclopropane in 81% yield with 95% D-content. From this result, it was concluded that magnesium cyclopropylidene (**94**) is stable at −60 °C for at least 3 h.

Interestingly, inversion of stereochemistry of the magnesium cyclopropylidenes was observed in this study (Scheme 5). Both diastereomers of α-chlorocyclopropyl phenyl sulfoxides (**93** and **97**) were treated with i-PrMgCl and the generated magnesium cyclopropylidenes (**94** and **98**) were allowed to stand at several different temperatures. As shown in Scheme 5, after quenching the carbenoid with water after 5 min at −78 °C, **96** and **99** were obtained without any contamination of their diastereomers. On the other hand, when the reaction mixture was kept at −60 °C for 1 to 3 h, a mixture of both chlorocyclopropanes (**96** and **99**) was obtained. From these results, it was concluded that magnesium cyclopropylidenes **94** and **98** are configurationally stable below −60 °C and pyramidal inversion of the magnesium cyclopropylidenes slowly takes place at temperature around −60 °C.

To extend this chemistry as a new method for the synthesis of allenes, temperatures of the reaction and the nature of the Grignard reagent were investigated. PhMgCl at 0 °C was found to give the optimal condition for the preparation of allenes in good isolated yields (equation 27). Large-membered ring olefin, cyclohexadecene and *trans*-stilbene were converted to α-chlorocyclopropyl phenyl sulfoxides **100** and **102**. When treated with 2.5 eq of PhMgCl at 0 °C for 10 min, the one-carbon ring-expansion occurs to lead to the allene **101** and diphenylallene **103** in good yields. This reaction provided a general method for the synthesis of allenes from olefins with one-carbon elongation.

B. Nucleophilic Reactions of Magnesium Cyclopropylidenes

As mentioned above, carbenoids have both a nucleophilic and an electrophilic nature. Generation of a magnesium cyclopropylidene and its nucleophilic behavior were reported (equation 28)[33]. Treatment of ethyl dibromocyclopropanecarboxylate (**104**) with i-PrMgCl in ether at −50 °C for 10 min lead, via stereoselective bromine–magnesium exchange, to the formation of magnesium cyclopropylidene (**105**). The generated carbenoid **105** was found to be stable at −50 °C due to the interaction between the ester group and the magnesium. Reaction of carbenoid **105** with iodine and allyl bromide gave the products **106**, both as single isomers, in good yields. The reaction of **105** with benzaldehyde and cyclopentanone gave γ-lactones **107** and **108** in 60% yield.

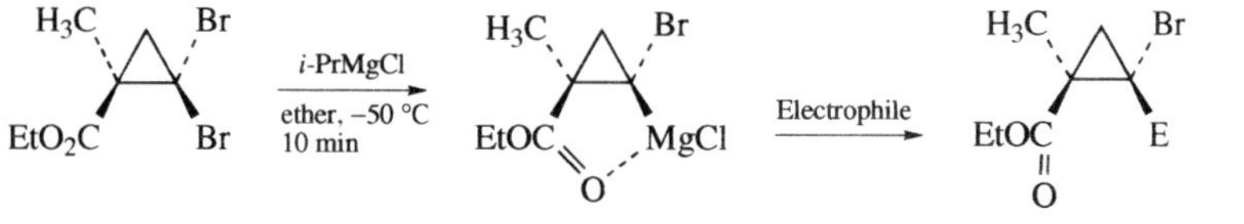

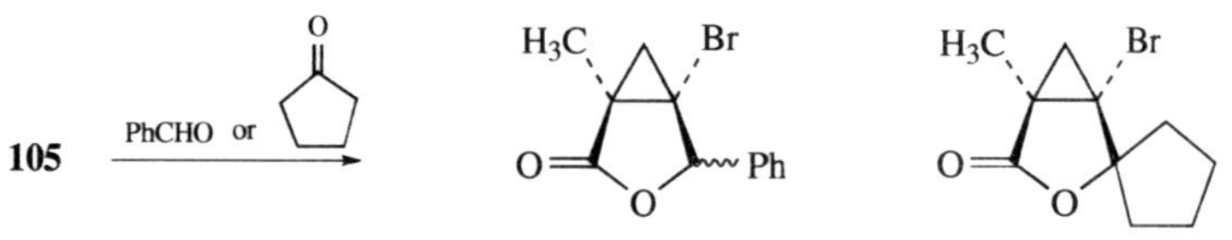

16. The chemistry of magnesium carbenoids

A similar chemistry was carried out from 2,2-dibromocyclopropanecarbonitrile (**109**) (equation 29)[34]. Thus, treatment of dibromide (**109**) with i-PrMgCl in a mixture of dichloromethane and ether (4:1) at $-50\,°C$ for 5 min resulted in the formation of magnesium cyclopropylidene (**110**) via a stereoselective bromine–magnesium exchange reaction. The nucleophilic reaction of carbenoid **110** with several electrophiles afforded the products **111** in about 80% yield. From the products **111**, highly substituted cyclopropanes could be synthesized using bromine- and sulfoxide–magnesium exchange reactions.

$$\underset{(\mathbf{109})}{\overset{H_3C\quad Br}{\underset{NC\quad Br}{\triangle}}} \xrightarrow[\text{CH}_2\text{Cl}_2\text{-ether,}\ -50\,°C,\ 5\ \text{min}]{i\text{-PrMgCl}} \underset{(\mathbf{110})}{\overset{H_3C\quad Br}{\underset{NC\quad MgCl}{\triangle}}} \xrightarrow{\text{Electrophile}} \underset{(\mathbf{111})}{\overset{H_3C\quad Br}{\underset{NC\quad E}{\triangle}}} \qquad (29)$$

Electrophile: H_2O (E = H, 76%), I_2 (E = I, 77%), $PhSSO_2Ph$ (E = SPh, 86%), allyl bromide (E = $H_2C=CHCH_2$, 78%)

C. Electrophilic Reactions of Magnesium Cyclopropylidenes

An interesting reaction using the electrophilic nature of magnesium carbenoids with α-sulfonyl lithium carbanion giving alkylidenecyclopropanes was reported by Satoh and Saito (equation 30)[35]. 1-Chlorocyclopropyl phenyl sulfoxide (**112**), synthesized from commercially available cyclopropyl phenyl sulfide in 93% overall yield, was treated with 2.5 eq of i-PrMgCl at $-78\,°C$. The sulfoxide–magnesium exchange reaction was found to take place instantaneously to give magnesium cyclopropylidene (**113**). To this carbenoid, three equivalents of an α-sulfonyl lithium carbanion was added and the reaction mixture was allowed to warm to $-50\,°C$ to give alkylidenecyclopropane (**115**) in moderate yield.

$$\underset{\text{SPh}}{\triangle} \longrightarrow \longrightarrow \underset{(\mathbf{112})}{\overset{Cl\quad S(O)Ph}{\triangle}} \xrightarrow[\text{THF, }-78\,°C]{i\text{-PrMgCl}} \underset{(\mathbf{113})}{\overset{Cl\quad MgCl}{\triangle}}$$

$$\underset{(\mathbf{115})}{\overset{R_1\quad R_2}{\underset{\triangle}{\rightthreetimes}}} \xleftarrow[-78\ \text{to}\ -50\,°C]{} \left[\underset{(\mathbf{114})}{\overset{R^1\quad R^2}{\underset{PhO_2S\quad MgCl}{\triangle}}}\right] \xleftarrow{} \underset{\text{Li}}{R^1R^2CSO_2Ph} \qquad (30)$$

$R^1 = H$, $R^2 = $ 1-Naphthyl (61%)

$R^1 = R^2 = Ph$ (48%)

The proposed mechanism of this reaction is composed by an initial S_N2-type nucleophilic substitution reaction of **113** with the nucleophilic α-sulfonyl lithium carbanion to give the alkylmagnesium species (**114**) having a sulfonyl group at the β-position. Then, a β-elimination reaction of magnesium sulfinate from the intermediate (**114**) occurs

TABLE 4. Synthesis of alkylidenecyclopropanes (**118**) from 1-chlorocyclopropyl phenyl sulfoxide (**116**) and α-sulfonyl lithium carbanions through magnesium cyclopropylidene (**117**)

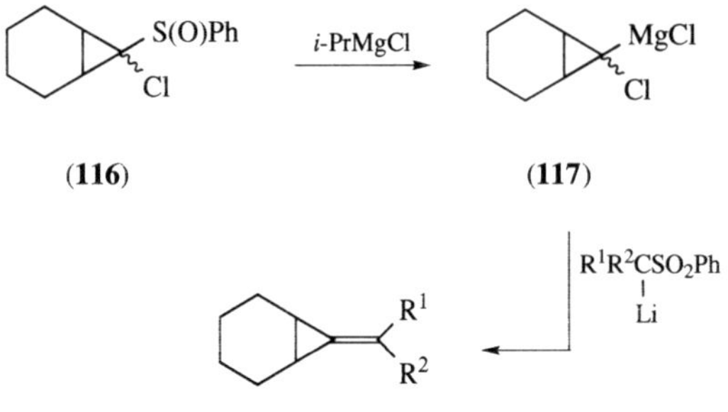

(**116**) (**117**) (**118**)

R^1	R^2	**118**/Yield(%)
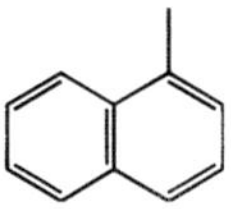	H	64
Ph	H	73
Ph	Ph	50
$C{\equiv}CC_5H_{11}$	H	25
$CH_2CH_2-C_6H_4-OCH_3$	H	32

to give the expected alkylidenecyclopropane (**115**). Examples for the synthesis of alkylidenecyclopropanes (**118**) from 1-chlorocyclopropyl phenyl sulfoxide (**116**) via magnesium cyclopropylidene (**117**) are summarized in Table 4.

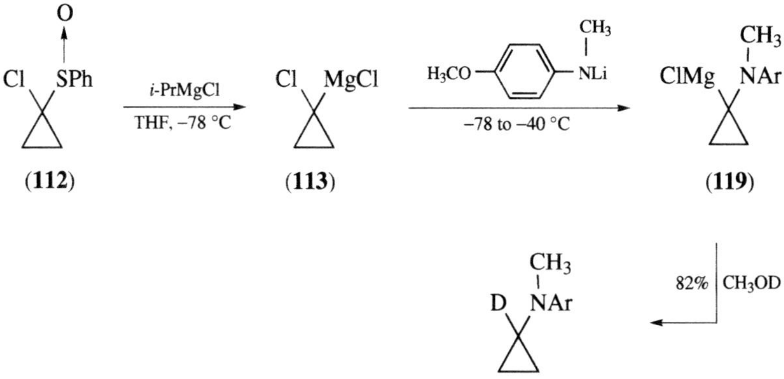

(**112**) (**113**) (**119**) (**120**)

D-content 98%

(31)

16. The chemistry of magnesium carbenoids   741

The electrophilic reaction of magnesium cyclopropylidene (**113**) with *N*-lithioarylamines was reported (equation 31)[36]. Thus, electrophilic reaction of magnesium cyclopropylidene (**113**) derived from **112** with *N*-lithio *N*-methyl *p*-anisidine resulted in the formation of α-amino-substituted cyclopropylmagnesium (**119**) in good yield. Methanolysis of the reaction mixture with CH_3OD gave α-deuteriated *N*-cyclopropyl-*N*-methyl-*p*-anisidine (**120**) in 82% yield with 98% D-content.

The reaction of **113** with several *N*-lithioamines was investigated and the results are summarized in Table 5. *N*-Methylaniline, *p*-chloro-*N*-methylaniline and *N*-benzyl-*p*-anisidine gave 60–67% yield of the desired *N*-cyclopropyl arylamines. Diphenylamine gave the desired product; however, the yield was not satisfactory. Interestingly, dibenzylamine did not afford the desired product at all. This result indicated that the magnesium cyclopropylidene (**113**) only reacts with *N*-lithio arylamines. The reaction of **113** with *N*-lithio nitrogen-containing heterocyclic compounds was also studied. From the results shown in Table 5, the yields of the reaction are variable as a function of the heterocyclic compounds used.

The reactivity of the formed α-amino-substituted cyclopropylmagnesium (**119**) with some electrophiles was investigated[36]. Cyclopropylmagnesium (**119**) was found to have a low nucleophilicity and, for example, reaction with benzaldehyde gave only 40% yield

TABLE 5. Reaction of magnesium cyclopropylidene (**113**) derived from 1-chlorocyclopropyl phenyl sulfoxide (**112**) with *N*-lithio amines

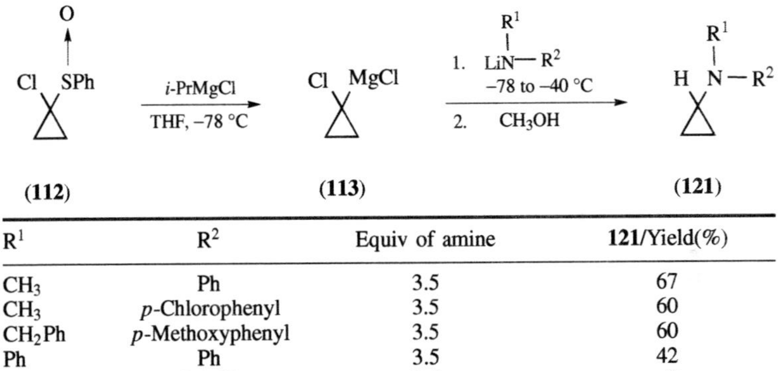

R^1	R^2	Equiv of amine	**121**/Yield(%)
CH_3	Ph	3.5	67
CH_3	*p*-Chlorophenyl	3.5	60
CH_2Ph	*p*-Methoxyphenyl	3.5	60
Ph	Ph	3.5	42
CH_2Ph	CH_2Ph	3.5	0
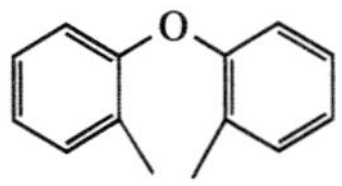 phenoxazine		2.0	87
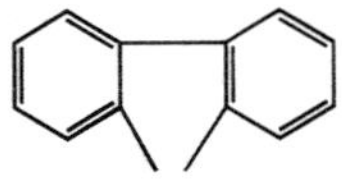 carbazole		2.0	43
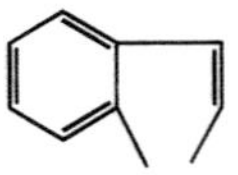 indoline		2.0	21
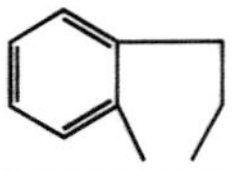 indole		2.0	75

of the adduct. The reaction of **119** with ethyl chloroformate gave maximum 20% yield of the desired ethoxycarbonylated product.

On the other hand, the reaction of **119** with carbon disulfide followed by iodomethane gave dithioester **122** in high yield (equation 32). Methanolysis of the dithioester **122** in methanol with excess $Hg(OCOCF_3)_2$ gave the cyclopropyl α-amino acid derivative **123** in high yield.

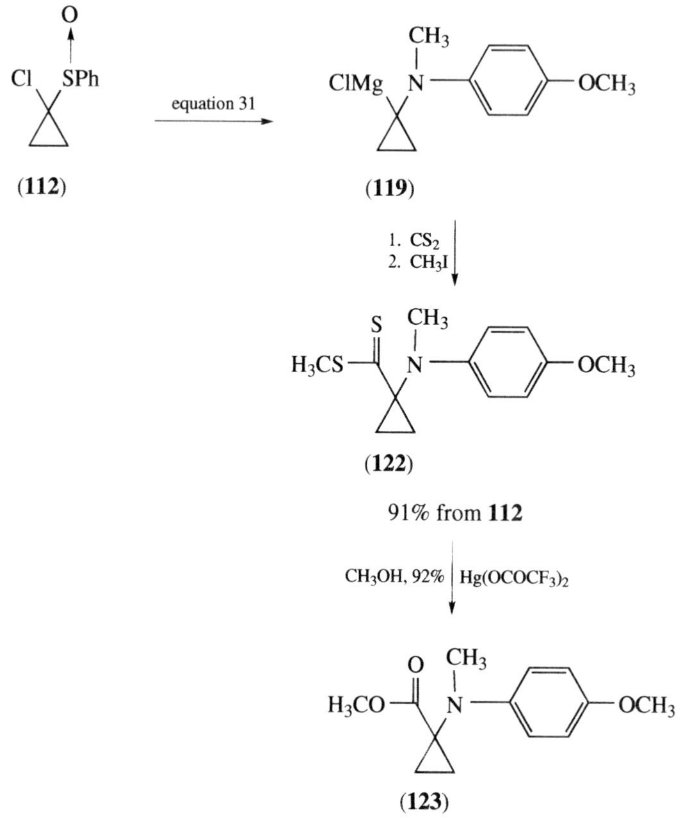

(32)

V. MAGNESIUM ALKYLIDENE CARBENOIDS

A. Generation and Nucleophilic Property of Magnesium Alkylidene Carbenoids, the Fritsch–Buttenberg–Wiechell Rearrangement

Alkylidene carbenoids are sp^2 carbenoids and are known to be very interesting reactive intermediates[37]. The most famous reaction of alkylidene carbenoids is the Fritsch–Buttenberg–Wiechell rearrangement (equation 33)[14].

Treatment of 1-haloalkene (**124**) with a strong base resulted in the formation of carbenoid **125** by the hydrogen–metal exchange reaction. The metal in **125** is usually Na, Li or K and the organometals were found to be highly unstable; elimination of Metal-X resulted in the formation of alkylidene carbene **126**. The Fritsch–Buttenberg–Wiechell rearrangement takes place either from the alkylidene carbenoid or carbene to afford

16. The chemistry of magnesium carbenoids

acetylene (**127**). As the alkylidene carbenoid (**125**) was usually generated at room temperature or higher, the behavior of the alkylidene carbenoid was unclear until recently.

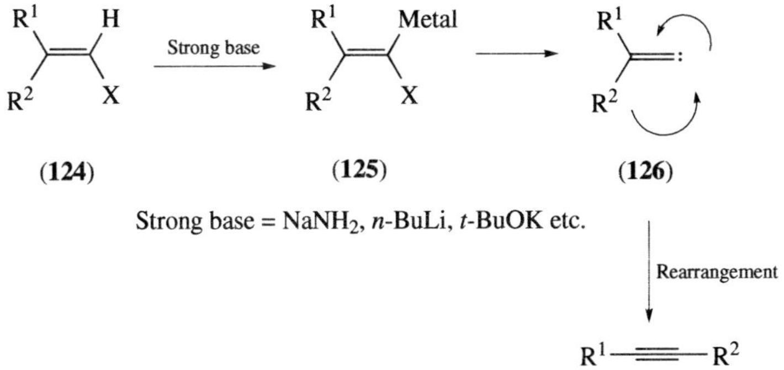

$$\underset{(124)}{\overset{R^1}{\underset{R^2}{>}}\!\!=\!\!\overset{H}{\underset{X}{<}}} \xrightarrow{\text{Strong base}} \underset{(125)}{\overset{R^1}{\underset{R^2}{>}}\!\!=\!\!\overset{\text{Metal}}{\underset{X}{<}}} \longrightarrow \underset{(126)}{\overset{R^1}{\underset{R^2}{>}}\!\!=\!\!:} \quad \xrightarrow{\text{Rearrangement}} \underset{(127)}{R^1\!\!-\!\!\!\equiv\!\!\!-R^2}$$

Strong base = NaNH$_2$, n-BuLi, t-BuOK etc.   (33)

In 1993, Satoh and coworkers reported the preparation of lithium- and magnesium-alkylidene carbenoids from 1-chlorovinyl phenyl sulfoxides by sulfoxide–metal exchange reaction at low temperature (Scheme 6)[38]. 1-Chlorovinyl phenyl sulfoxide (**128**) is easily synthesized from the corresponding aldehyde and chloromethyl phenyl sulfoxide in high yield. Sulfoxide **128** was treated with t-BuLi in THF at $-78\,°C$ to give the terminal alkyne **131**. Obviously, the intermediate of this reaction was the alkylidene carbenoid **129**.

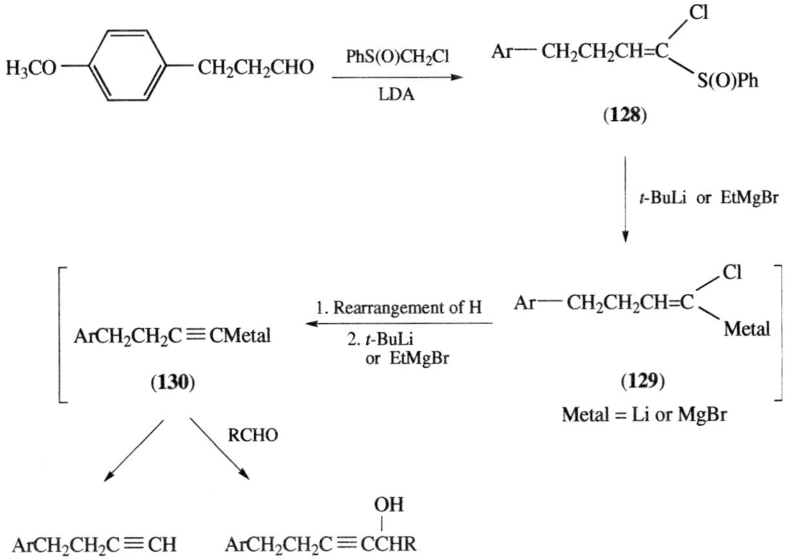

SCHEME 6

The Fritsch–Buttenberg–Wiechell rearrangement of the carbenoid gave the alkyne **131**, which was further metallated in the reaction mixture with the excess of the alkylmetal to give the acetylide **130**. Addition of an aldehyde gave the propargylic alcohol **132**. When this reaction was carried out with EtMgBr, the corresponding magnesium alkylidene carbenoid (**129**; Metal = MgBr) was generated; however, even at −78 °C the rearrangement took place to give the alkyne **131**.

On the other hand, magnesium alkylidene carbenoid **134** derived from 1-halovinyl *p*-tolyl sulfoxides **133**, easily prepared from ketones (cyclopentadecanone is shown as representative example) and halomethyl aryl sulfoxide in three steps in high overall yields, showed quite interesting properties (equation 34)[39]. For instance, treatment of 1-chlorovinyl *p*-tolyl sulfoxide (**133a**) with EtMgCl in THF at −78 °C resulted instantaneously in the formation of magnesium alkylidene carbenoid **134a**. The formation of carbenoid **134a** was confirmed by quenching the reaction mixture with CD_3OD to afford deuteriated chloroalkene (**135**) in high yield.

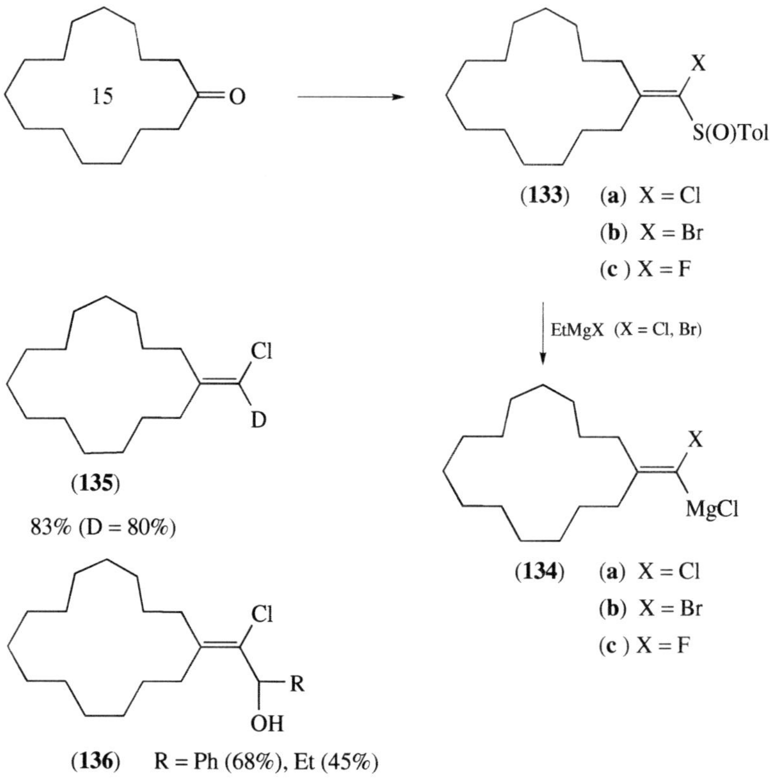

Moreover, magnesium carbenoid **134a** was found to be stable below −78 °C for at least 30 min. The carbenoid (**134b**; X = Br) showed similar properties compared with **134a**; however, **134c** (X = F) was found to be relatively unstable. The nucleophilicity of **134a** was examined by classical reaction with ketones and aldehydes and it was found that carbenoid **134a** has a rather low nucleophilicity; only aldehydes reacted to give the adduct **136** in about 60% yields. Interestingly, the Fritsch–Buttenberg–Wiechell rearrangement was rarely observed from the magnesium alkylidene carbenoids derived from ketones.

16. The chemistry of magnesium carbenoids 745

The configurational stability of the magnesium alkylidene carbenoids was examined (equation 35)[39b]. Thus, at first, (*E*)-1-chlorovinyl *p*-tolyl sulfoxide (**137**) and its (*Z*)-isomer (**138**) were synthesized from acetophenone. Sulfoxide **137** was treated with EtMgCl in THF at −78 °C for 5 min to give (*E*)-chloroalkene **141** in 95% as a single isomer after hydrolysis. Prolonging the reaction time to 60 min gave a similar result.

$$
\begin{array}{c}
\text{Ph}\diagdown\quad\diagup\text{S(O)Tol-}p \\
\text{H}_3\text{C}\diagup\quad\diagdown\text{Cl}
\end{array}
\xrightarrow[\text{THF, }-78\,°C]{\text{EtMgCl}}
\left[\begin{array}{c}
\text{Ph}\diagdown\quad\diagup\text{MgCl} \\
\text{H}_3\text{C}\diagup\quad\diagdown\text{Cl}
\end{array}\right]
\xrightarrow{\text{H}_2\text{O}}
\begin{array}{c}
\text{Ph}\diagdown\quad\diagup\text{H} \\
\text{H}_3\text{C}\diagup\quad\diagdown\text{Cl}
\end{array}
$$

(**137**)   (**139**)   (**141**)

(35)

$$
\begin{array}{c}
\text{Ph}\diagdown\quad\diagup\text{Cl} \\
\text{H}_3\text{C}\diagup\quad\diagdown\text{S(O)Tol-}p
\end{array}
\xrightarrow[\text{THF, }-78\,°C]{\text{EtMgCl}}
\left[\begin{array}{c}
\text{Ph}\diagdown\quad\diagup\text{Cl} \\
\text{H}_3\text{C}\diagup\quad\diagdown\text{MgCl}
\end{array}\right]
\xrightarrow{\text{H}_2\text{O}}
\begin{array}{c}
\text{Ph}\diagdown\quad\diagup\text{Cl} \\
\text{H}_3\text{C}\diagup\quad\diagdown\text{H}
\end{array}
+ \mathbf{141}
$$

(**138**)   (**140**)   (**142**)

5 min   **142** : **141** = 1.8 : 1
60 min   **142** : **141** = 1 : 1.1

On the other hand, treatment of (*Z*)-1-chlorovinyl *p*-tolyl sulfoxide (**138**) with EtMgCl for 5 min gave a mixture of *Z*-chloroalkene **142** accompanied by its *E*-isomer (**141**) in a ratio of 1.8:1. Prolonging the reaction time to 60 min gave an almost equimolar mixture of **142** and **141** in a 1:1.1 ratio. These results indicated that magnesium alkylidene carbenoid **140** isomerizes even at −78 °C. Magnesium alkylidene carbenoid **139** is therefore thermodynamically more stable than **140**.

$$
\begin{array}{c}
\text{O} \\
\text{Ph}\diagdown\!\!\diagup\text{COOC}_2\text{H}_5
\end{array}
\longrightarrow
\begin{array}{c}
\text{Br}\diagdown\quad\diagup\text{Br} \\
\text{Ph}\diagup\quad\diagdown\text{COOC}_2\text{H}_5
\end{array}
\xrightarrow[\substack{\text{ether, }-50\,°C \\ 15\,\text{min}}]{i\text{-PrMgCl}}
\left[\begin{array}{c}
\text{Br}\diagdown\quad\diagup\text{MgCl} \\
\text{Ph}\diagup\quad\diagdown\text{COOC}_2\text{H}_5
\end{array}\right]
$$

(**143**)   (**144**)

(36)

$$
\begin{array}{c}
\text{Br}\diagdown\quad\diagup\text{R} \\
\quad\diagup\quad\diagdown\text{R} \\
\text{Ph}\diagup\quad\diagdown\text{O} \\
\quad\quad\diagdown\text{O}
\end{array}
\qquad
\begin{array}{c}
\text{Br}\diagdown\quad\diagup\text{I} \\
\text{Ph}\diagup\quad\diagdown\text{COOC}_2\text{H}_5
\end{array}
\xleftarrow{\text{electrophile}}
$$

(**146**)   (a) R = Ph (79%)   (**145**) 89%
        (b) R = −(CH$_2$)$_4$− (71%)

The nucleophilic nature of magnesium alkylidene carbenoids was also reported (equation 36)[40]. Treatment of dibromoalkene **143**, derived from ethyl phenylglyoxylate, with i-PrMgCl in ether at $-50\,°C$ for 15 min resulted in the formation of the magnesium alkylidene carbenoid having an ethyl ester group in the molecule (**144**). Carbenoid **144** could be trapped with several electrophiles such as iodine, or benzophenone and cyclopentanone to afford the iodide **145** and lactones **146a** and **146b**, respectively, in good yields.

B. Electrophilic Reactions of Magnesium Alkylidene Carbenoids

1. One-pot synthesis of tetrasubstituted olefins

From a synthetic organic chemistry point of view, the electrophilic nature of magnesium alkylidene carbenoids is much more interesting than their nucleophilic nature. Satoh and coworkers found that treatment of 1-chlorovinyl p-tolyl sulfoxide (**147**) with an excess of PhMgBr in THF at -85 to $-50\,°C$ for 2 h followed by CD_3OD gave the deuterio styrenyl derivative **150** in 80% yield with a complete deuterium incorporation (equation 37)[39b].

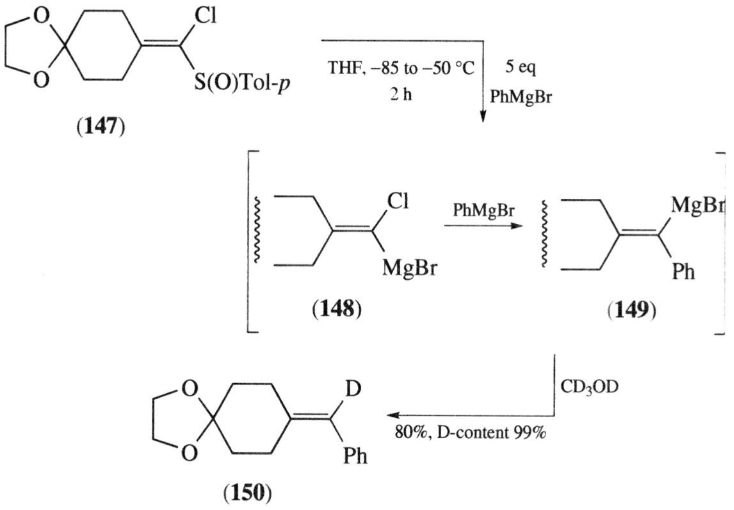

The reaction proceeds as follows. At first, the sulfoxide–magnesium exchange reaction of **147** gave the magnesium alkylidene carbenoid **148**. Based on the electrophilic nature of carbenoid **148**, nucleophilic substitution of **148** on the sp^2 carbon atom by PhMgBr resulted in the formation of alkenyl Grignard reagent **149**. Finally, the carbanion was quenched with CD_3OD to afford the deuteriated olefin **150**. This reaction resulted in an interesting double substitution of sulfinyl and chloro groups by phenyl and deuterio groups on the olefinic sp^2 carbon in a one-pot procedure.

Selected examples of this new method are summarized in Table 6. As shown in this Table, PhMgBr and p-methoxyphenylmagnesium bromide reacted well with **147** to give first the magnesium alkylidene carbenoid and, after nucleophilic substitution with an additional equivalent of Grignard, the corresponding alkenyl Grignard reagent. A large variety of electrophiles can be added to the reaction mixture and the corresponding tetrasubstituted olefins **151** are obtained in good yields (with the exception of the reaction of acetone).

16. The chemistry of magnesium carbenoids 747

TABLE 6. Synthesis of tetrasubstituted olefins (151) by the reaction of 1-chlorovinyl p-tolyl sulfoxide (147) with ArMgBr followed by addition of electrophiles

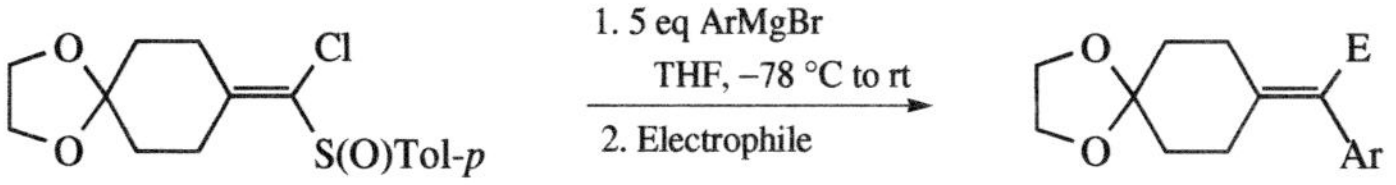

Ar	Electrophile	151	
		E	Yield (%)
Ph	CH_3CH_2CHO	CH_3CH_2CHOH	81
Ph	CH_3COCH_3	$CH_3CH(OH)CH_3$	21
Ph	ClCOOEt	EtOCO	65
Ph	I_2	I	53
Ph	 Ph⌇⌇O (styrene oxide)	$PhCH_2CH_2CH(OH)CH_2$	42
Ph	PhNCO	PhNHCO	87
p-Methoxyphenyl	PhCHO	PhCH(OH)	87
p-Methoxyphenyl	ClCOOEt	EtOCO	67
p-Methoxyphenyl	PhNCO	PhNHCO	87

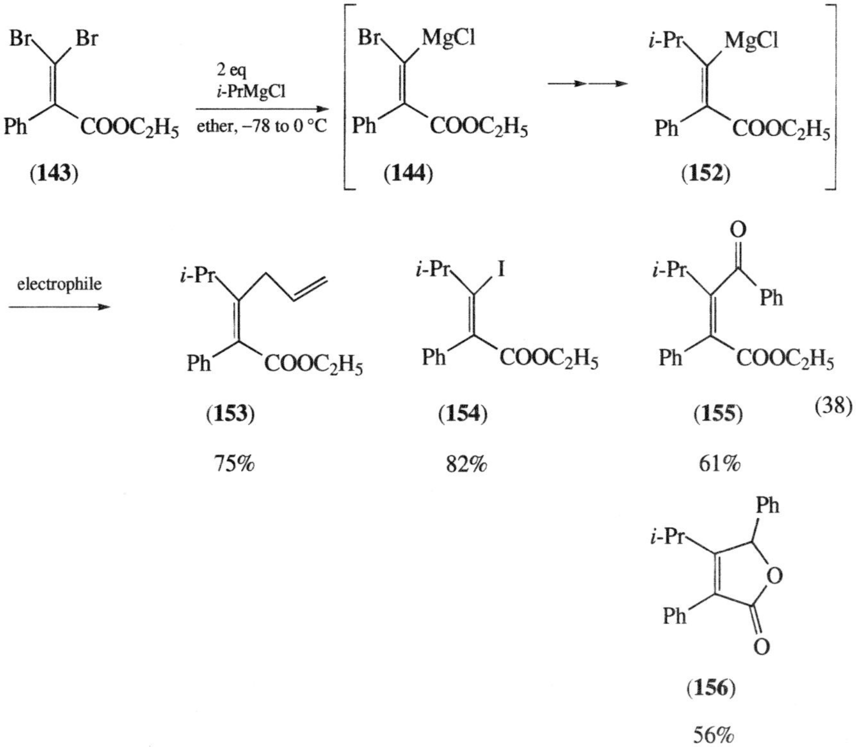

(38)

(153) 75%

(154) 82%

(155) 61%

(156) 56%

A similar reaction was reported by Knochel, Marek and coworkers (equation 38)[40]. Thus, dibromide **143** was treated with 2 equivalents of i-PrMgCl at $-78\,^\circ$C and the reaction mixture was warmed to $0\,^\circ$C to give the functionalized alkenyl Grignard reagent **152** through magnesium alkylidene carbenoid intermediate **144**. Trapping the alkenyl Grignard reagent **152** with different electrophiles such as allyl bromide, iodine, benzoyl chloride and benzaldehyde gave the corresponding olefins **153–156** respectively in moderate to good yields.

2. Synthesis of allenes by alkenylation of magnesium alkylidene carbenoids with α-sulfonyl lithium carbanions

The electrophilic reaction of magnesium alkylidene carbenoids with other nucleophiles than the original Grignard reagent can also be carried out. For example, treatment of magnesium alkylidene carbenoid **157**, derived from **147**, with α-sulfonyl lithium carbanion afforded allenes **159** in moderated yields (equation 39)[41].

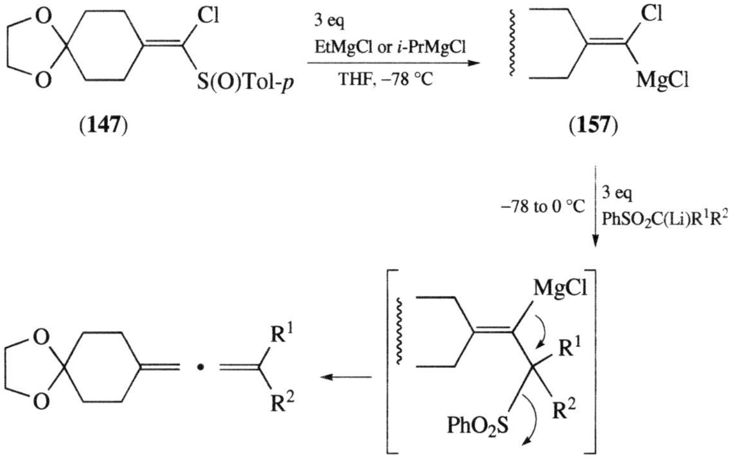

(39)

(**159**) R^1 = Ph, R^2 = H   65%
R^1 = 1-Naphthyl, R^2 = H   51%
R^1 = Vinyl, R^2 = H   51%
R^1 = CH_2CH_2Ar, R^2 = H   42%
R^1 = $C\equiv CC_5H_{11}$, R^2 = H   65%
R^1 = Ph, R^2 = Ph   41%

The proposed mechanism is as follows: First, the α-sulfonyl lithium carbanion attacks the electrophilic carbenoid carbon atom to give the vinylmagnesium intermediate (**158**). As the sulfonyl moiety is a good leaving group, a β-elimination takes place to afford the allenes (**159**).

3. Reaction of magnesium alkylidene carbenoids with N-lithio arylamines

A very interesting direct alkenylation of arylamines at the *ortho*-position by the reaction of magnesium alkylidene carbenoids with N-lithio arylamines to give **162** was reported by Satoh and coworkers (equation 40)[42]. Magnesium alkylidene carbenoid **157**, derived

16. The chemistry of magnesium carbenoids 749

from **147**, was treated with three equivalents of *N*-lithio aniline at −78 °C and the reaction mixture was gradually allowed to warm to −10 °C to give *ortho*-alkenylated aniline **160** in 49% yield. Toluene was found to be the best solvent for this reaction. The generality of this unprecedented reaction was investigated and selected results are summarized in Table 7.

(40)

Magnesium alkylidene carbenoid **157** reacted with *p*-anisidine in 44% yield; however, the yield decreased with *p*-chloroaniline. 2-Methylaniline gave only *ortho*-alkenylated product and 2,6-dimethylaniline gave no alkenylated product. These results indicate that this reaction only gives *ortho*-alkenylated products. Interestingly, the reaction with 1-aminonaphthalene and 1-aminoanthracene gave much better yields. Magnesium alkylidene carbenoid **161**, generated from 1-chlorovinyl *p*-tolyl sulfoxide synthesized from acetone, reacts similarly with *N*-lithio arylamines to give **162**. Especially, the reaction with 1-aminonaphthalene and 1-aminoanthracene gave about 80% yields of *ortho*-alkenylated arylamines.

Very interesting results were obtained from the reaction of magnesium alkylidene carbenoids with *meta*-substituted arylamines (Table 8)[42b]. The reaction of magnesium alkylidene carbenoids **157** and **161** with three *meta*-substituted anilines was carried out and the results are summarized in Table 8. The reaction of **157** with *meta*-anisidine gave two products **163** and **164** (in a 30:13 ratio) in 43% yield. The main product was found to have the alkenyl group at the more hindered position (**163**). As shown in the Table, all the other *meta*-substituted aniline derivatives also gave the more hindered alkenylated compounds as the main product in variable ratio.

To gain a better understanding of the regioselectivity of the reaction, theoretical study using the Gaussian 98 program was performed[42b]. Thus, electrostatic potential-derived charges using the CHelpG scheme of Breneman were calculated with the structures optimized at the MP2/6-31(+)G* level and the more negative charge was found to be on carbon-2 in the most stable conformer.

Stereochemistry of this reaction is also quite interesting. Thus, both geometrical isomers of 1-chlorovinyl *p*-tolyl sulfoxides (**165**–**167**) were synthesized from 2-cyclohexenone, methyl vinyl ketone and 2-heptanone respectively, and the corresponding magnesium alkylidene carbenoids were generated and treated with *N*-lithio aniline or *N*-lithio 1-aminonaphthalene. The results are summarized in Table 9.

Interestingly, the reaction of the magnesium alkylidene carbenoids derived from *E*-**165** and *Z*-**165** with *N*-lithio aniline gave *Z*-*ortho*-alkenylated aniline *Z*-**168** and *E*-*ortho*-alkenylated aniline *E*-**168**, respectively, with high stereospecificity (entries 1 and 2). The same results were obtained from *E*-**165** and *Z*-**165** with *N*-lithio 1-aminonaphthalene which gives *Z*-**169** and *E*-**169** (entries 3 and 4). Furthermore, the reaction of the magnesium alkylidene carbenoid derived from *E*-**166** and *Z*-**166** with *N*-lithio 1-aminonaphthalene (entries 5 and 6) again showed high stereospecificity in products *Z*-**170** and

TABLE 7. Synthesis of *ortho*-alkenylated arylamines (**162**) by the reaction of the magnesium alkylidene carbenoids with *N*-lithio arylamines

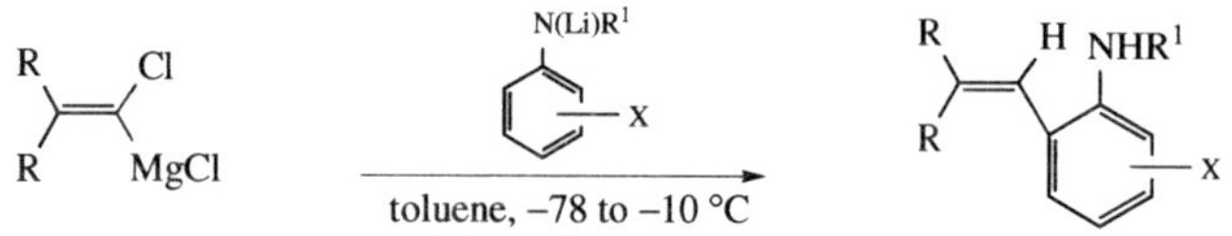

[Reaction scheme: (157) R₂C=C(Cl)(MgCl) + N(Li)R¹-aryl-X → (162) R₂C=CH–aryl(NHR¹)–X, toluene, −78 to −10 °C; (161) R = CH₃]

R	R	Arylamine	162/Yield (%)
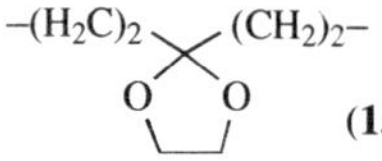 −(H₂C)₂–O–C(–O)–(CH₂)₂− (157)		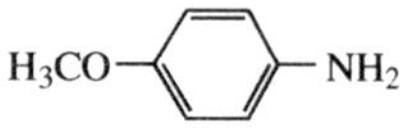 H₃CO–C₆H₄–NH₂	44
		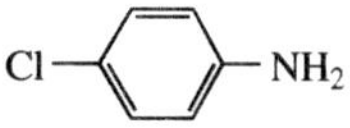 Cl–C₆H₄–NH₂	28
		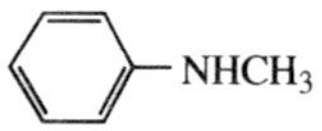 C₆H₅–NHCH₃	38
		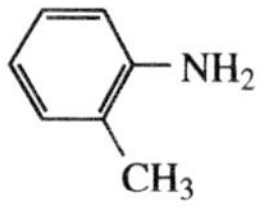 2-CH₃–C₆H₄–NH₂	32
		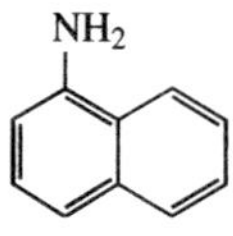 1-aminonaphthalene	66
		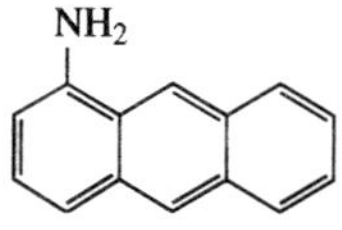 1-aminoanthracene	60
CH₃	CH₃ (161)	C₆H₅–NH₂	43
		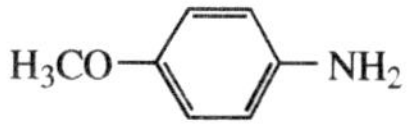 H₃CO–C₆H₄–NH₂	43
		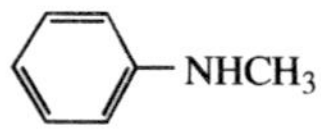 C₆H₅–NHCH₃	46
		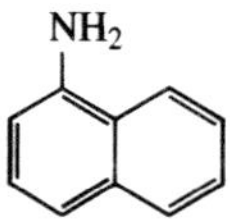 1-aminonaphthalene	81
		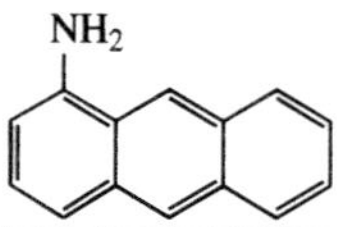 1-aminoanthracene	79

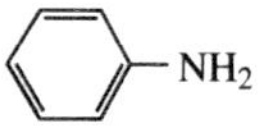

16. The chemistry of magnesium carbenoids   751

TABLE 8. Synthesis of *ortho*-alkenylated arylamines **163** and **164** by reaction of magnesium alkylidene carbenoids with *meta*-substituted *N*-lithio arylamines

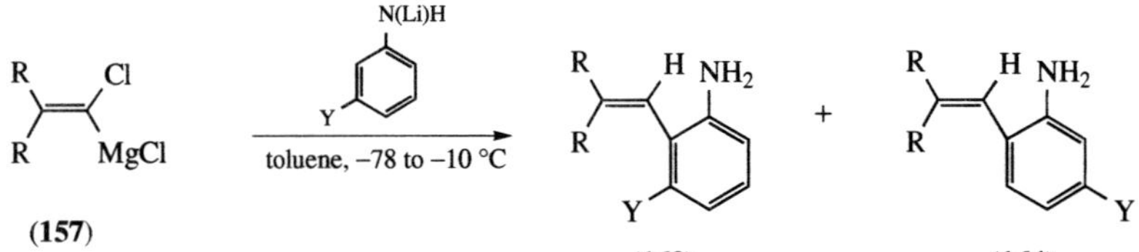

R	R	Arylamine	163 and 164	
			Yield (%)	163:164
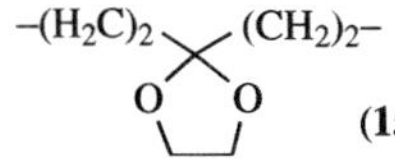 –(H₂C)₂ (CH₂)₂– O O (**157**)		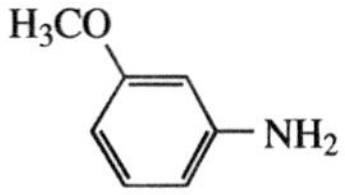 H₃CO–C₆H₄–NH₂	43	30:13
		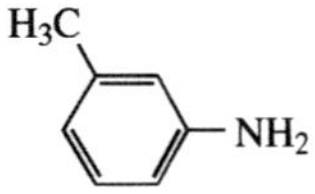 H₃C–C₆H₄–NH₂	33	19:14
		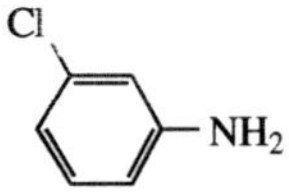 Cl–C₆H₄–NH₂	38	25:13
CH₃	CH₃ (**161**)	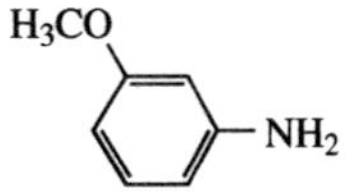 H₃CO–C₆H₄–NH₂	51	29:22

E-**170**. Obviously, the *N*-lithio arylamines attack backside to the chlorine atom to give the products stereospecifically with inversion of the configuration at the sp² carbon.

On the other hand, when this reaction was carried out with 1-chlorovinyl *p*-tolyl sulfoxide derived from unsymmetrical dialkyl ketone **167** with *N*-lithio 1-aminonaphthalene (entries 7 and 8), *Z*-*ortho*-alkenylated arylamine *Z*-**171** was obtained as the main product from both vinyl sulfoxides with low stereoselectivity. The stereospecificity and stereoselectivity mentioned above are explained from the high configurational stability of the magnesium carbenoids generated from 1-chlorovinyl *p*-tolyl sulfoxides derived from α,β-unsaturated ketones.

For a better understanding of the structure and the substitution reactions of magnesium alkylidene carbenoids, computational studies were performed. The result for the calculation of magnesium alkylidene carbenoid **161** is shown in Figure 1, and two different optimized structures were found depending on the basis set used for the theoretical calculations. It seems that the MP2 structure is closer to the reality.

The optimized structures for **173** and for *E*- and *Z*-1-chloro-2-methyl-3-butadienylmagnesium chloride (**172**), which are derived from *E*-**166** and *Z*-**166**, respectively, are shown in Figure 2. There are three characteristic differences between **161** and **172**. In **172**, the

TABLE 9. The reaction of magnesium alkylidene carbenoids derived from E- and Z-1-chlorovinyl p-tolyl sulfoxides with N-lithio aniline and N-lithio 1-aminonaphthalene

Entry	1-Chlorovinyl p-tolyl sulfoxide	Arylamine	Product	$E:Z$	Yield (%)
1	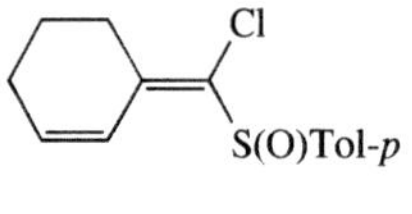 (E-165)	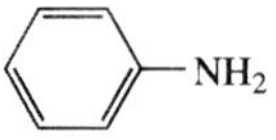 PhNH$_2$	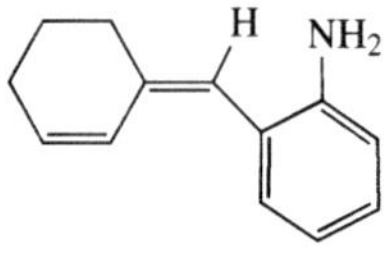 (Z-168)	6:94	53
2	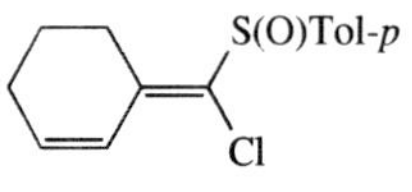 (Z-165)	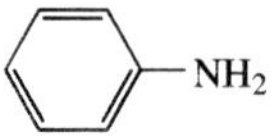 PhNH$_2$	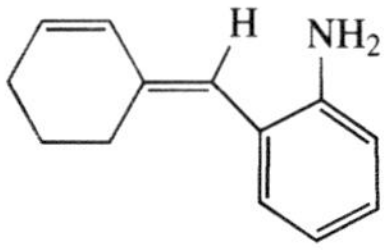 (E-168)	94:6	46
3	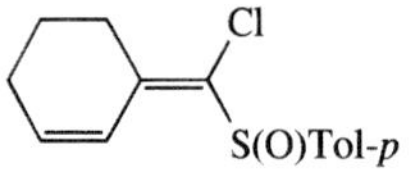 (E-165)	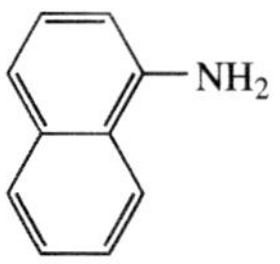 1-naphthyl-NH$_2$	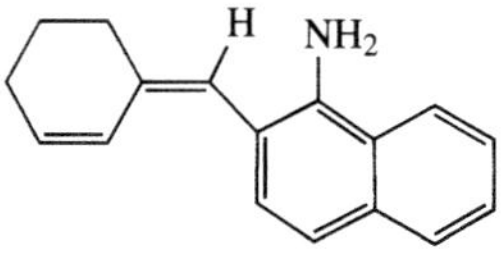 (Z-169)	3:97	65
4	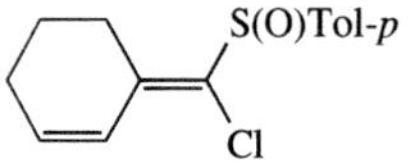 (Z-165)	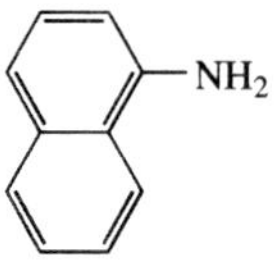 1-naphthyl-NH$_2$	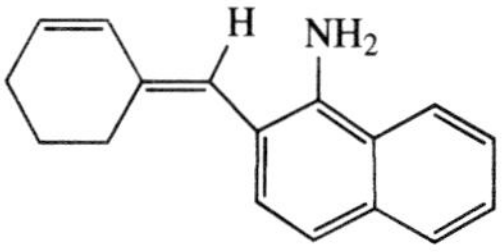 (E-169)	95:5	71
5	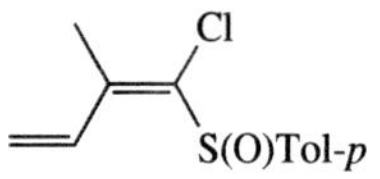 (E-166)	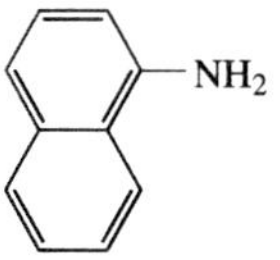 1-naphthyl-NH$_2$	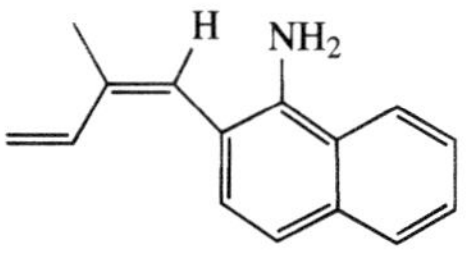 (Z-170)	4:96	68
6	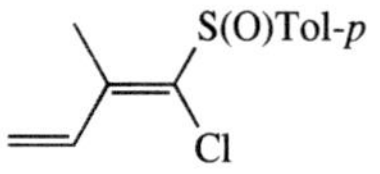 (Z-166)	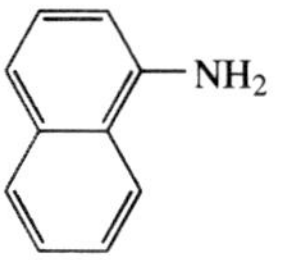 1-naphthyl-NH$_2$	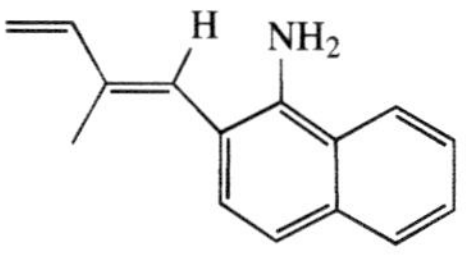 (E-170)	94:6	62

16. The chemistry of magnesium carbenoids 753

TABLE 9. (continued)

Entry	1-Chlorovinyl p-tolyl sulfoxide	Arylamine	Product	E:Z	Yield (%)
7	(E-167) n-H₁₁C₅ / Cl, S(O)Tol-p	1-naphthylamine-NH₂	(E-171) n-H₁₁C₅–C(Me)=CH–(2-(1-amino)naphthyl)	44:56	54
8	(Z-167) n-H₁₁C₅ / S(O)Tol-p, Cl	1-naphthylamine-NH₂	(Z-171)	34:66	55

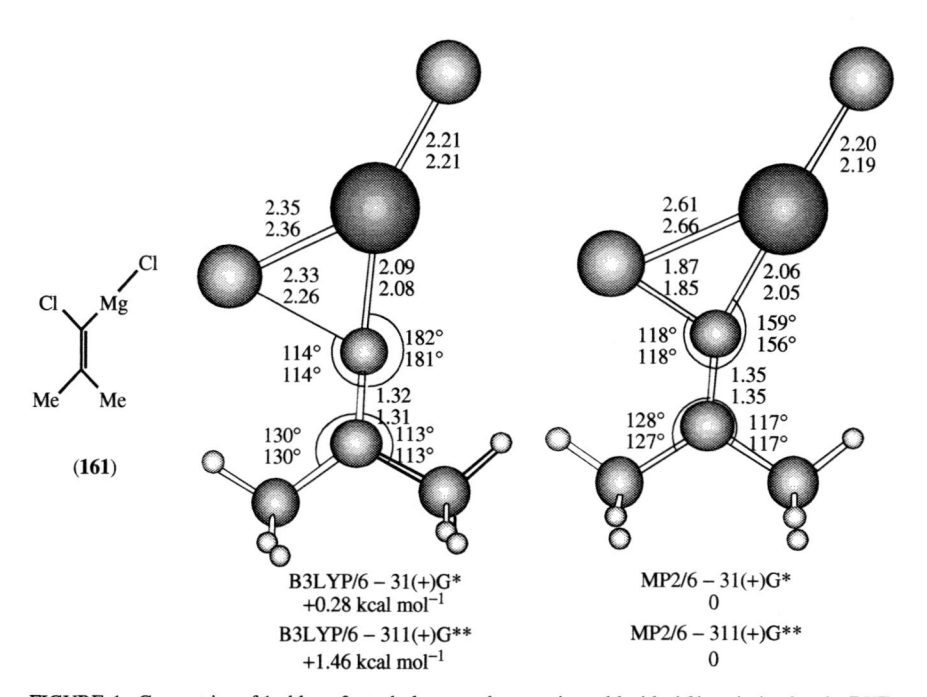

FIGURE 1. Geometries of 1-chloro-2-methylpropenylmagnesium chloride **161** optimized at the RHF, B3LYP and MP2 levels of theory with the 6-31(+)G* and 6-311(+)G** basis sets. The energies of these geometries were calculated by the CCSD(T) method with the corresponding basis set

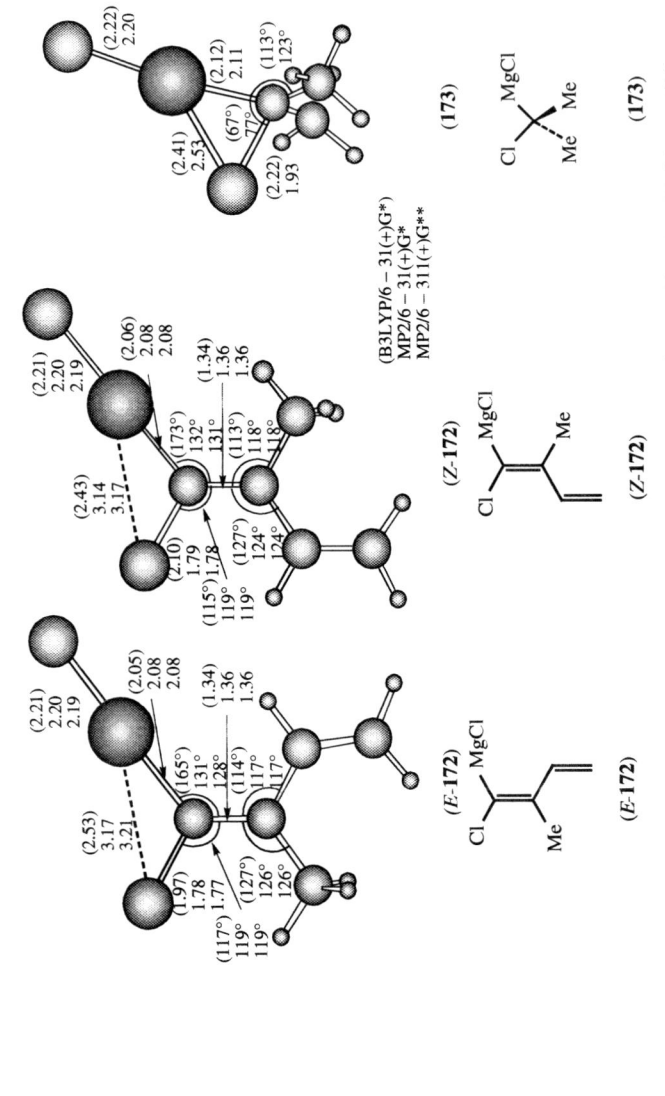

FIGURE 2. Geometries of *E*-**172**, *Z*-**172** and **173** optimized at B3LYP/6-31(+)G*, MP2/6-31(+)G* and MP2/6-311(+)G**

distance between the Mg and the vinyl–Cl is much longer, the C–Cl bond is shorter and the C=C–Mg angle is smaller as compared to **161**. Thus, the conjugated system is geometrically stabilized.

4. Reaction of magnesium alkylidene carbenoids with N-lithio nitrogen-containing heterocycles, lithium acetylides and lithium thiolates

Magnesium alkylidene carbenoids were found to be reactive with different nucleophiles to give new alkenylmagnesium compounds which could be trapped with electrophiles. As a whole, novel methods for the synthesis of tri- or tetrasubstituted olefins from the 1-chlorovinyl *p*-tolyl sulfoxides in a one-pot reaction are realized.

Thus, treatment of magnesium alkylidene carbenoid **157** with *N*-lithio indole in a toluene/ether mixture of solvents resulted in the formation of *N*-alkenylated indole **175** in 57% yield through the alkenylmagnesium intermediate **174** (equation 41)[43]. The generality of this reaction to give **176** was investigated and the results are shown in Table 10. Indazole gave the desired *N*-alkenylated product in 51% yield; however, pyrazole gave only 15% yield. Phenothiazine and phenoxazine gave good yields of the products. Interestingly, the simplest heterocycle pyrrole gave the *N*-alkenylated product in only 14% yield whereas the 2-alkenylated pyrrole (**177**) was obtained as a main product in 56% yield.

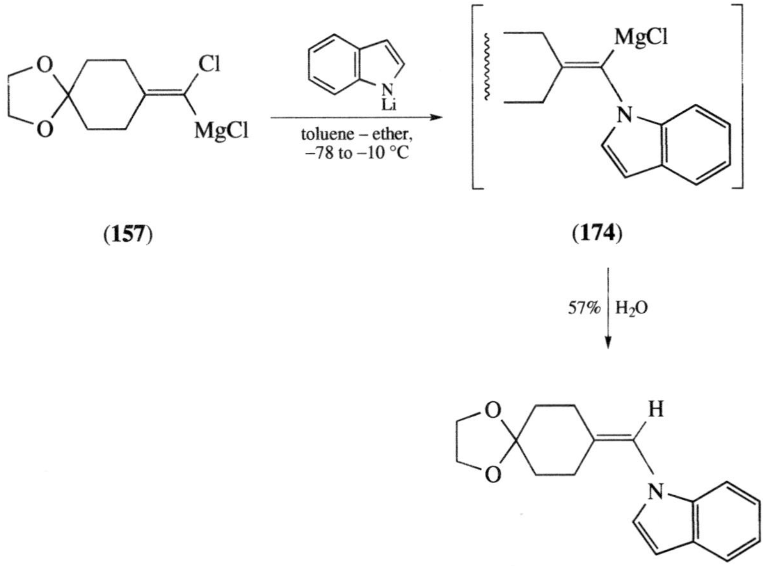

The stereochemistry of this reaction was also investigated (equation 42). Thus, geometrical isomers of 1-chlorovinyl *p*-tolyl sulfoxides Z-**178** and E-**178** were synthesized from 4-phenyl-2-butanone and they were treated with *i*-PrMgCl followed by *N*-lithio indole. As shown in equation 42, this reaction gave a mixture of isomers (E-**179** and Z-**179**) in relatively good yields but with low stereoselectivity.

TABLE 10. Reaction of magnesium alkylidene carbenoid **157** with N-lithio nitrogen-containing heterocycles

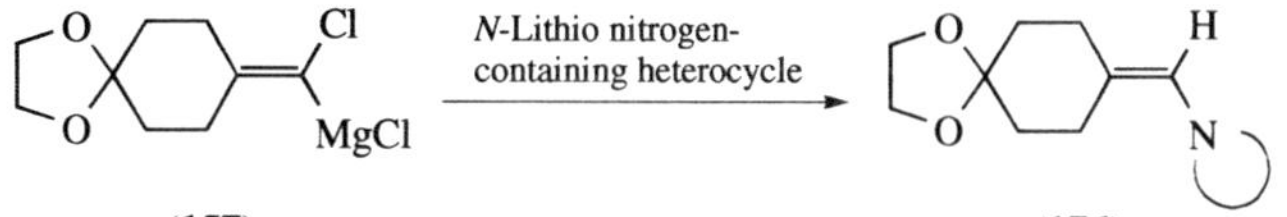

N-Lithio nitrogen-containing heterocycle	**176** Yield (%)
indazol-N-Li	51
pyrazol-N-Li	15
phenothiazin-N-Li	71
phenoxazin-N-Li	70
pyrrol-N-Li	14 (normal **176**) / 56 (product **177**, 2-substituted NH pyrrole)

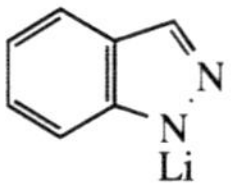

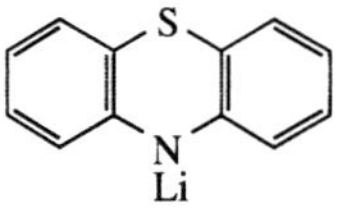

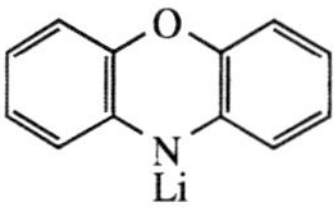

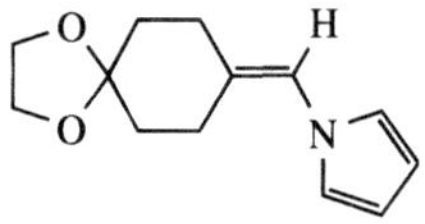

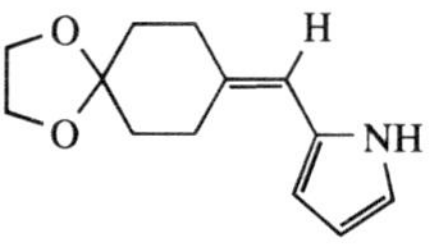

From a more synthetic perspective, reaction of the alkenylmagnesium derivatives with electrophiles is very interesting. Indeed, if the intermediates could be trapped with electrophiles, the reaction would provide a novel route to the preparation of nitrogen-containing heterocycles having a fully substituted enamine structure. This expectation proved to be possible (Table 11)[43].

The reaction of the magnesium alkylidene carbenoid **157** with N-lithio phenothiazine gave the alkenylmagnesium intermediate **180** in good yield. Quenching the reaction with deuteriomethanol gave the deuterio olefin (**181**, Table 11, E = D) in 71% yield with a deuterium incorporation of 98%. The reaction with iodomethane did not take place; however, using 5 mol% of CuI as a catalyst at room temperature resulted in the formation of the methylated olefin in 62% yield. The alkylation and allylation required CuI as a catalyst. Benzoyl chloride and phenyl isocyanate reacted with the alkenylmagnesium intermediate **180** to give the desired products (**181**).

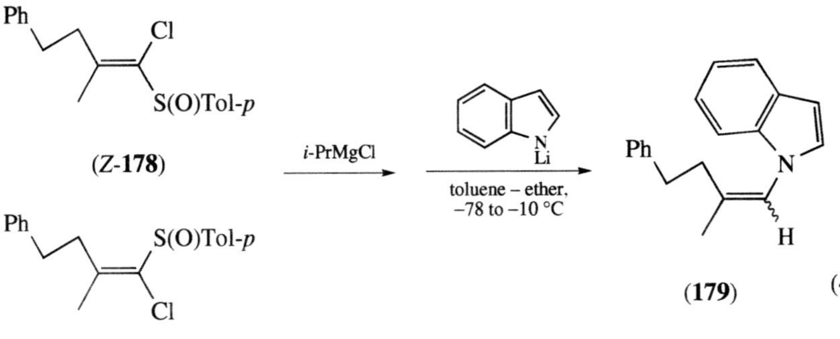

16. The chemistry of magnesium carbenoids

(Z-178) / (E-178) → (179) via i-PrMgCl, indole-Li, toluene–ether, −78 to −10 °C

(42)

178	179 / Yield (%)	E-179 : Z-179
Z-178	64	44 : 56
E-178	65	29 : 71

TABLE 11. Synthesis of phenothiazine having a fully substituted enamine structure (**181**)

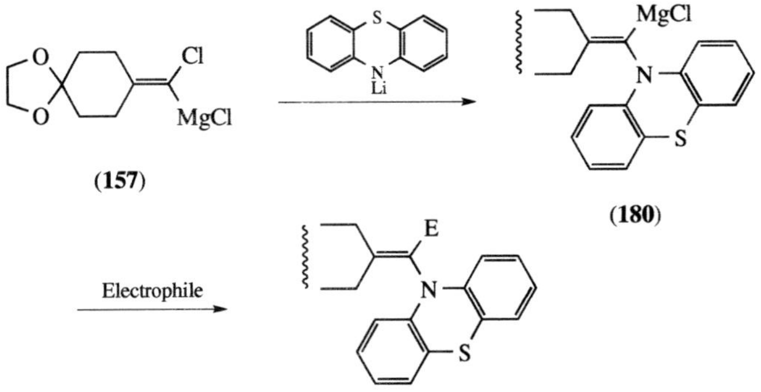

(**157**) + phenothiazine-Li → (**180**) → Electrophile → (**181**)

Electrophile	181	
	E	Yield (%)
CH_3OD	D	71[a]
CH_3I	CH_3	62[b]
CH_3CH_2I	CH_3CH_2	55[b]
$CH_2=CHCH_2I$	$CH_2=CHCH_2$	63[b]
$PhCH_2Br$	$PhCH_2$	30[b]
PhCOCl	PhCO	59
PhNCO	PhNHCO	39

[a] Deuterium content 98%.
[b] The reaction was carried out with CuI as a catalyst.

TABLE 12. Synthesis of enyne **183** from magnesium alkylidene carbenoids **134a** with lithium acetylides and electrophiles

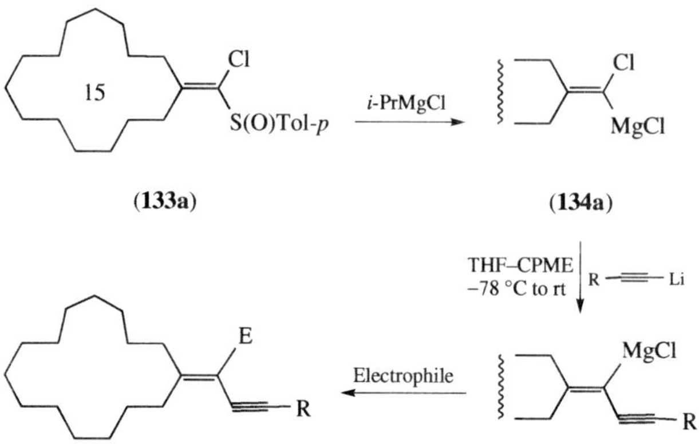

R	Electrophile	183	
		E	Yield (%)
$CH_3(CH_2)_3$	H_2O	H	63
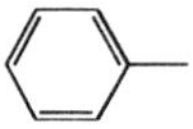 phenyl	H_2O	H	41
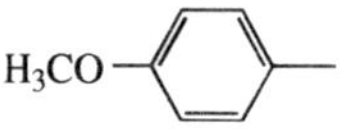 H_3CO-C$_6$H$_4$-	H_2O	H	49
 F-C$_6$H$_4$-	H_2O	H	38
$(CH_3)_3Si$	H_2O	H	16
$CH_3(CH_2)_3$	I_2	I	44

Lithium acetylides were found to react with magnesium alkylidene carbenoids to afford enynes (Table 12)[44]. Thus, magnesium alkylidene carbenoid **134a**, generated from 1-chlorovinyl p-tolyl sulfoxide **133a**, reacts with 1-hexynyl lithium (3 equivalents) to give, via **182**, the conjugated enyne **183** in 63% yield after hydrolysis. In this reaction, the presence of cyclopentylmethyl ether (CPME) as an additive was found to be effective. The scope of this reaction was investigated and the results are summarized in Table 12. Unfortunately, yields were moderate.

The reaction of magnesium alkylidene carbenoids with lithium thiolates gave trisubstituted alkenyl sulfides (**185**) in good yields through the alkenylmagnesium intermediates **184** (Table 13)[44]. Thus, to the magnesium alkylidene carbenoid **157**, prepared in toluene at −78°C, was added lithium p-toluenethiolate (3 equivalents) and alkenylsulfide **185** was obtained in 80% yield after hydrolysis. In this reaction, the presence of 1,2-dimethoxyethane (DME) as an additive was found to be effective. As shown in Table 13, the reaction with arenethiolates gave better yields as compared to the reaction with alkyl thiolates.

16. The chemistry of magnesium carbenoids

TABLE 13. Synthesis of vinyl sulfides (**185**) by the reaction of magnesium alkylidene carbenoids with lithium thiolates

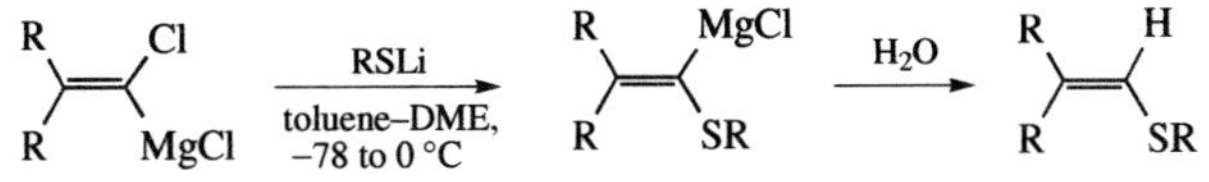

$$\underset{R}{\overset{R}{>}}\!\!=\!\!\underset{MgCl}{\overset{Cl}{<}} \xrightarrow[\text{toluene–DME,} \\ -78 \text{ to } 0\,°\text{C}]{RSLi} \underset{R}{\overset{R}{>}}\!\!=\!\!\underset{SR}{\overset{MgCl}{<}} \xrightarrow{H_2O} \underset{R}{\overset{R}{>}}\!\!=\!\!\underset{SR}{\overset{H}{<}}$$

(**184**) (**185**)

R	R	Lithium thiolate	**185**/Yield (%)
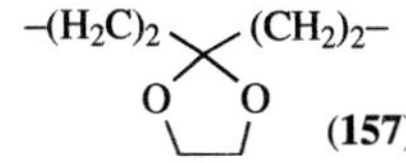 $-(H_2C)_2\underset{O\quad O}{\diagup}(CH_2)_2-$ (**157**)		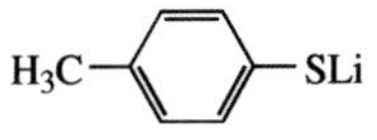 $H_3C-\!\!\bigcirc\!\!-SLi$	80
		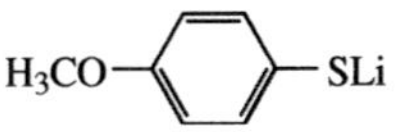 $H_3CO-\!\!\bigcirc\!\!-SLi$	82
		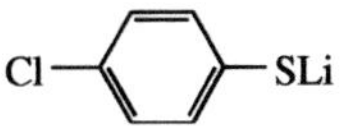 $Cl-\!\!\bigcirc\!\!-SLi$	72
		$n\text{-}C_{12}H_{25}SLi$	60
		$(CH_3)_3CSLi$	39
$-(CH_2)_{14}-$	(**134a**)	 $H_3C-\!\!\bigcirc\!\!-SLi$	51
CH_3	CH_3 (**161**)	 $H_3C-\!\!\bigcirc\!\!-SLi$	77
Ph	Ph	 $H_3C-\!\!\bigcirc\!\!-SLi$	68

TABLE 14. Synthesis of vinyl sulfides (**186**) by the reaction of magnesium alkylidene carbenoids with lithium *p*-toluenethiolate followed by some electrophiles

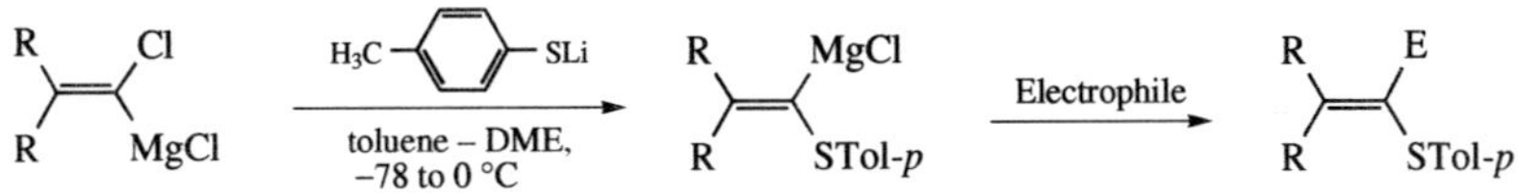

$$\underset{R}{\overset{R}{>}}\!\!=\!\!\underset{MgCl}{\overset{Cl}{<}} \xrightarrow[\text{toluene – DME,} \\ -78 \text{ to } 0\,°\text{C}]{H_3C-\bigcirc-SLi} \underset{R}{\overset{R}{>}}\!\!=\!\!\underset{STol\text{-}p}{\overset{MgCl}{<}} \xrightarrow{\text{Electrophile}} \underset{R}{\overset{R}{>}}\!\!=\!\!\underset{STol\text{-}p}{\overset{E}{<}}$$

(**184**) (**186**)

R	R	Electrophile	**186** E	Yield (%)
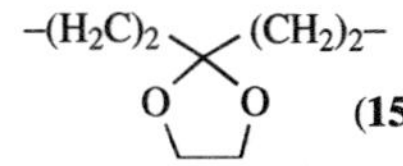 $-(H_2C)_2\underset{O\quad O}{\diagup}(CH_2)_2-$ (**157**)		D_2O	D	80[a]
		PhCHO	PhCHOH	64
		CH_3CH_2CHO	CH_3CH_2CHOH	39
		PhCOCl	PhCO	54
		ClCOOEt	EtOCO	11
		I_2	I	50
CH_3 (**161**)	CH_3	PhCHO	PhCHOH	58
		I_2	I	72

[a] Deuterium content 98%.

Reaction of the alkenylmagnesium intermediates **184** with several electrophiles led to trisubstituted alkenyl sulfides **186** (Table 14). Thus, when the reaction was quenched with D_2O, the deuteriated vinyl sulfide (**186**; E = D) was obtained in 80% yield with 98% deuterium incorporation. The reaction with aldehydes, benzoyl chloride and iodine gave moderate to good yields of the desired functionalized alkenyl sulfides (**186**); however, ethyl chloroformate did not give good result.

Development of new synthetic methods from aryl 1-chlorovinyl sulfoxides including the chemistry of magnesium alkylidene carbenoids has been reviewed by the author[45].

VI. MAGNESIUM β-OXIDO CARBENOIDS

A one-carbon homologation of carbonyl compounds is an important and extensively used method for the preparation of desired carbonyl compounds[46]. One-carbon ring-expansion[47] or one-carbon homologation of ketones or aldehydes via a β-oxido carbenoid is a representative example of the homologation, but few methods have been reported[48,49].

For example, as shown in equation 43, Taguchi, Nozaki and coworkers reported in 1974 a one-carbon ring enlargement of cyclododecanone (**187**) to cyclotridecanone (**190**) with dibromomethyllithium through β-oxido carbenoid (**188**)[48a,c]. This reaction was expected to proceed via a one-carbon expanded enolate (**189**). Cohen and coworkers used the bis(phenylthio)methyllithium[49] whereas Satoh and coworkers used α-sulfinyl lithium carbanion of 1-chloroalkyl aryl sulfoxides as the source of β-oxido carbenoids (equation 44)[50].

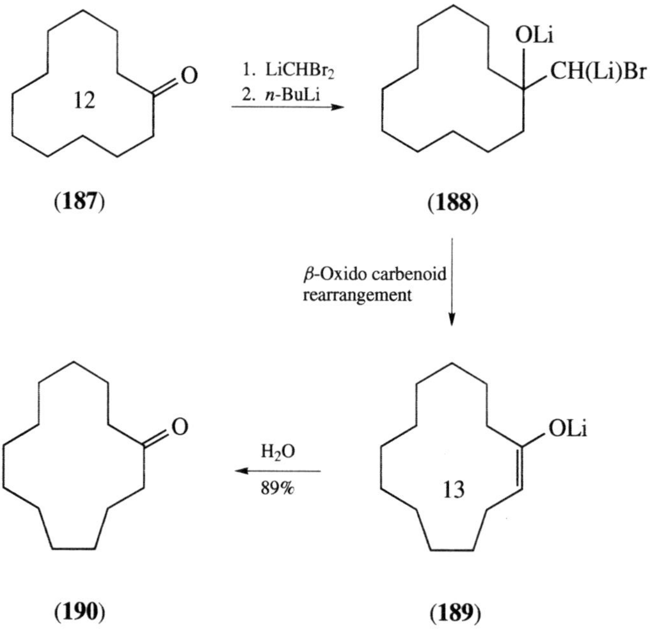

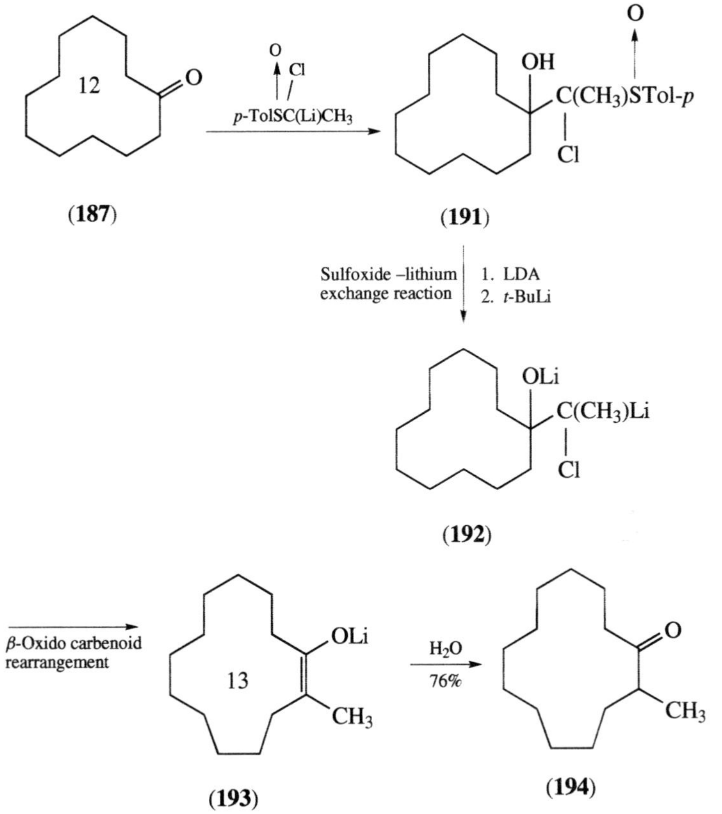

16. The chemistry of magnesium carbenoids

[Scheme showing conversion of **187** (cyclododecanone) via p-TolSC(Li)CH₃ to adduct **191**, then via sulfoxide–lithium exchange reaction (1. LDA, 2. t-BuLi) to β-oxido carbenoid **192**, then via β-oxido carbenoid rearrangement to one-carbon elongated enolate **193** (13-membered ring with OLi and CH₃), then H₂O (76%) to cyclotridecanone **194** with α-methyl group] (44)

Thus, treatment of α-sulfinyl lithium carbanion of 1-chloroethyl p-tolyl sulfoxide with cyclododecanone gave the adduct **191** in high yield. This adduct was further treated with LDA (to form the lithium alkoxide) followed by tert-butyllithium to give the β-oxido carbenoid **192** via the sulfoxide–lithium exchange reaction. The β-oxido carbenoid rearrangement then takes place to afford one-carbon elongated enolate **193**, which was finally treated with water to give a one-carbon homologated cyclotridecanone having a methyl group at the α-position (**194**) in 76% yield.

Satoh and coworkers further investigated this reaction and found that, in some cases, magnesium β-oxido carbenoids gave better results. Trapping of the enolate intermediates with several electrophiles was successfully carried out and a new method for the synthesis of one-carbon expanded cyclic α,α-disubstituted ketones from lower cyclic ketones was realized. An example using 1,4-cyclohexanedione mono ethylene ketal (**195**) as a representative cyclic ketone is shown in Table 15[51].

Thus, α-sulfinyl lithium carbanion of 1-chloroethyl p-tolyl sulfoxide was reacted with 1,4-cyclohexanedione mono ethylene ketal (**195**) to afford the adduct (**196**) in quantitative yield. The adduct was treated with tert-butylmagnesium chloride (magnesium alkoxide was initially formed) followed by isopropylmagnesium chloride to result in the formation of magnesium β-oxido carbenoid **197**. The β-oxido carbenoid rearrangement then takes place to give one-carbon expanded magnesium enolate **198**. Finally, an electrophile was

TABLE 15. Synthesis of 2-methyl-2-(substituted)cycloheptanones (**199**) from cyclohexanone derivative **195**, 1-chloroethyl *p*-tolyl sulfoxide, and electrophiles

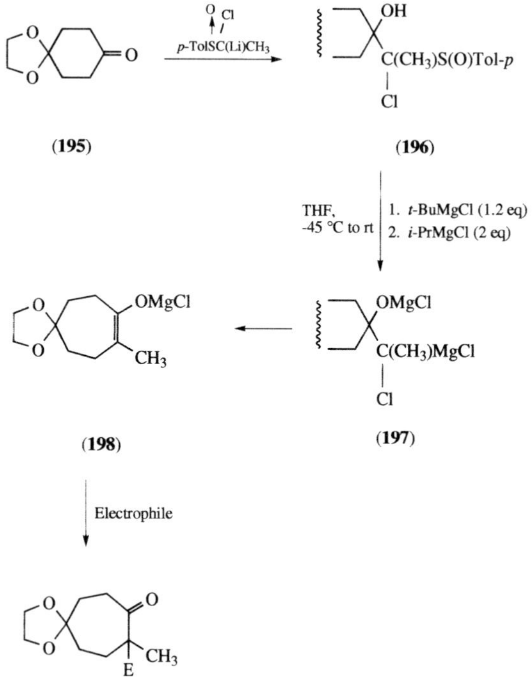

Electrophile	199	
	E	Yield (%)
CD_3OD	D	75[a]
CH_3CH_2CHO	CH_3CH_2CHOH	71
PhCHO	PhCHOH	64
PhCOCl	PhCO	63
CH_3I	CH_3	73[b]
$PhCH_2Br$	$PhCH_2$	75[b]
$CH_2=CHCH_2I$	$CH_2=CHCH_2$	59[b]
ClCOOEt	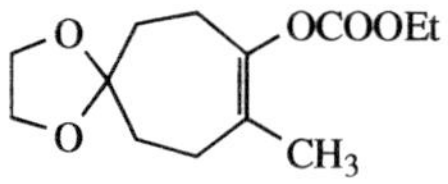 (**200**)	74
Et_3SiCl	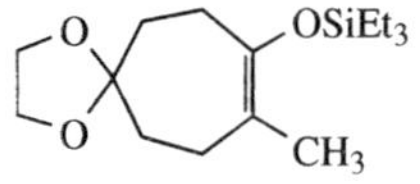 (**201**)	75[b]

[a] Deuterium content 95%.
[b] HMPA was added as an additive.

added to the reaction mixture to give one-carbon expanded ketone having a quaternary center at the α-position (**199**).

Quenching this reaction with deuteriomethanol gave 2-methylcycloheptanone having deuterium at the 2-position (**199**; E = D) in 75% yield with 95% deuterium incorporation. Aldehydes and benzoyl chloride gave the desired products in 60–70% yields. Alkylation of the enolate intermediate (**198**) was successfully carried out with alkyl halides in the presence of HMPA in good yields. The reaction with ethyl chloroformate and chlorotriethylsilane gave enol carbonate (**200**) and silyl enol ether (**201**) in 74 and 75% yield, respectively.

This chemistry was found to be applicable to large-membered cyclic ketones, for example cyclopentadecanone, and selected results are shown in Scheme 7. The reaction of α-sulfinyl lithium carbanion of 1-chloroethyl *p*-tolyl sulfoxide gave the adduct in almost quantitative yield. The adduct was treated with *tert*-butylmagnesium chloride followed by isopropylmagnesium chloride to result in the formation of magnesium β-oxido carbenoid **202**. The β-oxido carbenoid rearrangement then takes place to give one-carbon expanded magnesium enolate **203**. Quenching of this enolate intermediate with propanal afforded α,α-disubstituted cyclohexadecanone (**204**) in 73% yield. Benzoyl chloride gave α-benzoylated ketone (**204**; E = COPh) in 71% yield. Interestingly, the reaction with ethyl chloroformate gave not the enol carbonate but an α-ethoxycarbonylated product (**204**; E = COOEt), though the yield was not satisfactory.

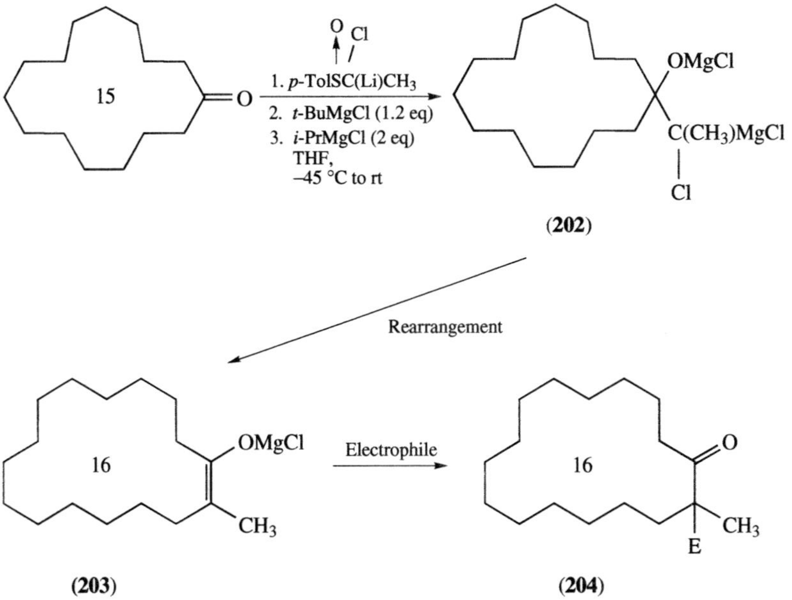

Electrophile: CH_3CH_2CHO E = CH_3CH_2CHOH 73%
PhCOCl E = PhCO 71%
ClCOOEt E = COOEt 33%

SCHEME 7

The chemistry mentioned above could also be applied to aldehydes. As an example, *p*-anisaldehyde and 1-chlorobutyl *p*-tolyl sulfoxide were used as shown in Table 16. Thus, treatment of α-sulfinyl lithium carbanion of 1-chlorobutyl *p*-tolyl sulfoxide with

TABLE 16. Synthesis of α-substituted ketones (**208**) from *p*-anisaldehyde, 1-chlorobutyl *p*-tolyl sulfoxide and electrophiles

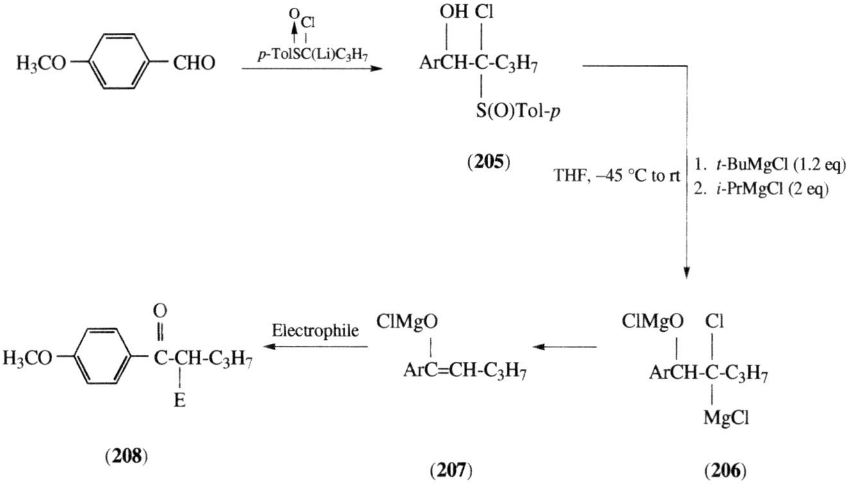

$H_3CO-C_6H_4-CHO$ $\xrightarrow{p\text{-TolSC(Li)}C_3H_7}$ ArCH(OH)-C(Cl)(S(O)Tol-*p*)-C$_3$H$_7$

(**205**)

THF, −45 °C to rt | 1. *t*-BuMgCl (1.2 eq)
2. *i*-PrMgCl (2 eq)

$H_3CO-C_6H_4-C(=O)-CH(E)-C_3H_7$ ⟵ Electrophile ⟵ ArC(OMgCl)=CH-C$_3$H$_7$ ⟵ ArCH(OMgCl)-C(Cl)(MgCl)-C$_3$H$_7$

(**208**)   (**207**)   (**206**)

Electrophile	208	
	E	Yield (%)
CH$_3$CH$_2$CHO	CH$_3$CH$_2$CHOH	75
CH$_3$I	CH$_3$	99[a]
PhCH$_2$Br	PhCH$_2$	85[a]
CH$_2$=CHCH$_2$I	CH$_2$=CHCH$_2$	96[a]
ClCOOEt	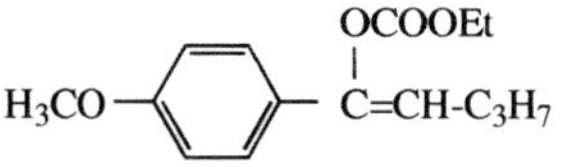 H$_3$CO-C$_6$H$_4$-C(OCOOEt)=CH-C$_3$H$_7$ (**209**)	71

[a] HMPA was added as an additive.

p-anisaldehyde gave a mixture of two diastereoisomers (**205**) in quantitative yield. Treatment of the main isomer with *tert*-butylmagnesium chloride followed by isopropylmagnesium chloride resulted in the formation of magnesium β-oxido carbenoid **206**. In this case, rearrangement of the hydrogen on the carbon bearing the oxygen took place to give magnesium enolate **207**. Quenching this enolate intermediate with propanal gave the desired α-substituted ketone (**208**) in 75% yield. Alkylation of the enolate **207** with iodomethane, benzyl bromide and allyl iodide gave α-alkylated ketones **208** in good to quantitative yields. The reaction of the enolate **207** with ethyl chloroformate gave again the enol carbonate **209** in 71% yield.

Application of the method described above to unsymmetrical cyclic ketones such as 2-substituted cyclohexanones gave 2,7-disubstituted and 2,2,7-trisubstituted cycloheptanones (Scheme 8)[52]. Treatment of α-sulfinyl lithium carbanion of 1-chloroethyl *p*-tolyl sulfoxide with 2-substituted cyclohexanones (**210a** and **210b**) afforded adducts as a mixture of two diastereomers. The main adducts were first treated with *t*-BuMgCl followed by *i*-PrMgCl (4 equiv) at 0°C to room temperature to give the magnesium β-oxido carbenoid **211**. The

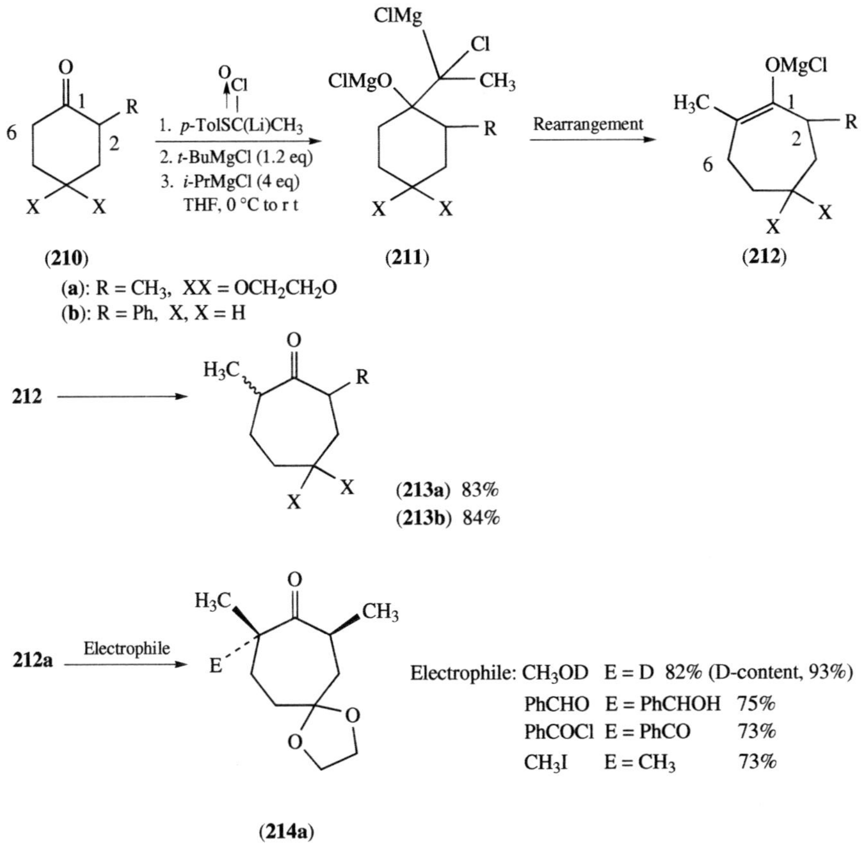

16. The chemistry of magnesium carbenoids

SCHEME 8

(210)
(a): R = CH$_3$, XX = OCH$_2$CH$_2$O
(b): R = Ph, X, X = H

Reagents: 1. *p*-TolSC(Li)CH$_3$; 2. *t*-BuMgCl (1.2 eq); 3. *i*-PrMgCl (4 eq); THF, 0 °C to rt → (211) → Rearrangement → (212)

212 → (213a) 83%; (213b) 84%

212a + Electrophile → (214a)

Electrophile: CH$_3$OD   E = D   82% (D-content, 93%)
PhCHO   E = PhCHOH   75%
PhCOCl   E = PhCO   73%
CH$_3$I   E = CH$_3$ 73%

β-oxido carbenoid rearrangement then took place to afford one-carbon ring-expanded magnesium enolates **212**. Hydrolysis of the magnesium enolate afforded 2,7-disubstituted cycloheptanone derivatives **213a** and **213b** in 83 and 84% yields, respectively.

Interestingly, the formation of **213** implies that the carbon–carbon insertion took place between the carbons C$_1$ and C$_6$ of the starting cyclohexanone (**210**). The migrating group is not the same as that usually reported in this type of rearrangement[48]. In addition, magnesium enolate intermediate **212** could be trapped with several electrophiles such as benzaldehyde, benzoyl chloride and iodomethane to give 2,2,7-trisubstituted cycloheptanones (**214a**) in good yields. This method is very useful for the synthesis of 2,7-disubstituted and 2,2,7-trisubstituted cycloheptanones from 2-substituted cyclohexanones with one-carbon ring-expansion in only two chemical steps.

Using dichloromethyl phenyl sulfoxide in this procedure as a one-carbon homologating agent gave interesting results (equation 45)[53]. Thus, treatment of the α-sulfinyl lithium carbanion of dichloromethyl phenyl sulfoxide at -60°C with cyclobutanone and cyclopentanone gave the adducts **215** in almost quantitative yields. The sulfoxide–magnesium exchange reaction of the adducts **215** with EtMgBr gave magnesium β-oxido carbenoids

216. The β-oxido carbenoid rearrangement then took place to give one-carbon expanded magnesium enolate having a chlorine atom (**217**), which was treated with water to afford α-chloroketone (**218**) in moderate yield. Unfortunately, this method could not be applied to larger cycloalkanones and acyclic ketones. Application of this method to aldehydes gave chloromethyl aryl ketones and chloromethyl alkyl ketones in moderate yields[53].

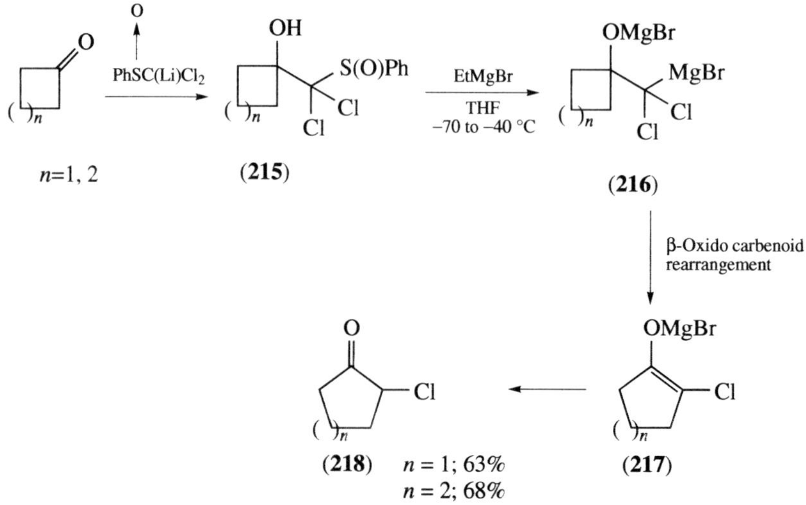

Finally, magnesium carbenoid α to carbonyl moiety was reported (equation 46)[53]. The α-sulfinyl lithium carbanion of chloromethyl phenyl sulfoxide was reacted with methyl esters to give α-chloro-α-sulfinylmethyl ketones (**219**) in 80–95% yields. Treatment of **219** with EtMgBr in THF at low temperature resulted in the formation of magnesium carbenoid **220** and α-chloroketones **222** in moderate yields after hydrolysis. Obviously, the sulfoxide–magnesium exchange reaction of **219** proceeded to give magnesium carbenoid **220**; however, Wolff-type rearrangement does not take place, but instead, magnesium enolates (**221**) were produced in this reaction.

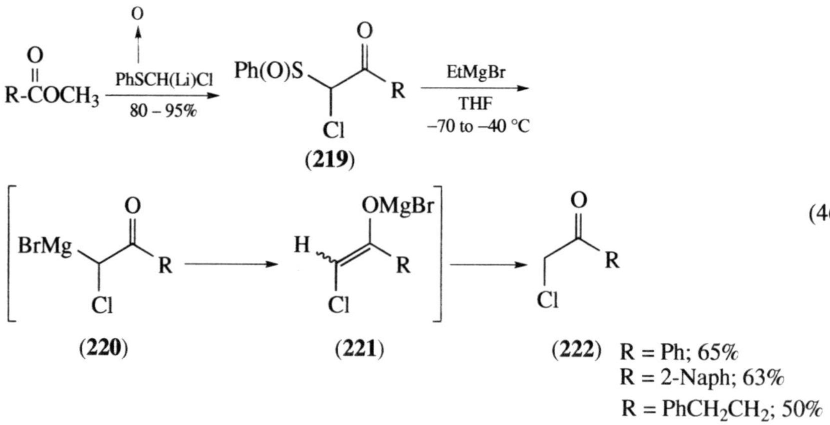

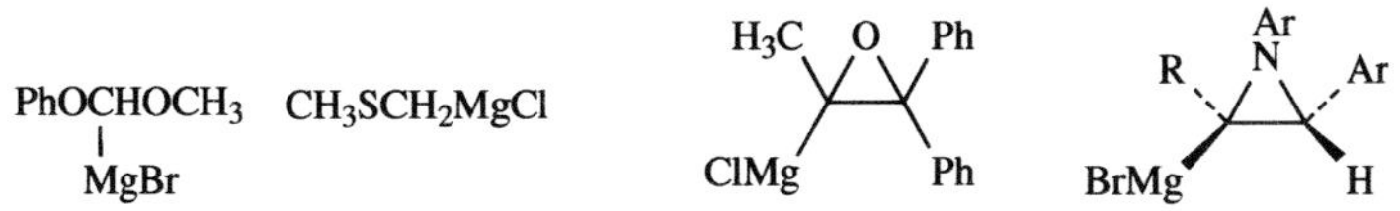

16. The chemistry of magnesium carbenoids

PhOCHOCH$_3$ | MgBr

(223)

CH$_3$SCH$_2$MgCl

(224)

[structure: H$_3$C, O, Ph, ClMg, Ph]

(225)

[structure: Ar, N, Ar, R, BrMg, H]

(226)

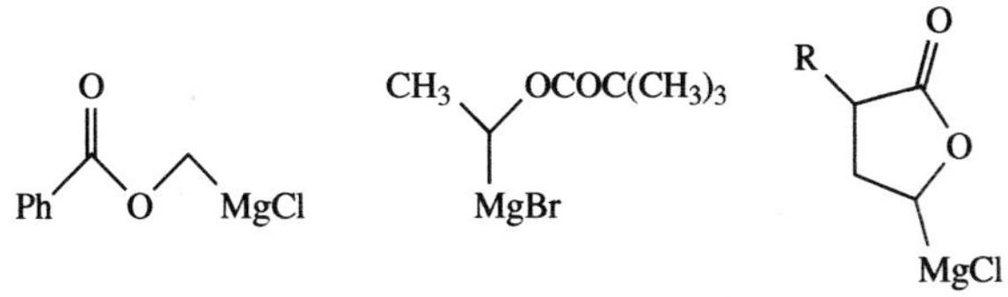

[structure: Ph–C(=O)–O–CH$_2$–MgCl]

(227)

[structure: CH$_3$–CH(OCOC(CH$_3$)$_3$)–MgBr]

(228)

[structure: R-substituted γ-butyrolactone with MgCl]

(229)

FIGURE 3

VII. OTHER MAGNESIUM CARBENOIDS

Some other α-heteroatom-substituted Grignard reagents, such as magnesioacetal (223)[54], methylthiomethyl Grignard reagent 224[55], oxiranyl Grignard reagent 225[56], aziridinylmagnesium halide (226)[57], functionalized magnesim carbenoids 227 and 228[58] and γ-magnesio-γ-butyrolactones (229)[59], have been reported (Figure 3).

In some cases, they act as magnesium carbenoids; however, they usually react as α-heterosubstituted Grignard reagents. Therefore, the author will not address the chemistry of these α-heteroatom-substituted Grignard reagents in detail.

VIII. CONCLUSIONS AND PERSPECTIVE

As outlined in this chapter, magnesium carbenoids are relatively stable compounds as compared to the corresponding lithium carbenoids. Therefore, we can prepare the corresponding carbenoids in a similar way to the usual preparation of more classical Grignard reagents. Generation of the magnesium carbenoids can be performed mainly in two ways: the halogen–magnesium exchange and the sulfoxide–magnesium exchange reactions at low temperature, usually at −78°C. As mentioned above, the preparation of magnesium carbenoids, from sulfoxides having a halogen on the α-position using the sulfoxide–magnesium exchange reaction, has a much higher versatility than the halogen–magnesium exchange reaction. The magnesium carbenoids show both nucleophilic and electrophilic properties; however, the electrophilic reaction of the magnesium carbenoids is far more interesting for synthetic purposes as mentioned above.

The chemistry of magnesium carbenoids started practically in the last 20 years; in other words, it is rather a new field in chemistry. Many new and interesting results will be forthcoming from this field.

IX. REFERENCES

1. In this chapter, the term 'carbenoid' is suggested for the description of intermediates which exhibit reactions qualitatively similar to those of free carbenes.
2. Some monographs and reviews concerning carbenes and carbenoids:
   (a) W. Kirmse, *Carbene Chemistry*, Academic Press, New York, 1971.
   (b) F. Z. Dorwald, *Metal Carbenes in Organic Synthesis*, Wiley-VCH, Weinheim, 1999.
   (c) G. Bertrand (Ed.), *Carbene Chemistry*, Marcel Dekker, New York, 2002.
   (d) G. Kobrich, *Angew. Chem., Int. Ed. Engl.*, **11**, 473 (1972).

(e) P. J. Stang, *Chem. Rev.*, **78**, 383 (1978).
(f) S. D. Burke and P. A. Grieco, *Org. React.*, **26**, 361 (1979).
(g) H. F. Schaefer III, *Acc. Chem. Res.*, **12**, 288 (1979).
(h) H. Wynberg and E. W. Meijer, *Org. React.*, **28**, 1 (1982).
(i) A. Oku and T. Harada, *J. Synth. Org. Chem. Jpn.*, **44**, 736 (1986).
(j) A. Oku, *J. Synth. Org. Chem. Jpn.*, **48**, 710 (1990).
(k) A. Padwa and K. E. Krumpe, *Tetrahedron*, **48**, 5385 (1992).
(l) A. Padwa and M. D. Weingarten, *Chem. Rev.*, **96**, 223 (1996).
(m) M. Braun, *Angew. Chem. Int. Ed.*, **37**, 430 (1998).
3. D. Seyferth, R. L. Lambert, Jr. and E. M. Hanson, *J. Organomet. Chem.*, **24**, 647 (1970).
4. R. W. Hoffmann and A. Kusche, *Chem. Ber.*, **127**, 1311 (1994).
5. V. Schulze, P. G. Nell, A. Burton and R. W. Hoffmann, *J. Org. Chem.*, **68**, 4546 (2003).
6. V. Schulze, M. Bronstrup, V. P. W. Bohm, P. Schwerdtfeger, M. Schimeczek and R. W. Hoffmann, *Angew. Chem. Int. Ed.*, **37**, 824 (1998).
7. V. Schulze, R. Lowe, S. Fau and R. W. Hoffmann, *J. Chem. Soc., Perkin Trans. 2*, 463 (1998).
8. (a) S. Oae, *Reviews on Heteroatom Chemistry*, MYU, Tokyo, **4**, 196 (1991).
(b) B. J. Wakefield, *Organomagnesium Methods in Organic Synthesis*, Academic Press, London, 1995, p. 58.
9. (a) H. Uno, F. Semba, T. Tasaka and H. Suzuki, *Chem. Lett.*, 309 (1989).
(b) H. Uno, K. Sakamoto, F. Semba and H. Suzuki, *Bull. Chem. Soc. Jpn.*, **65**, 210 (1992).
(c) H. Uno, K. Sakamoto, F. Semba and H. Suzuki, *Bull. Chem. Soc. Jpn.*, **65**, 218 (1992).
10. (a) T. Satoh and K. Takano, *Tetrahedron*, **52**, 2349 (1996).
(b) T. Satoh, *J. Synth. Org. Chem. Jpn.*, **61**, 98 (2003).
11. (a) T. Satoh, T. Oohara, Y. Ueda and K. Yamakawa, *Tetrahedron Lett.*, **29**, 313 (1988).
(b) T. Satoh, T. Oohara, Y. Ueda and K. Yamakawa, *J. Org. Chem.*, **54**, 3130 (1989).
12. R. W. Hoffmann and P. G. Nell, *Angew. Chem. Int. Ed.*, **38**, 338 (1999).
13. P. R. Blakemore, S. P. Marsden and H. D. Vater, *Org. Lett.*, **8**, 773 (2006).
14. (a) A. Hassner and C. Stumer, *Organic Synthesis Based on Name Reactions*, Second edition, Pergamon, Amsterdam, 2002.
(b) B. P. Mundy, M. G. Ellerd and F. G. Favaloro, Jr., *Name Reactions and Reagents in Organic Synthesis*, Second edition, Wiley-Interscience, New York, 2005.
(c) L. Kurti and B. Czako, *Strategic Applications of Named Reactions in Organic Synthesis*, Elsevier, Amsterdam, 2005.
15. C. Bolm and D. Pupowicz, *Tetrahedron Lett.*, **42**, 7349 (1997).
16. R. C. Hahn and J. Tompkins, *Tetrahedron Lett.*, **31**, 973 (1990).
17. R. W. Hoffmann, O. Knopff and A. Kusche, *Angew. Chem. Int. Ed.*, **39**, 1462 (2000).
18. T. Satoh, A. Kondo and J. Musashi, *Tetrahedron*, **60**, 5453 (2004).
19. M. Pohmakotr, W. Ieawsuwan, P. Tuchinda, P. Kongsaeree, S. Prabpai and V. Reutrakul, *Org. Lett.*, **6**, 4547 (2004).
20. R. W. Hoffmann, B. Holzer, O. Knopff and K. Harms, *Angew. Chem. Int. Ed.*, **39**, 3072 (2000).
21. B. Holzer and R. W. Hoffmann, *Chem. Commun.*, 732 (2003).
22. R. W. Hoffmann, *Chem. Soc. Rev.*, **32**, 225 (2003).
23. C. D. Lima, M. Julia and J-N. Verpeaux, *Synlett*, 133 (1992).
24. J. L. G. Ruano, S. A. Alonso de Diego, M. R. Martin, E. Torrente and A. M. M. Castro, *Org. Lett.*, **6**, 4945 (2004).
25. (a) T. Satoh, A. Osawa and A. Kondo, *Tetrahedron Lett.*, **45**, 6703 (2004).
(b) T. Satoh, A. Osawa, T. Ohbayashi and A. Kondo, *Tetrahedron*, **62**, 7892 (2006).
26. (a) S. V. Kessar and P. Singh, *Chem. Rev.*, **97**, 721 (1997).
(b) A. R. Katritzky and M. Qi, *Tetrahedron*, **54**, 2647 (1998).
(c) W. H. Pearson and P. Stoy, *Synlett*, 903 (2003).
27. T. Satoh, J. Musashi and A. Kondo, *Tetrahedron Lett.*, **46**, 599 (2005).
28. J. Clayden and M. Julia, *Synlett*, 103 (1995).
29. T. Satoh, S. Ogata and D. Wakasugi, *Tetrahedron Lett.*, **47**, 7249 (2006).
30. O. Knopff, C. Stiasny and R. W. Hoffmann, *Organometallics*, **23**, 705 (2004).
31. P. D. Gardner and M. Narayana, *J. Org. Chem.*, **26**, 3518 (1961).
32. T. Satoh, T. Kurihara and K. Fujita, *Tetrahedron*, **57**, 5369 (2001).
33. V. A. Vu, I. Marek, K. Polborn and P. Knochel, *Angew. Chem. Int. Ed.*, **41**, 351 (2002).
34. F. Kopp, G. Sklute, K. Polborn, I. Marek and P. Knochel, *Org. Lett.*, **7**, 3789 (2005).

16. The chemistry of magnesium carbenoids 769

35. T. Satoh and S. Saito, *Tetrahedron Lett.*, **45**, 347 (2004).
36. T. Satoh, M. Miura, K. Sakai and Y. Yokoyama, *Tetrahedron*, **62**, 4253 (2006).
37. P. J. Stang, *Chem. Rev.*, **78**, 383 (1978).
38. T. Satoh, Y. Hayashi and K. Yamakawa, *Bull. Chem. Soc. Jpn.*, **66**, 1866 (1993).
39. (a) T. Satoh, K. Takano, H. Someya and K. Matsuda, *Tetrahedron Lett.*, **36**, 7097 (1995).
    (b) T. Satoh, K. Takano, H. Ota, H. Someya, K. Matsuda and M. Koyama, *Tetrahedron*, **54**, 5557 (1998).
40. V. A. Vu, I. Marek and P. Knochel, *Synthesis*, 1797 (2003).
41. (a) T. Satoh, T. Sakamoto and M. Watanabe, *Tetrahedron Lett.*, **43**, 2043 (2002).
    (b) T. Satoh, T. Sakamoto, M. Watanabe and K. Takano, *Chem. Pharm. Bull.*, **51**, 966 (2003).
42. (a) T. Satoh, Y. Ogino and M. Nakamura, *Tetrahedron Lett.*, **45**, 5785 (2004).
    (b) T. Satoh, Y. Ogino and K. Ando, *Tetrahedron*, **61**, 10262 (2005).
43. T. Satoh, J. Sakurada and Y. Ogino, *Tetrahedron Lett.*, **46**, 4855 (2005).
44. M. Watanabe, M. Nakamura and T. Satoh, *Tetrahedron*, **61**, 4409 (2005).
45. T. Satoh, *The Chemical Record*, The Japan Chemical Journal Forum and Wiley Periodicals, Inc. **3**, 329 (2004).
46. (a) O. W. Lever, Jr., *Tetrahedron*, **32**, 1943 (1976).
    (b) S. F. Martin, *Synthesis*, 633 (1979).
    (c) J. C. Stowell, *Chem. Rev.*, **84**, 409 (1984).
    (d) N. F. Badham, *Tetrahedron*, **60**, 11 (2004).
47. (a) G. R. Krow, *Tetrahedron*, **43**, 3 (1987).
    (b) M. Hesse, *Ring Enlargement in Organic Chemistry*, Wiley-VCH, Weinheim, 1991.
    (c) P. Dowd and W. Zhang, *Chem. Rev.*, **93**, 2091 (1993).
48. (a) H. Taguchi, H. Yamamoto and H. Nozaki, *J. Am. Chem. Soc.*, **96**, 6510 (1974).
    (b) H. Taguchi, H. Yamamoto and H. Nozaki, *Tetrahedron Lett.*, 2617 (1976).
    (c) H. Taguchi, H. Yamamoto and H. Nozaki, *Bull. Chem. Soc. Jpn.*, **50**, 1592 (1977).
    (d) J. Villieras, P. Perriot and J. F. Normant, *Synthesis*, 968 (1979).
    (e) H. D. Ward, D. S. Teager and R. K. Murray, Jr., *J. Org. Chem.*, **57**, 1926 (1992).
49. W. D. Abraham, M. Bhupathy and T. Cohen, *Tetrahedron Lett.*, **28**, 2203 (1987).
50. (a) T. Satoh, N. Itoh, K. Gengyo and K. Yamakawa, *Tetrahedron Lett.*, **33**, 7543 (1992).
    (b) T. Satoh, N. Itoh, K. Gengyo, S. Takada, N. Asakawa, Y. Yamani and K. Yamakawa, *Tetrahedron*, **50**, 11839 (1994).
51. (a) T. Satoh and K. Miyashita, *Tetrahedron Lett.*, **45**, 4859 (2004).
    (b) K. Miyashita and T. Satoh, *Tetrahedron*, **61**, 5067 (2005).
52. T. Satoh, S. Tanaka and N. Asakawa, *Tetrahedron Lett.*, **47**, 6769 (2006).
53. T. Satoh, Y. Mizu, T. Kawashima and K. Yamakawa, *Tetrahedron*, **51**, 703 (1995).
54. I. Tabushi, K. Takagi and R. Oda, *Nippon Kagaku Zassi*, **85**, 302 (1964); *Chem. Abstr.*, **61**, 13226g (1964).
55. K. Ogura, M. Fujita, K. Takahashi and H. Iida, *Chem. Lett.*, 1697 (1982).
56. T. Satoh, S. Kobayashi, S. Nakanishi, K. Horiguchi and S. Irisa, *Tetrahedron*, **55**, 2515 (1999).
57. (a) T. Satoh, R. Matsue, T. Fujii and S. Morikawa, *Tetrahedron Lett.*, **41**, 6495 (2000).
    (b) T. Satoh, R. Matsue, T. Fujii and S. Morikawa, *Tetrahedron*, **57**, 3891 (2001).
58. S. Avolio, C. Malan, I. Marek and P. Knochel, *Synlett*, 1820 (1999).
59. S. Sugiyama, H. Shimizu and T. Satoh, *Tetrahedron Lett.*, **47**, 8771 (2006).

CHAPTER 17

Catalytic enantioselective conjugate addition and allylic alkylation reactions using Grignard reagents

FERNANDO LÓPEZ

Departamento de Química Orgánica, Facultad de Química, Universidad de Santiago de Compostela, Avda. de las ciencias, s/n, 15782, Santiago de Compostela, Spain
e-mail: qofer@usc.es

and

ADRIAAN J. MINNAARD and BEN L. FERINGA

Stratingh Institute for Chemistry, University of Groningen, Nijenborgh 4, 9747 AG Groningen, The Netherlands
Fax: ++31 503634296; e-mails: A.J.Minnaard@rug.nl and B.L.Feringa@rug.nl

I. INTRODUCTION	772
II. ENANTIOSELECTIVE CONJUGATE ADDITION TO CYCLIC ENONES	774
III. ENANTIOSELECTIVE CONJUGATE ADDITION TO ACYCLIC ENONES	779
IV. ENANTIOSELECTIVE CONJUGATE ADDITION TO α,β-UNSATURATED ESTERS AND THIOESTERS	779
V. APPLICATION OF THE CONJUGATE ADDITION OF GRIGNARD REAGENTS TO α,β-UNSATURATED THIOESTERS IN THE SYNTHESIS OF NATURAL PRODUCTS	786
VI. MECHANISTIC STUDIES	788
VII. ENANTIOSELECTIVE ALLYLIC ALKYLATION WITH GRIGNARD REAGENTS	791
VIII. CONCLUSIONS	798
IX. REFERENCES AND NOTES	800

The chemistry of organomagnesium compounds
Edited by Z. Rappoport and I. Marek © 2008 John Wiley & Sons, Ltd

I. INTRODUCTION

The conjugate addition of organometallic reagents to α,β-unsaturated compounds is one of the basic methods in our repertoire for the construction of carbon–carbon bonds[1]. These addition reactions have been used as key steps in the synthesis of numerous biologically active compounds, and show a broad scope due to the large variety of donor and acceptor compounds that can be employed. It is evident that a tremendous effort has been devoted over the last three decades to develop asymmetric variants of this reaction[2].

The first successful approaches were based on the, often Cu-mediated, conjugate addition of organolithium and organomagnesium (Grignard) reagents to α,β-unsaturated systems covalently modified with chiral auxiliaries (Scheme 1[3])[4].

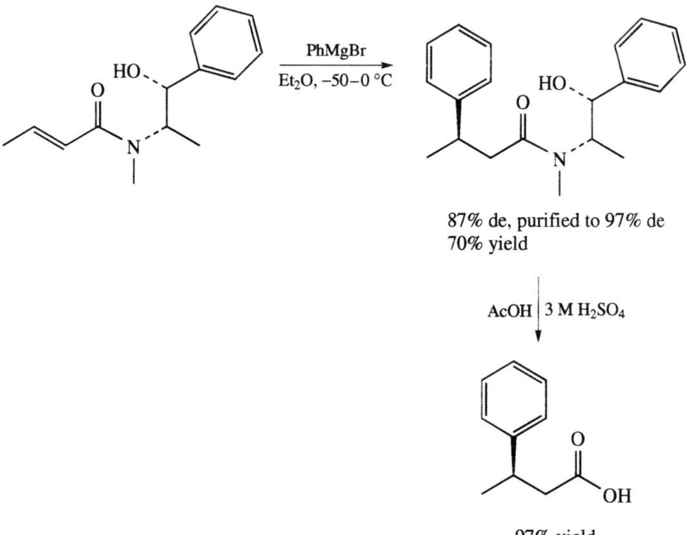

SCHEME 1. Asymmetric conjugate addition of Grignard reagents to substrates containing chiral auxiliaries

Other strategies made use of organocopper compounds with chiral nontransferable groups, such as chiral alkoxycuprates and amidocuprates[2, 4]. For instance, Corey and coworkers reported in 1986 enantioselectivities of over 90% by using a chiral ephedrine-derived alkoxycuprate[5]. The use of organolithium reagents in the presence of stoichiometric amounts of chiral ether **1** or amine **2** ligands was also explored, providing high enantioselectivities in the conjugate addition to α,β-unsaturated N-cyclohexylimines and sterically crowded esters (Scheme 2)[6].

Although some of these strategies provide high enantioselectivities with a number of substrates, the development of catalytic rather than stoichiometric processes is the main challenge in order to provide truly efficient synthetic methods.

It was not until the late 1980s that the feasibility of a catalytic ($\leqslant$ 10 mol% chiral catalyst) and enantioselective conjugate addition was demonstrated. Lippard and coworkers reported the first enantioselective conjugate addition of a Grignard reagent to an enone, using catalytic amounts of Cu–amide complex **3** (Figure 1)[7]. Subsequently, a variety of catalytic systems, based on, e.g., Cu thiolates **4–7**[8], and phosphine–oxazoline ligand **8**[9], was introduced for the conjugate addition of Grignard reagents. Although the

SCHEME 2. Asymmetric conjugate addition of organolithium reagents with stoichiometric chiral ligands. Adapted with permission from *Acc. Chem. Res.*, **40**, 179–188 (2007). Copyright 2007 American Chemical Society

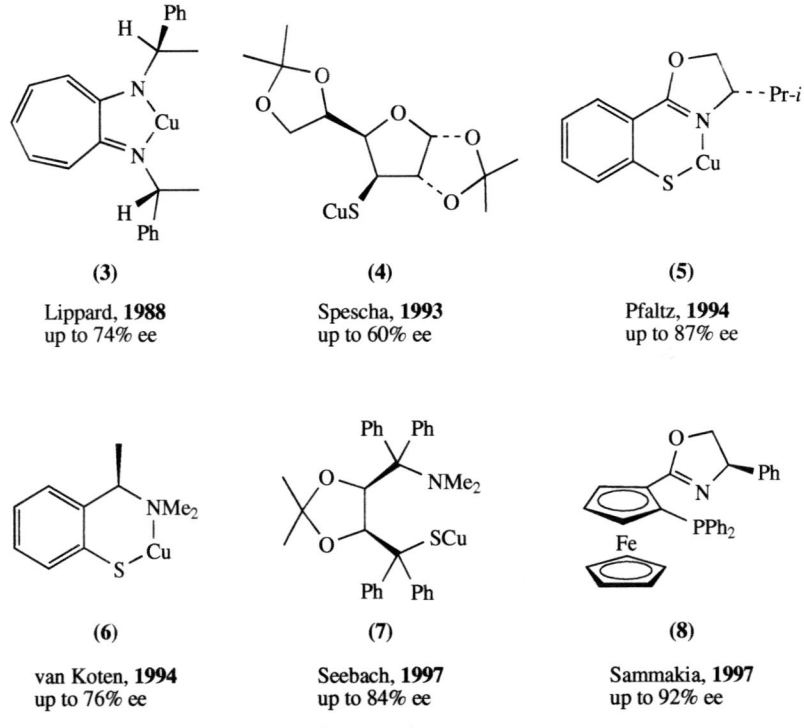

(3) Lippard, **1988** up to 74% ee

(4) Spescha, **1993** up to 60% ee

(5) Pfaltz, **1994** up to 87% ee

(6) van Koten, **1994** up to 76% ee

(7) Seebach, **1997** up to 84% ee

(8) Sammakia, **1997** up to 92% ee

FIGURE 1. Selected catalysts developed for the conjugate addition of Grignard reagents to enones. Adapted with permission from *Acc. Chem. Res.*, **40**, 179–188 (2007). Copyright 2007 American Chemical Society

scope remained limited and ee values infrequently reached the 90% level (Figure 1), high enantioselectivity (92%) was observed in two examples for ligand **8**. Despite the fact that the parameters governing the stereocontrol were not completely clear, these excellent contributions provided an important basis to allow the development of the catalytic methodology.

The development of a catalytic enantioselective method for the conjugate addition of Grignard reagents was for a long time hampered by the reactivity of the Grignard which causes a fast uncatalyzed reaction. In addition, organomagnesium reagents are 'hard' nucleophiles which as such prefer direct addition above conjugate addition[1,2,10]. Therefore, the development of catalytic conjugate addition reactions using less reactive organometallics such as organozinc, organocopper, organoaluminum, or arylboron and arylsilicon reagents dominated. These reagents are used with catalysts based on copper, rhodium, palladium, nickel and cobalt.

Early work by Soai and coworkers showed the viability of performing the conjugate addition of dialkylzinc reagents to enones with modest enantioselectivities using substoichiometric amounts of chiral complexes of Ni and Co[11]. Dialkylzinc reagents have distinct advantages compared to Grignard reagents, because they show low reactivity in the uncatalyzed reaction and a high tolerance of functional groups, both in the substrate and zinc reagent. A Cu-catalyzed conjugate addition of Et_2Zn to 2-cyclohexenone with 32% ee was subsequently reported[12]. The discovery by Feringa and coworkers in 1996 that chiral monodentate phosphoramidites are excellent ligands for the asymmetric Cu-catalyzed conjugate addition of R_2Zn reagents[13] led to a method for the highly enantioselective Cu-catalyzed conjugate addition of dialkylzinc reagents to enones[14,15]. This, in turn, stimulated the development of a broad range of efficient phosphorus-based catalysts for the Cu-catalyzed conjugate addition of dialkylzinc reagents. In addition, this methodology found application in natural product synthesis[16–20].

Complementary to the use of zinc reagents for the introduction of (functionalized) alkyl groups is the rhodium-catalyzed conjugate addition of aryl- and alkenylboron reagents. This method rapidly became popular, also because arylboron reagents are air and moisture stable and a large variety of them is commercially available[21].

The efficiency of dialkylzinc and boron reagents in the catalytic enantioselective conjugate addition clearly displaced for a number of years the use of Grignard reagents in this transformation[22]. Compared to dialkylzinc reagents, Grignard reagents are known to show lower functional group tolerance, although recently considerable advances have been made in the use of functionalized organomagnesium compounds[23]. Nevertheless, there are significant incentives to use Grignard reagents instead of dialkylzinc or organoboron compounds. Grignard reagents are cheap and readily available from all kinds of alkyl, alkenyl, aryl and alkynyl halides. This is in contrast to diorganozincs. In fact, as most zinc and boron reagents themselves originate from the corresponding Grignard, it serves atom economy to use the latter reagents directly. For large-scale synthesis, Grignard reagents are preferred over organozincs for environmental reasons and are commonly used in fine chemical and pharmaceutical industries. A disadvantage of the conjugate addition of boronic acids is the concomitant protonation of the resulting enolate that precludes its use in tandem reactions[24].

II. ENANTIOSELECTIVE CONJUGATE ADDITION TO CYCLIC ENONES

Most ligands used so far in the field of the copper-catalyzed conjugate addition of Grignard reagents combine phosphorus, sulfur or selenium with nitrogen or oxygen donor atoms in their structure, to coordinate selectively with copper and magnesium of the organometallic species, respectively[8,9,25]. The fact that free Cu salts show high activity in the conjugate addition of Grignard reagents, even at low temperature, makes tight binding of Cu ions by bidentate ligands probably essential to avoid a nonselective background reaction. Interestingly, although chiral diphosphine ligands have dominated the field of asymmetric catalysis in the last 30 years[26], until very recently none of these ligands was reported to be effective in the conjugate addition of Grignard reagents. *A priori*, diphosphines would not match with the metal-differentiating coordination concept, although in several of these diphosphine ligands the two phosphorus atoms have very different electronic and steric properties[27].

In 2004, our group reported the highly regio- and enantioselective conjugate addition of Grignard reagents using copper catalysts with ferrocenyl-based diphosphine ligands. Where bidentate phosphines such as BINAP, Trost ligand and DuPHOS led to poor enantioselectivities in the model reaction (5–28% ee) (Scheme 3, Figure 2), promising enantioselectivities (45–70% ee) were obtained with Mandyphos, Walphos and Josiphos. Among the ferrocenyl ligands, Taniaphos[28] (Figure 2). provided in the preliminary screening the highest enantioselectivity (95% ee), although with modest regioselectivity (conjugate addition *versus* direct addition = 60:40).

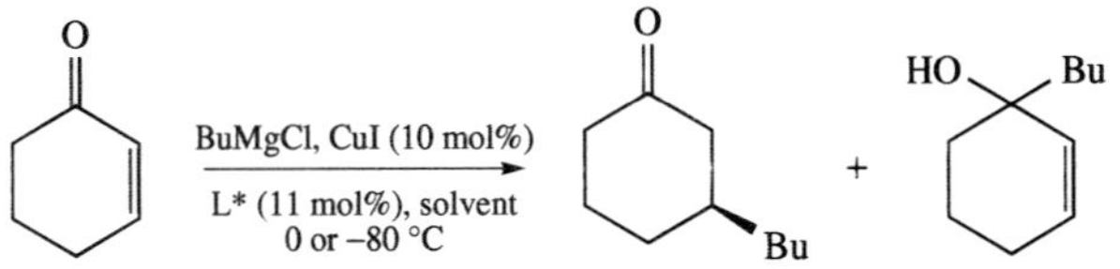

SCHEME 3. Enantioselective copper-catalyzed addition of butylmagnesium chloride to 2-cyclohexenone

Optimization of the reaction parameters led to conditions using 5 mol% of CuCl, 6 mol% of Taniaphos and 1.15 equiv of EtMgBr in Et_2O at 0 °C, which afforded full conversion in 15 min with a regioselectivity of 95% and an excellent 96% ee[29]. The results with a variety of Grignard reagents using these optimal conditions are shown in Table 1.

The products were obtained with 90–96% ee using RMgBr reagents with linear alkyl chains (R = Me, *n*-Pr, *n*-Bu). Employing Grignard reagents with branched alkyl chains, a strong influence of the substitution pattern on the enantioselectivity was observed. In particular, the incorporation of isopropyl and isobutyl fragments resulted in poor ee values, although isoamylmagnesium bromide afforded the corresponding 1,4-addition product with 95% ee. Noteworthy is that, with the α- and β-branched Grignard reagents *i*-PrMgBr and *i*-BuMgBr, Josiphos provides excellent regiocontrol (99%) with moderate to high (54%–92%) enantioselectivities. Contrary to Taniaphos, Josiphos is more effective at low temperatures (e.g. −60 °C *vs* 0 °C) and in combination with CuBr•SMe_2 instead of CuCl.

Therefore, the proper selection of Taniaphos or Josiphos in a complementary way allows the use of a broad range of Grignard reagents. Moreover, the reaction turned out to be not limited to cyclohexenone, as other cyclic enones as well as lactones provided high levels of regio- and enantioselectivity (Figure 3)[29].

Recently, diaminocarbenes, also called NHCs (for *N*-heterocyclic carbenes), have been shown to be a viable alternative to phosphorus ligands. Almost simultaneously, both the copper/NHC-catalyzed enantioselective conjugate addition of dialkylzincs[30] and Grignard reagents to 3-substituted enones was reported. For the Grignard addition reactions, the groups of Mauduit, Alexakis and coworkers[31] studied a series of chiral imidazolium salts as procatalysts, whereupon **9** and **10** turned out to be the most efficient (Figure 4). Good yields and high to excellent enantioselectivities were obtained (Table 2) apparently in the complete absence of direct addition product.

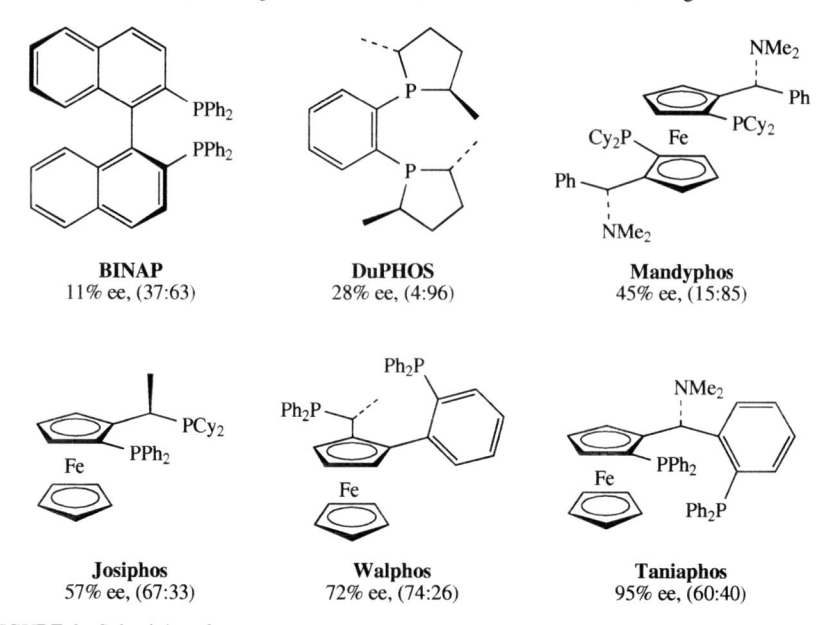

FIGURE 2. Selectivity of several diphosphine ligands in the model reaction. In parentheses, regioselectivity: (conjugate addition versus direct addition). Adapted with permission from *Acc. Chem. Res.*, **40**, 179–188 (2007). Copyright 2007 American Chemical Society

Ligands shown:
- **BINAP** — 11% ee, (37:63)
- **DuPHOS** — 28% ee, (4:96)
- **Mandyphos** — 45% ee, (15:85)
- **Josiphos** — 57% ee, (67:33)
- **Walphos** — 72% ee, (74:26)
- **Taniaphos** — 95% ee, (60:40)

TABLE 1. Enantioselective copper-catalyzed conjugate addition of Grignard reagents to 2-cyclohexenone [a,b]

RMgBr	L*	1,4:1,2	ee (%) (1,4)
EtMgBr	Taniaphos	95:5	96
MeMgBr	Taniaphos	83:17	90
n-PrMgBr	Taniaphos	81:19	94
n-BuMgBr	Taniaphos	88:12	96
i-PrMgBr	Taniaphos	78:22	1
i-BuMgBr	Taniaphos	62:38	33
(isopentyl)MgBr	Taniaphos	76:24	95
i-PrMgBr [c]	Josiphos	99:1	54
i-BuMgBr [c]	Josiphos	99:1	92
EtMgBr [c]	Josiphos	99:1	56

[a] >98% conversion after 15 min at 0 °C using CuCl.
[b] >98% conversion after 2 h at −60 °C using CuBr•SMe$_2$.
[c] Adapted with permission from *Acc. Chem. Res.*, **40**, 179–188 (2007). Copyright 2007 American Chemical Society.

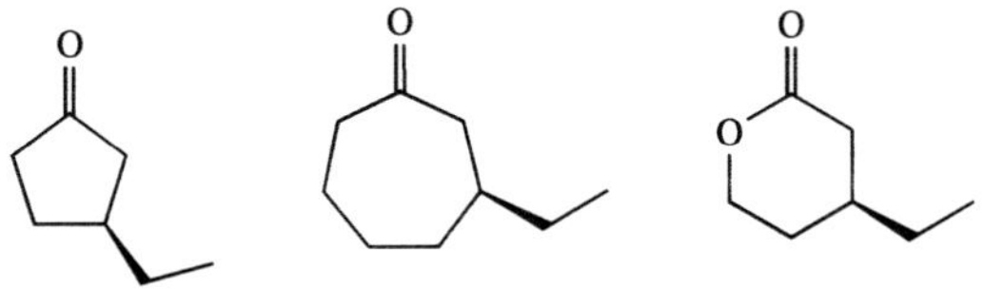

92% ee[a]
(99:1)[c]

87% ee[b]
(80:20)[c]

82% ee[a]
(99:1)[c]

[a]Josiphos, CuBr·SMe$_2$, −60 °C
[b]Taniaphos, CuCl, 0 °C
[c]1,4:1,2 ratio

FIGURE 3. Representative examples of conjugate addition products using Cu/ferrocenyl diphosphine catalysts. Adapted with permission from *Acc. Chem. Res.*, **40**, 179–188 (2007). Copyright 2007 American Chemical Society

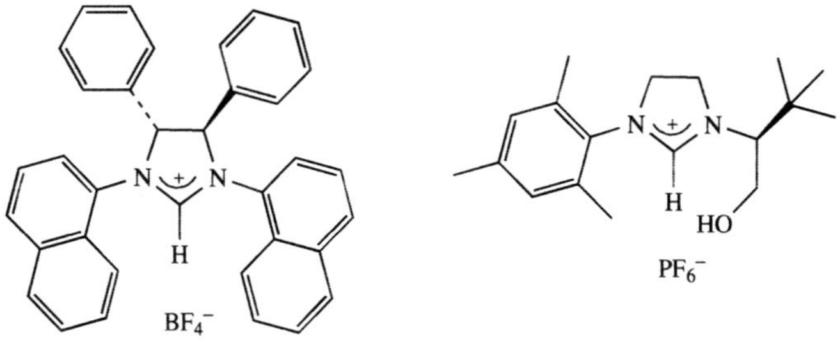

(9) BF$_4^-$

(10) PF$_6^-$

FIGURE 4. Chiral NHC ligands for the enantioselective copper-catalyzed conjugate addition of Grignard reagents

TABLE 2. Cu/NHC-catalyzed addition of Grignard reagents to 3-substituted 2-cyclohexenones

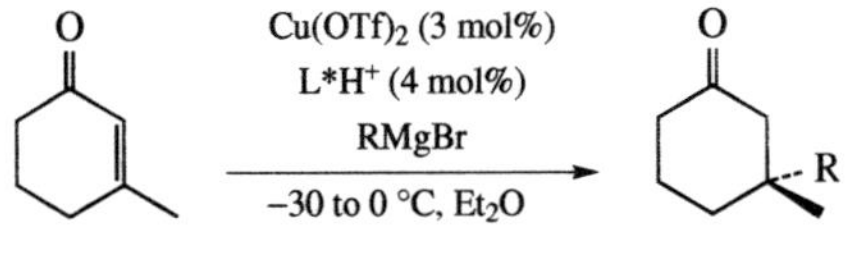

Cu(OTf)$_2$ (3 mol%)
L*H$^+$ (4 mol%)
RMgBr
−30 to 0 °C, Et$_2$O

RMgBr	L*	Yield (%)	ee (%)
EtMgBr	9	90	73 (S)
EtMgBr	10	81	80 (R)
BuMgBr	10	100 (conv)	77 (R)
ButenylMgBr	10	80	90 (S)
i-BuMgBr	10	72	96 (S)
i-PrMgBr	10	77	77 (R)
c-pentMgBr	10	80	85 (R)
c-hexMgBr	10	79	74 (R)
t-BuMgBr	10	0	
PhMgBr	10	61	66 (R)

TABLE 3. Cu/NHC-catalyzed addition of Grignard reagents to 3-substituted-2-cycloalkenones

RMgBr	L*	Enone	Yield (%)	ee (%)	Product
EtMgBr	9	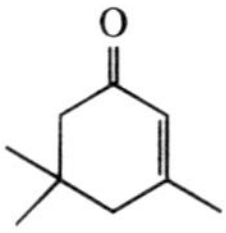	57	71 (S)	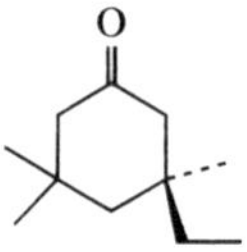
EtMgBr	10	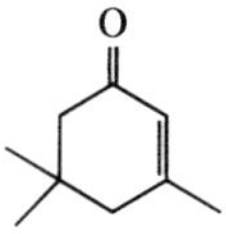	85	82 (R)	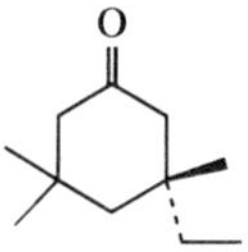
MeMgBr	10	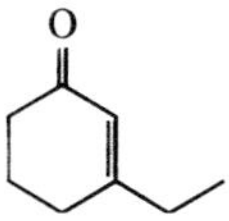	67	68 (S)	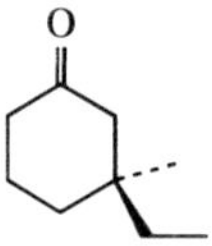
EtMgBr	10	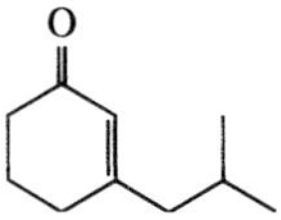	69	81 (S)	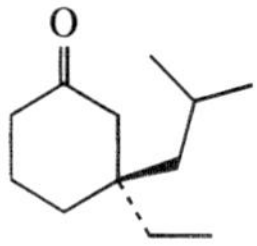
EtMgBr	10	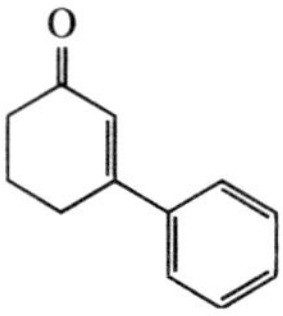	87	72 (S)	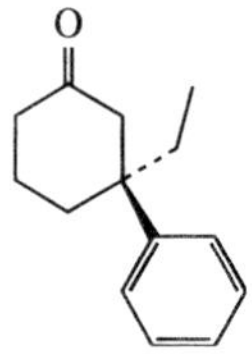
EtMgBr	10	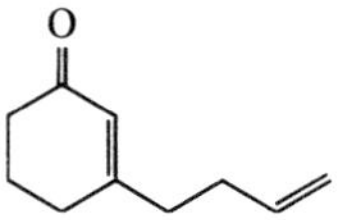	84	69 (R)	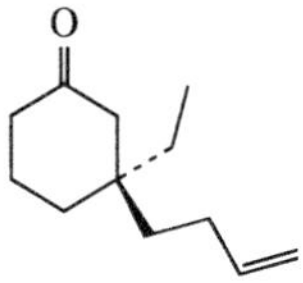
EtMgBr	10	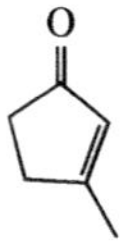	90	46 (R)	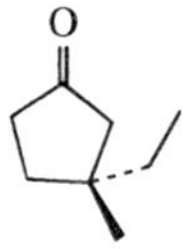
EtMgBr	10	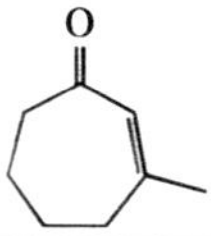	99 (conv)	82 (R)	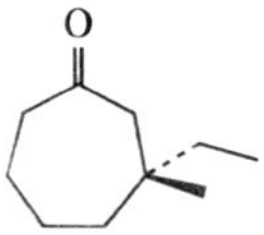

A series of differently substituted cyclic enones was used as substrate. Using a seven-membered ring enone the enantioselectivity remained high, but for a five-membered ring analogue the ee dropped (Table 3).

III. ENANTIOSELECTIVE CONJUGATE ADDITION TO ACYCLIC ENONES

The method developed for the conjugate addition of Grignard reagents using Cu complexes of ferrocenyl-based diphosphines was subsequently expanded to linear enones[32]. The β-substituted ketones resulting from the conjugate addition to linear enones are common subunits in natural products and important building blocks for the synthesis of physiologically active molecules. A number of procedures for their enantioselective preparation has been reported to date although the enantioselectivities are usually substrate and ligand dependent[19a, 20, 22, 33].

Initially, the addition of EtMgBr to (E)-3-nonen-2-one was investigated, catalyzed by CuCl and Taniaphos (Table 4). The product was obtained with good regioselectivity at 0 °C, but surprisingly with complete lack of enantioselectivity. Performing the conjugate addition at low temperature and in particular using Josiphos dramatically enhanced the selectivity up to 86% ee. Further improvement could be obtained by using the less coordinating solvent t-BuOMe instead of Et$_2$O.

These conditions resulted also in high selectivities when Grignard reagents with different linear alkyl chains were used, whereas the substrate scope included a variety of aliphatic linear enones (Scheme 4). Particularly noteworthy is the addition of MeMgBr (e.g. to octenone), which provides the corresponding ketones with 97–98% ee, even when only 1 mol% of catalyst is employed[34].

Both β-substituted aliphatic and aromatic enones can be used. For instance, benzylideneacetone, β-thienyl- and β-furyl-substituted enones reacted smoothly in t-BuOMe at −75 °C with RMgBr reagents to give the corresponding ketones with high yields, regioselectivities and enantioselectivities of 90–97% (Scheme 4). In contrast, the conjugate addition of α-branched and aryl Grignard reagents or the use of sterically hindered enones provided only moderate enantioselectivities[32].

IV. ENANTIOSELECTIVE CONJUGATE ADDITION TO α,β-UNSATURATED ESTERS AND THIOESTERS

The conjugate addition of Grignard reagents to α,β-unsaturated acid derivatives, in particular to esters, is highly attractive. Despite the enormous synthetic potential of the resulting

TABLE 4. Enantioselective conjugate addition of EtMgBr to (E)-3-nonen-2-one [a,b,c]

L*	CuX	Solvent	T (°C)	1,4:1,2	ee (%) (1,4)
Taniaphos	CuCl	Et$_2$O	0	84:16	1
Taniaphos	CuCl	Et$_2$O	−75	70:30	48
Josiphos	CuBr•SMe$_2$	Et$_2$O	−75	91:9	86
Josiphos	CuBr•SMe$_2$	t-BuOMe	−75	99:1	90

[a] EtMgBr added to a solution of (E)-3-nonen-2-one, 5 mol% CuX and 6 mol% ligand.
[b] All conversions are >98%.
[c] Adapted with permission from *Acc. Chem. Res.*, **40**, 179–188 (2007). Copyright 2007 American Chemical Society.

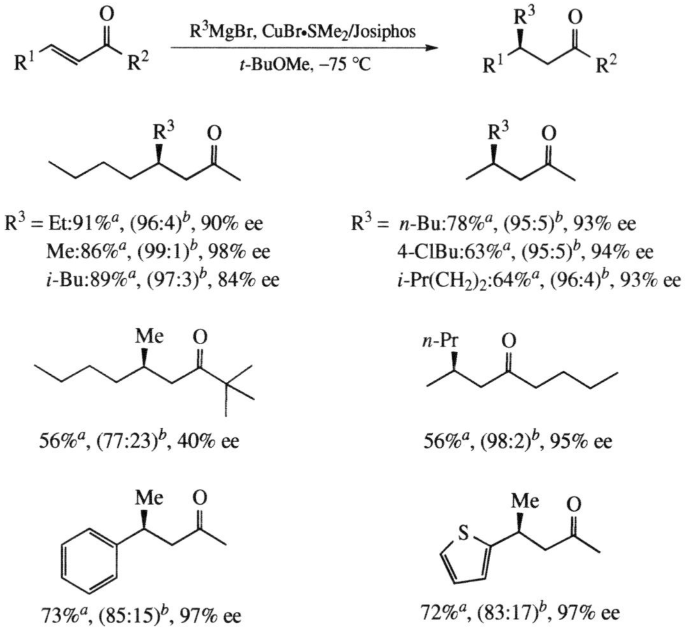

$R^1\!\!-\!\!CH=CH\!-\!C(O)R^2 \xrightarrow[\text{t-BuOMe, $-75\,^\circ$C}]{R^3MgBr,\ CuBr\cdot SMe_2/Josiphos} R^1\!\!-\!\!CH(R^3)\!-\!CH_2\!-\!C(O)R^2$

R^3 = Et: 91%^{a}, (96:4)^b, 90% ee
Me: 86%^{a}, (99:1)^b, 98% ee
i-Bu: 89%^{a}, (97:3)^b, 84% ee

R^3 = n-Bu: 78%^{a}, (95:5)^b, 93% ee
4-ClBu: 63%^{a}, (95:5)^b, 94% ee
i-Pr(CH$_2$)$_2$: 64%^{a}, (96:4)^b, 93% ee

56%^{a}, (77:23)^b, 40% ee

56%^{a}, (98:2)^b, 95% ee

73%^{a}, (85:15)^b, 97% ee

72%^{a}, (83:17)^b, 97% ee

^{a} Isolated yield ^{b} 1,4:1,2 ratio.

SCHEME 4. Enantioselective conjugate addition of RMgBr reagents to linear enones. Adapted with permission from *Acc. Chem. Res.*, **40**, 179–188 (2007). Copyright 2007 American Chemical Society

β-substituted esters as chiral building blocks for natural product synthesis, progress during the last decades in the enantioselective conjugate addition of organometallic reagents to unsaturated esters has been limited[35]. The lower intrinsic reactivity of α,β-unsaturated esters compared to that of enones may account for this paucity of methodologies. Indeed, no combinations of catalysts and alkyl organometallic reagents had previously been shown to be successful for these conjugate additions, although an enantioselective conjugate addition of dialkylzinc reagents to the more reactive α,β-unsaturated N-acyloxazolidinones has been reported by Hird and Hoveyda [36,37].

An initial study demonstrated that Josiphos and its 'inverted' analogue **11** were very effective in promoting the Cu-catalyzed conjugate addition of EtMgBr to unsaturated crotonates (Scheme 5). It is noteworthy that the use of sterically hindered esters, which usually helps to avoid undesired 1,2-additions, or alternatives for esters such as an oxazolidinone, are not required. Indeed, the highest conversions and stereoselectivities are obtained with methyl crotonate[38].

Interestingly, the dinuclear Cu complexes **12** and **13** (Scheme 6) could be recovered from the crude reaction mixtures or, alternatively, prepared independently by mixing equimolar amounts of ligands and CuBr·SMe$_2$ in an appropriate solvent. It was established that these Cu complexes participate in the catalytic cycle, as the reaction of methyl crotonate and EtMgBr with the independently prepared (or recovered) complexes (0.5 mol%) afforded the product with the same yields and enantioselectivities as previously obtained with the complexes prepared *in situ*.

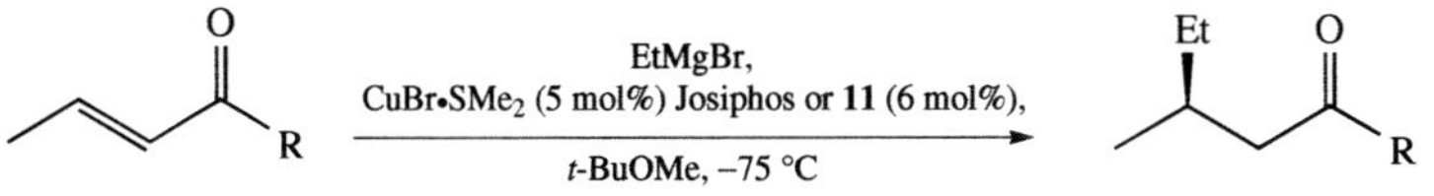

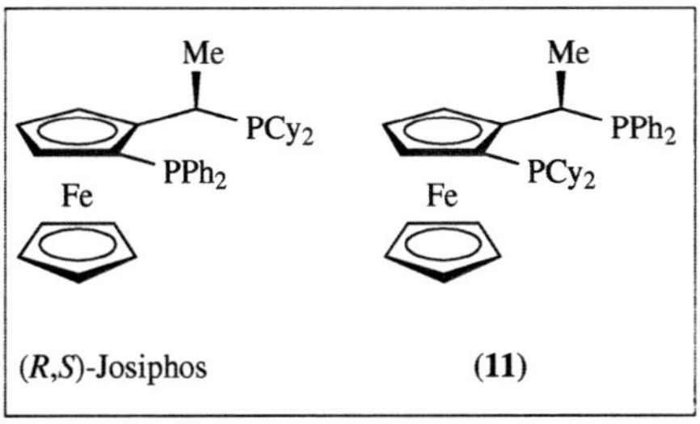

17. Catalytic enantioselective conjugate addition   781

SCHEME 5. Screening of catalysts and crotonic acid derivatives in the copper-catalyzed Grignard addition. Adapted with permission from *Acc. Chem. Res.*, **40**, 179–188 (2007). Copyright 2007 American Chemical Society

	R	conv(%)	ee(%)
a	OMe	99 (96)[a]	95 (92)[a]
b	OEt	99 (96)[a]	90 (78)[a]
c	OPr-i	80 (98)[a]	53 (54)[a]
d	(oxazolidinone)	99	50

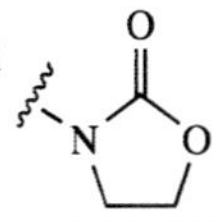

[a] Results with **11** in parentheses

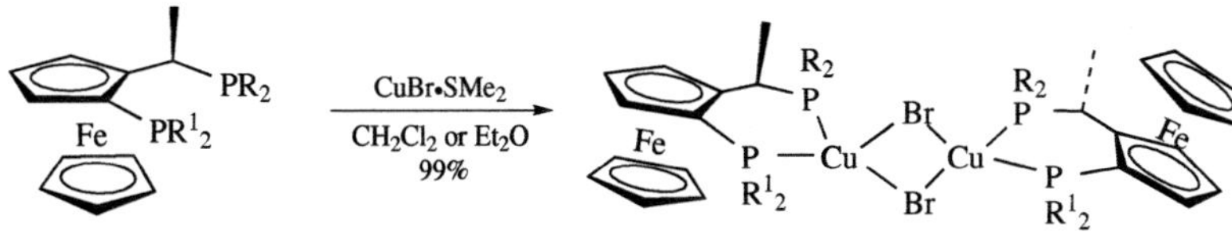

Josiphos (R = Cy, R^1 = Ph)
11 (R = Ph, R^1 = Cy)

SCHEME 6. Preparation of the Cu complexes **12** and **13**. Adapted with permission from *Acc. Chem. Res.*, **40**, 179–188 (2007). Copyright 2007 American Chemical Society

Table 5 summarizes the scope of the reaction. As a general trend, linear aliphatic Grignard reagents provided excellent results in the conjugate addition to methyl crotonate, affording the products with excellent regio- and enantioselectivities and complete conversions using 0.5 mol% of catalyst. With regard to the substrates, less hindered α,β-unsaturated esters, without branching at the γ position, afford better results with the Cu complex of Josiphos (**12**). However, for substrates with bulky groups or aromatic rings at the double bond, a superior efficiency is observed when catalyst **13** is used instead.

The conjugate addition of Grignard reagents can also be performed with the corresponding *Z*-enoates, leading to the products with opposite absolute configurations. However, lower ee values were consistently obtained in these reactions. In the reaction with *Z*-methyl cinnamate, analysis of the reaction mixture at different times revealed that an isomerization of the *Z*-enoate to the corresponding *E*-enoate occurred during the reaction, causing the decrease in ee.

From the perspective of potential applications to the synthesis of biologically active compounds, the introduction of a methyl group via the conjugate addition of MeMgBr to α,β-unsaturated esters is a particularly relevant goal. Unfortunately, the addition of MeMgBr to methyl-2-hexenoate showed the limitation of the methodology. Although the product was formed with high enantioselectivity (93% ee) the reaction rate was prohibitively low due to the decreased reactivity of MeMgBr.

TABLE 5. Enantioselective Cu/ferrocenyl diphosphine-catalyzed conjugate addition to α,β-unsaturated esters [a,b]

Substrate	RMgBr	Product	Cat (mol%)	Yield (%)[f]	ee (%)
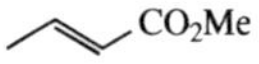 ⌒⌒CO₂Me (crotonate)	n-BuMgBr	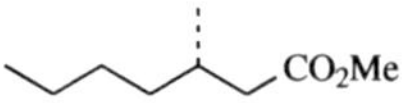 n-Bu adduct, CO₂Me	12 (0.5)	92	95
 ⌒⌒CO₂Me	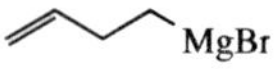 allyl-MgBr	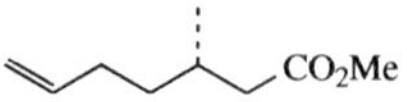 allyl adduct, CO₂Me	12 (0.5)	67	85
 ⌒⌒CO₂Me	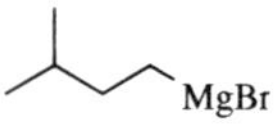 i-Bu-MgBr	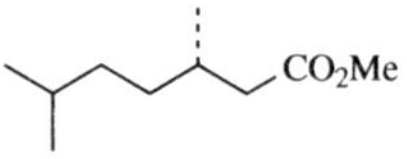 i-Bu adduct, CO₂Me	12 (0.5)	90	96
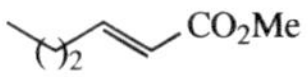 (C₂)⌒⌒CO₂Me	EtMgBr	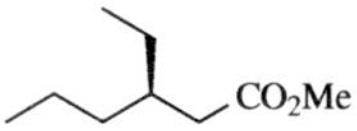 Et adduct, CO₂Me	12 (0.5)	99[c]	93
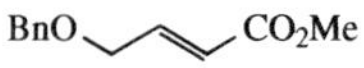 BnO⌒⌒CO₂Me	EtMgBr	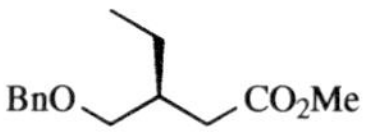 BnO, Et adduct, CO₂Me	12 (2.5)	85	86
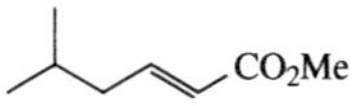 i-Pr⌒⌒CO₂Me	n-BuMgBr	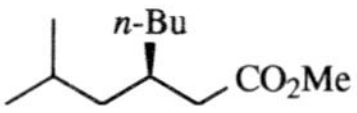 n-Bu adduct, CO₂Me	12 (2.5)	99[c]	92
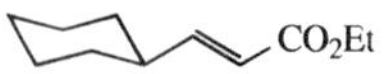 Cy-CH=CH-CO₂Et	EtMgBr	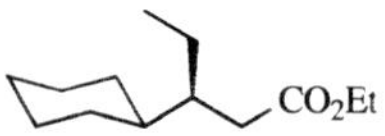 Cy, Et adduct, CO₂Et	13 (2.5)	86[d]	98
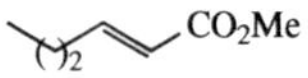 (C₂)⌒⌒CO₂Me	MeMgBr	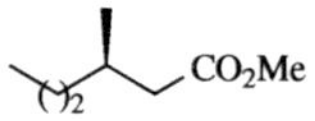 Me adduct, CO₂Me	12 (2.5)	19[c]	93
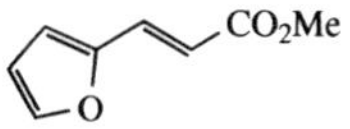 furyl-CH=CH-CO₂Me	EtMgBr	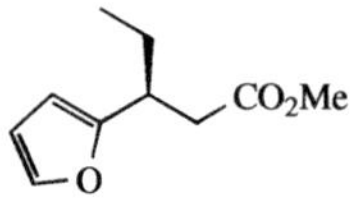 furyl, Et adduct, CO₂Me	13 (0.5)	90[d]	95
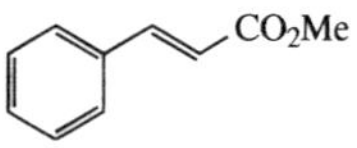 Ph-CH=CH-CO₂Me (E)	EtMgBr	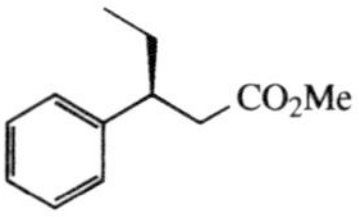 Ph, Et adduct, CO₂Me	13 (1.5)	94[d,e]	98 (S)
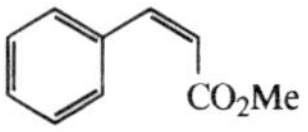 Ph-CH=CH-CO₂Me (Z)	EtMgBr	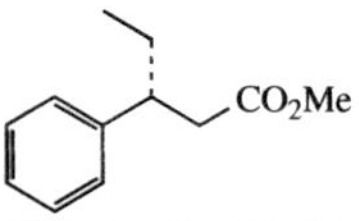 Ph, Et adduct, CO₂Me	13 (1.5)	100[c−e]	53 (R)

[a] Cu complex (see table), 1.15 equiv. of RMgBr, t-BuOMe, −75 °C.
[b] Adapted with permission from *Acc. Chem. Res.*, **40**, 179–188 (2007). Copyright 2007 American Chemical Society.
[c] Conversation(GC)
[d] 2.5 equiv. of RMgBr employed.
[e] Carried out in CH₂Cl₂.
[f] Isolated yeild.

The above-mentioned study was followed by a report of the groups of Ji, Loh and coworkers, who reported the application of a catalyst based on cuprous iodide and Tol-BINAP for the same purpose[39]. Noteworthy is that the effective use of a C_2-symmetric ligand in this reaction marks the end of the aforementioned metal-differentiating coordination concept. It was shown that a variety of Grignard reagents could be used for the

TABLE 6. CuI/Tol-BINAP-catalyzed conjugate addition of Grignard reagents

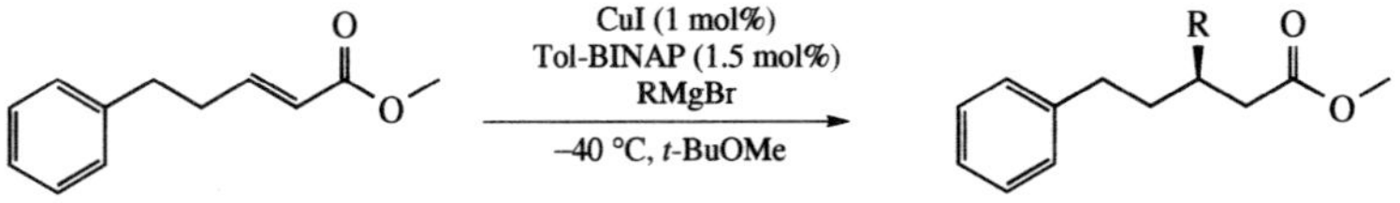

RMgBr	Yield (%)	ee (%)
EtMgBr	88	93
PrMgBr	90	92
i-PrMgBr	89	91
BuMgBr	90	92
PentMgBr	86	90
HeptMgBr	89	92
ButenylMgBr	90	94
i-BuMgBr	91	86
MeMgBr	20	98

TABLE 7. CuI/Tol-BINAP-catalyzed conjugate addition of EtMgBr

Unsaturated ester	Yield (%)	ee (%)	Product
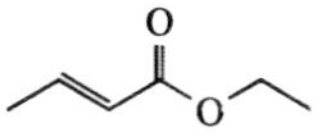	83	74	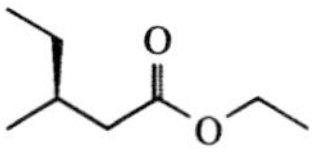
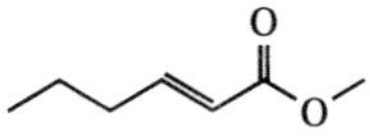	85	87	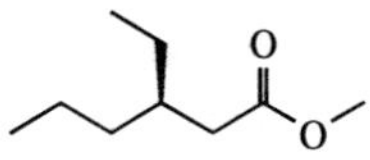
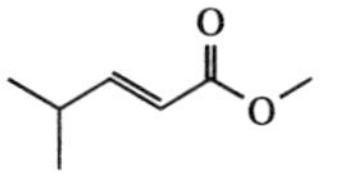	67	68	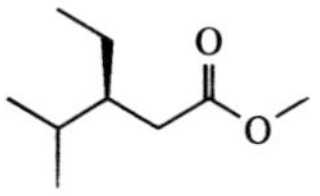
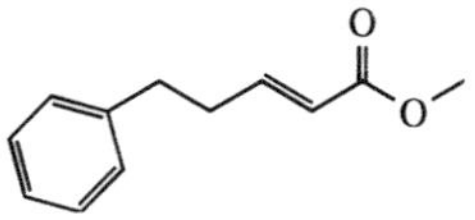	88	93	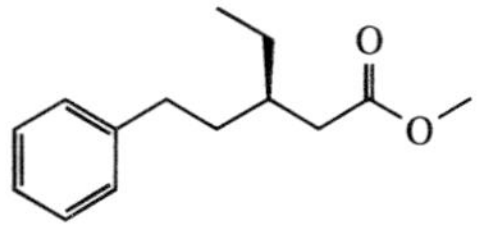
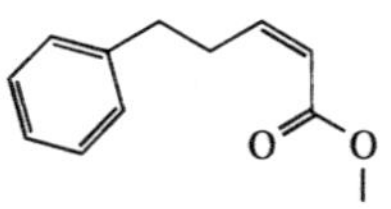	86	94 (S)	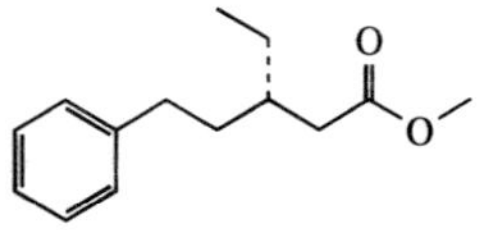
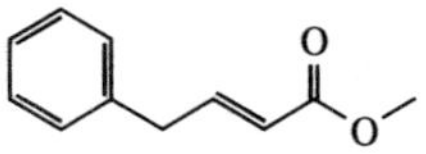	85	94	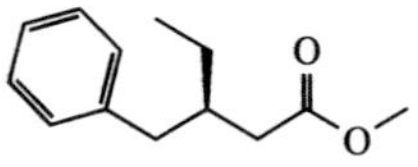
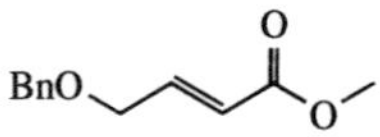	83	73	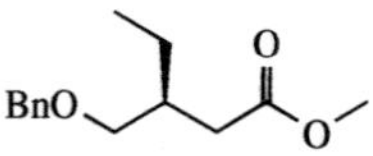
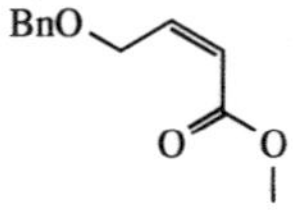	86	87	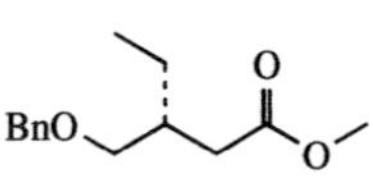

(*continued overleaf*)

TABLE 7. (continued)

Unsaturated ester	Yield (%)	ee (%)	Product
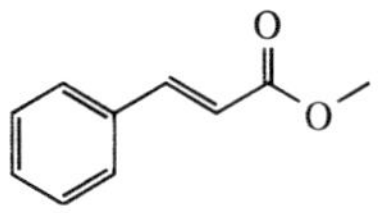 (methyl cinnamate)	90[a]	93	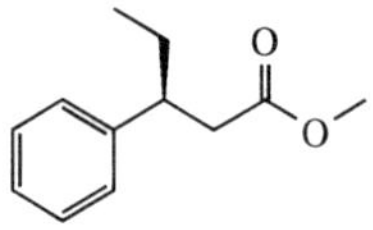 (β-ethyl phenyl product)
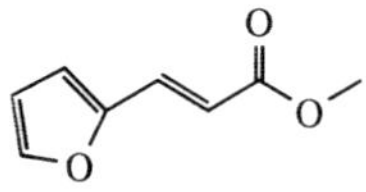 (methyl 3-(furan-2-yl)acrylate)	80[a]	85	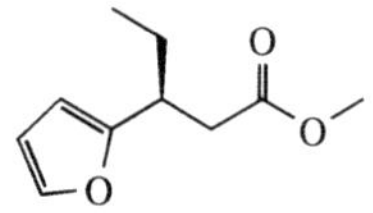 (β-ethyl furyl product)
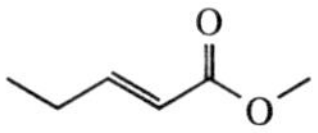 (methyl pent-2-enoate)	80[b]	74	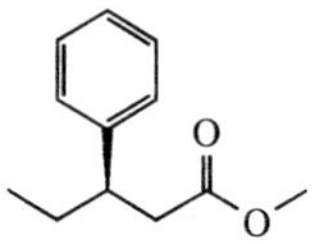 (β-phenyl ethyl product)

[a] 5 mol% CuI and 7.5 mol% Tol-BINAP were used.
[b] PhMgBr was used.

highly enantioselective addition to unsaturated methyl esters (Table 6). Table 7 shows the scope of the reaction using EtMgBr as the nucleophile. As in the case of Cu/Josiphos and Cu/Taniaphos, Z-enoates give the opposite configuration of the product, in this case apparently without concomitant isomerization of the substrate as the ee is not compromised. As for the Cu/ferrocenyl diphosphine catalysts, enantioselectivities drop considerably when PhMgBr is used as the nucleophile and conversions are significantly lower with MeMgBr.

To address the lack of reactivity in the methyl Grignard additions, we focussed our attention on the more reactive but equally readily accessible α,β-unsaturated thioesters[40]. The addition of MeMgBr to a series of unsaturated thioesters revealed the success of this approach[41]. The complex prepared *in situ* from CuBr•SMe$_2$ (1.0 mol%) and Josiphos (1.1 mol%) catalyzed the conjugate addition of MeMgBr providing the corresponding β-methyl-substituted thioesters with complete regioselectivity and excellent enantioselectivities (95–96% ee) (Table 8). The drastically higher yields obtained for the methyl adducts from α,β-unsaturated thioesters, compared to the oxoester analogs, are most probably due to their inherent electronic properties, which are closer to those of enones.

TABLE 8. Enantioselective conjugate addition of MeMgBr to α,β-unsaturated thioesters[a]

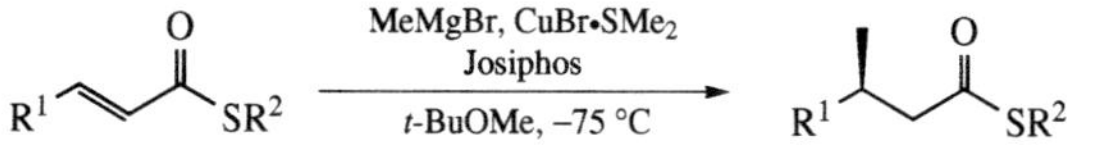

$$R^1\text{-CH=CH-C(O)-SR}^2 \xrightarrow[\text{t-BuOMe, }-75\,°\text{C}]{\text{MeMgBr, CuBr•SMe}_2,\text{ Josiphos}} R^1\text{-CH(Me)-CH}_2\text{-C(O)-SR}^2$$

R^1	R^2	Yield (%)	ee (%)
n-Pent	Et	90	96
n-Pent	Me	93	96
n-Pr	Et	92	96
BnO(CH$_2$)$_3$	Et	94	95
Ph	Et	88	95

[a] MeMgBr, CuBr•SMe$_2$ (1.0 mol%), Josiphos (1.2 mol%), t-BuOMe, −75 °C.

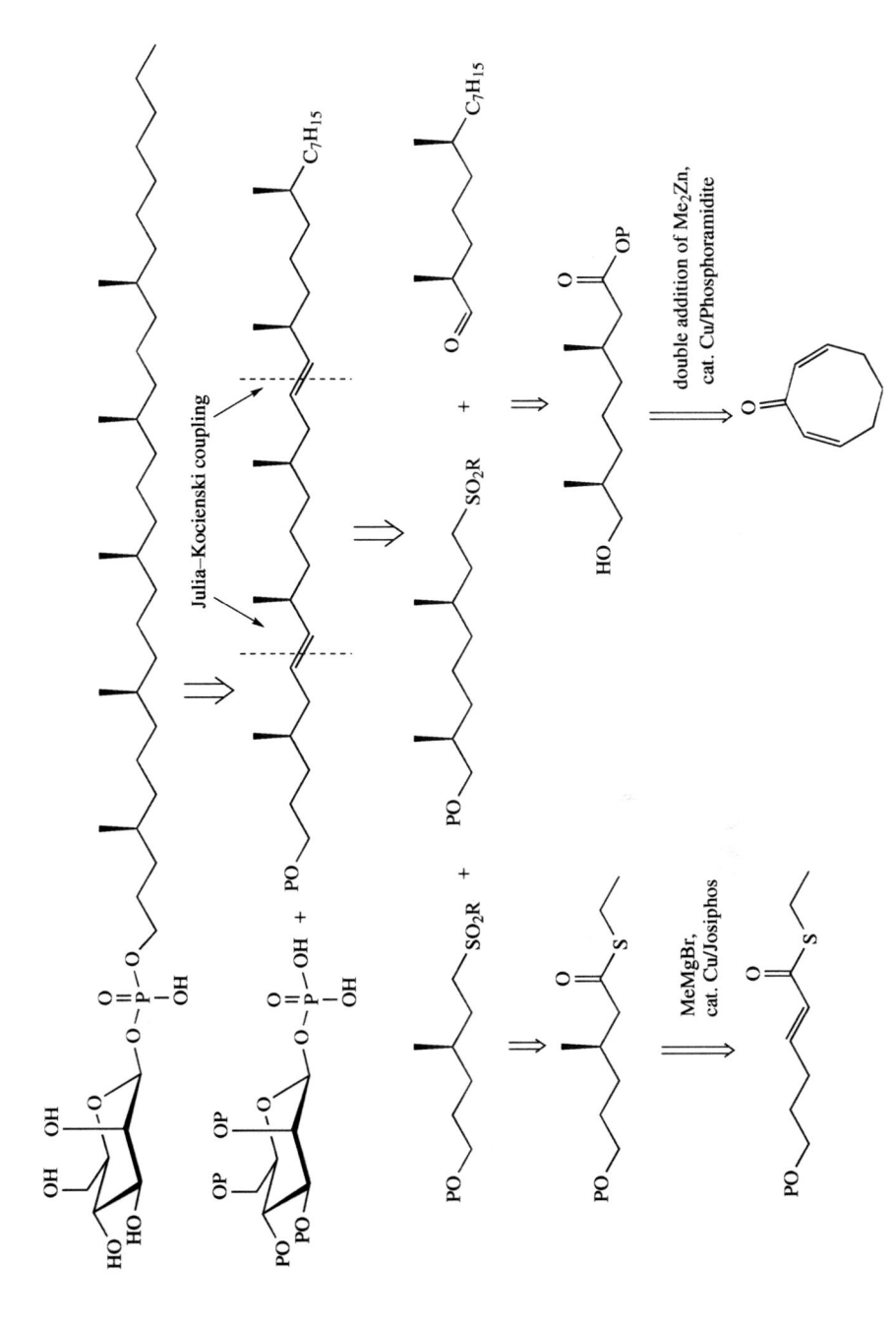

FIGURE 5. A β-D-mannosyl phosphomycoketide from *Mycobacterium tuberculosis*

V. APPLICATION OF THE CONJUGATE ADDITION OF GRIGNARD REAGENTS TO α,β-UNSATURATED THIOESTERS IN THE SYNTHESIS OF NATURAL PRODUCTS

An illustration of the use of β-methyl-substituted thioesters in the synthesis of natural products is present in the total synthesis of a β-D-mannosyl phosphomycoketide from *Mycobacterium tuberculosis* (Figure 5)[42]. Addition of MeMgBr to ethyl 6-benzyloxy-2-hexene thioate catalyzed by Cu/Josiphos (92% yield, 93% ee) furnished one of the building blocks. The other four methyl groups were introduced using copper/phosphoramidite-catalyzed dimethylzinc addition.

Moreover, the enantioselective conjugate addition of Grignard reagents to α,β-unsaturated thioesters allows access to enantiopure *syn*- and *anti*-1,3-dimethyl arrays by way of an iterative procedure[41,43]. The approach relies on sequential enantioselective conjugate additions, the protocol of which is shown in Scheme 7. The first stereogenic center is created by the addition of MeMgBr, using Josiphos (95% ee). The resulting thioester is converted in one step into the corresponding aldehyde, which subsequently undergoes a

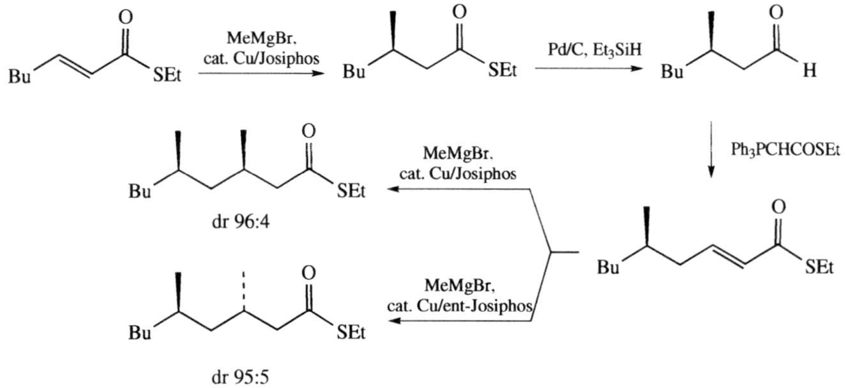

SCHEME 7. An iterative catalytic route to enantiopure *syn*- and *anti*-1,3-dimethyl arrays. Adapted with permission from *Acc. Chem. Res.*, **40**, 179–188 (2007). Copyright 2007 American Chemical Society

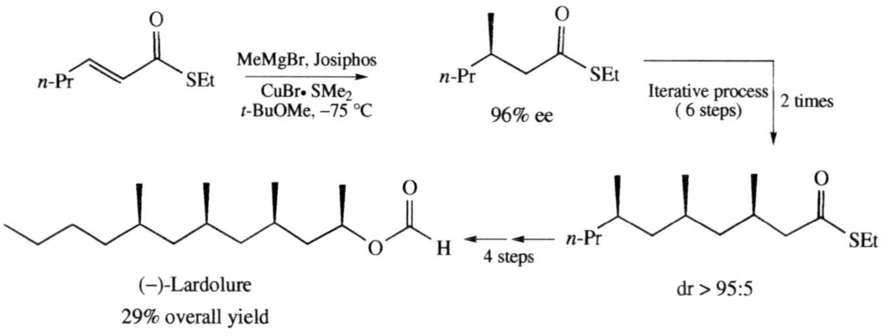

SCHEME 8. Application of the iterative conjugate addition in the synthesis of (−)-Lardolure. Adapted with permission from *Acc. Chem. Res.*, **40**, 179–188 (2007). Copyright 2007 American Chemical Society

(−)-Mycocerosic acid

FIGURE 6. Mycocerosic acid from *M. tuberculosis*

Wittig reaction to give the desired Michael acceptor. A second catalytic conjugate addition using Josiphos or its enantiomer affords with excellent diastereoselectivities the *syn*- or *anti*-1,3-dimethyl derivative.

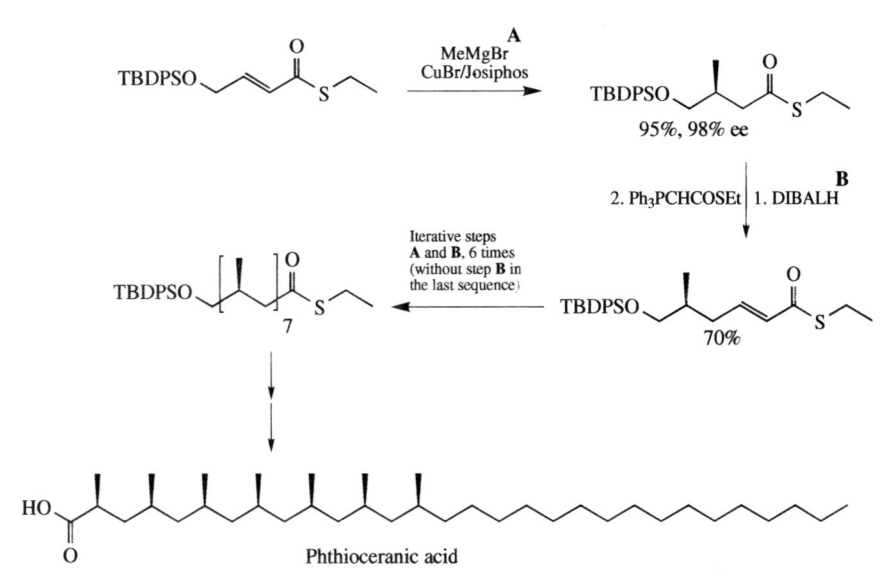

SCHEME 9. Catalytic asymmetric synthesis of Phthioceranic acid

The synthetic utility of this iterative catalytic protocol is especially apparent in the preparation of natural products containing so-called deoxypropionate units[44]. This was demonstrated in the asymmetric total synthesis of (−)-Lardolure, the aggregation pheromone of the acarid mite *Lardoglyphus konoi* (Scheme 8)[41]. Furthermore, as a convincing illustration of the high efficiency of this iterative protocol, this strategy has been applied in the synthesis of Mycocerosic acid[45] (Figure 6) and Phthioceranic acid[46], lipids present in the cell wall of *M. tuberculosis* (Scheme 9).

As the product of the conjugate addition of a Grignard reagent is a magnesium enolate, it is tempting to use this species in a subsequent diastereoselective reaction. The development of a tandem conjugate addition–aldol reaction, starting with the Cu/Josiphos-catalyzed addition of methylmagnesium bromide to unsaturated thioesters, turned out to be very successful[47]. A fast reaction of the magnesium enolate at low temperature with a suitable aldehyde afforded the corresponding *syn,syn* aldol product predominantly. The excellent diastereoselectivity of this acyclic three-component reaction is remarkable as exemplified in Scheme 10. This strategy was used for an efficient synthesis of Phaseolinic acid (Scheme 11).

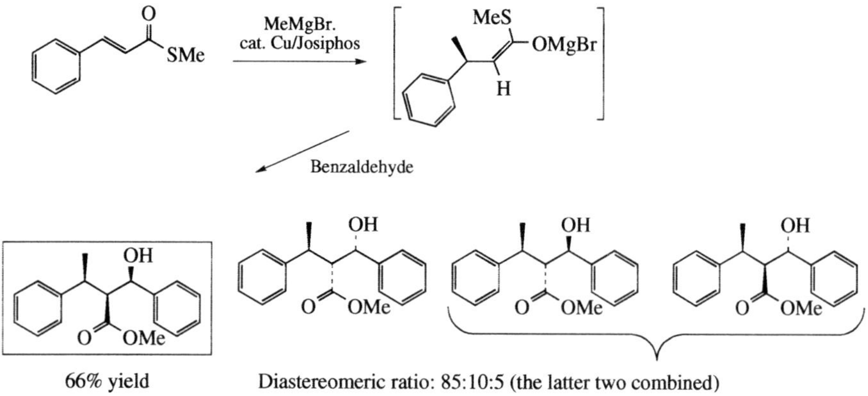

SCHEME 10. A tandem conjugate addition–aldol reaction

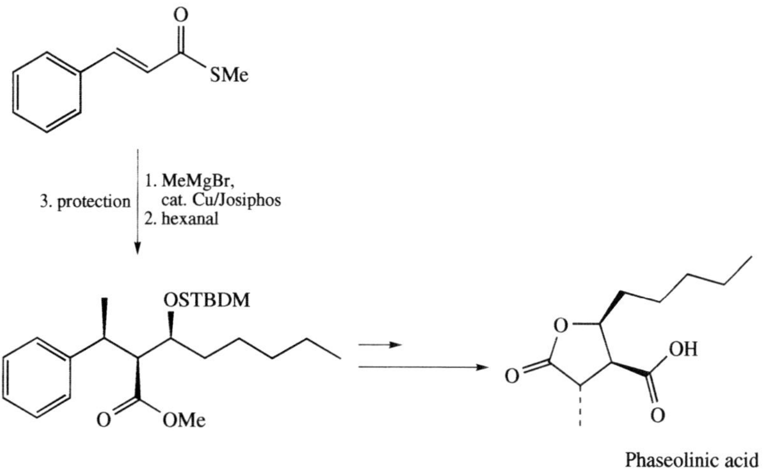

SCHEME 11. Total synthesis of Phaseolinic acid

VI. MECHANISTIC STUDIES

An extensive spectroscopic and mechanistic study on the enantioselective Cu/ferrocenyl bisphosphine-catalyzed conjugate addition has been performed[48]. Several parameters such as solvent, nature of the halide present in the Grignard reagent and Cu(I) source, and additives (i.e. dioxane and crown ethers) were identified. These factors directly affect the formation and nature of the intermediate active species, and therefore the selectivity, rate and overall outcome of the reaction. Importantly, the presence of Mg^{2+} and Br^- ions in the reaction are essential in order to achieve high selectivity and efficiency.

Kinetic studies carried out on a model reaction, the addition of EtMgBr to methyl crotonate catalyzed by Cu/Josiphos, indicated that the rate of the conjugate addition reaction is dependent on catalyst, Grignard reagent and substrate. Although the determination of the reaction order in methyl crotonate and Grignard reagent was impeded due to side reactions and inhomogeneity, the observation that the reaction rate increases with their

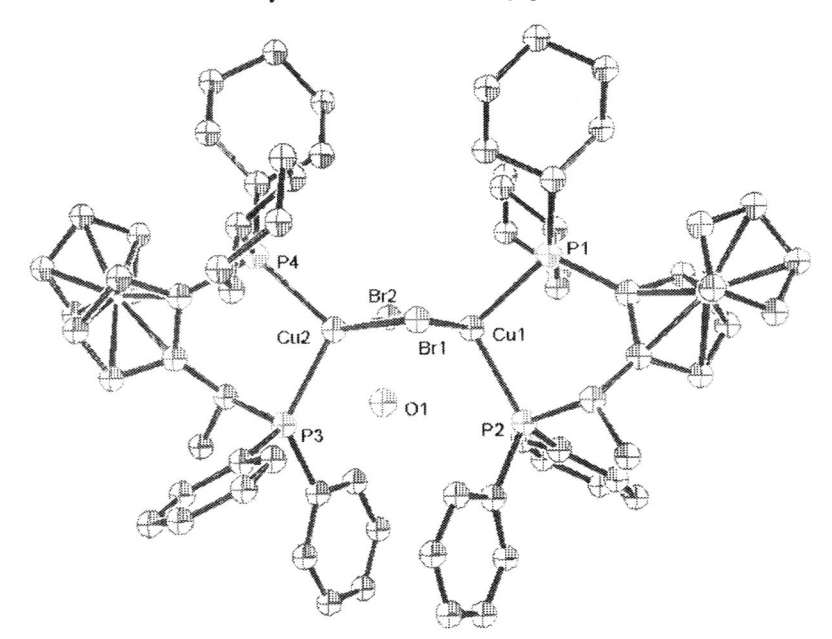

FIGURE 7. The X-ray structure of **13** (hydrogen atoms are omitted for clarity). Adapted with permission from *Acc. Chem. Res.*, **40**, 179–188 (2007). Copyright 2007 American Chemical Society

concentrations suggests that both reactants are involved in the rate-determining step. On the other hand, the order of the reaction (1.10) with respect to the catalyst suggests that a mononuclear species is involved. This was also supported by the observation that the ee of the product shows a linear dependency on the ee of the catalyst. The structure of the initial dinuclear Cu–Josiphos complex **13** was established by X-ray analysis (Figure 7).

A reaction pathway which is consistent with the experimental, kinetic and spectroscopic results is proposed in Scheme 12.

Alkyl transfer from the magnesium halide to the chiral Cu complexes generates the Cu complex **A**, as deduced from NMR experiments. Very likely, this complex functions in a similar manner as previously postulated for organocuprate additions[49].

The second intermediate proposed is a Cu–olefin π-complex with an additional interaction of Mg^{2+} with the carbonyl oxygen of the enone (enoate). The formation of a π-complex is presumably followed by intramolecular rearrangement to a Cu(III) intermediate, where Cu forms a σ-bond with the β-carbon of the enone (enoate), in fast equilibrium with the π-complex.

This catalytic cycle gives an explanation for the observed isomerization of Z-enoates to their E-isomers, which occurs within the time scale of the reaction. Indeed, these isomerization experiments provide evidence for the presence of a fast equilibrium between a π-complex and a Cu(III) species (σ-complex), which should be followed by the rate limiting, reductive elimination step and the formation of complex **A** again.

The proposed catalytic cycle is in accordance with the results of the kinetic studies. The dependence of the reaction rate on the substrate and Grignard reagent indicates that both reactants are involved in the rate-limiting step. This step is preceded by fast equilibria between complexes, for example, a substrate-bound σ-complex and π-complex and substrate-unbound complex **A**.

SCHEME 12. Proposed catalytic cycle. Adapted with permission from *Acc. Chem. Res.*, **40**, 179–188 (2007). Copyright 2007 American Chemical Society.

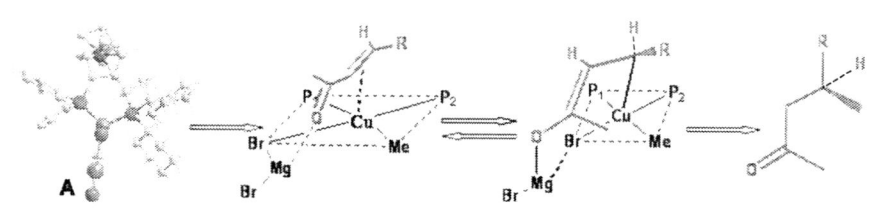

FIGURE 8. Model for the enantioselective conjugate addition of Grignard reagents (P_1: PPh_2 moiety P_2-: PCy_2). Adapted with permission from *Acc. Chem. Res.*, **40**, 179–188 (2007). Copyright 2007 American Chemical Society

Semiempirical [PM3(tm)] calculations indicated that Cu complex **A** adopts a distorted tetrahedral structure with the positioning of the Grignard reagent at the bottom face of the complex (Figure 8).

In the proposed model it can be envisioned that the enone approaches the alkylcopper complex **A** from the least hindered side and binds to the top apical position. This forces the complex to adopt a square pyramidal geometry, which is stabilized via π-complexation of the alkene moiety to the Cu and, importantly, through the interactions between Mg and the carbonyl moiety of the skewed enone. Formation of a transition structure with the chair-like seven-membered ring conformation is proposed in the next step, where Cu forms a σ-bond approaching from the bottom side of the β-carbon leading to the Cu(III) intermediate with the absolute configuration shown (Figure 8). Up to this stage in the catalytic cycle, complex formation is reversible, but in the subsequent rate-determining step, the alkyl transfer step, the product stereochemistry is established. To avoid steric interactions with the dicyclohexyl moieties at the nearby phosphorus, the final transfer of the alkyl group occurs as shown in Figure 8. Although this model predicts the correct sense of asymmetric induction, it is nevertheless a model and further mechanistic studies and DFT calculations need to be performed to shed light on the factors that determine the origin of the enantioselectivity.

VII. ENANTIOSELECTIVE ALLYLIC ALKYLATION WITH GRIGNARD REAGENTS

As recently highlighted by Woodward, enantioselective S_N2' allylic substitution reactions are mechanistically related to conjugate addition reactions[50]. Theoretical studies carried out by Nakamura and coworkers for the conjugate addition and allylic alkylation using Gilman's cuprates revealed profound mechanistic similarities between these two processes[49, 51].

Compared to the enantioselective allylic alkylation using soft nucleophiles[52], the reaction with Grignard reagents has received much less attention. The first enantioselective copper-catalyzed allylic alkylation with alkylmagnesium reagents was reported in 1995 by the groups of Bäckvall, van Koten and coworkers (Scheme 13)[53].

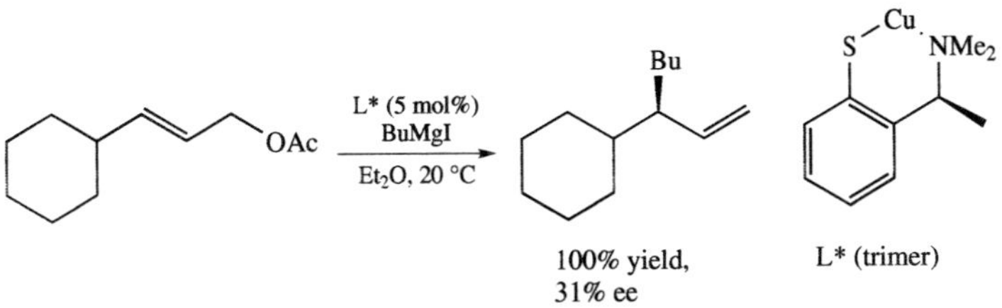

SCHEME 13. Copper-catalyzed allylic substitution using an arenethiolate ligand

In contrast, the first highly enantioselective version of this reaction used bulky dialkylzincs and was reported a few years later by Dübner and Knochel[54]. For linear dialkylzincs, highly efficient catalysts were reported soon after this disclosure by Hoveyda and coworkers[55] and Feringa and coworkers[56]. As in the field of conjugate addition reactions, organozinc reagents dominated the field until recently[50]. Grignard reagents regained attention, however, after two reports of the Alexakis group on the highly enantioselective allylic substitution of cinnamyl chlorides catalyzed by a Cu/phosphoramidite (Table 9)[57].

In a subsequent report, the scope of this method was expanded to cyclic and linear β-substituted allyl chlorides using a slightly different ligand (Table 10)[58].

Similar to Hoveyda's work, which describes NHC ligands for the allylic alkylation with dialkylzinc reagents[59], Okamoto and coworkers reported the enantioselective allylic alkylation with Grignard reagents using an α-methyl naphthylamine-based NHC complex (Scheme 14)[60].

Surprisingly, also the use of N-heterocyclic carbenes as such, e.g. without copper, is a valuable approach for the allylic alkylation using Grignard reagents. The NHC acts as a Lewis base that activates the reagent and modifies its reactivity. This was used in a versatile preparation of esters containing quaternary stereocenters (Table 11)[61]. The formation of the corresponding cyclopropyl-containing side product, which is formed in the absence of ligand, could rather effectively be suppressed, although reaction times were rather long (24–60 h).

Next to the successful conjugate addition of Grignard reagents, the catalyst generated from Josiphos and CuBr•SMe$_2$ also effectively promotes the allylic alkylation of cinnamyl bromide with MeMgBr, affording the corresponding products with good regioselectivity (85:15) and high ee (85% ee)[62]. Under the same conditions, the allylic alkylation of

TABLE 9. Asymmetric allylic substitution using a Cu/phosphoramidite ligand

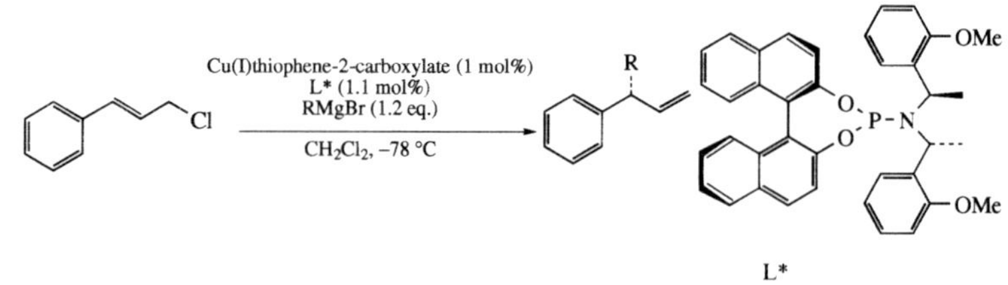

Substrate	R	Product	Yield (%)	S_N2'/S_N2	ee (%)
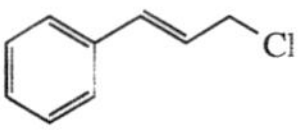 cinnamyl-Cl	Et	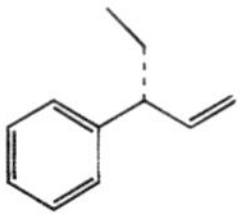	86	99/1	96 (R)
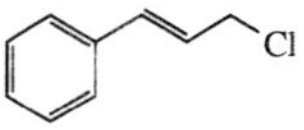 cinnamyl-Cl	Me	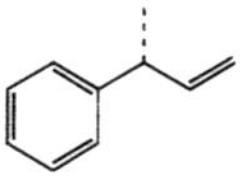	100 (conv.)	89/11	96
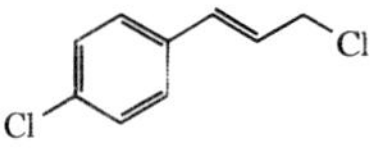 4-Cl-cinnamyl-Cl	Me	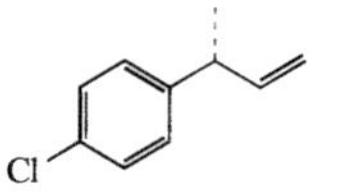	90 (conv.)	90/10	95
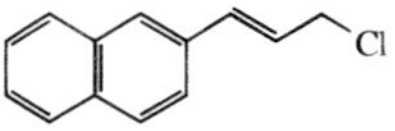 2-naphthyl-allyl-Cl	Me	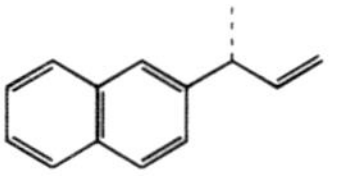	100 (conv.)	84/16	93
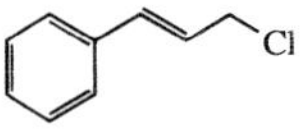 cinnamyl-Cl	Butenyl	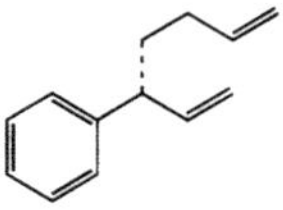	83	96/4	92 (R)
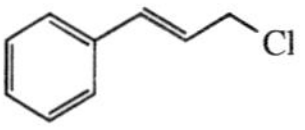 cinnamyl-Cl	Pentenyl	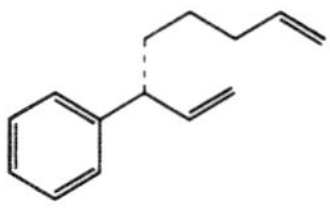	81	91/1	96 (R)
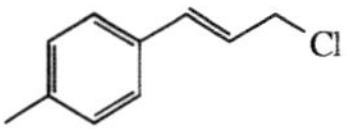 4-Me-cinnamyl-Cl	Et	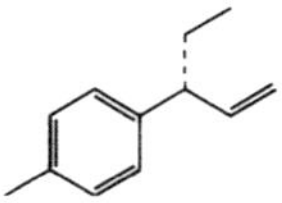	85	99/1	96 (R)
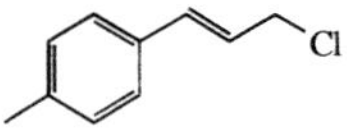 4-Me-cinnamyl-Cl	Butenyl	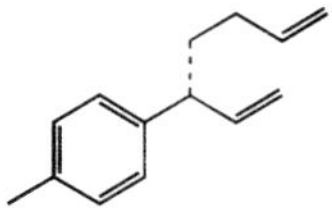	83	97/3	93 (R)
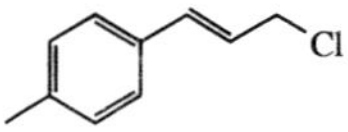 4-Me-cinnamyl-Cl	Pentenyl	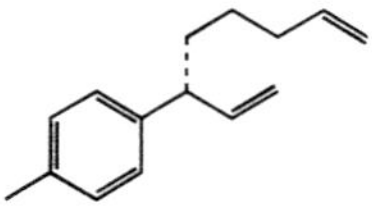	86	91/1	94 (R)
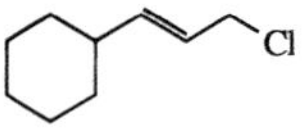 cyclohexyl-allyl-Cl	Et	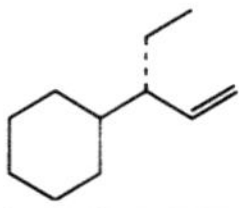	82	99/1	91 (−)

TABLE 10. Asymmetric allylic substitution of disubstituted allylic chlorides using a Cu/phosphoramidite ligand

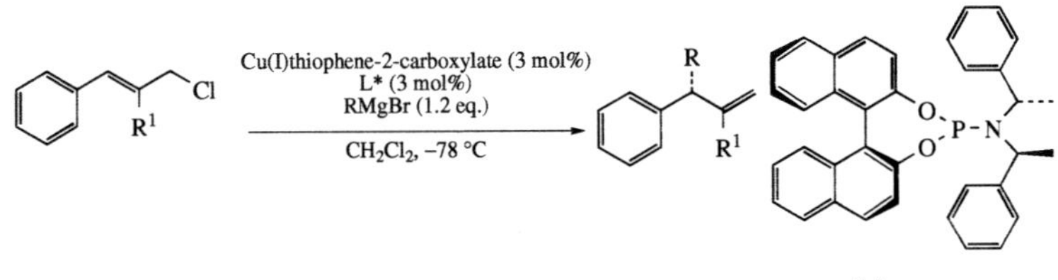

Substrate	R	Product	Yield (%)	S_N2'/S_N2	ee (%)
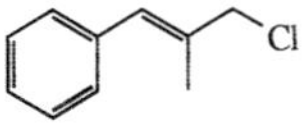	Et	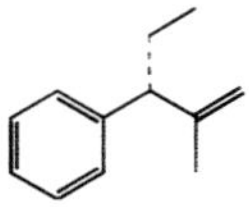	87	92/8	98 (+)
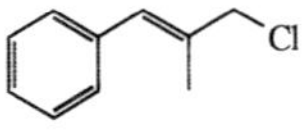	Pr	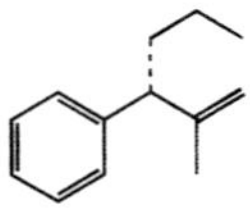	85	84/16	97 (+)
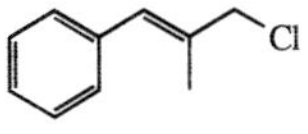	Pent	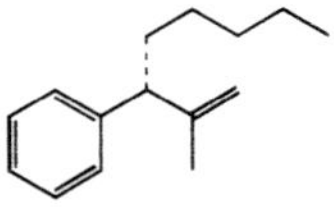	83	83/17	96 (+)
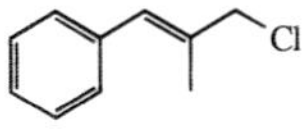	Butenyl	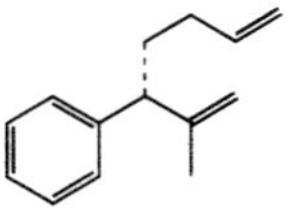	84	89/11	97 (+)
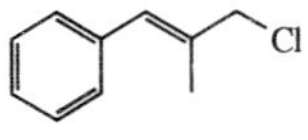	Pentenyl	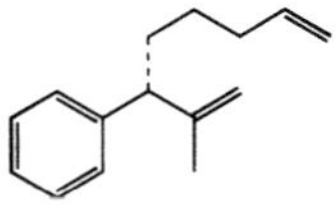	87	87/13	96 (+)
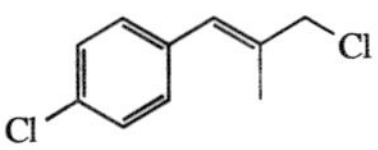	Et	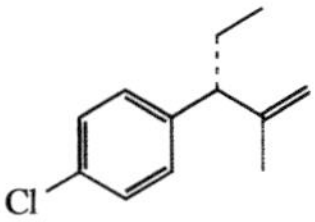	87	92/8	96 (+)
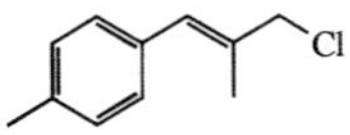	Et	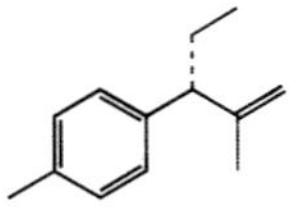	85	84/16	96 (+)
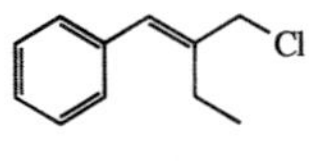	Et	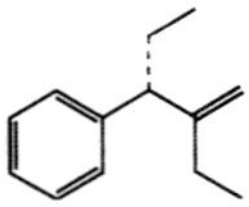	83	83/17	92 (+)

TABLE 10. (continued)

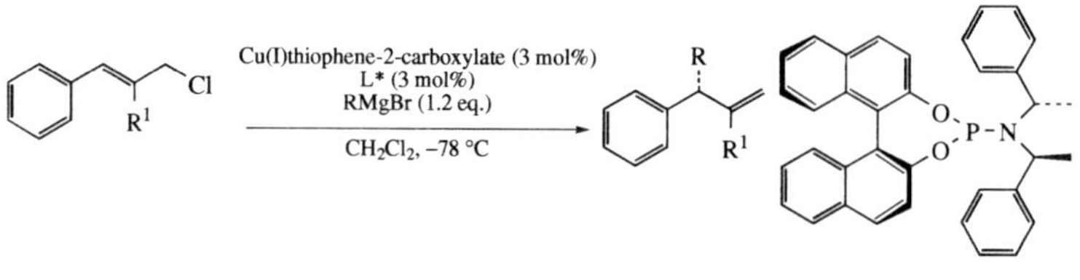

Substrate	R	Product	Yield (%)	$S_N 2'/S_N 2$	ee (%)
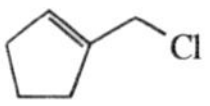 cyclopentenyl-CH₂Cl	Bu	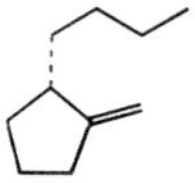	99 (conv)	96/4	98
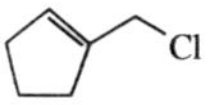 cyclopentenyl-CH₂Cl	Hex	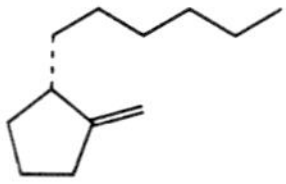	91	98/2	98
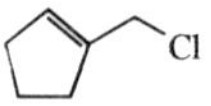 cyclopentenyl-CH₂Cl	phenethyl	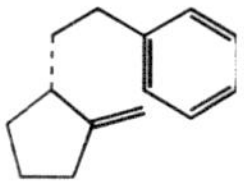	99 (conv)	97/3	98
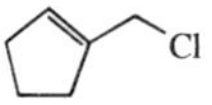 cyclopentenyl-CH₂Cl	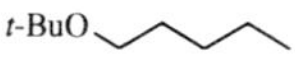 t-BuO~~~		60	98/2	98
cyclohexenyl-CH₂Cl	Bu		73	81/19	98
cyclohexenyl-CH₂Cl	Butenyl		83	97/3	99
cyclohexenyl-CH₂Cl	Hex		67	97/3	98
cyclohexenyl-CH₂Cl	phenethyl		78	85/15	99
cyclohexenyl-CH₂Cl	t-BuO~~~	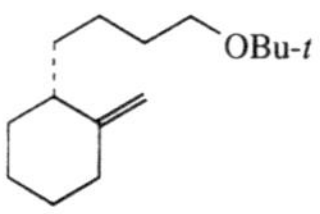	99 (conv)	91/9	99

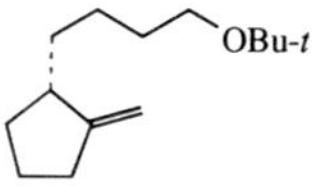

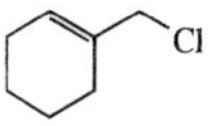

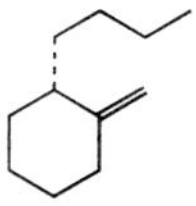

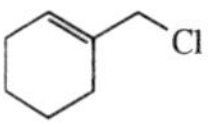

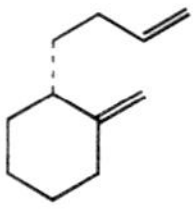

Reaction conditions: Cu(I)thiophene-2-carboxylate (3 mol%), L* (3 mol%), RMgBr (1.2 eq.), CH₂Cl₂, −78 °C.

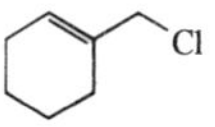

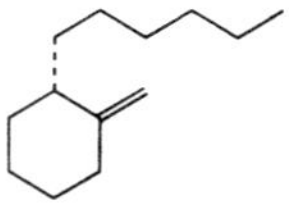

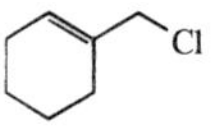

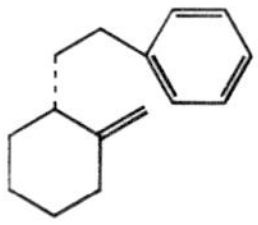

L* = BINOL-derived phosphoramidite with bis(α-methylbenzyl)amine.

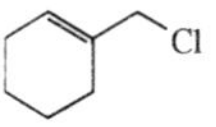

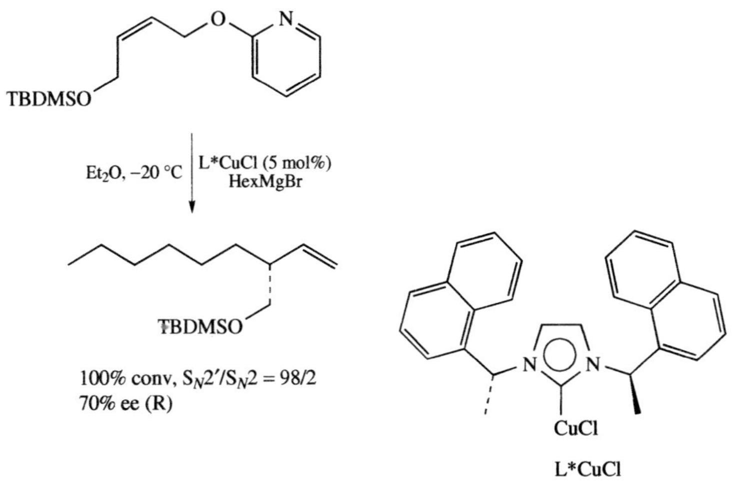

SCHEME 14. Synthesis of secondary allylic alcohols by Cu/NHC-catalyzed allylic substitution

TABLE 11. Allylic alkylation of γ-chloro-α,β-unsaturated esters

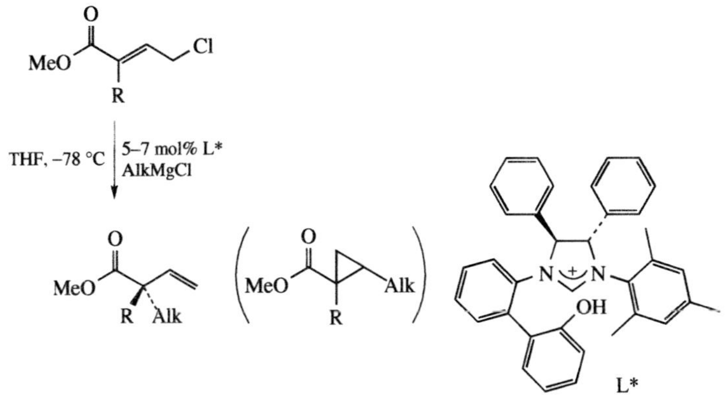

R	Alkyl	S_N2'/S_N2	Cyclopropyl product (%)	S_N2' Yield (%)	ee (%)
Me	i-Pr	90/10	9	80	97
Me	c-Pent	81/19	12	57	75
Me	c-Hex	92/8	27	63	94
Me	n-Bu	86/14	28	34	63
Et	i-Pr	91/9	7	73	97
Bu	i-Pr	91/9	7	75	98
Et	c-Pent	88/12	8	66	90
Bu	c-Pent	78/22	13	59	85
Et	c-Hex	93/7	19	60	96
Bu	c-Hex	92/8	13	57	96
Et	n-Bu	88/12	26	35	79

cinnamyl bromide with EtMgBr provided a modest regioselectivity (38:62) and a disappointing 56% ee. However, a dramatic improvement was observed using Taniaphos as the ligand and CH_2Cl_2 instead of t-BuOMe as the solvent. The desired product was obtained with a good regioselectivity (82:18) and an excellent ee (96%).

The scope of the method turned out to be particularly broad (Table 12). The allylic substitution of cinnamyl bromide could also be performed with other linear alkyl Grignard

TABLE 12. Cu/Taniaphos-catalysed enantioselective allylic alkylation with Grignard reagents

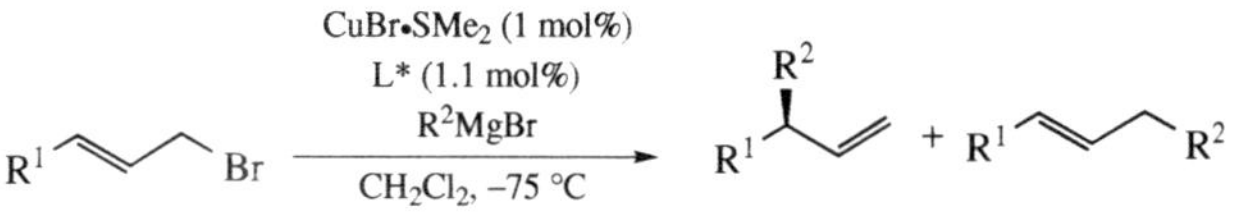

R^1	R^2	S_N2'/S_N2	Product	Yield (%)	ee (%)
Ph	Et	82/18	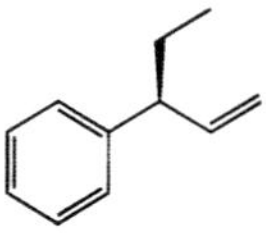	92	95
Ph	Bu	87/13	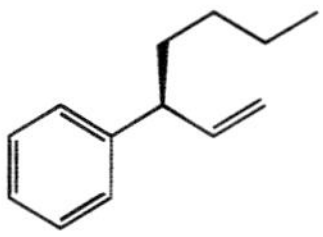	92	94
Ph	Butenyl	91/9	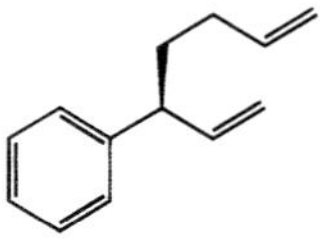	93	95
Ph	Me	97/3	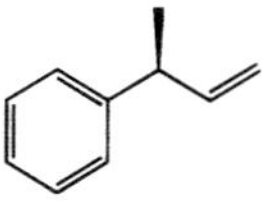	91	98
1-Naph	Me	100/0	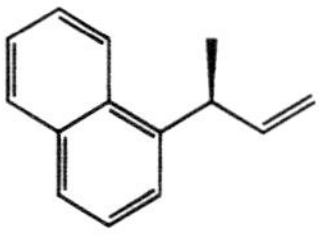	87	96
p-ClC$_6$H$_4$	Me	99/1	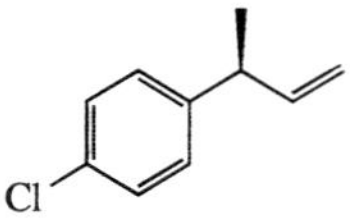	95	97
p-MeOOC$_6$H$_4$	Me	98/2	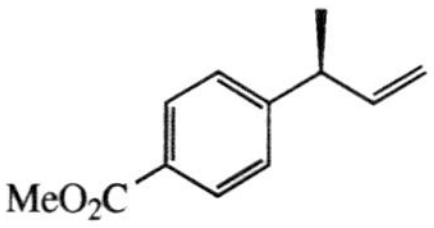	94	97
BnOCH$_2$	Me	100/0	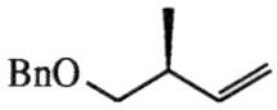	93	92
Bu	Me	100/0	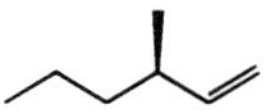	99	92

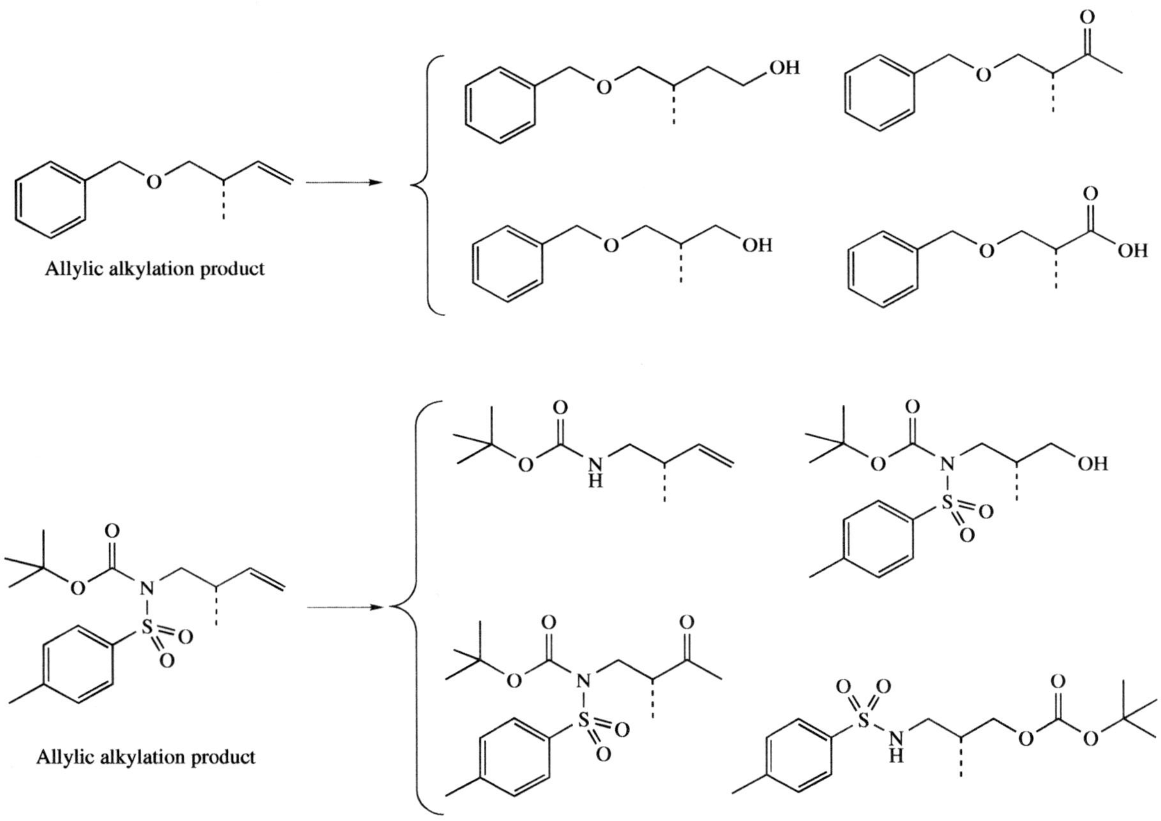

SCHEME 15. Synthesis of bifunctional building blocks through copper/Taniaphos-catalyzed allylic alkylation

reagents. Most important, the alkylations with MeMgBr afforded the products with almost complete control of regioselectivity and enantioselectivity ($\geq 96\%$) and also linear allylic bromides turned out to be excellent substrates.

The utility of the copper/Taniaphos-catalyzed allylic alkylation was further illustrated in two subsequent reports. The application of aliphatic allylic bromides containing protected alcohols and amines leads to the efficient synthesis of bifunctional building blocks (Scheme 15)[63].

A recent and entirely novel application of this reaction to readily available 3-bromopropenyl esters, leads to virtually enantiopure allylic esters, e.g. alcohols[64]. The reaction is chemo-, regio- and enantioselective as illustrated in Scheme 16, which also shows an application of this method for the synthesis of a naturally occurring butenolide. Known catalytic methods for the preparation of allylic alcohol derivatives involve metal-catalyzed allylic substitution using oxygen nucleophiles. Most of these methods provide ethers[65], although the use of carboxylic acids[66] and the preparation of amines[67] have also been reported. The synthesis of allylic esters using catalytic enantioselective carbon–carbon bond formation is therefore a new and versatile addition. In addition, the reaction is a nice complement to the kinetic resolution of allylic alcohols using Sharpless' asymmetric epoxidation. The scope of the method is large as shown in Table 13.

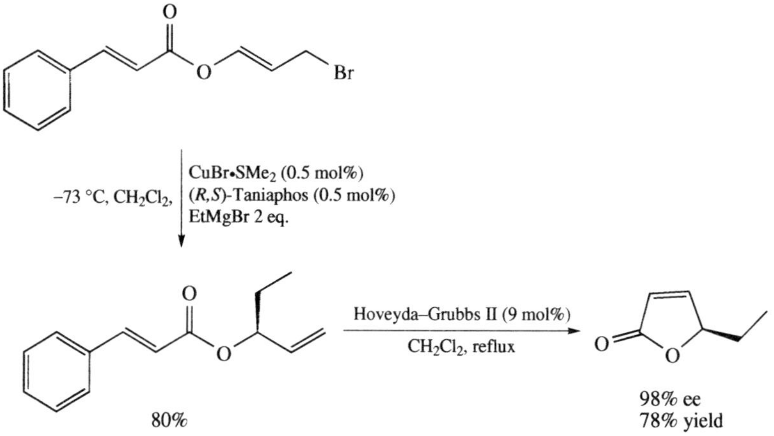

SCHEME 16. Copper-catalyzed asymmetric synthesis of chiral allylic esters; synthesis of a natural butenolide using allylic alkylation followed by ring-closing metathesis

VIII. CONCLUSIONS

Although already studied for a long time, the development of efficient and versatile catalysts for the conjugate addition and allylic alkylation using Grignard reagents is a recent development. In the conjugate addition reactions, the high enantiomeric excesses, the versatile asymmetric conjugate addition to α,β-unsaturated esters and thioesters and the formation of quaternary stereocenters are particularly noteworthy features. In addition, an iterative and catalytic approach to deoxypropionate subunits has been developed and applied to the synthesis of multimethyl branched natural products. In the allylic alkylation, the different substitution patterns of the substrate and the alkylation of 3-bromopropenyl esters have strongly broadened the synthetic utility of the reaction.

TABLE 13. The allylic substitution of 3-bromopropenyl esters using Cu/Taniaphos

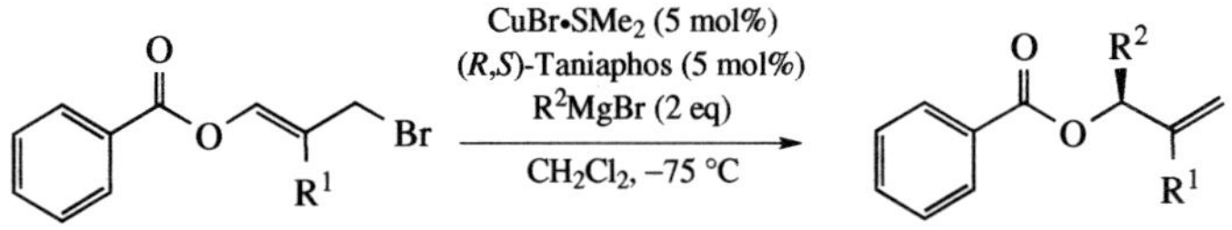

Reaction conditions: CuBr·SMe$_2$ (5 mol%), (R,S)-Taniaphos (5 mol%), R^{2}MgBr (2 eq), CH$_2$Cl$_2$, −75 °C

R^1	R^2	Product	Yield (%)	ee (%)
H	Et	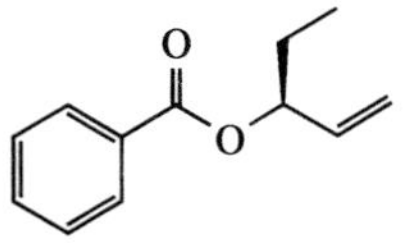	87	(+)98
H	Pent	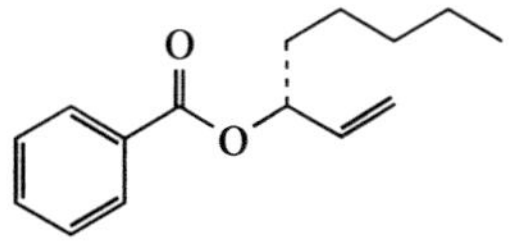	99	(−)97[a]
H	i-Bu	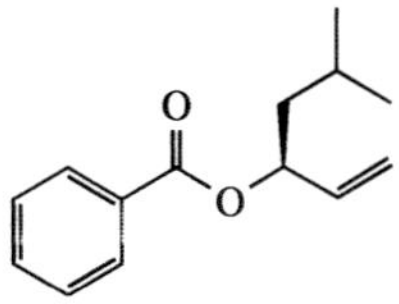	—	—
H	Butenyl	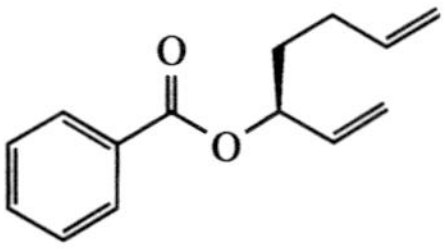	96	(+)97
H	Phenethyl	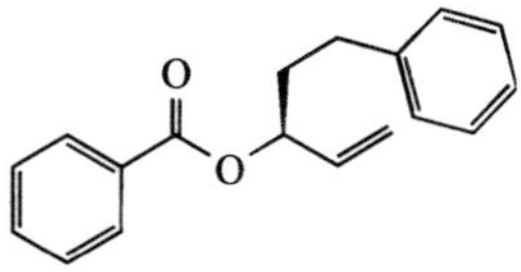	93	(+)93
H	Octadecyl	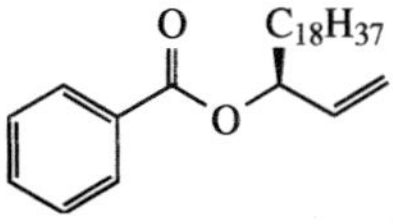	93	(+)95
Me	Et	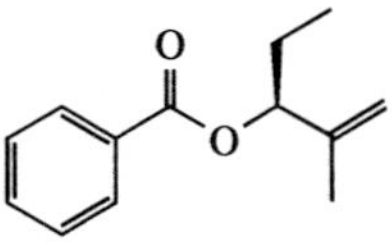	97[b]	97
Me	Pent	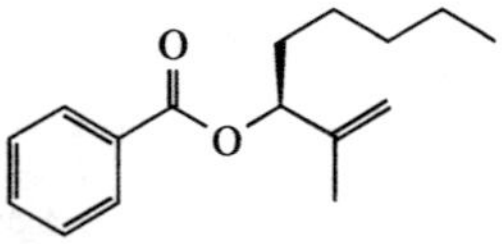	96[b]	98

[a] The enantiomer of the ligand was used.
[b] A mixture of regioisomers was isolated.

In view of this, it is evident that in the coming years, expansion of the catalytic toolbox for the conjugate addition and allylic substitution of these readily accessible organometallic reagents will be reported.

IX. REFERENCES AND NOTES

1. (a) P. Perlmutter, *Conjugate Addition Reactions in Organic Synthesis*, Tetrahedron Organic Chemistry Series 9, Pergamon, Oxford, 1992.
   (b) V. Caprio, 'Recent Advances in Organocopper Chemistry', *Lett. Org. Chem.*, **3**, 339 (2006).
2. (a) K. Tomioka and Y. Nagaoka, in *Comprehensive Asymmetric Catalysis* (Eds. E. N. Jacobsen, A. Pfaltz and H. Yamamoto), Vol. 3, Springer-Verlag, New York, 1999, pp. 1105–1120.
   (b) B. L. Feringa and A. H. M. de Vries, in *Advances in Catalytic Processes* (Ed. M. Doyle), Vol. 1, JAI Press, Connecticut, 1995, pp 151–192.
3. (a) T. Kogure and E. L. Eliel, *J. Org. Chem.*, **49**, 578 (1984).
   (b) T. Mukaiyama and N. Iwasawa, *Chem. Lett.*, 913 (1981).
4. B. E. Rossiter and N. M. Swingle, *Chem. Rev.*, **92**, 771 (1992).
5. E. J. Corey, R. Naef and F. J. Hannon, *J. Am. Chem. Soc.*, **108**, 7114 (1986).
6. Y. Asano, A. Iida and K. Tomioka, *Tetrahedron Lett.*, **38**, 8973 (1997) and references cited therein.
7. (a) G. M. Villacorta, C. P. Rao and S. J. Lippard, *J. Am. Chem. Soc.*, **110**, 3175 (1988).
   (b) K-H. Ahn, B. Klassen and S. J. Lippard, *Organometallics*, **9**, 3178 (1990).
8. (a) M. Spescha and G. Rihs, *Helv. Chim. Acta*, **76**, 1219 (1993).
   (b) Q-L. Zhou and A. Pfaltz, *Tetrahedron*, **50**, 4467 (1994).
   (c) M. van Klaveren, F. Lambert, D. J. F. M. Eijkelkamp, D. M. Grove and G. van Koten, *Tetrahedron Lett.*, **35**, 6135 (1994).
   (d) D. Seebach, G. Jaeschke, A. Pichota and L. Audergon, *Helv. Chim. Acta*, **80**, 2515 (1997).
9. (a) E. L. Stangeland and T. Sammakia, *Tetrahedron*, **53**, 16503 (1997).
   (b) M. Kanai and K. Tomioka, *Tetrahedron Lett.*, **36**, 4275 (1995).
10. J. F. G. A. Jansen and B. L. Feringa, *J. Org. Chem.*, **55**, 4168 (1990).
11. K. Soai, T. Hayasaka and S. Ugajin, *J. Chem. Soc., Chem. Commun.*, 516 (1989).
12. A. Alexakis, J. Frutos and P. Mangeney, *Tetrahedron: Asymmetry*, **4**, 2427 (1993).
13. A. H. M. de Vries, A. Meetsma and B. L. Feringa, *Angew. Chem., Int. Ed. Engl.*, **35**, 2374 (1996).
14. B. L. Feringa, M. Pineschi, L. A. Arnold, R. Imbos and A. H. M. de Vries, *Angew. Chem., Int. Ed. Engl.*, **36**, 2620 (1997).
15. L. A. Arnold, R. Naasz, A. J. Minnaard and B. L. Feringa, *J. Am. Chem. Soc.*, **123**, 5841 (2001).
16. B. L. Feringa, R. Naasz, R. Imbos and L. A. Arnold, in *Modern Organocopper Chemistry* (Ed. N. Krause), Wiley-VCH, Weinheim, 2002, pp. 224.
17. N. Krause and A. Hoffmann-Röder, *Synthesis*, 171 (2001).
18. A. H. Hoveyda, A. W. Hird and M. A. Kacprzynski, *Chem. Commun.*, 1779 (2004).
19. (a) A. Alexakis and C. Benhaim, *Eur. J. Org. Chem.*, 3221 (2002).
    (b) M. P. Sibi and S. Manyem, *Tetrahedron*, **56**, 8033 (2000).
20. For relevant recent examples, see:
    (a) M. Shi, C-J. Wang and W. Zhang, *Chem. Eur. J.*, **10**, 5507 (2004).
    (b) A. P. Duncan, and J. L. Leighton, *Org. Lett.*, **6**, 4117 (2004).
21. (a) Y. Takaya, M. Ogasawara, T. Hayashi, M. Sakai and N. Miyaura, *J. Am. Chem. Soc.*, **120**, 5579 (1998).
    (b) T. Hayashi and K. Yamasaki, *Chem. Rev.*, **103**, 2829 (2003).
    (c) J. G. Boiteau, A. J. Minnaard and B. L. Feringa, *Org. Lett.*, **5**, 681 (2003).
22. K. Tomioka, in *Comprehensive Asymmetric Catalysis, Suppl. 2* (Eds. E. N. Jacobsen, A. Pfaltz and H. Yamamoto), Springer-Verlag, New York, 2004, pp. 109–124.
23. W. W. Lin, O. Baron and P. Knochel, *Org. Lett.*, **8**, 5673 (2006) and references cited therein.
24. For an exception to the rule, see: K. Yoshida, M. Ogasawara and T. Hayashi *J. Org. Chem.*, **68**, 1901 (2003).
25. (a) D. M. Knotter, D. M. Grove, W. J. J. Smeets, A. L. Spek and G. van Koten, *J. Am. Chem. Soc.*, **114**, 3400 (1992).

(b) A. L. Braga, S. J. N. Silva, D. S. Lüdtke, R. L. Drekener, C. C. Silveira, J. B. T. Rocha and L. A. Wessjohann, *Tetrahedron Lett.*, **43**, 7329 (2002).
(c) M. Kanai, Y. Nakagawa and K. Tomioka, *Tetrahedron*, **55**, 3843 (1999) and references cited therein.
26. R. Noyori, *Asymmetric Catalysis in Organic Synthesis*, Wiley, New York, 1994.
27. (a) T. J. Colacot, *Chem. Rev.*, **103**, 3101 (2003).
(b) A. Togni, C. Breutel, A. Schnyder, F. Spindler, H. Landert and A. Tijani, *J. Am. Chem. Soc.*, **116**, 4062 (1994).
28. T. Ireland, G. Grossheimann, C. Wieser-Jeunesse and P. Knochel, *Angew. Chem., Int. Ed.*, **38**, 3212 (1999).
29. B. L. Feringa, R. Badorrey, D. Peña, S. R. Harutyunyan and A. J. Minnaard, *Proc. Natl. Acad. Sci. U.S.A.*, **101**, 5834 (2004).
30. K.S. Lee, M. K. Brown, A. W. Hird and A. H. Hoveyda, *J. Am. Chem. Soc.*, **128**, 7182 (2006).
31. D. Martin, S. Kehrli, M. d'Augustin, H. Clavier, M. Mauduit and A. Alexakis, *J. Am. Chem. Soc.*, **128**, 8416 (2006).
32. F. López, S. R. Harutyunyan, A. J. Minnaard and B. L. Feringa, *J. Am. Chem. Soc.*, **126**, 12784 (2004).
33. For selected examples, see:
(a) H. Mizutani, S. J. Degrado and A. H. Hoveyda, *J. Am. Chem. Soc.*, **124**, 779 (2002).
(b) B. H. Lipshutz and J. M. Servesko, *Angew. Chem., Int. Ed.*, **42**, 4789 (2003).
(c) T. Hayashi, K. Ueyama, N. Tokunaga and K. A. Yoshida, *J. Am. Chem. Soc.*, **125**, 11508 (2003).
(d) P. K. Fraser and S. Woodward, *Chem. Eur. J.*, **9**, 776 (2003).
34. With 1 mol% of catalyst the products were obtained with equal enantioselectivity and slightly lower regioselectivity.
35. For a Rh-catalyzed asymmetric conjugate addition of arylboronic reagents to α,β-unsaturated esters, see: S. Sakuma, M. Sakai, R. Itooka and N. Miyaura, *J. Org. Chem.*, **65**, 5951 (2000) and references cited therein.
36. A. W. Hird and A. H. Hoveyda, *Angew. Chem., Int. Ed.*, **42**, 1276 (2003).
37. For an alternative strategy, see: J. Schuppan, A. J. Minnaard and B. L. Feringa, *Chem. Commun.*, 792 (2004).
38. F. López, S. R. Harutyunyan, A. Meetsma, A. J. Minnaard and B. L. Feringa, *Angew. Chem., Int. Ed.*, **44**, 2752 (2005).
39. S.-Y. Wang, S.-J. Ji and T.-P. Loh, *J. Am. Chem. Soc.*, **129**, 276 (2007).
40. The reduced electron delocalization in the thioester moiety, compared to oxoesters, results in a higher reactivity towards conjugate addition reactions; see Reference 41.
41. R. Des Mazery, M. Pullez, F. López, S. R. Harutyunyan, A. J. Minnaard and B. L. Feringa, *J. Am. Chem. Soc.*, **127**, 9966 (2005).
42. R. P. Van Summeren, D. B. Moody, B. L. Feringa and A. J. Minnaard, *J. Am. Chem. Soc.*, **128**, 4546 (2006).
43. The occurrence of polydeoxypropionate chains in numerous biologically relevant compounds and the paucity of iterative catalytic asymmetric procedures is a stimulus for the development of new methodologies.
44. For alternative iterative approaches, see:
(a) B. Liang, T. Novak, Z. Tan and E. Negishi, *J. Am. Chem. Soc.*, **128**, 2770 (2006).
(b) T. Novak, Z. Tan, B. Liang and E. Negishi, *J. Am. Chem. Soc.*, **127**, 2838 (2005) and references cited therein.
45. B. ter Horst, B. L. Feringa and A. J. Minnaard, *Chem. Commun.*, 489 (2007).
46. B. ter Horst, B. L. Feringa and A. J. Minnaard, *Org. Lett.*, **9**, 3013 (2007).
47. G. P. Howell, S. P. Fletcher, K. Geurts, B. ter Horst and B. L. Feringa, *J. Am. Chem. Soc.*, **128**, 14977 (2006).
48. S. R. Harutyunyan, F. López, W. R. Browne, A. Correa, D. Peña, R. Badorrey, A. Meetsma, A. J. Minnaard and B. L. Feringa, *J. Am. Chem. Soc.*, **128**, 9103 (2006).
49. E. Nakamura and S. Mori, *Angew. Chem., Int. Ed.*, **39**, 3751 (2000).
50. (a) S. Woodward, *Angew. Chem., Int. Ed.*, **44**, 5560 (2005).
For recent reviews on copper-catalyzed asymmetric allylic substitution, see:
(b) H. Yorimitsu and K. Oshima, *Angew. Chem., Int. Ed.*, **44**, 4435 (2005).

(c) A. Alexakis, C. Malan, L. Lea, K. Tissot-Croset, D. Polet and C. Falciola, *Chimia*, **60**, 124 (2006).
51. M. Yamanaka, S. Kato and E. Nakamura, *J. Am. Chem. Soc.*, **126**, 6287 (2004).
52. B. M. Trost and M. L. Crawley, *Chem. Rev.*, **103**, 2921 (2003).
53. (a) M. van Klaveren, E. S. M. Persson, A. del Villar, D. M. Grove, J. E. Bäckvall and G. van Koten, *Tetrahedron Lett.*, **36**, 3059 (1995).
 (b) J. Meuzelaar, A. S. E. Karlström, M. Van Klaveren, E. S. M. Persson, A. del Villar, G. Van Koten and J. E. Bäckvall, *Tetrahedron*, **56**, 2895 (2000).
 (c) A. S. E. Karlström, F. F. Huerta, G. J. Meuzelaar and J. E. Bäckvall, *Synlett*, 923 (2001).
 (d) H. K. Cotton, J. Norinder and J. E. Bäckvall, *Tetrahedron*, **62**, 5632 (2006).
 Although not enantioselective but stereospecific, the related copper-mediated and diphenylphosphanylbenzoyl-directed *syn*-allylic substitution with Grignard reagents is mentioned here:
 (e) C. Herber and B. Breit, *Angew. Chem., Int. Ed.*, **44**, 5267 (2005).
54. (a) F. Dübner and P. Knochel, *Angew. Chem., Int. Ed.*, **38**, 379 (1999).
 (b) F. Dübner and P. Knochel, *Tetrahedron Lett.*, **41**, 9233 (2000).
55. C. A. Luchaco-Cullis, H. Mizutani, K. E. Murphy and A. H. Hoveyda, *Angew. Chem., Int. Ed.*, **40**, 1456 (2001).
56. H. Malda, A. W. van Zijl, L. A. Arnold and B. L. Feringa, *Org. Lett.*, **3**, 1169 (2001).
57. (a) K. Tissot-Croset, D. Polet and A. Alexakis, *Angew. Chem., Int. Ed.*, **43**, 2426 (2004).
 (b) K. Tissot-Croset and A. Alexakis, *Tetrahedron Lett.*, **45**, 7375 (2004).
 (c) A. Alexakis, C. Malan, L. Lea, C. Benhaim and X. Fournioux, *Synlett*, 927 (2001).
 (d) A. Alexakis and K. Croset, *Org. Lett.*, **4**, 4147 (2002).
 (e) K. Tissot-Croset, D. Polet, S. Gille, C. Hawner and A. Alexakis, *Synthesis*, 2586 (2004).
58. C. A. Falciola, K. Tissot-Croset and A. Alexakis, *Angew. Chem., Int. Ed.*, **45**, 5995 (2006).
59. (a) A. O. Larsen, W. Leu, C. N. Oberhuber, J. E. Campbell and A. H. Hoveyda, *J. Am. Chem. Soc.*, **126**, 11130 (2004).
 (b) J. J. van Veldhuizen, J. E. Campbell, R. E. Giudici and A. H. Hoveyda, *J. Am. Chem. Soc.*, **127**, 6877 (2005).
60. (a) S. Tominaga, Y. Oi, T. Kato, D. K. An and S. Okamoto, *Tetrahedron Lett.*, **45**, 5585 (2004).
 (b) S. Okamoto, S. Tominaga, N. Saino, K. Kase and K. Shimoda, *J. Organomet. Chem.*, **690**, 6001 (2005). Note that due to a drawing error in a Table it erroneously looks as if silyl enol ethers are used as substrates.
61. Y. Lee and A. H. Hoveyda, *J. Am. Chem. Soc.*, **128**, 15604 (2006).
62. F. López, A. W. van Zijl, A. J. Minnaard and B. L. Feringa, *Chem. Commun.*, 409 (2006).
63. A. W. Van Zijl, F. López, A. J. Minnaard and B. L. Feringa, *J. Org. Chem.*, **72**, 2558 (2007).
64. K. Geurts, S. P. Fletcher and B. L. Feringa, *J. Am. Chem. Soc.*, **128**, 15572 (2006).
65. (a) B. M. Trost and A. Aponick, *J. Am. Chem. Soc.*, **128**, 3931 (2006) and references cited therein.
 (b) C. Shu and J. F. Hartwig, *Angew. Chem., Int. Ed.*, **43**, 4794 (2004) and references cited therein.
66. S. F. Kirsch and L. E. Overman, *J. Am. Chem. Soc.*, **127**, 2866 (2005).
67. C. E. Anderson and L. E. Overman, *J. Am. Chem. Soc.*, **125**, 12412 (2003).

Author Index

This author index is designed to enable the reader to locate an author's name and work with the aid of the reference numbers appearing in the text. The page numbers are printed in normal type in ascending numerical order, followed by the reference numbers in parentheses. The numbers in *italics* refer to the pages on which the references are actually listed.

Aavula, B. R. 289 (264), *311*
Abarbri, M. 518 (44, 46), 519 (51), 525 (41), *585*
Abarca, A. 278 (208), *310*
Abarca, B. 689 (26, 27), 713 (24), *714, 715*
Abate, Y. 171 (92), *186*
Abbondondola, J. A. 171 (103), *186*
Abd El-Hady, D. 277 (200), *310*
Abd El-Hamid, M. 277 (200), *310*
Abdel-Fattah, A. A. A. 559 (232), *590*
Abdel-Hadi, M. 550 (167), *589*
Abegg, R. 2 (10), *92*, 107 (25), *127*
Abel, E. W. 58 (3, 4), *92*, 512 (4), *583*, 657 (90), *678*
Abeles, R. H. 348 (245), *364*
Abeln, D. 50 (166), *96*
Abo El-Maali, N. 277 (200), *310*
Abraham, W. D. 760 (49), *769*
Abrahams, J. P. 329 (94, 102), *361*
Abrams, S. 288 (257), *311*
Abreu, J. B. 172 (108), *186*
Acar, O. 277 (206), *310*
Ackermann, L. 550 (171, 172), *589*
Adamo, C. 384 (30), *402*
Adamowicz, L. 323 (55), *360*
Adams, F. C. 302 (340, 341), 304 (319), *313*
Adams, J. A. 347 (233), 348 (221, 234, 240, 243), *364*, 560 (234), *590*, 673 (148, 149), *679*
Adderley, C. J. R. 476 (91), *508*
Adhikari, R. 495 (111), *509*
Adzuma, K. 355 (311), *366*
Afeefy, H. Y. 113 (56), *128*
Afghan, B. K. 300 (320), *313*

Aggarwal, S. K. 303 (344, 350), 304 (354), *313, 314*
Agostiano, A. 212 (115), *217*
Agrawal, Y. K. 282 (233), *311*
Agus, Z. S. 268 (87), *307*
Ahmadi, M. A. 246 (91), *262*
Ahmed, H. M. 280 (220), *310*
Ahmed, S. 305 (364), *314*
Ahn, K-H. 772 (7), *800*
Ahn, N. G. 347 (228), *364*
Ahr, M. 554 (191), *589*, 626 (76), *630*
Aimetti, J. A. 499 (117), *509*
Ainai, T. 558 (217), *590*
Akamine, P. 348 (241), *364*
Akao, A. 524 (63), *585*, 699 (31), 702 (39), *715*
Akihiro, Y. 538 (105), *587*
Akishin, P. A. 183 (154), 184 (152), *187*, 391 (15), *402*
Akiyama, M. 194 (29), 213 (119), *215, 217*
Akkerman, O. S. 3 (15), 34 (106, 116), 35 (117, 118), 38 (83), 40 (140), 43 (146), 44 (142), 61 (198), 66 (213–216), 86 (269), *92, 94, 95, 97, 99*, 116 (58), 120 (67), 121 (71), *128*, 183 (140), *187*, 318 (22), *359*
Akola, J. 331 (120), *362*
Akutagawa, S. 657 (88), *678*
Alam, H. 318 (21), *359*
Alami, M. 555 (195–198), *589*, 607 (48), 608 (51), *629*
Albachi, R. 516 (29), *584*
Alberti, A. 259 (144), *263*, 580 (313), *593*
Alcaraz, M. 302 (337), 304 (331), *313*
Aleandri, L. E. 610 (60), *630*
Aleksiuk, O. 471 (93), *508*

The chemistry of organomagnesium compounds
Edited by Z. Rappoport and I. Marek © 2008 John Wiley & Sons, Ltd

Alemparte, C. 583 (319), *593*
Alexakis, A. 558 (219, 220), 564 (255), 565 (257), *590*, *591*, 633 (1), 635 (16), *676*, 774 (12), 775 (31), 779 (19), 791 (50, 57, 58), *800–802*
Alexopoulos, E. 74 (245), *98*, 413 (43), *435*
Al-Fekri, D. M. 103 (7), *126*, 519 (53), *585*
Alhambra, C. 350 (263), *365*
Ali, A. M. M. 276 (198), *310*
Al-Juaid, S. S. 23 (77, 78), 63 (112, 206, 208), *94*, *97*
Al-Laham, M. A. 384 (30), *402*
Allan, A. 391 (16), *402*
Allan, J. F. 422 (59), 424 (61), *435*, 469 (79), *508*
Allan, J. R. 425 (70), *436*
Allan, L. M. 302 (338), *313*
Allen, A. 303 (345), *313*
Almo, S. C. 357 (342), *366*
Alonso de Diego, S. A. 727 (24), *768*
Alonso, T. 46 (158), *96*, 123 (89), *129*
Alonso-Rodríguez, E. 271 (164), *309*
Alpaugh, H. B. 268 (13), *306*
Altarejos, J. 440 (7), *506*
Althammer, A. 550 (171), *589*
Altink, R. M. 34 (106), *94*
Altman, R. B. 270 (160), *309*
Altman, S. 335 (150), *362*
Altura, B. M. 268 (20, 97, 98), 270 (153), *306*, *307*, *309*
Altura, B. T. 268 (19, 20, 97, 98), 270 (153), 276 (18), *306*, *307*, *309*
Alvarez, E. 608 (54, 55), *629*
Alves, J. 355 (320), *366*
Amao, Y. 214 (135), *218*
Ammann, D. 276 (14), *306*
Amstrong, D. R. 18 (67), *93*
Amzel, L. M. 329 (93), *361*
An, D. K. 791 (60), *802*
An, X. 356 (326), *366*
Ananvoranich, S. 336 (160), *362*
Anastassopoulou, J. 334 (142), *362*
Anctil, E. J.-G. 537 (102), *586*
Andersen, A. 124 (97), 125 (105), *129*, 161 (56), *185*
Andersen, R. A. 26 (94), 29 (102), 91 (277), *94*, *99*, 123 (85), 125 (102), *129*, 183 (147, 149), *187*
Anderson, C. E. 798 (67), *802*
Anderson, J. D. 424 (66), *436*, 473 (85), *508*
Anderson, R. A. 23 (75), *93*, 150 (12), *153*
Anderson, V. E. 348 (238), *364*
Andersson, I. 357 (343), 359 (354), *366*, *367*
Andersson, P. 519 (53), *585*
Andersson, P. G. 564 (252), *591*
Ando, K. 749 (42), *769*
André, I. 323 (54, 57), *360*

Andrés, J. 357 (347), 359 (356, 357), *366*, *367*, 393 (25), *402*
Andres, J. L. 369 (2), 384 (30), *401*, *402*
Andrews, L. 157 (20), 158 (28, 31), 159 (33, 34), *184*, *185*
Andrews, P. C. 19 (68), 64 (209), *93*, *97*
Andrews, T. J. 358 (348–350), 359 (339, 351, 353), *366*, *367*
Andrikopoulos, P. C. 9 (52), 17 (66), 20 (70), 22 (72, 74), *93*, 419 (50), *435*
Androulakis, G. 278 (207), *310*
Angermund, K. 33 (114), 45 (157), 86 (266), *94*, *96*, *99*
Anisimov, A. V. 299 (306), *312*
Anteunis, M. 292 (282), *312*
Antolini, F. 62 (201), *97*
Antonucci, V. 293 (267), *311*
Aoyama, T. 556 (207), *590*
Apeloig, Y. 156 (1), *184*
Aponick, A. 798 (65), *802*
Appleby, I. C. 103 (6), *126*
Arai, M. 107 (21), *126*
Araki, S. 285 (237), *311*
Araki, Y. 201 (41, 42, 70), *216*
Arayama, K. 572 (287), *592*
Arduengo, A. J. 53 (175), *96*
Argade, N. P. 557 (209), *590*
Ariese, F. 269 (151), *309*
Arkhangelsky, S. E. 330 (106, 107), *361*
Armentrout, P. B. 124 (97), 125 (105), *129*, 160 (39), 161 (56), *185*
Armstrong, D. R. 9 (52), 17 (66), 20 (70), 22 (72, 74), 57 (186), *93*, *96*, 124 (91), *129*, 413 (42), 419 (50), *435*
Arnaud, M. J. 288 (254, 255), *311*
Arnold, J. J. 352 (301), *365*
Arnold, L. A. 774 (14–16), 791 (56), *800*, *802*
Arpe, H. J. 214 (124), *218*
Arrayás, R. G. 659 (95), *678*
Arrebola, F. J. 302 (335), *313*
Arruda, M. A. Z. 271 (167), *309*
Arsenian, M. A. 268 (103), *308*
Arseniyadis, S. 468 (83), 469 (84), *508*, 528 (85), *586*
Artacho, R. 271 (168), *309*
Asahi, T. 194 (30), 201 (76), *215*, *216*
Asai, F. 579 (311), *592*
Asakawa, N. 760 (50), 764 (52), *769*
Asano, Y. 426 (71), *436*, 498 (41), *507*, 772 (6), *800*
Asaoko, M. 563 (248), *591*
Ashby, E. C. 23 (76), 58 (11), *92*, *93*, 103 (4, 7), 107 (29), 108 (31, 32, 40), *126*, *127*, 132 (1, 2), 134 (13), 138 (21), 139 (9), 140 (5), 146 (40), 147 (42), *153*, *154*, 183 (148), *187*, 389 (3), 391 (8, 10), *401*, 411 (40), 424 (68), 428 (83), *435*, *436*, 513 (14), 519 (53), *584*, *585*, 683 (16), *714*

Ashkenazi, V. 248 (113), 262
Ashley-Facey, A. 172 (108), 186
Aso, Y. 290 (274), 312
Atas, M. 278 (214), 310
Athar, T. 16 (63), 93
Atsmon, J. 268 (144), 308
Atwood, J. L. 28 (100), 47 (47), 93, 94
Audergon, L. 774 (8), 800
Audrain, H. 646 (37), 677
Ault, B. S. 157 (23), 185, 391 (17), 402
Aultman, J. F. 512 (7), 583
Aurbach, D. 247 (106, 108), 248 (96, 107, 113), 250 (115), 251 (112), 252 (114, 116), 253 (117–119), 258 (109–111), 262
Avedissian, H. 555 (198), 589, 626 (29), 629
Avent, A. G. 62 (120), 63 (112), 94, 95
Avola, S. 550 (167), 589
Avolio, S. 534 (81), 586, 767 (58), 769
Awad, H. 44 (44), 93, 689 (26–28), 713 (24, 25), 714, 715
Axten, J. 391 (22), 402
Ayala, P. Y. 384 (30), 402
Baal-Zedaka, I. 211 (110), 217
Baba, T. 341 (201), 363
Babcock, G. T. 201 (72), 216
Babu, C. S. 322 (44), 360
Bac, P. 267 (1), 268 (50, 54), 306, 307
Baccolini, G. 674 (151), 679
Bach, T. 525 (65), 585
Bachain, P. 301 (358), 314
Bachand, C. 500 (118), 509
Bachert, D. 103 (8), 126, 519 (52), 585
Bäckvall, J. E. 791 (53), 802
Bäckvall, J.-E. 544 (131), 555 (201, 204), 557 (210), 558 (218), 588, 590
Baddeley, G. V. 476 (91), 508
Badham, N. F. 760 (46), 769
Badorrey, R. 775 (29), 788 (48), 801
Baer, R. 347 (237), 364
Baev, A. K. 122 (77), 128
Baeyer, A. 2 (6), 92, 107 (24), 127
Bai, Y. 277 (205), 310
Bailey, P. J. 74 (124, 243, 244), 76 (249), 95, 98, 150 (43), 154, 406 (19), 407 (21, 22), 414 (45), 422 (15, 20), 423 (17), 434, 435
Bailey, S. M. 113 (2), 126
Bailey, W. F. 395 (4), 401, 527 (28), 584
Baily, P. J. 81 (255, 256), 98
Baird, M. S. 528 (82), 586
Baizer, M. M. 219 (8), 233 (2), 238 (6), 239 (78), 260, 261
Bajaj, A. V. 3 (24), 92
Bakale, R. P. 395 (35), 402
Baker, D. C. 478 (65), 508
Baker, J. 369 (2), 401
Baker, T. A. 352 (275), 365
Bakhtina, M. 352 (290), 365
Bakos, J. T. 339 (178), 363

Balaban, T. S. 356 (330), 366
Baldwin, J. E. 547 (149), 588
Balenkova, E. S. 547 (148), 588
Balkovec, J. M. 443 (13), 506
Ballard, S. G. 193 (17), 215
Ballem, K. H. D. 74 (246), 98
Ballesteros, R. 689 (26, 27), 713 (24), 714, 715
Baltzer, G. 268 (39), 306
Baluja-Santos, C. 268 (36), 306
Ban, N. 336 (161), 362
Banas, K. 299 (311), 312
Banatao, D. R. 270 (160), 309
Banon-Tenne, D. 633 (1), 676
Bara, M. 267 (1), 268 (50, 54, 102), 306–308
Baranov, V. 124 (96), 129, 180 (134), 187
Baranov, V. I. 180 (135), 187
Barber, J. 355 (321), 366
Barber, J. J. 259 (149), 263
Barbier, P. 2 (1), 92
Barbot, F. 657 (89), 678
Barbour, H. 268 (86), 307
Barbour, H. M. 282 (158), 309
Bard, A. J. 219 (3, 5), 260
Bare, W. D. 158 (28, 31), 185
Barford, D. 347 (226), 364
Barish, R. A. 268 (83), 307
Barkovskii, G. B. 171 (5), 184
Barley, H. R. L. 419 (50), 435
Barlow, S. E. 118 (61), 128
Barmasov, A. V. 198 (59), 216
Barnes, K. 219 (1), 260
Barnett, N. D. R. 15 (62), 38 (126), 93, 95, 149 (47), 154
Baron, O. 268 (10, 11), 306, 525 (66), 534 (99), 540 (114), 585–587, 774 (23), 800
Barone, R. 260 (147), 263
Barone, V. 384 (30), 402
Barran, P. E. 175 (122), 187, 318 (23), 359
Barrett, J. 208 (88), 217
Barrientos, C. 106 (16), 126
Barth, I. 357 (336), 366
Barthelat, J. C. 331 (114–116), 361, 362
Bartmann, E. 45 (154), 95
Bartoli, G. 566 (263), 591, 674 (151), 679
Barton, T. L. 268 (59), 307
Baruah, M. 548 (161), 588
Bassindale, M. J. 424 (64), 436, 473 (86), 508
Batey, R. T. 334 (140), 362
Batra, V. K. 352 (299), 365
Bätting, K. 656 (84), 678
Battioni, P. 634 (12), 676
Bau, R. 86 (268), 99
Bauer, C. E. 193 (28), 214 (139), 215, 218
Bäuerle, W. 287 (253), 311
Baum, E. 26 (97), 94
Baumann, F. 52 (173), 53 (174), 96
Baumeister, D. 301 (359), 314

Baumgartner, J. 38 (130), *95*
Baxter, D. C. 304 (351), *313*
Baxter, J. T. 512 (5), *583*
Bayh, O. 44 (44), *93*, 689 (26-28), *715*
Bazureau, J. P. 44 (44), *93*, 689 (28), *715*
Beale, S. I. 208 (90-92), *217*
Beals, B. J. 259 (146), *263*
Beard, W. A. 352 (288, 299), *365*
Beardah, M. S. 184 (156), *187*
Beattie, T. 499 (116), *509*
Bebenek, K. 355 (315), *366*
Beceiro-González, E. 271 (164), *309*
Becerra, E. 283 (236), *311*
Beck, K. R. 141 (28), *153*
Becke, A. D. 384 (28), *402*
Becker, J. S. 287 (251), *311*
Becker, W. E. 107 (29), *127*, 389 (3), *401*
Becker, W. F. 108 (33, 34, 38), *127*
Beckmann, J. 514 (24), *584*
Bedford, R. B. 546 (144), *588*, 615 (68-70), *630*
Beeby, A. 196 (43), *216*
Beese, L. S. 354 (310), *366*
Begtrup, M. 517 (38), 518 (49), 525 (64), 574 (297), *584, 585, 592*
Begum, S. A. 545 (136), *588*
Begun, G. M. 124 (92), *129*
Behrend, C. 285 (156), *309*
Behrens, U. 10 (54), 42 (144), 48 (163), *93, 95, 96*, 122 (81), *128*
Beigelman, L. 340 (187), 341 (188), *363*
Bejun, G. M. 182 (4), *184*
Belchior, J. C. 270 (152), *309*
Beletskaya, I. P. 171 (5, 100), 175 (101), *184, 186*, 268 (5), *306*
Belfiore, F. 268 (49), *307*
Bell, C. E. 351 (284), *365*
Bell, L. 673 (140, 141), *679*
Bell, S. P. 352 (275), *365*
Beller, M. 544 (129, 130), *588*, 596 (27), *629*
Bello, Z. I. 259 (146), *263*
Belmonte Vega, A. 302 (335), *313*
Belov, S. A. 65 (211), *97*
Benafoux, D. 539 (111), *587*
Benaglia, M. 259 (144), *263*
Bencze, L. 653 (69, 70), *677*
Bendicho, C. 272 (170), *309*
Benech, H. 268 (43), *306*
Benhaim, C. 558 (219), 564 (255), *590, 591*, 779 (19), 791 (57), *800, 802*
Benjamini, E. 347 (230), *364*
Benkeser, R. A. 141 (28), *153*
Benkovic, S. J. 355 (274), *365*
Benn, R. 71 (227), *97*, 123 (80, 86), *128, 129*, 143 (32, 37), 151 (15), 152 (50, 51), *153, 154*
Bension, R. M. 659 (97), *678*
Berdinsky, V. L. 330 (109), *361*

Berezin, B. D. 214 (138), *218*
Berezin, M. B. 214 (138), *218*
Berg, A. 206 (78), *217*
Bergauer, M. 518 (49), *585*
Bergbreiter, D. E. 289 (265), *311*
Bergenhem, N. C. H. 344 (217), *364*
Berger, I. 335 (31), *360*
Bergkamp, M. A 198 (57), *216*
Bergman, R. G. 91 (277), *99*
BergStresser, G. L. 44 (147), *95*
Bérillon, L. 518 (43), 525 (41, 67), 534 (81, 98), *585, 586*
Bermejo-Martinez, F. 268 (36), *306*
Bernard, D. 517 (34), *584*
Bernardi, A. 484 (101), *509*
Bernardi, P. 580 (313), *593*
Berner, H. J. 419 (55), *435*
Bernhard, L. 289 (271), *312*
Bertagnolli, H. 3 (16), *92*
Berthelot, J. 125 (101), *129*, 182 (138), *187*
Bertini, F. 258 (133), *263*
Bertrand, G. 718 (2), *767*
Bertrand, J. 458 (50), 478 (97), *507, 508*
Bestman, H. J. 537 (104), *586*
Betham, M. 615 (70), *630*
Bhagi, A. K. 428 (86), *436*
Bhakat, K. K. 353 (307), *366*
Bhatia, A. V. 559 (227), *590*
Bhatt, R. K. 558 (218), *590*
Bhattacharya, A. 595 (8), *628*
Bhupathy, M. 760 (49), *769*
Biali, S. E. 471 (93), *508*
Bianchet, M. A. 329 (93), *361*
Bianciotto, M. 331 (114), *361*
Bica, K. 618 (74), *630*
Bickelhaupt, F. 3 (15, 28), 34 (106, 116), 35 (117, 118), 38 (83), 39 (135), 40 (137, 138, 140), 43 (146), 44 (142), 58 (187), 61 (136, 198), 62 (199), 66 (213-216), 86 (269), *92, 94-97, 99*, 116 (58), 120 (66-68), 121 (69, 71), *128*, 145 (27), *153*, 183 (140), *187*, 268 (3, 4), *306*, 318 (22), *359*, 384 (31), *402*, 512 (6), *583*
Bicker, G. R. 293 (266), *311*
Bierbaum, V. M. 118 (61), *128*
Billing, D. G. 538 (109), *587*
Bilodeau, M. T. 503 (123), *509*
Bindels, R. J. M. 268 (66), *307*
Bindra, R. 634 (12), *676*
Binkley, J. S. 369 (1, 2), *401*
Biran, C. 467 (77), 472 (76), 473 (74), *508*, 539 (111), *587*
Bischoff, L. 689 (27), *715*
Bivens, L. 144 (29), *153*
Bjerregaard, T. 518 (49), *585*
Björklund, E. 304 (351), *313*
Black, C. B. 320 (7), *359*
Blackmore, I. J. 71 (231), *98*, 411 (33), *435*

Blackwell, H. E. 707 (48), *715*
Blades, A. T. 175 (115, 123), *186, 187*
Blagouev, B. 503 (121), *509*
Blair, S. 84 (265), *99*, 298 (304), *312*, 427 (73), *436*
Blakemore, P. R. 722 (13), *768*
Blanco, F. 689 (26, 27), 713 (24), *714, 715*
Blankenship, R. E. 193 (28), 214 (139), *215, 218*, 356 (327), *366*
Blart, E. 198 (66), *216*
Bläser, D. 71 (230), *98*, 405 (12), *434*
Blau, W. J. 212 (113), *217*
Blessing, R. H. 328 (82), *361*
Bloch, R. 571 (285), *592*
Blom, R. 29 (102), *94*, 123 (85), *129*, 183 (147, 149), *187*
Blomberg, C. 3 (18), 40 (137, 138), 62 (199), *92, 95, 97*, 107 (30), 120 (66, 68), *127, 128*, 145 (27), *153*, 258 (132), *263*, 384 (31), *402*
Blount, K. F. 341 (180), *363*
Bobbio, C. 702 (38), *715*
Bobkowski, W. 268 (110), *308*
Bobrovskii, A. P. 196 (51), *216*
Boche, G. 59 (192), 62 (204), *97*, 395 (34), *402*, 516 (32), *584*
Bochkarev, A. 327 (67), *360*
Bochman, M. 74 (247), *98*
Bochmann, M. 421 (18), *434*
Bocian, D. F. 201 (38, 39, 45, 46), *215, 216*
Bock, C. W. 319 (25), 324 (35), *360*, 391 (22), *402*
Bock, H. 62 (203), *97*
Boero, M. 341 (204), *363*
Boersma, J. 69 (29), 77 (251), *92, 98*
Boese, R. 71 (230), *98*, 405 (12), *434*
Boff, J. M. 211 (111), *217*
Bogdanović, B. 33 (114, 115), 44 (113, 149, 150, 152, 153), 45 (154, 156), 47 (159), *94–96*, 513 (17), 514 (22), 540 (118), *584, 587*, 610 (60), 621 (79), *630*
Böhland, H. 419 (55), *435*
Böhm, H.-J. 300 (317), *313*
Böhm, V. P. W. 135 (11), *153*, 550 (166, 168), *589*, 720 (6), *768*
Bohme, D. K. 124 (94, 96), 125 (95), *129*, 156 (13), 160 (54), 161 (55), 180 (46, 134, 135), *184, 185, 187*
Böhner, C. 711 (49), *715*
Bohr, V. A. 353 (305), *366*
Boiteau, J. G. 774 (21), *800*
Boldea, A. 285 (242), *311*
Bolesov, I. G. 528 (82), *586*
Bollewein, T. 81 (261), *98*
Bollwein, T. 302 (302), *312*, 409 (30), 410 (37), *435*
Bolm, C. 517 (34), *584*, 596 (30), *629*, 722 (15), *768*
Bols, M. 548 (161), *588*

Boman, X. 268 (138), *308*
Bommer, J. C. 193 (21), 214 (133), *215, 218*
Bonafoux, D. 467 (77), 472 (76), *508*
Boncella, J. M. 183 (147), *187*
Bonde, S. E. 604 (46), *629*
Bongini, A. 564 (253), *591*
Bonini, B. F. 259 (144), *263*, 559 (228), *590*
Bonne, F. 574 (295), *592*
Bonnekessel, M. 554 (189), *589*, 608 (57), *629*
Bonnet, V. 524 (60), 525 (41), 537 (103), 553 (186), *585, 586, 589*
Bonnett, R. 214 (134), *218*
Bons, P. 540 (118), 587, 610 (60), *630*
Bordeau, M. 467 (77), 472 (76), 473 (74), *508*, 539 (111), *587*
Borg, D. C. 196 (47), *216*
Born, R. 550 (172), *589*
Borodin, V. A. 121 (70), *128*
Borzak, S. 268 (106), *308*
Borzov, M. V. 65 (211), *97*
Bosco, M. 566 (263), *591*, 674 (151), *679*
Boudet, N. 524 (62), *585*
Boudier, A. 525 (41), *585*
Boukherroub, R. 243 (85), *262*
Boulain, J. C. 344 (215), *364*
Boutevin, B. 711 (50), *715*
Bowen, B. P. 356 (327), *366*
Bowen, M. E. 289 (264), *311*
Bowyer, W. J. 259 (146), *263*, 513 (12), *583*
Boyd, W. A. 144 (29), *153*
Boyer, P. D. 329 (91), *361*
Boyles, H. B. 391 (7), *401*
Boymond, L. 517 (35), 534 (81, 100), *584, 586*, 619 (45), *629*, 690 (30), *715*
Brade, K. 576 (301), *592*
Braga, A. L. 774 (25), *801*
Braithwaite, D. 233 (22), 234 (23), 244 (21), *260*
Braithwaite, D. G. 239 (70), 240 (64–69), *261*
Brancaleoni, D. 580 (313), *593*
Brand, R. A. 610 (60), *630*
Brändén, C. I. 359 (355), *367*
Bräse, S. 531 (93), *586*
Brasuel, M. 285 (156), *309*
Braun, M. 718 (2), *768*
Brautigam, C. A. 352 (295), *365*
Breaker, R. R. 336 (153), *362*
Breckenridge, W. H. 157 (16, 19), 159 (18), *184*
Brehon, B. 555 (195), *589*
Breit, B. 455 (42), *507*, 557 (213), *590*, 791 (53), *802*
Bremer, T. H. 235 (50), *261*
Brenek, S. J. 103 (10), *126*
Breneman, C. M. 113 (56), *128*
Brenner, A. 246 (99), *262*
Brereton, R. G. 211 (109), *217*
Breton, G. 524 (60), *585*

Brettreich, M. 675 (154), *679*
Breutel, C. 775 (27), *801*
Brewer, J. M. 320 (26), 349 (249, 251, 254, 256–258), *360*, *364*, *365*
Brian, C. 539 (111), *587*
Brick, P. 351 (286), *365*
Brintzinger, H. H. 125 (102), *129*
Brintzinger, H.-H. 48 (164), 49 (165), *96*
Brion, J.-D. 555 (197, 198), *589*, 607 (48), 608 (51), *629*
Britt, R. D. 341 (198), *363*
Brochette, P. 198 (60), *216*
Brockerhoff, H. 301 (356), *314*
Brodbelt, J. S. 178 (130), *187*
Brodilova, J. 193 (20), *215*
Broeke, J. 519 (53), *585*
Brombrun, A. 557 (210), *590*
Bronstrup, M. 135 (11), *153*, 720 (6), *768*
Brookings, D. C. 673 (141), *679*
Brosche, T. 537 (104), *586*
Brown, J. D. 514 (19), *584*
Brown, M. 302 (324), *313*
Brown, M. K. 775 (30), *801*
Brown, R. F. C. 540 (115), *587*
Brown, S. B. 208 (97), 210 (100), 211 (108), *217*
Brown, S. N. 469 (78), *508*
Brown, T. L. 4 (35, 36), *92*, 682 (2), *714*
Browne, W. R. 788 (48), *801*
Bruce, D. W. 546 (144), *588*, 615 (68–70), *630*
Bruening, G. 339 (178), *363*
Bruhn, C. 38 (127), 42 (143), *95*, 136 (16), *153*
Bruice, T. C. 341 (202), *363*
Brunschwig, B. S. 196 (48), *216*, 259 (143), *263*
Brunvoll, J. 183 (146), *187*
Bruzzoniti, M. C. 273 (182), *309*
Bryce-Smith, D. 674 (150), *679*
Brym, M. 64 (209), *97*
Bučević-Popović, V. 345 (218), *364*
Buchachenko, A. L. 330 (106–110), 331 (104, 111), *361*
Bucher, A. 44 (44), *93*, 689 (28), *715*
Buchwald, S. L. 550 (170), *589*
Buck, P. 576 (303), *592*
Buckingham, M. J. 286 (246), *311*
Budzelaar, P. H. M. 77 (251), *98*
Buechler, J. A. 347 (231), *364*
Buell, G. R. 661 (105), *678*
Bugg, C. E. 318 (16), *359*
Bühler, R. 517 (39), *585*
Buhlmayer, P. 443 (13), *506*
Bulgakov, O. V. 344 (215), *364*
Bulgakov, R. G. 235 (51), *261*
Bullock, T. H. 51 (51), *93*
Bulska, E. 287 (192), *310*
Bulygin, V. V. 329 (92), *361*
Bumagin, N. A. 553 (187), *589*
Bünder, W. 25 (90), *94*
Bunlaksananusorn, T. 525 (41), *585*
Bunnage, M. E. 454 (40), *507*
Burack, J. L. 268 (19), *306*
Burant, J. C. 384 (30), *402*
Burdon, J. 654 (77), *678*
Bures, F. 559 (223), *590*, 601 (44), *629*
Burke, D. H. 341 (193), *363*
Burke, J. M. 341 (195, 197), *363*
Burke, S. D. 718 (2), *768*
Burla, M. C. 328 (83), *361*
Burns, C. J. 183 (147), *187*
Burns, D. H. 545 (134), *588*
Buron, F. 702 (40), *715*
Burt, S. K. 158 (27), *185*, 331 (117, 118), *362*
Burton, A. 720 (5), *768*
Burton, D. J. 516 (30), 517 (34), *584*
Burton, R. F. 287 (250), *311*
Busacca, C. A. 555 (203), *590*
Busch, D. H. 342 (207), *363*
Busch, F. R. 499 (115), *509*
Bush, J. E. 283 (234), *311*
Bushnell, D. A. 355 (313), *366*
Bustamante, C. 336 (163), *362*
Buttrus, N. H. 63 (207), *97*
Butts, C. P. 556 (206), *590*
Buzayan, J. M. 339 (178), *363*
Cabrol, N. 478 (97), *508*
Cachau, R. E. 331 (118), *362*
Cadman, M. L. F. 493 (106), *509*
Cahiez, G. 517 (34, 35), 518 (43), 519 (51), 534 (81, 98, 100), 544 (133), 545 (137), 546 (143), 552 (176), 554 (191), 555 (195, 196, 198), 559 (223), 563 (245), *584–586*, *588–591*, 601 (43, 44), 607 (47), 618 (72), 619 (45), 626 (29, 76), *629*, *630*, 635 (16), *676*, 690 (30), *715*
Cai, M. 540 (121, 122), *587*
Calabrese, G. 268 (91), *307*
Calabrese, J. C. 527 (28), *584*
Calaza, M. I. 525 (41), *585*
Caldecott, K. W. 351 (283), *365*
Calderon, R. L. 268 (130), *308*
Cali, P. 517 (38), 574 (297), *584*, *592*
Calingaert, G. 238 (77), *261*
Calvin, M. 207 (85–87), *217*
Cameron, C. E. 352 (301), *365*
Cameron, I. L. 268 (12), *306*
Camici, L. 475 (81), *508*
Caminiti, R. 318 (17), *359*
Cammi, R. 384 (30), *402*
Campbell, J. E. 791 (59), *802*
Campbell, M. J. 579 (310), *592*
Campos Pedrosa, L. de F. 268 (96), *307*
Camus, C. 348 (246), 350 (261), *364*, *365*
Canè, F. 580 (313), *593*

Canfranc, E. 278 (208), *310*
Cannon, K. C. 55 (178), *96*, 225 (40), *261*
Cannone, J. J. 337 (169), *363*
Cao, Z. 275 (193), *310*
Capelo, J. L. 272 (170), *309*
Capozzi, M. A. M. 583 (318), *593*
Caprio, V. 774 (1), *800*
Carda, M. 391 (26), 393 (25), *402*
Cardani, S. 484 (101), *509*
Cardellicchio, C. 583 (317, 318), *593*, 601 (41, 42), *629*
Cardillo, G. 564 (253), *591*
Carey, F. A. 387 (32), *402*
Carmichael, I. 206 (77), *217*
Caro, C. F. 32 (111), 37 (121), 62 (120), *94*, *95*, 416 (49), *435*
Caro, Y. 301 (358), *314*
Carpentier, J.-F. 69 (223), *97*
Carreira, L. A. 349 (258), *365*
Carrera, A. 303 (348), *313*
Carrera, S. 659 (95), *678*
Carretero, J. C. 518 (48), *585*, 659 (95), *678*
Carroll, G. L. 253 (124), *263*
Carroll, J. D. 293 (266), *311*
Carson, A. S. 109 (46, 47), *127*
Carswell, E. L. 424 (63), *435*, 473 (89), *508*
Carter, J. P. 323 (57), *360*
Carvalho, S. 270 (152), *309*
Carver, J. P. 323 (54), *360*
Casareno, R. 322 (44), *360*
Casarini, A. 580 (313), *592*
Casellato, U. 268 (148), *308*
Casey, W. H. 257 (129), *263*
Cason, J. 600 (37), 601 (38, 39), *629*
Castedo, L. 540 (115), *587*
Castellato, U. 64 (210), *97*
Castillo, A. 268 (119), *308*
Castillo, R. 393 (25), *402*
Castro, A. M. M. 727 (24), *768*
Castro, B. 442 (10), 484 (100), *506*, *508*
Castro, C. 352 (301), *365*
Castro, J. M. 440 (7), *506*
Cazeau, P. 472 (76), 473 (74), *508*, 539 (111), *587*
Ceccarelli, S. M. 625 (93), *630*
Cech, T. R. 320 (29), 335 (149), 337 (154, 170, 171), *360*, *362*, *363*
Cerdà, V. 283 (236), *311*
Cernak, I. 268 (145), *308*
Černohorský, T. 279 (217), *310*
Ceulemans, M. 302 (340, 341), *313*
Cevette, M. J. 268 (73), *307*
Chaboche, C. 544 (133), 554 (191), *588*, *589*, 626 (76), *630*
Chacon, D. R. 82 (162), *96*, 417 (13), *434*
Chai, G. 349 (251), *364*
Challacombe, M. 384 (30), *402*
Chamberlin, S. A. 559 (227), *590*

Chambers, R. D. 516 (30), 517 (34), *584*
Champ, M. A. 302 (330), *313*
Chan, H.-K. 545 (134), *588*
Chan, T. H. 515 (26), *584*
Chandler, D. 329 (90), *361*
Chandra, T. 196 (44), *216*
Chang, C.-C. 81 (228), *98*, 149 (46), *154*, 422 (25), *435*
Chang, C. H. 270 (159), *309*
Chang, C. K. 198 (57), 201 (71–73), *216*
Chang, J. 272 (177), *309*
Chang, W. 356 (326), *366*
Chanon, M. 157 (15), *184*, 260 (147), *263*, 391 (20), *402*, 513 (15), *584*
Chao, L.-C. 411 (40), *435*, 683 (16), *714*
Chapdelaine, M. J. 451 (33), *507*
Chapleur, Y. 484 (100), *508*
Chasid, O. 253 (119), *262*
Chau, A. S. Y. 300 (320), *313*
Chau, Y. K. 300 (320, 321), 302 (324, 338), 303 (318), *313*
Chazalviel, J.-N. 241 (79, 80), 243 (81–83, 85), *262*
Chechik-Lankin, H. 642 (29), 643 (31), *677*
Cheeseman, J. R. 384 (30), *402*
Chemla, F. 516 (32), *584*
Chen, A. 244 (89), *262*
Chen, B. 213 (118), *217*
Chen, C. 547 (154), *588*
Chen, C.-D. 547 (155), *588*
Chen, G. 540 (121, 122), *587*
Chen, J. 162 (59, 60), 163 (62), 164 (63), 171 (105), *185*, *186*
Chen, K. 123 (90), *129*
Chen, M. 159 (36), *185*
Chen, M. L. 277 (205), *310*
Chen, S. J. 335 (11), *359*
Chen, T.-A. 514 (19), *584*
Chen, W. 384 (30), *402*
Chen, W.-C. 566 (262), *591*
Chen, X. 162 (61), *185*
Chen, Y. 291 (280), *312*, 325 (73), *361*
Chen, Y.-H. 547 (154), *588*
Chenal, T. 69 (223), *97*
Cheney, M. C. 277 (201), *310*
Cheng, D. 514 (23), *584*, 619 (78), *630*
Cheng, F.-C. 278 (209, 211, 212), *310*
Cheng, W.-L. 547 (154), *588*
Cheng, Y. 348 (244), *364*
Cheng, Y. C. 162 (60), 163 (62), 171 (105), *185*, *186*
Chérest, M. 666 (114), *678*
Chernomorsky, S. 214 (131), *218*
Chernow, B. 268 (81), *307*
Chernyh, I. N. 219 (7), *260*
Cherry, D. A. 667 (118), *678*
Chevrot, C. 236 (60), 258 (38, 130), *261*, *263*, 291 (279), *312*

Chiang, M. Y. 81 (228), *98*, 149 (46), *154*, 422 (25), *435*
Chiavarini, S. 302 (323), *313*
Chiba, N. 563 (248), *591*
Chiba, S. 579 (309), *592*
Chickos, J. 121 (70), *128*
Chiechi, R. C. 531 (94), *586*
Childs, B. J. 479 (82), *508*
Chinkov, N. 563 (249), *591*, 633 (1), *676*
Chipperfield, J. R. 122 (76), *128*
Chisholm, M. H. 73 (240), *98*, 418 (11), *434*
Chittleborough, G. 268 (35), *306*
Chiu, T. K. 335 (144), *362*
Chivers, T. 78 (253), *98*, 244 (89), *262*, 421 (34), *435*
Cho, C.-H. 553 (178), *589*
Cho, D.-G. 557 (215), *590*
Cho, Y.-H. 554 (194), *589*
Chong, J. M. 54 (177), 83 (176), *96*, 426 (31), *435*
Choplin, A. 174 (114), *186*
Chou, S. K. 621 (80), *630*
Chowdhury, F. A. 560 (235), *590*
Chrisman, W. 554 (193), *589*
Christensen, B. G. 499 (116), *509*
Christensen, C. 576 (303), *592*
Christian, E. L. 337 (173), *363*
Christian, G. D. 280 (223), *310*
Christian, W. 348 (247), *364*
Christiansen, T. F. 283 (234), *311*
Christophersen, C. 525 (64), *585*
Chrost, A. 25 (82), *94*, 299 (307), *312*
Chu, B. 305 (362, 363), *314*
Chu, C.-M. 566 (262), *591*
Chuang, M.-C. 566 (262), *591*
Chubanov, V. 268 (56), *307*
Chubb, J. E. 713 (52), *715*
Chung, K.-G. 558 (219), *590*
Churakov, A. V. 65 (211), *97*
Churney, K. L. 113 (2), *126*
Chusid, O. 247 (106), 248 (113), 251 (112), 252 (114, 116), 258 (111), *262*
Cini, R. 328 (83), *361*
Cioslowski, J. 384 (30), *402*
Citra, A. 158 (31), *185*
Cittadini, A. 268 (58), *307*, 317 (2), *359*
Citterio, D. 285 (237), *311*
Cladera, A. 283 (236), *311*
Clark, D. R. 395 (4), *401*
Clausen, T. 268 (125), *308*
Clavier, H. 565 (257), *591*, 775 (31), *801*
Clay, M. D. 647 (41), *677*
Clayden, J. 553 (179), *589*, 730 (28), *768*
Clegg, W. 15 (62), 20 (70), 38 (126), 83 (263), 84 (265), *93*, *95*, *98*, *99*, 149 (47), *154*, 298 (304), *312*, 413 (42), 419 (50), 422 (41, 59), 427 (73), *435*, *436*

Cleland, W. W. 319 (24), 349 (250), 359 (339), *359*, *364*, *366*
Clemens, M. R. 214 (130), *218*
Clifford, P. J. 103 (10), *126*
Clifford, S. 384 (30), *402*
Cloke, F. G. N. 51 (168), *96*
Cloutier, R. 285 (241), *311*
Coates, G. E. 4 (34), *92*, 410 (39), 428 (87), *435*, *436*, 682 (12), *714*
Cobb, M. H. 347 (237), *364*
Coburn, E. R. 398 (39), *402*
Coelho, L. M. 271 (167), *309*
Cogan, D. A. 575 (298, 299), *592*, 673 (145), *679*
Cogdell, R. J. 324 (58), 356 (325, 331), *360*, *366*
Cohen, H. 250 (115), *262*
Cohen, T. 514 (23), *584*, 760 (49), *769*
Cohen, Y. 248 (96, 113), 251 (112), 258 (109, 110), *262*
Cohn, M. 329 (89), 331 (112), *361*
Cola, S. Y. 303 (343), *313*
Colacot, T. J. 53 (174), *96*, 775 (27), *801*
Colasson, B. 576 (303), *592*
Cole, P. E. 335 (143), *362*
Cole, S. C. 70 (225), *97*, 430 (77), *436*
Coleman, G. H. 578 (307), *592*
Coles, M. P. 70 (225), *97*, 430 (77), *436*
Colin, C. L. 190 (1), *214*
Collett, J. R. 337 (169), *363*
Collins, J. D. 52 (173), 53 (174), *96*
Collman, J. P. 438 (2), *506*
Colombo, L. 484 (101), *509*
Colonge, J. 439 (3), *506*
Combellas, C. 513 (15), *584*
Comes-Franchini, M. 559 (228), *590*
Commeyras, A. 516 (29), *584*
Compton, R. N. 124 (92), *129*, 182 (4), *184*
Compton, T. R. 288 (259), *311*
Conlon, D. A. 648 (47), *677*
Conn, R. S. E. 640 (25, 26), *677*
Conroski, K. 119 (64), *128*
Conway, B. 81 (224), *97*, 414 (35), 430 (80), *435*, *436*
Cook, N. C. 599 (36), *629*
Cook, P. F. 347 (236), *364*
Cooney, J. J. A. 553 (179), *589*
Cooper, G. D. 112 (52), *127*
Corbel, B. 462 (59), *507*
Corbet, J. P. 596 (23), *629*
Corchado, J. C. 350 (263), *365*
Cordray, K. 214 (136), *218*
Corey, E. J. 484 (102), *509*, 582 (316), *593*, 772 (5), *800*
Corich, M. 500 (55), *507*
Cornejo, J. 208 (90–92), *217*
Cornelis, R. 269 (151), *309*
Correa, A. 788 (48), *801*

Correll, C. C. 335 (147), *362*
Corriu, R. 661 (105), *678*
Cosma, P. 212 (115), *217*
Cossi, M. 384 (30), *402*
Costello, R. B. 268 (136), *308*
Cotterell, C. 148 (48), *154*
Cottet, F. 702 (38), *715*
Cotton, F. 25 (87), *94*
Cotton, H. K. 791 (53), *802*
Coudray, C. 288 (256–258), *311*
Coutrot, P. 489 (8, 9), 491 (105), *506, 509*
Cowan, J. A. 268 (51), *307*, 318 (4), 320 (7, 8), 321 (37), 322 (44), 328 (1), 347 (6), 354 (27), *359, 360*
Cowan, P. J. 462 (58), 494 (109), *507, 509*
Cowden, C. J. 648 (47), *677*
Cox, J. D. 110 (50), *127*
Coxall, R. A. 74 (124, 243), 81 (255), *95, 98*, 406 (19), 407 (21), 423 (17), *434*
Coxhall, R. A. 150 (43), *154*
Crabtree, R. H. 58 (5), *92*, 633 (1), *676*
Cracco, R. Q. 268 (19), *306*
Cram, D. J. 395 (5), *401*
Cramer, R. E. 68 (219), *97*
Crandall, J. K. 634 (12), *676*
Craun, F. F. 268 (130), *308*
Crauste, C. 491 (105), *509*
Crawford, H. M. 653 (176), *676*
Crawford, J. J. 424 (64), *436*, 473 (86), *508*
Crawford, K. S. K. 113 (55), *127*
Crawley, M. L. 791 (52), *802*
Creighton, J. R. 122 (78), *128*
Crimmins, M. T. 468 (83), *508*
Crombie, L. 462 (57), 493 (106, 107), 494 (108), *507, 509*
Croset, K. 791 (57), *802*
Cross, R. L. 329 (92), *361*
Crossland, I. 3 (17), *92*, 391 (9), *401*
Crossland, L. 451 (30), *507*
Croteau, D. L. 351 (280), *365*
Crothers, D. M. 335 (143), *362*
Cuddihy, K. P. 259 (146), *263*
Cudzilo, S. 103 (5), *126*
Cui, C. 74 (248), *98*, 424 (16), *434*
Cui, Q. 384 (30), *402*
Cui, X.-L. 547 (158), *588*
Cukier, R. I. 201 (73), *216*
Cundy, D. J. 495 (111), *509*
Cunniff, P. 283 (166), *309*
Čurdová, E. 279 (217), *310*
Curran, D. J. 277 (201), *310*
Curri, M. L. 212 (115), *217*
Cutshall, N. S. 667 (120), *678*
Cuvigny, T. 608 (54, 55), *629*
Cvetovich, R. 103 (8), *126*, 519 (52), *585*
Czako, B. 742 (14), *768*
D'Augustin, M. 565 (257), *591*, 775 (31), *801*
D'Hollander, R. 292 (282), *312*

D' Sa, B. 472 (94), *508*
D'Souza, F. 201 (41, 42), *216*
D'Souza, L. M. 337 (169), *363*
Daasbjerg, K. 256 (28), 259 (136, 137), *260, 263*
Dakternieks, D. 514 (24), 550 (164), *584, 588*
Dalai, S. 452 (36), *507*
Dale, S. H. 419 (50), *435*
Dalton, J. 198 (57), *216*
Damen, H. 655 (64), *677*
Damin, A. 125 (108), *129*
Damrau, H.-R. 49 (165), *96*
Damrauer, R. 118 (61), *128*
Daniels, A. D. 384 (30), *402*
Danielson, N. D. 305 (364), *314*
Danielsson, L.-G. 269 (151), *309*
Dankwardt, J. W. 433 (90), *436*, 553 (182), *589*
Danly, D. E. 239 (78), *261*
Danopoulos, A. A. 615 (70), *630*
Dapprich, S. 384 (30), *402*
Daris, J. P. 500 (118), *509*
Darki, A. 144 (29), *153*
Daskapan, T. 579 (309), *592*
Datta, A. 571 (282), *592*
Daugaard, G. 268 (141), *308*
Davanzo, C. U. 295 (288), *312*
Davidson, F. 53 (175), *96*
Davidson, W. 282 (158), *309*
Davidsson, Ö. 59 (145), *95*
Davies, I. W. 5 (43), *93*
Davies, S. G. 454 (40), *507*
Davis, D. A. 559 (227), *590*
Davis, F. A. 575 (299), *592*
Davis, I. W. 703 (41), *715*
Davis, S. R. 158 (26), *185*, 391 (18, 19), *402*
Dawson, G. J. 673 (141), *679*
de Armas, G. 283 (236), *311*
de Armas, J. 561 (240), *591*
de Diego, J. E. 518 (48), *585*
de Groot, A. 474 (47), *507*
de Groot, H. J. M. 356 (332), *366*
de Jong, J. C. 268 (66), *307*
de Kanter, F. J. J. 86 (269), *99*
De, L. 331 (119), *362*
de la Calle-Guntiñas, M. B. 302 (323), 304 (319), *313*
de la Fuente Blanco, J. A. 482 (6), *506*
de la Peña, M. 341 (191), *363*
De Luca, L. 559 (226), *590*
de Meijere, A. 452 (36), *507*, 531 (93), 547 (153), 550 (126), 582 (315), *586–588, 593*, 596 (28), 629, 633 (1), *676*
de Pablos, F. 302 (334), *313*
de Valk, H. W. 268 (95), *307*
de Vries, A. H. M. 774 (2, 13, 14), *800*
Deakyne, C. A. 107 (20), *126*
DeAntonis, D. M. 103 (10), *126*

Debal, A. 487 (22), *506*
Decambox, P. 275 (189), *310*
Dechert, S. 54 (95), *94*
Deelman, B.-J. 519 (53), *585*
Defrees, D. J. 369 (1, 2), *401*
Degrado, S. J. 779 (33), *801*
Deguchi, J. 206 (82), *217*
Dehmel, F. 518 (46), 519 (51), 525 (41), *585*
DeHoff, B. S. 645 (34), *677*
Dei Cas, L. 268 (105), *308*
Deiderich, F. 633 (1), *676*
Dekker, J. 77 (251), *98*
del Amo, V. 557 (208), 580 (314), *590*, *593*
del Prado, M. 518 (48), *585*
del Rosario, J. D. 395 (36), *402*
Del Villar, A. 558 (218), *590*, 791 (53), *802*
Delacroix, T. 534 (98), *586*
Delagoutte, E. 351 (270, 271), *365*
Delaney, M. O. 545 (134), *588*
Delbaere, L. T. J. 332 (125), *362*
Demay, S. 525 (41), *585*
Dembech, P. 475 (81), *508*, 580 (313), *592*, *593*
Demel, P. 455 (42), *507*, 557 (213), *590*
Demiroglu, A. 278 (215), *310*
Demissie, T. 53 (174), *96*
Demtschuk, J. 54 (95), *94*
Denisko, O. V. 547 (156), *588*
Denmark, S. E. 571 (284, 286), *592*
DePuy, C. H. 118 (61), *128*
Derkacheva, V. M. 213 (120, 122), *217*
DeRose, V. J. 336 (168), 341 (198), *363*
Des Mazery, R. 457 (45), *507*, 787 (41), *801*
Desai, H. 144 (29), *153*
Dessy, R. E. 107 (28), *127*, 244 (89), 258 (36, 88), *261*, *262*, 633 (10), *676*
Dettwiler, J. E. 563 (246), *591*
DeWald, H. A. 633 (3), *676*
DeWolfe, R. H. 141 (33), *154*
Dhupa, N. 268 (85), *307*
Di Carro, M. 302 (332), *313*
Di Furia, F. 500 (55), *507*
Di Ninno, F. 499 (116), *509*
Diaz, N. 348 (242), *364*
Dick, C. M. 74 (124, 243), 81 (255, 256), *95*, *98*, 150 (43), *154*, 406 (19), 407 (21, 22), 422 (15), 423 (17), *434*
Dick, C. M. E. 74 (244), *98*
Dickerson, R. E. 335 (144), *362*
Dickey, J. B. 289 (270), *312*
Didierjean, C. 491 (105), *509*
Didiluk, M. T. 672 (137), *679*
Didiuk, M. T. 557 (212), *590*, 673 (143, 144), *679*
Die, M. 555 (202), *590*
Diebler, H. M. 318 (18), *359*
Dieckmann, R. 345 (218), *364*
Diederich, F. 550 (126), *587*, 633 (1), *676*

Diers, J. R. 201 (39), *215*
Dieter, R. K. 474 (95), *508*, 559 (222), *590*
Dieterich, F. 325 (68), *360*
Dietze, H.-J. 287 (251), *311*
Dietzek, B. 357 (337), *366*
Digeser, M. H. 427 (72), *436*
Digiovanni, S. 295 (286), *312*
Dilsiz, N. 278 (214), *310*
Ding, L. N. 162 (60), *185*
Ding, Y. 74 (248), *98*, 424 (16), *434*
Dinsmore, A. 538 (109), *587*
Dirkx, W. M. R. 302 (341), *313*
Dirnbach, E. 344 (216), *364*
Dismukes, G. C. 322 (42), *360*
Dix, D. T. 138 (13), *153*
Dobrovetsky, E. 327 (67), *360*
Dobson, M. P. 175 (122), 176 (119–121), *186*, *187*, 318 (23), *359*
Dohle, W. 268 (8), *306*, 518 (36, 42, 47), 519 (51), 525 (41, 68), 555 (198), 577 (305), *584*, *585*, *589*, *592*, 607 (47), *629*
Dohmeier, C. 68 (218), *97*
Dol, I. 278 (213), *310*
Dolev, E. 268 (144), *308*
Dolling, U. 103 (8), *126*, 519 (52), *585*
Dolling, U.-H. 648 (47), *677*
Dolman, S. J. 703 (41), *715*
Dolphin, D. 193 (14), 196 (47), *215*, *216*
Dombovári, J. 287 (251), *311*
Domcke, W. 332 (131), *362*
Donato, F. 268 (130), *308*
Dong, C.-G. 550 (165), *588*
Dong, F. 184 (160), *187*
Donkervoort, J. G. 545 (137), *588*
Donohoe, T. J. 485 (24), *507*
Dooley, C. A. 303 (343), *313*
Dorea, J. G. 268 (127), *308*
Dorigo, A. E. 422 (47), *435*
Dorough, G. D. 207 (85), *217*
Dorup, I. 268 (125), *308*
Dorwald, F. Z. 718 (2), *767*
Dorziotis, I. 654 (72), *678*, 699 (31), *715*
Dos Santos, M. 555 (198), *589*, 608 (51), *629*
Dotson, L. 268 (44), *306*
Doublié, S. 352 (298), *365*
Doudna, J. A. 334 (140), 337 (154), *362*
Douglas, R. H. 191 (9), *215*
Dove, A. P. 76 (242), 77 (250), *98*, 424 (44), *435*, 484 (99), *508*
Dowd, P. 760 (47), *769*
Downs, A. J. 157 (17, 20), *184*
Doyle, D. A. 327 (67), *360*
Doyle, M. 774 (2), *800*
Draeger, M. 514 (24), *584*
Drane, M. 238 (63), *261*
Dranka, M. 410 (36), *435*
Draper, D. E. 334 (139), 335 (10), 336 (164, 166), *359*, *362*, *363*

Drekener, R. L. 774 (25), *801*
Dreosti, I. E. 268 (123), *308*
Driessens, F. C. 268 (68, 69), *307*
Dror, Y. 211 (110), *217*
Du, Y. 337 (169), *363*
Dubbaka, S. R. 576 (301), 580 (314), *592*, *593*
Dube, H. 531 (92), *586*
Dübner, F. 791 (54), *802*
Dubois, J. E. 458 (51), *507*, 482 (98), *508*
Dubois, L. H. 172 (109), *186*
Dubois, T. 241 (79), *262*
Duboudin, J.-G. 635 (13–15), 648 (49, 50), *676*, *677*
Ducros, V. 288 (257), *311*
Dudek, J. 268 (107), *308*
Dudev, M. 321 (39), *360*
Dudev, T. 320 (28), 321 (37–39), 322 (43, 44), *360*
Dufford, R. T. 235 (49), *261*
Duke, R. B. 391 (8), *401*
Dulai, K. 191 (9), *215*
Dumouchel, S. 700 (33, 34), *715*
Dunbar, R. C. 125 (99, 106), *129*, 156 (11, 14), 160 (48, 52), 162 (53), *184*, *185*
Duncan, A. P. 779 (20), *800*
Duncan, J. L. 391 (16), *402*
Duncan, M. A. 124 (98), *129*, 168 (75), 184 (158, 159), *186*, *187*
Duncan, T. M. 329 (92), *361*
Dunn, K. 550 (164), *588*
Dunogues, J. 467 (77), 472 (76), *508*, 539 (111), *587*
Duplais, C. 546 (143), 559 (223), *588*, *590*, 601 (44), 618 (72), *629*, *630*
Duquerroy, S. 348 (246), 350 (261), *364*, *365*
Durlach, J. 267 (1), 268 (33, 50, 54, 102, 110), *306–308*
Durlach, V. 268 (33, 102), *306*, *308*
Dürr, C. 610 (60), *630*
Duthie, A. 514 (24), *584*
Dvorak, D. 518 (45), *585*
Dvoráková, H. 611 (61, 62), *630*
Dykstra, C. E. 171 (99), *186*
Dyrda, G. 196 (49), *216*
Dzhemilev, U. M. 123 (87), *129*, 421 (53), *435*, 671 (131–133, 135), *679*
Eaborn, C. 23 (77, 78), 32 (110), 63 (112, 206, 208), *94*, *97*
Eaborn, C. A. 63 (207), *97*
Eastham, J. F. 288 (261), *311*, 682 (15), *714*
Eaton, P. E. 419 (54, 56, 57), *435*, 466 (66), *508*, 538 (105, 106), *587*
Ebdrup, S. 525 (64), *585*
Ebel, G. E. 513 (13), *583*
Ebersole, M. H. 196 (50), *216*
Eberson, L. 258 (56), 259 (61), *261*
Eckert, J. 246 (92), *262*
Edmondson, D. L. 358 (348), *367*

Edwards, A. M. 327 (67), *360*
Edwards, G. L. 479 (82), *508*
Edwards, P. G. 86 (268), *99*
Efthimiou, E. 282 (232), *311*
Egekeze, J. O. 293 (266), *311*
Egli, M. 335 (9, 31, 145, 146), *359*, *360*, *362*
Egorov, A. M. 299 (306), *312*
Eguchi, K. 300 (314), *313*
Eicher, T. 451 (29), *507*
Eigen, G. 318 (18), *359*
Eijkelkamp, D. J. F. M. 774 (8), *800*
Eilerts, N. W. 73 (240), *98*, 418 (11), *434*
Einspahr, H. 318 (16), *359*
Eisch, J. J. 268 (2), *306*, 664 (110–113), *678*
Ek, F. 611 (64), *630*
Ekane, S. 268 (117), *308*
Eklov, B. M. 300 (300), *312*
El-Demerdash, S. H. 176 (124), *187*
El-Hamruni, S. M. 63 (112), *94*
El-Kaderi, H. M. 78 (252), *98*, 122 (77), *128*, 422 (28, 29), *435*
El-Kaderi, M. 72 (234), *98*
El-Kheli, M. N. A. 63 (207), *97*
El-Khouly, M. E. 201 (41, 42), *216*
Elhanine, M. 169 (81), *186*
Eliel, E. L. 395 (4), *401*, 772 (3), *800*
Elin, R. J. 268 (41, 135), 270 (29, 77), 275 (193), *306–308*, *310*
Ellenberger, T. 352 (298), 355 (316), *365*, *366*
Eller, K. 160 (40, 41), *185*
Ellerd, M. G. 742 (14), *768*
Ellis, A. M. 123 (84), *129*, 184 (156, 157), *187*
Ellison, J. J. 62 (202), *97*
Ellman, J. A. 575 (298–300), *592*
El-Nahas, A. M. 176 (124), *187*
El-Shereefy, E.-S. E. 176 (124), *187*
Elving, P. J. 300 (315), *313*
Emanuel', V. L. 268 (118), *308*
Emel'ianov, A. B. 268 (118), *308*
Emerson, K. 699 (31), *715*
Emteborg, H. 304 (351), *313*
Ende, D. J. 103 (10), *126*
Engel, N. 210 (102), *217*
Engelbrecht, S. 329 (95), *361*
Engelhardt, F. C. 648 (47), *677*
Engelhardt, L. M. 58 (155), 81 (260), *96*, *98*, 405 (5), 427 (74), *434*, *436*
Englert, U. 29 (103), *94*
Englich, U. 71 (229), *98*, 407 (23), *434*
Erdik, E. 579 (309), *592*
Erickson, W. E. 667 (54), *677*
Erikkson, M. C. 555 (203), *590*
Erikson, H. 38 (129), *95*
Eriksson, J. 59 (61), 60 (185), *93*, *96*
Erkelens, C. 356 (332), *366*
Ermini, G. 512 (10), *583*
Ershov, V. V. 330 (108), *361*
Ertel, T. S. 3 (16), *92*

Eshaghi, S. 325 (70), *361*
Esipov, S. E. 171 (5), 175 (101), *184*, *186*
Eskins, M. 493 (107), *509*
Estela, J. M. 283 (236), *311*
Etcheberry, A. 243 (83), *262*
Etkin, N. 419 (26), *435*
Evans, D. A. 503 (62, 123, 124), 504 (90), 505 (61), *508*, *509*, 539 (112), *587*
Evans, D. E. 108 (41), *127*
Evans, D. F. 139 (26), 140 (19), 150 (45), *153*, *154*
Evans, M. C. W. 356 (329), *366*
Evans, S. 124 (93), *129*
Evans, W. H. 113 (2), *126*
Evans, W. V. 220 (12), 228 (16), 231 (20), 233 (22), 234 (23), 235 (18), 238 (17), 244 (19, 21), 258 (15), *260*
Evers, E. A. I. M. 116 (58), *128*
Evstigneev, V. B. 198 (58), *216*
Evstigneeva, E. V. 110 (48), *127*
Ewbank, P. C. 286 (244), *311*
Ezhov, Y. S. 183 (153), 184 (155), *187*
Ezhova, M. B. 52 (173), 53 (174), *96*
Fabbri, G. 580 (313), *593*
Faber, T. 661 (104), *678*
Fabre, J. L. 608 (52, 53), *629*
Fabre, S. 74 (124, 243, 244), 81 (255, 256), *95*, *98*, 150 (43), *154*, 406 (19), 407 (21, 22), 422 (15), 423 (17), *434*
Fahey, D. W. 156 (8), 160 (45), *184*, *185*
Faid, K. 285 (238, 241), *311*
Fajer, J. 196 (47), *216*
Fajert, J. 201 (71), *216*
Fakhfakh, M. A. 608 (49, 50), *629*
Falciola, C. 791 (50), *802*
Falciola, C. A. 558 (220), *590*, 791 (58), *802*
Falck, J. R. 558 (218), *590*
Faller, L. D. 329 (101), *361*
Fallis, A. G. 567 (268), *591*, 633 (1), 646 (37, 38), 647 (41–43), 648 (44, 45, 48), *676*, *677*
Fang, Q. K. 583 (319), *593*
Fantucci, P. 331 (119), *362*
Farády, L. 653 (69–71), *677*, *678*
Farghaly, O. A. 276 (198, 199), *310*
Farias, P. A. M. 276 (197), *310*
Farkas, J. 684 (18), 713 (53), *714*, *715*
Farkas, O. 384 (30), *402*
Farnham, W. B. 527 (28), *584*
Farquhar, G. D. 359 (353), *367*
Farrar, J. M. 166 (76), *186*
Farrell, R. P. 552 (175), *589*
Farthing, C. N. 547 (149), *588*
Farwell, J. D. 305 (305), *312*
Fatiadi, A. J. 528 (86), *586*
Fau, S. 137 (10), *153*, 720 (7), *768*
Faust, R. 196 (43), *216*
Fauvarque, J. F. 258 (130), *263*
Favaloro, F. G. 742 (14), *768*

Fazakerley, G. V. 108 (41), *127*, 140 (19), *153*
Fazekas, M. 212 (113), *217*
Fedorchuk, C. 78 (253), *98*, 421 (34), *435*
Fedorov, A. A. 357 (342), *366*
Fedorov, E. V. 357 (342), *366*
Fedoseev, G. B. 268 (118), *308*
Fedushkin, I. L. 79 (254), *98*, 414 (32), 426 (48), *435*
Fehsenfeld, F. C. 156 (8), 160 (45), *184*, *185*
Feillet-Coudray, C. 288 (256–258), *311*
Feitelson, J. 198 (55), *216*
Fejfarová, K. 86 (270), *99*
Feldberg, S. W. 196 (48), *216*
Felding, J. 518 (49), *585*
Feldman, A. K. 576 (303), *592*
Felkin, H. 656 (83), 665 (115), 666 (114), *678*
Fellah, S. 243 (81–83, 85), *262*
Fellmann, P. 458 (51), 482 (98), *507*, *508*
Felton, R. H. 196 (47), *216*
Fendler, J. H. 206 (79), *217*
Feng, B. 337 (169), *363*
Feng, Z. 253 (120), *262*
Ferguson, E. E. 156 (8), 160 (45), *184*, *185*
Ferguson, H. D. 103 (10), *126*
Feringa, B. L. 457 (44, 45), 500 (46), *507*, 558 (220, 221), 563 (244), 564 (255), *590*, *591*, 774 (2, 10, 13–16, 21), 775 (29), 779 (32), 780 (37, 38), 786 (42), 787 (41, 45–47), 788 (48), 791 (56, 62), 798 (63, 64), *800–802*
Ferment, O. 268 (40), *306*
Fernandez-Lazaro, F. 201 (40), *216*
Fernholt, L. 23 (76), *93*, 183 (148), *187*
Ferracutti, N. 442 (10), *506*
Ferrão, M. F. 295 (288), *312*
Fiammengo, R. 336 (162), *362*
Fiandanese, V. 601 (40–42), *629*
Fiaschisi, R. 555 (203), *590*
Fidélis, A. 241 (80), 243 (82), *262*
Field, E. 231 (20), 233 (22), 239 (74), 244 (19), *260*, *261*
Field, M. J. 348 (242), *364*
Fields, E. K. 595 (3), *628*
Figadère, B. 554, 555 (190, 197, 198), *589*, 607 (48), 608 (49–51), 614 (65), *629*, *630*
Filgueiras, A. V. 272 (170), *309*
Fillebeen, T. 73 (239), *98*, 424 (10), *434*
Finch, J. T. 340 (186), *363*
Findl, E. 246 (91), *262*
Finkbeiner, H. L. 112 (52), *127*
Fischer, A. K. 30 (108), *94*, 243 (86), *262*
Fischer, E. O. 88 (88), 89 (89), *94*
Fischer, M. 540 (116), 566 (266), *587*, *591*
Fischer, R. 39 (133), *95*
Fishel, W. P. 289 (269), *312*
Fitzgerald, R. 303 (344), *313*
FitzGerald, S. 196 (43), *216*
Fjellvaag, H. 107 (19), *126*

Author Index

Flaherty, K. M. 340 (181), *363*
Flatman, P. W. 268 (60), *307*
Fleck, O. 355 (317), *366*
Fleischmann, M. 241 (37), *261*
Fleming, F. F. 432 (88), *436*, 528 (85, 87), 563 (250), 565 (258–261), *586*, *591*, 647 (39, 40), 667 (121, 122), 670 (123–125), 671 (126–130), 677–679
Fleming, I. 439 (5), 485 (4), *506*, 540 (115), *587*
Fleming, M. J. 547 (151), *588*
Fletcher, K. S. 277 (201), *310*
Fletcher, S. P. 500 (46), *507*, 558 (221), *590*, 787 (47), 798 (64), *801*, *802*
Florencio, M. H. 302 (322), *313*
Florer, J. B. 325 (69), *361*
Flores, R. 341 (191), *363*
Florián, J. 352 (300), *365*
Fochi, M. 259 (144), *263*, 559 (228), 576 (301), *590*, *592*
Focia, P. J. 318 (21), *359*
Fogh-Andersen, N. 276 (16), *306*
Fokin, V. V. 576 (303), *592*
Folest, J. C. 236 (60), 258 (38), *261*, 291 (279), *312*
Fondekar, K. P. 473 (85), *508*
Fontani, M. 30 (108), *94*, 243 (86), *262*
Forbes, G. C. 17 (65), *93*
Ford, W. T. 140 (14), *153*
Foresman, J. B. 369 (2), 384 (30), *401*, *402*
Forgione, P. 567 (268), *591*, 633 (1), 646 (38), 647 (43), 648 (44), *676*, *677*
Forman, A. 196 (47), *216*
Forster, A. C. 339 (179), *363*
Foster, W. E. 653 (57), *677*
Fournier, A. 608 (50), *629*
Fournioux, X. 558 (219), *590*, 791 (57), *802*
Fox, D. J. 369 (2), 384 (30), *401*, *402*
Fox, J. M. 560 (233), *590*, 659 (98, 99, 100), *678*
Fraenkel, G. 116 (59), *128*, 138 (13), 148 (48), *153*, *154*
Frajerman, C. 666 (114), *678*
France, M. R. 168 (75), *186*
Francis, C. L. 495 (111), *509*
Franck, X. 554 (190), 555 (197, 198), *589*, 607 (48), 608 (49–51), 614 (65), *629*, *630*
Frandt, J. 653 (62), *677*
Franks, M. A. 648 (46), *677*
Franz, K. 268 (73), *307*
Franz, T. 573 (290), *592*
Franzen, R. G. 534 (101), *586*
Franzini, E. 331 (119), *362*
Franzke, C. 301 (357), *314*
Fraser, P. K. 779 (33), *801*
Frash, M. V. 124 (94), *129*, 180 (46), *185*
Frech, W. 304 (351), *313*
Frederiksen, M. U. 707 (47), *715*

Fredj, A. B 357 (335), *366*
Freeborn, B. 335 (147), *362*
Freeman, S. 493 (106), *509*
Freer, A. A. 324 (58), 356 (325, 331), *360*, *366*
Freijee, F. J. M. 120 (67), 121 (71), *128*
Freiser, B. S. 160 (43), 162 (42, 57), *185*
French, H. E. 238 (63), 246 (14), *260*, *261*
Frenzen, G. 516 (32), *584*
Freycon, M. T. 268 (116), *308*
Freyer, W. 196 (54), *216*
Freymann, D. M. 318 (21), *359*
Friedberg, E. C. 355 (316), *366*
Friedman, H. 235 (50), *261*
Friedrichsen, W. 517 (40), *585*
Friesner, R. A. 350 (265), *365*
Frisch, A. C. 544 (129, 130), *588*, 596 (27), *629*
Frisch, M. J. 369 (2), 384 (30), *401*, *402*
Fromme, P. 356 (36, 322, 324, 330), *360*, *366*
Froschauer, E. M. 325 (68), *360*
Frost, R. M. 546 (144), *588*, 615 (68–70), *630*
Fruchier, A. 491 (105), *509*
Frutos, J. 774 (12), *800*
Fry, A. J. 255 (127, 128), *263*
Frydman, B. 211 (103), *217*
Frydman, L. 328 (84), *361*
Frydman, V. 328 (84), *361*
Frye, S. V. 395 (4), *401*
Fu, G. C. 571 (283), *592*, 618 (66), *630*
Fu, Q. 169 (80), *186*
Fuchimoto, Y. 468 (80), *508*
Fuchs, C. F. 604 (7), *628*
Fuhrhop, J.-H. 208 (93–95), *217*
Fuji, K. 480 (92), *508*
Fujii, T. 767 (57), *769*
Fujii, Y. 561 (239), *591*
Fujikura, S. 661 (103), *678*
Fujimori, K. 701 (36), *715*
Fujino, O. 279 (218), *310*
Fujisawa, H. 484 (28), *507*
Fujisawa, T. 486 (25), *507*
Fujita, H. 329 (99), *361*
Fujita, K. 737 (32), *768*
Fujita, M. 767 (55), *769*
Fujiwara, K. 303 (349), *313*
Fujiwara, Y. 513 (14), *583*
Fujiyoshi, Y. 356 (328), *366*
Fukaya, F. 193 (27), *215*
Fukin, G. K. 426 (48), *435*
Fukuhara, K. 625 (91), *630*, 700 (32), *715*
Fukui, H. 279 (218), *310*
Fukumoto, H. 123 (75), *128*
Fukuzumi, S. 200 (69), 201 (40, 70), *216*
Funabashi, Y. 570 (280), *591*
Funseth-Smotzer, J. 323 (51), *360*
Furcola, N. C. 289 (271), *312*

Furois, J. M. 198 (60), *216*
Fürstner, A. 514 (19), 546 (145), 554 (188, 189), 559 (231), *584*, *588–590*, 596 (32), 608 (57), 614 (58, 59), 616 (73), 624 (35), 626 (83, 94–96), *629*, *630*
Furukawa, N. 517 (34), *584*
Furuya, A. 163 (66, 67), *185*
Furuyama, T. 419 (51), *435*
Fuson, R. C. 299 (309), *312*, 633 (2–5), *676*
Futai, M. 329 (97), *361*
Fuxreiter, M. 355 (319), *366*
Gaboyard, M. 711 (50), *715*
Gabrielsson, A. 200 (67), *216*
Gadde, S. 201 (41, 42), *216*
Gaddum, L. W. 235 (49), 246 (14), *260*, *261*
Gaderbauer, W. 38 (130), *95*
Gaertner, P. 618 (74), *630*
Gaertner, R. 633 (3), *676*
Gafni, A. 344 (216, 217), *364*
Gagas, D. 119 (64), *128*
Gago, S. 341 (191), *363*
Gahan, L. R. 343 (210), *364*
Gaidies, A. 610 (60), *630*
Gait, M. J. 339 (184), *363*
Gajda, T. 547 (147), *588*
Gaje, J. R. 503 (124), *509*
Gal, J.-Y. 290 (273), 292 (281), *312*
Galili, T. 201 (71), 206 (78), *216*, *217*
Galimov, E. M. 330 (108), *361*
Galizzi, A. 327 (77), *361*
Gallagher, D. J. 22 (71), *93*
Galland, L. 268 (19), *306*
Galle, J. E. 664 (113), *678*
Galli, C. 198 (56), *216*
Gallucci, J. C. 50 (98), *94*, 123 (83), *129*
Games, D. E. 493 (107), *509*
Games, D. F. 462 (57), *507*
Gan, Z. 357 (333), *366*
Gao, H.-W. 282 (228), *310*
Gao, J. 350 (263), *365*
Gao, K. 163 (69), *186*
Gao, Y. 540 (120), *587*
Gapeev, A. 125 (99), *129*
Garavaglia, S. 327 (77), *361*
García Alonso, J. I. 273 (181), *309*, 303 (346), *313*
García, F. 51 (51), *93*
García-Fernández, R. 273 (181), *309*
García García, P. 424 (66), *436* 473 (85) *508*,
Garcia Ruano, J. L. 583 (319), *593*
Gardiner, K. 335 (150), *362*
Gardiner, M. G. 26 (92), 51 (168), *94*, *96*
Gardner, P. D. 735 (31), *768*
Garland, H. O. 268 (92), *307*
Garrido Frenich, A. 302 (335), *313*
Garst, J. F. 55 (180, 182), *96*, 172 (106–108), *186*, 512 (5), 513 (11, 14), *583*
Gasser, R. 268 (26), *306*

Gasyna, Z. 196 (52), *216*
Gaucheron, F. 268 (129), *308*
Gauld, J. W. 336 (160), *362*
Gault, Y. 173 (110–113), 174 (114), *186*
Gautheron, B. 50 (98), *94*, 123 (83), *129*
Gavrilova, V. A. 198 (58), *216*
Gavryushin, A. 633 (1), *676*
Gawaz, M. 268 (62), *307*
Gawley, R. E. 395 (37), *402*
Gebara-Coghlan, M. 495 (111), *509*
Geetha, V. 576 (303), *592*
Geissler, M. 682 (10), *714*
Gellene, G. I. 159 (38), *185*
Genchel, V. G. 110 (48), *127*
Genders, J. D. 246 (98), *262*
Genet, J.-P. 559 (224), *590*
Gengyo, K. 760 (50), *769*
Gennari, C. 625 (93), *630*
Gentemann, S. 201 (46), *216*
Gentil, S. 50 (98), *94*, 123 (83), *129*
George, P. 319 (25), *360*
Gerber, B. 276 (195, 196), *310*
Gerbhardt, P. 39 (133), *95*
Gerlt, J. A. 357 (342), *366*
Geurts, K. 500 (46), *507*, 558 (221), *590*, 787 (47), 798 (64), *801*, *802*
Geven, W. B. 287 (248), *311*
Geyer, A. 49 (165), *96*
Ghahramani, M. 268 (34), *306*
Ghandour, M. A. 276 (198), *310*
Ghozland, F. 458 (50), *507*
Giacomelli, G. 559 (226), *590*
Giannotti, C. 198 (62), *216*
Gianopoulos, J. G. 268 (135), *308*
Gibbs, C. S. 347 (235), *364*
Gibson, V. C. 71 (231), 74 (241), 76 (242), 77 (250), *98*, 411 (33), 418 (14), 424 (44), *434*, *435*, 484 (99), *508*
Gilardi, R. 419 (57), *435*, 538 (105), *587*
Gilbert, T. M. 91 (278), *99*
Gilfillan, C. J. 20 (70), *93*
Gilje, J. W. 68 (219), *97*
Gill, P. M. W. 369 (2), 384 (30), *401*, *402*
Gille, S. 791 (57), *802*
Gillespie, S. E. 328 (85), *361*
Gilman, H. 289 (269, 270), 295 (289–295), 296 (296, 298), 297 (297), *312*, 619 (75), *630*, 633 (7), 653 (6, 8), *676*
Giraitis, A. P. 228 (43, 44), *261*
Girard, J. E. 157 (22), *185*
Girgis, S. I. 282 (232), *311*
Girijavallabhan, V. M. 574 (296), *592*
Giudici, R. E. 791 (59), *802*
Gizbar, C. 253 (118, 119), *262*
Gizbar, H. 247 (106), 248 (96, 113), 251 (112), 252 (114, 116), *262*
Glanz, M. 54 (95), *94*
Glaus, T. M. 276 (195, 196), *310*

Gless, R. D. 451 (34), *507*
Glover, C. V. C. 349 (257), *364*
Glusker, J. P. 319 (25), 324 (35), *360*
Gmeiner, P. 518 (49), *585*
Gneupel, K. 246 (92), *262*
Gobley, O. 50 (98), *94*, 123 (83), *129*
Godlewska-Żyłkievicz, B. 273 (184), *309*
Goedheer, J. C. 208 (98, 99), *217*
Goel, A. B. 138 (21), 139 (9), *153*
Goel, B. 147 (42), *154*
Gofer, Y. 247 (106), 248 (96, 113), 250 (115), 251 (112), 252 (114, 116), 253 (117–119), 258 (111), *262*
Goff, J. P. 268 (126), *308*
Göhr, H. 246 (97), *262*
Gokel, G. W. 121 (70), *128*
Gold, H. 513 (18), *584*
Gold, M. G. 347 (226), *364*
Goldberg, I. 252 (114), *262*
Goldberg, N. 156 (1), *184*
Golden, J. T. 91 (277), *99*
Goldie, H. 332 (125), *362*
Gol'dshleger, U. I. 125 (103), *129*
Goldsmith, E. J. 347 (237), *364*
Golen, J. A. 659 (99), *678*
Gomez, J. 198 (66), *216*
Gommermann, N. 268 (8), *306*, 518 (42), 548 (160), 550 (126), *585*, *587*, *588*
Gomperts, R. 369 (2), 384 (30), *401*, *402*
Goncharova, V. A. 268 (118), *308*
Gonella, M. 268 (91), *307*
Gong, Z. 123 (90), *129*
Gonzalez, C. 369 (2), 384 (30), *401*, *402*
González, F. 391 (26), 393 (25), *402*
Gonzalez-Portal, A. 268 (36), *306*
González-Rodríguez, M. J. 302 (335), *313*
González-Soto, E. 271 (164), *309*
Good, M. C. 347 (229), *364*
Goodby, J. W. 546 (144), *588*, 615 (68), *630*
Goodman, M. F. 352 (300), *365*
Goodwin, C. J. 454 (40), *507*
Görbing, M. 517 (34), *584*
Goris, J. 268 (37), *306*
Görls, H. 39 (133), *95*
Görög, S. 292 (283), *312*
Gorrichon-Guigon, L. 458 (50), 478 (97), *507*, *508*
Gosh, P. 335 (148), *362*
Gossage, R. A. 545 (137), *588*
Gossauer, A. 210 (102), *217*
Gosselin, F. 703 (41), *715*
Goto, H. 675 (158), *679*
Götte, M. 352 (301), *365*
Gottfriedsen, J. 54 (95), *94*
Gottlieb, H. E. 252 (114, 116), *262*
Gottschling, D. E. 335 (149), *362*
Gouin, L. 652 (55), 653 (56), *677*
Goursot, A. 157 (15), *184*, 391 (20), *402*

Gouterman, M. 193 (14), *215*
Gozzi, C. 596 (26), *629*
Grabowski, P. J. 335 (149), *362*
Graening, T. 517 (40), *585*
Graham, D. V. 13 (59), 22 (72), *93*
Graille, J. 301 (355, 358, 361), *314*
Grant, B. 347 (233), *364*
Grant, C. V. 328 (84), *361*
Graves, D. J. 347 (230), *364*
Gray, T. C. 52 (173), 53 (174), *96*
Gray, T. G. 51 (171, 172), *96*
Gready, J. E. 359 (351), *367*
Greaves, J. 302 (330), *313*
Green, L. G. 576 (303), *592*
Green, M. L. H. 124 (93), *129*
Greenbaum, E. 214 (136), *218*
Greenberg, A. 113 (56), *128*
Greene, T. M. 157 (17, 20), *184*
Gregory, T. D. 247 (100, 101), *262*
Greiser, T. 6 (48), 38 (122), 42 (49), *93*, *95*
Grenet, S. 439 (3), *506*
Gretsova, N. S. 213 (120), *217*
Gribov, E. 125 (108), *129*
Grieco, P. A. 718 (2), *768*
Griesser, R. 332 (123), *362*
Griffioen, S. 66 (213), *97*
Grignard, V. 2 (2, 7), 92, 107 (23), *127*, 268 (2), *306*, 389 (3), *401*, 512 (1), *583*
Grigorenko, B. L. 331 (117, 118), *362*
Grigsby, W. J. 82 (162), *96*, 417 (13), *434*
Grilley, D. 336 (164), *362*
Grimes, R. N. 51 (169), *96*
Grimm, B. 191 (8), *215*
Grison, C. 489 (8, 9), 491 (105), *506*, *509*
Gritzo, H. 48 (164), *96*
Grizard, D. 288 (256), *311*
Grognet, J. M. 268 (43), *306*
Gromada, J. 69 (223), *97*
Gromek, J. M. 65 (212), *97*, 395 (37), *402*
Gronert, S. 118 (61), *128*
Grosche, M. 519 (53), 550 (168), *585*, *589*
Gröser, T. 675 (156), *679*
Grossheimann, G. 775 (28), *801*
Grossie, D. A. 458 (49), *507*
Grove, D. M. 774 (8, 25), 791 (53), *800*, *802*
Grover, P. 583 (319), *593*
Grubbs, R. D. 268 (52), *307*, 325 (61), 328 (5), *359*, *360*
Gruber, P. R. 191 (7), *215*
Grulich, P. F. 325 (74), *361*
Gruter, G.-J. M. 66 (215), *97*
Grutzner, J. B. 140 (14), 141 (28), *153*
Gryko, D. T. 201 (45), *216*
Gstöttmayr, C. W. K. 550 (166, 168), *589*
Guallar, V. 350 (265), *365*
Guay-Woodford, L. M. 268 (64), *307*
Güçer, Ş 271 (161), *309*
Guddat, L. W. 343 (210), *364*

Gudermann, T. 268 (55, 56), *307*
Gudipati, V. 432 (88), *436*, 565 (259), *591*, 647 (39, 40), *677*
Guenter, F. 574 (296), *592*
Guerin, C. 661 (105), *678*
Guerrier-Takada, C. 335 (150), *362*
Gueux, E. 288 (256–258), *311*
Guggenberger, L. J. 58 (20, 190), *92*, *96*, 391 (14), *402*
Gui, L. 356 (326), *366*
Guida-Pietrasanta, F. 711 (50), *715*
Guiet-Bara, A. 267 (1), 268 (50, 54, 102), *306–308*
Guijarro, A. 514 (20), *584*
Guijarro, D. 519 (53), *585*
Guilard, R. 193 (11, 13, 15), 194 (32), 195 (16), *215*
Guillamet, B. 550 (173), *589*
Guillaume, T. 268 (138), *308*
Guitian, E. 540 (115), *587*
Gunther, T. 268 (63), *307*
Guo, H. 554 (192), *589*
Guo, J. 667 (121), *678*
Guo, W. 162 (61), 163 (71), 166 (77), 167 (82–85), 168 (73, 86), 169 (80, 87), *185*, *186*
Gupta, P. 268 (24), *306*
Gupta, R. K. 268 (24), *306*
Gupta, S. 286 (247), *311*, 428 (85), *436*
Gurinovich, V. V. 211 (107), *217*
Guss, J. M. 30 (105), *94*
Gust, D. 194 (35), 195 (16), *215*
Gustafsson, B. 59 (145), *95*
Gutell, R. R. 337 (169), *363*
Gutierrez, A. M. 201 (40), *216*
Gutman, E. L. 289 (271), *312*
Gutmann, V. 226 (41), *261*
Gutteridge, S. 357 (338, 346), 359 (339), *366*
Gyepes, R. 86 (270), 89 (273), *99*
Haaland, A. 23 (76), 26 (91), 29 (102), *93*, *94*, 123 (85), *129*, 183 (139, 146, 148, 149), *187*
Haas, O. 246 (95), *262*
Habiak, V. 546 (143), *588*, 618 (72), *630*
Hackmann, C. 142 (30), *154*
Hadei, N. 550 (167), *589*
Haelters, J. P. 462 (59), *507*
Hafner, W. 88 (88), *94*
Haga, M. 675 (158), *679*
Hagaman, E. 656 (83), *678*
Hagopian, R. A. 579 (309), *592*
Hahn, R. C. 723 (16), *768*
Hairault, L. 711 (50), *715*
Håkansson, M. 38 (129), 59 (61, 145), 60 (185), *93*, *95*, *96*
Halasa, A. F. 19 (46), *93*
Haley, M. M. 531 (93, 94), *586*
Halford, S. E. 355 (311), *366*
Halow, I. 113 (2), *126*

Hamada, Y. 370 (112), 458 (52), *507*, *509*
Hamdouchi, C. 259 (145), 260 (150), *263*, 512 (5), 518 (48), *583*, *585*
Hamilton, C. L. 119 (62), *128*, 654 (76), *678*
Hamlin, R. 351 (286), *365*
Hamman, C. 340 (183), *363*
Hammann, C. 341 (190), *363*
Hammerich, O. 219 (9), 253 (122, 123), *260*, *263*
Hammes, W. 517 (39), *585*
Hampton, E. M. 268 (84), *307*
Han, C. 282 (229), *310*
Han, J. 341 (197), *363*
Han, K.-L. 169 (94, 95), 170 (97, 98), *186*
Han, R. 73 (235–238), *98*, 148 (17), *153*, 418 (7), 419 (6, 52), 422 (60), 423 (8, 9), *434*, *435*
Han, W. 193 (19), *215*
Han, Z. 395 (35), *402*, 583 (319), *593*
Hanafusa, T. 259 (134), *263*, 391 (11), *402*
Hanawalt, E. M. 711 (51), 713 (53), *715*
Hanay, W. 419 (55), *435*
Hand, E. S. 478 (65), *508*
Handler, G. S. 107 (28), *127*, 258 (88), *262*
Handwerker, S. M. 268 (19), *306*
Handy, S. T. 515 (25), *584*
Hanks, S. K. 347 (224), *364*
Hannon, F. J. 772 (5), *800*
Hansen, J. 65 (212), *97*, 336 (161), *362*, 395 (37), *402*
Hansen, M. U. 514 (19), *584*
Hansford, K. A. 563 (246), *591*
Hanson, E. M. 719 (3), *768*
Hanson, W. 514 (19), *584*
Hanusa, T. P. 58 (5), *92*
Hao, H. 74 (248), *98*, 424 (16), *434*
Hao, W. 540 (122), *587*
Hara, M. 194 (29), 213 (119), *215*, *217*
Harada, T. 718 (2), *768*
Harding, M. M. 321 (33, 34), *360*
Hardy, J. A. 344 (211), *364*
Hargittai, I. 183 (142), *187*
Hargittai, M. 183 (143–145, 150, 151), *187*
Harino, H. 302 (333), *313*
Harkins, P. C. 347 (237), *364*
Harmat, N. 180 (137), *187*, 425 (69), *436*
Harms, K. 59 (192), 62 (204), *97*, 395 (34), *402*, 516 (32), 527 (75), *584*, *586*, 725 (20), *768*
Harriman, A. 198 (63), *216*
Harrington, D. J. 320 (29), *360*
Harrington, P. E. 707 (47), *715*
Harrington, R. W. 84 (265), *99*, 298 (304), *312*, 427 (73), *436*
Harris, M. E. 348 (238), *364*
Harris, R. K. 151 (49), *154*
Harrison, R. M. 303 (345), *313*
Harrity, J. P. A. 673 (147), *679*

Hartig, J. R. 68 (220), *97*
Hartl, F. 200 (67), *216*
Hartl, F. T. 347 (236), *364*
Hartley, F. R. 156 (2), *184*
Hartman, F. C. 359 (339), *366*
Hartman, R. J. 398 (39), *402*
Hartmann, N. 540 (117), *587*
Hartwich, G. 193 (28), *215*
Hartwig, A. 351 (278), *365*
Hartwig, J. F. 553 (178, 180), *589*, 798 (65), *802*
Hartwig, T. 610 (60), *630*
Harutyunyan, S. R. 457 (44, 45), *507*, 564 (255), *591*, 775 (29), 779 (32), 780 (38), 787 (41), 788 (48), *801*
Harvey, J. N. 333 (133), *362*
Harvey, P. C. 190 (1), *214*
Harvey, S. 46 (158), 58 (155), *96*, 123 (89), *129*
Hascall, T. 73 (239), 82 (162), *96*, *98*, 417 (13), 424 (10), *434*
Hase, T. A. 288 (263), *311*
Hasegawa, T. 214 (132), *218*
Hashimoto, W. 327 (78), *361*
Hasobe, T. 201 (70), *216*
Hassan, J. 596 (26), *629*
Hässig, M. 276 (195, 196), *310*
Hassner, A. 742 (14), *768*
Hatada, M. 349 (256), *364*
Hatano, M. 19 (45), *93*, 569 (274), 570 (278, 279), *591*, 684 (19), *714*
Hauk, D. 520 (56), *585*, 682 (9), *714*
Hawkes, S. A. 63 (112), *94*
Hawner, C. 791 (57), *802*
Hawthornthwaite-Lawless, A. M. 324 (58), 356 (325, 331), *360*, *366*
Hayasaka, T. 774 (11), *800*
Hayashi, K. 458 (52), 468 (80), *507*, *508*
Hayashi, T. 546 (143), 547 (146), 553 (185), 554 (191, 194), 568 (272), *588*, *589*, *591*, 615 (67), 619 (77), 622 (88), 623 (89), *630*, 637 (23), 639 (24), *677*, 774 (21, 24), 779 (33), *800*, *801*
Hayashi, Y. 623 (90), *630*, 743 (38), *769*
Haycock, J. W. 347 (237), *364*
Hayes, D. 424 (63, 66), *435*, *436*, 473 (85, 89), *508*
Hayes, P. C. 224 (29), *260*
Hayes, T. K. 559 (227), *590*
Hazell, T. 268 (121), *308*
Hazimeh, H. 260 (147), *263*, 513 (15), *584*
Hazra, T. K. 353 (307), *366*
He, X. 469 (78, 79), *508*
Head-Gordon, M. 369 (2), 384 (30), *401*, *402*
Healy, E. M. 259 (146), *263*
Heathcock, C. H. 468 (83), 469 (84), 478 (65), 484 (103), 485 (4), *506*, *508*, *509*
Heathcote, P. 356 (329), *366*

Heck, L. L. 295 (293–295), 296 (296), *312*
Heckman, J. E. 341 (195), *363*
Heckmann, B. 558 (218), *590*
Heckmann, G. 525 (65), *585*
Heeg, J. 50 (101), *94*
Heeg, M. J. 47 (160, 161), 72 (234), 78 (252), 83 (233), 84 (264), *96*, *98*, 122 (77, 79), *128*, 303 (303), *312*, 419 (27), 422 (28, 29), 433 (89), *435*, *436*
Hefter, G. 322 (47), *360*
Hegedus, L. S. 563 (251), *591*
Hein, J. W. 214 (128), *218*
Hein, M. 573 (290), *592*
Heinemann, F. W. 136 (16), *153*
Heinemann, W. 42 (143), *95*
Helfert, S. C. 51 (171, 172), 52 (173), 53 (174), *96*
Helmy, R. 291 (280), *312*
Henderson, K. W. 17 (64), 19 (69), 69 (222), 83 (263), *93*, *97*, *98*, 150 (41), *154*, 413 (42), 422 (41, 47, 59), 424 (61–67), 425 (70), 429 (76), *435*, *436*, 460 (56), 469 (78, 79), 473 (85–89), *507*, *508*, 539 (110), *587*
Henderson, L. C. 74 (124), *95*, 150 (43), *154*, 423 (17), *434*
Henderson, M. J. 31 (109), *94*
Hendersson, K. W. 539 (112), *587*
Hendrikx, C. J. 474 (47), *507*
Hendry, G. A. 210 (100), *217*
Hendry, L. 625 (92), *630*
Hengge, A. C. 319 (24), *359*
Henneberg, D. 653 (66), 655 (63–65), *677*
Henningsen, I. 351 (287), *365*
Hennion, G. F. 621 (80), *630*
Henrotte, J. G. 268 (61), *307*
Henry, D. J. 550 (164), *588*
Herber, C. 74 (124), *95*, 150 (43), *154*, 423 (17), *434*, 791 (53), *802*
Herberich, G. E. 29 (103), *94*
Herbst-Irmer, R. 57 (186), *96*, 124 (91), *129*
Hermann, G. 357 (337), *366*
Hermes Sales, C. 268 (96), *307*
Hernandez, D. 457 (48), *507*
Herndon, D. N. 268 (134), *308*
Herold, D. A. 303 (344, 350), 304 (354), *313*, *314*
Heron, N. M. 560 (234), *590*, 673 (148, 149), *679*
Herrmann, W. A. 519 (53), 550 (166, 168), *585*, *589*
Herschlag, D. 338 (176), 341 (188), *363*
Hervé du Penhoat, C. 608 (54, 55), *629*
Herzog, R. 59 (132), *95*, 149 (6), *153*
Heslop, J. A. 4 (34), *92*, 428 (87), *436*, 682 (12), *714*
Hesse, M. 760 (47), *769*
Hettinger, W. P. 240 (72), *261*

Hevia, E. 9 (52), 13 (59), 17 (64), 20 (70), 22 (71, 72, 74), 73 (73), 81 (224, 259), *93, 97, 98*, 414 (35, 46), 419 (50), 430 (80), 431 (82), *435, 436*, 460 (56), *507*
Hewgill, F. R. 476 (91), *508*
Hey, E. 427 (74), *436*
Heyn, A. S. 667 (54), *677*
Hibbert, H. 300 (313), *313*
Hideki, Y. 545 (139), *588*
Hideyuki, H. 544 (128), *588*
Higashi, S. 555 (199), *589*, 608 (56), *629*
Higashiyama, K. 547 (159), *588*
Higuchi, T. 707 (46), *715*
Hilbers, C. W. 287 (248), *311*
Hill, C. L. 513 (14), *583, 584*
Hill, E. A. 119 (63, 64), *128*, 144 (29), *153*, 654 (73, 75), *678*
Hill, M. S. 63 (112, 208), *94, 97*
Hiller, J. 87 (271), *99*
Hills, I. D. 618 (66), *630*
Himeshima, N. 214 (135), *218*
Himmel, H.-J. 157 (17), *184*
Himo, F. 341 (202), *363*
Hingorani, M. M. 353 (272), *365*
Hinkley, W. 667 (117), *678*
Hino, N. 559 (225), *590*
Hioki, T. 452 (37), 502 (119, 120), *507, 509*
Hirai, A. 622 (86), *630*, 658 (92), *678*
Hiraishi, A. 193 (22–26), *215*
Hirano, T. 107 (22), *127*
Hird, A. W. 774 (18), 775 (30), 780 (36), *800, 801*
Hird, M. 546 (144), *588*, 615 (68–70), *630*
Hirsch, A. 675 (154–156), *679*
Hisamoto, H. 285 (237), *311*
Hitchcock, P. B. 23 (77, 78), 24 (80), 32 (110, 111), 37 (121), 51 (168), 62 (120, 201), 63 (112, 206–208), 70 (225), 71 (231), *94–98*, 411 (33), 416 (49), 430 (77), *435, 436*
Hitomi, K. 353 (306), *366*
Hiyama, T. 285 (243), *311*, 487 (68, 104), *508, 509*
Hoang, T. 699 (31), *715*
Hoarau, C. 689 (26, 27), *715*
Hocek, M. 611 (61–63), *630*
Hochstrasser, R. M. 198 (56), *216*
Hocková, D. 611 (62), *630*
Hocquemiller, R. 554 (190), 555 (197, 198), *589*, 607 (48), 608 (49–51), 614 (65), *629, 630*
Hodgson, D. M. 547 (151), *588*
Hodoscek, M. J. 125 (104), *129*
Hoeijmakers, J. H. J. 351 (279), *365*
Hoelzer, B. 550 (163), *588*
Hoffman, R. 86 (86), *94*
Hoffman, R. J. 247 (100, 101), *262*
Hoffmann, R. W. 135 (11), 137 (10), *153*, 516 (32), 527 (75, 76, 78, 79), 528 (80), 534 (81), 540 (115), 550 (163), *584, 586–588*, 661 (104), *678*, 719 (4), 720 (5–7), 721 (12), 724 (17), 725 (20, 22), 727 (21), 734 (30), *768*
Hoffmann-Röder, A. 774 (17), *800*
Hofmann, F. R. 419 (55), *435*
Hogenbirk, M. 61 (198), 86 (269), *97, 99*
Holland, M. J. 349 (257), *364*
Holland, P. L. 91 (277), *99*
Hollingsworth, C. 107 (28), *127*
Holloway, C. E. 3 (26), 25 (25), *92*
Hollstein, E. 301 (357), *314*
Holm, T. 3 (17), *92*, 108 (36), 109 (37, 42, 43), 115 (45), 116 (13), 119 (14), *126, 127*, 235 (54, 55), 258 (52, 53), 259 (135), *261, 263*, 389 (3), 391 (9), 398 (39), *401, 402*, 451 (30), *507*
Holmes, A. B. 537 (104), *586*
Holmes, R. R. 30 (107), *94*
Holten, D. 201 (38, 39, 45, 46), *215, 216*
Holtkamp, H. C. 40 (137, 138), *95*, 120 (66, 68), *128*
Holton, R. A. 473 (75), *508*, 597 (33), 598 (34), *629*
Holzer, B. 527 (75), *586*, 725 (20), 727 (21), *768*
Holzwarth, A. R. 356 (330), *366*
Honda, K. 270 (157), *309*
Honeyman, G. W. 69 (222), 73 (73), *93, 97*, 150 (41), *154*, 419 (50), *435*
Honeymoon, G. W. 429 (76), *436*
Hong, Q. 572 (288), *592*
Hong, S. 330 (103), *361*
Hoogstraten, C. G. 341 (198), *363*
Hope, G. A. 107 (20), *126*
Hopkinson, A. C. 124 (94, 96), 125 (95), *129*, 160 (54), 176 (129), 180 (46, 134, 135), *185, 187*
Hopman, M. 63 (112), *94*
Horáček, M. 86 (270), 87 (271), 88 (272), 89 (273), *99*
Horan, N. R. 672 (137), *679*
Hordyjewska, A. 268 (109), *308*
Horibe, H. 556 (207), *590*
Horiguchi, K. 767 (56), *769*
Hormnirum, P. 424 (44), *435*
Hormnirun, P. 76 (242), *98*
Horn, E. 40 (140), *95*
Horrobin, D. F. 301 (360), *314*
Horsburgh, L. 422 (59), *435*
Hosmane, N. S. 51 (170–172), 52 (173), 53 (174), *96*
Hosmane, S. N. 53 (174), *96*
Hou, X. 279 (216), *310*
Houghton, J. D. 210 (100), *217*
Houk, K. N. 391 (24), *402*
Houpis, I. N. 654 (72), *678*, 699 (31), *715*
Houri, A. F. 672 (137–139), 673 (145), *679*

House, D. 485 (24), *507*
House, H. O. 137 (7), *153*, 156 (3), *184*, 295 (287), *312*
Housecroft, C. E. 512 (4), *583*
Houte, H. 268 (7), *306*
Hoveyda, A. H. 557 (212), 560 (234), 561 (240), *590*, *591*, 671 (136), 672 (137–139), 673 (142–149), *679*, 774 (18), 775 (30), 779 (33), 780 (36), 791 (55, 59, 61), *800–802*
Howard, J. A. K. 24 (80), 65 (211), *94*, *97*
Howden, M. E. H. 654 (77), *678*
Howell, G. P. 500 (46), *507*, 787 (47), *801*
Howson, W. 547 (147), *588*
Hoye, T. R. 300 (300), *312*
Hoylaerts, M. F. 345 (220), *364*
Hruby, V. J. 455 (43), *507*
Hsieh, K. 119 (64), *128*
Hsieh, Y.-T. 547 (154), *588*
Hsu, S.-F. 16 (63), *93*
Hsueh, M.-L. 16 (63), *93*
Hu, G. G. 335 (30), *360*
Hu, Q.-S. 550 (165), *588*
Hu, S. 167 (84), *186*
Hu, Y. 163 (65), 164 (64), 167 (83), 169 (79, 80, 93, 96), 170 (98), *185*, *186*
Huang, C.-H. 547 (153), *588*
Huang, H. W. 320 (7), *359*
Huang, J. 550 (166), *588*
Huang, J.-W. 547 (155), *588*
Huang, J.-Y. 81 (228), *98*, 149 (46), *154*, 422 (25), *435*
Huang, L.-F. 547 (153), *588*
Huang, P.-Y. 398 (39), *402*
Huang, Q. 107 (18), *126*
Huang, W. 213 (118), *217*
Huang, X. 644 (32), *677*
Huang, Y.-L. 278 (209, 211, 212), *310*
Huang, Z. 159 (36), *185*
Huckett, S. C. 610 (60), *630*
Huennekens, F. M. 207 (86, 87), *217*
Huerta, F. F. 791 (53), *802*
Huffman, J. C. 73 (240), *98*, 196 (44), *216*, 418 (11), *434*
Huffman, J. L. 353 (303), *365*
Huffman, M. 293 (267), *311*
Hug, G. 206 (77), *217*
Hug, G. L. 190 (2, 3), *214*
Hugerat, M. 201 (71), *216*
Huggett, R. J. 302 (330), *313*
Hughes, D. L. 74 (247), *98*, 421 (18), *434*
Hughes, G. 5 (43), *93*
Hughes, T. R. 329 (89), *361*
Huijgen, H. J. 268 (88), *307*
Huisgen, R. 540 (115), *587*
Hulanicki, A. 273 (184), *309*
Hulce, M. 451 (33), *507*
Hull, H. S. 122 (75), *128*
Hullihen, J. 329 (93), *361*

Hulsbergen, F. B. 356 (332), *366*
Hummert, M. 79 (254), *98*, 414 (32), *435*
Humphrey, S. M. 51 (51), 74 (247), *93*, *98*, 421 (18), *434*
Hung, I. 152 (52), *154*
Hunt, D. 191 (9), *215*
Hunt, J. C. A. 574 (294), *592*
Hunter, T. 347 (222, 224), *364*
Huong Giang, L. T. 302 (333), *313*
Hupe, E. 268 (7), *306*, 525 (41), *585*
Hurd, P. W. 458 (49), *507*
Hurley, P. T. 259 (143), *263*
Hurry, W. H. 599 (5), *628*
Husk, G. R. 664 (110), *678*
Hutcheon, M. L. 329 (92), *361*
Hutchins, C. J. 339 (179), *363*
Hutchison, D. A. 141 (28), *153*
Huth, A. 538 (107), *587*
Huth, J. A. 305 (364), *314*
Hutjgen, H. J. 275 (191), *310*
Hwang, C.-S. 59 (191), *97*
Hynninen, P. H. 191 (9), 211 (106), 214 (112), *215*, *217*
Iacuone, A. 583 (317), *593*
Iannello, S. 268 (49), *307*
Ibragimov, A. A. 123 (87), *129*
Iché-Tarrat, N. 331 (115, 116), *362*
Ichikawa, K. 145 (39), *154*
Ichikawa, T. 579 (309), *592*
Ichimura, A. 259 (138), *263*
Ida, R. 357 (333), *366*
Ide, M. 686 (22), 688 (23), *714*
Idriss, K. A. 280 (220), *310*
Ieawsuwan, W. 552 (177), *589*, 725 (19), *768*
Iida, A. 772 (6), *800*
Iida, H. 767 (55), *769*
Iida, T. 524 (63), *585*, 699 (31), 702 (39), *715*
Iitsuka, D. 466 (71), *508*
Iizuka, T. 90 (276), *99*
Ijare, O. B. 286 (247), *311*
Ikeda, H. 335 (31), *360*
Ikeda, T. 285 (237, 243), *311*
Iko, Y. 329 (97), *361*
Ikonomou, M. G. 175 (115), *186*
Ikumi, A. 544 (128), 545 (135), *588*
Ila, H. 268 (10, 11), *306*, 525 (66), *585*
Ilani, A. 206 (80), *217*
Ilgen, F. 540 (116), *587*
Ilgenfritz, G. 318 (18), *359*
Ilim, M. 277 (206), *310*
Imahori, H. 200 (68, 69), 201 (70), *216*
Imamoto, T. 63 (60), *93*, 182 (6), *184*, 570 (280), *591*
Imanishi, M. 270 (157), *309*
Imbos, R. 774 (14, 16), *800*
Imhof, R. 246 (95), *262*
Imi, K. 661 (103), *678*
Imizu, Y. 171 (24), *185*

Imker, H. J. 357 (342), *366*
Inagaki, K. 193 (27), *215*
Inagaki, S. 125 (109), *129*
Inganäs, O. 285 (240), *311*
Ingrosso, C. 212 (115), *217*
Inhoffen, H. H. 211 (104), *217*
Inoue, A. 5 (42), 41 (41), *93*, 341 (199–201), *363*, 523 (59), 528 (77, 83), *585*, *586*, 633 (1), *676*, 690 (29), 699 (11), 703 (43), 706 (42, 44), *714*, *715*
Inoue, K. 194 (29), 208 (96), 213 (119), *215*, *217*
Inoue, R. 567 (270), *591*, 636 (18), *676*
Inoue, S. 196 (37), *215*, 274 (186, 187), *309*, *310*
Inoue, T. 214 (137), *218*, 622 (84), *630*, 658 (93), *678*
Ireland, R. E. 494 (110), *509*
Ireland, T. 775 (28), *801*
Irisa, S. 767 (56), *769*
Isaacs, N. W. 324 (58), 356 (325, 331), *360*, *366*
Isago, H. 192 (12), *215*
Isaka, K. 517 (34), 519 (54), *584*, *585*
Iseki, Y. 196 (37), *215*
Iseri, L. T. 268 (100), *307*
Ishifune, M. 254 (121), *263*
Ishihara, K. 19 (45), *93*, 569 (274), 570 (278, 279), *591*, 684 (19), *714*
Ishii, D. 300 (312), *313*
Ishii, T. 463 (63), *508*
Isoyama, N. 193 (25), *215*
Itagaki, M. 284 (178), *309*
Itami, K. 544 (131), 555 (199, 204), *588–590*, 608 (56), *629*, 641 (27, 28), 662 (109), 664 (108), *677*, *678*
Ito, M. 193 (26), *215*, 558 (217), *590*
Ito, O. 201 (41, 42, 70), *216*
Ito, S. 546 (142), *588*, 616 (71), *630*, 701 (36), *715*
Itoh, H. 329 (100), *361*
Itoh, K. 517 (34), 519 (54), *584*, *585*, 675 (157, 158), *679*
Itoh, N. 479 (14, 16), *506*, 760 (50), *769*
Itoh, S. 193 (24, 25), *215*
Itoh, T. 699 (31), *715*
Itono, S. 107 (22), *127*
Itooka, R. 780 (35), *801*
Iturraspe, J. 211 (103), *217*
Ivanov, D. 503 (53, 121), *507*, *509*
Iwahara, T. 662 (106), *678*
Iwai, S. 353 (306), *366*
Iwaki, M. 193 (24), *215*
Iwaki, S. 193 (25), *215*
Iwamoto-Kihara, A. 329 (97), *361*
Iwasawa, N. 772 (3), *800*
Iyer, S. S. 73 (240), *98*, 418 (11), *434*
Izatt, R. M. 328 (85, 86), *361*

Izod, K. 84 (265), *99*, 298 (304), *312*, 427 (73), *436*
Izumi, T. 353 (307), *366*
Jablonski, A. 295 (288), *312*
Jackson, P. R. 268 (142), *308*
Jacob, A. P. 182 (133), *187*
Jacob, K. 30 (108), *94*, 243 (86), *262*
Jacob Morris, J. 469 (78), *508*
Jacobsen, E. N. 774 (2), 779 (22), *800*
Jaenschke, A. 48 (163), *96*, 122 (81), *128*
Jaeschke, G. 774 (8), *800*
Jäger, V. 573 (290), *592*
Jaggar, A. J. 63 (206), *97*
Jaidane, N. 357 (335), *366*
Jain, N. 421 (58), *435*, 700 (35), *715*
Jaiswal, A. 353 (307), *366*
James, A. W. G. 462 (57), *507*
James, P. F. 181 (131), 182 (133), *187*
Jamison, W. C. L. 667 (118), *678*
Janin, J. 348 (246), 350 (261), *364*, *365*
Janke, N. 44 (152), 45 (154, 156), 47 (159), *95*, *96*
Jansen, B. J. M. 474 (47), *507*
Jansen, J. F. G. A. 774 (10), *800*
Jaramillo, C. 518 (48), *585*
Jarvis, J. A. J. 30 (104), *94*
Jäschke, A. 336 (162), *362*
Jasien, P. G. 171 (99, 103), *186*
Jastrzebski, J. T. B. H. 69 (29), 79 (257), *92*, *98*, 545 (137), *588*
Jayasena, S. D. 341 (192, 194), *363*
Jayaweera, P. 175 (115), *186*
Jeltsch, A. 355 (318, 320), *366*
Jenkins, H. D. B. 106 (15), *126*
Jensen, A. 270 (154), *309*
Jensen, A. E. 518 (36), 519 (51), *584*, *585*
Jensen, C. M. 71 (232), *98*
Jewitt, B. 124 (93), *129*
Jezequel, M. 544 (133), *588*
Ji, S. 565 (256), *591*
Ji, S.-J. 782 (39), *801*
Jia, J. 293 (285), *312*
Jiang, G. 302 (327), *313*
Jiang, G.-b. 302 (336), *313*
Jiang, H. 123 (90), *129*
Jiang, L. 272 (176), *309*, 329 (87), *361*
Jiang, P. 391 (22), *402*
Jiwoua Ngounou, C. 283 (162), *309*
Joachimiak, A. 327 (67), *360*
Job, A. 653 (68), *677*
Jockusch, R. A. 318 (19, 20), *359*
Johansen, T. N. 519 (50), *585*
John, N. B. St. 295 (293), *312*
Johnson, B. G. 369 (2), 384 (30), *401*, *402*
Johnson, C. 67 (217), *97*
Johnson, F. B. 268 (13), *306*
Johnson, J. S. 579 (310), *592*
Johnson, K. A. 352 (289, 291, 292), *365*

Johnson, L. N. 348 (239), *364*
Johnson, S. C. 478 (65), *508*
Jolibois, P. 107 (27), *127*, 220 (10, 11), *260*
Jolly, B. S. 81 (260), *98*, 405 (5), *434*
Jońca, Z. 277 (203, 204), *310*
Jones, B. T. 279 (216), *310*
Jones, C. 64 (209), *97*
Jones, E. G. 290 (275), *312*
Jones, M. C. 519 (53), *585*
Jones, P. 534 (96), *586*
Jones, R. 133 (4), *153*, 301 (301), *312*
Jones, R. G. 295 (291), *312*
Jones, R. O. 331 (120), *362*
Jones, R. V. H. 673 (140, 141), *679*
Joordens, J. J. M. 287 (248), *311*
Jordan, K. D. 391 (12), *402*
Jordan, P. 356 (36, 322), *360, 366*
Jousseaume, B. 635 (13–15), 648 (49, 50), *676, 677*
Joyce, C. M. 352 (297), 355 (274), *365*
Joyce, G. F. 336 (151), *362*
Juan, D. 268 (76), *307*
Julia, M. 553 (179), *589*, 608 (52–55), *629*, 727 (23), 730 (28), *768*
Julius, M. 516 (32), *584*
Juneau, K. 320 (29), *360*
Junge, W. 329 (95), *361*
Jungreis, E. 280 (221), *310*
Jungwirth, P. 322 (47), *360*
Junk, C. L. 190 (1), *214*
Junk, P. C. 46 (158), 64 (209), 81 (260), *96–98*, 123 (89), *129*, 405 (5), *434*
Justyniak, I. 410 (36), *435*
Jutzi, P. 67 (93), 68 (221), *94, 97*
Kabeláč, M. 323 (50), *360*
Kacprzynski, M. A. 774 (18), *800*
Kadish, K. M. 193 (11, 13, 15), 194 (32), 195 (16), *215*
Kaeseberg, C. 665 (115), *678*
Kageyama, H. 60 (194–196), 61 (197), *97*
Kahn, L. R. 369 (1), *401*
Kai, Y. 51 (167), 60 (194, 196), 61 (197), *96, 97*, 121 (74), *128*, 254 (126), *263*, 540 (119), *587*
Kaim, W. 52 (173), 53 (174), 79 (258), *96, 98*
Kaliya, O. L. 213 (120, 122), *217*
Kalojanoff, A. 214 (126, 127), *218*
Kalsbeck, W. A. 201 (46), *216*
Kalscheuer, R. 301 (359), *314*
Kambe, N. 544 (127, 128), 545 (135, 136), 560 (235), 561 (239, 241, 242), *588, 590, 591*
Kamei, K. 699 (31), *715*
Kamei, T. 641 (27, 28), *677*
Kamikawa, T. 553 (185), *589*
Kamiya, N. 356 (323), *366*
Kamm, B. 191 (7), *215*
Kamm, M. 191 (7), *215*

Kamm, O. 2 (9), *92*
Kanai, M. 564 (254), *591*, 774 (9, 25), *800, 801*
Kanakamma, P. P. 547 (154), *588*
Kanatani, R. 661 (101, 107), 662 (106), *678*
Kanazawa, A. 285 (243), *311*
Kandil, S. A. 633 (10), *676*
Kane, H. J. 358 (348), *367*
Kanegisa, N. 540 (119), *587*
Kanehisa, N. 51 (167), *96*, 121 (74), *128*, 254 (126), *263*
Kanemasa, S. 452 (35), 474 (96), *507, 508*
Kang, S.-K. 557 (215), *590*
Kanno, K.-I. 554 (192), *589*
Kanoufi, F. 513 (15), *584*
Kao, K.-H. 566 (262), *591*
Kaplan, F. 138 (22), *153*
Kapoor, P. N. 428 (86), *436*
Kapoor, R. N. 428 (86), *436*
Kapur, G. N. 395 (4), *401*
Kar, A. 557 (209), *590*
Karaghiosoff, K. 27 (99), *94*
Karakyriakos, E. 159 (35), *185*
Karam, J. D. 352 (294), *365*
Karbstein, K. 341 (188), *363*
Karen, P. 107 (18, 19), *126*
Karen, V. L. 107 (18), *126*
Karfunkel, H. R. 675 (155), *679*
Kargin, Yu. M. 219 (7), *260*
Karlson, L. 304 (351), *313*
Karlsson, R. 347 (232), *364*
Karlström, A. S. E. 555 (201, 204), *590*, 558 (218), *590*, 791 (53), *802*
Karlström, E. 544 (131), *588*
Karnachi, T. 214 (137), *218*
Karpeisky, A. 340 (187), *363*
Karplus, M. 341 (203), *363*
Karpov, A. S. 547 (148), *588*
Karppanen, H. 268 (21), 270 (155), *306, 309*
Karr, P. A. 201 (41, 42), *216*
Karras, M. 640 (25, 26), *677*
Kasai, N. 51 (167), 60 (194–196), 61 (197), *96, 97*, 121 (74), *128*, 254 (126), *263*, 540 (119), *587*
Kasai, Y. 341 (199, 201), *363*
Kaschube, W. 86 (266), *99*
Kase, K. 791 (60), *802*
Kashimura, S. 254 (121), *263*
Kasho, V. N. 329 (101), *361*
Kasparov, V. V. 183 (153), 184 (155), *187*
Kasper, B. 213 (123), *217*
Kasting, J. F. 355 (13), *359*
Katagiri, T. 389 (3), *401*
Katayama, Y. 193 (23), *215*
Kathuria, P. 268 (44), *306*
Kato, S. 198 (64), *216*, 531 (90), *586*, 699 (31), *715*, 791 (51), *802*
Kato, T. 791 (60), *802*

Kato, Y. 675 (158), *679*, 699 (31), *715*
Kato-Yamada, Y. 329 (99), *361*
Katritzky, A. R. 547 (156–158), 559 (232), 572 (288, 289), *588*, *590*, *592*, 729 (26), *768*
Kattnig, E. 626 (96), *630*
Katz, A. K. 324 (35), *360*
Katz, J. J. 211 (105), *217*
Kauffman, S. L. 289 (271), *312*
Kaul, F. A. R. 519 (53), *585*
Kaupp, M. 79 (258), *98*
Kawabata, T. 480 (92), *508*
Kawai, R. 348 (243), *364*
Kawai, S. 327 (78), *361*
Kawamura, S.-I. 452 (37), *507*
Kawasaki, M. 699 (31), *715*
Kawashima, T. 766 (53), *769*
Kazakov, V. P. 235 (51), *261*
Kazantsev, A. V. 336 (157), *362*
Kazmaier, U. 487 (19, 23), *506*, *507*
Kazuta, K. 556 (207), *590*
Kealy, T. J. 25 (84), *94*
Kebarle, P. 175 (115, 123), *186*, *187*
Kehres, D. G. 325 (3), *359*
Kehrli, S. 565 (257), *591*, 775 (31), *801*
Keith, T. 384 (30), *402*
Keizer, S. P. 193 (19), *215*
Kellog, M. S. 499 (117), *509*
Kelly, J. 293 (267), *311*
Kemp, B. E. 347 (230), *364*
Kennard, C. H. L. 26 (92), *94*
Kennedy, A. L. 9 (52), *93*
Kennedy, A. R. 13 (59), 15 (62), 17 (64–66), 18 (67), 19 (68, 69), 20 (70), 22 (71, 72, 74), 69 (222), 73 (73), 81 (224, 259), *93*, *97*, *98*, 150 (41), *154*, 414 (35, 46), 419 (50), 422 (59), 424 (61), 425 (70), 429 (76), 430 (80), 431 (82), *435*, *436*, 460 (56), 469 (79), *507*, *508*
Kephart, J. C. 214 (125), *218*
Kern, D. 344 (212), *364*
Kern, J. 356 (324), *366*
Kerr, W. J. 424 (63–67), *435*, *436*, 473 (85–89), *508*, 539 (110, 112), *587*
Kerwin, S. M. 478 (65), *508*
Kessar, S. V. 540 (115), *587*, 729 (26), *768*
Keyes, W. R. 288 (254), *311*
Khairallah, G. N. 181 (132), *187*
Khalilov, A. M. 671 (132), *679*
Khalilov, L. M. 123 (87), *129*
Kham, K. 258 (130), *263*, 291 (279), *312*
Khan, M. A. 81 (81), *94*
Khan, M. S. 139 (26), 150 (45), *153*, *154*
Khan, S. I. 86 (268), *99*
Kharasch, M. S. 288 (146), *308*, 389 (3), *401*, 512 (3), *583*, 595 (1, 3, 8), 599 (5), 604 (7), 624 (4), 625 (2, 6), *628*, 653 (67), *677*
Khawaja, J. A. 268 (21), 270 (155), *306*, *309*
Khetrapala, C. L. 286 (247), *311*

Kholmogorov, V. E. 196 (51), *216*
Kholmogorov, V. Ye. 198 (59), *216*
Khutoreskaya, G. 327 (67), *360*
Khvoinova, N. M. 426 (48), *435*
Khvorova, A. 341 (192, 194), *363*
Kibune, N. 302 (326), *313*
Kiefer, W. 357 (337), *366*
Kienle, M. 576 (301), 580 (314), *592*, *593*
Kii, S. 524 (63), *585*, 702 (39), *715*
Kikuchi, M. 563 (248), *591*
Kikuchi, W. 549 (162), *588*
Kilbourn, B. T. 30 (104), *94*
Kilic, E. 278 (215), *310*
Kiliç, Z. 277 (206), *310*
Kiljunen, H. 288 (263), *311*
Kim, C.-B. 553 (178), *589*
Kim, C. G. 70 (226), *97*, 432 (79), *436*
Kim, D. 196 (53), *216*
Kim, E. E. 354 (309), *366*
Kim, J. 70 (226), *97*, 432 (79), *436*
Kim, J.-H. 270 (159), *309*
Kim, K. M. 268 (13), *306*
Kim, T.-S. 667 (120), *678*
Kim, Y. 70 (226), *97*, 432 (79), *436*
Kimball, D. B. 531 (93), *586*
Kimura, E. 675 (157), *679*
Kimura, M. 270 (157), *309*, 395 (38), *402*, 558 (218), *590*
Kimura, T. 622 (88), *630*, 637 (23), *677*
Kina, A. 554 (194), *589*
King, A. O. 293 (266), *311*
King, B. A. 11 (56), *93*, 686 (17), *714*
King, W. A. 359 (351), *367*
Kinosita, K. 329 (96, 100), *361*
Kinter, M. 303 (350), 304 (354), *313*, *314*
Kintopf, S. 123 (88), *129*
Kinugasa, H. 254 (121), *263*
Kinzelmann, H.-G. 44 (152), *95*
Kirby, S. P. 116 (1), *126*
Kirchner, F. 578 (307), *592*
Kirmse, W. 718 (2), *767*
Kirsch, S. F. 798 (66), *802*
Kirschleger, B. 288 (262), *311*, 516 (31), *584*
Kisanga, P. 472 (94), *508*
Kise, H. 193 (25), 194 (29), 213 (119), *215*, *217*
Kishimoto, N. 193 (27), *215*
Kisker, C. 351 (280), *365*
Kisko, J. L. 73 (239), *98*, 424 (10), *434*
Kitagawa, K. 5 (42), 41 (41), *93*, 523 (59), *585*, 699 (11), *714*
Kitagawa, T. 259 (138), *263*
Kitamura, M. 579 (309), *592*
Kitamura, T. 453 (38, 39), *507*
Kitamura, Y. 285 (237), *311*
Kitazume, T. 558 (218), *590*
Kitching, W. 244 (89), *262*
Kitoh, Y. 479 (16), 485 (15, 17), *506*

Kitsunai, T. 285 (243), *311*
Kjekshus, Å. 107 (18), *126*
Klabunde, K. J. 171 (24, 25), *185*
Klabunde, T. 322 (49), *360*
Klages, F. 578 (307), *592*
Klassen, B. 772 (7), *800*
Kleiber, P. D. 162 (59, 60), 163 (62), 164 (63), 169 (88–91), 171 (92, 105), *185*, *186*
Kleiger, S. C. 595 (1), *628*
Kleimeier, J. 68 (221), *97*
Klein, D. J. 334 (141), *362*
Klein, G. L. 268 (134), *308*
Klein, T. E. 270 (160), *309*
Klein, W. R. 514 (19), *584*
Kleinrock, N. S. 159 (38), *185*
Klenow, H. 351 (287), *365*
Klix, R. C. 559 (227), *590*
Kloth, M. 64 (209), *97*
Klug, A. 340 (186), *363*
Kluge, S. 332 (14), 350 (268), *359*, *365*
Kluger, R. 331 (113), *361*
Klukas, O. 356 (36), *360*
Knauf, W. 537 (104), *586*
Kneisel, F. F. 268 (8), *306*, 518 (42), 534 (96), *585*, *586*
Knighton, D. R. 347 (232), *364*
Knizek, J. 27 (99), *94*
Knochel, P. 268 (7, 8, 10, 11), 291 (276, 277), *306*, *312*, 517 (35), 518 (36, 42–44, 46, 47), 519 (51), 520 (55), 524 (61, 62), 525 (41, 66–69), 526 (72), 527 (28), 528 (83, 84, 87), 529 (88), 531 (57, 89, 92), 533 (95), 534 (81, 96-99, 100), 539 (113), 540 (114, 116), 543 (125), 545 (138), 548 (160), 550 (126), 553 (186), 555 (198), 557 (208), 559 (223), 563 (245), 566 (266), 569 (277), 576 (301, 304), 577 (305, 306), 578 (308), 580 (314), *584–593*, 601 (44), 607 (47), 619 (45), *629*, 633 (1), *676*, 690 (29, 30), 714 (7, 8), *714*, *715*, 738 (33), 739 (34), 748 (40), 767 (58), *768*, *769*, 774 (23), 775 (28), 791 (54), *800–802*
Knochen, M. 278 (213), *310*
Knoers, N. V. 268 (65), *307*
Knoers, N. V. A. M. 268 (66), *307*
Knopff, O. 527 (75, 78, 79), *586*, 661 (104), *678*, 724 (17), 725 (20), 734 (30), *768*
Knotter, D. M. 774 (25), *800*
Knox, J. E. 47 (161), *96*, 122 (79), *128*
Ko, B.-T. 16 (63), *93*
Ko, K-Y. 395 (4), *401*
Ko, Y. H. 330 (103), *361*
Kobayashi, F. 201 (76), *216*
Kobayashi, H. 474 (96), *508*
Kobayashi, K. 107 (21), *126*, 453 (38, 39), 466 (71), 468 (80), 487 (67, 68, 72, 104), *507–509*

Kobayashi, M. 193 (24, 25), 194 (29), 213 (119), *215*, *217*
Kobayashi, S. 767 (56), *769*
Kobayashi, T. 198 (64), *216*
Kobayashi, Y. 480 (11, 12), *506*, 558 (217), *590*
Kobe, K. A. 231 (45), *261*
Kober, R. 517 (39), *585*
Kobetz, P. 222 (26), *260*
Kobrich, G. 718 (2), *767*
Kobyshev, G. I. 198 (65), *216*
Koch, E-C. 103 (5), *126*
Kochi, J. K. 258 (133), *263*, 596 (11, 17), 604 (9, 10, 13–15), 615 (12, 16), *628*, *629*
Kocienski, P. 666 (116), *678*
Koebrich, G. 576 (303), *592*
Kofink, C. 531 (89), *586*
Kofink, C. C. 545 (138), *588*
Koga, N. 387 (33), *402*
Kogure, T. 772 (3), *800*
Kohl, A. 325 (70), *361*
Kohler, E. P. 299 (309, 310), *312*
Koide, N. 285 (243), *311*
Koizumi, M. 339 (185), *363*
Kojima, C. 341 (199), *363*
Kojima, K. 157 (21), *185*
Kok, G. L. 219 (5), *260*
Kok, W. T. 275 (191), *310*
Kolisek, M. 325 (66, 68), *360*
Kolodner, R. D. 351 (282), *365*
Kolonits, M. 183 (151), *187*
Kolotuchin, S. V. 566 (264), *591*
Kolthoff, I. M. 300 (315), *313*
Komander, D. 347 (226), *364*
Komaromi, I. 384 (30), *402*
Komatsu, H. 285 (237), *311*
Komatsu, K. 125 (100), *129*
Komiya, N. 419 (57), *435*, 538 (105), *587*
Kondo, A. 527 (74), *586*, 727 (18), 728 (25), 730 (27), *768*
Kondo, J. 528 (77, 83), *586*, 703 (43), 706 (42, 44), *715*
Kondo, K. 556 (207), *590*
Kondo, Y. 341 (199), *363*, 404 (2), *434*, 538 (105, 108), *587*, 713 (55), *715*
Kondyrew, N. W. 224 (31, 32), 244 (87), 246 (13, 30), *260–262*
Kongsaeree, P. 725 (19), *768*
Konigsberg, W. 352 (301), *365*
Konigsberg, W. H. 352 (294), *365*
Konishi, H. 453 (38, 39), 466 (71), 468 (80), 487 (72), *507*, *508*
Konrad, M. 268 (67), *307*
Konstantinović, S. 540 (118), *587*
Koo, W. W. 268 (70), *307*
Kooijman, H. 40 (140), 66 (215), 86 (269), *95*, *97*, *99*
Koon-Church, S. E. 259 (146), *263*

Koop, U. 505 (26), *507*
Koops, R. W. 559 (227), *590*
Kooriyama, Y. 486 (25), *507*
Kopecky, K. R. 395 (5), *401*
Kopelman, R. 285 (156), *309*
Kopf, J. 6 (48), 38 (122), 42 (49), *93*, *95*, 682 (10), *714*
Kopnyshev, S. B. 121 (70), *128*
Kopp, F. 268 (8), 291 (277), *306*, *312*, 525 (41), 528 (84), 531 (57), 555 (198), 569 (277), 576 (304), *585*, *586*, *589*, *591*, *592*, 607 (47), *629*, 739 (34), *768*
Koppel, H. 268 (26), *306*, 332 (131), *362*
Koppetsch, G. 33 (114, 115), *94*
Koradin, C. 548 (160), *588*
Korn, T. 268 (8), *306*, 518 (42), *585*
Korn, T. J. 559 (223), *590*, 601 (44), *629*
Kornberg, A. 327 (76), *361*
Kornberg, R. D. 355 (313), *366*
Korneeva, V. S. 352 (301), *365*
Korotkov, V. I. 198 (59), *216*
Korpics, C. J. 398 (39), *402*
Korppi-Tommola, J. 357 (334), *366*
Kosar, W. 258 (131), *263*
Koschatzky, K. H. 537 (104), *586*
Koss, A.-M. 673 (149), *679*
Kossa, W. C. 654 (80), *678*
Kost, G. J. 276 (15), *306*
Kostas, I. D. 66 (215), *97*
Koth, C. M. 327 (67), *360*
Kotoku, M. 556 (207), *590*
Koudsi, Y. 458 (50), *507*
Kouznetsov, D. A. 330 (106, 107), *361*
Kowalewska, A. 32 (110), *94*
Koyama, M. 526 (70), *586*, 746 (39), *769*
Koyanagi, G. K. 161 (55), *185*
Kozhushkov, S. I. 582 (315), *593*
Krafczyk, R. 53 (175), *96*
Krafft, M. E. 473 (75), *508*, 597 (33), 598 (34), *629*
Kraft, B. J. 196 (44), *216*
Kramer, J. G. 289 (271), *312*
Krasovskaya, V. 539 (113), *587*
Krasovskiy, A. 291 (276, 277), *312*, 520 (55), 526 (72), 527 (28), 529 (88), 531 (57), 539 (113), 543 (125), 557 (208), 569 (277), 580 (314), *584–587*, *590*, *591*, *593*, 633 (1), *676*, 690 (29), 714 (7, 8), *714*, *715*
Kraszewska, I. 410 (36), *435*
Kraus, K. W. 600 (37), 601 (39), *629*
Krause, H. 554 (189), *589*, 608 (57), 614 (59), 624 (35), *629*
Krause, N. 455 (42), 474 (95), 505 (26), *507*, *508*, 596 (21), *629*, 633 (9), *676*, 774 (16, 17), *800*
Kräutler, B. 210 (101), *217*
Krauß, N. 356 (36, 322, 324, 330), *360*, *366*
Kravchuk, A. V. 338 (176), *363*

Krebs, B. 322 (49), *360*
Krebs, E. G. 347 (230), *364*
Krejčová, A. 279 (217), *310*
Krisanov, R. S. 65 (211), *97*
Krishnamurthy, D. 521 (58), 583 (319), *585*, *593*
Krishnan, R. 369 (1), *401*
Krishnan, S. 707 (48), *715*
Kristensen, J. 518 (49), *585*
Krivan, V. 287 (253), *311*
Krogh, S. C. 283 (234), *311*
Krogsgaard-Larsen, P. 370 (113), *509*, 519 (50), *585*
Kroll, J. 301 (357), *314*
Kromann, H. 519 (50), *585*
Kromov, V. I. 212 (117), *217*
Kropf, H. 213 (123), *217*
Kroth, H. J. 26 (96), *94*
Krow, G. R. 55 (178), *96*, 225 (40), *261*, 760 (47), *769*
Krueger, C. 123 (80), *128*
Kruger, C. 152 (51), *154*
Krüger, C. 33 (114, 115), 45 (156, 157), 47 (159), 67 (93), 71 (227), 86 (266), *94*, *96*, *97*, *99*
Kruger, K. 335 (149), *362*
Krumpe, K. E. 718 (2), *768*
Krut'ko, D. P. 65 (211), *97*
Krysan, D. J. 497 (60), *507*
Krywult, B. 495 (111), *509*
Krzesinski, J. M. 268 (138), *308*
Kuang, T. 356 (326), *366*
Kubišta, J. 86 (270), 89 (273), *99*
Kubo, T. 701 (36), *715*
Kubota, K. 566 (265), *591*
Kubota, T. 285 (237), *311*, 558 (218), *590*
Kucharski, L. M. 325 (74, 75), *361*
Kudin, K. N. 384 (30), *402*
Kühlbrandt, W. 356 (328), *366*
Kuhn, A. 57 (186), *96*, 124 (91), *129*
Kuhn, N. 71 (230), *98*, 405 (12), *434*
Kuimelis, R. G. 337 (159), *362*
Kukovinets, A. G. 671 (132), *679*
Kulinkovich, O. 582 (315), *593*
Kulinkovich, O. G. 268 (6), *306*, 582 (315), *593*
Kulpmann, W. R. 268 (30), *306*
Kumada, M. 661 (101, 107), 662 (106), *678*
Kumamaru, T. 303 (349), 304 (352, 353), *313*, *314*
Kumar, S. N. 214 (137), *218*
Kumazawa, K. 484 (69), *508*
Kundu, K. 23 (78), *94*
Kundu, S. 201 (70), *216*
Kuniyasu, H. 544 (127, 128), 545 (135, 136), 561 (239), *588*, *591*
Kunkel, T. A. 351 (273), 355 (315), *365*, *366*
Küntzel, H. 237 (62), *261*

Kunz, W. 322 (47), *360*
Kupfer, V. 88 (272), *99*
Kurihara, T. 737 (32), *768*
Kuriyan, J. 347 (227), *364*
Kuroda, A. 480 (92), *508*
Kurti, L. 742 (14), *768*
Kusche, A. 527 (78), 534 (81), *586*, 719 (4), 724 (17), *768*
Kushlan, D. M. 682 (5), *714*
Kuwabara, T. 341 (189), *363*
Kuwahara, Y. 206 (81), *217*
Kuwata, K. 212 (114), *217*
Kuyper, J. 291 (278), *312*
Kuznetsov, D. A. 331 (104), *361*
Kuznetsova, N. A. 213 (120, 122), *217*
Kuznetsova, S. V. 299 (306), *312*
Kwan, C. L. 604 (15), *629*
Kyler, K. S. 528 (85), *586*
Laaziri, H. 534 (81), *586*
Labbauf, A. 120 (65), *128*
Labeeuw, M. 268 (114), *308*
Labeeuw, O. 559 (224), *590*
LaBelle, J. T. 356 (327), *366*
Ladbury, J. E. 300 (317), *313*
Ladlow, M. 711 (49), *715*
Laemmle, J. 108 (32), *127*, 389 (3), *401*, 683 (16), *714*
Lagarden, M. 610 (60), *630*
Lahiri, S. 341 (198), *363*
Lai, J.-J. 208 (97), *217*
Lajer, H. 268 (141), *308*
Lajeunesse, E. 344 (215), *364*
Lakhdar, Z. B. 357 (335), *366*
Lakowicz, J. R. 283 (235), *311*
Lamarche, B. J. 352 (290), *365*
Lamberet, G. 301 (355), *314*
Lambert, D. 341 (195), *363*
Lambert, F. 774 (8), *800*
Lambert, R. L. 517 (34), *584*, 719 (3), *768*
Lamm, G. 320 (32), 323 (52), *360*
Lammi, R. K. 201 (39, 45), *215, 216*
Lamoure, C. 344 (215), *364*
Lan, X. 547 (156), *588*
Lander, E. S. 347 (223), *364*
Lander, P. A. 563 (251), *591*
Landert, H. 775 (27), *801*
Lang, F. 699 (31), *715*
Lang, K. 193 (20), *215*
Lang, S. 520 (56), *585*, 682 (9), *714*
Lange, G. 4 (31), *92*, 686 (1), *714*
Lanzisera, D. V. 157 (20), 159 (34), *184, 185*
Lappert, M. F. 24 (80), 30 (104), 31 (109), 32 (111), 37 (121), 40 (139), 62 (120, 201), *94, 95, 97*, 305 (305), *312*, 416 (49), *435*
Lapsker, I. 211 (110), *217*
Larchevêque, M. 487 (22), *506*
Largo, A. 106 (16), *126*
Larhed, M. 513 (18), *584*

Larsen, A. O. 791 (59), *802*
Larsen, C. D. 328 (86), *361*
Larsen, R. D. 293 (266), *311*
Larsen, T. M. 349 (252, 255, 259), *364, 365*
Larson, G. L. 457 (48), *507*
Laszlo, P. 86 (86), *94*
Latham, R. A. 156 (3), *184*
Latham, R. L. 137 (7), *153*
Lau, S. Y. W. 5 (43), *93*
Laurent, A. 647 (42), *677*
Laurich, D. 554 (189), *589*, 626 (95), *630*
Lautens, M. 658 (94), *678*
Laux, M. 505 (26), *507*
Lavilla, I. 272 (170), *309*
Law, M. C. 515 (26), *584*
Lawler, R. G. 138 (20), *153*
Lawrence, L. M. 513 (14), *584*
Lay, J. O. 176 (125), *187*
Laye, P. G. 109 (47), *127*
Layfield, R. A. 51 (51), *93*
Layh, M. 37 (121), *95*
Lazzari, D. 580 (313), *592, 593*
Le Du, M. H. 344 (215), *364*
Le Paith, J. 596 (30), *629*
Lea, L. 558 (219), *590*, 791 (50, 57), *802*
Leach, C. S. 268 (93), *307*
Leader, G. R. 300 (316), *313*
Leal-Granadillo, I. A. 303 (346), *313*
Leazer, J. L. 103 (8), *126*, 519 (52), *585*
Lebioda, L. 349 (251, 256–258), *364, 365*
Lebret, B. 711 (50), *715*
Leclerc, F. 341 (203), *363*
Leclerc, M. 285 (238, 239, 241, 242), *311*
Lecomte, F. 328 (82), *361*
LeCours, S. M. 198 (56), *216*
Lee, C. 384 (28), *402*
Lee, C.-F. 559 (230), *590*
Lee, C. H. 235 (18), *260*, 466 (66), *508*, 538 (105), *587*
Lee, C.-H. 419 (54), *435*
Lee, C.-P. 278 (211), *310*
Lee, F. H. 228 (16), 235 (18), 238 (17), *260*
Lee, G.-H. 81 (228), *98*, 149 (46), *154*, 422 (25), *435*, 547 (154), *588*
Lee, H. 103 (9), *126*, 517 (37), *584*
Lee, J. 514 (20), *584*, 654 (72), *678*, 699 (31), *715*
Lee, J.-B. 278 (209, 212), *310*
Lee, J. I. 166 (76), *186*
Lee, J. Y. 355 (312), *366*
Lee, K.-S. 268 (135), *308*
Lee, K.S. 775 (30), *801*
Lee, Y. 791 (61), *802*
Lefrancois, M. 173 (111), *186*
Legros, J. 596 (30), *629*
Lehmann, C. W. 624 (35), *629*
Lehmkuhl, H. 45 (157), 71 (227), *96, 97*, 123 (80, 86, 88), *128, 129*, 143 (37), *154*, 221

(24, 25), 222 (27), 233 (2), *260*, 653 (62, 66), 655 (59–61, 63–65, 81), 656 (87), *677, 678*
Lehmkul, H. 151 (15), 152 (51), *153, 154*
Lehner, R. S. 499 (115), *509*
Leibfritz, D. 136 (8), *153*
Leiboda, L. 320 (26), 349 (248, 254), *360, 364*
Leier, C. V. 268 (105), *308*
Leighton, J. L. 779 (20), *800*
Leinweber, C. M. 289 (271), *312*
Leising, F. 69 (223), *97*
Leitner, A. 554 (189), *589*, 614 (58, 59), 626 (94), *629, 630*
Lemaire, M. 596 (26), *629*
Lentz, D. 26 (96), *94*
Leontis, N. B. 335 (148), *362*
Leotta, G. J. 563 (247), *591*
Lepage, O. 626 (96), *630*
Lepifre, F. 552 (176), *589*
Leprêtre, A. 518 (43), 534 (81), *585, 586*
Leroi, G. E. 201 (73), *216*
Lesar, A. 125 (104), *129*
Lescoute, A. 341 (192), *363*
Lesley, S. A. 325 (70), *361*
Leslie, A. G. W. 329 (94, 102), *361*
Leśniewska, B. 273 (184), *309*
Lessène, G. 473 (74), *508*
Lesseue, G. 539 (111), *587*
Leu, W. 791 (59), *802*
Leung, M.-K. 547 (155), *588*
Leung, W.-P. 24 (80), *94*
Leupold, D. 196 (54), *216*
Leuser, H. 633 (1), *676*
Levanon, H. 201 (71), 206 (78), *216, 217*
Lever, O. W. 760 (46), *769*
Levesque, I. 285 (242), *311*
Levi, E. 247 (106), 248 (96, 113), *262*
Levi, M. D. 253 (117), *262*
Lewandowski, W. 277 (203, 204), *310*
Lewenstam, A. 268 (21), 270 (155), *306, 309*
Lewiński, J. 25 (82), *94*, 299 (307), *312*, 410 (36), *435*
Lewinski, K. 320 (26), *360*
Lewis, L. D. 268 (112), *308*
Lewis, N. S. 243 (84), 259 (143), *262, 263*
Lewis, R. 531 (92), *586*
Lex, J. 517 (40), *585*
L'Hostis-Kervella, I. 462 (59), *507*
Li, F. 201 (46), *216*
Li, G. 455 (43), *507*
Li, G. Y. 550 (171, 174), *589*
Li, L. 107 (20), *126*, 272 (176), *309*
Li, N. 395 (35), *402*
Li, P. 272 (169), *309*
Li, W. 484 (102), *509*
Li, X. 293 (285), *312*
Li, Y. 272 (169), *309*
Li, Z. 62 (120), *95*

Liang, B. 787 (44), *801*
Liang, Y.-J. 278 (212), *310*
Liao, J. C. 329 (90), *361*
Liao, L.-a. 659 (98, 99), *678*
Liao, S. 44 (113, 153), 45 (154), *94, 95*
Liashenko, A. 384 (30), *402*
Licheri, G. 318 (17), *359*
Lichtenwalter, M. 619 (75), *630*
Licini, G. 500 (55), *507*
Liddle, S. T. 74 (124), 76 (249), *95, 98*, 150 (43), *154*, 422 (20), 423 (17), *434*
Liebenow, C. 247 (35, 102–105), *261, 262*, 305 (365), *314*
Liebman, J. F. 107 (20), 110 (49), 111 (51), 113 (55, 56), *126–128*
Liebscher, D.-E. 268 (90), *307*
Liebscher, D.-H. 268 (90), *307*
Lieff, M. 300 (313), *313*
Lilley, D. M. 340 (183), *363*
Lilley, D. M. J. 341 (194), *363*
Lim, C. 320 (28), 321 (37–39), 322 (43, 44), *360*
Lim, W. A. 347 (229), *364*
Lima, C. D. 727 (23), *768*
Limbach, M. 452 (36), *507*
Lin, C.-C. 16 (63), 81 (228), *93, 98*, 149 (46), *154*, 422 (25), *435*
Lin, L. 272 (174), *309*
Lin, M.-C. 278 (209, 211, 212), *310*
Lin, N. 337 (169), *363*
Lin, P. 352 (299), *365*
Lin, T. 272 (176), *309*
Lin, W. 540 (114, 116), 566 (266), *587, 591*
Lin, W. W. 774 (23), *800*
Lin, W.-W. 566 (262), *591*
Lin, X. 301 (360), *314*
Lin, Y. L. 321 (39), *360*
Lin, Y.-M. 566 (262), *591*
Linares-Palomino, P. J. 440 (7), *506*
Lindahl, T. 353 (302), *365*
Lindqvist, O. 271 (165), *309*
Lindqvist, Y. 359 (355), *367*
Lindsay, D. M. 518 (36), 525 (41), 531 (89), *584–586*
Lindsay Smith, J. R. 200 (67), *216*
Lindsell, W. E. 58 (3, 4), *92*, 512 (4), 513 (13), *583*
Lindsey, J. S. 201 (38, 39, 45, 46), *215, 216*
Linnanto, J. 357 (334), *366*
Linsk, J. 239 (73, 74), 240 (42, 75), *261*
Linstrom, P. J. 116 (3), *126*
Lion, C. 666 (114), *678*
Liotta, D. C. 667 (117–119), *678*
Liparini, A. 270 (152), *309*
Lipkowski, J. 25 (82), *94*, 299 (307), *312*
Lippard, S. J. 322 (41), *360*, 772 (7), *800*
Lipscomb, W. N. 53 (174), *96*, 322 (49), *360*

Lipshutz, B. H. 142 (30), *154*, 512 (2), 554 (193), *583*, *589*, 596 (20), *629*, 779 (33), *801*
Little, B. F. 4 (37), *92*
Little, R. D. 253 (124), *263*
Liu, C. 282 (229), *310*
Liu, C.-Y. 533 (95), 534 (97), *586*
Liu, D.-K. 162 (60), *185*
Liu, G. 293 (285), *312*, 384 (30), *402*, 575 (298, 299), *592*
Liu, H. 163 (68, 69), 166 (77), 167 (82, 83, 85), 168 (73, 86), 169 (78–80, 87, 93–96), 170 (97, 98), *185*, *186*, 336 (160), 350 (262), *362*, *365*
Liu, H.-C. 163 (70, 71), 168 (72, 74), *186*
Liu, H.-J. 572 (276), *591*
Liu, H.-W. 278 (209, 211, 212), *310*
Liu, J.-Y. 566 (262), *591*
Liu, L. 272 (169), *309*, 391 (19), *402*, 644 (33), *677*
Liu, Q. 159 (36), *185*
Liu, T.-J. 566 (262), *591*
Liu, W. 528 (87), *586*
Liu, X. 560 (233), *590*
Liu, X-K. 463 (63), *508*
Liu, Y. 268 (72), 302 (337), 304 (331), *307*, *313*
Liu, Z. 356 (326), *366*
Livshin, S. 642 (29), *677*
Llewellyn, C. A. 211 (109), *217*
Lloyd, C. 574 (294), *592*
Lloyd-Jones, G. C. 556 (206), *590*
Loader, C. 493 (107), *509*
Łobiński, R. 269 (151), 302 (341), *309*, *313*
Lobitz, P. 247 (35, 103), *261*, *262*
LoBrutto, R. 291 (280), *312*
Locos, O. B. 212 (113), *217*
Loeb, L. A. 332 (128), 351 (276, 277), *362*, *365*
Loewe, R. S. 286 (244), *311*
Loh, T. 565 (256), *591*
Loh, T.-P. 782 (39), *801*
Lohrenz, J. C. W. 516 (32), *584*
London, R. E. 268 (25), *306*
Lone, S. 352 (293), *365*
Long, A. M. 352 (298), *365*
Lonnerdal, B. 268 (124), *308*
Looney, A. 73 (235), *98*, 419 (6), *434*
Loos, D. 68 (218), *97*
Lopez-Avila, V. 302 (337), 304 (331), *313*
López, C. 271 (168), *309*
López, F. 457 (44, 45), *507*, 563 (244), 564 (255), *591*, 779 (32), 780 (38), 787 (41), 788 (48), 791 (62), 798 (63), *801*, *802*
López-Mahía, P. 271 (164), *309*
Lopez, X. 322 (45), *360*
Lorimer, G. H. 357 (344, 345), 358 (352), 359 (339), *366*, *367*

Loroño-González, D. 74 (124), *95*, 150 (43), *154*, 414 (45), 423 (17), *434*, *435*
Lou, B-S. 455 (43), *507*
Love, B. E. 290 (275), *312*
Love, C. 666 (116), *678*
Loveland, L. L. 349 (251), *364*
Lovell, T. 341 (202), *363*
Lovey, R. G. 574 (296), *592*
Lowe, E. D. 348 (239), *364*
Lowe, G. 348 (239), *364*
Lowe, R. 137 (10), *153*, 720 (7), *768*
Lowther, W. T. 343 (209), *364*
Lu, B. Z. 395 (35), *402*
Lu, G. 277 (202), *310*
Lu, K.-J. 52 (173), 53 (174), *96*
Lu, T. 318 (21), *359*
Lu, W. Y. 169 (88–91), *186*
Lu, X. 162 (61), 167 (83, 84), *185*, *186*
Lu, Z. 248 (96, 107, 113), *262*, 567 (271), *591*, 648 (51, 52), *677*
Lu, Z.-H. 583 (319), *593*
Lu, Z.-R. 32 (110), *94*
Lubbe, W. J. 325 (75), *361*
Lubell, W. D. 563 (246), *591*
Lubin, C. 495 (111), *509*
Lucast, L. 334 (140), *362*
Luchaco-Cullis, C. A. 791 (55), *802*
Lüdtke, D. S. 774 (25), *801*
Luftmann, H. 301 (359), *314*
Luh, T.-Y. 547 (152–154), 559 (230), *588*, *590*
Lui, K. 246 (91), *262*
Lui, X. 659 (100), *678*
Lukaski, H. C. 268 (57), *307*
Lukyanets, E. A. 213 (120, 122), *217*
Lukzen, N. N. 331 (111), *361*
Luna, A. 125 (101), *129*, 182 (138), *187*
Lund, H. 219 (5, 8, 9), 238 (6), 239 (78), 253 (122, 123), 256 (28), 259 (136, 137), *260*, *261*, *263*
Lund, T. 259 (136, 137), *263*
Lung, F-D. 455 (43), *507*
Lunin, V. V. 327 (67), *360*
Luo, Q. 213 (118), *217*
Luo, X. 123 (90), *129*
Lusztyk, J. 26 (91), *94*, 183 (146), *187*
Lutter, R. 329 (94, 102), *361*
Lutz, H. 276 (195), *310*
Lutz, M. 36 (119), *95*
Luzikova, E. V. 553 (187), *589*
Lyalin, G. N. 198 (65), *216*
Lynch, J. 699 (31), *715*
Ma, S. 567 (271), *591*, 648 (51, 52), 658 (94), *677*, *678*
Maass, G. 318 (18), 355 (320), *359*, *366*
Macaev, F. Z. 474 (47), *507*
Macciantelli, D. 259 (144), *263*
MacGregor, M. 413 (42), *435*

Mach, K. 86 (270), 87 (271), 88 (272), 89 (273-275), 99
Mack, J. 193 (15, 18), 215
Mackey, O. N. D. 428 (84), 436
Macklin, T. K. 553 (184), 589
MacMahon, T. J. 162 (57), 185
Madabusi, L. V. 337 (169), 363
Madigan, M. T. 193 (28), 214 (139), 215, 218
Maeda, H. 559 (225), 590
Maekawa, Y. 302 (326), 313
Maercker, A. 654 (77, 78), 678
Magerlein, B. J. 289 (268), 311
Magi, E. 302 (332), 313
Magiera, D. 583 (319), 593
Magnuson, V. R. 83 (262), 98, 417 (4), 434
Maguire, J. A. 51 (170-172), 52 (173), 53 (174), 96
Maguire, M. E. 268 (51, 53), 307, 325 (3, 61, 64, 69, 74, 75), 326 (60), 327 (67, 72), 328 (1), 359-361
Maguire, R. J. 302 (338), 313
Mahiuddin, S. 322 (47), 360
Mahmoud, J. S. 276 (197), 310
Mahrwald, R. 180 (137), 187, 425 (69), 436
Mahuteau-Betzer, F. 554 (191), 589, 626 (76), 630
Maibaum, J. 370 (114), 509
Maier, S. 487 (19), 506
Mair, J. H. 539 (112), 587
Mairanovskii, S. G. 219 (4), 260
Maj-Zurawska, M. 287 (249), 311
Maj-Żurawska, M. 275 (190), 310
Maj-Żurawska, M. 287 (192), 310
Makropoulos, N. 27 (99), 81 (261), 94, 98, 302 (302), 312, 409 (30), 410 (37), 435
Maksimchuk, K. R. 352 (301), 365
Maksimenka, R. 357 (337), 366
Malakauskas, K. K. 268 (118), 308
Malan, C. 534 (81), 558 (219), 586, 590, 767 (58), 769, 791 (50, 57), 802
Malda, H. 791 (56), 802
Malick, D. K. 384 (30), 402
Maligres, P. 699 (31), 715
Mallard, W. G. 116 (3), 126
Mallory, J. 253 (124), 263
Malmgren, S. M. 438 (1), 506
Malon, A. 287 (192, 249), 310, 311
Maloney, L. 340 (187), 363
Malpass, D. B. 682 (15), 714
Manabe, Y. 145 (38), 154
Manas, M. 268 (48), 307
Mandy, K. 538 (109), 587
Manes, T. 345 (220), 364
Mangeney, P. 574 (293), 592, 774 (12), 800
Mann, B. E. 151 (49), 154
Mann, C. K. 219 (1), 260
Manning, G. S. 334 (137), 362
Mano, M. 468 (80), 487 (72), 508

Manojew, D. P. 246 (30), 261
Mansoorabadi, S. O. 349 (252), 364
Mantoura, R. F. C. 211 (109), 217
Manyem, S. 779 (19), 800
Manz, J. 357 (336), 366
Mao, M. K. T. 443 (13), 506
Mao, X. A. 329 (87), 361
Maqanda, W. 213 (121), 217
Marcaccio, M. 259 (144), 263
Marcantoni, E. 566 (263), 591
Marchese, G. 601 (40-42), 629
Marchi-Delapierre, C. 260 (147), 263, 513 (15), 584
Marciniak, B. 190 (2, 3), 214
Marco, J. A. 391 (26), 402
Marder, L. 295 (288), 312
Maree, M. D. 213 (121), 217
Marek, I. 41 (141), 69 (29), 92, 95, 110 (49), 127, 196 (4), 215, 297 (260), 311, 525 (69), 528 (83, 84), 534 (81), 560 (233), 563 (249), 566 (265), 585, 586, 590, 591, 633 (1), 642 (29), 643 (31), 676, 677, 738 (33), 739 (34), 748 (40), 767 (58), 768, 769
Marier, J. R. 268 (122), 308
Marina, M. L. 278 (208), 310
Marini, E. 580 (313), 592
Markarian, A. A. 330 (106, 107), 361
Markell, M. S. 268 (19), 306
Markham, G. D. 319 (25), 324 (35), 360
Markies, P. R. 3 (15), 34 (106, 116), 35 (117, 118), 38 (83), 43 (146), 44 (142), 66 (213, 214, 216), 92, 94, 95, 97, 183 (140), 187, 318 (22), 359
Markó, L. 653 (69-71), 677, 678
Marks, V. 252 (116), 262
Maroni-Barnaud, Y. 458 (50), 478 (97), 507, 508
Maroni, P. 458 (50), 507
Marquez, V. E. 335 (31), 360
Marr, I. L. 302 (334), 313
Marsais, F. 524 (60), 585, 689 (26, 27), 702 (40), 713 (24, 25), 714, 715
Marsch, M. 59 (192), 62 (204), 97, 395 (34), 402, 516 (32), 584
Marschner, C. 38 (130), 95, 305 (305), 312
Marsden, C. J. 183 (150), 187
Marsden, S. P. 722 (13), 768
Marsh, T. 335 (150), 362
Marshal, J. A. 645 (34), 677
Marshall, E. L. 76 (242), 77 (250), 98, 424 (44), 435, 484 (99), 508
Marshall, G. D. 280 (222), 310
Marshall, J. A. 494 (110), 509
Marshall, W. F. 53 (175), 96, 550 (174), 589
Marsischky, G. T. 351 (282), 365
Marsoner, H. J. 268 (34), 306
Martel, A. 500 (118), 509
Marti, T. M. 355 (317), 366

Martick, M. 341 (196), *363*
Martin, D. 565 (257), *591*, 775 (31), *801*
Martin, J. A. 595 (1), *628*
Martin-Landa, I. 302 (334), *313*
Martin, M. R. 727 (24), *768*
Martin, R. 550 (170), 554 (188), *589*, 596 (32), 616 (73), *629*, *630*
Martin, R. L. 369 (2), 384 (30), *401*, *402*
Martin, R. M. 538 (105), *587*
Martin, S. F. 760 (46), *769*
Martin, T. R. 40 (139), *95*
Martina, V. 601 (40), *629*
Martinek, R. G. 269 (149), *308*
Martinez-Isla, A. 282 (232), *311*
Martínez Vidal, J. L. 302 (335), *313*
Martinho Simões, J. A. 111 (51), *127*
Martinot, L. 246 (46, 59), *261*
Marumo, S. 201 (74, 75), *216*
Maruoka, K. 580 (312), *592*
Maruyama, K. 194 (30), 201 (75, 76), *215*, *216*, 389 (3), 391 (10), *401*
Maruyama, T. 285 (243), *311*
Marvi, M. 512 (8), *583*
Maryanoff, C. A. 667 (117–119), *678*
Marzi, E. 702 (38), *715*
Marzilli, T. A. 138 (20), *153*
Masalov, N. 474 (47), *507*
Mase, T. 524 (63), 531 (90), *585*, *586*, 699 (31), 702 (39), *715*
Mash, E. A. 289 (264), *311*
Mashima, H. 550 (169), 555 (200), *589*
Mashima, K. 51 (167), *96*, 121 (74), 123 (73), *128*, 254 (126), *263*, 540 (119), *587*
Masironi, R. 268 (120), *308*
Mason, R. 30 (105), *94*
Mason, R. S. 302 (328), *313*
Mason, S. A. 24 (80), *94*
Massa, W. 62 (204), *97*, 395 (34), *402*
Massey, L. K. 268 (128), *308*
Masson, E. 702 (38), *715*
Mataga, N. 194 (30), 201 (74–76), *215*, *216*
Mateos, A. F. 482 (6), *506*
Mathers, F. C. 246 (90), *262*
Mathiasson, L. 304 (351), *313*
Matile, P. 210 (101), *217*
Matoba, T. 302 (333), *313*
Matsubara, K. 90 (276), *99*
Matsuda, K. 526 (70), *586*, 746 (39), *769*
Matsue, R. 767 (57), *769*
Matsui, S. 485 (20), *506*
Matsui, Y. 701 (36), *715*
Matsukawa, S. 570 (280), *591*
Matsumoto, K. 517 (34), 519 (54), *584*, *585*
Matsumoto, T. 475 (73), *508*
Matsumoto, Y. 419 (51), *435*, 686 (22), *714*
Matsumura, T. 19 (45), *93*, 570 (279), *591*, 684 (19), *714*

Matsuo, K. 259 (139–142), *263*, 546 (142), *588*, 616 (71), 622 (84), *630*, 658 (93), *678*
Matsuo, T. 212 (114), *217*
Matsuo, Y. 123 (73), *128*
Matsuumi, M. 558 (217), *590*
Matsuura, K. 193 (23), *215*
Matsuura, T. 208 (96), *217*
Matsuyama, T. 259 (134), *263*, 391 (11), *402*
Mattalia, J.-M. 260 (147), *263*, 513 (15), *584*
Matte, A. 332 (125), *362*
Matthews, B. W. 343 (209), *364*
Mauduit, M. 565 (257), *591*, 775 (31), *801*
Mauser, H. 359 (351), *367*
Mauskop, A. 268 (19), *306*
Mauzerall, D. 193 (17), 206 (80), 208 (89, 93), *215*, *217*
Mauzerall, D. C. 198 (55), *216*
Mayer, A. 246 (93, 94), *262*
Mayer, P. 409 (30), *435*
Mayer, V. A. 289 (271), *312*
Mayerle, E. A. 240 (42), *261*
Mayo, F. R. 595 (1), *628*
Mayr, H. 557 (208), *590*
Mazur, H. 288 (257, 258), *311*
Mbofung, C. M. F. 283 (162), *309*
McBee, E. T. 516 (29), *584*
McCaffrey, J. G. 157 (19), *184*
McCammon, J. A. 348 (244), *364*
McCarthy, W. J. 323 (55), *360*
McCarty, A. L. 201 (41, 42), *216*
McCord, J. 268 (106), *308*
McCormick, M. 667 (118), *678*
McCoull, W. 575 (299), *592*
McCullough, R. D. 286 (244), *311*
McDermott, G. 324 (58), 356 (325, 331), *360*, *366*
McDonald, J. E. 51 (171, 172), *96*
McDonald, J. H. 395 (4), *401*
McFail-Isom, L. 335 (30), *360*
McGarvey, G. J. 395 (38), *402*
McGeary, C. A. 23 (77, 78), *94*
McKay, D. B. 340 (181), *363*
McKean, D. C. 391 (16), *402*
McKinley, A. J. 159 (35), *185*
McLaughlin, L. W. 337 (159), *362*
McLeod, D. 472 (94), *508*
McMahon, J. P. 575 (300), *592*
McMahon, R. J. 106 (17), *126*
McNamara, J. M. 648 (47), *677*
McTyre, R. B. 268 (107), *308*
Meetsma, A. 774 (13), 780 (38), 788 (48), *800*, *801*
Mehler, K. 45 (157), 71 (227), *96*, *97*, 123 (80, 86, 88), *128*, *129*, 143 (37), 151 (15), 152 (51), *153*, *154*
Mehrotra, R. 268 (44), *306*
Meij, I. C. 268 (65, 66), *307*
Meijer, E. W. 718 (2), *768*

Meiners, A. F. 516 (29), *584*
Meinwald, J. 625 (92), *630*
Mekelburger, H. B. 485 (4), *506*
Meldal, M. 576 (303), *592*
Melnik, M. 3 (26), 25 (25), *92*
Memon, Z. S. 268 (19), *306*
Menard, M. 500 (118), *509*
Méndez, M. 554 (189), 559 (231), *589, 590*, 614 (59), 626 (83), *629, 630*
Mendler, B. 487 (23), *507*
Menefee, A. L. 349 (252), *364*
Ménez, A. 344 (215), *364*
Mennucci, B. 384 (30), *402*
Mentasti, E. 273 (182), *309*
Menunier, L. 404 (3), *434*
Mercero, J. M. 322 (45), *360*
Mergelsberg, I. 574 (296), *592*
Merkley, J. H. 664 (111–113), *678*
Merlevede, W. 268 (37), *306*
Mertinat-Perrier, O. 292 (281), *312*
Mervaala, E. 268 (21), 270 (155), *306, 309*
Mesquita, R. B. R. 280 (225), *310*
Metra, M. 268 (105), *308*
Meulemans, T. M. 474 (47), *507*
Meunier, P. 50 (98), *94*, 123 (83), *129*
Meuzelaar, G. J. 558 (218), *590*, 791 (53), *802*
Meuzelaar, J. 791 (53), *802*
Meyer, D. 550 (172), *589*
Meyer, F. J. 4 (31), *92*, 686 (1), *714*
Meyer, R. 458 (50), *507*
Meyers, A. I. 566 (264), *591*
Meyers, C. H. 289 (269), *312*
Meyers, R. A. 280 (221), *310*
Michael, J. P. 538 (109), *587*
Michaelis, K. 537 (104), *586*
Michrin, N. 211 (110), *217*
Miginiac, P. 657 (89), *678*
Mignani, G. 596 (23), *629*
Mihalios, D. 519 (53), *585*
Miki, H. 527 (73), 540 (122), *586, 587*
Miki, K. 51 (167), 60 (194–196), 61 (197), *96, 97*, 121 (74), *128*, 254 (126), *263*, 540 (119), *587*
Milburn, R. K. 124 (94, 96), 125 (95), *129*, 160 (54), 180 (46, 134, 135), *185, 187*
Milburn, R. R. 553 (183), *589*
Mildvan, A. S. 328 (88), *361*
Miles, W. H. 395 (36), *402*
Millam, J. M. 384 (30), *402*
Millán, J. L. 345 (220), *364*
Millart, H. 268 (33), *306*
Miller, C. G. 325 (69), *361*
Miller, J. 552 (175), *589*
Miller, J. A. 433 (90), *436*, 544 (132), 553 (181), 555 (205), *588–590*
Miller, J. D. 545 (134), *588*
Miller, M. A. 201 (38, 39), *215*
Miller, M. J. 659 (96), *678*

Miller, R. B. 257 (129), *263*
Miller, R. E. 184 (160–162), *187*
Miller, S. A. 25 (85), *94*
Miller, T. A. 123 (84), *129*
Milligan, W. O. 458 (49), *507*
Mills, J. E. 667 (118), *678*
Mimouni, F. 268 (133), *308*
Min, D. B. 211 (111), *217*
Minaard, A. J. 564 (255), *591*
Minasov, G. 335 (145, 146), *362*
Mineno, M. 555 (199), *589*, 608 (56), *629*
Mingardi, A. 564 (253), *591*
Mingos, D. M. P. 633 (1), *676*
Mingos, M. D. P. 58 (5), *92*
Minnaard, A. J. 457 (44, 45), *507*, 563 (244), *591*, 774 (15, 21), 775 (29), 779 (32), 780 (37, 38), 786 (42), 787 (41, 45, 46), 788 (48), 791 (62), 798 (63), *800–802*
Minofar, B. 322 (47), *360*
Minor, W. 349 (258), *365*
Mioskowski, C. 558 (218), *590*
Misaizu, F. 163 (66, 67), *185*
Mishima, M. 466 (70), *508*
Misra, V. K. 334 (139), 336 (166), *362, 363*
Mistry, J. 493 (106), *509*
Mitchell, H. L. 259 (148), *263*, 513 (14), *583*
Miti, N. 343 (210), *364*
Mitra, S. 353 (307), *366*
Mitsudo, K. 664 (108), *678*
Mitsuhashi, S. 475 (73), *508*
Mittendorf, R. 268 (135), *308*
Mityaev, V. I. 212 (117), *217*
Mitzel, F. 196 (43), *216*
Miura, M. 741 (36), *769*
Miura, T. 475 (73), *508*
Miwa, T. 274 (186, 187), *309, 310*, 463 (63), *508*
Miyake, Y. 558 (219), *590*
Miyamoto, T. 569 (274), *591*
Miyashita, K. 480 (18), *506*, 761 (51), *769*
Miyaura, N. 596 (22), *629*, 774 (21), 780 (35), *800, 801*
Mizu, Y. 766 (53), *769*
Mizumoto, A. 212 (116), *217*
Mizutani, H. 779 (33), 791 (55), *801, 802*
Mizutani, K. 557 (216), 560 (236), *590*
Mizuuchi, K. 355 (311), *366*
Mo, X. 357 (333), *366*
Mochida, K. 157 (21), *185*
Mochizuki, H. 285 (243), *311*
Mochizuki, S. 341 (199), *363*
Modena, G. 500 (55), *507*
Moeder, C. W. 293 (266), *311*
Mogano, D. 538 (109), *587*
Moir, J. H. 424 (65–67), *436*, 473 (85, 87, 88), *508*
Mojovic, L. 550 (173), *589*

Mokhallalati, M. K. 574 (292), *592*, 674 (152, 153), *679*
Molander, G. A. 604 (46), *629*
Molina, A. 654 (72), *678*, 699 (31), *715*
Molina, D. M. 325 (70), *361*
Moliner, V. 393 (25), *402*
Mollenhauer, D. 350 (269), *365*
Molnar, J. 183 (150), *187*
Monarca, S. 268 (130), *308*
Moncrieff, D. 57 (186), *96*, 124 (91), *129*
Mongin, F. 44 (44), *93*, 404 (2), *434*, 524 (60), 537 (103), 550 (173), 553 (186), *585, 586, 589*, 689 (26–28), 700 (33, 34), 713 (24, 25), *714, 715*
Monnens, L. A. H. 287 (248), *311*
Monnereau, C. 198 (66), *216*
Montaño, G. A. 356 (327), *366*
Montes de Lopez-Cepero, I. 457 (48), *507*
Montet, D. 301 (355), *314*
Montgomery, J. A. 384 (30), *402*
Montreux, A. 69 (223), *97*
Moody, C. J. 574 (294), *592*
Moody, D. B. 786 (42), *801*
Moore, A. L. 194 (35), *215*
Moore, D. T. 184 (161, 162), *187*
Moore, G. J. 516 (33), *584*
Moore, P. B. 334 (141), 335 (147, 148), 336 (161), *362*
Moore, R. D. 268 (24), *306*
Moore, T. A. 194 (35), 195 (16), *215*
Morabito, R. 302 (323), *313*
Morales-Morales, D. 71 (232), *98*
Morales Sánchez, M. C. 302 (335), *313*
Moreau, P. 516 (29), *584*
Moreda-Piñeiro, J. 271 (164), *309*
Moreno-Torres, R. 271 (168), *309*
Morgat, J. L. 235 (47, 48), *261*
Mori, S. 327 (78), *361*, 791 (49), *801*
Morikawa, O. 453 (38, 39), 466 (71), 468 (80), 487 (72), *507, 508*
Morikawa, S. 767 (57), *769*
Morita, E. H. 341 (199), *363*
Morita, K. 214 (132), *218*
Morita, N. 701 (36), *715*
Moritani, T. 305 (362, 363), *314*
Morken, J. P. 557 (212), *590*, 672 (138, 139), 673 (142–144), *679*
Morlender-Vais, N. 563 (249), *591*
Morley, C. P. 67 (93), *94*, 428 (84), *436*
Mornet, R. 652 (55), 653 (56), *677*
Morokuma, K. 384 (30), 387 (33), *402*, 419 (51), *435*
Morris, J. J. 424 (62), *435*
Morris, M. D. 219 (3, 5), *260*
Morris, M. E. 268 (71), *307*
Morris, M. L. 268 (112), *308*
Morrison, C. A. 76 (249), *98*, 422 (20), *434*
Morrison, J. D. 486 (54), *507*

Morrison, R. T. 599 (5), *628*
Moryganov, A. P. 214 (138), *218*
Moser-Veillon, P. B. 268 (136), *308*
Mosher, H. S. 108 (39), *127*
Moshkovich, M. 247 (108), 248 (96, 107, 113), 251 (112), 258 (109, 110), *262*
Most, K. 74 (245), *98*, 413 (43), *435*
Motellier, S. 275 (189), *310*
Mountokalakis, T. D. 268 (115), *308*
Moyano, N. 211 (103), *217*
Moyeux, A. 546 (143), *588*, 618 (72), *630*
Mozzanti, G. 559 (228), *590*
Mukai, T. 327 (78), *361*
Mukaiyama, T. 484 (28), *507*, 549 (162), *588*, 772 (3), *800*
Mulholland, A. J. 333 (133), *362*
Müller, A. 59 (192), *97*, 516 (32), *584*
Muller, B. H. 344 (215), *364*
Müller, G. 36 (119), *95*
Müller, I. 74 (245), *98*, 413 (43), *435*
Müller, K. M. 337 (169), *363*
Muller, M. 135 (11), *153*
Müller, M. D. 302 (325), *313*
Mullica, D. F. 458 (49), *507*
Mullineaux, C. W. 191 (9), *215*
Mulvey, R. E. 9 (52), 13 (59), 15 (62), 17 (64–66), 18 (67), 19 (68, 69), 20 (70), 22 (71, 72, 74), 38 (126), 39 (39), 40 (40), 69 (222), 73 (73), 81 (224, 259), 83 (263), *93, 95, 97, 98*, 149 (47), 150 (41), *154*, 404 (1, 2), 413 (42), 414 (35, 46), 419 (50), 422 (41, 47), 429 (76), 430 (80), 431 (82), *434–436*, 460 (56), *507*, 682 (6), 713 (54), *714, 715*
Mulvihill, M. J. 659 (96), *678*
Mulzer, J. 538 (107), *587*
Mumford, F. E. 633 (4), *676*
Munch-Petersen, J. 398 (39), *402*
Mundy, B. P. 742 (14), *768*
Muniategui-Lorenzo, S. 271 (164), *309*
Munoz, R. 268 (81), *307*
Muntau, H. 269 (151), *309*
Muntean, F. 124 (97), 125 (105), *129*, 161 (56), *185*
Münzel, H. 287 (253), *311*
Murafuji, T. 701 (36), *715*
Murai, T. 579 (311), *592*
Murakami, K. 568 (273), *591*, 636 (22), *677*
Murakami, Y. 556 (207), *590*
Muralidharan, K. R. 574 (292), *592*, 674 (153), *679*
Murata, K. 327 (78), *361*
Murayama, T. 475 (73), *508*
Murdoch, J. R. 579 (309), *592*
Murov, S. 206 (77), *217*
Murphy, E. 268 (32), *306*
Murphy, K. E. 791 (55), *802*
Murray, J. B. 340 (187), *363*

Murray, P. T. 268 (89), *307*
Murray, R. K. 765 (48), *769*
Murrell, J. N. 175 (122), *187*, 318 (23), *359*
Murso, A. 520 (56), *585*, 682 (9), *714*
Murthy, K. V. S. 272 (173), *309*
Murugesh, M. G. 558 (217), *590*
Musashi, J. 527 (74), *586*, 727 (18), 730 (27), *768*
Musgrave, W. K. R. 516 (30), 517 (34), *584*
Muslukhov, R. R. 123 (87), *129*
Myers, A. G. 559 (229), *590*
Mynott, R. 33 (114, 115), 44 (153), 45 (156), *94–96*
Naasz, R. 774 (15, 16), *800*
Nabeshima, Y. 285 (243), *311*
Nackashi, J. 103 (4), *126*, 146 (40), *154*, 428 (83), *436*
Naef, R. 772 (5), *800*
Nagana Gowda, G. A. 286 (247), *311*
Nagano, T. 546 (143), 547 (146), 554 (191), *588*, *589*, 615 (67), 619 (77), *630*
Nagao, Y. 463 (63), *508*
Nagaoka, Y. 774 (2), *800*
Nagase, H. 517 (34), 519 (54), *584*, *585*
Nagase, S. 391 (23), *402*
Nagashima, H. 675 (157), *679*
Nagashima, K. V. P. 193 (23), *215*
Nagashima, U. 107 (22), *127*
Nagel, U. 39 (131), *95*
Nair, S. K. 547 (157), *588*
Naitoh, Y. 544 (127), *588*
Nakagawa, J. 285 (237), *311*
Nakagawa, Y. 774 (25), *801*
Nakahashi, R. 453 (39), *507*
Nakajama, N. 571 (284), *592*
Nakajima, H. 675 (158), *679*
Nakajima, K. 675 (157), *679*
Nakajima, S. 194 (30), *215*
Nakamura, A. 51 (167), *96*, 121 (74), 123 (73), *128*, 254 (126), *263*, 540 (119), 560 (235), *587*, *590*
Nakamura, E. 387 (33), *402*, 546 (142), 550 (169), 555 (200), 566 (265), *588*, *589*, *591*, 616 (71), 622 (84, 86), *630*, 658 (92, 93), *678*, 791 (49, 51), *801*, *802*
Nakamura, M. 302 (326), *313*, 387 (33), *402*, 526 (71), 546 (142), *586*, *588*, 616 (71), 622 (84, 86), *630*, 658 (92, 93), *678*, 749 (42), 758 (44), *769*
Nakamura, T. 707 (45), *715*
Nakanishi, S. 767 (56), *769*
Nakashima, T. 329 (99), *361*, 487 (72), *508*
Nakata, K. 558 (217), *590*
Nakata, M. 555 (202), *590*, 686 (22), 688 (23), *714*
Namkoong, E.-Y. 557 (215), *590*
Nanayakkara, A. 384 (30), *402*
Napieraj, A. 547 (147), *588*

Narasaka, K. 579 (309), *592*
Narayana, M. 735 (31), *768*
Narula, A. K. 428 (85), *436*
Nasielski, J. 550 (167), *589*
Naso, F. 583 (317, 318), *593*
Nassar, R. 424 (62), *435*
Nasser, D. 282 (233), *311*
Näther, C. 62 (203), *97*
Naumov, V. A. 183 (154), *187*
Navarro, M. 271 (168), *309*
Naylor, R. D. 116 (1), *126*
Ndjouenkeu, R. 283 (162), *309*
Neber, G. 578 (307), *592*
Nedden, G. 39 (131), *95*
Negishi, E. 671 (134), *679*, 787 (44), *801*
Négrel, J.-C. 391 (20), *402*
Negrel, J.-C. 157 (15), *184*
Nell, P. G. 527 (76), *586*, 720 (5), 721 (12), *768*
Nelson, B. 325 (63), *360*
Nelson, J. M. 220 (12), *260*
Nelson, S. G. 504 (90), *508*, 539 (112), 547 (150), *587*, *588*
Nemanick, E. J. 259 (143), *263*
Nemoto, T. 699 (31), *715*
Nemukhin, A. V. 121 (72), *128*, 158 (27), 160 (32), *185*, 268 (147), *308*, 331 (117, 118), *362*
Nemutlu, E. 274 (188), *310*
Nenajdenko, V. G. 547 (148), *588*
Nerín, C. 303 (348), *313*
Nesterova, T. N. 122 (82), *128*
Netherton, N. R. 618 (66), *630*
Netzel, T. L. 198 (57), *216*
Neumann, B. 68 (221), *97*
Neumann, H. M. 108 (32), *127*, 391 (8), *401*, 411 (40), *435*
Neumann, S. M. 596 (11), 604 (13), *628*, *629*
Neves, A. 343 (210), *364*
Neville, M. E. 347 (236), *364*
Newman, J. 357 (346), *366*
Newman, K. 302 (328), *313*
Newmann, H. M. 389 (3), *401*
Nguyen, B. T. 512 (9), *583*
Nguyen, N. 647 (41), *677*
Ni, H.-R. 119 (63), *128*
Ni Mhuircheartaigh, E. M. 212 (113), *217*
Nicaise, O. J.-C. 571 (284, 286), *592*
Nichols, G. 121 (70), *128*
Nicholson, J. 304 (354), *314*
Nicholson, J. K. 286 (246), *311*
Nicoloff, N. 503 (53), *507*
Nie, L.-H. 273 (183), *309*
Niegowski, D. 325 (70), *361*
Nielsen, F. H. 268 (57), *307*
Nielsen, M. F. 253 (123), *263*
Niemczyk, M. P. 194 (31), *215*
Niemeyer, M. 540 (117), *587*

Nightingale, D. 235 (49), *261*
Nii, S. 560 (235), 561 (241), *590*, *591*
Niikura, S. 563 (248), *591*
Nikiforov, G. A. 330 (108), *361*
Nikolaev, E. N. 171 (5), *184*
Nikolaev, S. A. 175 (101), *186*
Nilsson, P. 513 (18), *584*
Nilsson, R. 285 (240), *311*
Ninomiya, Y. 452 (35), *507*
Nishi, S. 502 (119), *509*
Nishide, K. 487 (104), *509*
Nishii, Y. 622 (85), *630*
Nishikawa, T. 636 (20, 21), *677*
Nishimae, S. 567 (270), *591*, 636 (18), *676*
Nishimura, Y. 194 (30), 201 (75, 76), *215*, *216*
Nishiyama, H. 517 (34), 519 (54), *584*, *585*
Nissen, P. 336 (161), *362*
Niwaki, K. 259 (140), *263*
Nizovtsev, A. V. 528 (82), *586*
Njock, M. G. K. 357 (335), *366*
Nobbs, T. J. 355 (311), *366*
Noble, M. E. M. 348 (239), *364*
Nobuto, D. 419 (51), *435*
Nocera, D. G. 201 (73), *216*
Noe, M. C. 582 (316), *593*
Nogueira, J. M. F. 302 (322), *313*
Nogueras, M. 440 (7), *506*
Noji, H. 329 (96, 98, 99, 100), *361*
Nolan, S. P. 550 (166), *588*
Noll, B. C. 469 (78, 79), *508*
Nolph, K. D. 268 (44), *306*
Noltemeyer, M. 74 (248), *98*, 424 (16), *434*
Noltes, J. G. 30 (105), *94*
Nomoto, T. 35 (117, 118), 43 (146), *95*
Nomura, N. 557 (211), *590*, 621 (81, 82), *630*
Nonoyama, N. 531 (90), *586*
Noodleman, L. 341 (202), *363*
Nordahl Larsen, A. L. 370 (113), *509*
Nordlander, J. E. 141 (34), 142 (35), *154*, 654 (74), *678*
Nordlund, P. 325 (70), *361*
Norinder, J. 791 (53), *802*
Normant, J. 633 (1), *676*
Normant, J. F. 517 (34), *584*, 633 (1), 635 (16), *676*, 765 (48), *769*
Normant, J.-F. 566 (265), *591*
Noske, H. J. 301 (357), *314*
Notaras, E. G. A. 212 (113), *217*
Nöth, H. 27 (99), *94*, 409 (30), 427 (72), *435*, *436*
Noubi, L. 283 (162), *309*
Novak, D. P. 26 (91), *94*, 183 (146), *187*
Novák, P. 246 (95), *262*
Novak, T. 293 (267), *311*, 787 (44), *801*
Novoderezhkin, V. I. 194 (34), *215*
Nowak, A. 268 (110), *308*
Nowak, T. 350 (260), *365*
Nowotny, M. 355 (312), *366*
Noyori, R. 775 (26), *801*
Nozaki, H. 484 (69), *508*, 661 (103), *678*, 765 (48), *769*
Nozaki, K. 194 (30), *215*
Nsango, M. 357 (335), *366*
Nudelman, N. S. 260 (147), *263*
Nudenberg, W. 595 (8), *628*
Nugent, W. A. 258 (133), *263*, 615 (16), *629*
NuLi, Y. 253 (120), *262*
Nummy, L. 517 (37), *584*
Nummy, L. J. 103 (9), *126*
Nunzi, A. 328 (83), *361*
Nuske, H. 531 (93), *586*
Nuttall, R. L. 113 (2), *126*
Nuzzo, R. G. 172 (109), *186*
Nyman, E. S. 214 (112), *217*
Nyokong, T. 192 (12), 213 (121), *215*, *217*
Oae, S. 720 (8), *768*
Oberhuber, C. N. 791 (59), *802*
O'Brian, P. J. 354 (285), *365*
O'Brien, C. 550 (167), *589*
O'Hair, R. A. J. 156 (2), 180 (136), 181 (131, 132), 182 (133), *184*, *187*
O'Hara, C. T. 9 (52), 13 (59), 17 (66), 19 (69), 20 (70), 22 (71, 72), *93*
O'Hare, P. A. G. 107 (20), *126*
Occhialini, D. 259 (136), *263*
Ochterski, J. 384 (30), *402*
Oda, R. 767 (54), *769*
Ödman, F. 304 (351), *313*
Odobel, F. 198 (66), *216*
O'Donnel, M. 353 (272), *365*
Oei, T. O. 268 (79), *307*
Oesch, U. 276 (14), *306*
Oesterlund, K. 564 (252), *591*
Ogasawara, M. 774 (21, 24), *800*
Ogata, M. 214 (132), *218*
Ogata, S. 734 (29), *768*
Ogino, Y. 749 (42), 756 (43), *769*
Ogura, F. 290 (274), *312*
Ogura, K. 767 (55), *769*
Ohashi, M. 90 (276), *99*
Ohbayashi, T. 728 (25), *768*
Ohkohchi, M. 201 (75), *216*
Ohkubo, K. 201 (40), *216*
Ohmiya, H. 545 (139, 140), 546 (141), 560 (237, 238), 568 (273), *588*, *591*, 636 (22), *677*
Ohmori, H. 559 (225), *590*
Ohmori, K. 707 (46), *715*
Ohno, K. 163 (66, 67), *185*, 517 (34), 519 (54), *584*, *585*
Ohno, M. 303 (349), *313*
Ohno, T. 194 (30), *215*
Ohta, A. 701 (36), *715*
Ohtani, H. 198 (64), *216*
Ohtsuka, E. 339 (185), *363*
Oi, Y. 791 (60), *802*

Oikonomakos, N. G. 348 (239), *364*
Ok, E. 278 (215), *310*
Oka, K. 285 (237), *311*
Okada, K. 567 (269), *591*, 636 (17, 19), *676*
Okada, S. 699 (31), *715*
Okada, T. 201 (70, 74–76), *216*, 284 (178), *309*
Okamoto, K. 125 (100), *129*
Okamoto, S. 791 (60), *802*
Okamoto, Y. 60 (194–196), 61 (197), *97*, 303 (349), 304 (353), *313*, *314*
Oku, A. 718 (2), *768*
Okubo, M. 259 (138–142), *263*, 391 (10), *401*
Okunchikov, V. V. 213 (122), *217*
Okuno, H. 556 (207), *590*
Okura, I. 214 (137), *218*
Olbrysch, O. 653 (66), *677*
Olcucu, A. 278 (214), *310*
Olerich, M. A. 276 (17), *306*
Oliva, M. 357 (347), 359 (356, 357), *366*, *367*, 393 (25), *402*
Oliveira, C. C. 280 (224), *310*
Ollis, D. L. 351 (286), *365*
Olmstead, M. M. 82 (162), *96*, 417 (13), *434*
Olsson, R. 611 (64), *630*
Olszewski, T. K. 491 (105), *509*
Omote, H. 329 (97), *361*
Onan, K. D. 485 (21), *506*
Onda, K-I. 479 (14, 16), 485 (15, 17), *506*
O'Neil, B. T. 499 (115), *509*
O'Neil, P. 83 (263), *98*
O'Neil, P. A. 38 (126), *95*, 149 (47), *154*, 413 (42), 422 (41), *435*
Ong, C. M. 419 (26), *435*
Onishi, S. 268 (139), *308*
Ono, A. 576 (302), *592*
Ono, N. 531 (91), *586*
Ono, T. 198 (64), *216*
Onsager, L. 384 (29), *402*
Oohara, T. 721 (11), *768*
Operti, L. 162 (44, 57), *185*
Oppolzer, W. 656 (82, 84–86), *678*
Orchard, A. F. 124 (93), *129*
Ordóñez, L. A. 268 (119), *308*
Orf, J. 121 (70), *128*
Organ, M. G. 550 (167), *589*
Orhanović, S. 344 (219), *364*
Orita, M. 341 (189), *363*
Orlorva, M. A. 330 (107), *361*
Orlova, M. A. 330 (106), *361*
Orr, M. 214 (136), *218*
Ortega-Blake, I. 323 (55, 56), *360*
Örtendahl, M. 38 (129), *95*
Orth, P. 356 (324), *366*
Ortiz, J. 201 (40), *216*
Ortiz, J. V. 384 (30), *402*
Os, C. W. v. 287 (248), *311*
Osada, Y. 212 (116), *217*

Osawa, A. 728 (25), *768*
Osawa, E. 675 (158), *679*
Osborn, H. M. I. 547 (147), *588*
Osborne, C. G. 268 (107), *308*
Oscarson, J. L. 328 (85, 86), *361*
O'Shea, P. D. 5 (43), *93*, 703 (41), *715*
Oshima, K. 5 (42), 41 (41), *93*, 276 (9), *306*, 523 (59), 527 (73), 528 (77, 83), 540 (122), 545 (139, 140), 546 (141), 557 (216), 560 (236–238), 561 (243), 567 (269, 270), 568 (273), *585–588*, *590*, *591*, 596 (31), 623 (90), *629*, *630*, 633 (1), 636 (17–22), 643 (30), *676*, *677*, 690 (29), 699 (11), 703 (43), 706 (42, 44), 707 (45), *714*, *715*, 791 (50), *801*
Oshima, Koichiro 675 (158), *714*
Oshiyama, A. 341 (204), *363*
Osman, A. 224 (29), *260*
Osowska-Pacewicka, K. 547 (147), *588*
Ossola, F. 64 (210), *97*, 268 (148), *308*
Oster, G. 327 (81), 329 (90), *361*
Osuka, A. 194 (30), 201 (74–76), *215*, *216*
Oswald, J. 513 (14), *584*
Ota, H. 526 (70), *586*, 746 (39), *769*
Otsubo, T. 290 (274), *312*
Otsuka, S. 657 (88), *678*
Ottesen, L. K. 611 (64), *630*
Ottow, E. 538 (107), *587*
Ou, X. 272 (176), *309*
Ough, E. 196 (52), *216*
Ouyang, J. M. 277 (205), *310*
Overcash, D. M. 246 (90), *262*
Overman, L. E. 563 (247), *591*, 798 (66, 67), *802*
Owen, D. J. 348 (239), *364*
Owens, J. M. 259 (146), *263*
Owens, T. D. 575 (298), *592*
Özaltin, N. 274 (188), *310*
Ozanam, F. 241 (79, 80), 243 (81–83, 85), *262*
Ozenne, C. 301 (355), *314*
Ozin, G. A. 157 (19), *184*
Paap, J. 48 (163), *96*, 122 (81), *128*
Pace, N. 335 (150), *362*
Pace, N. R. 336 (157), *362*
Pack, G. R. 320 (32), 323 (51, 52), *360*
Pacold, M. 73 (240), *98*, 418 (11), *434*
Padwa, A. 718 (2), *768*
Pages, N. 267 (1), 268 (50, 54), *306*, *307*
Pahira, J. J. 268 (113), *308*
Pai, E. F. 327 (71), *361*
Pai Fondekar, K. 424 (66), *436*
Pajerski, A. D. 39 (134), 44 (147), 56 (57), 69 (183), *93*, *95*, *96*, 410 (38), 429 (78), *435*, *436*, 682 (5, 14), 713 (53), *714*, *715*
Pallaud, R. 235 (47, 48), *261*
Pan, J. 337 (172), *363*
Pan, T. 336 (167), *363*
Panasenko, A. A. 123 (87), *129*

Pande, N. 337 (169), *363*
Panov, D. 512 (8, 9), *583*
Paolucci, F. 259 (144), *263*
Papageorgiou, T. 278 (207), *310*
Papasergio, R. I. 31 (109), *94*
Papazachariou, I. M. 282 (232), *311*
Pape, A. 36 (119), *95*
Papillon, J. P. N. 707 (47), *715*
Papiz, M. Z. 324 (58), 356 (325, 331), *360, 366*
Paquette, L. A. 50 (98), *94*, 123 (83), *129*
Paradies, H. H. 517 (34), *584*
Paris, G. 134 (18), *153*
Paris, G. E. 140 (5), *153*, 428 (83), *436*
Park, C.-H. 557 (215), *590*
Park, E. J. 285 (156), *309*
Park, K. 553 (178), *589*
Parker, R. D. 289 (272), 292 (284), *312*
Parker, V. B. 113 (2), *126*
Parkin, A. 74 (124), *95*, 150 (43), *154*, 423 (17), *434*
Parkin, G. 73 (235–239), *98*, 148 (17), *153*, 344 (213), *364*, 418 (7), 419 (6, 52), 422 (60), 423 (8, 9), 424 (10), *434, 435*
Parkinson, J. A. 20 (70), 69 (222), *93, 97*, 150 (41), *154*, 429 (76), *436*
Parnis, J. M. 157 (19), *184*
Parr, R. G. 384 (28), *402*
Parris, G. E. 108 (40), *127*, 146 (40), *154*
Parrish, J. D. 253 (124), *263*
Parsons, S. 74 (124, 243, 244), 76 (249), 81 (255, 256), *95, 98*, 150 (43), *154*, 406 (19), 407 (21, 22), 414 (45), 422 (15, 20), 423 (17), *434, 435*
Partridge, J. C. 191 (9), *215*
Parts, E. O. 116 (57), *128*
Parves, M. 44 (147), *95*
Parvez, M. 39 (134), 56 (57, 184), 69 (183), 78 (253), *93, 95, 96, 98*, 421 (34), 429 (78), *435, 436*, 682 (5, 14), *714*
Pasquinet, E. 711 (50), *715*
Pasternak, K. 268 (109), *308*
Pasto, D. J. 621 (80), *630*
Patai, S. 113 (55), *127*, 451 (29), *507*, 528 (86), *586*
Patel, D. J. 119 (62), *128*, 654 (76), *678*
Patel, P. H. 332 (128), *362*
Patel, S. S. 352 (289, 292, 296), *365*
Patil, S. 538 (109), *587*
Patricia, J. J. 527 (28), *584*
Patrick, B. O. 74 (246), *98*
Patterson, D. B. 19 (46), *93*
Patterson, E. V. 106 (17), *126*
Paul, A. 196 (54), *216*
Paul, A. G. 478 (65), *508*
Paunier, L. 268 (42), *306*
Pauson, P. L. 25 (84), *94*
Pavela-Vrančić, M. 345 (218), *364*

Pavela-Vrani, M. 344 (219), *364*
Paver, M. A. 57 (186), *96*, 124 (91), *129*
Pavlenko, S. A. 333 (132), *362*
Payandeh, J. 327 (71), *361*
Pearce, F. G. 357 (340), 358 (349, 350), *366, 367*
Pearce, R. 30 (104), *94*
Pearson, R. 234 (23), 258 (15), *260*
Pearson, T. H. 228 (43), *261*
Pearson, W. H. 729 (26), *768*
Pedersen, E. B. 701 (37), *715*
Pedersen, J. B. 331 (111), *361*
Pedersen, L. C. 352 (299), *365*
Pedersen, L. G. 352 (299), *365*
Pedersen, P. L. 329 (93), 330 (103), *361*
Pedersen, S. U. 256 (28), 259 (136, 137), *260, 263*
Pedley, J. B. 109 (47), 116 (1), *126, 127*
Pedrares, A. S. 324 (59), *360*
Pei, W. 619 (78), *630*
Peled, A. 211 (110), *217*
Pelissier, H. 540 (115), *587*
Pellicena, P. 347 (227), *364*
Peña, D. 775 (29), 788 (48), *801*
Pendergrass, E. 289 (265), *311*
Penedo, J. C. 341 (194), *363*
Peng, C. Y. 384 (30), *402*
Pennella, D. R. 648 (46), *677*
Pentikäinen, U. 333 (133), *362*
Penwell, A. J. 567 (268), *591*, 648 (48), *677*
Peracchi, A. 341 (188), *363*
Péralez, E. 157 (15), *184*, 391 (20), *402*
Percival, W. C. 599 (36), *629*
Pereira-Filho, E. R. 271 (167), *309*
Périchon, J. 236 (60), 258 (38, 130), *261, 263*, 291 (279), *312*
Perlmutter, P. 774 (1), *800*
Perpall, H. J. 293 (266), *311*
Perrier, O. 290 (273), *312*
Perriot, P. 765 (48), *769*
Pershin, A. P. 330 (108), *361*
Persson, E. 558 (218), *590*
Persson, E. S. M. 557 (210), *590*, 791 (53), *802*
Perutz, R. N. 200 (67), *216*
Peschek, G. A. 193 (24), *215*
Peschke, M. 175 (123), *187*
Petek, S. 489 (8, 9), *506*
Peter, K. 285 (240), *311*
Peters, E.-M. 573 (290), *592*
Peters, K. 573 (290), *592*
Petersen, H. 525 (64), *585*
Peterson, J. M. 653 (8), *676*
Peterson, T. H. 91 (277), *99*
Petersson, G. A. 384 (30), *402*
Petit, S. 275 (189), *310*
Petrella, A. 212 (115), *217*

Petrie, S. 125 (106, 107), *129*, 156 (9-14), 160 (52), 162 (53), *184*, *185*
Petrov, A. S. 320 (32), 323 (51, 52), *360*
Petrova, N. V. 110 (48), *127*
Petrusová, L. 88 (272), 89 (273), *99*
Peyrat, J. F. 607 (48), 608 (51), *629*
Peyrat, J.-F. 555 (197, 198), *589*
Pfaltz, A. 774 (2, 8), 779 (22), *800*
Pfitzner, A. 409 (30), 427 (72, 75), *435*, *436*
Phansavath, P. 559 (224), *590*
Philbert, M. A. 285 (156), *309*
Phillips, N. H. 527 (28), *584*
Phomphrai, K. 73 (240), 98, 418 (11), *434*
Piarulli, U. 625 (93), *630*
Piccaluga, G. 318 (17), *359*
Piccirilli, J. A. 338 (176), *363*
Pichota, A. 774 (8), *800*
Pierce, J. 357 (338, 345), *366*
Pierce, O. R. 516 (29), *584*
Pietsch, E. C. 648 (46), *677*
Pietzsch, C. 30 (108), *94*, 243 (86), *262*
Pilard, J.-F. 253 (122), *263*
Pilcher, G. 110 (50), *127*
Pileni, M. P. 198 (60), *216*
Pillai, C. K. S. 323 (53), *360*
Pimerzin, A. A. 122 (82), *128*
Pina, M. 301 (355, 358, 361), *314*
Pineschi, M. 774 (14), *800*
Pingoud, A. 355 (318-320), *366*
Pinho, P. 519 (53), *585*
Pinkerton, R. C. 222 (26), 228 (43), *260*, *261*
Pinkus, A. G. 458 (32, 49), *507*
Pinna, G. 318 (17), *359*
Piotroski, H. 81 (261), *98*
Piotrowski, H. 302 (302), *312*, 409 (30), 410 (37), *435*
Pipik, B. 648 (47), *677*
Piskorz, P. 384 (30), *402*
Pistón, M. 278 (213), *310*
Pitteloud, R. 656 (85, 86), *678*
Pizziconi, V. B. 356 (327), *366*
Platz, H. 537 (104), *586*
Plé, N. 518 (43), *585*, 702 (40), *715*
Pletcher, D. 241 (37), 246 (98), *261*, *262*
Pley, H. W. 340 (181), *363*
Plieninger, H. 517 (39), *585*
Ploch, D. 301 (361), *314*
Ploom, L. R. 116 (57), *128*
Po, P. L. 159 (37), *185*
Podall, H. E. 653 (57), *677*
Podell, E. 320 (29), *360*
Podkovyrov, A. I. 122 (77), *128*
Poerwono, H. 547 (159), *588*
Poh, J. 357 (333), *366*
Pohl, R. 611 (63), *630*
Pohmakotr, M. 725 (19), *768*
Polášek, M. 87 (271), *99*

Polborn, K. 409 (30), *435*, 528 (84), *586*, 738 (33), 739 (34), *768*
Polet, D. 558 (220), *590*, 791 (50, 57), *802*
Poli, G. 484 (101), *509*
Polidori, G. P. 328 (83), *361*
Polikarpov, E. V. 121 (72), *128*, 160 (32), *185*
Polyakova, E. V. 279 (219), *310*
Pomelli, C. 384 (30), *402*
Poncini, L. 537 (104), *587*
Pons, B. 303 (348), *313*
Pont, F. 288 (254, 255), *311*
Poonia, N. S. 3 (24), *92*
Pople, J. A. 369 (1, 2), 384 (30), *401*, *402*
Popp, J. 357 (337), *366*
Porat, A. 211 (110), *217*
Porcheddu, A. 559 (226), *590*
Poretz, R. D. 214 (131), *218*
Porra, R. J. 191 (8), *215*
Pörschke, K.-R. 86 (266), *99*
Porsev, V. V. 158 (30), 171 (104), *185*, *186*
Porter, H. D. 633 (2), *676*
Porter, R. F. 159 (37, 38), *185*
Posner, G. H. 596 (18, 19), *629*, 633 (9), *676*
Potapov, D. A. 171 (102), 175 (101), *186*
Poter, G. 198 (63), *216*
Powell, L. A. 272 (175), *309*
Power, P. P. 58 (58), 59 (191), 62 (202), 79 (79), 82 (162), *93*, *94*, *96*, *97*, 417 (13), *434*, 682 (4), *714*
Poyner, R. R. 349 (250, 253, 259), *364*, *365*
Pozet, N. 268 (114), *308*
Pozniak, B. P. 160 (48), *185*
Prabpai, S. 725 (19), *768*
Prada-Rodríguez, D. 271 (164), *309*
Prathapan, S. 201 (38, 39), *215*
Prebil, S. 125 (104), *129*
Preis, C. 411 (24), *434*
Premo, J. G. 240 (71), *261*
Preuss, H. 79 (258), *98*
Prévost, C. 515 (27), *584*
Price, S. M. 356 (325), *366*
Pridgen, L. N. 574 (292), *592*, 674 (152, 153), *679*
Prince, S. M. 324 (58), 356 (331), *360*, *366*
Procter, G. 503 (122), *509*
Prody, G. A. 339 (178), *363*
Prokhorenko, V. I. 356 (330), *366*
Prosenc, M.-H. 49 (165), *96*
Proulx, J. 268 (85), *307*
Prout, F. S. 398 (39), *402*
Provot, O. 555 (197, 198), *589*, 607 (48), 608 (51), *629*
Prust, J. 74 (245), *98*, 413 (43), *435*
Pryde, P. G. 268 (135), *308*
Przybilski, R. 341 (190), *363*
Psarras, T. 244 (89), 258 (36), *261*, *262*
Puar, M. S. 574 (296), *592*
Puchta, G. T. 519 (53), *585*

Puga, Y. M. 103 (10), *126*
Pullez, M. 457 (45), *507*, 787 (41), *801*
Pullins, S. H. 168 (75), *186*
Pupowicz, D. 517 (34), *584*, 722 (15), *768*
Pygall, C. F. 124 (93), *129*
Pyle, A. M. 334 (136), 337 (12), *359, 362*
Qi, D. 302 (327), *313*
Qi, M. 729 (26), *768*
Qin, J. 349 (251), 355 (311), *364, 366*
Qiu, G. 547 (157), *588*
Quamme, G. A. 268 (31, 74, 75), 288 (252), *306, 307, 311*
Quéguiner, G. 518 (43), 524 (60), 525 (41), 534 (81), 537 (103), *586*, 550 (173), 553 (186), *585, 586, 589*, 689 (26, 27), 700 (33, 34), 713 (24, 25), *714, 715*
Quevauviller, P. 302 (323), *313*
Quintin, J. 554 (190), *589*, 614 (65), *630*
Rabezzana, R. 162 (44), *185*
Rabinowitch, E. 207 (83, 84), *217*
Rabuck, A. D. 384 (30), *402*
Rachamim, M. 206 (78), *217*
Rachtan, M. 299 (311), *312*
Radojević, M. 303 (345), *313*
Radzio-Andezelm, E. 347 (225), *364*
Rafailovich, M. 305 (362), *314*
Raghavachari, K. 369 (2), 384 (30), *401, 402*
Raghavendra, A. S. 194 (33), *215*
Ragsdale, S. W. 342 (208), *363*
Rague, B. 484 (100), *508*
Rahman, A. U. 540 (115), *587*
Rahn, B. J. 604 (46), *629*
Rakita, P. E. 3 (18, 27), 55 (178), 58 (13), *92, 96*, 225 (40), 258 (131, 132), 259 (145), *261, 263*, 389 (3), *401*, 512 (3, 5–7), *583*
Ralph, W. C. 239 (74), *261*
Rambaud, M. 288 (262), *311*, 516 (31), *584*
Rambidi, N. G. 183 (153), 184 (155), *187*
Rambo, R. P. 334 (140), *362*
Ramesh, V. 286 (247), *311*
Ramiandrasoa, P. 552 (176), 555 (195, 196), *589*
Ramírez-Solís, A. 323 (56), *360*
Ramirez, U. D. 318 (21), *359*
Ramsay, L. E. 268 (142), *308*
Ramsden, H. E. 44 (148), *95*
Ranade, C. R. 208 (96), *217*
Rangel, A. O. S. S. 280 (225), *310*
Rankin, D. W. H. 183 (141), *187*
Rao, C. P. 772 (7), *800*
Rao, J. K. 272 (172), *309*
Rao, S. A. 582 (316), *593*
Rapoport, H. 451 (34), *507*
Rappoport, Z. 41 (141), 69 (29), *92, 95*, 110 (49), 113 (55), 118 (60), *127, 128*, 156 (1), *184*, 196 (4), *215*, 297 (260), *311*, 471 (93), *508*, 528 (86), *586*
Rapsomanikis, S. 303 (345), *313*

Rardin, R. L. 322 (41), *360*
Rasmussen, H. S. 268 (104), *308*
Raston, C. L. 19 (68), 26 (92), 31 (109), 40 (139), 44 (151), 46 (158), 51 (168), 58 (155), 81 (260), *93–96, 98*, 123 (89), *129*, 405 (5), 427 (74), *434, 436*
Rataboul, F. 544 (130), *588*
Rathjen, P. D. 339 (179), *363*
Rathke, M. W. 462 (58), 494 (109), *507, 509*
Ravi Kumar, G. 272 (172, 173), *309*
Ray, J. 148 (48), *154*
Rayment, I. 349 (253, 255, 259), *364, 365*
Rayssiguier, Y. 268 (102), 288 (256–258), *308, 311*
Ready, J. M. 622 (87), *630*, 652 (53), *677*
Reamer, B. 654 (72), *678*
Reamer, R. 699 (31), *715*
Record, M. R. 334 (138), *362*
Redden, P. R. 301 (360), *314*
Reddinger, J. 286 (244), *311*
Reddy, C. K. 534 (96), *586*
Reddy, G. S. 357 (345), *366*
Redeker, T. 68 (221), *97*
Redondo, P. 106 (16), *126*
Redwood, M. E. 428 (87), *436*
Reed, D. 38 (126), *95*, 149 (47), *154*
Reed, D. P. 395 (4), *401*
Reed, G. H. 349 (250, 252, 253, 255, 259), *364, 365*
Rees, C. W. 71 (231), *98*, 411 (33), *435*
Rees, T. C. 654 (75, 79, 80), *678*
Reetz, M. F. 425 (69), *436*
Reetz, M. T. 180 (137), *187*
Reeves, J. T. 517 (37), *584*
Reeves, T. 103 (9), *126*
Reffy, B. 183 (151), *187*
Regev, A. 201 (71), *216*
Reginato, G. 580 (313), *592*
Reich, H. J. 527 (28), *584*
Reich, I. L. 527 (28), *584*
Reich, R. 653 (68), *677*
Reichard, G. A. 484 (102), *509*
Reid, A. F. 122 (75), *128*
Reider, P. J. 654 (72), *678*, 699 (31), *715*
Reimuth, O. 288 (146), *308*
Reinehr, D. 653 (62, 66), 655 (59–61, 63–65), 656 (87), *677, 678*
Reinhart, R. A. 268 (38), *306*
Reinmuth, O. 389 (3), *401*, 512 (3), *583*
Reisinger, C.-P. 550 (168), *589*
Reist, E. J. 601 (38), *629*
Reiter, S. M. 121 (70), *128*
Reményi, A. 347 (229), *364*
Ren, H. 524 (61), 526 (72), 534 (97), *585, 586*
Replogle, E. S. 369 (2), 384 (30), *401, 402*
Resing, K. A. 347 (228), *364*
Resnick, L. M. 268 (19, 101), *306, 308*
Respess, W. L. 295 (287), *312*

Reusch, C. E. 276 (195, 196), *310*
Reutrakul, V. 725 (19), *768*
Rezabal, E. 322 (45), *360*
Rha, R. 334 (140), *362*
Riber, E. 270 (154), *309*
Ricchiardi, G. 125 (108), *129*
Ricci, A. 475 (81), *508*, 519 (51), 531 (89), 534 (81), 559 (228), 563 (245), 576 (301), 580 (313), *585*, *586*, *590–593*
Rich, D. H. 370 (114), *509*
Richard, O. 268 (116), *308*
Richardson, C. C. 352 (298), *365*
Richey, G. H. 172 (107), *186*
Richey, H. G. 3 (16, 17, 28), 11 (56), 39 (134), 44 (147, 151), 55 (182), 56 (55, 57, 184), 58 (14), 69 (183), *92*, *93*, *95*, *96*, 268 (4, 5), *306*, 389 (3), *401*, 410 (38), 429 (78), 431 (81), *435*, *436*, 451 (30), *507*, 512 (3, 5, 6), 543 (124), *583*, *587*, 634 (11), 645 (35, 36), 654 (75, 79, 80), 657 (91), 659 (97), 667 (54), *676–678*, 682 (5, 14), 684 (13, 18), 686 (17), 711 (51), 713 (52, 53), *714*, *715*
Richmann, P. N. 68 (219), *97*
Richoux, M.-C. 198 (63), *216*
Richter, J. 191 (7), *215*
Richtmyer, N. K. 299 (310), *312*
Ridley, D. 410 (39), 428 (87), *435*, *436*
Riech, I. 247 (106), *262*
Rieger, D. L. 503 (123), *509*
Rieke, R. D. 254 (125), *263*, 514 (19, 20), 540 (119), *584*, *587*
Rihs, G. 774 (8), *800*
Rijksen, G. 287 (248), *311*
Rinaldi, R. 331 (115), *362*
Ritter, C. 268 (34), *306*
Rivaro, P. 302 (332), *313*
Rivera, S. L. 395 (36), *402*
Rizzi, M. 327 (77), *361*
Robb, M. A. 369 (2), 384 (30), *401*, *402*
Robbins, D. L. 124 (98), *129*
Roberts, B. A. 19 (68), *93*
Roberts, C. W. 516 (29), *584*
Roberts, D. A. 666 (116), *678*
Roberts, J. D. 119 (62), *128*, 136 (8), 138 (22–25), 141 (34), 142 (35, 36), *153*, *154*, 654 (74, 76–78), *678*
Robertson, H. E. 183 (141), *187*
Robinet, J. J. 336 (160), *362*
Robinson, G. C. 240 (76), *261*
Robinson, M. J. 347 (237), *364*
Robl, C. 68 (218), *97*
Robles, E. S. J. 123 (84), *129*
Rocha, J. B. T. 774 (25), *801*
Roche, K. E. 268 (107), *308*
Rock, P. A. 257 (129), *263*
Rodger, P. J. A. 17 (65), *93*
Rodgers, J. D. 667 (118), *678*
Rodigari, F. 303 (342), *313*

Rodriguez-Cruz, S. E. 318 (19, 20), *359*
Rodríguez, S. 391 (26), *402*
Roenn, M. 555 (201), *590*
Roesky, H. 74 (245), *98*
Roesky, H. W. 74 (248), *98*, 413 (43), 424 (16), *434*, *435*
Roffia, S. 259 (144), *263*
Rogers, H. R. 259 (148), *263*, 513 (14), *583*
Rogers, R. D. 50 (98), *94*, 123 (83), *129*
Rogers, R. J. 259 (148), *263*, 513 (14), *583*
Roithova, J. 162 (58), *185*
Rojas, F. G. 559 (227), *590*
Roland, S. 574 (293), *592*
Roldán, J. L. 518 (48), *585*
Rollick, K. L. 615 (16), *629*
Romani, A. M. 268 (47), *307*
Romani, A. M. P. 268 (53), *307*, 325 (64, 65), 342 (62), *360*
Romano, L. J. 352 (293), *365*
Rondelez, Y. 329 (99), *361*
Ronzini, L. 601 (40–42), *629*
Root, K. S. 513 (14), *584*
Rosenplänter, J. 29 (103), *94*
Rosito, V. 583 (318), *593*
Roskoski, J. R. 347 (236), *364*
Ross, M. J. 413 (42), *435*
Rossana, D. M. 485 (21), *506*
Rossini, F. D. 120 (65), *128*
Rossiter, B. E. 772 (4), *800*
Rostovtsev, V. V. 576 (303), *592*
Roth, K. 537 (104), *586*
Rothman, A. M. 634 (11), *676*
Rottländer, M. 517 (35), 525 (41), 534 (81, 100), *584–586*, 690 (30), 701 (37), *715*
Röttländer, M. 619 (45), *629*
Rousset, C. J. 671 (134), *679*
Roussi, G. 666 (114), *678*
Rowe, B. R. 156 (8), 160 (45), *184*, *185*
Rowlings, R. B. 17 (65, 66), 18 (67), 19 (68, 69), *93*
Roy, A. H. 553 (178), *589*
Roy, A. M. 553 (180), *589*
Roy, G. 353 (307), *366*
Roy, R. 353 (307), *366*
Roychowdhury-Saha, M. 341 (193), *363*
Rozhnov, A. N. 122 (82), *128*
Ruano, J. L. G. 727 (24), *768*
Ruben, M. 546 (145), *588*
Rüchardt, C. 654 (74), *678*
Rude, R. K. 268 (78), 276 (17), *306*, *307*
Rüdiger, W. 191 (8), *215*
Rue, C. 124 (97), 125 (105), *129*, 161 (56), *185*
Rueping, M. 552 (177), *589*
Rüffer, T. 38 (127), 136 (16), 42 (143), 59 (132), *95*, 149 (6)*153*
Ruffner, D. E. 339 (182), *363*

Rufińska, A. 71 (227), *97*, 123 (80, 86), *128, 129*, 143 (32, 37), 151 (15), 152 (50, 51), *153, 154*
Ruhland, T. 516 (32), *584*
Ruhlandt-Senge, K. 71 (229), *98*, 324 (59), *360*, 407 (23), *434*
Ruiz-Lopez, M. 331 (116), *362*
Ruiz-López, M. D. 271 (168), *309*
Ruiz-Vicent, L. E. 323 (56), *360*
Rumyantseva, S. V. 214 (138), *218*
Rundle, R. E. 58 (19, 20, 188, 190), *92, 96*, 391 (14), *402*
Russel, C. A. 124 (91), *129*
Russell, A. T. 547 (149), *588*
Russell, C. A. 57 (186), *96*
Russell, C. J. 424 (63), *435*, 473 (89), *508*
Russell, J. 148 (48), *154*
Ruszczycky, M. W. 348 (238), *364*
Ruzicka, J. 280 (222), *310*
Ryan, M. F. 268 (82, 86), *307*
Ryan, M. P. 268 (140), *308*
Ryan, R. R. 91 (278), *99*
Ryde, K. W. 268 (79), *307*
Ryde, U. 322 (40), *360*
Rylander, R. 268 (111), *308*
Ryppa, C. 191 (8), *215*
Ryzen, E. 268 (78), *307*
Ryznar, J. W. 240 (71), *261*
Sabatier, M. 288 (254, 255), *311*
Sabbert, D. 329 (95), *361*
Sachdev, H. 411 (24), *434*
Sadler, P. J. 286 (246), *311*
Saenger, W. 356 (36, 324), *360, 366*
Safont, V. S. 357 (347), 359 (356, 357), *366, 367*, 393 (25), *402*
Sahni, J. 325 (63), *360*
Sahrawat, K. L. 272 (172, 173), *309*
Saino, N. 791 (60), *802*
Saint-Martin, H. 323 (55, 56), *360*
Saito, I. 208 (96), *217*
Saito, M. 675 (158), *679*
Saito, N. 285 (237), *311*
Saito, S. 739 (35), *769*
Sakabe, Y. 145 (39), *154*
Sakai, K. 741 (36), *769*
Sakai, M. 774 (21), 780 (35), *800, 801*
Sakai, S. 125 (109), *129*, 391 (12), *402*
Sakamoto, F. 302 (329), *313*
Sakamoto, K. 720 (9), *768*
Sakamoto, S. 63 (60), *93*, 182 (6), *184*
Sakamoto, T. 538 (105, 108), *587*, 713 (55), *715*, 748 (41), *769*
Sakata, G. 566 (265), *591*
Sakata, Y. 201 (70), *216*
Saksena, A. K. 574 (296), *592*
Sakuma, S. 780 (35), *801*
Sakurada, J. 756 (43), *769*
Sakurai, Y. 193 (25, 26), *215*

Salem, G. 190 (1), *214*
Salem, M. 268 (81), *307*
Sales, Z. S. 424 (62), *435*
Salido, S. 440 (7), *506*
Salih, B. 303 (347), *313*
Salinger, R. 244 (89), *262*
Salinger, R. M. 108 (39), *127*
Salkeld, A. A. 268 (107), *308*
Salvucci, M. E. 358 (341), *366*
Samaj, J. 325 (66), *360*
Sambongi, Y. 329 (97), *361*
Sammakia, T. 774 (9), *800*
Sancar, A. 351 (281), *365*
Sànchez, A. 440 (7), *506*
Sánchez-Barbara, L. F. 74 (247), *98*, L. F., 421 (18)*434*
Sánchez-Martínez, C. 518 (48), *585*
Sander, L. 518 (49), *585*
Sander, W. 540 (115), *587*
Sanders, G. T. 268 (88), *307*
Sanders, G. T. B. 275 (191), *310*
Sanders, R. 268 (88), *307*
Sands, J. 335 (149), *362*
Sane, D. C. 279 (216), *310*
Sano, S. 463 (63), *508*
Santabarbara, S. 356 (329), *366*
SantaMaria, C. 103 (10), *126*
Santelli, M. 540 (115), *587*
Sanz-Medel, A. 273 (181), 303 (346), *309, 313*
Sapountzis, I. 268 (8), *306*, 518 (36, 42), 525 (41, 68), 531 (92), 540 (116), 543 (125), 550 (126), 559 (223), 566 (266), 576 (301, 304), 577 (306), *584–587, 590–592*, 601 (44), 629, 690 (29), 714 (8), *714, 715*
Saraymen, R. 278 (215), *310*
Saris, N. E. 268 (21), *306*
Saris, N.-E. L. 270 (155), *309*
Sartini, R. P. 280 (224), *310*
Sarvestani, M. 103 (9), *126*, 517 (37), *584*
Sarzanini, C. 273 (182), *309*
Sasahara, T. 145 (39), *154*
Sasaki, C. 302 (329), *313*
Sasaki, H. 576 (302), *592*
Sasaki, Y. 484 (28), *507*
Sassian, M. 512 (8), *583*
Sastre-Santos, A. 201 (40), *216*
Satar, A. K. M. A. 352 (294), *365*
Sato, F. 540 (120), 543 (124), *587*, 700 (32), *715*
Sato, T. 572 (287), *592*
Satoh, T. 479 (14, 16), 480 (18), 485 (15, 17), *506*, 526 (70, 71), 527 (74), *586*, 720 (10), 721 (11), 727 (18), 728 (25), 730 (27), 734 (29), 737 (32), 739 (35), 741 (36), 743 (38), 746 (39), 748 (41), 749 (42), 756 (43), 758 (44), 760 (45, 50), 761 (51), 764 (52), 766 (53), 767 (56, 57, 59), *768, 769*
Satomi, H. 452 (37), *507*

Sattelberger, A. P. 91 (278), *99*
Sauer, A. 540 (115), *587*
Sauvé, G. 286 (244), *311*
Savchenko, A. I. 582 (315), *593*
Savèant, J. M. 235 (57), 259 (58), *261*
Savory, J. 516 (30), 517 (34), *584*
Sayles, D. C. 624 (4), *628*
Scarpa, A. 268 (47), *307*, 342 (62), *360*
Scerbo, R. 302 (323), *313*
Schaefer, H. F. 718 (2), *768*
Schäfer, H. 237 (62), *261*
Schaper, F. 48 (164), 49 (165), *96*
Scharenberg, A. M. 325 (63), *360*
Schat, G. 34 (116), 35 (118), 38 (83), 40 (138, 140), 61 (198), 62 (199), 66 (213), 86 (269), *94*, *95*, *97*, *99*, 116 (58), 120 (66–68), 121 (71), *128*, 384 (31), *402*
Schechter, A. 247 (108), 248 (96, 107, 113), 251 (112), 252 (114), 253 (118), 258 (110), *262*
Scheer, H. 191 (5, 8), 193 (28), 211 (105, 106), 214 (133), *215*, *217*, *218*
Scheinman, S. J. 268 (64), *307*
Scheiper, B. 554 (189), *589*, 608 (57), *629*
Schenk, G. 343 (210), *364*
Scheuplein, R. 268 (107), *308*
Schiesser, C. H. 550 (164), *588*
Schilling, B. E. R. 29 (102), *94*, 123 (85), *129*, 183 (149), *187*
Schimeczek, M. 720 (6), *768*
Schimperna, G. 484 (101), *509*
Schink, H. E. 564 (252), *591*
Schlecker, W. 538 (107), *587*
Schlegel, H. 122 (79), *128*
Schlegel, H. B. 47 (161), *96*, 369 (1, 2), 384 (30), *401*, *402*
Schlenk, W. 3 (12), *92*, 107 (26), *127*, 132 (3), *153*, 225 (39), *261*, 391 (21), *402*
Schlichte, K. 44 (153), 45 (154, 156), 47 (159), *95*, *96*
Schlingmann, K. P. 268 (55, 56, 67), *307*
Schloss, J. D. 50 (98), *94*, 123 (83), *129*
Schloss, J. V. 358 (352), *367*
Schlosser, M. 142 (31), *154*, 702 (38), *715*
Schmalz, H.-G. 517 (40), *585*
Schmid, G. 89 (274, 275), *99*
Schmidt, E. 201 (72), *216*
Schmidt, H. 59 (132), *95*, 149 (6), *153*
Schmidt, H.-G. 74 (248), *98*, 424 (16), *434*
Schmidt, M. 357 (337), *366*
Schnare, M. N. 337 (169), *363*
Schneider, B. 323 (50), *360*
Schneider, G. 300 (317), *313*, 359 (355), *367*
Schneider, H. 519 (53), *585*
Schneider, I. R. 339 (178), *363*
Schneider, W. P. 289 (268), *311*
Schneiderbauer, S. 409 (30), *435*
Schnöckel, H. 26 (97), 68 (218, 220), *94*, *97*

Schnyder, A. 775 (27), *801*
Schofield, C. J. 547 (149), *588*
Schomburg, G. 653 (66), 655 (64, 65), *677*
Schormann, M. 74 (248), *98*, 424 (16), *434*
Schrader, E. A. 648 (46), *677*
Schrader, U. 208 (95), *217*
Schreiber, S. L. 707 (48), *715*
Schröder, F. A. 69 (189), *96*
Schroeder, D. 162 (58), *185*
Schroth, G. 653 (62), 655 (63–65), *677*
Schubert, B. 14 (50), 42 (144), *93*, *95*
Schuessler, H. A. 160 (50, 51), *185*
Schüler, P. 51 (51), *93*
Schulman, C. C. 268 (117), *308*
Schulten, M. 71 (230), *98*, 405 (12), *434*
Schultz, E. 596 (26), *629*
Schultz, M. 26 (94), *94*, 125 (102), *129*
Schultz, R. A. 355 (316), *366*
Schulz, C. P. 169 (81), *186*
Schulze, F. 295 (289, 290), *312*
Schulze, V. 135 (11), 137 (10), *153*, 720 (5–7), *768*
Schumann, H. 26 (96), 54 (95), 79 (254), *94*, *98*, 414 (32), *435*
Schumm, R. H. 113 (2), *126*
Schuppan, J. 780 (37), *801*
Schurko, R. W. 152 (52), *154*
Schutte, S. 26 (96), *94*
Schwarz, H. 156 (1), 160 (40), *184*, *185*
Schweigel, M. 325 (66, 68), *360*
Schwenk-Kircher, H. 27 (99), *94*
Schwenke, D. C. 279 (216), *310*
Schwerdtfeger, P. 720 (6), *768*
Schweyen, R. J. 325 (66, 68), *360*
Schwickardi, M. 33 (114), 44 (113), *94*, 513 (17), 540 (118), *584*, *587*, 621 (79), *630*
Scolastico, C. 484 (101), *509*
Scott, W. G. 340 (186, 187), 341 (196), *363*
Scuseria, G. E. 384 (30), *402*
Seburg, R. A. 106 (17), *126*
Seck, M. 555 (197), *589*, 607 (48), *629*
Seconi, G. 475 (81), *508*, 580 (313), *592*, *593*
Sedaira, H. 280 (220), *310*
Seebach, D. 65 (212), *97*, 395 (37), *402*, 571 (281), *591*, 774 (8), *800*
Seeger, R. 369 (1), *401*
Seelig, M. S. 268 (131), *308*
Segal, J. A. 74 (241), 76 (242), *98*, 418 (14), 424 (44), *434*, *435*
Segelmann, A. 214 (131), *218*
Seidel, G. 554 (189), *589*, 626 (95), *630*
Seidel, N. 30 (108), *94*, 243 (86), *262*, 525 (41), *585*
Seifert, F. 227 (34), *261*
Seifert, T. 27 (99), *94*, 427 (72), *436*
Seiler, A. 246 (97), *262*
Seiler, H. G. 269 (150), *308*
Seiler, P. 65 (212), *97*, 395 (37), *402*

Seip, R. 23 (76), *93*, 183 (148), *187*
Seitz, L. M. 4 (35–37), *92*, 682 (2), *714*
Selg, P. 125 (102), *129*
Seliem, M. M. 277 (200), *310*
Seligman, P. F. 302 (330), *313*
Sell, M. S. 514 (19), *584*
Semba, F. 720 (9), *768*
Sembri, L. 566 (263), *591*
Semmelhack, H. F. 540 (115), *587*
Senanayake, C. 395 (35), *402*
Senanayake, C. H. 103 (9), *126*, 517 (37), 521 (58), 583 (319), *584*, *585*, *593*
Senge, M. O. 191 (6–8), 192 (10), 196 (4), 212 (113), 214 (139), *215*, *217*, *218*
Sengupta, S. 512 (2), *583*, 596 (20), *629*
Senior, A. E. 330 (105), *361*
Senn, H. M. 350 (264), *365*
Seo, Y.-S. 553 (178), *589*
Štěpnička, P. 86 (270), *99*
Sergeev, G. B. 121 (72), *128*, 158 (27), 160 (32), *185*
Sergeeva, N. N. 196 (4), *215*
Seriaga, M. P. 513 (11), *583*
Servesko, J. M. 779 (33), *801*
Servis, K. L. 268 (78), *307*
Seth, J. 201 (38, 39, 46), *215*, *216*
Seudeal, N. 224 (29), *260*
Sévignon, M. 596 (26), *629*
Sevvana, M. 452 (36), *507*
Seyberth, H. W. 268 (67), *307*
Seyferth, D. 517 (34), *584*, 661 (102), *678*, 719 (3), *768*
Shadrina, M. S. 331 (117), *362*
Shafer, P. R. 654 (74), *678*
Shafirovich, E. Ya. 125 (103), *129*
Shah, R. D. 667 (118), *678*
Shaikh, N. 544 (129), *588*
Shakoor, A. 45 (157), *96*
Shan, S. 338 (176), *363*
Shang, X. 572 (276), *591*
Shang, Z. 337 (169), *363*
Shaper., A. G. 268 (120), *308*
Shapiro, H. 238 (77), *261*
Sharma, H. K. 428 (86), *436*
Sharma, S. 428 (85), *436*
Sharpless, K. B. 576 (303), *592*
Shaw, J. T. 503 (62), 505 (61), *508*
Shaw, Y.-J. 547 (154), *588*
Shen, H. 123 (90), *129*
Shen, J. R. 356 (323), *366*
Shen, Y. 567 (267), *591*
Sheng, Y. 169 (88, 89), *186*
Shepherd, L. H. 653 (58), *677*
Sherrington, D. C. 69 (222), 73 (73), *93*, *97*, 150 (41), *154*, 429 (76), *436*
Shi, L. 325 (74), *361*
Shi, M. 779 (20), *800*
Shi, T. 176 (129), *187*
Shi, Y.-J. 648 (47), *677*
Shi, Z. 347 (228), *364*
Shia, K.-S. 572 (276), *591*
Shiba, T. 193 (26), *215*
Shibutani, T. 517 (34), *584*
Shieu, J.-C. 547 (154), *588*
Shigeru, K. 514 (21), *584*
Shigeta, Y. 357 (336), *366*
Shikanai, D. 572 (287), *592*
Shilai, M. 538 (105), *587*
Shimada, K. 193 (22–25), *215*
Shimada, T. 554 (194), *589*
Shimizu, A. 453 (39), *507*
Shimizu, H. 767 (59), *769*
Shimizu, M. 466 (70), 486 (25), *507*, *508*
Shimoda, K. 791 (60), *802*
Shimokubo, H. 567 (270), *591*
Shimomura, A. 201 (70), *216*
Shin, C. S. 270 (159), *309*
Shin, H. 667 (120), *678*
Shin, J.-S. 557 (215), *590*
Shin, S. 707 (47), *715*
Shingu, H. 125 (100), *129*
Shinokubo, H. 5 (42), 41 (41), *93*, 276 (9), *306*, 523 (59), 527 (73), 528 (77, 83), 540 (122), 560 (236), 561 (243), 567 (269), *585*–*587*, *590*, *591*, 596 (31), 623 (90), *629*, *630*, 636 (17, 18, 20, 21), *676*, *677*, 699 (11), 703 (43), 706 (42, 44), 707 (45), *714*, *715*
Shintani, R. 571 (283), *592*, 623 (89), *630*, 639 (24), *677*
Shintou, T. 549 (162), *588*
Shioiri, T. 370 (112), 458 (52), *507*, *509*
Shioji, H. 302 (333), *313*
Shiono, T. 285 (243), *311*
Shiragami, H. 661 (103), *678*
Shirai, M. 713 (55), *715*
Shirakawa, E. 568 (272), *591*, 622 (88), 623 (89), *630*, 637 (23), 639 (24), *677*
Shiraki, R. 259 (141), *263*
Shirazi, A. R. 271 (165), *309*
Shireman, B. T. 707 (47), *715*
Shirey, T. L. 268 (108), *308*
Shiro, M. 463 (63), *508*
Shiue, J.-L. 566 (262), *591*
Shkel, I. A. 334 (138), *362*
Shono, T. 254 (121), *263*
Shook, B. C. 528 (85), *586*
Showalter, A. K. 352 (290), *365*
Shu, C. 798 (65), *802*
Shubert, D. C. 604 (46), *629*
Shui, X. 335 (30), *360*
Shults, R. H. 621 (80), *630*
Shumaker, J. B. 633 (7), *676*
Shuman, H. 268 (22), *306*
Shumov, Y. S. 212 (117), *217*
Shuvaeva, O. V. 279 (219), *310*

Shvartsburg, A. A. 176 (125–128), *187*
Sibi, M. P. 426 (71), *436*, 498 (41), *507*, 779 (19), *800*
Sidorskikh, P. F. 212 (117), *217*
Siede, W. 355 (316), *366*
Siefert, J. L. 355 (13), *359*
Siero, J. P. J. 208 (98), *217*
Sierra, I. 278 (208), *310*
Sigel, H. 327 (79), 331 (121), 332 (122–124, 126), *361*, *362*
Sigel, R. K. O. 334 (136), 336 (156), *362*
Siggaard-Andersen, O. 276 (16), *306*
Sikorsky, P. 44 (113), *94*
Sillen, L. G. 286 (245), *311*
Silva, S. J. N. 774 (25), *801*
Silveira, C. C. 774 (25), *801*
Silver, B. B. 268 (27), *306*
Silver, M. S. 654 (74), *678*
Silverman, G. S. 3 (18, 27), 55 (178), 58 (13), 92, 96, 225 (40), 258 (131, 132), 259 (145), *261*, *263*, 512 (3, 5, 6), *583*
Silvermann, G. S. 389 (3), *401*
Silverstein, B. 268 (107), *308*
Simaan, S. 560 (233), *590*
Simon, J. 268 (117), *308*
Simon, W. 276 (14), *306*
Simonet, J. 253 (122), *263*
Simpson, G. W. 495 (111), *509*
Sims, P. A. 349 (252), *364*
Simuste, H. 512 (9), *583*
Singh, J. 268 (48), *307*
Singh, P. 729 (26), *768*
Singh, S. K. 572 (289), *592*
Sinha, P. 577 (306), 578 (308), *592*
Sinitsyna, T. M. 268 (118), *308*
Sirover, M. A. 351 (276, 277), *365*
Sisido, K. 484 (69), *508*
Sitzmann, H. 26 (94), *94*
Siu, K. W. M. 176 (125, 129), *187*
Skalitzky, D. 119 (64), *128*
Skamnaki, V. T. 348 (239), *364*
Skatova, A. A. 79 (254), *98*, 414 (32), 426 (48), *435*
Skell, P. S. 157 (22), *185*
Skelton, B. 123 (89), *129*
Skelton, B. W. 40 (139), 46 (158), 81 (260), 95, 96, 98, 405 (5), *434*
Skiebe, A. 675 (156), *679*
Skinner, H. A. 109 (46), *127*
Sklute, G. 528 (84), *586*, 739 (34), *768*
Skov, K. 259 (137), *263*
Slawin, A. M. Z. 574 (294), *592*
Slayden, S. W. 110 (49), 111 (51), 113 (55, 56), *127*, *128*
Slim, G. 339 (184), *363*
Sliwiński, W. 410 (36), *435*
Slok, F. A. 519 (50), *585*
Slota, R. 196 (49), *216*

Smalley, J. F. 196 (48), *216*
Smeets, W. J. J. 3 (15), 34 (106, 116), 35 (117, 118), 38 (83), 40 (140), 95, 43 (146), 44 (142), 66 (213–216), 79 (258), 92, 94, 95, 97, 98, 183 (140), *187*, 318 (22), *359*, 774 (25), *800*
Smeets, W. J. J. ,
Smidansky, E. 352 (301), *365*
Smil, D. V. 648 (45), *677*
Smirnov, V. V. 171 (5, 100, 102), 175 (101), *184*, *186*, 268 (5), *306*
Smirnova, I. N. 329 (101), *361*
Smith, D. M. A. 323 (55), *360*
Smith, J. D. 23 (77, 78), 32 (110), 63 (112, 206–208), 94, 97
Smith, K. D. 28 (100), *94*
Smith, K. M. 74 (246), 98, 193 (11, 13, 15), 194 (32), 195 (16), 208 (97), 211 (108), 214 (139), *215*, *217*, *218*
Smith, M. B. 108 (33, 34, 38), *127*
Smith, N. K. 268 (12), *306*
Smith, R. S. 23 (76), 93, 183 (148), *187*, 391 (10), *401*, 604 (14), *629*
Smith, S. G. 108 (35), *127*, 389 (3), *401*
Smith, S. J. 343 (210), *364*
Smith, W. B. 108 (31), *127*
Smitrovich, J. H. 557 (214), *590*
Snavely, M. D. 325 (69), *361*
Sneyd, J. C. R. 122 (76), *128*
Snider, B. B. 640 (25, 26), *677*
Snieckus, V. 537 (102), 553 (183, 184), *586*, *589*
Soai, K. 774 (11), *800*
Sobolev, G. A. 183 (154), *187*
Soep, B. 169 (81), *186*
Sofia, A. 544 (131), *588*
Sofield, C. D. 26 (94), *94*
Sofoniou, M. C. 281 (227), *310*
Soi, A. 675 (155, 156), *679*
Sokolov, V. V. 452 (36), *507*
Sölch, O. 278 (210), *310*
Soler-López, M. 334 (135), *362*
Solladié-Cavallo, A. 574 (295), *592*
Solladié, G. 486 (54), *507*
Solomon, M. 667 (118), *678*
Solov'ev, V. N. 121 (72), *128*, 158 (27), 160 (32), *185*
Somashekar, B. S. 286 (247), *311*
Someya, H. 526 (70), 560 (238), *586*, *591*, 746 (39), *769*
Somlyo, A. P. 268 (22), *306*
Song, J. J. 103 (9), *126*, 517 (37), *584*
Song, J. Z. 699 (31), *715*
Sonnentag, H. 119 (64), *128*
Soriaga, M. P. 172 (106, 108), *186*, 512 (5), *583*
Sosnick, T. R. 336 (167), *363*
Sosnicki, J. G. 686 (20, 21), *714*

Soto, A. M. 336 (164), *362*
Soto, J. E. 172 (108), *186*
Søtofte, I. 30 (105), *94*
Souhassou, M. 328 (82), *361*
Souza, F. E. S. 567 (268), *591*, 648 (45, 48), *677*
Sowadski, J. M. 347 (232), 348 (234), *364*
Spatling, L. 268 (132), *308*
Spatz, J. H. 550 (172), *589*
Spek, A. L. 3 (15), 34 (106, 116), 35 (117, 118), 38 (83), 40 (138, 140), 43 (146), 44 (142), 62 (199), 66 (213–216), 77 (251), 79 (258), 86 (269), *92*, *94*, *95*, *97–99*, 120 (68), *128*, 183 (140), *187*, 318 (22), *359*, 384 (31), *402*, 774 (25), *800*
Spero, D. M. 573 (291), *592*
Sperry, D. C. 166 (76), *186*
Spescha, M. 774 (8), *800*
Spialter, L. 661 (105), *678*
Spichiger, U. 268 (26), *306*
Spiess, A. 525 (65), *585*
Spikes, J. D. 193 (21), 214 (133), *215*, *218*
Spina, G. 583 (318), *593*
Spindler, F. 775 (27), *801*
Spiridonov, V. P. 183 (154), 184 (152), *187*, 391 (15), *402*
Spivey, A. C. 547 (149), *588*
Spliethoff, B. 44 (113), 45 (154), *94*, *95*, 554 (193), *589*
Sporikou, C. N. 537 (104), *586*
Spoto, G. 125 (108), *129*
Spreitzer, R. J. 358 (341), *366*
Spring, D. R. 707 (48), 711 (49), *715*
Squiller, E. P. 56 (55), 69 (183), *93*, *96*, 410 (38), 429 (78), 431 (81), *435*, *436*, 684 (13), *714*
Sreedhara, A. 320 (8), *359*
Ssusi, A. K. 224 (31), *261*
Stace, A. 175 (117), *186*
Stace, A. J. 175 (116, 118, 122), 176 (119–121), *186*, *187*, 318 (23), *359*
Stadnikov, G. L. 2 (8), *92*
Stahl, T. 79 (258), *98*
Stahle, M. 142 (31), *154*
Stahley, M. R. 339 (174, 177), *363*
Staley, R. H. 160 (47), *185*
Stalke, D. 41 (141), 57 (186), *95*, *96*, 124 (91), *129*
Stallard, M. O. 303 (343), *313*
Stammler, H.-G. 68 (221), *97*
Stan, M. A. 547 (150), *588*
Stańczyk, W. A. 32 (110), *94*
Standen, M. C. H. 673 (140, 141), *679*
Stanford, S. 596 (25), *629*
Stang, P. J. 596 (28), *629*, 633 (1), *676*, 718 (2), 742 (37), *768*, *769*
Stangeland, E. L. 774 (9), *800*
Stanton, J. F. 106 (17), *126*

Stanway, S. J. 547 (151), *588*
Stapleton, L. 512 (7), *583*
Starowieyski, K. B. 25 (82), *94*, 183 (146), *187*, 299 (307), *312*
Stasch, A. 74 (245), *98*, 413 (43), *435*
Staubitz, A. 518 (47), 577 (305), *585*, *592*
Stec, B. 349 (248, 254), *364*
Steel, D. G. 344 (216, 217), *364*
Steel, P. J. 547 (158), *588*
Steele, I. 419 (57), *435*, 538 (105), *587*
Stefan, R. I. 280 (226), *310*
Stefanov, B. B. 384 (30), *402*
Steglich, W. 517 (39), *585*
Stegmann, W. 288 (252), *311*
Steichen, J. J. 268 (70), *307*
Steinborn, D. 38 (127), 42 (143), 59 (132), *95*, 136 (16), 149 (6), *153*
Steinbüchel, A. 301 (359), *314*
Steiner, A. 57 (186), *96*, 124 (91), *129*
Steinig, A. G. 573 (291), *592*
Steitz, J. A. 355 (175), *363*
Steitz, T. A. 334 (141), 335 (147), 336 (161), 351 (286), 352 (127, 297), 354 (310), 355 (175, 314), *362*, *363*, *365*, *366*
Steitz, T. J. 352 (294, 295), *365*
Stengelin, M. 329 (101), *361*
Stephan, D. W. 419 (26), *435*
Stephenson, N. A. 342 (207), *363*
Stergiades, I. A. 450 (27), *507*
Sternbach, D. D. 485 (21), *506*
Steward, O. W. 432 (88), *436*, 563 (250), 565 (259–261), *591*, 647 (39, 40), 667 (122), 670 (123, 125), 671 (126–130), *677*, *679*
Stewart, J. J. P. 369 (2), *401*
Stey, T. 41 (141), *95*
Stiasny, C. 734 (30), *768*
Stiasny, H. C. 527 (79), *586*
Stickney, J. L. 172 (108), *186*
Stiel, H. 196 (54), *216*
Stile, M. 476 (64), *508*
Stiles, P. L. 184 (162), *187*
Still, W. C. 395 (4), *401*
Stillman, M. J. 193 (15, 18, 19), 196 (52), *215*, *216*
Stivers, J. T. 353 (308), *366*
Stöckl, D. 273 (236), *309*
Stoll, A. H. 529 (88), *586*
Stoll, H. 79 (258), *98*
Stone, F. G. A. 58 (3, 4), *92*, 512 (4), *583*, 657 (90), *678*
Stone, F. G. S. 513 (13), *583*
Stone, J. F. 299 (309), *312*
Stork, G. A. 474 (47), *507*
Stormo, G. D. 339 (182), *363*
Stoudt, S. J. 713 (53), *715*
Stowasser, R. 342 (205), *363*
Stowell, J. C. 760 (46), *769*
Stoy, P. 729 (26), *768*

Strain, M. C. 384 (30), *402*
Sträter, N. 322 (49), *360*
Stratmann, R. E. 384 (30), *402*
Straub, B. F. 527 (28), *584*, 714 (7), *714*
Strauss, H. F. 656 (85), *678*
Strell, M. 214 (126, 127), *218*
Striccoli, M. 212 (115), *217*
Strissel, C. 305 (305), *312*
Strobel, S. A. 339 (174, 177), *363*
Strohmeier, W. 226 (33), 227 (34), *261*
Stryer, L. 327 (80), *361*
Stubbe, J. 348 (245), *364*
Stuckey, G. D. 38 (128), *95*
Stucky, D. 62 (200), *97*
Stucky, G. 58 (188), *96*
Stucky, G. D. 47 (47), 58 (19), 62 (205), 67 (217), 83 (262), *92*, *93*, *97*, *98*, 384 (31), *402*, 417 (4), *434*
Studte, C. 557 (213), *590*
Stulgies, B. 531 (93), 547 (153), *586*, *588*
Stumer, C. 742 (14), *768*
Su, G. 108 (35), *127*, 389 (3), *401*
Su, M. I. 352 (290), *365*
Su, X. 583 (319), *593*
Subba Reddy, B. V. 576 (303), *592*
Subirana, J. A. 334 (135), *362*
Subramaniam, V. 344 (217), *364*
Subramanian, J. 208 (95), *217*
Subramanian, S. 337 (169), *363*
Sudakin, D. L. 214 (129), *218*
Suess, J. 537 (104), *586*
Suga, S. 219 (9), *260*
Suga, T. 579 (309), *592*
Sugihara, Y. 701 (36), *715*
Sugimoto, F. 272 (171), *309*
Sugimoto, O. 514 (21), *584*
Suginome, H. 487 (67), *508*
Sugio, T. 193 (27), *215*
Sugita, T. 145 (39), *154*
Sugiyama, S. 767 (59), *769*
Sui, S. F. 325 (73), *361*
Sui, Z. 421 (58), *435*, 700 (35), *715*
Sullivan, A. C. 63 (207), *97*
Sultanov, R. M. 671 (131, 132), *679*
Sun, J. 169 (78), *186*
Sun, J.-L. 169 (94, 95), 170 (97, 98), *186*
Sun, L. 272 (169), *309*
Sun, M. 553 (178), *589*
Sun, S. 329 (90), *361*
Sun, W. 293 (285), *312*
Sun, X. 521 (58), *585*
Sun, Z. H. 325 (73), *361*
Sundaresan, N. 323 (53), *360*
Sundberg, R. J. 387 (32), *402*
Sundheim, O. 353 (303), *365*
Sung, M. M. 70 (226), *97*, 432 (79), *436*
Suresh, C. H. 323 (53), *360*
Surman, M. D. 659 (96), *678*

Suter, M. 409 (30), *435*
Suzmura, K. 341 (189), *363*
Suzuki, A. 596 (22), *629*
Suzuki, H. 90 (276), *99*, 720 (9), *768*
Suzuki, K. 285 (237), *311*, 707 (46), *715*
Suzuki, N. 671 (134), *679*
Suzuki, S. 327 (78), *361*, 570 (278), *591*
Suzuki, Y. 285 (237), *311*
Sviridov, S. V. 582 (315), *593*
Svith, H. 256 (28), *260*
Swain, C. G. 391 (7), *401*
Sweeney, J. B. 547 (147), *588*
Sweeney, O. R. 296 (296), *312*
Swierczewski, G. 666 (114), *678*
Swingle, N. M. 772 (4), *800*
Swiss, J. 297 (297), *312*
Swiss, K. A. 667 (117, 119), *678*
Syfrig, M. A. 395 (37), *402*
Sykes, P. 391 (6), *401*
Symons, R. H. 336 (158), 339 (179), *362*, *363*
Szepesi, G. 292 (283), *312*
Szmacinski, H. 283 (235), *311*
Szulbinski, W. S. 198 (61), *216*
Tabor, S. 352 (298), *365*
Tabushi, I. 767 (54), *769*
Taddei, M. 475 (81), *508*
Taguchi, H. 765 (48), *769*
Tainer, J. A. 353 (303, 306), *365*, *366*
Taira, K. 337 (152), 341 (189, 199–201), *362*, *363*
Tajiri, A. 701 (36), *715*
Takabatake, H. 453 (38), *507*
Takada, S. 760 (50), *769*
Takafumi, K. 568 (272), *591*
Takagi, K. 767 (54), *769*
Takagi, Y. 341 (199–201), *363*
Takahashi, H. 547 (159), *588*
Takahashi, K. 559 (225), *590*, 767 (55), *769*
Takahashi, T. 554 (192), *589*, 671 (134), *679*
Takai, K. 484 (103), *509*
Takaichi, S. 193 (23–25), 194 (29), 213 (119), *215*, *217*
Takano, K. 107 (22), *127*, 485 (17), *506*, 526 (70), *586*, 720 (10), 746 (39), 748 (41), *768*, *769*
Takaya, Y. 774 (21), *800*
Takayama, Y. 700 (32), *715*
Takeda, A. 270 (157), *309*
Takeda, T. 270 (157), *309*
Takehisa, T. 463 (63), *508*
Takeuchi, S. 329 (99), *361*
Takeuchi, T. 274 (186, 187), *309*, *310*
Taki, M. 341 (199), *363*
Takle, A. K. 574 (294), *592*
Talmard, C. 13 (59), 22 (71, 72), *93*
Tamaki, K. 201 (70), *216*
Tamao, K. 661 (101, 107), 662 (106), *678*
Tamaru, Y. 452 (37), *507*, 596 (24), *629*

Tamborski, C. 516 (33), *584*
Tamm, M. 53 (175), *96*, 452 (36), *507*
Tamura, M. 206 (81, 82), *217*, 604 (9, 10), *628*
Tamura, Y. 502 (119, 120), *509*
Tan, Z. 103 (9), *126*, 517 (37), *584*, 787 (44), *801*
Tan, Z. J. 335 (11), *359*
Tanabe, M. 329 (97), *361*
Tanabe, Y. 622 (85), *630*
Tanaka, H. 193 (27), *215*, 300 (314), *313*
Tanaka, J. 452 (35), 474 (96), *507, 508*
Tanaka, M. 259 (140), *263*, 302 (333), *313*
Tanaka, N. 60 (195), *97*
Tanaka, R. 572 (287), *592*
Tanaka, S. 764 (52), *769*
Tanaka, Y. 341 (189, 199), *363*
Tang, H. 56 (184), *96*
Tang, J. 567 (269), *591*, 636 (17), *676*
Tang, K. H. 352 (290), *365*
Tang, T. P. 575 (298), *592*
Tani, K. 123 (73), *128*
Taniguchi, S. 201 (74, 75), *216*
Tanji, K. 514 (21), *584*
Tanmatsu, M. 468 (80), *508*
Tanno, T. 198 (64), *216*
Tano, T. 193 (27), *215*
Tao, S. 304 (352, 353), *314*
Tapia, O. 357 (347), 359 (356, 357), *366, 367*
Targrove, M. A. 305 (364), *314*
Tarhouni, R. 516 (31), *584*
Tari, L. W. 332 (125), *362*
Tasaka, T. 720 (9), *768*
Tasler, S. 554 (193), *589*
Tateno, M. 341 (204), *363*
Tauber, A. Y. 191 (9), *215*
Tavakkoli, K. 63 (207), *97*
Tawney, P. O. 625 (2, 6), *628*
Taylor, N. J. 83 (176), *96*
Taylor, S. S. 347 (225, 231, 232), 348 (234, 241), *364*
Taylor, T. C. 357 (343), 359 (354), *366, 367*
Tcherkez, G. G. B. 359 (353), *367*
Teager, D. S. 765 (48), *769*
Tease, R. L. 272 (175), *309*
Teat, S. J. 424 (61), *435*
Tebboth, J. A. 25 (85), *94*
Tedrow, J. S. 503 (62), 505 (61), *508*
Teerlinck, C. E. 259 (146), *263*, 513 (12), *583*
Teixeira, P. 302 (322), *313*
Templeton, D. M. 269 (151), *309*
Ten Eyck, L. F. 347 (232), 348 (234), *364*
Teng, W. 324 (59), *360*
ter Horst, B. 500 (46), *507*, 787 (45–47), *801*
Terakura, K. 341 (204), *363*
Terao, J. 544 (127, 128), 545 (135, 136), 560 (235), 561 (239, 241, 242), *588, 590, 591*
Terasaki, H. 675 (157), *679*
Terauchi, N. 563 (248), *591*

Terenin, A. N. 198 (65), *216*
Teresa Aranda, M. 583 (319), *593*
Tereshko, V. 335 (31, 145, 146), *360, 362*
Terpstra, R. A. 268 (69), *307*
Terwey, D. P. 340 (187), *363*
Tesche, B. 554 (193), *589*
Tesfaldet, Z. O. 280 (226), *310*
Tessier, P. E. 567 (268), *591*, 647 (41), 648 (48), *677*
Teuchner, K. 196 (54), *216*
Tews, E. C. 162 (57), *185*
Teyssot, A. 243 (82, 83), *262*
Thakker, R. V. 268 (64), *307*
Thatcher, G. R. J. 331 (113), *361*
Themelis, D. G. 281 (227), *310*
Therien, M. J. 198 (56), *216*, 579 (309), *592*
Therkelsen, F. D. 701 (37), *715*
Thewalt, U. 88 (272), 89 (274, 275), *99*
Thewalt, Y. 87 (271), *99*
Thibonnet, J. 525 (41, 67), *585*
Thiel, W. 350 (264), *365*
Thienpont, L. M. 273 (180), *309*
Thirumlalai, D. 337 (172), *363*
Thivet, A. 555 (195), *589*
Thoennes, D. 6 (48), 9 (53), 38 (122, 123), *93, 95*, 682 (3), *714*
Thomas, C. J. 52 (173), *96*
Thomas, G. L. 711 (49), *715*
Thomas, M. J. 279 (216), *310*
Thompson, C. A. 159 (33), *185*
Thompson, R. I. 160 (49–51), *185*
Thorarensen, A. 555 (201), *590*
Thorp, L. 2 (9), *92*
Thorup, N. 701 (37), *715*
Thurston, C. E. 231 (45), *261*
Thyssen, K. 370 (113), *509*
Tian, H. 213 (118), *217*
Tichelaar, G. R. 398 (39), *402*
Tiekink, E. R. 550 (164), *588*
Tijani, A. 775 (27), *801*
Tilley, T. D. 38 (130), *95*, 305 (305), *312*
Tilstam, U. 513 (16), *584*
Tinga, M. A. G. M. 40 (140), *95*
Tinoco, I. 336 (163), *362*
Tishkov, A. 557 (208), *590*
Tissot-Croset, K. 558 (220), *590*, 791 (50, 57, 58), *802*
Tius, M. A. 450 (27), *507*
Tjepkema, M. W. 646 (37), *677*
Tjurina, L. A. 171 (5, 100, 102), 175 (101), *184, 186*, 268 (5), *306*
Tobrman, T. 518 (45), *585*
Tochtermann, W. 4 (33), *92*
Todo, H. 545 (136), *588*
Togni, A. 775 (27), *801*
Tohdo, K. 370 (112), *509*
Tokoro, N. 300 (312), *313*
Tokunaga, N. 779 (33), *801*

Tolmann, W. B. 322 (41), *360*
Tolstikov, G. A. 235 (51), *261*
Tomasi, J. 384 (30), *402*
Tomasini, C. 564 (253), *591*
Tomaszkiewicz, I. 107 (20), *126*
Tomedi, P. 295 (288), *312*
Tomilov, A. P. 219 (7), *260*
Tomimoto, K. 531 (90), *586*, 699 (31), *715*
Tominaga, S. 791 (60), *802*
Tomioka, K. 564 (254), *591*, 772 (6), 774 (2, 9, 25), 779 (22), *800*, *801*
Tompkins, J. 723 (16), *768*
Tonami, K. 302 (329), *313*
Toney, J. 38 (128), 62 (200, 205), 67 (217), *95*, *97*, 384 (31), *402*
Tongate, C. 275 (193), *310*
Tooke, D. M. 20 (70), *93*
Topf, J. M. 268 (89), *307*
Topiol, S. 369 (1), *401*
Topol, I. A. 158 (27), *185*, 268 (147), *308*, 331 (117, 118), *362*
Topolski, M. 513 (12), *583*
Tornøe, C. W. 576 (303), *592*
Torrente, E. 727 (24), *768*
Torres, L. E. 457 (48), *507*
Torres, R. A. 341 (202), *363*
Tortajada, J. 125 (101), *129*, 182 (138), *187*
Torti, S. V. 648 (46), *677*
Tortorella, P. 583 (317, 318), *593*
Tosiello, L. 268 (94), *307*
Touitou, Y. 268 (40), *306*
Touster, J. 255 (127, 128), *263*
Towrie, M. 200 (67), *216*
Toyozawa, A. 341 (199), *363*
Trachtman, M. 319 (25), *360*, 391 (22), *402*
Trafny, E. A. 348 (234), *364*
Traxler, M. D. 503 (53), *507*
Treber, J. 45 (154), 47 (159), *95*, *96*
Trécourt, F. 44 (44), *93*, 524 (60), 537 (103), 550 (173), 553 (186), *585*, *586*, *589*, 689 (26–28), 700 (33, 34), 713 (24, 25), *714*, *715*
Trellopoulos, A. V. 281 (227), *310*
Tremaine, J. F. 25 (85), *94*
Tresset, G. 329 (99), *361*
Tressol, J. C. 288 (256), *311*
Tribolet, R. 332 (126), *362*
Trindle, C. 158 (31), *185*
Tripoli, R. 473 (74), *508*, 539 (111), *587*
Trost, B. M. 439 (5), 443 (13), 485 (4), *506*, 540 (115), *587*, 707 (47), *715*, 791 (52), 798 (65), *802*
Troupel, M. 236 (60), 258 (38), *261*, 291 (279), *312*
Troxler, R. E. 208 (97), *217*
Troxler, R. F. 211 (108), *217*
Troy, J. 391 (22), *402*
Trucks, G. W. 369 (2), 384 (30), *401*, *402*

Truglio, J. J. 351 (280), *365*
Truhlar, D. G. 350 (263), *365*
Trzcinski, W. A. 103 (5), *126*
Tsai, M. D. 352 (290, 296), *365*
Tsang, R. C. 268 (133), *308*
Tsay, F.-R. 103 (8), *126*, 519 (52), *585*
Tsay, Y.-H. 33 (114), 45 (157), *94*, *96*
Tschaen, D. 699 (31), *715*
Tschelinzeff, W. 104 (11, 12), *126*
Tseng, H.-R. 559 (230), *590*
Tshelintsev, V. 104 (12), *126*
Tsien, R. Y. 268 (23), *306*
Tso, E. L. 268 (83), *307*
Tsodikov, O. V. 334 (138), *362*
Tsuda, T. 300 (312), *313*
Tsuge, O. 452 (35), 474 (96), *507*, *508*
Tsuji, M. 274 (187), *310*
Tsuji, T. 545 (139), 546 (141), *588*, 707 (45), *715*
Tsukada, M. 145 (39), *154*
Tsunoi, S. 302 (333), *313*
Tsurusaki, N. 259 (140), *263*
Tsuruta, H. 212 (116), *217*
Tsutsui, H. 579 (309), *592*
Tsutsumi, T. 259 (138, 139), *263*
Tsvirko, M. P. 211 (107), *217*
Tuchinda, P. 725 (19), *768*
Tuck, D. G. 224 (29), *260*
Tucker, B. J. 336 (153), *362*
Tulub, A. A. 158 (29, 30), *185*, 332 (130), 333 (129, 134), *362*
Tulub, A. V. 158 (30), 171 (104), *185*, *186*
Tunuli, S. 206 (79), *217*
Turck, A. 518 (43), *585*, 702 (40), *715*
Turgeman, R. 247 (108), 248 (96, 113), 250 (115), 251 (112), 258 (111), *262*
Turnbull, A. G. 122 (75), *128*
Turnlund, J. R. 288 (254, 255), *311*
Turon, F. 301 (358), *314*
Tuulmets, A. 512 (8, 9), *583*
Tuulmets, A. V. 116 (57), *128*
Tvaroska, I. 323 (54, 57), *360*
Twamley, B. 81 (81), *94*
Tzanavaras, P. D. 281 (227), *310*
Uchibori, Y. 391 (23), *402*
Uchida, M. 304 (353), *314*
Uchiyama, M. 404 (2), 419 (51), *434*, *435*, 713 (55), *715*
Ueda, I. 329 (97), *361*
Ueda, Y. 721 (11), *768*
Uemura, S. 558 (219), *590*
Ueyama, K. 779 (33), *801*
Ugajin, S. 774 (11), *800*
Ugalde, J. M. 322 (45), *360*
Uhlenbeck, O. C. 339 (182), 341 (180), *363*
Uhm, H. L. 3 (27), *92*, 512 (6), *583*
Ulaşan, M. 277 (206), *310*
Umemoto, H. 159 (18), *184*

Author Index

Umetani, S. 279 (218), *310*
Umland, F. 279 (185), *309*
Umpleby, J. D. 656 (83), *678*
Unger, M. A. 302 (330), *313*
Ungváry, F. 55 (182), *96*, 512 (5), , *583*, 172 (107), *186*, 512 (5), *583*
Uno, H. 720 (9), *768*
Unterer, S. 276 (195, 196), *310*
Uppal, J. S. 160 (47), *185*
Urabe, H. 572 (287), *592*, 625 (91), *630*
Urpi, F. J. 503 (123), *509*
Usher, D. A. 342 (205), *363*
Usman, N. 340 (187), *363*
Usón, I. 74 (245), *98*, 413 (43), *435*
Utimoto, K. 527 (73), 540 (122), *586*, *587*, 636 (19), 661 (103), *676*, *678*
Utley, J. H. P. 253 (123), *263*
Uzal Barbeito, L. A. 302 (339), *313*
Valente, C. 550 (167), *589*
Valiev, M. 348 (243), *364*
Vallino, M. 59 (193), *97*
van den Heuvel, L. P. 268 (65), *307*
van den Heuvel, L. P. W. J. 268 (66), *307*
van der Kerk, G. J. M. 77 (251), *98*
van der Sluis, P. 38 (83), *94*
van der Wal, G. 116 (44), 121 (71), *127*, *128*
Van Draanen, N. A. 468 (83), *508*
van Gammeren, A. J. 356 (332), *366*
van Grondelle, R. 194 (34), *215*
Van Hook, J. W. 268 (80), *307*
Van Houten, B. 351 (280), *365*
van Ingen, H. E. 275 (191), *310*
Van Klaveren, M. 558 (218), *590*, 774 (8), 791 (53), *800*, *802*
van Koten, G. 30 (105), 69 (29), 79 (257, 258), *92*, *94*, *98*, 519 (53), 545 (137), 558 (218), *585*, *588*, *590*, 774 (8, 25), 791 (53), *800*, *802*
van Leeuwen, H. P. 269 (151), *309*
van Lier, J. E. 193 (21), *215*
Van Nuwenborg, J. E. 273 (180), *309*
van Staden, J. F. 280 (226), *310*
Van Summeren, R. P. 786 (42), *801*
van Veldhuizen, J. J. 791 (59), *802*
van Willigen, H. 196 (50), *216*
van Zijl, A. W. 791 (56, 62), 798 (63), *802*
Vance, C. J. 241 (37), *261*
Vandenheede, J. R. 268 (37), *306*
Vander, R. J. 295 (292), *312*
Vaquero, M. P. 268 (137), *308*
Varchi, G. 519 (51), 531 (89), 534 (81), 559 (228), 563 (245), *585*, *586*, *590*, *591*
Varga, V. 89 (274, 275), *99*
Vargas, W. 71 (229), *98*, 407 (23), *434*
Vasil'ev, V. P. 121 (70), *128*
Vasilevski, D. A. 582 (315), *593*
Vater, H. D. 722 (13), *768*
Vedsø, P. 518 (49), 525 (64), *585*

Veenstra, A. 325 (74), *361*
Veith, U. 573 (290), *592*
Velarde-Ortiz, R. 514 (20), *584*
Verkade, J. 472 (94), *508*
Verma, D. K. 302 (338), *313*
Verpeaux, J-N. 727 (23), *768*
Verpeaux, J. N. 608 (52, 53), *629*
Verresha, G. 571 (282), *592*
Vestergren, M. 59 (61, 145), 60 (185), *93*, *95*, *96*
Vestfrid, Y. 247 (106), 252 (116), *262*
Vicario, J. L. 545 (137), *588*
Vicens, Q. 337 (171), *363*
Vickery, T. 103 (8), *126*, 519 (52), *585*
Victorov, N. B. 540 (123), *587*
Viebrock, H. 10 (54), 41 (125), 50 (166), *93*, *95*, *96*
Vierling, W. 268 (46), *307*
Viestfrid, Yu. 253 (117), *262*
Vigneron, J. 243 (83), *262*
Vigroux, A. 331 (114–116), *361*, *362*
Villà, J. 350 (263), *365*
Villacorta, G. M. 772 (7), *800*
Villalva-Servín, N. P. 647 (42), *677*
Villena, A. 34 (106, 116), 66 (213, 216), *94*, *95*, *97*
Villieras, J. 288 (262), *311*, 442 (10), *506*, 765 (48), *769*
Villiéras, J. 516 (31), *584*
Villiger, V. 2 (6), *92*, 107 (24), *127*
Vinarov, D. A. 350 (260), *365*
Vink, R. 268 (145), *308*
Visser, M. S. 673 (144, 146, 147), *679*
Vlcek, A. 200 (67), *216*
Vlismas, T. 289 (272), 292 (284), *312*
Vogels-Mentink, G. M. 287 (248), *311*
Vogt, M. 341 (198), *363*
Voitjuk, A. A. 333 (132), *362*
Volante, R. P. 654 (72), *678*, 699 (31), *715*
Volden, H. 29 (102), *94*, 123 (85), *129*
Volden, H. V. 183 (147, 149), *187*
Vollet, J. 26 (97), 68 (220), *94*, *97*
Voloshin, M. 300 (300), *312*
von der Brück, D. 517 (39), *585*
von Hippel, P. H. 351 (270, 271), *365*
Von Rein, F. W. 645 (35, 36), *677*
von Schnering, H. G. 573 (290), *592*
von Zezschwitz, P. 531 (93), *586*
Voorbergen, P. 62 (199), *97*, 384 (31), *402*
Vormann, J. 268 (73), *307*
Vostrikova, O. S. 421 (53), *435*, 671 (131–133), *679*
Vostrowsky, O. 537 (104), *586*
Vrade, H. 543 (124), *587*
Vrana, K. E. 347 (236), *364*
Vreugdenhill, A. D. 107 (30), *127*
Vrieze, C. 79 (257), *98*
Vrieze, K. 291 (278), *312*

Vrkic, A. K. 181 (131), *187*
Vu, V. A. 268 (8), *306*, 518 (36, 42), 525 (41, 67, 69), 528 (83), *584–586*, 738 (33), 748 (40), *768, 769*
Wacker, W. E. 268 (28), *306*
Wada, T. 661 (102), *678*, 699 (31), *715*
Wada, Y. 329 (97), *361*
Wade Downey, C. 503 (62), 505 (61), *508*
Waggoner, K. M. 58 (58), *93*, 682 (4), *714*
Wagman, D. D. 113 (2), *126*
Wagner, A. J. 268 (10, 11), *306*, 525 (66), *585*
Wagner, B. 287 (192), *310*
Wagner, B. O. 136 (8), *153*
Wagner, C. 59 (132), *95*, 149 (6), *153*
Wagner, I. 350 (267), *365*
Wagner, R. C. 599 (36), *629*
Wagnerova, D. M. 193 (20), *215*
Wahab, A. 322 (47), *360*
Wakabayashi, K. 560 (237), *591*
Wakao, N. 193 (23–26), 194 (29), 213 (119), *215, 217*
Wakasugi, D. 734 (29), *768*
Wakasugi, K. 622 (85), *630*
Wakefield, B. J. 389 (3), *401*, 451 (31), *507*, 512 (3), *583*, 633 (1), 657 (90), 674 (150), *676, 678, 679*, 690 (29), *715*, 720 (8), *768*
Wakisaka, A. 341 (199), *363*
Waladkhani, A. R. 214 (130), *218*
Walborsky, H. M. 55 (179), *96*, 259 (145), 260 (150), *263*, 512 (5), 513 (12, 14), *583*
Wald, S. A. 395 (35), *402*
Walder, L. 219 (8), *260*
Walfort, B. 59 (132), *95*, 149 (6), *153*
Walker, F. 134 (18), *153*
Walker, G. C. 355 (316), *366*
Walker, J. E. 329 (94, 102), *361*
Walker, N. 175 (122), *187*, 318 (23), *359*
Walker, N. R. 184 (159), *187*
Walling, C. 55 (181), *96*, 513 (14), *583*
Wallis, J. M. 67 (93), *94*
Walter, D. 124 (97), 125 (105), *129*, 161 (56), *185*
Wältermann, M. 301 (359), *314*
Walters, I. A. S. 454 (40), *507*
Walters, R. S. 184 (159), *187*
Walther, D. 39 (133), *95*
Walther, H. 160 (49–51), *185*
Walther, L. E. 278 (210), *310*
Wan, W. B. 531 (94), *586*
Wan, Z. 547 (150), *588*
Wang, C-J. 779 (20), *800*
Wang, C. 168 (73, 74), *186*
Wang, C. C. 352 (294), *365*
Wang, C.-S. 163 (71), *186*
Wang, D. N. 356 (328), *366*
Wang, G. T. 122 (78), *128*
Wang, H. 574 (296), *592*
Wang, H.-M. 170 (97), *186*

Wang, J. 253 (120), *262*, 276 (197), *310*, 352 (294), *365*, 644 (33), *677*
Wang, J.-Q. 51 (170), *96*
Wang, K. 356 (326), *366*
Wang, L. 169 (80), *186*, 277 (202), *310*
Wang, L. Y. 322 (46), *360*
Wang, M. 559 (232), *590*
Wang, P. 272 (176), *309*, 328 (85, 86), *361*
Wang, Q. 565 (258–261), *591*, 667 (121, 122), 670 (123–125), 671 (126, 127), *678, 679*
Wang, S. 341 (188), *363*, 565 (256), *591*, 644 (33), *677*
Wang, S.-Y. 782 (39), *801*
Wang, S. Z. 325 (73), *361*
Wang, T. 291 (280), *312*, 595 (8), *628*
Wang, X.-J. 521 (58), *585*
Wang, Y. 52 (173), 53 (174), 81 (228), *96, 98,* 149 (46), *154*, 293 (285), *312*, 422 (25), *435*, 547 (154), 566 (262), *588, 591*
Wanklyn, J. A. 4 (30), *92*
Warashina, M. 341 (189), *363*
Warburg, O. 348 (247), *364*
Ward, H. D. 765 (48), *769*
Ward, H. R. 138 (20), *153*
Wardell, J. L. 302 (339), *313*
Warnock, D. G. 268 (64), *307*
Warren, M. A. 325 (74), *361*
Warshel, A. 352 (300), *365*
Wasielewski, M. R. 194 (31), 195 (36), *215*
Wasser, P. K. W. 208 (95), *217*
Watabe, H. 561 (242), *591*
Watanabe, K. 284 (178), *309*
Watanabe, M. 526 (71), *586*, 748 (41), 758 (44), *769*
Watanabe, T. 194 (29), 213 (119), *215, 217*, 701 (36), *715*
Waterhouse, A. 621 (80), *630*
Watkins, E. K. 657 (91), *678*
Watson, S. C. 288 (261), *311*
Watt, D. S. 528 (85), *586*
Watt, G. D. 328 (86), *361*
Weakley, T. J. R. 531 (94), *586*
Weare, J. H. 348 (243), *364*
Weatherstone, S. 15 (62), 17 (66), 81 (224, 259), *93, 97, 98*, 414 (35, 46), 430 (80), 431 (82), *435, 436*
Weaver, D. 667 (121), *678*
Webb, L. J. 243 (84), *262*
Weber, B. 571 (281), *591*
Weber, F. 26 (94), *94*
Weber, G. 349 (249), *364*
Weber, J. 330 (105), *361*
Webster, D. E. 122 (76), *128*
Wedekind, J. E. 349 (253, 255, 259), *364, 365*
Weeber, A. 49 (165), *96*
Weghuber, J. 325 (66), *360*
Wehlacz, J. T. 634 (12), *676*
Wehmschulte, R. J. 79 (79), 81 (81), *94*

Wei, G. 563 (250), *591*, 671 (128–130), *679*
Wei, X.-H. 62 (120, 201), *95*, *97*
Weigmann, R. 579 (309), *592*
Weinberg, M. S. 268 (107), *308*
Weiner, M. 595 (8), *628*
Weingarten, M. D. 718 (2), *768*
Weinhold, F. 268 (147), *308*
Weinmann, H. 513 (16), *584*
Weiss, E. 5 (38), 6 (48), 9 (53), 10 (54), 14 (50), 25 (21, 22, 90), 38 (122, 123), 41 (125), 42 (49, 144), 50 (166), 89 (89), *92–96*, 682 (3, 10), *714*
Weiss, F. T. 300 (315), *313*
Weiss, J. 207 (83, 84), *217*
Welker, M. E. 648 (46), *677*
Weller, D. D. 451 (34), *507*
Welling, M. 160 (49–51), *185*
Wells, J. A. 344 (211), *364*
Welmaker, G. S. 563 (247), *591*
Welsby, A. M. 109 (47), *127*
Wende, W. 355 (319), *366*
Wenkert, E. 656 (83), *678*
Weskamp, T. 550 (166, 168), *589*
Wessjohann, L. A. 774 (25), *801*
Westeppe, U. 44 (152, 153), 45 (154, 156), *95*, *96*, 540 (118), *587*
Westera, G. 145 (27), *153*
Westerhausen, M. 27 (99), 81 (261), *94*, *98*, 302 (302), *312*, 409 (30), 410 (37), 427 (72, 75), *435*, *436*
Westhof, E. 341 (192), *363*
Weston, J. 332 (14), 344 (214), 350 (266), *359*, *364*, *365*
Westover, K. D. 355 (313), *366*
Whang, D. D. 268 (79, 84), *307*
Whang, R. 268 (45, 79, 84, 99), *306*, *307*
Whetten, A. 171 (25), *185*
Whitby, R. J. 673 (140, 141), *679*
White, A. H. 31 (109), 40 (139), 46 (158), 58 (155), 81 (260), *94–96*, *98*, 123 (89), *129*, 405 (5), 427 (74), *434*, *436*
White, A. J. P. 71 (231), 74 (241), 76 (242), 77 (250), *98*, 411 (33), 418 (14), 424 (44), *434*, *435*, 484 (99), *508*
White, D. A. 238 (6), *260*
White, J. D. 667 (120), *678*
Whiteside, R. A. 369 (1), *401*
Whitesides, G. M. 137 (7), 138 (22–24), 141 (34), *153*, *154*, 156 (3), *184*, 259 (148, 149), *263*, 513 (14), *583*, *584*
Whittle, R. R. 56 (55), 69 (183), *93*, *96*, 410 (38), 429 (78), 431 (81), *435*, *436*, 684 (13), *714*
Wiederhold, L. R. 353 (307), *366*
Wiehe, A. 191 (8), *215*
Wieneke, B. 27 (99), *94*
Wieser-Jeunesse, C. 775 (28), *801*
Wilcox, C. S. 485 (4), *506*

Wilcox, D. E. 343 (48), *360*
Wilczok, U. 45 (154), *95*, 610 (60), *630*
Wildschutz, T. 268 (117), *308*
Wilke, G. 86 (266, 267), *99*
Wilkes, J. G. 176 (125, 126, 128), *187*
Wilkinson, G. 23 (75), 25 (87), 58 (3, 4), *92–94*, 150 (12), *153*, 512 (4), 513 (13), *583*, 657 (90), *678*
Wilkinson, H. S. 583 (319), *593*
Wilkinson, P. D. 289 (269), *312*
Willard, G. F. 424 (68), *436*
Willems, J. L 287 (248), *311*
Willey, K. F. 124 (98), *129*
William, A. D. 480 (11, 12), *506*
Williams, D. J. 71 (231), 74 (241), 76 (242), 77 (250), *98*, 411 (33), 418 (14), 424 (44), *434*, *435*, 484 (99), *508*
Williams, E. R. 318 (19, 20), *359*
Williams, L. D. 335 (30), *360*
Williams, M. A. 300 (317), *313*
Williamson, R. C. N. 282 (232), *311*
Williard, P. G. 439 (5), *506*
Wilson, D. M. 353 (305), *366*
Wilson, P. D. 646 (37), 647 (43), 648 (44), *677*
Wilson, S. H. 352 (288, 299), *365*
Wilson, T. J. 341 (194), *363*
Winkler, R. 318 (18), *359*
Winnefeld, K. 278 (210), *310*
Winter, C. H. 47 (160, 161), 50 (101), 72 (234), 78 (252), 83 (233), 84 (264), *94*, *96*, *98*, 122 (77, 79), *128*, 303 (303), *312*, 419 (27), 422 (28, 29), 433 (89), *435*, *436*
Winterton, R. C. 247 (100, 101), *262*
Wiss, J. 512 (10), *583*
Wissing, E. 79 (258), *98*
Witanowski, M. 138 (23, 25), *153*
Witt, H. T. 356 (36, 324), *360*, *366*
Wittig, G. 4 (31, 32), *92*, 686 (1), *714*
Woerpel, K. A. 557 (214), *590*
Wolf, F. I. 268 (58), *307*, 317 (2), *359*
Wolfes, H. 355 (320), *366*
Wong, A. 357 (333), *366*
Wong, I. 352 (289, 292), *365*
Wong, K.-Y. 515 (26), *584*
Wong, M. W. 369 (2), 384 (30), *401*, *402*
Wong, P. T. S. 300 (320), *313*
Wong, T. 646 (37), *677*
Wong, T. H. 162 (60), 164 (63), 169 (88, 89), *185*, *186*
Wood, R. D. 353 (304), 355 (316), *365*, *366*
Woodbury, N. W. 356 (327), *366*
Woodle, M. 206 (80), *217*
Woods, K. L. 268 (143), *308*
Woods, L. A. 296 (298), *312*
Woodson, S. A. 336 (155, 165), 337 (172), *362*, *363*
Woodward, C. A. 176 (120, 121), *186*

Woodward, J. 214 (136), *218*
Woodward, S. 779 (33), 791 (50), *801*
Woon, D. E. 107 (22), *126*
Wotiz, H. 107 (28), *127*
Wowk, A. 295 (286), *312*
Wozniak, R. 25 (82), *94*, 299 (307), *312*
Wright, D. S. 57 (186), *96*, 124 (91), *129*
Wright, G. F. 300 (313), *313*
Wright, L. 293 (267), *311*
Wright, R. R. 175 (122), *187*, 318 (23), *359*
Wu, A-B. 458 (32), *507*
Wu, G. 357 (333), *366*
Wu, H.-F. 178 (130), *187*
Wu, M.-J. 674 (152), *679*
Wu, S. 162 (61), *185*
Wu, T.-M. 16 (63), *93*
Wu, W. 201 (72, 73), *216*
Wu, Y. 168 (73), *186*
Wu, Y.-D. 163 (70, 71), 168 (72, 74), *186*, 391 (24), *402*
Wurst, J. R. 514 (20), *584*
Würthwein, E.-U. 579 (309), *592*
Wyckoff, H. W. 354 (309), *366*
Wynberg, H. 718 (2), *768*
Wynne, K. 198 (56), *216*
Xenos, D. 278 (207), *310*
Xia, A. 47 (160, 161), 50 (101), 72 (234), 78 (252), 83 (233), 84 (264), *94*, *96*, *98*, 122 (77, 79), *128*, 303 (303), *312*, 419 (27), 422 (28, 29), 433 (89), *435*, *436*
Xia, J. 540 (121, 122), *587*
Xie, M. 644 (32, 33), *677*
Xing, X. 351 (284), *365*
Xiong, H. 254 (125), *263*, 540 (119), *587*
Xiong, Y. 419 (54), *435*, 466 (66), *508*, 538 (105), *587*
Xu, D. 107 (20), *126*
Xu, J. 421 (58), *435*, 700 (35), *715*
Xu, X. 619 (78), *630*
Xu, Y. 521 (58), *585*
Xu, Z. 671 (136), 672 (137–139), 673 (145), *679*
Xuong, N. G. 351 (286), *365*
Xuong, N. H. 347 (232), 348 (234, 241), *364*
Yadav, J. S. 576 (303), *592*
Yagi, K. 357 (336), *366*
Yaginuma, F. 576 (302), *592*
Yago, M. D. 268 (48), *307*
Yahara, K. 194 (29), 213 (119), *215*, *217*
Yamabe, S. 384 (27), *402*, 569 (275), *591*
Yamada, A. 198 (64), *216*
Yamada, S. 212 (114), *217*, 514 (21), *584*
Yamada, T. 327 (78), *361*
Yamagami, T. 568 (272), *591*, 622 (88), 623 (89), *630*, 637 (23), 639 (24), *677*
Yamaguchi, K. 63 (60), *93*, 156 (7), 182 (6), *184*, 419 (51), *435*

Yamaguchi, S. 568 (272), *591*, 622 (88), *630*, 637 (23), *677*
Yamakawa, K. 479 (14, 16), 485 (15, 17), *506*, 721 (11), 743 (38), 760 (50), 766 (53), *768*, *769*
Yamamoto, C. 302 (326), *313*
Yamamoto, H. 557 (211), 580 (312), *590*, *592*, 621 (81, 82), *630*, 633 (1), *676*, 690 (29), *715*, 765 (48), *769*, 774 (2), 779 (22), *800*
Yamamoto, M. 274 (187), *310*
Yamamoto, T. 285 (243), *311*
Yamamura, M. 194 (29), 213 (119), *215*, *217*
Yamanaka, M. 791 (51), *802*
Yamani, Y. 760 (50), *769*
Yamasaki, K. 341 (199), *363*, 774 (21), *800*
Yamashita, H. 290 (274), *312*
Yamataka, H. 259 (134), *263*, 391 (11), *402*, 466 (70), 471 (93), *508*
Yamauchi, Y. 559 (225), *590*
Yamazaki, I. 194 (30), 201 (75, 76), 206 (81, 82), *215–217*
Yamazaki, S. 384 (27), *402*, 569 (275), *591*
Yamazaki, T. 558 (218), *590*
Yan, H. 356 (326), *366*
Yan, M.-C. 566 (262), *591*
Yan, N. 659 (99), *678*
Yanagida, T. 329 (97), *361*
Yanagisawa, A. 557 (211), *590*, 621 (81, 82), *630*
Yang, B. 547 (158), *588*
Yang, C. G. 419 (57), *435*
Yang, C.-G. 538 (105), *587*
Yang, D.-Y. 278 (209, 211, 212), *310*
Yang, F. 302 (324, 338), 303 (318), *313*
Yang, H. 572 (289), *592*
Yang, J. 253 (120), *262*
Yang, J.-X. 282 (228), *310*
Yang, J. Z. 547 (156), *588*
Yang, K.-C. 81 (228), *98*, 149 (46), *154*, 422 (25), *435*
Yang, L.-L. 278 (211), *310*
Yang, L.-M. 559 (230), *590*
Yang, N. C. 595 (8), *628*
Yang, Qi 495 (111), *509*
Yang, S. 163 (65, 68–71), 164 (64), 166 (77), 167 (82–85), 168 (72–74, 86), 169 (78–80, 87), 93–96), 170 (97, 98), *185*, *186*
Yang, S. D. 268 (37), *306*
Yang, S. I. 201 (38, 39), *215*
Yang, S. K. 335 (143), *362*
Yang, W. 350 (262), 355 (312), *365*, *366*, 384 (28), *402*
Yang, X. 163 (65, 68, 69), 164 (64), *185*, *186*
Yang, Y. L. 353 (308), *366*
Yang, Z. 247 (35, 103), *261*, *262*, 279 (216), *310*, 572 (288), *592*
Yang, Z. Y. 516 (30), 517 (34), *584*
Yao, C.-F. 566 (262), *591*

Author Index

Yao, J. 567 (267), *591*
Yao, S.-Z. 273 (183), *309*
Yap, G. P. A. 647 (42, 43), *677*
Yarkony, D. 332 (131), *362*
Yarus, M. 337 (173), *363*
Yasar, S. B. 271 (161), *309*
Yasuda, H. 51 (167), 96, 121 (74), 123 (73), *128*, 254 (126), *263*, 540 (119), *587*
Yasuda, M. 686 (22), 688 (23), *714*
Yasuda, N. 524 (63), *585*, 702 (39), *715*
Yasuda, R. 329 (96, 100), *361*
Yasui, H. 557 (216), *590*, 643 (30), *677*
Yazawa, T. 300 (314), *313*
Yee, N. K. 103 (9), *126*, 517 (37), *584*
Yee, R. 253 (124), *263*
Yeh, C. S. 124 (98), *129*
Yeh, S.-M. 547 (154), *588*
Yellowlees, L. J. 81 (255, 256), *98*, 407 (21, 22), *434*
Yeo, W. W. 268 (142), *308*
Yiin, S.-J. 547 (154), *588*
Yin, H.-M. 169 (95), *186*
Yin, Y. W. 355 (314), *366*
Yokoi, N. 193 (25), *215*
Yokoo, T. 527 (73), 540 (122), *586*, *587*
Yokoyama, Y. 741 (36), *769*
Yonemoto, W. 347 (231), *364*
Yonemura, H. 212 (114), *217*
Yong, K. H. 54 (177), 83 (176), *96*, 426 (31), *435*
Yoon, K. 305 (362, 363), *314*
Yoon, T. 559 (229), *590*
Yorimitsu, H. 545 (139, 140), 546 (141), 557 (216), 560 (237, 238), 567 (269), 568 (273), *588*, *590*, *591*, 636 (17, 22), 643 (30), *676*, *677*, 707 (45), *715*, 791 (50), *801*
Yorimitsu, Hideki 675 (158), *714*
York, O. 633 (5), *676*
Yoshida, A. 538 (108), *587*
Yoshida, J. 219 (9), *260*, 608 (56), *629*, 641 (27, 28), 662 (109), 664 (108), *677*, *678*
Yoshida, J.-I. 555 (199), *589*
Yoshida, K. 774 (24), *800*
Yoshida, K. A. 779 (33), *801*
Yoshida, M. 329 (96, 98, 100), *361*
Yoshida, Y. 157 (21), *185*
Yoshida, Z. I. 502 (119, 120), *509*
Yoshida, Z.-I. 452 (37), *507*
Yoshikai, N. 550 (169), 555 (200), *589*
Yoshinari, K. 341 (189), *363*
Yoshino, S. 268 (139), *308*
Yoshino, T. 145 (38), *154*
Yoshiwara, H. 517 (34), 519 (54), *584*, *585*
Young, M. A. 171 (105), *186*
Young, W. G. 141 (33), *154*
Youngblood, W. J. 201 (45), *216*
Yousef, R. I. 59 (132), *95*, 149 (6), *153*
Yu, B.-S. 273 (183), *309*

Yu, C. 293 (285), *312*
Yu, N. 337 (169), *363*
Yu, S. 103 (4), *126*
Yu, S. H. 116 (59), *128*
Yu, Z. 514 (23), *584*
Yuan, H. S. H. 86 (268), *99*
Yuan, Q.-G. 273 (183), *309*
Yuan, T. 162 (61), *185*
Yuan, T.-M. 547 (154), *588*
Yuki, H. 60 (194–196), 61 (197), *97*
Yun, H.-S. 553 (178), *589*
Yurkowski, M. 301 (356), *314*
Yvernault, T. 290 (273), 292 (281), *312*
Zabicky, J. 297 (260), *311*
Zacharoulis, D. 278 (207), *310*
Zagatto, E. A. G. 280 (224), *310*
Zakrzewski, J. 198 (62), *216*
Zakrzewski, V. G. 384 (30), *402*
Zaleski, J. M. 196 (44), 201 (73), *216*
Zanazzi, P. F. 328 (83), *361*
Zandler, M. E. 201 (41, 42), *216*
Zanello, P. 30 (108), *94*, 243 (86), *262*
Zani, L. 596 (30), *629*
Zapf, A. 544 (129, 130), *588*
Zarzuelo, M. M. 583 (319), *593*
Zaug, A. J. 335 (149), *362*
Zawadzka, M. 212 (113), *217*
Zawadzki, S. 547 (147), *588*
Zecchina, A. 125 (108), *129*
Žemva, B. 74 (248), *98*, 424 (16), *434*
Zeng, Y. 277 (202), *310*
Zerbini, I. 268 (130), *308*
Zerevitinov, T. 299 (308), *312*
Zewge, D. 699 (31), *715*
Zhai, L. 286 (244), *311*
Zhang, D. 622 (87), *630*, 652 (53), *677*
Zhang, E. 320 (26), 349 (256, 258), *360*, *364*, *365*
Zhang, F. 659 (99), *678*
Zhang, H. 51 (171, 172), 52 (173), 53 (174), *96*, 200 (67), 201 (72), *216*
Zhang, J. 268 (72), *307*, 347 (237), 356 (326), *364*, *366*
Zhang, L. 521 (58), *585*
Zhang, M.-X. 419 (56, 57), *435*, 538 (105, 106), *587*
Zhang, P. 395 (37), *402*
Zhang, R. 327 (67), *360*
Zhang, W. 760 (47), *769*, 779 (20), *800*
Zhang, X. 168 (73), *186*
Zhang, X.-H. 163 (70), 168 (72, 74), *186*
Zhang, Y. 348 (244), 350 (262), *364*, *365*
Zhang, Y. H. 322 (46), *360*
Zhang, Z. 528 (87), 563 (250), 565 (260, 261), *586*, *591*, 670 (123, 125), 671 (127–130), *679*
Zhao, H. 540 (122), *587*
Zhao, J.-F. 282 (228), *310*

Zhao, L. 162 (61), 167 (83), *185*, *186*, 452 (36), *507*
Zhao, L. J. 322 (46), *360*
Zhao, X. 161 (55), *185*
Zhel'vis, A. I. 224 (32), *261*
Zhen, P. 277 (202), *310*
Zheng, C. 51 (170), *96*
Zheng, J. 347 (232), *364*
Zheng, W. 107 (20), *126*
Zheng, X. 29 (103), *94*
Zhong, X. 352 (296), *365*
Zhonghua, L. 253 (118), *262*
Zhou, D. M. 337 (152), 341 (201), *362*, *363*
Zhou, G. 648 (47), *677*
Zhou, G. X. 291 (280), *312*
Zhou, J. 348 (240), *364*
Zhou, J. M. 341 (201), *363*
Zhou, M. 159 (36), *185*
Zhou, Q-L. 774 (8), *800*
Zhou, Q. 302 (327), *313*, 325 (73), *361*
Zhou, Q.-f. 302 (336), *313*
Zhou, X. 277 (202), *310*

Zhou, Y. 329 (92), *361*
Zhou, Z.-Z. 282 (228), *310*
Zhu, B.-Y. 572 (276), *591*
Zhu, D. 51 (171, 172), 53 (174), *96*
Zhu, L. 272 (174), *309*
Zhu, S. 514 (23), *584*
Zhu, W. 123 (90), *129*
Zieger, H. E. 142 (36), *154*
Ziegler, K. 221 (24, 25), 222 (27), *260*
Ziemer, K. 62 (203), *97*
Ziller, J. W. 69 (223), *97*
Zimmerman, H. E. 503 (53), *507*
Zirngast, M. 38 (130), *95*
Zisbrod, Z. 268 (19), *306*
Zoellner, E. A. 289 (270), *312*
Zoller, M. J. 347 (235), *364*
Zouni, A. 356 (324), *366*
Zsurka, G. 325 (66), *360*
Zubritskii, L. M. 540 (123), *587*
Zuiderweg, E. R. P. 344 (212), *364*
Zureck, J. 333 (133), *362*
Zuther, F. 214 (126, 127), *218*
Zwierzak, A. 547 (147), *588*

Subject Index

Entries are in letter-by-letter alphabetical order ignoring spaces and punctuation marks. Page numbers in *italic* refer to Figures and Tables not included in the relevant page ranges.

AAS *see* Atomic absorption spectrometry
Ab initio calculations, modified Schlenk
   equilibrium, 378–380
Absorption spectra, tetrapyrroles, 192–193
Acetaldehyde, Mg cation complex, 169
Acetamide, Mg cation fragmentation, 176, 177
Acetates, aldol stereochemistry, *447*
Acetic acid
   Mg cation complex, 169
   organomagnesate anion decarboxylation, 181
Acetolysis, enthalpies of reaction, 120
Acetone
   enthalpy of formation, 115
   Mg cation binding energy, 175
Acetonitrile, Mg cation fragmentation, 176, 177
Acetophenone
   butylation, 684–685
   potentiometric titration, 292
Acetylenes
   halogen–Mg exchange, 526
   hydromagnesiation, 540, 543
Acetylenic fragments, PET, 196, 197
Acetylenic organomagnesium amides, structures, 81–82
Acetylides, donor–acceptor complexes, 42–43
Acid–base reactions, organomagnesium anions, 180
Acid Chrome Blue K, UV–visible spectrophotometry, 282
L-Acosamine, aminosugar, 487, 488
Active hydrogen compounds
   gas chromatography, 295
   Grignard analytical reagents, 299–300
Acyclic enones, enantioselective conjugate addition, 774, 779, 780

α-Acyl aminoesters, preparation, 462, 497–498
Acylation, Fe-catalyzed, 559, 599–603
C-Acylation, dialkyl malonate, 462, 494–496
Acyl chlorides, organometallic compound acylation, 559
N-Acyloxazolidinones
   aldol-type reaction, 503
   amination, 504
   preparation, 462, *463*
Acyl silanes, α,β-unsaturated, 450
N-Acylthiazolidinethiones
   aldol-type reaction, 503
   preparation, 462, *463*
Addition reactions
   1,2-addition, 17, 450–451, 569
   1,4-addition, 4–5, 450–451, 478, 543, 547, 563–565
   asymmetric, 454–457, 564
   addition–elimination, 471
   conjugated unsaturated carbonyl compounds, 624–625, 626, 628
   enantioselective addition, 571
   conjugate addition, 771–791
   functionalized Grignard reagents, 543, 559–575
   Grignard carbonyl additions
      experimental background, 370–374
      polar mechanisms, 370, 387–388, 389, 391, 394, 396, 400–401
      theoretical studies, 369–370, 374–401
      transition-state structures, 375–376, *377*, 380–382, 388–389, 391, *393*, 394–396
      without solvent molecules, 384–389, *390*, *391*

The chemistry of organomagnesium compounds
Edited by Z. Rappoport and I. Marek © 2008 John Wiley & Sons, Ltd

Addition reactions (*continued*)
  with solvent molecules, 389, 391, *392*, *393*, 401
  nucleophilic, 256–257, 683–686
  organomagnesiates
    1,2-addition, 17
    1,4-addition, 4–5
    oxidative, 512–515
  see also Carbomagnesiation; Conjugate addition
Adduct-forming reactions
  Mg cations, 160
  radiative associative, 160
Adenosine diphosphate *see* ADP
Adenosine monophosphate *see* AMP
Adenosine triphosphate *see* ATP
ADP, glycolytic cycle, 345, *346*
AES *see* Atomic emission spectrometry
Agostic interactions
  diorganomagnesium compounds, 24
  heteroleptic organomagnesiates, 18
Air exclusion, calorimetry, 104
Alcohols
  active hydrogen determination, 299
  Mg cation reactions, 161, 165–166
Aldehydes
  one-carbon ring expansion, 763–764
  organomagnesium amide reactions, 423–426
Aldol-type reactions
  α-bromoketones, 440, 482
  diketones, 493–495
  β-hydroxyacid synthesis, 486
  β-lactam, 499–500
  lactone synthesis, 486
  magnesium enolate preparation, 447–450, 463
  oxazolidinones, 503
  silyl enol ether, 472–473, 484, 486
  α-sulfinyl magnesium carbanion enolates, 459, 486, 487, 500
  sulfoxides, 486, 487
  tandem conjugate addition, 453, *788*, 789
  thioamides, 500–501, *502*
Alkali-activated PTFE tube, UV–visible fluorometry, 283
Alkali alkoxides, heteroleptic organomagnesiate production, 15–16
Alkaline phosphatase (AP), allosteric behavior, 344
Alkalinity, hydrolysis, 289
Alkanes
  Mg atom reactions, 157
  Mg cation reactions, 160, 162–163, 180
Alkenes
  carbomagnesiation reactions, 633, 653–674
    allyl/homoallyl alcohols, 664–671

    allylmagnesiation, 655–657
    intermolecular, 655–656, 662, 663, 665–666, 671–674
    intramolecular, 654–655, 656, 661–662, 665–666
    simple alkenes, 653–655
    strained alkenes, 657–661
    vinylsilanes, 661–664
    Zr-catalyzed ethylmagnesiation, 671–674
  Mg cation reactions, 160
ortho-Alkenylated products, formation, 749, *750–751*
Alkenylation
  arylamines, 748–749, *750–751*
  Fe-catalyzed, 604–608, 626
  magnesium alkylidene carbenoids, 748
Alkenylboron reagents, Rh-catalyzed conjugate addition, 774
Alkenyl bromides
  Br_2–Mg exchange, 527
  Fe-catalyzed cross-coupling, 604–605, *606*, 607–608
Alkenyl chlorides, Fe-catalyzed cross-coupling, 605, *606*, 607–608
Alkenyl derivatives, Fe-catalyzed cross-coupling, 608–610
Alkenyl halides
  Fe-catalyzed cross-coupling, 605–608
  halogen–Mg exchange, 525–526
Alkenyl iodides, I_2–Mg exchange, 526–527
Alkenyl sulfides, trisubstituted, 758–760
Alkoxides
  cubane structure, 70–71, 145, *146*
  heteroleptic monoorganomagnesium compounds, 69–71
  NMR spectra, 145–147
  organomagnesium-group 16-bonded complexes, 428–432
  solution-state species, 132, *133*
Alkoxydialkylmagnesate, addition reactions, 711, 713
Alkoxymagnesium bromides, enthalpies of formation, 114–115, *116*
Alkylation
  enantioselective allylic, 792–799
  enolate intermediate, *762*, 763
  Fe-catalyzed, 615–618
Alkyl bromides
  cross-coupling reactions
    Fe-catalyzed, 617
    transition-metal-catalyzed, 544–546
n-Alkyl bromides, transition-metal-catalyzed cross-coupling, 544
Alkyl chlorides, primary, 544
Alkyl fluorides, transition-metal-catalyzed cross-coupling, 545
Alkyl Grignard reagents

Subject Index

Fe-catalyzed arylation and heteroarylation, 610–614
oxidation at inert electrodes, 232–233
distribution of anodic products, *232*
Alkyl halides
  halogen–Mg exchange, 528
  Mg atom reactions, 157–158
  Mg cation reactions, 160–161
  Mg cluster reactivity, 157
  multi-carbon homologation, 723–724
  Schlenk equilibrium, 108–109, 134
Alkylidene carbenoids *see* Magnesium alkylidene carbenoids
Alkylidene cyclopropanes, synthesis, 739–740
Alkylmagnesium amides, weak acid deprotonation, 419
Alkylmagnesium bromides, methylene increment, 110–111
Alkylmagnesium chlorides
  lithium chloride complexes, 682
  Schlenk equilibrium, 107
  transition-metal-catalyzed cross-coupling, 545
Alkylmagnesium compounds, NMR spectra, 133–138, 141
Alkylmagnesium fluorides, Schlenk equilibrium, 107
Alkylphosphonium ylides, addition, 567
Alkyl transfer, catalytic enantioselective conjugate addition, 790
Alkyl trihaloacetate, halogen–metal exchange, 442
Alkynes
  carbomagnesiation reactions, 633–653
    alkynylsilanes, 639–641, 642
    intermolecular, 635–639
    intramolecular, 633–635
    N-, O- and S-attached alkynes, 641–644
    propargyl alcohols, 645–652
    propargyl amines, 652–653
  Mg film self-hydrogenation, 173
  unactivated, 567, 568
Alkynylsilanes, carbomagnesiation, 639–641, 642
All-carbon quaternary chiral centers, 565
Allenes
  Doering–LaFlamme synthesis intermediates, 735–738
  magnesium alkylidene carbenoid alkenylation, 748
2,-Allenols, synthesis by substitution, 559, 560
β-Allenyl ester magnesium enolates, asymmetric synthesis, 449, 450, 506
Allenyl iodides, I_2–Mg exchange, 527, 528
Allenyl ketones, α,β-enone formation, 563
Allosteric enzymes, protein-based, 343–345
Allyl alcohols
  carbomagnesiation, 664–671

cyclopropanation, 722
substitution, 557
Allyl amines, functionalized Grignard reactions, 548, 549
Allyl carbamates, substitution, 557
Allyl ethers, substitution, 557
Allylic alkylation, enantioselective, 792–799
Allylic isomers, allylmagnesium compounds, 142–143, 144
Allylic N-protected aminoesters, magnesium chelate, 446
Allylic substitutions, functionalized Grignard reagents, 543, 557–558
Allylmagnesiation, alkenes, 655–657
Allylmagnesium bromide, DME complex, 59
Allylmagnesium chloride
  di-Grignard reagents, 62
  tetraorganomagnesiate production, 7
Allylmagnesium compounds
  allylic exchange, 142, *143*
  dynamic equilibria, 143, *144*
  $^n J_{HH}$ spin coupling constants, 143–144, *145*
  NMR spectra, 141–145, 151
  Schlenk equilibrium, 107, *140*, 142, 144–145
Allylmagnesium β-diketiminate, structure, 76, 77
Allylmagnesium halides, transition-metal-catalyzed cross-coupling, 545–546
Allylsilanes, catalyzed addition to C=C bonds, 561
Aluminum compounds, organic derivatives, 103
Amide enolates, reactions with electrophiles, 499–500
Amides
  heteroleptic monoorganomagnesium compounds, 54, 71–83
  mixed Mg/Li, 539
  secondary, 81, *82*
  Weinreb, 559
  also Organomagnesium amides
Amidocuprates, oxidative coupling, 580–581
Amination
  N-acyloxazolidinones, 504
  electrophilic, 575–580
Amines
  active hydrogen determination, 299
  carbomagnesiation, 641–642, 652–653, 667
  Mg cation complexes, 167–169
  sterically hindered, 457, 464, 538
α-Amino acids
  electrophilic reactions, 729, *730*
  functionalized Grignard reagent synthesis, 517–518
β-Amino acids, chiral derivatives, 454

β-Amino-α,β-diesters, Michael addition preparation, 489
Aminoester enolates, 487, 489
Aminoesters, allylic N-protected, 446
γ-Amino-β-ketoester derivatives, ethyl hydrogen malonate, 497
α-Amino-substituted carbanions, non-stabilized, 728–729
α-Amino-substituted cyclopropylmagnesium, 740, 741–742
Aminosugar, L-acosamine, 487, 488
Ammonia, organomagnesium ion reactivity, 180
AMP, cAMP, 347–348
Analytical aspects, 265–314
  acronyms, 266–267
  elemental analysis of Mg, 268, 269–288
  Grignard analytical reagents, 269, 299–305
  speciation analysis, 268, 269, 288–299
Angle strain, magnesacycloalkanes, 120–121
Anhydrides, enantioselective desymmetrization, 571
Aniline, potentiometric titration, 291
Anilines, functionalized Grignard reagent protection, 530–531
Anion abstraction, Mg cation–organohalogen complexes, 163
Anionic ligands, Mg cation fragmentation, 176–178
Anions
  anthracene radical anions, 45, 46
  bimolecular reactions, 180–182
  bis-anions, 15
  chiral oxazolidinones, 503–505
  β-diketiminate, 74
  indenyl, 29
  magnesate complex fragmentation, 178–179
  speciation analysis of Mg, 269
  tris[aryl]magnesiate, 35
ANN (artificial neural networks), 269–270
Anodic oxidation, 227–244
  diorganomagnesium compounds, 227–228
  Grignard reagents, 228–243
    addition to olefins, 237
    inert electrodes, 229–236
    reactivity, 235–236, 258–260
    sacrificial anodes, 237–241
    semiconductor anodes, 241–243
Antenna pigments, chlorophylls, 191
Anthracene, magnesium compounds, 44–47, 514
9-Anthraldehyde, liquid chromatography, 293
Anti-carbomagnesiation
  alkenes, 672–673
  alkynes, 644, 645–651
AP (alkaline phosphatase), 342
Arene metalation, heteroleptic organomagnesiate production, 22

Aromatic compounds, carbomagnesiation reactions, 674–675, 676
Aromatic Grignard reagents, Fe-catalyzed homocoupling, 619–621
Arrays, porphyrins, 201, 204–205
Arsenic compounds, derivatization, 303
Artificial neural networks (ANN), speciation analysis, 269–270
Aryl acetic acids, Ivanov reaction, 458, 502, 503
Aryl amidocuprates, oxidative coupling, 580–581
Aryl amines, direct alkenylation, 748–749
Arylation, Fe-catalyzed, 610–614
Arylazo tosylates, electrophilic amination, 577–578
Arylboron reagents, Rh-catalyzed conjugate addition, 774
Aryl chlorides
  dechlorination, 554
  Fe-catalyzed cross-coupling, 611, 612
  functionalized Grignard reagent preparation, 513
  Kumada–Tamao-Corriu reactions, 550
Aryl fluorides, transition-metal-catalyzed cross-coupling, 550, 551
Aryl glycines, synthesis, 574–575
Aryl Grignard reagents
  Fe-catalyzed reactions
    addition, 624–625
    alkylation, 615–618
    heteroarylation, 614–615
    homocoupling, 619–621
  functionalized, 601, 603
  oxidation at inert electrodes, 233–235
    distribution of anodic products, 233
Arylmagnesium compounds
  dynamic equilibria, 140
  NMR spectra, 134, 138–140, 151
Aryloxides
  NMR spectra, 146, 147
  organomagnesium-group 16-bonded complexes, 428–432
  solution-state species, 132, 133
Aryl sulfoxide, halogen–Mg exchange, 526
Aryl tosylates, Fe-catalyzed cross-coupling, 611, 612
Aryl triflates, Fe-catalyzed cross-coupling, 611, 612
Arynes, functionalized Grignard reagent synthesis, 540, 542
Association, Grignard reagents, 108
Asymmetric 1,4-addition
  magnesium enolate preparation, 454–457
  Michael acceptors, 564
Asymmetric aldol-type reactions, α-sulfinyl magnesium carbanion enolates, 459, 486, 487, 500

Subject Index

Asymmetric homologation, chiral magnesium carbenoids, 722
Asymmetric synthesis
  cyclopropane derivatives, 734
  magnesium enolates, 454–457
    β-allenyl ester magnesium enolates, 449, 450, 506
    β-sulfinyl ester magnesium enolates, 448, 449
Ate complexes, 681–715
Atomic absorption spectrometry (AAS)
  elemental analysis, 272, 277–278
  Mg in body fluids, 270
Atomic emission spectrometry (AES)
  elemental analysis, 278–279
  ICP–AES, 279, 284
  Mg in body fluids, 270
Atomization see Enthalpies of atomization
ATP
  biosynthesis, 329–331
  hydrolysis, 331–333
  isotope effect, 330
  MgATP, 328–329
  Mg^{2+} binding, 317, 331–332
  NMR spectroscopy, 286–287
ATP synthases
  biosynthesis of ATP, 329–330
  PET, 194–195
Aza-allyl type bonding, diorganomagnesium compounds, 32
Aza-crowns, monoorganomagnesium cations, 56
Aziridines, ring opening, 547

Bacteria, photosynthesis, 194
Bacteriochlorins, photosynthesis, 192–193
Bacteriochlorophylls, photosynthesis, 191
Barbier conditions, functionalized Grignard reagent synthesis, 514
Base excision repair, DNA, 353–354
Bases, acid–base reactions, 180
Benzalacetomesitylene, Michael addition, 451
Benzene complexes, photodissociation spectra, 163
Benzimidazoles, synthesis, 577
Benzophenone, potentiometric titration, 292
Benzotriazole, substitution leaving group, 547, 548
O-Benzoyl-N,N-dialkylhydroxylamines, electrophilic amination, 579
Benzyl-n-chloroamines, electrophilic amination, 578–579
Benzylic magnesium reagents, S–Mg exchange, 529–530
Benzyne, radical cations, 163–164
Biaryl compounds, transition-metal-catalyzed cross-coupling, 554

Bicyclo[n.1.0]alkanes, 1,3-C–H insertion of magnesium carbenoids, 732, 733–734
Bidentate coordination
  Mg cation–alcohol complexes, 166
  Mg^{2+}, 322, 324
Bifunctional organomagnesium compounds, structure, 39–40
1,1-Bimetallic species, addition reactions, 566–567
Bimolecular reactions, gas-phase organomagnesium ions, 179–182
BINAP ligand, catalytic enantioselective conjugate addition, 775, 776, 782–784
Binary magnesium salts, enthalpies of atomization, 115
Binding energies, Mg cations, 175
Binding mechanisms, Mg, 320–324, 331–332
Biochemistry, Mg, 315–367
Biochromism, polythiophene, 285
Biological issues, Mg analysis, 268–269, 270, 275–276, 278–279, 282, 287–288
Biomedical issues, Mg analysis, 268–269, 278
Biosynthesis, ATP, 329–331
2,2'-Biquinoline, indicator, 288, 289
Bis(allyl)magnesium compounds, NMR spectra, 143, 144–145, 151
Bis(amide) species
  deprotonation of arenes, 419, 421
  disproportionation reactions, 422
  formation, 412
1,2-Bis(2-aminophenoxy)ethane-N,N,N',N'-tetraacetic acid, UV–visible spectrophotometry, 282, 283
Bis-anions, heteroleptic organomagnesiates, 15
Bis($ortho$-anisyl)magnesium, structure, 34, 35
Bis(benzene)chromium, heteroleptic organomagnesiate production, 22
Bis[1,3-bis{(dimethoxy)methyl}phenyl] magnesium, structure, 34, 35
Bis[1,3-bis{(dimethylphosphino)methyl} phenyl]magnesium, structure, 36
Bis $tert$-butylmagnesium, structure, 24–25
Bis(4-$tert$-butylphenyl)magnesium, donor–acceptor complexes, 41–42
Bis(cyclopentadienyl)magnesium see Magnesocene
Bis-deprotonation, ethyl hydrogen malonate, 458
Bis[3-(dialkyl)aminobutyl]magnesium, structure, 32–33
Bis[4-(dialkylamino)butyl]magnesium, structure, 32–33
1,5-Bis(3,5-dichloro-2-hydroxy)-3-cyanoformazan, UV–visible spectrophotometry, 282
Bis[2,6-diethylphenyl]magnesium, structure, 24, 25

1,2-Bis[(2,6-diisopropylphenyl)imino]
  acenaphthene, radical anion, 79, *80*
Bis[2-(*N*,*N*-dimethylamino)ethyl]ether,
  halogen–Mg exchange, 521, 523
Bis([2-(dimethylamino)methyl]ferrocenyl)
  magnesium, structure, 30, *31*
2,6-Bis(imino)pyridines, dialkylmagnesium
  compound reactions, 71–72
Bis(indenyl)magnesium, structure, 28–29, 48
exo,exo-Bis(iso-dicyclopentadienyl)magnesium,
  structure, 27
Bis(1-methylboratabenzene)magnesium,
  structure, 29
Bis(2-methylbutyl)magnesium, sparteine
  complex, 148
Bis(neopentyl)magnesium, structure, 23
Bis(pentamethylcyclopentadienyl)magnesium,
  carbene complex, 53–54
1,4-Bis(phenyl)-2-butene-1,4-diylmagnesium
  tris-THF complex, structure, 51, *52*
Bis[α-(2-pyridyl)-α,α-bis(trimethylsilyl)
  methyl]magnesium, structure, 31–32
N,*N*'-Bis(salicylidene)2,3-diaminobenzofuran,
  UV–visible fluorometry, 283, 284
Bis(2-thienyl)magnesium, donor–acceptor
  complexes, 43
Bis(thiomethyl)magnesium compounds,
  structure, 42, *43*
Bis(toluene)chromium, heteroleptic
  organomagnesiate production, 22
Bis(*p*-tolyl)magnesium, structure, 37–39
Bis[2,4,6-tri-*t*-butylphenyl]magnesium,
  structure, 23–24
9,10-Bis(trimethylsilyl)anthracene magnesium,
  structure, 45–46
1,4-Bis(trimethylsilyl)-2-butene-1,4-
  diylmagnesium, TMEDA complex, 51,
  52
Bis[(trimethylsilyl)methyl]magnesium,
  structure, 23–24
Bis(trimethylsilyl)peroxide, hydroxycarbonyl
  compound synthesis, 475
Blood
  AAS, 278
  AES, 279
  electrolyte determination, 275–276
  mass spectrometry, 288
BMDA (bromomagnesium diisopropylamide),
  472, 473
Boc-protected 2-substituted pyrroles,
  magnesium enolate, 447, 449
Body fluids, elemental analysis, 270, 278–279
Bond dissociation energies, 102
  magnesium cationic sandwiches, 124
Bond energies
  cyclopentadienyl magnesium, 123
  heteroatom organomagnesium bromides,
    115

  magnesium cationic sandwiches, 124–125
Bonds
  C=C addition, 471, 559–568
  C–halogen bond strength, 516
  C–H bond activation, 158
  C=N addition, 571–575
  C–O bond activation, 159, 161, 167
  C=O addition, 569–571
  C–X bond activation, 48, 156, 158, 159
  electron-deficient, 5
  Mg–F, 103
Bone, mass spectrometry, 288
Boronic acids, conjugate addition, 774
Boron reagents, conjugate addition, 774
Borosilicate glass, silanol groups, 300
Breast cancer, elemental analysis in hair, 278
Bridge bidentate coordination, Mg^{2+}, 322
Bridging amide nitrogen atoms, heteroleptic
  organomagnesiates, 21
Britton–Robinson buffer, UV–visible
  spectrophotometry, 282
Bromine–magnesium exchange
  functionalized Grignard reagent synthesis,
    515, 519–526, 527, 528, 529
  lithium magnesates, 692–694, 696–700,
    702, 703–707, 711, 712–713
o-Bromoarylmagnesium, functionalized
  Grignard reagent synthesis, 520, 521
9-Bromo-9-[(bromomagnesium)methylene]
  fluorene, THF complex, 59
Bromocresol Green, UV–visible
  spectrophotometry, 280, 281
α-Bromoketones
  Mg-mediated aldol-type reaction, 440, 482
  permutational halogen/metal
    interconversions, 437–440, 482
  Polysantol preparation, 440
2-(Bromomagnesio)-1,3-phenylene-16-crown-5,
  66, *67*
2-(Bromomagnesio)-1,3-xylyl-15-crown-4, 66,
  67
2-(Bromomagnesio)-1,3-xylyl-18-crown-5, 66,
  67
Bromomagnesium diisopropylamide (BMDA),
  472, 473
Bromomagnesium enolates, 438–440, 449
3-Bromopropenyl esters, enantioselective
  allylic alkylation, 799
Bromopyrogallol Red, elemental analysis of
  Mg, 277
Bromothiophenes, Br_2–Mg exchange, 525
Bronsted acids, magnesium bromide salts, 113,
  115
Bruceantin tetrahydrofuran ring, 478
1,3-Butadiene, transition-metal-catalyzed
  cross-coupling, 544
Butanol
  gas chromatography, 295

titration, 288, 291, 293
Butenolide, enantioselectic allylic alkylationchela, 799
Buttenberg, Fritsch–Buttenberg–Wiechell rearrangement, 742–744
t-Butyl acetate, metalation, 457, 466
tert-Butylallylmagnesium chloride, Schlenk equilibrium, 145
Butylmagnesium, nucleophilic addition, 684–685
t-Butylmagnesium chloride, sparteine complex, 60
2-[(4-*tert*-Butylphenyl)magnesio]-1,3-xylene-18-crown-5, 35–36

C_2-symmetric ligand, catalytic enantioselective conjugate addition, 782–784
Cadmium compounds, derivatization, *304*
Calmagite, UV–visible spectrophotometry, 282
Calorimeters, 104–105
Calorimetry, 104–106
cAMP, protein kinase interactions, 347–348
Cancer, pathogenesis indicators, 278
Capillary electrophoresis, tube coating, 301, 305
Carbanions
  cycloalkylmagnesium halides, 118
  non-stabilized α-amino-substituted, 728–729
  α-sulfinyl lithium, 760–766
  α-sulfinyl magnesium enolates, 459, 486, 487, 500
  α-sulfonyl lithium, 727–728, 748
Carbazoles, functionalized, 533, 534
Carbenacyclopropanes *see* Magnesium cyclopropylidenes
Carbenes
  bis(pentamethylcyclopentadienyl)magnesium complexes, 53–54
  synthesis intermediates, 718, 735
Carbenoids *see* Magnesium carbenoids
Carbocations, Mg atom reactions, 159
Carbocyclic compounds, thermochemistry, 117
Carbomagnesiation reactions, 631–679
  alkenes, 633, 653–674
  alkynes, 633–653
  anti-carbomagnesiation, 644, 645–651
  aromatic compounds, 674–675, 676
  functionalized Grignard reagents, 543
Carbometalation, Fe-catalyzed, 622–623
Carbon, Mg compounds, 106–107
Carbon–carbon bonds
  C=C addition, 471, 559–568
    1,4-addition to Michael acceptors, 563–565

activated alkenes, 565–567
catalyzed, 559–562
triple C–C bonds, 567–568
Carbon dioxide
  liquid chromatography, 293
  organomagnesium ion reactivity, 180
Carbon dioxide fixing, rubisco enzyme, 357–359
Carbon–halogen bonds
  reactivity order, 516
  reductive metal insertion, 438–441
Carbon–hydrogen bond activation, 158
1,2-Carbon–hydrogen insertion reactions, 732
1,3-Carbon–hydrogen insertion reactions, 729–734
1,5-Carbon–hydrogen insertion reactions, 734
Carbon–magnesium bonds, reactivity, 517
Carbon monoxide
  Mg complexes, 125
  organomagnesium ion reactivity, 180
Carbon–nitrogen bonds, C=N addition, 571–575
Carbon–oxygen bonds
  C–O activation, 159, 161, 167
  C=O addition, 569–571
Carbon tetrachloride, titration, 289
Carbon–X bond activation, 156, 158, 159, 162
Carbonyl compounds
  conjugated unsaturated, 624–625, 626, 628
  Grignard additions
    experimental background, 370–374
    rare-earth metal salts, 569–570
    theoretical studies, 369–370, 374–401, 569
  one-carbon homologation, 760–766
  synthesis by substitution, 559
Carboxylate ligands, magnesate anions, 178–179
Carboxylates, Mg^{2+} coordination, 321–322
Carboxylic acid derivatives, α,β-unsaturated, 451
Carboxylic acid dianions, 503
Carboxylic esters, magnesium enolate, 446
Catalysts
  addition to C=C bonds, 559–562
  carbomagnesiation
    alkenes, 657–660, 662, 663, 664–665, 671–674
    alkynes, 634–644, 648, 651–652
  cooperative, 567, 568
  cross-coupling reactions, 543–547, 550–556
  enantioselective conjugate addition, 771–791
Catalyzed Claisen rearrangment, 444
Catalyzed Mukaiyama reaction, 456, 457, 474
Cathodic reduction, 244–253
  general mechanism, 244–245

Cathodic reduction, (*continued*)
  Mg deposition and reverse process, 245–253
    kinetics, 252–253
    organomagnesium cations, 246
Cationic sandwiches, thermochemistry, 124–125
Cations
  dimagnesium, 10
  heteroleptic monoorganomagnesium, 56–57
  Mg biochemistry, 316–327
  organic substrate reactions, 160–170
Cellulose, matrix obliteration, 272
Chalcogenides, heavy group 15 complexes, 433–434
Chalcone, 1,4-addition, 478
Charcoal, Ni catalyst immobilization, 554
Charge reduction, ligated Mg cations, 175
Charge transfer
  carbocation–Mg atom reactions, 159
  Mg cation complexes, 163
Chelation
  allylic N-protected aminoesters, 446
  carbomagnesiation, 636, 642, 647, 667–670, 671–672
  β-diketones, 461, 463, 493–494
  Grignard addition calculations, 380–384, 391, 393–396, *397*
  β-ketoesters, 445, 493–494
  stereoselectivity, 440, 441, 491, 492, 493
Chemical shifts
  alkylmagnesium compounds, 134, *135–137*, 138–139, 141
  ^{25}Mg NMR spectroscopy, 151–152, *153*
Chemical vapor deposition, organomagnesium alkoxides, 432
Chemisorption, Mg(0001) single-crystal surface, 172, 173
Chiral acetates, aldol stereochemistry, *447*
Chiral additives, Grignard reagents, 60–61
Chiral β-amino acid derivatives, 454
Chiral auxiliaries, 1,4-addition to Michael acceptors, 563–564
Chiral centers, all-carbon quaternary, 565
Chiral dioxolones, aldol stereochemistry, *448*
Chiral enolates, 505–506
Chiral ferrocenyl-based diphospines, asymmetric 1,4-addition, 456
Chiral Grignard reagents, synthesis, 514–515, 527–528, 725–727
Chiral magnesium amides, 454, 498
Chiral magnesium carbenoids, asymmetric homologation, 722
Chiral magnesium enolates, 500
Chiral oxazolidinone anions, 503–505
Chiral oxime ethers, addition to C=N bonds, 574

Chiral sulfinimines, addition to C=N bonds, 575
Chiral sulfoxides, synthesis, 582–583
Chloranil, oxidative coupling of amidocuprates, 580, 581
Chlorins
  photooxidation, 207, 210–211
  photosynthetic pigments, 125, 191–193
α-Chloroalkyl aryl sulfoxides, preparation, 721
Chloroaluminate complexes, cathodic reduction, 248–249
2-Chloro-4,6-dimethoxy-1,3,5-triazine, carbonyl compound synthesis, 559
α-Chloroesters, preparation, 444
α-Chloroglycidic ester, 440, 441
α-Chloro-β-hydroxy sulfoxides, desulfinylation, 444
Chlorophyll-bound magnesium, speciation analysis, 270
Chlorophyllin, 213–214
Chlorophylls
  applied photochemistry, 213–214
  Mg^{2+} coordination, 324
  photochemical reactions, 209–212
  photosynthesis, 191, 356–357
1-Chlorovinyl p-tolyl sulfoxide
  geometrical isomers, 749, 751, *752*
  magnesium alkylidene carbenoid formation, 744, 745, 746–747
Chromatic chemosensors, elemental analysis of Mg, 285–286
Chromatography
  elemental analysis of Mg, 273–274
  speciation analysis, 292–295
Chromism, polythiophene, 285
Chromium-catalyzed reactions
  addition to triple C–C bonds, 567, 568
  carbomagnesiation, 636–637, 638
Chromophores, absorption spectra, 193
CID *see* Collision-induced dissociation
Claisen rearrangement
  ester enolates, 446, 487, 491
  magnesium dihalide, 444
Clerodane, synthesis, 456, 457, 474
Cluster Grignard reagents, 171, 172
Clusters *see* Magnesium clusters
CND (conductivity measurement), 273
^{13}C NMR spectroscopy
  alkoxides, 145, *146–147*, 147
  alkylmagnesium compounds, 134, *135–137*
  allylmagnesium compounds, 141, *142*, 143, 151
  bis(cyclopentadienyl) magnesium, 152, *153*
  co-ordination complexes, *149–150*
  Evans-type 4-phenyloxazolidinone auxiliary, 455
  magnesium carbenoids, 720
  speciation analysis, 297

Cobalt-*bis*-(1,3-diphenylphosphino)propane complex, 545–546
Cobalt-catalyzed reactions
  addition to C=C bonds, 560–561
  cross-coupling reactions, 545–546
Cobalt chloride, cyclization reactions, 545–546
Coldspray ionization, 156
Collision-induced dissociation (CID)
  Mg cation reactions, 161, 182
  magnesium cationic sandwiches, 124
Colorimetry, elemental analysis of Mg, 279–283
Color tests, speciation analysis, 295–296
Column separation methods, elemental analysis of Mg, 273–275
Complexation energies, 102
Complexes
  ate complexes, 681–715
  electrochemistry, 248–252
  Grignard reagents, 58–66, 108
  ligated Mg cation photoactivation, 162–170
  Mg–CO thermochemistry, 125
  NMR spectra, 148, *149–150*
  organomagnesium-group 15-bonded, 403–427, 428
  organomagnesium-group 16-bonded, 427–434
  see also Donor–acceptor complexes
Complexometric titration
  elemental analysis, 282–283
  Mg in body fluids, 270
Concentration, NMR solutions, 132–133
Condensation reactions, Mg cation–alcohols, 161
Conductivity measurement (CND), ion chromatography, 273
Configurational stability
  functionalized Grignard reagents, 514–515
  magnesium alkylidene carbenoids, 745, 751
Conglomerates, monoorganomagnesium cations, 57
Conjugate addition
  asymmetric, 454–457, 565
  carbomagnesiation, 654, 667–671, 674–675, 676
  magnesium enolate preparation, 450–457
  see also Enantioselectivity, conjugate addition
Conjugated unsaturated carbonyl compounds, Fe-catalyzed addition, 624–625, 626, 628
Cooperative catalysis, addition to triple C–C bonds, 567, 568
Coordination
  intramolecularly coordinating substituents, 30–36
  Mg, 318–324
    bidentate, 322, 324

binding modes, 320–324, 331–332
bridge bidentate, 322
coordination sphere, 318–324
inner sphere, 320–321, 323–324
mixed sphere, 320–321
monodentate, 322, 324
outer sphere, 320–321, 324
template effect, 342
NMR spectra, 148, *149–150*
Copper, amidocuprate oxidative coupling, 580–581
Copper(I) salts, catalysts, 557–558
Copper(II) chloride lithium chloride, 544
Copper-catalyzed reactions
  addition to triple C–C bonds, 567, 568
  allylic substitution, 557–558
  carbomagnesiation, 634–635, 636, 637–639, 641–644, 648, 651–652, 659, 660, 662, 663
  conjugate addition, 455, 457, 772–791
  cross-coupling reactions, 544–545
  enantioselective
    allylic alkylation, 792–799
    conjugate addition, 772–791
    olefin π-complex, 790
    semiempirical [PM3(tm)] calculations, 790–791
Copper–magnesium cluster compound, structure, 86, *87*
Corriu, Kumada–Tamao-Corriu reactions, 550
Coumarin 343, UV–visible fluorometry, 285
Coupling reactions
  Kumada coupling reaction, 513
  oxidative, 580–581
    homocoupling, 547
  see also Cross-coupling reactions
Covalency, ferrocene, 122
Covalently bound magnesium, speciation analysis, 270
o-Cresolphthalein, UV–visible spectrophotometry, 280, 281
Cross-coupling reactions
  Fe-catalyzed, 604–618, 626–628
    alkenylation, 604–608, 626
    alkylation, 615–618
    arylation and heteroarylation, 610–615, 626, 627–628
  lithium tripyridylmagnesate, 688
  transition-metal-catalyzed, 543–547, 550–556
  transition-metal-free, 556–557
Crown ethers
  12-Crown-4, 178
  donor–acceptor complexes, 34–36, 42, 44
  Grignard reagents, 66–67
  Mg cation fragmentation, 178
  organomagnesiate production, 11–12, 14
Cryoscopy, speciation analysis, 299

2,1,1-Cryptand, triorganomagnesate
   production, 12–13
Cryptands, disproportionation reactions, 11,
   12–13
Crystallography, magnesium carbides, 107
CSD database, organomagnesium compounds,
   3
Cubane structure
   alkoxides, 70–71, 145, *146*
   diorganomagnesium compounds, 29, 50, *51*
   see also Heterocubane structure
Cumene, Grignard reagent coupling, 295
Cumulene, 106
Cyanoacetylene
   Mg cation reactions, 160
   magnesium cationic sandwiches, 125
Cyanohydrins, functionalized Grignard reagent
   protection, 534, 535–536
Cyanopolyenes, Mg cation reactions, 160
Cyclic compounds
   cAMP, 347–348
   ring-opening of small cycles, 543, 547, 548
   thermochemistry, 117–121
Cyclic dodecamers, ethylmagnesium amides,
   82, *83*
Cyclic enones, enantioselective conjugate
   addition, 774–779
Cyclization, radical, 623–624
Cycloalkanes, enthalpies of formation,
   117–118
Cycloalkylmagnesium halides,
   thermochemistry, *110*, 117–120
Cyclobutanes, substituted, 547
Cyclobutyl Grignard reagents,
   thermochemistry, 118–119
Cycloheptanones, 2,2,7-trisubstituted, 765
2-Cyclohexen-1-one, diethylmagnesium
   reactions, 684, *685*
1,5-Cyclooctadiene, NMR spectroscopy, 296
Cyclooctene, NMR spectroscopy, 296
Cyclopentadienyl anion, formation, 182
Cyclopentadienylmagnesium, lowest excited
   electronic state, 123
Cyclopentadienylmagnesium amides, 83, *84*
Cyclopentadienylmagnesium amidinate
   complex, 72–73
Cyclopentadienylmagnesium
   bromide(tetraethylethylenediamine), 67
Cyclopentadienylmagnesium *tert*-butylthiolate,
   84, *85*
Cyclopentadienylmagnesium chloride, dimer,
   67–68
Cyclopentadienylmagnesium β-diketiminate,
   78
Cyclopentadienylmagnesium ethoxide, *70*, 71
Cyclopentadienylmagnesium halide complexes,
   67–69
Cyclopentadienyl(neopentyl)magnesium, 29

Cyclopropanation
   allylic alcohols, 722
   Simmons–Smith-type, 722
Cyclopropanes
   magnesium carbenoid reactions, 730–731,
      734, 735
   substituted, 547
   synthesis, 582
Cyclopropanols, synthesis, 582
Cyclopropenes, addition to C=C bonds, 560
Cyclopropenyl derivatives, magnesium
   cationic sandwiches, 125
Cyclopropylamines, synthesis, 582
Cyclopropyl α-amino acid, synthesis, 742
Cyclopropyl bromides, Br_2–Mg exchange, 528
Cyclopropyl Grignard reagents,
   thermochemistry, 118–119
Cyclopropylidenes see Magnesium
   cyclopropylidenes
Cyclopropyl iodides, I_2–Mg exchange, 528
Cyclopropylmagnesium, α-amino-substituted,
   740, 741–742
Cyclopropyne, 106
Cystosine, Mg cation complex, 170

DABCO
   donor–acceptor complexes, 39, *40*
   tetraorganomagnesiate production, 9
Danofloxacin, 499
DCTA, ion chromatography, 273
Decarboxylation, magnesate anions, 178–179,
   180, 181
Dechlorination, aryl chlorides, 554
Decomposition, organomagnesium films, 173
Dehydration, Mg cation–acetic acid complex,
   169
Dehydrogenation, Mg film reactions, 173–174
Dehydropyridine, radical cation formation,
   164–165
Density functional theory (DFT), Grignard
   carbonyl additions, 369–370, 376
Deoxypropionate units, enantioselective
   conjugate addition, 786, 788, 799
Deposition–dissolution of magnesium
   cathodic reduction, 245–253
   kinetics, 252–253
Deprotonation
   bis-deprotonation, 458
   1,1'-di-deprotonation, 19
   enantioselective, 424, 469–470, 473
   epoxides, 547
   hindered Hauser bases, 464, 465
   ketones, 424, 425, 596–599
   toluene, 19
   triorganomagnesates, 686–690, 713
   see also Protonation

Subject Index

Deprotonative metalation, heteroleptic organomagnesiate production, 18–22
Derivatization reagents, Grignard reagents, 300–301, *302–304*
Desulfinylation
  α-chloro-β-hydroxy sulfoxides, 444
  α,α-disulfenylated lactones, 443–444
  α-halo-β-ketosulfoxide, 444, 479
  isopropylmagnesium chloride, 445
  β-keto-sulfinylcarboxylic acid derivatives, 444–445
  β-keto-sulfoxide, 444, 479
Desymmetrization, enantioselective, 571
Deuteriation, magnesium enolates, 441, 479, 489
Deuterium oxide, Mg film desorption, 173
DFT (density functional theory), 369–370
α,β-Diacylglycerols, derivatization, 301
Dialkylmagnesium compounds
  2,6-bis(imino)pyridine reactions, 71–72
  β-diketimine reactions, 73–78
  enthalpies of formation, 116–117
  sparteine complexes, 148, *150*
Dialkyl malonate C-acylation, 462, 494–496
N,N-Dialkyl-α-methacrylthioamides, Michael addition, 452
Dialkylphosphine oxides, transition-metal-catalyzed cross-coupling, 550, 552
N,N-Dialkyl sulfamate, *ortho*-directing group, 553
Dialkylzinc reagents, conjugate addition, 774
Diaminocarbenes, catalytic enantioselective conjugate addition, 775, *777*, *778*
Dianions
  carboxylic acid, 503
  tetraorganomagnesium, 8–9
1,4-Diarylbutadienes, functionalized Grignard reagent synthesis, 540, 543
Diarylmagnesium compounds, ^{1}H NMR spectra, 138
Diarylmagnesium reagents, Br$_2$–Mg exchange, 521, 522, 523
Diastereoselectivity, carbomagnesiation, 671–672
Diazines, transition-metal-catalyzed cross-coupling, 553
DIBAH (diisobutylaluminum hydride), 513
Dibenzofuranic derivatives, 476, *477*
Dibromo-β-lactam, halogen–metal exchange, 443, 446
Dibutylmagnesium, 2,2′-ethylidenebis(2,4-di-*tert*-butylphenol) reaction, 16
Dicarbonyl enolates, reactions with electrophiles, 489–499
Dichloromethyl phenyl sulfoxide, one-carbon homologation, 765–766

Dickerson–Drew dodecamer, 335
Dicyclopentadienyl titanium dichloride, functionalized Grignard reagent synthesis, 540, 543
1,1′-Di-deprotonation, heteroleptic organomagnesiate production, 19
Dienes, Mg film self-hydrogenation, 173
1,3-Dienyl triflates, transition-metal-catalyzed cross-coupling, 555
Diethyl N-Boc-iminomalonate, functionalized Grignard reagent synthesis, 517–518
Diethyl ether, chemisorbed film solubility, 173
Diethylmagnesium
  18-crown-6 complex, 44
  nucleophilic addition, 684, *685*
  structure, 25
Diffuse reflectance infrared Fourier transform spectroscopy (DRIFTS), 295
Diffusively bound magnesium(II) ions, in RNA, 270
Diglyme, donor–acceptor complexes, 41–42
1,1-Di-Grignard reagents, structure, 61–62
1,1-Dihalocyclopropanes, Doering–LaFlamme allene synthesis, 735–738
Dihydroquinoline-2*H*-3-carboxylic acid derivatives, 453
Diisobutylaluminum hydride (DIBAH), Mg activation, 513
Diisopropylamine, Mg cation complex, 168
β-Diketiminates
  anion chelation, 74
  enolization, 424
  metalation, 419, 421
  O$_2$ reactions, 422–423
  structure, 417–418
  synthesis, 411
β-Diketimines, dialkylmagnesium compound reactions, 73–78
β-Diketones
  aldol-type reactions, 493–495
  chelated magnesium enolate, 461, 463
Dilithium tetrabutylmagnesate, deprotonation, 689–690
Dilution *see* Enthalpies of dilution
Dimagnesacycloalkanes, thermochemistry, 120–121
1,7-Dimagnesiocyclododecane tetra THF complex, 40
Dimagnesium cations, tetraorganomagnesiates, 10
Dimerization
  alkoxy- and aryloxymagnesium hydrides, 147
  PET, 206
  see also Enthalpies of dimerization
Dimethoxyethane *see* DME complexes
1,2-Dimethoxyethane, Mg cation complex photofragmentation, 166

1,3-Dimethylallylmagnesium chloride, Schlenk equilibrium, 145
Dimethylamine, Mg cation complex, 168
(Dimethylamino)dimethylmagnesium chloride, structure, 64
1-[2-(Dimethylamino)ethyl]-2,3,4,5-tetramethylcyclopentadienylmagnesium bromide, 68–69
2-[(Dimethylamino)methyl]ferrocenyl groups, 30, *31*
Dimethyl ether, chemisorption, 172
N,N-Dimethylformamide, Mg cation complex, 169
Dimethylmagnesium
  donor–acceptor complexes, 39, *40*
  structure, 25, *26*
Dimethylsilyl chloride, Grignard reagent coupling, 295
Dimethyl sulfoxide, Mg cation fragmentation, 176
Diorganomagnesium compounds
  anodic oxidation, 227–228
  dioxane complexes, 39, *40*
  disproportionation reactions, 11
  electrochemical synthesis, 221–224, 253–256
  nucleophilic addition, 256–257
  heteroleptic organomagnesiate production, 15–16
  structural chemistry, 23–54
    donor–acceptor complexes, 36–44, 47–54
    donor-base-free compounds, 23–25
    intramolecularly coordinating substituents, 30–36
    magnesium anthracene compounds, 44–47
    multi-hapto-bonded groups, 25–29, 47–54
    σ-bonded compounds, 36–44
  thermochemistry, 116–117
Dioxanate, Schlenk equilibrium, 107
Dioxane, diorganomagnesium complexes, 39, *40*
Dioxolones, aldol stereochemistry, *448*
Diphenyl ditelluride, titration, 290
2,2-Diphenylethyl-2,4,6-trimethylphenyl ketone (Kohler's ketone), 457, 458
Diphenylmagnesium
  donor–acceptor complexes, 43–44
  structure, 25, *26*
cis-Diphenylvinylenemagnesium THF complex, 40–41
Diphosphine ligands
  asymmetric 1,4-addition, 456
  conjugate addition of Grignard reagents, 775
Dipole moments, Grignard reagent calculations, *375*

Dipropylamine, Mg cation complex, 168
Direct alkenylation, arylamines, 748–749
Direct oxidative addition, Mg to organic halides, 512–515
Displacement reactions, Mg cation–alkyl halides, 160–161
Disproportionation reactions
  cryptands, 11, 12–13
  Lewis base donor solvent effects, 412
  organomagnesium amide complexes, 422
Dissolution, electrochemical deposition–dissolution of Mg, 245–253
α,α-Disulfenylated lactones, desulfinylation, 443–444
Dithioacetals, ring opening, 547, 548
1,3,16,18-Dixylylene-30-crown-8, donor–acceptor complexes, 42
DME complexes, Grignard reagents, 59–60
DNA
  capillary electrophoresis, 305
  Mg^{2+} interactions, 320–321, 333–342
  ribozymes, 335–342
  replication and repair, 350–355
    base excision repair, 353–354
    DNA replicases, 351–353
    generic two-ion mechanism, 354–355
DNA replicases, 351–353
Doering–LaFlamme allene synthesis, magnesium cyclopropylidene intermediates, 735–738
Donor–acceptor complexes
  multi-hapto-bonded groups, 47–54
  σ-bonded diorganomagnesium compounds, 36–44
  see also Complexes
Donor–acceptor transfer compounds, photochemistry, 196–201
Donor-base-free diorganomagnesium compounds, structures, 23–25
Double bonds, addition reactions, 11, 35, 559–567
Double titration, organolithium compounds, 289
DRIFTS (diffuse reflectance infrared Fourier transform spectroscopy), 295
Dry ashing, matrix obliteration, 271, 272
DuPHOS ligand, catalytic enantioselective conjugate addition, 775, 776
Dynamic equilibria
  allylmagnesium compounds, 143, *144*
  arylmagnesium compounds, 140

Eclipsed structure, magnesocene, 183
EDTA
  AAS, 278
  complexometric analysis, 282–283

ion chromatography, 273
NMR spectroscopy, 286
EGTA, UV–visible spectrophotometry, 280, 281
Electric conductivity
 organomagnesium solutions, 224–227
 solvent effects, 226–227
Electroanalytical determination, elemental analysis of Mg, 276–277
Electrochemistry, 219–263
 anodic oxidation, 227–244
  diorganomagnesium compounds, 227–228
  Grignard reagents, 228–243
  methyl group grafting, 241–243
 cathodic reduction, 244–253
  general mechanism, 244–245
  Mg deposition and reverse process, 245–253
 diorganomagnesium compound synthesis, 221–224, 253–256
 nucleophilic addition, 256–257
 elemental analysis of Mg, 275–277
 Grignard reagents, 305
 organomagnesium intermediates, 253–257
 solution conductivity, 224–227
Electrochromism, polythiophene, 285
Electrodes
 inert, 229–236
 ion-selective, 275–276
Electrolytes, blood/plasma/serum, 275–276
Electron-deficient bonds, organomagnesiate structures, 5
Electron diffraction, gas-phase, 183–184
Electron donor–acceptor transfer compounds, 196–201
Electronegativity, methyl deviation correlation, 111–112, 118
Electronic absorption spectra, tetrapyrroles, 192–193
Electron impact, ligated Mg cations, 175
Electron spin resonance (ESR) spectroscopy
 monomethyl magnesium radical, 159
 speciation analysis, 299
Electron transfer
 ligated Mg cations, 175–176
 Mg cation–amine complexes, 168
 magnesium anthracene compounds, 45
 photochemistry, 191, 194–206
  donor–acceptor compounds, 196–201
  heteroligand systems, 201, 204–206
 see also Single-electron transfer
Electrooxidation
 Grignard reagents, 237–241
 organoelemental compounds, 240
Electrophilic reactions
 amination, 575–580
 carboxylic acid dianions, 503

chiral oxazolidinone anions, 503–505
magnesium alkylidene carbenoids
 alkenylation, 748
 N-lithio arylamines, 748–755
 N-lithio nitrogen-containing heterocycles, 755–757
 lithium acetylides, 758
 lithium thiolates, 758–760
 tetrasubstituted olefin one-pot synthesis, 746–748
magnesium amide enolates, 499–500
magnesium carbenoids
 1,3-C–H insertion reaction, 729–734
 Grignard reactions, 723–725
 N-lithio arylamine reactions, 728–729, 730
 secondary chiral Grignard reagent synthesis, 725–727
 α-sulfonyl lithium carbanion reactions, 727–728, 748
magnesium cyclopropylidenes, 739–742
magnesium dicarbonyl enolates, 489–499
magnesium ester enolates, 484–489
magnesium ketone enolates, 472–484
magnesium lactam enolates, 499–500
magnesium lactone enolates, 484–489
magnesium thioamide enolates, 500–502
magnesium thioesters, 500–502
Electrophoresis
 capillary, 301, 305
 elemental analysis of Mg, 273–274
Electropositivity, Mg, 102–103, 125
Electrospray ionization (ESI), gas-phase Mg ion fragmentation, 174–175, 178
Electrothermal atomic absorption spectrometry (ETAAS), 278
Elemental analysis of magnesium, 268, 269–288
 column separation methods, 273–275
 electrochemical methods, 275–277
 sample preparation, 271–273
 spectral methods, 277–288
Elimination reactions
 addition–elimination, 471
 β-hydride, 544
 Mg cation–organic substrates, 160–161
Emission spectroscopy, tetraphenylporphyrins, 199
Enamides, 1,4-addition to Michael acceptors, 563–564
Enantioselectivity
 carbomagnesiation, 659–661, 673, 674
 conjugate addition, 771–791
  acyclic enones, 774, 779, 780
  cyclic enones, 774–779
  magnesium enolate preparation, 454
  mechanistic studies, 789–791
  organomagnesium amides, 426

Enantioselectivity (*continued*)
  polar mechanism, 370
  tandem aldol reaction, *788*, 789
  α,β-unsaturated esters, 779–784
  α,β-unsaturated thioesters, 784
  natural product synthesis, *785*, 786–789
deprotonation, 424, 425, 469–470, 473
desymmetrization of anhydrides, 571
Grignard reagents
  addition to ketones, 571
  allylic alkylation, 792–799
  chiral additives, 60–61
  ketone deprotonation, 424, 425
  protonation, 480, 481
Endocyclic compounds, thermochemistry, 117
Energy currency of life, Mg biochemistry, 327–333
Energy thresholds, magnesium cationic sandwiches, 124
Enolases, metabolic enzymes, 348–350
Enolate intermediate, alkylation, *762*, 763–765
Enolates *see* Magnesium enolates
Enolization
  hindered ketones, 478
  organomagnesate anions, 180–181
  solvent-free Grignard reagents, 157
Enol phosphates
  magnesium enolate reactions, 480, 481, 489–491
  transition-metal-catalyzed cross-coupling, 544
Enol triflates, transition-metal-catalyzed cross-coupling, 555
Enones
  acyclic, 774, 779
  cyclic, 774–779
α,β-Enones, 1,4-addition, 563
Enthalpies of atomization, binary magnesium salts, 115
Enthalpies of dilution, Grignard reagents, 107–108
Enthalpies of dimerization, magnesacycloalkanes, 120
Enthalpies of formation
  acetone, 115
  alkoxymagnesium bromides, 114–115, *116*
  cycloalkanes, 117–118
  cycloalkylmagnesium halides, *110*, 117–120
  dialkylmagnesium compounds, 116–117
  difference, 112, 114, 119
  Grignard reagents, 115–116
  hydrocarbylmagnesium bromides, 109, *110*, 114
  magnesium sandwich species
    magnesocene, 122
    neutral magnesium half-sandwiches, 123
  Mg–fluorocarbon reactions, 103
  organomagnesium bromides, 111, 112, 113–114, 115
  parent hydrocarbons, 112
  see also Molar standard enthalpies of formation
Enthalpies of hydrogenation, unsaturated organomagnesium bromides, 113, 119
Enthalpies of hydrolysis
  butylmagnesium chloride, 110
  calorimetry, 104
  magnesacycles, 121
Enthalpies of protonation, unsaturated organomagnesium halides, 112
Enthalpies of reaction
  acetolysis, 120
  Bronsted acids, 113, 115
  calorimetry, 105–106
  Grignard reagents, *116*
  hydrocarbylmagnesium bromides, 109, *110*
  magnesium sandwich species
    cationic sandwiches, 124
    neutral magnesium half-sandwiches, 123
  organomagnesium bromides, 111
  phenylethynylmagnesium bromide, 112
  see also Molar standard enthalpies of reaction
Enthalpies of solution, enthalpy of formation calculation, 116
Enthalpies of solvation, Schlenk equilibrium, 109, 117
Enthalpies of vaporization, calorimetry, 105–106
Entropy
  cycloalkylmagnesium halides, 120
  magnesacycloalkanes, 120
Environmental aspects, Mg metal, 102–103
Environmental pollution control, organometallic compounds, 300
Enzymes (protein-based), 342–359
  allosteric systems, 343–345
  DNA replication and repair, 350–355
  enzymatic modes of action, 342–345
  metabolic enzymes, 345–350
  photosynthesis, 355–359
  sequential systems, 343, 348–350
  template enzymes, 342–343
  see also ATP synthases
Epimers, functionalized Grignard reagents, 514–515
Epoxides, deprotonation, 547
Equilibrium constant, Schlenk equilibrium, 108
ESI (electrospray ionization), 174–175, 178
Esophageal cancer, elemental analysis in hair, 278
ESR spectroscopy *see* Electron spin resonance spectroscopy

Subject Index

Ester enolates
  Claisen rearrangement, 446, 487, 491
  reactions with electrophiles, 484–489
Esters
  quaternary stereocenters, 792, 796
  α,β-unsaturated, 779–784
Ether-functionalization, diorganomagnesium compounds, 33
Ethyl hydrogen malonate
  γ-amino-β-ketoester derivatives, 497
  bis-deprotonation, 458
2,2'-Ethylidenebis(2,4-di-*tert*-butylphenol), 16
Ethyl isocyanate, Mg cation complex, 169
Ethyl isothiocyanate, Mg cation complex, 169
Ethylmagnesiation, Zr-catalyzed, 671–674
Ethylmagnesium amides, cyclic dodecamer, 82, 83
Ethylmagnesium bromide
  ethoxide structure, 58–59
  sparteine complexes, 60–61
Ethylmagnesium tris(3-*tert*-butylpyrazolyl) phenylborate, 73, 74
O-Ethyl *S*-(tetrahydro-2-oxo-3-furanyl) dithiocarbonate, magnesium enolate, 447
Evans-type 4-phenyloxazolidinone auxiliary, NMR spectroscopy, 447
Evaporation, solvent, 15
Evening primrose oil, analysis, 301
Exchange reactions
  alkyl–halide, 108–109, 134
  allylmagnesium compounds, 142, *143*
  halogen–metal, 439, 442, 443, 446, 718
  halogen–Mg, 515–537, 690–711, 712–713, 718–720
  H–metal, 742–743
  see also Ligand exchange
Excited states
  cyclopentadienyl magnesium, 123
  Mg atoms, 157, 158, 159
Exo,exo-bis(iso-dicyclopentadienyl)magnesium, structure, 27
Exothermic reactions, Mg fluorocarbon reactions, 103

FAAS (flame atomic absorption spectrometry), 271–272, 277–278, 288
FAES (flame atomic emission spectrometry), 278
Fecal endogenous excretion (FEE), mass spectrometry, 288
FEE (fecal endogenous excretion), 288
Ferrocene
  nomenclature, 122
  structure, 25
Ferrocenylacetylene, NMR spectroscopy, 297
Ferrocenyl-based diphosphines
  asymmetric 1,4-addition, 456
  conjugate addition of Grignard reagents, 775
Ferrocenyl dianions, heteroleptic organomagnesiate production, 20, *21*
Fertilizers, complexometric analysis, 283
FIA (flow injection analysis) systems, 280
Films, formation of organomagnesium species, 172–174
Flame atomic absorption spectrometry (FAAS), 271–272, 277–278, 288
Flame atomic emission spectrometry (FAES), 278
Flow injection analysis (FIA) systems, UV–visible spectrophotometry, 280
Fluorescence, Mg(II) tetrapyrroles, 193
Fluoroarylmagnesium compounds, ^{19}F NMR spectra, 139–140
Fluorocarbons, pyrolants, 103
Fluorometry, elemental analysis of Mg, 270, 283–285, 288
2-Fluoropyridine, photodissociation spectra, 164
3-Fluoropyridine, deprotonation, 688
^{19}F NMR spectroscopy, fluoroarylmagnesium compounds, 139–140
Food additives, chlorophyll, 213–214
Force constants, cyclopentadienyl magnesium, 123
Forensic analysis, organometallic compounds, 300
Formaldehyde, Mg cation complex, 169
Formation
  Grignard reagent calculations, 376, 378
  solvent-free environments, 156–179
  ligated Mg ion fragmentation, 166, 174–179
  Mg atom reactions, 157–159
  Mg cation reactions, 160–170
  Mg cluster reactions, 171–172
  Mg surfaces and films, 172–174
  see also Enthalpies of formation; Molar standard enthalpies of formation
Fragmentation of organomagnesium ions
  ligated, 166, 174–179
  unimolecular reactions, 182
Free energy, universal energy currency of life, 327–333
Free-energy difference, cycloalkylmagnesium halides, 119
Free hydrated magnesium ions, speciation analysis, 269
Free radicals *see* Radicals
Fritsch–Buttenberg–Wiechell rearrangement, alkylidene carbenoids, 742–744
FT–ICR (Fourier transform–ion cyclotron resonance), 160
Fullerenes
  carbomagnesiation, 674–675, 676
  PET, 196, 200–201

Fulvenes, cyclopentadienylmagnesium
   compounds, 27–28
Functionalized Grignard reagents
   acylation, 559, 601, 603
   addition, 543
      1,4-addition, 543, 547, 563–565
      C=C bonds, 559–568
      C=N bonds, 571–575
      C=O bonds, 569–571
      multiple bonds, 559–575
   amination, 575–581
      electrophilic, 575–580
      oxidative coupling, 580–581
   chiral sulfoxide synthesis, 582–583
   cyclopropane synthesis, 582
   reactivity, 543–583
   substitution
      Sp-center, 559
      Sp^2-center, 544, 550–559
      Sp^3-center, 543–550
   synthesis, 512–543
      direct oxidative addition, 512–515
      functional group tolerance, 530–537
      halogen–Mg exchange, 515–537
      metalation, 537–540, 541

Gallium compounds, derivatization, *304*
Gas chromatography
   organometallic compounds, 300–301,
      302–304
   speciation analysis, 293, 295
Gas phase
   ligated Mg ion fragmentation, 174–179
   Mg atoms, 157–159
   organomagnesium ions
      bimolecular reactions, 179–182
      unimolecular reactions, 182
      structure determination, 183–184
Gas-phase electron diffraction, structure
      determination, 183–184
Geometries
   Grignard reagent calculations, 374–375,
      376, *377*, 380, *381*
   magnesium alkylidene carbenoids, 751,
      753–754, 755
GFAAS (graphite furnace atomic absorption
      spectroscopy), 272, 273, 277–278
Gibbs energies, 102
   cycloalkylmagnesium halides, 119
Gilman's color tests, speciation analysis,
      295–296
Gilman's cuprates, enantioselective allylic
      alkylation, 792
Glaucyrone, β-ketoester reactions, 494
Glucidic α-ketoacids, 492
Glucidic phosphoenolpyruvic acid derivatives,
      491

Glycerides, analysis, 301
Glycol ethers, oxidative addition solvents, 512
Glycolytic cycle, metabolic enzymes, 345, *346*
Glymes, Mg cation fragmentation, 178
Glyoxal, 1,2-bismimine addition reaction, 574
Grafting, semiconductor anodes, 241–243
Graphite furnace atomic absorption
      spectroscopy (GFAAS), 272, 273,
      277–278
Grignard, Victor, 2, 268
Grignard reactions
   chemisorption, 173
   computational studies, 384–399
      carbonyl addition without solvent
         molecules, 384–389, *390*, *391*
      carbonyl addition with solvent molecules,
         389, 391, *392*, *393*, 401
      chelation-controlled addition, 380–384,
         391, 393–396, *397*
      computational methods, 384
      Grignard reagent structures, 384–387
      polar mechanisms, 370, 387–388, 389,
         391, 394, 396, 400–401
      single-electron transfer, 370, 372,
         396–399
      transition-state structures, 375–376, *377*,
         380–382, 388–389, 391, *393*,
         394–396
   model reactions, 380–384
   organomagnesate anions, 180–181
   photochemistry, 190
Grignard reagents
   addition–elimination, 471
   alkyl, 232–233
   analytical reagents and aids, 269, 299–305
      active hydrogen determination, 299–300
      derivatization reagents, 300–301,
         302–304
   anodic oxidation, 228–243
      addition to olefins, 237
      inert electrodes, 229–236
      reactivity, 235–236, 258–260
      sacrificial anodes, 237–241
      semiconductor anodes, 241–243
   aryl, 601, 603
   association, 108
   benzylic, 529–530
   calculations, 374–380
      computational studies, 384–399
      model Grignard reactions, 380–384
   carbonyl additions
      experimental background, 370–374
      Fe-catalyzed reactions, 624–625, 626,
         628
      theoretical studies, 369–370, 374–401
   cathodic reduction
      general mechanism, 244–245

Mg deposition and reverse process, 245–248
chiral synthesis, 514–515, 527–528, 725–727
cluster, 171
complexation, 108
conjugate addition, 450–452, 772–791
crown ethers, 66–67
derivatization reagents, 300–301, *302–304*
diaryl, 521, 522, 523
1,1-di-Grignard reagents, 61
DME complexes, 59–60
electric conductivity, 224–227
electrochemical behavior, 305
electrophilic amination, 575–581
enantioselectivity
  allylic alkylation, 792–799
  chiral additives, 60–61
  conjugate addition, 772–791
enthalpies of dilution, 107–108
enthalpies of reaction, 115–116
Fe-catalyzed reactions, 595–630
  acylation, 559, 599–603
  addition to conjugated unsaturated carbonyls, 624–625, 626, 628
  alkenylation, 604–608, 626
  alkylation, 615–618
  arylation and heteroarylation, 610–615
  carbometalation, 622–623
  cross-coupling reactions, 604–618, 626, 627–628
  deprotonation of ketones, 596–599
  homocoupling, 619–621, 626, 627
  organic synthesis, 625–628
  radical cyclization, 623–624
  substitution, 621–622
formation
  mechanisms, 55–56, 376, 378
  Mg surfaces and films, 1974
fragmentation reactions, 182
functionalized
  aryl, 601, 603
  reactivity, 543–583
  synthesis, 512–543
α-heteroatom-substituted, 767
history, 2–3
magnesium carbenoid electrophilic reactions, 723–725
NMR spectra, 131–132, 144–145
optically enriched, 721
oxonium structure, 107
perfluorinated synthesis, 516
rate constant–oxidation potential correlations, 258–260
reactive halide coupling, 295
Schlenk equilibrium, 132, 140–141, 144–145, 151
secondary chiral, 725–727

solvent-free, 157
structural chemistry, 58–69, 384–387
α-sulfur-stabilized, 725, 726
surface conditioning, 301, 305
thermochemistry, 103
see also Alkyl Grignard reagents; Aryl Grignard reagents; Methyl Grignard reagents
Group 1 introns, self-splicing, 337–339
Group 15, organomagnesium complexes, 403–427, 428
Group 16, organomagnesium complexes, 427–434

Haber–Weiss process, photooxidation, 207
Hair analysis, pathogenesis indicators for cancer, 278
Half-sandwiches, thermochemistry, 123, 124
Halides
  exchange reactions, 108–109, 134
  gas chromatography, 295
Halobenzene complexes, photodissociation spectra, 163
Halogen compounds
  heteroleptic monoorganomagnesium compounds, 58–69
  see also Grignard reagents
Halogen–lithium exchange
  functionalized Grignard reagent synthesis, 530, 532
  mechanism, 516
Halogen–magnesium exchange
  carbenoid generation, 718–720
  functionalized Grignard reagent synthesis, 515–537
  magnesium enolate preparation, 442–443
  triorganomagnesates, 690–711, 712–713
Halogen–metal exchange reactions, 439, 442, 443, 446, 718
α-Haloketones, preparation, *444*, 479
α-Halo-β-ketosulfoxide, desulfinylation, 444, 479
Hammerhead ribozymes, 339–342
Hapticity, magnesocene, 122
Heat of formation, 102
Heat of reaction, 102
Heteroanionic enolates, 460
Heteroaryl amidocuprates, oxidative coupling, 580–581
Heteroarylation
  alkyl Grignard reagents, 610–614, 626, 627–628
  aryl Grignard reagents, 614–615
Heteroaryl bromides, Br_2–Mg exchange, 524, 525
Heteroaryl chlorides
  Fe-catalyzed cross-coupling, 611, *612*
  Kumada–Tamao-Corriu reactions, 550

Heteroaryl Grignard reagents, Fe-catalyzed homocoupling, 619–621
Heteroaryl iodides, functionalized Grignard reagent synthesis, 518–519
Heteroatoms
  α-heteroatom-substituted Grignard reagents, 767
  monoorganomagnesium compounds, 54–85
  organomagnesium bromides, 113–116
  permutational heteroatom/metal enolate preparation, 441–445
Heterocubane structure
  cyclopentadienylmagnesium *tert*-butylthiolate, 84, *85*
  cyclopentadienylmagnesium ethoxide, *70*, 71
  diorganomagnesium compounds, 29, 50, *51*
  methylmagnesium *tert*-butoxide, 70
  see also Cubane structure
Heterocycles, *N*-lithio nitrogen-containing, 755–757
N-Heterocyclic carbenes (NHCs)
  catalytic enantioselective reactions
    allylic alkylation, 792, *796*
    conjugate addition, 775, *777*, *778*
  transition-metal-catalyzed cross-coupling, 550, 551
Heterogeneous processes, Grignard reagent formation, 172
Heteroleptic compounds
  diorganomagnesium compounds, 29
  monoorganomagnesium compounds, 54–85
  organomagnesiates, 5, 14–22
    bis-anions, 15
Heteroligand systems, PET, 201, 204–206
Hexaaquomagnesium ion, coordination, 318–320
Hexamethylenetetramine (HMTA), Fe-catalyzed alkylation, 617
High-performance liquid chromatography (HPLC), elemental analysis of Mg, 274
Hindered amines, 457, 464, 538
Hindered Hauser bases
  deprotonation
    regioselectivity, 464
    stereoselectivity, 465
  metalation, 464
Hindered ketones, enolization, 478
Hindered magnesium amides, metalation, 464–469
^{1}H NMR spectroscopy
  alkoxides, 145, *146–147*
  alkylmagnesium compounds, 134–138, 141
  allylmagnesium compounds, 141–142, 143, 144–145, 151
  arylmagnesium compounds, 138–139
  co-ordination complexes, 148, *149–150*
  elemental analysis of Mg, 286

Evans-type 4-phenyloxazolidinone auxiliary, 455
peroxides, 147
speciation analysis, 296–298
  No-D NMR spectroscopy, 296
Homoallyl alcohols, carbomagnesiation, 664–671
Homocoupling
  Fe-catalyzed, 554, 619–621, 626, 627
  oxidative, 547
  transition-metal-free, 557
Homoleptic iron carbonyls, 125
Homoleptic organomagnesiates
  structural chemistry, 5
  thermochemistry, 103
Homologation
  alkyl halides, 723–724
  asymmetric, 722
  one-carbon, 760–766
Homologous series, organomagnesium halides, 110–112, 117
Homopropargylic alcohols, I$_2$–Mg exchange, 527, 528
Horseradish peroxidases, porphyrin photooxidation, 206–207
HPLC (high-performance liquid chromatography), 274
β-Hydride elimination, transition-metal-catalyzed cross-coupling, 544
β-Hydride transfer, ketone reduction, 425–426
Hydrocarbyllithium compounds, 103
Hydrocarbylmagnesium bromides, enthalpies of formation, 109, *110*, 114
Hydrocarbylmetals, 103
Hydrogen
  C–H bond activation, 158
  organomagnesium ion reactivity, 180
  see also Active hydrogen compounds
Hydrogenation
  self-hydrogenation, 173
  see also Enthalpies of hydrogenation
Hydrogen atom abstraction
  Mg–alkane reactions, 157
  Mg cation–formaldehyde complex, 169
Hydrogen compounds
  gas chromatography, 295
  Grignard analytical reagents, 299–300
Hydrogen–metal exchange, magnesium alkylidene carbenoids, 742–743
Hydrolysis
  alkalinity titration, 289
  ATP, 331–333
  pancreatic lipase, 301
  see also Enthalpies of hydrolysis
Hydromagnesiation, acetylenes, 540, 543
β-Hydroxyacids, aldol-type reactions, 486

Hydroxycarbonyl compounds,
    bis(trimethylsilyl)peroxide synthesis, 475
Hydroxyl-assisted carbomagnesiation, 665,
    666, 667–671

IC (ion chromatography), 273
ICP–AES (inductively coupled plasma–atomic
    emission spectrometry), 279, 284
ICP–MS (inductively coupled plasma–mass
    spectrometry), 287–288
IE (ionization energy), 174–175
Imines, addition to C=N bonds, 571–572,
    573–574
Iminium triflates, functionalized Grignard
    reactions, 548, 549
Immonium ion, Mg cation–amine complex,
    168
Indenyl anions, bis(indenyl)magnesium, 29
Indicators, speciation analysis, 288, 289, 290
Indoles, magnesiation, 538
Induction period, oxidative addition, 513
Inductive effects, hydrocarbylmagnesium
    bromides, 114
Inductively coupled plasma
    AES, 279, 284
    mass spectrometry, 287–288
Inert electrodes
    anodic oxidation of Grignard reagents,
        229–236
        distribution of anodic products, *230, 232,
            233*
Infrared (IR) spectroscopy
    Mg atom reactions, 157, 158
    magnesium enolate, 451
    near-IR spectroscopy, 512
    speciation analysis, 295
Insertion reactions
    C–halogen bonds, 438–441
    Mg atoms with organic substrates, 157, 158,
        159
    magnesium carbenoids
        1,2-C–H insertion, 732
        1,3-C–H insertion, 729–734
        1,5-C–H insertion, 734
    Mg cation complexes, 166–167, 168
Interconversion
    permutation of H_2/metal, 457–470
    permutation of metal/metal salts, 445–450
Interfacial chemistry, Grignard reagent
    formation, 172–174
Intermediate compounds
    chiral sulfoxides, 582–583
    electrode reactions, 253–257
Intermolecular agostic interactions,
    diorganomagnesium compounds, 24
Intermolecular carbomagnesiation

alkenes, 655–656, 662, 663, 665–666,
    671–674
alkynes, 635–639
Intramolecular carbomagnesiation
    alkenes, 654–655, 656, 661–662, 665–666
    alkynes, 633–635, 654–655
Intramolecular coordination
    diorganomagnesium compounds
        substituents, 30–36
        oxygen atom, 66
Inverse crown structures, heteroleptic
    organomagnesiates, 5, 14, 19, 20–21
Inversion, pyramidal, 738
Iodine, titration, 290–291
Iodine–lithium exchange, carbenoid
    generation, 718
Iodine–magnesium exchange
    carbenoid generation, 719–720
    functionalized Grignard reagent synthesis,
        517–519, 526–527, 528, 531, 532
    lithium tributylmagnesate, 690–692, 694,
        695, 697, 700–703, 707, 708–710
β-Iodo-α-ketoester magnesium enolate, 440,
    441, 491, 492
Iodoketones
    enol phosphate preparation, 480, 481
    halogen–metal exchange, 439
Iodomagnesium enolates, 438–441, 449, 491,
    492
Ion chromatography (IC), elemental analysis
    of Mg, 273
Ionic liquids, functionalized Grignard reagent
    synthesis, 515
Ionic ring–metal interactions, magnesocene,
    122
Ionization energy (IE), ligated Mg ions,
    174–175
Ionization reactions, conductivity
    measurements, 225–227
Ionochromism, polythiophene, 285
Ion-selective electrodes (ISE), elemental
    analysis of Mg, 275–276, 283
Iridium, organomagnesium compounds, *90*, 91
Iron, CO complexes, 125
Iron(III) catalysts
    acetylacetonate, 546, 601–603, 605–609,
        615–617, 621–622, 623
    chloride, 546, 596, 599–601, 604, 615–617,
        619–620, 623
    cross-coupling reactions, 546
Iron-catalyzed reactions
    carbomagnesiation, 637–639, 652, 657–659
    Grignard reagents, 595–630
        acylation, 559, 599–603
        addition to conjugated unsaturated
            carbonyl compounds, 624–625,
            626, 628
        addition to triple C–C bonds, 567, 568

Iron-catalyzed reactions (*continued*)
  alkenylation, 604–608, 626
  alkylation, 615–618
  allylic substitution, 557
  arylation and heteroarylation, 610–615
  carbometalation, 622–623
  cross-coupling reactions, 553–554, 604–618, 626, 627–628
  deprotonation of ketones, 596–599
  homocoupling, 554, 619–621, 626, 627
  organic synthesis, 625–628
  radical cyclization, 623–624
  substitution, 621–622
IR spectroscopy *see* Infrared spectroscopy
ISE (ion-selective electrodes), 275–276, 283
(Isodicyclopentadienyl)(butyl)magnesium, TMEDA complex, 50, *51*
Isomers, allylmagnesium compounds, 142–143, *144*
Isopropylamine, Mg cation complex, 168
Isopropylmagnesium chloride
  desulfinylation, 445
  halogen–metal exchange, 442
  I_2–Mg exchange, 531, 532
  lithium chloride reagent, 531, 533–537, 714
Isopropylmagnesium β-diketiminate, structure, 77
Isopropylmagnesium tris(3-*tert*-butylpyrazolyl)hydroborate, 73, *74*
Isotopically labelled magnesium
  biosynthesis of ATP, 330
  reverse isotope dilution technique, 287–288
  Schlenk equilibrium, 107
Isovanillin, active hydrogen determination, 300
Ivanov reaction, arylacetic acids, 458, 502, 503

Jojoba wax, analysis, 301
Josiphos ligand
  catalytic enantioselective conjugate addition, 775, *776*, *777*, 780, 781, 786, 788–789
  catalyzed 1,4-addition, 564
  magnesium thioester reactions, 501
$^N J_{HH}$ spin coupling constants, allylmagnesium compounds, 143–144, *145*

Ketenes
  Grignard reagents, 471
  Mg cation–acetic acid dehydration, 169
β-Ketoesters
  glaucyrone, 494
  magnesium chelate reactions, 445, 493–494
  transition-metal-catalyzed cross-coupling, 555
  xanthyrone, 494
Ketone enolates, reactions with electrophiles, 472–484

Ketones
  deprotonation, 424, 425, 596–599
  enantioselective addition, 571
  hindered, 478
  Kohler's ketone, 457, 458
  Michler's ketone, 295, 296
  one-carbon ring expansion, *762*, 763, 764–765
  organomagnesium amide reactions, 423–426
  tetrasubstituted, 457
β-Ketophosphonates, preparation, 462
β-Ketosilanes, preparation, 457
β-Keto-sulfinylamides, magnesium enolate preparation, 444
β-Keto-sulfinylcarboxylic acid derivatives, desulfinylation, 444–445
β-Keto-sulfoxide, desulfinylation, 444, 479
Kinetics
  electrochemical Mg deposition, 252–253
  ligated Mg cation fragmentation, 176
Klenow fragment, DNA replicase, 351–352
KMG-20-AM/27-AM, UV–visible fluorometry, 285
Kohler's ketone, metalation, 457, 458
Krasnovskii reduction, chlorophyll, 211, *212*
Kumada coupling reaction, microwave activation, 513
Kumada–Tamao-Corriu reactions, aryl electrophiles, 550

β-Lactam
  aldol-type reactions, 499–500
  halogen–metal exchange, 442, 443
  synthesis, 475, 476
Lactones
  aldol-type reactions, 486
  magnesium enolate, 484–489
LaFlamme, Doering–LaFlamme allene synthesis intermediates, 735–738
Lanthanum(III), FAAS analysis, 272
Lanthanum chloride lithium chloride, addition to multiple bonds, 569, 572
(–)-Lardolure, enantioselective conjugate addition, *786*, 788
Large-angle X-ray scattering, structural studies, 3
Laser ablation, Mg atom formation, 157, 159
Lewis bases, titration, 288
Ligand exchange
  Mg^{2+}, 318–321
  cationic sandwiches, 124
  sulfoxides, 444, 720
Ligand loss, Mg cation–alcohol reactions, 161
Ligand redistribution, organomagnesium amides, 411
Ligands
  anionic, 176–178

carboxylate, 178–179
enantioselective conjugate addition, 774–777, 780–784
Mg^{2+} coordination
  carboxylates, 321–322
  nitrogen compounds, 324
  phosphate functionalities, 322–324
  water, 318–320
organomagnesium cation reactivity, 180
Ligand switching, Mg cation–alcohol reactions, 161
Ligated magnesium ions
  adduct-forming reactions, 160
  electron impact, 175
  gas-phase fragmentation, 174–179
  ligands, 174–175
  photoactivation reactions, 162–170
Linear sweep cathodic stripping voltammetry (LSCSV), sodium pentothal, 276
Liquid chromatography
  high-performance, 274
  speciation analysis, 292–293
N-Lithio amines, magnesium carbenoid reactions, 741
N-Lithio arylamines
  magnesium alkylidene carbenoid reactions, 748–755
  magnesium carbenoid reactions, 728–729, 730
N-Lithio nitrogen-containing heterocycles, 755–757
Lithium acetylides, magnesium alkylidene carbenoid reactions, 526, 758
Lithium aluminum hydride, titration, 288, 290
Lithium butyldimethylmagnesium, I_2–Mg exchange, 692, 695
Lithium chloride
  addition to multiple bonds, 569–570, 572
  alkylmagnesium chloride complexes, 682, 714
  Cu(II) chloride, 544
  functionalized Grignard reagent synthesis, 519–521, 531, 533–537
  lanthanum chloride, 569
  2,2,6,6-tetramethylpiperidine magnesium chloride, 539–540, 541
  transition-metal-free homocoupling, 556, 557
Lithium dialkylamides, metalation, 537
Lithium dibutylisopropylmagnesate, halogen-Mg exchange, 692, 694, 697, 699, 700
Lithium magnesiates, halogen–Mg exchange reagents, 523
Lithium tetramethylpiperidide, epoxide deprotonation, 547
Lithium thiolates, magnesium alkylidene carbenoid reactions, 758–760

Lithium tributylmagnesate
  deprotonation, 688
  halogen–Mg exchange, 690–692, 693, 696–698, 707
Lithium trimethylmagnesate, 681, 683–684, 706, 707
Lithium triorganomagnesates, 682
Lithium triphenylmagnesate, 681, 683
Lithium tris(sec-butyl)magnesate, Br_2–Mg exchange, 706, 707
Lithium tris(2,2,6,6-tetramethylpiperidino) magnesate, 713
Lithospermic acid, liquid chromatography, 293, 294
Lowest excited electronic state, cyclopentadienyl magnesium, 123
Low-temperature ashing (LTA), matrix obliteration, 271
LSCSV (linear sweep cathodic stripping voltammetry), 276
LTA (low-temperature ashing), 271
Lung cancer, elemental analysis in hair, 278
Lymphoblastic leukemia, elemental analysis in hair, 278
Lymphoma, elemental analysis in hair, 278

Mag-fura-2/5/Red, UV–visible fluorometry, 283, 284, 288
Mag-indo-1, UV–visible fluorometry, 283, 284
Magnesacarboranes, molecular geometry, 51–53
Magnesacycles
  magnesium anthracene compounds, 46–47
  thermochemistry, 117, 120–121
Magnesacycloalkanes, thermochemistry, 120–121
Magnesacyclohexane, structure, 39–40
Magnesate anions
  bimolecular reactions, 180–182
  fragmentation, 178–179
Magnesiation
  hydromagnesiation of acetylenes, 540, 543
  magnesium amide bases, 537–540
  see also Carbomagnesiation; Metalation
Magnesium
  activation, 513
  in adult human body, 270
  biochemistry, 315–367
    cation transporters, 324–327
    DNA and RNA interactions, 320–321, 333–342
    isotope effect, 330
    protein-based enzymes, 342–359
    universal energy currency of life, 327–333
  bound to DNA, 270
  CO complexes, thermochemistry, 125

Magnesium (*continued*)
  coordination, 318–324
    binding modes, 320–324, 331–332
    carboxylate ligands, 321–322
    hexaaquomagnesium ion, 318–320
    nitrogen ligands, 324
    phosphate functionalities, 322–324
  direct oxidative addition to organic halides, 512–515
  electrochemical deposition, 245–253
    kinetics, 252–253
  electropositivity, 102–103, 125
  elemental analysis, 268, 269–288
  films, 172–174
  free hydrated ions, 269
  isotopically labelled, 107
  properties, 102–103
  Rieke magnesium, 514
  second ionization energy, 175
  surfaces, 172–174
  vaporization, 157
^{24,25,26}Magnesium, reverse isotope dilution technique, 287–288
Magnesium acetylides, structure, 42–43
Magnesium–adenosine triphosphate (MgATP), 328–329
Magnesium alcoholates, solubility, 521
Magnesium alkoxide, metalation, 461, 493
Magnesium alkylidene carbenoids
  electrophilic reactions
    alkenylation, 748
    N-lithio arylamines, 748–755
    N-lithio nitrogen-containing heterocycles, 755–757
    lithium acetylides, 758
    lithium thiolates, 758–760
    tetrasubstituted olefin one-pot synthesis, 746–748
  generation, 742–745
  nucleophilic property, 744–746
  structure, 751, *753–754*, 755
Magnesium allenyl enolates, 449, 450, 506
Magnesium amide bases
  chiral, *470*
  magnesium enolate preparation, 464–465, *466–469*
  metalation agents, 537–543
Magnesium amide enolates, 499–500
Magnesium amides
  chiral, 454, 498
  hindered, 464–469
Magnesium β-aminoenolate, 454
Magnesium aminoester enolates, 487, 489, 497
Magnesium–anthracene compounds
  functionalized Grignard reagent synthesis, 514
  structural chemistry, 44–47

Magnesium atoms
  excited state, 157, 158, 159
  formation, 157
  organic substrate reactions, 157–159
    alkanes, 157
    alkyl halides, 157–158
    unsaturated substrates, 158–159
  reactivity, 171
  triplet states, 158
Magnesium bicarbonate, as-received Mg surfaces, 172
Magnesium bisamides, metalation, 539
Magnesium carbenoids, 717–769
  chiral, 722
  cyclopropanation of allylic alcohols, 722
  electrophilic reactions
    1,3-C–H insertion reaction, 729–734
    Grignard reagent reactions, 723–725
    N-lithio arylamine reactions, 728–729, *730*
    secondary chiral Grignard reagent synthesis, 725–727
    α-sulfonyl lithium carbanion reactions, 727–728, 748
  generation, 718–722
    halogen–Mg exchange, 718–720
    sulfoxide–Mg exchange, 718, 720–722
  halogen–Mg exchange, 516
  α-heteroatom-substituted Grignard reagents, 767
  magnesium alkylidene carbenoids, 742–760
  magnesium cyclopropylidenes, 735–742
  magnesium β-oxido carbenoids, 479, 760–766
Magnesium carbides, 106–107
Magnesium cations
  bimolecular reactions, 179–180
  organic substrate reactions, 160–170
    adduct-forming reactions, 160
    alcohols, 161, 165–166
    alkane complexes, 162–163
    alkyl halides, 160–161
    amine complexes, 167–169
    C=X bonds, 169–170
    ether complexes, 166–167
    MgX$^+$, 162
    organohalogen complexes, 163–165
    photoactivation, 162–170
    radicals, 160–170
    reactivity, 160
  speciation analysis, 271
Magnesium chiral enolates, 505–506
Magnesium chloride, cation reactions, 162
Magnesium clusters
  Cu–Mg cluster structure, 86, *87*
  solvent-free environments
    alkyl halide reactivity, 157
    Mg cation–ether complexes, 167

Subject Index 877

organic substrate reactions, 171–172
  reactivity, 171
Magnesium cyclopropylidenes
  electrophilic reactions, 739–742
  intermediates in Doering–LaFlamme allene synthesis, 735–738
  nucleophilic reactions, 738–739
  pyramidal inversion, 738
Magnesium dialkylmalonate, 462, 494–499
Magnesium dicarbonyl enolates, 489–499
Magnesium dihalide
  catalyzed Claisen rearrangement, 444
  transmetalation, 491, 492
Magnesium enolates
  chiral, 500
  deuteriation, 441, 479, 489
  enantioselective conjugate addition, 789
  IR spectroscopy, 451
  one-carbon ring expansion, 762, 763–765
  preparation, 438–472
    conjugate addition, 450–457, 789
    permutational H_2/metal interconversions, 457–470
    permutational metal/metal salts interconversion, 445–450
    permutation of heteroatom/metal, 441–445
    reductive metal insertion, 438–441
  reactivity, 472–506
    carboxylic acid dianions, 503
    chiral oxazolidinone anions, 503–505
    electrophiles, 472–505
    magnesium amide enolates, 499–500
    magnesium dicarbonyl enolates, 489–499
    magnesium ester enolates, 484–489
    magnesium ketone enolates, 472–484
    magnesium lactam enolates, 499–500
    magnesium lactone enolates, 484–489
    magnesium thioamide enolates, 500–502
    magnesium thioesters, 500–502
Magnesium ester enolates, 484–489
Magnesium ethyl carbonate (MMC, Stile's reagent), 463, 476
Magnesium–fluorine bonds, 103
Magnesium Green, UV–visible fluorometry, 283, 284, 285
Magnesium halides, gas-phase structures, 3084
Magnesium hydride
  functionalized Grignard reagent synthesis, 540, 543
  titanocene hydride–magnesium hydride complex, 88–89
Magnesium hydrogen alkylmalonate, 494–499
Magnesium hydroxide, as-received Mg surfaces, 172
Magnesium ionophores, ion-selective electrodes, 275

Magnesium α-ketoester enolates, 489–493
Magnesium ketone enolates, 472–484
Magnesium lactam enolates, 442, 443, 499–500
Magnesium lactone enolates, 484–489
Magnesium lithospermate, liquid chromatography, 293, 294
Magnesium Orange, UV–visible fluorometry, 283, 284
Magnesium organohaloaluminate salts, cathodic reduction, 248–251
Magnesium oxide, cation reactions, 162
Magnesium β-oxido carbenoids, 479, 760–766
Magnesium phthalocyanine, PET, 198
Magnesium sandwich species
  cationic sandwiches, 124–125
  half-sandwiches, 123, 124
  thermochemistry, 115, 122–125
  triple decker (club) sandwiches, 123–124
Magnesium(II)-selective fluorophores, UV–visible fluorometry, 283–284
Magnesium thioamide enolates, 500–502
Magnesium thioesters, 500–502
Magnesocene
  ^{25}Mg NMR spectra, 152, *153*
  multi-hapto-bonded complexes, 25–26, 47–50
  organomagnesium cation formation, 180, 182
  structure, 183
  thermochemistry, 122–123
ansa-Magnesocene complexes, structure, 48–50
Magon, UV–visible spectrophotometry, 282, 283
Mag-quin-1/2, UV–visible fluorometry, 283, 284
MALDI (matrix assisted laser desorption ionization), 156, 171
Mandyphos ligand, catalytic enantioselective conjugate addition, 775, 776
Manganese, catalyzed carbomagnesiation, 636, 637
Manganese salts, transmetalation, 552
Mass spectrometers, EI sources, 156
Mass spectrometry
  elemental analysis of Mg, 287–288
  FT–ICR, 160
  ICP–MS, 287–288
  solvent-free environments, 156, 159, 160, 162
Matrix assisted laser desorption ionization (MALDI), 156, 171
Matrix isolation, Mg atom reactions, 157–159, 171
Matrix obliteration, elemental analysis sample preparation, 271–272

Subject Index

Mechanisms
  enantioselective conjugate addition, 789–791
  semiempirical [PM3(tm)] calculations, 790–791
Medicinal applications, chlorophyll, 191, 211, 214
Metabolic enzymes (protein-based), 345–350
  enolases, 348–350
  glycolytic cycle, 345, *346*
  protein kinases, 345, 347–348
α-Metalated nitriles, functionalized Grignard reagent synthesis, 528, 529
Metalation
  arene, 22
  deprotonative, 18–22
  Fe-catalyzed carbometalation, 622–623
  hindered Hauser bases, 464
  hindered magnesium amides, 464–469
  Kohler's ketone, 457, 458
  magnesium alkoxide, 461, 493
  magnesium enolate preparation, 457–470
  organomagnesium amides, 419–421, 537–540, 541
  thioamides, 452
  see also Transmetalation
Metal-bonded butyl groups, heteroleptic organomagnesiates, 22
Metal carbonyls, 125
Metal insertion, C–halogen bonds, 438–441
Metalla-cyclobutane, Grignard reaction, 86
Metallocycles, 106
Metalloids, CO complexes, 125
Metallo-olefins, 106
Metalloproteins, speciation analysis, 269
Metallotetrapyrroles, photochemistry, 212–213
Metal powders, water determination, 299
Metastable conditions, organomagnesium cation fragmentation, 182
Methane, active hydrogen determination, 299–300
Methanol, Mg cation fragmentation, 176
2-Methoxyethanol, Mg cation complex photofragmentation, 166
2-(Methoxymethyl)phenylmagnesium bromide, 65, 66
N-Methylacetamide, Mg cation binding energy, 175
(2-Methylallyl)magnesium, allylic exchange, 143
Methylamine, Mg cation complex, 168
N-Methylation, enthalpies of formation, 113–114
Methyl bromide, enthalpy of reaction, 105–106
Methyl crotonate, enantioselective conjugate addition, 781, 789

Methyl deviations, alkylmagnesium bromides, 111
Methylene increment, organomagnesium bromides, 110–111, 113–114
9-Methylfluorene, titration, 288, 289
Methyl Grignard reagents
  oxidation at inert electrodes, 229–231
  distribution of anodic products, *230*
Methyl groups, electrochemical grafting, 241–243
Methyl halide, Mg atom reactions, 158
Methyl 2-halogeno-2-cyclopropylideneacetate, Michael addition, 452
Methyl 4-iodobenzoate, functionalized Grignard reagent synthesis, 517
Methylmagnesium alkoxides, ^{1}H NMR spectra, 145
Methylmagnesium *tert*-butoxide, 70
Methylmagnesium tris(3-*tert*-butylpyrazolyl)hydroborate, 73, *74*
Methylmagnesium tris(3-*tert*-butylpyrazolyl)phenylborate, 73, *74*
Methyl methacrylate, polymerization, 16, 484
N-Methylpyrrolidinone (NMP)
  alkenylation, 604–605
  functionalized Grignard reagent synthesis, 534, 535
Methylthymol Blue, UV–visible spectrophotometry, 280, 281
Methyl 2-(trimethylsilyl)propenoate, Michael addition, 452
MgATP (Mg–adenosine triphosphate), 328–329
^{25}Mg NMR spectroscopy, 143, 151–152, *153*
Michael addition
  β-amino-α,β-diester preparation, 489
  benzalacetomesitylene, 451
  C=C bonds, 563–565
  N,N-dialkyl-α-methacrylthioamides, 452
  magnesium aminoester enolates, 497
  magnesium enolate preparation, 451–454
  methyl 2-halogeno-2-cyclopropylideneacetate, 452
  methyl 2-(trimethylsilyl)propenoate, 452
  α,β-unsaturated carboxylic acid derivatives, 451
  α,β-unsaturated thioesters, 501
Michler's ketone, Gilman's color test, 295, 296
Microfluorimetry, elemental analysis of Mg, 288
Microwave-aided extraction, elemental analysis of Mg, 271
Microwave digestion, FAAS analysis, 272
Microwave irradiation, aryl chloride/bromide activation, 513

Subject Index

Microwave spectroscopy, Mg film reactions, 174
Milk
  complexometric analysis, 283
  mineralization, 271, 279
Mineralization, matrix obliteration, 271
Mixed magnesate reagents, nucleophilic addition, 685–686
Mixed magnesium/lithium amides, metalation, 539
Mixed metal alkyl–amido base, triorganomagnesate production, 13–14
Mixed organomagnesium transition-metal compounds, 85–91
MMC (magnesium ethyl carbonate (Stile's reagent)), 463, 476
Model reactions, Grignard calculations, 380–384
Modified Schlenk equilibrium, *ab initio* calculations, 378–380
Moisture exclusion, calorimetry, 104
Molar standard enthalpies of formation, 102
  see also Enthalpies of formation
Molar standard enthalpies of reaction, 102
  see also Enthalpies of reaction
Molecular geometry
  slipped geometry, 53–54
  see also Structural chemistry
Monensin, synthesis, 482, 483
β-Monoacylglycerols, derivatization, 301
Monodentate coordination, Mg^{2+}, 322, 324
Monoorganomagnesium β-diketiminates, 73–78
Monoorganomagnesium heteroleptic compounds, 54–85
  amides, 71–83
  cations, 56–57
  halogen compounds, 58–69
  oxygen compounds, 69–71
  phosphorus compounds, 84–85
  sulfur compounds, 84, 85
Mukaiyama reaction, 450, 456, 457, 474
Multi-carbon homologation, alkyl halides, 723–724
Multi-hapto-bonded groups, diorganomagnesium compounds, 25–29, 47–54
Multiple bonds, organomagnesium reagent addition, 471, 559–575
Multivariate PLS regression, IR spectroscopy, 295
Mycocerosic acid, enantioselective conjugate addition, 787, 788

1,8-Naphthalene diylmagnesium THF complex, 40–41

Nasopharyngeal cancer, elemental analysis in hair, 278
Natural products
  catalytic enantioselective synthesis, 786–789, 799
  Fe-catalyzed Grignard reactions, 625–628
Near-IR spectroscopy, concentration monitoring, 512
Neodymium trichloride, addition to C=O bonds, 570
Neopentylmagnesium bromide, functionalized Grignard reagent synthesis, 534, 535
NHCs *see* N-Heterocyclic carbenes
Nickel
  catalyst immobilization on charcoal, 554
  organomagnesium compounds, 86
Nickel-catalyzed reactions
  addition to C=C bonds, 560
  carbomagnesiation, 635, 640, 653–654, 664–665
  cross-coupling reactions, 543–544, 550–552, 554
Nitriles
  α-metalated, 528, 529
  α,β-unsaturated, 565–566
Nitroalkenes, addition, 566
Nitroarenes, electrophilic amination, 576
Nitro compounds, I_2–Mg exchange, 531, 533
Nitrogen, organomagnesium ion reactivity, 180
Nitrogen compounds
  carbomagnesiation, 641–642
  N-lithio nitrogen-containing heterocycles, 755–757
  Mg^{2+} coordination, 324
Nitrogen oxides, organomagnesium ion reactivity, 180
Nitrosoarenes, electrophilic amination, 576–577
NMP (N-methylpyrrolidinone), 534, 535, 604–605
NMR spectroscopy, 131–153
  alkoxides, 145–147
  alkylmagnesium compounds, 133–138, 141
  allylmagnesium compounds, 141–145
  arylmagnesium compounds, 134, 138–140
  aryloxides, *146*, 147
  co-ordination complexes, 148, *149–150*
  elemental analysis of Mg, 286–287
  peroxides, 147–148
  solution concentration, 132–133
  speciation analysis, 296–298
  vinylmagnesium compounds, 145
  see also ^{13}C NMR spectroscopy; ^{19}F NMR spectroscopy; ^{1}H NMR spectroscopy; ^{25}Mg NMR spectroscopy; ^{17}O NMR spectroscopy
No-D NMR spectroscopy, speciation analysis, 296

Non-stabilized α-amino-substituted carbanions, 728–729
Nuclear magnetic resonance spectroscopy *see* NMR spectroscopy
Nucleobases, tautomeric forms, 170
Nucleophilicity, magnesium alkylidene carbenoids, 744–746
Nucleophilic reactions
  diorganomagnesium compounds, 256–257
  magnesium cyclopropylidenes, 738–739
  triorganomagnesates, 683–686, 711, 713

Olefination, Peterson, 474–475, 487, 488
Olefins
  anodic addition of Grignard reagents, 237
  Cu–olefin π-complex, 790
  tetrasubstituted, 746–748
One-carbon homologation, magnesium β-oxido carbenoids, 760–766
One-pot synthesis, tetrasubstituted olefins, 746–748
^{17}O NMR spectroscopy, peroxides, 147–148
Optically active compounds
  chloroalkyl aryl sulfoxides, 721, 725–727
  magnesium carbenoid generation, 721–722
Organic alkali compounds, stability, 103
Organic chromophores, absorption spectra, 193
Organic halides
  direct oxidative addition of Mg, 512–515
  Mg cation reactions, 162
Organic substrates
  Mg atom reactions, 157–159
  unsaturated substrates, 158–159
Organoaluminum compounds, 103
Organoantimony compounds, derivatization, 300, *304*
Organoelemental compounds, electrooxidation of Grignard reagents, 228, 237–241
Organohaloaluminate salts, cathodic reduction, 248–251
Organohalogens, Mg cation complexes, 163–165
Organolanthanides, titration with I$_2$, 290–291
Organolead compounds, derivatization, 300, *303*
Organolithium compounds, titration, 288, 289
Organolithium reagents, lithium magnesiate preparation, 523
Organomagnesates, 681–715
  anion reactivity, 178–179, 180
Organomagnesia and rings, thermochemistry, 117–121
Organomagnesiates
  structural chemistry, 4–22
    heteroleptic, 5, 14–22
    homoleptic, 5
    tetraorganomagnesiates, 5–12

  triorganomagnesiates, 12–14
Organomagnesium alkoxides, 428–432
Organomagnesium amides
  metalation, 537–540, 541
  reactivity, 419–426
  structure, 71–83, *405–410*, 412–419
  synthesis, 54, 71–83, 403, 404–412, *413–416*
Organomagnesium aryloxides, 428–432
Organomagnesium ate complexes, 681–715
Organomagnesium bromides
  heteroatoms, 113–116
  methylene increment, 110–111
  unsaturated
    enthalpies of hydrogenation, 113, 119
    formal protonation reactions, 112
Organomagnesium cations *see* Magnesium cations
Organomagnesium chloroaluminate complexes, cathodic reduction, 248–249
Organomagnesium compounds
  analytical aspects, 265–314
    speciation analysis, 268, 269, 288–299
  bifunctional, 39–40
  film decomposition, 173
  functionalized
    reactivity, 543–583
    synthesis, 512–543
  gas-phase structures, 183–184
  solvent-free environments, 155–187
Organomagnesium fluorides, 103
Organomagnesium-group 15-bonded complexes, 403–427, 428
Organomagnesium-group 16-bonded complexes, 427–434
Organomagnesium halides
  homologous series, 110–112, 117
  thermochemistry, 109–116
  unsaturated compounds, 112–113
Organomagnesium ions
  gas-phase bimolecular reactions, 179–182
  gas-phase unimolecular reactions, 182
  insertion reactions, 168
  ligated Mg ion fragmentation, 174–179
Organomagnesium phosphanide, 427, 428
Organomagnesium sulfides, 433–434
Organomagnesium transition-metal compounds, structures, 85–91
Organomercury compounds, derivatization, *304*
Organometallic compounds
  derivatization, 300–301, *302–304*
  electrooxidation of Grignard reagents, 228, 237–241
Organotin compounds, gas chromatography, 300, *302–303*, 304
Organozinc compounds, titration with I$_2$, 290–291

Osmometric measurements, Schlenk equilibrium, 107
Oxazolidinones
  1,4-addition to Michael acceptors, 563–564
  chiral, 503–505
Oxidation
  Mg cation reactions, 160–161, 163
  photoinduced ring-opening, 207–209, 210, 211
  see also Anodic oxidation
Oxidation potentials, Grignard reagent correlations, 258–260
Oxidative addition
  induction period, 513
  Mg to organic halides, 512–515
Oxidative coupling
  polyfunctional aryl/heteroaryl amidocuprates, 580–581
  silver tosylate homocoupling, 547
β-Oxido carbenoids, 479, 760–766
Oxime ethers, chiral, 574
Oxine
  high-performance liquid chromatography, 274
  UV–visible spectrophotometry, 279, 280–281
Oxonium structure, Grignard reagents, 107
Oxononium ions, stability, 104
Oxygen
  organomagnesium amide reactions, 422–423
  organomagnesium ion reactivity, 180
Oxygen compounds
  carbomagnesiation, 643
  C–O bond activation, 159, 161, 167
  heteroleptic monoorganomagnesium compounds, 69–71
  intramolecular coordination, 66

Palladium-catalyzed reactions
  addition to C=C bonds, 561
  carbomagnesiation, 641, 642
  cross-coupling reactions, 543–544, 550, 552, 553
Pancreatic lipase hydrolysis, glyceride analysis, 301
Particulate matter, Mg speciation analysis, 270–271
Pathogenesis indicators for cancer, AAS, 278
PDT (photodynamic therapy), 191, 211, 214
Pearl, ICP–AES Mg analysis, 279
PEBBLEs, UV–visible fluorometry, 285
PEPPSI, transition-metal-catalyzed cross-coupling, 550, 551
Perfluorinated Grignard reagents, synthesis, 516
Perfluoroalkylmagnesium halides, halogen–Mg exchange synthesis, 516

Perfluorophenylmagnesium bromide, synthesis, 516
Periodic table, bond energies, 115
Permutational hydrogen/metal interconversions, 437–440, 457–470, 482
Permutational metal/metal salts interconversions, 445–450
Permutation of heteroatom/metal, 441–445
Peroxides, NMR spectra, 147–148
Perturbative quantum chromodynamics (PQCD), ion chromatography, 273
Perylene-linked systems, PET, 201, 202–203
PET (photoinduced electron transfer), 191, 194–206
Peterson olefination, 474–475, 487, 488
Pharmacological issues, Mg analysis, 268–269
(−)-Phaseolinic acid, synthesis, 500, 501
1,10-Phenanthroline, indicator, 288, 289
Phenylacetylene, NMR spectroscopy, 297
4-(Phenylazo)diphenylamine, indicator, 289
1,2-Phenylenemagnesium THF complex, structure, 40–41
Phenylethynylmagnesium bromide, enthalpy of reaction, 112
2-(Phenylmagnesio)-1,3-xylene-15-crown-4, 34–36
Phenylmagnesium bromide ethoxide, structure, 58
N-Phenyl-1-naphthylamine, titration, 288, 289
4-Phenylnipecotic acid, preparation, 451
4-Phenyloxazolidinone, Evans-type auxiliary, 447
1-Phenylpropyne, transition-metal-catalyzed cross-coupling, 545
Pheophytinization, porphyrins, 194
Phosphates, Mg^{2+} coordination, 322–324
Phosphines, monomagnesium compounds, 84–85
Phosphomycoketides, enantioselective conjugate addition, 785, 786
Phosphoramidites
  allylic substitution, 558
  Cu-catalyzed enantioselective reactions
    allylic alkylation, 792, 793, 794
    conjugate addition, 774, 792, 793
Phosphorus compounds, heteroleptic monoorganomagnesium compounds, 84–85
Photoactivation
  excited-state Mg atoms, 157
  ligated Mg cation complexes
    alcohols, 165–166
    alkanes, 162–163
    amines, 167–169
    C=X bonds, 169–170
    ethers, 166–167
    organohalogens, 163–165

Photochemical reactions
  chlorophyll, 209–212, 213–214
  porphyrins, 206–209
  ring-opening reactions, 207–209, 210, 211
Photochemistry, 189–218
  applied, 212–214
  electron transfer, 191, 194–206
  Mg cation–alkane complexes, 163
  phthalocyanines, 190–191, 198
    applied photochemistry, 213
    electron transfer systems, 196, 198
    porphyrin dyads, 196
  porphyrins, 190–191, 192–194
    applied photochemistry, 212–214
    reactions, 206–209
    stability, 193–194
Photochromism, polythiophene, 285
Photodissociation spectroscopy, Mg cation complexes, 162–164, 165–166, 168–169, 171
Photodynamic therapy (PDT), chlorophylls, 191, 211, 214
Photofragmentation, Mg cation complexes, 166–170
Photoinduced electron transfer (PET), 191, 194–206
Photoionization, Mg cation–ether clusters, 167
Photooxidation
  phthalocyanines, 213
  ring-opening reactions, 207–209, 210, 211
Photooxygenation, chlorophylls, 211, *212*
Photopigments, porphyrins, 193–194
Photosensitizers, chemical transformations, 206, 211
Photosynthesis, 191–196
  chlorophylls, 191, 356–357
  photochemical reaction center, 194, *195*
  photosystems I and II, 355–356
  protein-based enzymes, 355–359
  rubisco, 357–359
  Z-scheme, 194, *195*
Photosystems I and II, photosynthesis, 355–356
Phthalocyanines
  photochemistry, 190–191, 198
    applied photochemistry, 213
    electron transfer systems, 196, 198
    porphyrin dyads, 196
Phthioceranic acid, enantioselective conjugate addition, 788
Phytochlorin, photosynthesis, 192
Piezochromism, polythiophene, 285
Piezoelectric detection, ion chromatography, 273
Pigments, photosynthesis, 191–192
Pinacolborates, functionalized synthesis, 534, 537
Plasma
  AAS, 278
  electrolyte determination, 275–276
  electrophoresis, 274
  mass spectrometry, 288
  speciation analysis, 269–270
Platinum terpyridine acetylide complex, PET, 198
PLS (partial least squares) regression, IR spectroscopy, 295
Pnictogenides, heavy group 15 complexes, 427
^{31}P NMR spectroscopy
  elemental analysis of Mg, 286–287
  speciation analysis, 297–298
Polar bonds, Mg, 103, 118
Polar mechanisms, Grignard carbonyl additions, 370, 387–388, 389, 391, 394, 396, 400–401
Polycenter electron-deficient bonds, magnesium carbides, 107
Poly(chlorotrifluoroethylene), surface conditioning, 305
Polycyclic aromatic hydrocarbons, Mg cation reactions, 160
Polyenes, Mg cation reactions, 160
Polyfluorinated pyridines, Mg cation complexes, 164
Polyfunctional aryl/heteroaryl amidocuprates, oxidative coupling, 580–581
Polymerization
  methyl methacrylate, 16, 484
  organomagnesium-group 15-bonded complex catalysts, 421
Polymers, chromatic chemosensors, 285
Polymetallic host, deprotonative metalation, 19
Polyphenol-bound magnesium, speciation analysis, 271
Polysaccharide-bound magnesium, speciation analysis, 271
Polysantol, preparation, 440
Polythiophene, chromatic chemosensors, 285
Polyvinyl alcohol, surface conditioning, 305
Porphyrin photochemistry, 125, 190–191, 192–194
  applied photochemistry, 212–214
  arrays, 201, 204–205
  dimerization, 206
  electron transfer systems, 196–206
  photochemical reactions, 206–209
  phthalocyanine dyads, 196
  stability, 193–194
Portland cement, voltametric analysis, 277
Potassium triiodide, potentiometric titration, 291–292
Potential energy profile, Grignard reagent calculations, 376, *377*, *381*
Potentiometric titration, speciation analysis, 291–292

PQCD (perturbative quantum chromodynamics), 273
Preconcentration, elemental analysis sample preparation, 272–273
Primary alkyl chlorides, transition-metal-catalyzed cross-coupling, 544
Proazaphosphatrane, magnesium enolate, 472
Probe encapsulated by biologically localized embedding (PEBBLE), 285
Propargyl alcohols
  carbomagnesiation, 645–652
  Fritsch–Buttenberg–Wiechell rearrangement, 744
Propargyl amines
  carbomagnesiation, 652–653
  functionalized Grignard reactions, 548, 549
β-Propiolactones, ring opening, 547
Propylamine, Mg cation complex, 168
Propylmagnesium bromide, DME complexes, 59
Prostatic carcinoma, elemental analysis in hair, 278
Protecting groups, functionalized Grignard reagents, 530–531, 534
Protein-based enzymes
  Mg biochemistry, 342–359
    allosteric systems, 343–345
    DNA replication and repair, 350–355
    enzymatic modes of action, 342–345
    metabolic enzymes, 345–350
    photosynthesis, 355–359
    sequential systems, 343, 348–350
    template enzymes, 342–343
Protein kinases, 345, 347–348
Proteins, speciation analysis, 269
Protonation
  enantioselective, 480, 481
  see also Deprotonation; Enthalpies of protonation
Proton transfer
  carbocation–Mg atom reactions, 159
  Mg cation reactions, 162, 175
Protoporphyrins, photooxidation, 207–208
PTFE, alkali-activated, 283
Pulsed photoacoustic spectroscopy, PET, 198
Purpurin, UV–visible spectrophotometry, 279–280
Pyramidal inversion, magnesium cyclopropylidenes, 738
1-Pyreneacetic acid, titration, 288, 289
1-Pyrenemethanol, titration, 288, 289
Pyridine-enhanced precatalyst preparation stabilization and initiation (PEPPSI), 550, 551
Pyridines
  diethylmagnesium reactions, 684
  Mg cation complex, 168
  metalation, 537
  transition-metal-catalyzed cross-coupling, 553
Pyridylmagnesium, Br_2–Mg exchange, 524
2-Pyridylmagnesium bromide, THF complex, 65
2-Pyridylsilylmethyl tris(trimethylsilyl)methylmagnesium bromide, THF complex, 63, 64
Pyroangelensolide, preparation, 482, 483
Pyrolants, fluorocarbons, 103
Pyrophosphates, Mg^{2+} coordination, 323–324
Pyrroles
  Boc-protected 2-substituted, 447, 449
  magnesiation, 538, 539

Quadrupole ion trap mass spectrometers, organomagnesium ions, 180
Quality control, organometallic compounds, 300
Quaternary chiral centers, all-carbon, 565
Quaternary stereocenters
  allylic alkylation of esters, 792, 796
  one-carbon expanded ketone, 762, 763
Quinalizarin, UV–visible spectrophotometry, 279–280, 282
Quinolines, transition-metal-catalyzed cross-coupling, 553
Quinones, PET, 194–196, 197–198, 207

Racemic magnesium carbenoids, generation, 720–721
Radiative association kinetics, bond dissociation energies, 124
Radiative associative adduct formation, Mg cation reactions, 160
Radical anions
  anthracene, 45, 46
  1,2-bis[(2,6-diisopropylphenyl)imino] acenaphthene, 79, 80
Radical cations, 160–170
Radical cyclization, Fe-catalyzed, 623–624
Radicals
  anodic oxidation of Grignard reagents, 229–237, 238, 258–259
  2-thenyl, 299
Rare-earth metal salts, addition to C=O bonds, 569–570
Rate constants, Grignard reagent correlations, 258–260
Reaction see Enthalpies of reaction; Molar standard enthalpies of reaction
Reactive halides, gas chromatography, 295
Reactivity
  Grignard reagents, 235–236, 258–260
  functionalized, 543–583

Reactivity (*continued*)
  Mg cations, 160–170
  magnesium enolates, 472–506
  organomagnesium amides, 419–426
Rearrangement reactions
  Claisen, 444, 446, 487, 491
  Fritsch–Buttenberg–Wiechell, 742–744
Redistribution reactions
  heteroleptic monoorganomagnesium compounds, 55
  ligand redistribution, 411
Reduction
  Krasnovskii reduction, 211, *212*
  see also Cathodic reduction
Reductive metal insertion, C–halogen bonds, 438–441
Reflectance, IR spectroscopy, 295
Reformatski reagent, enolate bridge-bonding, 77
Regioselectivity
  carbomagnesiation, 636–638, 641–644, 652–653, 675
  hindered Hauser base deprotonation, 464
Resonance effects, hydrocarbylmagnesium bromides, 114
Resorcinol, formation, 493, 505
Retro-Ritter reactions, Mg cation fragmentation, 177
Reverse isotope dilution technique, elemental analysis of Mg, 287–288
Rharasch reaction, natural product synthesis, 625–626
Rhenium complexes, tetraphenylporphyrins, 199–200
Ribozymes
  Mg^{2+} DNA/RNA interactions, 335–342
    group 1 introns, 337–339
    hammerhead ribozymes, 339–342
Rieke magnesium, functionalized Grignard reagent synthesis, 514
Ring contraction, cyclohexane, 476
Ring expansion, one-carbon, 760–766
Ring opening
  photoinduced, 207–209, 210, 211
  small cycles, 543, 547, 548
Ring strain
  cycloalkylmagnesium halides, 117
  magnesacycloalkanes, 120–121
Ritter reactions, retro-Ritter, 177
RMSEP parameter, IR spectroscopy, 295
RNA
  Mg^{2+} interactions, 333–342
    5S rRNA domain, 335
    Loop E motif, 335
    ribozymes, 335–342
  speciation analysis of bound Mg, 270
Root mean square error of prediction (RMSEP), IR spectroscopy, 295

Rosmarinic acid, liquid chromatography, 293, 294
Rotaxane structure, donor–acceptor complexes, 43–44
Rubisco, CO_2 fixing enzyme, 357–359
Ruthenium, organomagnesium compounds, 90–91
σ-bonded diorganomagnesium compounds, donor–acceptor complexes, 36–44

Sacrificial anodes, electrooxidation of Grignard reagents, 237–241
Salicylaldehyde phenylhydrazone, titration, 290
Sample preparation
  elemental analysis of Mg, 271–273
  matrix obliteration, 271–272
  preconcentration, 272–273
Sandwich species *see* Magnesium sandwich species
Scaffold effect, template enzymes, 342
Schlenck dimer, chelation-controlled addition, 393–395
Schlenk equilibrium, 2, 422
  ab initio calculations, 378–380
  alkylmagnesium compounds, 107, 134, 137
  allylmagnesium compounds, *140*, 142, 144–145
  Grignard reagents
    solutions, 132, 140–141, 144–145, 151
    structures, 384–387
  heteroleptic monoorganomagnesium compounds, 55
  thermochemistry, 107–109, 116
Scymmnol, synthesis, 495, 496
Secondary amides, organomagnesium amide structures, 81, *82*
Secondary chiral Grignard reagents, synthesis, 725–727
Second ionization energy, Mg, 175
SEFT (spin-echo Fourier transform), 286
Selected ion flow tube (SIFT), Mg cation reactions, 160
Self-assembly processes, heteroleptic organomagnesiate production, 18–19
Self-hydrogenation, alkyne/diene Mg films, 173
Semiconductor anodes, electrochemical oxidation of Grignard reagents, 241–243
Semiempirical calculations
  magnesium chelates, 504
  [PM3(tm)] calculations, 790–791
Sensitizers, PET, 206, 211
Sequential enzymes, protein-based, 343, 348–350
Sequential injection analysis (SIA) systems, UV–visible spectrophotometry, 280

Subject Index

Serines, α-substituted, 463, 505
Serum
  electrolyte determination, 275–276
  liquid chromatography, 293
  Mg speciation, 273
  UV–visible spectrophotometry, 282
SET see Single-electron transfer
Shell/shellfish, ICP–AES Mg analysis, 279
SIA (sequential injection analysis) systems, 280
SIFT (selected ion flow tube), 160
Silanol groups
  borosilicate glass, 300
  fused silica capillary tubes, 301, 305
Silicon anodes, electrochemical oxidation of Grignard reagents, 241–243
Silicon–magnesium exchange, silylenol ethers, 450
Silver perchlorate, potentiometric titration, 291
Silver tosylate, oxidative homocoupling, 547
Silyl dienol ethers, preparation, 597–599
Silyl enol ethers
  aldol-type reactions, 472–473, 484, 486
  preparation, 472–474
  Si–Mg exchange, 450
  transmetalation, 450
Simmons–Smith-type cyclopropanation, magnesium carbenoids, 722
Single-electron transfer (SET)
  Grignard carbonyl additions, 370, 372, 396–399, 400–401
  magnesium anthracene compounds, 45
Site-bound magnesium(II) ions, in RNA, 270
Small cycles, ring opening, 543, 547, 548
S_N2'-substitution, Cu(I) catalyst, 557
Sodium bis(2-methoxyethoxy)aluminum hydride, titration, 290
Solubility, lithium chloride effects, 521
Solutions
  Grignard reagent formation, 172–174
  NMR spectroscopy, 131–133
  see also Enthalpies of solution
Solvation see Enthalpies of solvation
Solvation energies, 104
Solvent effects
  electric conductivity, 226–227
  Grignard carbonyl additions, 387–391, 392, 393, 401
Solvent evaporation, Mg cation–amine complexes, 15
Solvent-free environments, 155–187
  bimolecular reactions, 179–182
  formation of organomagnesium species, 156–179
  Grignard reagents, 157, 384–389, 390, 391
  structures of organomagnesiums/magnesium halides, 183–184

unimolecular reactions, 182
Sparteine complexes
  dialkylmagnesium compounds, 148, 150
  Grignard reagents, 60–61
Sp-centers
  Sp^2-center, 544, 550–559
  Sp^3-center, 543–550
  substitution, 543–559
Speciation analysis, 268, 269, 288–299
  chromatographic methods, 273, 292–295
  cryoscopy, 299
  free hydrated Mg ions, 269, 271
  Mg bound to low molecular mass anions, 269
  Mg strongly bound in metalloproteins, 269
  Mg weakly bound to proteins, 269
  spectral methods, 295–299
  tiration methods, 288–292
  total Mg, 269, 270
Spectral methods
  elemental analysis of Mg, 277–288
  speciation analysis, 295–299
Spectrophotometry, Mg in body fluids, 270
Spin coupling constants, allylmagnesium compounds, 143–144, 145
Spin-echo Fourier transform (SEFT), NMR spectroscopy, 286
Spiro-organomagnesium compound, Tebbe-type, 86, 87
Square wave adsorptive stripping voltammetry (SWAdSV), elemental analysis of Mg, 276
Stability
  functionalized Grignard reagents, 514–515
  magnesium alkylidene carbenoids, 744–745, 751
  Mg^{2+}–DNA/RNA interactions, 320–321, 334–335
  organic alkali compounds, 103
  oxononium ions, 104
  porphyrin photochemistry, 193–194
Staggered structure, magnesocene, 183
Statine, enantioselective addition, 571
Stereoselectivity
  carbomagnesiation, 636–638, 641–644, 652–653, 670, 671–672, 675
  Grignard addition calculations, 380, 381
  hindered Hauser base deprotonation, 465
  magnesium chelate, 440, 441, 491, 492, 493, 504
  magnesium enolate preparation, 446–450, 455–456, 457
Sterically hindered amines
  magnesium amide formation, 538
  magnesium enolate preparation, 457, 464
Sterically hindered Hauser bases deprotonation
  regioselectivity, 464

Sterically hindered Hauser bases (*continued*)
    stereoselectivity, 465
    metalation, 464
Sterically hindered ketones, enolization, 478
Sterically hindered magnesium amides,
    metalation, 464–469
Stile's reagent (magnesium ethyl carbonate),
    463, 476
Strained alkenes, carbomagnesiation, 657–661
Strain energies, magnesacycloalkanes, 121
Strong Lewis bases, titration, 288
Structural chemistry, 1–99
    cubane structure, 29, 50, *51*, 145, *146*
    diorganomagnesium compounds, 23–54
    gas phase, 183–184
    heteroleptic monoorganomagnesium
        compounds, 54–85
    magnesium alkylidene carbenoids, 751,
        753–754, 755
    organomagnesiates, 4–22
    organomagnesium alkoxides/aryloxides,
        428–432
    organomagnesium amides, 71–83, *405–410*,
        412–419
    rotaxane structure, 43–44
    slipped geometry, 53–54
Substitution
    α-substituted serine preparation, 463, 505
    allylic, 543, 557–558
    Boc-protected 2-substituted pyrroles, 447,
        449
    Fe-catalyzed, 621–622
    α-heteroatom-substituted Grignard reagents,
        767
    $S_N 2'$-substitution, 557
    Sp-centers, 543–559
        Sp^2-center, 544, 550–559
        Sp^3-center, 543–550
    tetrasubstitution
        ketone preparation, 457
        olefin synthesis, 746–748
    trisubstitution
        alkenyl sulfides, 758–760
        2,2,7-trisubstituted cycloheptanones, 765
Sulfinimines, chiral, 575
β-Sulfinyl ester magnesium enolates,
    asymmetric synthesis, 448, *449*
α-Sulfinyl lithium carbanions, magnesium
    β-oxido carbenoids, 760–766
α-Sulfinyl magnesium carbanion enolates,
    asymmetric aldol-type addition, 459, 486,
    487, 500
Sulfonamides, substitution leaving groups, 553
Sulfonates, substitution leaving groups, 553
Sulfones, substitution leaving groups, 553
α-Sulfonyl lithium carbanions
    magnesium alkylidene carbenoid
        alkenylation, 748

magnesium carbenoid reactions, 727–728
Sulfoxide–magnesium exchange
    magnesium carbenoids
        generation, 718, 720–722
        Grignard reagent reactions, 725
        optically active, 721–722
        racemic, 720–721
    magnesium enolate preparation, 444
Sulfoxide–metal exchange reactions, 720–721
Sulfoxides
    aldol-type reactions, 486, 487
    chiral intermediates, 582–583
    I_2–Mg exchange, 527, 528
Sulfur compounds
    carbomagnesiation, 642–643, 644
    heteroleptic monoorganomagnesium
        compounds, 84, *85*
    magnesium enolate preparation, 443–444
Sulfur–magnesium exchange, benzylic
    magnesium reagent synthesis, 529–530
α-Sulfur-stabilized Grignard reagent, 725, 726
Surface chemistry
    Grignard reagent formation, 172–174
    Mg(0001) single-crystal chemisorption, 172,
        173
    surface modification, 172
Surface conditioning, Grignard reagents, 301,
    305
SWAdSV (square wave adsorptive stripping
    voltammetry), 276

TAEN (N,N′,N″-trimethyltriazacyclononane),
    9–11
TAG (triacylglycerols), 301
Tamao, Kumada–Tamao-Corriu reactions, 550
Tandem reactions, conjugate addition–aldol,
    453, 457, *788*, 789
Taniaphos ligand
    catalytic enantioselective reactions
        allylic alkylation, 792, 797–798
        conjugate addition, 775, 776, 777
Tautomeric forms, nucleobases, 170
Tautomerization, Mg cation–acetamide
    complex, 177
Tebbe-type spiro-organomagnesium
    compound, 86, *87*
Tellurium compounds, derivatization, *304*
3,3′,5,5′-Tetra-*tert*-butyldiphenoquionone,
    transition-metal-free homocoupling, 556,
    557
1,3,5,7-Tetracyanocyclooctatetraene,
    magnesium cationic sandwiches, 125
Tetrahydrocannabinols, preparation, 481
Tetrahydrofuran (THF) complexes
    diorganomagnesium compounds, 40–41,
        48–53
    donor–acceptor, 39–43

Grignard reagents, 58, 59, 63–66
Mg cation fragmentation, 176
4,4,5,5-Tetramethyl-1,3-dioxolan-2-one
O-phenylsulfoxime, electrophilic
amination, 579
2,2,6,6-Tetramethylpiperidine magnesium
chloride, lithium chloride complex,
539–540, 541
Tetraorganomagnesiates
dianions, 8–9
structural chemistry, 5–12
Tetraphenylporphyrins, PET, 199–200
Tetrapyrroles, photochemistry, 192–193, 212–213
Tetrasubstituted compounds
ketone preparation, 457
olefin synthesis, 746–748
Texas Red–dextran, UV–visible fluorometry, 285
Thallocene, magnesocene reactions, 123
Thenyl bromide, ESR spectroscopy, 299
2-Thenyl free radical, ESR spectroscopy, 299
Theoretical calculations
Grignard carbonyl additions, 369–370, 374–401
magnesium alkylidene carbenoid structure, 751, 753–754, 755
Mg–organic substrate reactions, 157–158
Thermochemistry, 101–130
calorimetry, 104–106
cyclic compounds, 117–121
diorganomagnesium compounds, 116–117
magnesium carbides, 106–107
Mg–CO complexes, 125
magnesium sandwich species, 115, 122–125
organomagnesia and rings, 117–121
organomagnesium halides, 109–116
Schlenk equilibrium, 107–109, 116
Thermochromism, polythiophene, 285
Thermometric titration, speciation analysis, 292
THF see Tetrahydrofuran complexes
Thiazoles, Br$_2$–Mg exchange, 525
Thienamycin, preparation, 442
(2-Thienyl)magnesium bromide, DME complex, 59
Thioamide enolates, reactions with electrophiles, 500–502
Thioamides
aldol-type reaction, 500–501, 502
metalation, 452
Thioesters
reactions with electrophiles, 500–502
α,β-unsaturated, 501, 784
natural product synthesis, 785, 786–789

Thioformamides, electrophilic amination, 579, 580
Thorium, organomagnesium compounds, 90, 91
Thymine, Mg cation complex, 170
Tight ion pairs, thermochemistry, 123
Titanium, organomagnesium compounds, 87–90
Titanocene dichlorides, catalyzed addition to C=C bonds, 561, 562
Titanocene hydride–magnesium hydride complex, 88–89
Titration
complexometric, 270, 282–283
speciation analysis, 288–292
indicators, 288, 289, 290
potentiometric titration, 291–292
thermometric titration, 292
visual endpoint, 288–291
TMEDA complexes
Grignard reagents, 59, 62
Fe-catalyzed reactions, 615–618, 622
multi-hapto-bonded groups, 50–52
Tol-BINAP ligand, catalytic enantioselective conjugate addition, 782–784
Toluene, deprotonation, 19
p-Tolylmagnesium bromide, DME complex, 60
Tosylates, transition-metal-catalyzed cross-coupling, 544
Transition metal carbonyls, 125
Transition-metal-catalyzed reactions
addition to C=C bonds, 559–562
allylic substitution, 557–558
cross-coupling, 543–547, 550–556
Transition metal compounds, organomagnesium moieties, 85–91
Transition-state structures, Grignard reagent calculations, 375–376, 377, 380–382, 388–389, 391, 393, 394–396
Transmetalation
cluster Grignard reagents, 171
functionalized Grignard reagents, 550, 552, 557
magnesium dihalide, 491, 492
magnesium enolate preparation, 445–450
silyl enol ethers, 450
see also Metalation
Triacylglycerols (TAG), derivatization, 301
Trialkylphosphines, transition-metal-catalyzed cross-coupling, 550
Trialkyltin halides, transmetalation, 550
Triarylmagnesiates, structural chemistry, 13
Triazenes, halogen–Mg exchange, 531, 533, 534
Triethylamine, Mg cation complex, 168
Triisopropylsilylamine, NMR spectroscopy, 297

Trimethylsilyl dienol ethers, preparation, 598, 599
Trimethylsilyl enol ether, preparation, 597
Trimethylsilylmethylmagnesium tris(3,5-dimethylpyrazolyl)hydroborate, 73, *74*
3-(Trimethylsilyl)propionic acid-d_4, NMR spectroscopy, 286
N,N',N''-Trimethyltriazacyclononane (TAEN), tetraorganomagnesiate production, 9–11
Triorganomagnesates
 deprotonation, 686–690, 713
 halogen–Mg exchange, 690–711, 712–713
 nucleophilic addition, 683–686, 711, 713
 preparation, 682
 structural chemistry, 12–14
Triple bonds, C–C bond addition reactions, 567–568
Triple decker (club) sandwiches, thermochemistry, 123–124
Triplet states
 Mg atoms, 158
 Mg(II) tetrapyrroles, 193
Tris[aryl]magnesiate anions, diorganomagnesium compounds, 35
Tris(ethylene)nickel(0), diorganomagnesium compounds, 86
Tris(pyrazolyl)borato alkylmagnesium derivatives, 73–74
(η^3-Tris(pyrazolyl)borato) derivatives, 411, 417–418
η^3-Tris(pyrazolyl)-hydroborate
 ketone reactions, 424, 425
 metalation, 419, 420
Tris(trimethylsilyl)methylmagnesium bromide, THF complex, 63, *64*
Trisubstituted compounds
 alkenyl sulfides, 758–760
 2,2,7-trisubstituted cycloheptanones, 765
Tropylium salts, magnesium cationic sandwiches, 125
Trost ligand, catalytic enantioselective conjugate addition, 775

Ultrasonic extraction, FAAS analysis, 272
Ultraviolet–visible fluorometry, elemental analysis of Mg, 283–285
Ultraviolet–visible spectrophotometry, elemental analysis of Mg, 279–283
Ultraviolet–visible spectroscopy
 Mg–organic substrate reactions, 171
 unsaturated substrates, 158
Unactivated alkynes, addition to triple C–C bonds, 567, 568
Unimolecular reactions, gas-phase organomagnesium ions, 182

Universal energy currency of life, 327–333
Universal methylene increment, 110–111
Unsaturated compounds
 conjugated carbonyls, 624–625, 626, 628
 Mg cation reactions, 160
 organomagnesium halides
  enthalpies of hydrogenation, 113, 119
  formal protonation reactions, 112
α,β-Unsaturated compounds
 acyl silanes, 450
 enantioselective conjugate addition
  esters, 779–784
  natural product synthesis, *785*, 786–789
  thioesters, 784
 Michael addition, 451, 501
 nitrile addition, 565–566
Unsaturated organic substrates, Mg atom reactions, 158–159
Uracil
 Br$_2$–Mg exchange, 524
 Mg cation complex, 170
Urine, mass spectrometry, 288

Vacuum tight equipment, calorimetry, 104
Vanillin, active hydrogen determination, 300
Vaporization *see* Enthalpies of vaporization
Vetraldehyde, active hydrogen determination, 300
Vibrational frequencies, Grignard reagent calculations, 375
Vinylmagnesium bromide
 DME complex, 59
 surface conditioning, 301, 305
Vinylmagnesium compounds, NMR spectra, 145
Vinylsilanes, carbomagnesiation, 661–664
Vinyl sulfides, magnesium alkylidene carbenoid reactions, *759*, 760
Viologen, PET, 198, 199
Visible spectroscopy *see* Ultraviolet–visible fluorometry; Ultraviolet–visible spectrophotometry; Ultraviolet–visible spectroscopy
Visual endpoint, titrations, 288–291
Voltametric methods, elemental analysis of Mg, 276

Walphos ligand, catalytic enantioselective conjugate addition, 775, *776*
Water
 determination in metal powders, 299
 gas chromatography, 295
 Mg cation binding energy, 175

organomagnesium ion reactivity, 180
titrations, 290
Waxes, analysis, 301
Weinreb amides, acylation reagents, 559
Wet digestion, mineralization of milk, 271, 279
Wiechell, Fritsch–Buttenberg–Wiechell rearrangement, 742–744

Xanthyrone, β-ketoester reactions, 494
X-ray absorption spectroscopy, structural studies, 3
X-ray photoelectron spectroscopy, Mg surfaces, 172
X-ray powder diffraction, diorganomagnesium compounds, 25
X-ray scattering, large-angle, 3
o-Xylidenemagnesium bis THF complex, 40

Yeast enolase, sequential behavior, 348–350
Yttrium trichloride, addition to C=O bonds, 570

Zerevitinov's method, active hydrogen determination, 299
Zinc chloride lithium chloride, addition to C=O bonds, 570
Zinc reagents, conjugate addition, 774
Zirconium
  catalyzed alkene ethylmagnesiation, 671–674
  organomagnesium compounds, 86–87
Zirconium-catalyzed reactions, addition to C=C bonds, 560, 561, 562
Zirconocene dichlorides, catalyzed addition to C=C bonds, 561, 562
Z-scheme, photosynthesis, 194, *195*

With kind thanks to Caroline Barlow for the creation of the Subject Index.